International Scientific Conference "Construction and Architecture: Theory and Practice of Innovative Development" (CATPID 2020)

IOP Conference Series: Materials Science and Engineering
Volume 1083

Nalchik, Russia
16 - 17 December 2020

Editors:

Batyr Yazyev
Stepan Litvinov
Anastasia Lapina
Anastasia Kotesova
Oksana Akay

ISBN: 978-1-7138-4586-7
ISSN: 1757-8981

Printed from e-media with permission by:

Curran Associates, Inc.
57 Morehouse Lane
Red Hook, NY 12571

Some format issues inherent in the e-media version may also appear in this print version.

Institute of Physics
Dirac House, Temple Back
Bristol BS1 6BE UK

Phone: 44 1 17 929 7481
Fax: 44 1 17 920 0979

techtracking@iop.org

Additional copies of this publication are available from:

Curran Associates, Inc.
57 Morehouse Lane
Red Hook, NY 12571 USA
Phone: 845-758-0400
Fax: 845-758-2633
Email: curran@proceedings.com
Web: www.proceedings.com

TABLE OF CONTENTS

PREFACE

The International Scientific Conference "Construction and Architecture: Theory and Practice for the innovation Development" (CATPID-2020 part II) is held from December 16th- 17th 2020 remotely. CATPID 2020 part II will be held in a separate conference format for the third time.

The work of the conference is aimed at providing an opportunity for scientists, graduate students and representatives of the construction industry to generalize the recent achievements results in the field of industrial and civil construction (design, organization of all types of work, construction materials, etc.), discussing promising directions for the construction industry development and establishing useful partner relationships for future interaction.

The conference is held on numerous topics:

- Structural mechanics and construction theory, civil and industrial engineering.
- Architecture, design and reconstruction of architectural heritage.
- Building materials and technologies of binders, concrete and building ceramics IOP.
- Mechanics of a deformable solid.
- Computational methods, mathematical models and information technologies in construction.
- Organization and technology of construction production. Surveying, cartography and photogrammetry.
- Innovations in utilities: heating, gas supply, ventilation and air conditioning, water supply and sanitation.
- Operation and reliability of transport and technological machines and complexes.
- Construction of roads, bridges, tunnels and airfields.
- Economics in construction.
- Reliability, efficiency and safety of fuel and energy facilities and construction complexes.
- Technosphere safety.
- Materials and nanotechnology in chemistry.

572 participants from 32 regions of the Russian Federation and 14 participants from 5 countries of the near and far abroad (Uzbekistan, Armenia, Vietnam, Taiwan, Belarus) were registered for the conference in 2020.

The important problems of construction and architecture (from ideas and projects to immediate implementation) are raised and the ways of their solution and optimization are suggested in the reports presented in frameworks of these topics. Many reports were prepared as a benefit of the state support of science in forms of grants, federal target programs, resolutions of the Government of the Russian Federation and other programs.

Particular attention is paid to young researchers, whose works are evaluated separately. The general impression associated with the organization and holding of the conference is very positive.

Annual holding of such conferences is planned to be organized on the main topics of the construction industry.

Editors:

Batyr Yazyev	ibm2225101@gmail.com
Stepan Litvinov	litvstep@gmail.com
Anastasia Lapina	nastya-L1@yandex.ru
Anastasia Kotesova	a.kotesova@mail.ru
Oksana Akay	oksanaakay@gmail.com

COMMITTEES

Conference Chairman

Batyr Yazyev	Don State Technical University

Deputy Chairman of the Organizing Committee

Pavel Akimov	Russian Academy of Architecture and Building Sciences

Organizing Committee Co-Chairs

Besarion Meskhi	Don State Technical University
Nikolay Karpenko	Department of Building Sciences
Yuri Konoplev	Academy of Sciences of the Republic of Tatarstan
Yuri Altudov	Kabardino-Balkarian Federal University
Svetlana Khashirova	Kabardino-Balkarian Federal University
Zaven Ter-Martirosyan	Moscow State University of Civil Engineering
Boris Sokolov	Kazan Giproniiaviaprom JSC
Iftikhar Abbasov	Southern Federal University
Valery Azarov	Volgograd State Technical University
Sergey Fedosov	Moscow State University of Civil Engineering
Vladimir Erofeev	Ogarev Mordovia State University
Levon Mayilyan	Self-regulatory organization Association "Association of builders of the South and North Caucasus districts"
Vitaliy Kolchunov	Russian Academy of Architecture and Building Sciences
Vladimir Travush	Russian Academy of Architecture and Building Sciences

Nail Kashapov	Kazan Federal University
Irina Akhmetova	Kazan State Energy University
Vladimir Andreev	Moscow State University of Civil Engineering
Aleksey Beskopylny	Don State Technical University
Sergey Kalashnikov	Volgograd State Technical University
Ashot Tamrazyan	Moscow State University of Civil Engineering
Armen Ter-Martirosyan	Moscow State University of Civil Engineering
Stepan Litvinov	Don State Technical University
Linar Sabitov	Kazan Federal University

Honorary Members of the Organizing Committe

Alexey Adamtsevich	Moscow State University of Civil Engineering
Friel Akhmetov	Kazan Federal University
Mukhtar Bekkiev	Mountain Geophysical Institute
Iskander Gilmanshin	Kazan Federal University
Georgy Nikitin	Kazan Giproniiaviaprom JSC
Oleg Radaykin	Kazan State University of Architecture and Engineering
Stanislav Roshchupkin	Sevastopol State University
Emer Sibgatullin	Kazan Federal University
Victor Taratuta	Kuban State Agrarian University
Svetlana Yazyeva	Don State Technical University

International Program Committee

Gagik Galstyan	National University of Architecture and Construction of Armenia (Yerevan)
Shan Yue	Shandong Jiaotong University
Narine Pirumyan	National University of Architecture and Construction of Armenia (Yerevan)
Lyu Chunguan	Shandong Jiaotong University
Amansahatov Chariyar	Deputy Director of the Academy of Sciences of Turkmenistan
Erkin Nafasov	Chief expert of Gosstroy of Turkmenistan
Abbas Hodzhaev	Tashkent architecture and civil engineering institute (TACII)
Ján Bujňák	Head of Department of Structures and Bridges, FCE UŽ, SK
Josef Vičan	Dean of FCE UŽ, SK
Zaynalobuddin Kobuliev	Institute of Water Resources and Ecology of the Academy of Sciences of the Tajikistan Republic
Zbigniew Wójcicki	FCE WrUT, PL
Stanisław Jemioło	FCE WUT, PL
Artur Zbiciak	FCE WUT, PL
Alejandro Valerio Pai Yang	México, Querétaro
Dietmar Wiegand	TU Wien, Austria
Vera Steiner	FH Wien, Austria
Milan Radosavljevic	Paisley, Renfrewshire, United Kingdom
Hellmich Christian	TU Wien, Austria
David Grigoryan	National University of Architecture and Construction of Armenia

Organising committee of the conference (editorial board)

IOP Conf. Series: Materials Science and Engineering 1083 (2021) 011002 doi:10.1088/1757-899X/1083/1/011002

Peer review declaration

All papers published in this volume of *IOP Conference Series: Materials Science and Engineering* have been peer reviewed through processes administered by the Editors. Reviews were conducted by expert referees to the professional and scientific standards expected of a proceedings journal published by IOP Publishing.

The International Scientific Conference "Construction and Architecture: Theory and Practice for the innovation Development" (CATPID-2020 Part II)

- **Type of peer review:** Single-anonymous
- **Describe criteria used by Reviewers when accepting/declining papers. Was there the opportunity to resubmit articles after revisions?** The main criterion used by reviewers is the scientific significance of the submitted materials. The authors had the opportunity to correct the article according to the comments of the reviewers.
- **Conference submission management system:** used own systems
- **Number of submissions received:** 158 submissions were received
- **Number of submissions sent for review:** 146 submissions were sent for review
- **Number of submissions accepted:** 114 submissions were accepted
- **Acceptance Rate (Number of Submissions Accepted / Number of Submissions Received X 100):** 72.2
- **Average number of reviews per paper:** 2 reviews per paper
- **Total number of reviewers involved:** 18 reviewers were involved
- **Any additional info on review process (ie plagiarism check system):**
- **Contact person for queries:**

Professor Batyr Yazyev
Don State Technical University, Rostov-on-Don, Russia
e-mail: ibm2225101@gmail.com

Experimental assessment of solar radiation effect on the stress-strain state of a span with an orthotropic slab

S Yu Gridnev*, I S Podlesnykh, N A Barchenkova

Voronezh State Technical University, 14 Moskovsky ave., Voronezh, 394000, Russia

E-mail: gridnev_s_y@rambler.ru

Abstract. The article is devoted to the experimental assessment of the solar radiation effect on the stress-strain state of a span with an orthotropic slab. Preliminary, comprehensive instrumental measurements of the road surface temperature distribution and the span elements were carried out at different times of the year. A detailed picture of the temperature distribution over the cross-section height was determined at different air temperatures and the Sun positions. To assess the solar radiation effect, instrumental measurements of deflections and deformations of the span's main beams under the influence of solar radiation were carried out and the nature of the change in the stress-strain state of the span during the day was determined. The measurements were carried out using the deflection meters and strain gauges mounted on the main beams that experience uneven heating throughout the day. Using the developed and previously tested finite element model, the calculated values of the span bearing elements' stress-strain state parameters for each case of exposure to solar radiation have been determined, taking into account the experimentally obtained temperatures. Comparison of the calculation results obtained on the basis of field measurements and calculation in the program of the finite element method has been carried out. The most unfavorable cases of the solar radiation effect on the main beams' displacement have been revealed.

1. Introduction

In the construction industry, it is mandatory to conduct comprehensive tests of building structures when inspecting buildings and structures, including bridge spans. A complex system of loads acts on the suer. In addition to transport, wind and self-weight loads, the span perceives temperature effects, which contribute to uneven heating of the elements. At the same time, this research area is poorly understood, but it can affect the stress-strain state (SSS), due to the constantly changing temperature field during the day and year. Therefore, in order to identify the features of the span operation under solar radiation conditions, it is necessary to carry out the field measurements and accumulate the data to form a complete picture of the temperature effect nature.

A review of the scientific articles devoted to the solar radiation effects' analysis has been carried out. In the articles [1-2], the temperature influence on the stress-strain state of the span has been noted and the calculation of temperature fields and stresses in the beams of the span has been made on the basis of the field measurements. In the works [3-5], the most unfavorable cases of exposure to solar radiation have been identified and its effect on the strength of the structure has been studied. The numerical results of the temperature distribution at different times of the year have been compared with the experimental measurements. The main problems and promising research of this topic have been noted.

The authors of the article carried out a number of works devoted to the solar radiation effects' analysis. In the articles [6-8], the temperature distribution along the steel-reinforced concrete span cross-section height has been studied and its effect on the strength of the asphalt concrete pavement has been identified. Based on the results of the works [9-11], a technique for performing full-scale temperature measurements on a span with an orthotropic slab at different times of the year has been developed and tested, a complete picture of the temperature distribution over the height of the cross-section has been compiled at various air temperatures and sun positions. The calculation of the stress-strain state has been carried out using the field measurements' results in various cases of exposure to solar radiation. Comparison of the calculation results obtained according to the recommendations of regulatory documents and on the field measurements basis has been made.

2. Research Objectives

2.1. To perform instrumental measurements of deflections and deformations of the main span beams when exposed to solar radiation

2.2. To study the change nature in SSS during the day based on the results' analysis of the in-situ temperature measurements, deflections and deformations of the elements

2.3. Using the developed finite element model of the span, to calculate the stress-strain state of the span for each case of exposure to solar radiation

2.4. To compare the results obtained on the basis of instrumental measurements and theoretical calculations.

3. The instrumental measurements' description

Instrumental measurements were carried out on a span with an orthotropic slab L = 43,1m road bridge across the river Vorona near Borisoglebsk in Voronezh region. A detailed description of the span is given in [10].

Site tests were carried out during daylight hours on July 21, 2020. The temperature measurements were taken in the span elements and instrumental measurements of the main beams' deflections were carried out at various air temperatures and sun position.

Temperature measurement was performed using a non-contact pyrometer "Megeon 16400". To verify the data, a contact thermometer "multimeter PM838" was used.

The deflections of the main beams were measured with the 6-PAO deflection meters of the Maximov's system with an absolute error ± 0.01 mm. The deflection meters 6-PAO are designed to measure linear displacements of individual points in structures. Deflection meters are used when testing building structures, engineering and construction surveys and when inspecting buildings, structures and bridge spans.

The deformations of the longitudinal fibers of the lower chord in the main beams were measured by the deformometers using dial gauges with a scale of 1 μm.

The installation diagram of the devices is shown in Figure 1.

Temperature measurements were carried out at 3 points of each span element (asphalt concrete pavement, main beam in the sun, main beam in the shadow). Deflection meters and deformometers were installed on the lower chord of the two main beams, 1/4 of the span length. Figure 2 shows the instruments' location during the field measurements on the span.

IOP Conf. Series: Materials Science and Engineering 1083 (2021) 012001 doi:10.1088/1757-899X/1083/1/012001

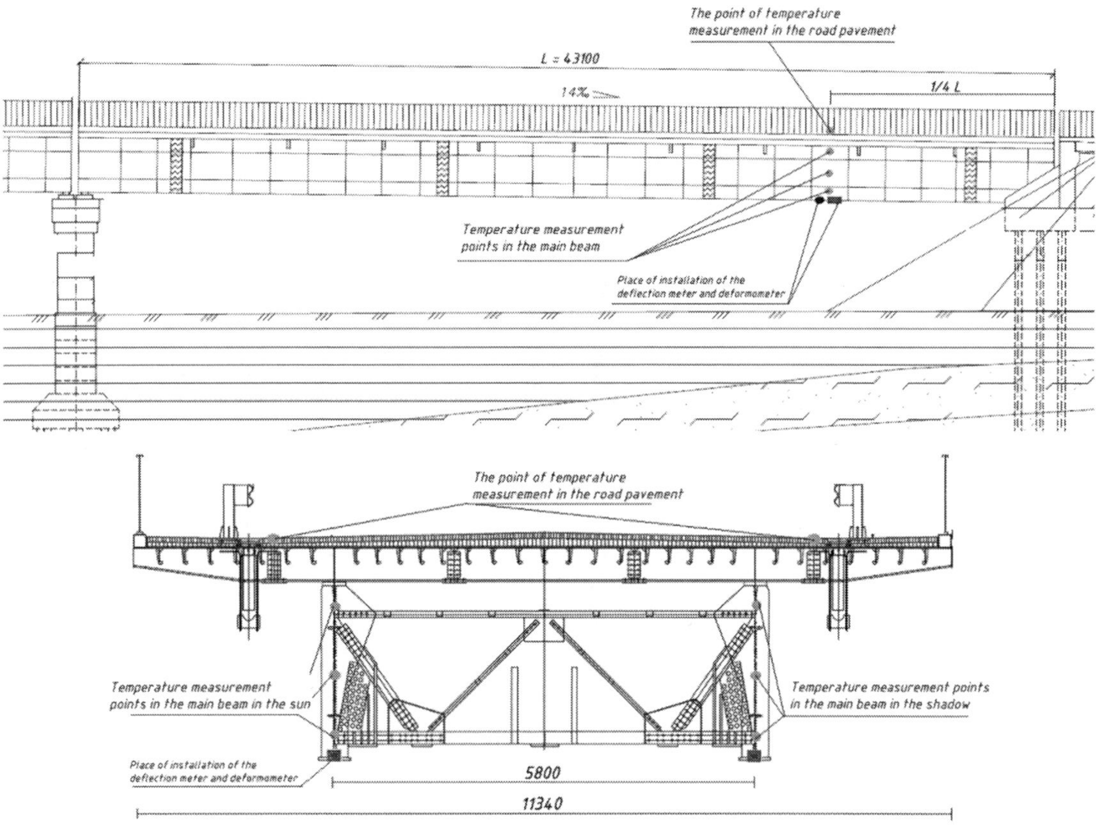

Figure 1. Installation diagram of measuring devices

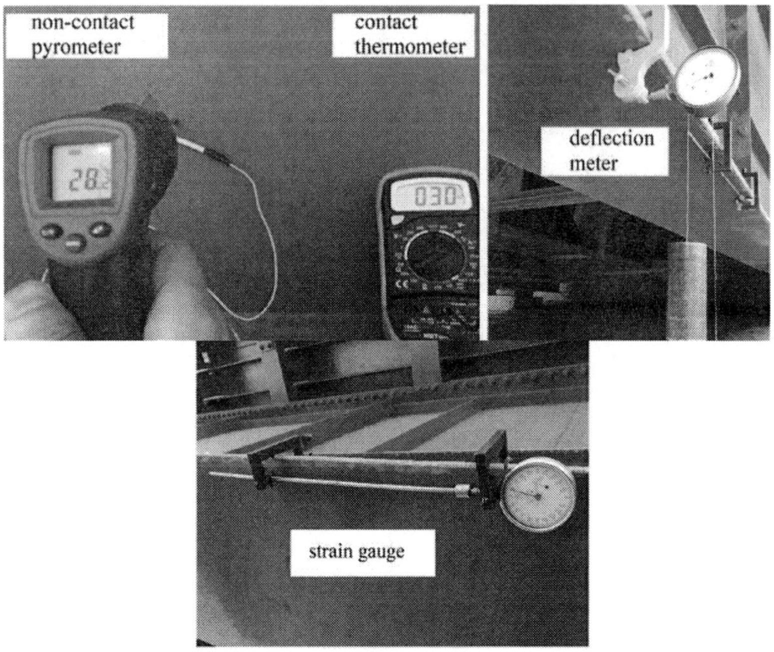

Figure 2. Devices for measuring the temperature of the span elements, deflections and deformations of the main beams

4. Results of instrumental measurements

The air temperature on the day of the study at the object changed from 22°C to 29 °C. The temperature comparison of the main beams in the sun and in the shade is shown in Figure 3.

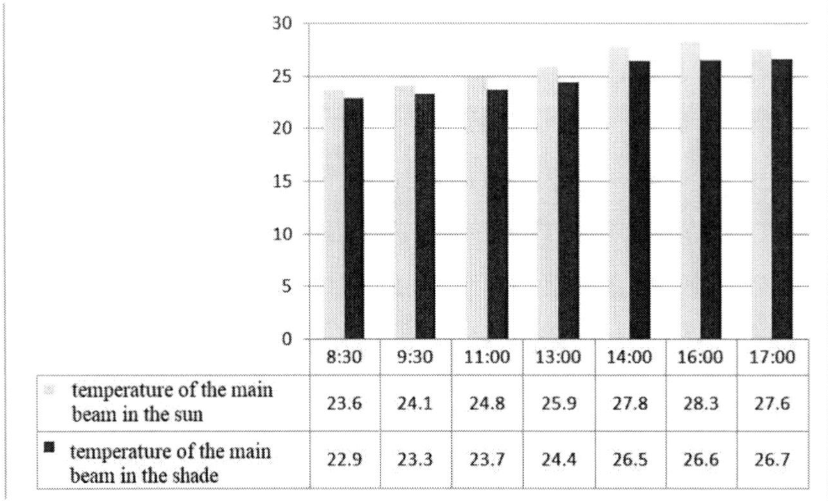

Figure 3. The temperature comparison of the main beams in the sun and in the shade

The maximum temperature of the main beams on the day of field measurements was 28.3°C. During the day, the temperature on the main beam under the influence of sunlight changed by 19.9%.

The deflection meter indicator with timekeeping in the process of field measurements is shown in Figure 4. The deflections and deformations' values obtained on the field measurements' basis are presented in Table 1 and in the diagram in Figure 5.

Table 1. Deflections, mm

	8-30	9-30	11-00	13-00	14-00	16-00	17-00
beam in the sun	0	-0.12	-0.63	-1.96	-1.99	-2.72	-2.36
beam in the shade	0	-0.09	-0.43	-0.14	-0.17	-0.68	-0.8

Deformation, mm

	8-30	9-30	11-00	13-00	14-00	16-00	17-00
beam in the sun	0	0.03	0.03	0.07	0.08	0.08	0.09
beam in the shade	0	0.01	0.01	0.01	0.02	0.02	0.02

Figure 4. Deflection meter indicator with time sheet during the field measurements

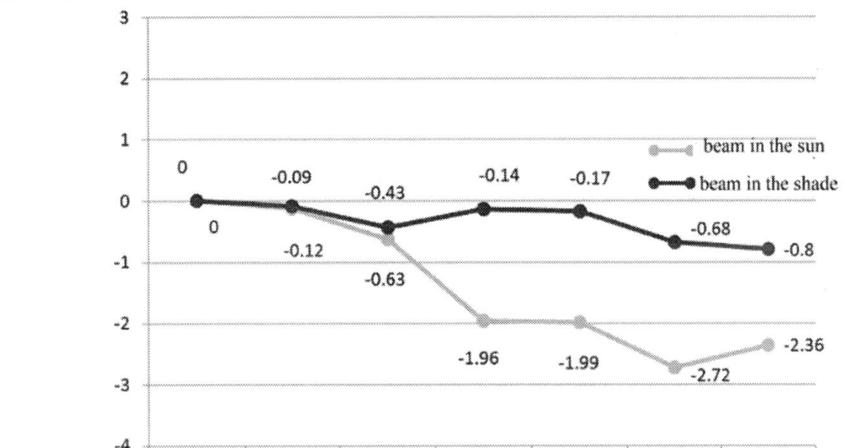

	8:30	9:30	11:00	13:00	14:00	16:00	17:00
beam in the sun	0	-0.12	-0.63	-1.96	-1.99	-2.72	-2.36
beam in the shade	0	-0.09	-0.43	-0.14	-0.17	-0.68	-0.8

Figure 5. Diagram of the main beams' displacement during the day

As a result of field measurements, it can be noted that, under the influence of solar radiation, the displacements of the main beams have negative values, i.e. the lower fibers are compressed. The maximum movement of the main beam during the day in the sun was -2.72 mm, of the main beam in the shade -0,8. In the first half of the day, when heating intensity has not reached its maximum values, the difference in the displacements of the two main beams does not exceed 0.2 mm. However, in the afternoon, as the temperature difference of the main beams increases, the beam movement in the sun is much higher than that of the beam in the shade. The movement of the beam in the sun at the same time is 2.04 mm higher than the movement of the beam in the shade.

5. The numerical studies' results analysis

One of the main criteria for evaluating the span work under the load action is the correspondence of the values obtained in field studies to the calculation results. To determine the bridge span bearing elements stress-strain state parameters' calculated values, a spatial finite element model has been developed. Spatial finite element model of the span is implemented in the complex "PC Lira CAD".

A detailed description of the finite element model was presented in [10]. The nature of the temperature distribution in the pavement layers was determined using the ELCUT Student software package for modeling thermal fields [7].

When performing the calculation of the finite element method (FEM), the displacements and stresses of the span elements were obtained. The comparative analysis of calculation results has been performed.

Figure 6 shows the main beam displacements' iso-fields, obtained in the FEM complex. A comparison of the obtained values of the main beams' displacement, obtained on the basis of instrumental measurements and theoretical calculations for each case of exposure to solar radiation, is shown in Figures 7 and 8.

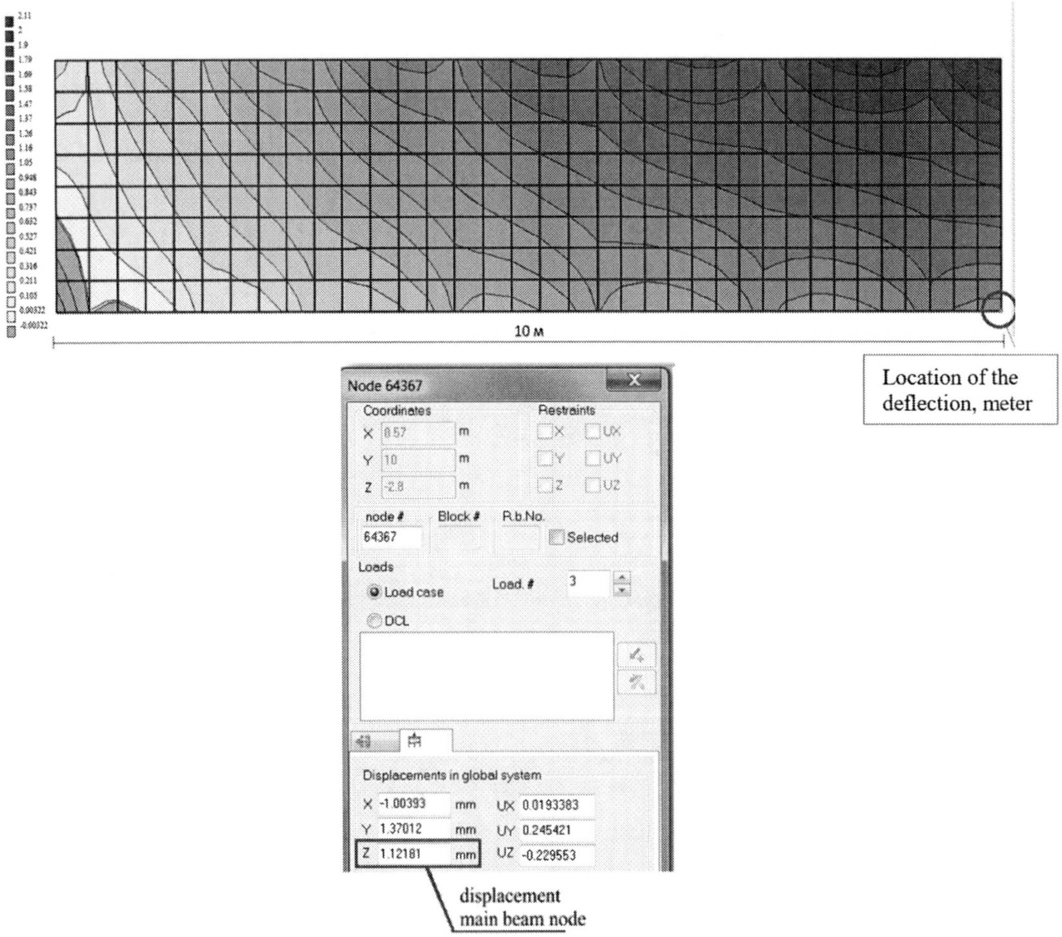

Figure 6. Iso-fields of the main beam's displacements obtained in the FEM complex

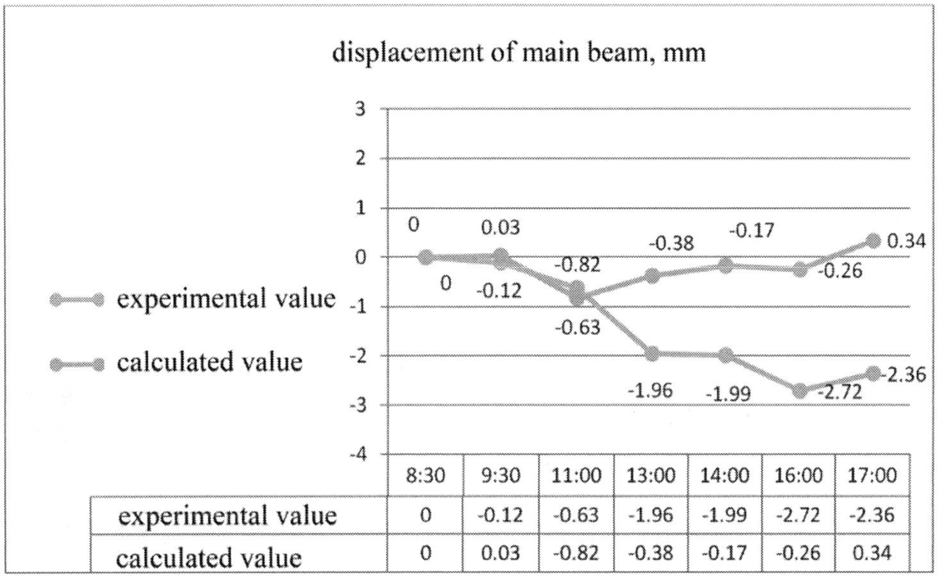

	8:30	9:30	11:00	13:00	14:00	16:00	17:00
experimental value	0	-0.12	-0.63	-1.96	-1.99	-2.72	-2.36
calculated value	0	0.03	-0.82	-0.38	-0.17	-0.26	0.34

Figure 7. Comparison of experimental and calculated deflections of the main beam in the sun

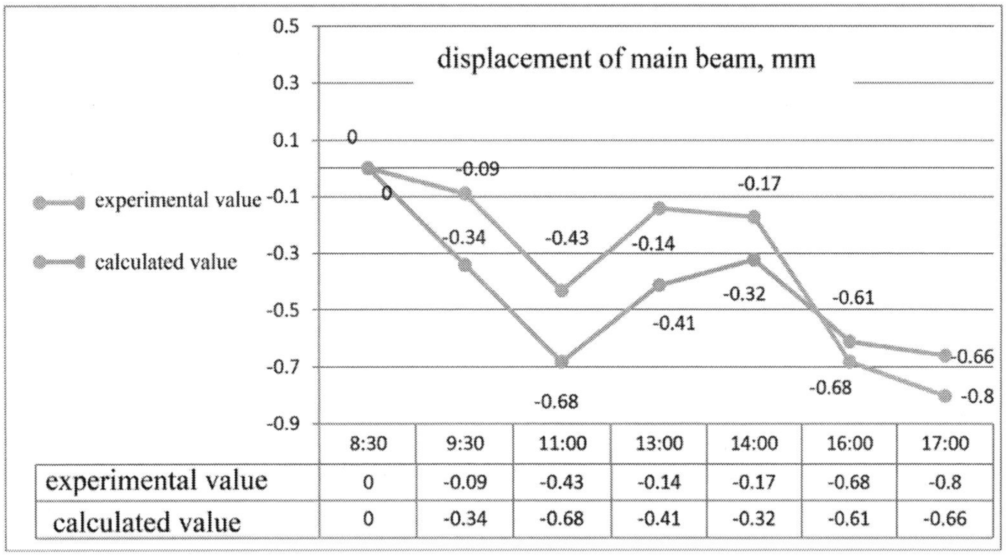

	8:30	9:30	11:00	13:00	14:00	16:00	17:00
experimental value	0	-0.09	-0.43	-0.14	-0.17	-0.68	-0.8
calculated value	0	-0.34	-0.68	-0.41	-0.32	-0.61	-0.66

Figure 8. Comparison of experimental and calculated deflections of the main beam in the shade

As it can be seen from the graph, the experimental deflections of the beam in the sun in the first half of the day are close to the calculated values. In the afternoon, with an increase in the solar radiation intensity, the experimental movement of the beam in the sun is much higher than the calculated values. In the measuring range from 13h to 17h, the discrepancy is 10% greater.

Experimental and calculated displacements of the main beam in the shade have a similar change pattern during the day. Moreover, the calculated values are predominantly higher than the experimental ones. Throughout the day, the discrepancy in the results varies from 0.7 mm to 0.27 mm.

Thus, the analysis of the results shows that the effect of solar radiation has a significant effect on the main beams' movement.

Summary

1. Under the solar radiation influence in the observed time range, the movements of the main beams decrease.
2. The movement of the beam in the sun at a fixed point in time is 2.04 mm higher than the movement of the beam in the shade.
3. The results of numerical studies made it possible to establish that, as the intensity of solar radiation increases, there is a decrease in the convergence of the main beams' displacements results according to the results of instrumental measurements and FEM calculation.

Conclusion

Experimental studies of the span main beams' deflections and deformations under the influence of solar radiation have been carried out, and the nature of the change in the stress-strain state of the span during the day has been determined. With the help of the developed and previously tested finite element model, the SSS was calculated for each case of exposure to solar radiation, taking into account the experimentally obtained temperatures of the span elements. A comparison of the results obtained on the basis of experimental studies and theoretical calculations showed in most cases their satisfactory coincidence. The most unfavorable cases of the solar radiation effect on the main beams' displacement have been revealed.

References

[1] Mishutin V O 2014 The need to take into account temperature deformations in the design and construction of bridge structures *Modern technologies. System analysis. Modeling* **3 (43)** 203-208.

[2] Bezbabicheva O I, Bilchenko A.V, Kislov A G 2010 Prediction of temperature stresses in facade beams of bridge structures *Science and progress of transport. Bulletin of the Dnipropetrovsk National University of Railway Transport* **33** 28-31.

[3] Linren Zhou, Yong Xia, James M W, Brownjohn Ki, Young Koo 2016 Temperature Analysis of a Long-Span Suspension Bridge Based on Field Monitoring and Numerical Simulation *Journal of Bridge Engineering* **21 (1).**

[4] XuMing Song, Hani Melhem, Jun Li, QingYuan Xu, LiJun Cheng 2016 Effects of Solar Temperature Gradient on Long-Span Concrete Box Girder during Cantilever Construction *Journal of Bridge Engineering* 21 (3).

[5] Bo Chen, Yu-zhou Sun, Gan-jun Wang, Ling-yan Duan 2014 Hindawi Publishing Corporation *Mathematical Problems in Engineering* **639867**.

[6] Podlesnykh I S, Gridnev S Yu, Barchenkova N A 2018 Analysis of the solar radiation influence on the stress-strain state of the steel-reinforced concrete span of the highway bridge *Internet journal "Transport facilities"* **2**. Title from the screen. Language. Russian, English. DOI: 10.15862/12SATS218

[7] Gridnev S Yu, Podlesnykh I S 2018 Influence of solar radiation on the safety of the upper layer of the pavement of the steel-reinforced concrete span *Collection of materials of the XIX International Scientific and Technical Conference "Actual problems of construction, construction industry and industry"* 34-36.

[8] Gridnev S Yu, Podlesnykh I S 2018 Temperature distribution over the highway bridge steel-reinforced concrete span cross-section height *Collection of materials of the XIX International Scientific and Technical Conference «Actual problems of construction, construction industry and industry»* 36.

[9] Podlesnykh I S, Gridnev S Yu, Barchenkova N A 2019 Verifying Regulatory Documents for Solar Radiation Level Control When Estimating Stress-Strain State of the Spans Based on the Orthotropic Slab *Proceedings of 24th International Conference "MECHANIKA 2019", Kaunas University of Technology* 116-121.

[10] Gridnev S.Yu., Podlesnykh, I.S. The influence of uneven heating on the stress distribution in the elements of the span with an orthotropic slab / Materials collection of the XX International Scientific and Technical Conference "Actual problems of construction, construction industry and architecture". – 2019. – 108-110 Pp.
[11] Gridnev S.Yu., Podlesnykh, I.S. Features of the temperature field in the elements of the span of a road bridge at negative temperatures close to zero / Materials collection of the XX International Scientific and Technical Conference "Actual problems of construction, construction industry and architecture". – 2019. – 110-112 Pp.
[12] Gridnev S.Yu., Podlesnykh, I.S. Method of measuring the temperature field of a span with an orthotropic slab under the influence of solar radiation / Materials collection of the XXI International Scientific and Technical Conference "Actual problems of construction and construction industry". – 2020. – 62-68 Pp.

IOP Conf. Series: Materials Science and Engineering 1083 (2021) 012002 doi:10.1088/1757-899X/1083/1/012002

Review and comparative analysis of computational algorithms for modeling the dynamics of elastically supported structural systems

Yu I Skalko[1], S Yu Gridnev[2*]

[1] Department of Computational Physics, Moscow Institute of Physics and Technology, 9 Institutskiy lane, Dolgoprudny, 141700, Russia
[2] Structural Mechanics Department ,Voronezh State Technical University, 14 Moskovskiy avenue, Voronezh, 394026, Russia

E-mail: gridnev_s_y@rambler.ru

Abstract. The work is devoted to the comparative analysis of computational algorithms for modeling the dynamics of elastically supported structural systems with intermediate and limiting supports in the presence of point forces applied to the superstructure. The considered algorithms are based on the model formulation in the form of a system of partial differential equations for generalized functions. This approach avoids the need to subordinate the basic functions to the boundary conditions and include the presence of sources of the forces, applied at a certain point, in the model.

1. Introduction

The construction of computational algorithms for calculating the dynamics of elastically supported load-bearing structures (structural systems) is associated with a number of specific difficulties. These difficulties are caused by the fact that the mathematical model of these structures' behavior contains the derivatives with respect to spatial variables of the fourth-fifth orders and higher. Structures can contain one-way connections, on and off switching connections, mechanisms with gaps, structures with elastic connections, which are pre-compressed by their own weight, structures with restraining arms, etc. The mathematical models of such systems, as a rule, have the form of partial differential equations' system with uneven nonlinear conditions at the internal and external boundaries. Quite often, the situations when an elastically supported structure is under the influence of localized forces, i.e. the forces, where the contact area with the superstructure, is negligible less than the size of the supporting structure, are considered. Such situations lead to the need to use mathematical models with point sources of acting forces.

In [1] - [4], a technique for constructing an approximate solution, in which the original problem is associated with an equation for generalized functions, the solution of which will also be a solution to the original problem, is proposed. The problem representation in the form of an equation for generalized functions makes it possible to formulate the problem and construct an approximate solution in the presence of point sources. Since the boundary conditions are taken into account in the equations for generalized functions (in a weak form), then when constructing an approximate solution there is no need to subject the basis functions to the boundary conditions. Later, for the hyperbolic systems of linear differential equations of the first order [5] - [8], this approach was developed for the case when it is possible to construct a fundamental solution of the problem operator for generalized functions. In this work, using the example of the elastic thin-walled rod transverse vibrations equations, a technique for applying the fundamental solution of the problem operator to construct an approximate solution of the initial-boundary value problem of this equation is presented.

2. Problem statement

The methods analysis for numerical modeling of the elastically supported structural systems' dynamics will be carried out using the example of the span structure's constructively nonlinear oscillations of a floating bridge with restrictive rigid supports. We take the model of an elastic bar with free ends of length $2l$ and linear density μ and full weight M as a mechanical model under the action of the forces combination from the dynamic pressures of car tires $\mathbb{R}_r(t,x)$, the interaction forces with floatation jackets $\mathbb{S}_s(t,x)=\frac{Mg}{N}-\eta_s u(t,x_s)$, transmitted by the means of elastic restraints (η_s - stiffness of this jacket, $u(t,x_s)$ - superstructure profile) and interaction forces with the transitional parts of the bridge $\mathbb{P}_p(t,x)$. The additional forces from the side of the restraint supports start acting when the gaps are closed on the elastic rod, $\mathbb{Q}_q^*=-\theta(u^*-u(t,x_q))\eta_q(u(t,x_q)-u^*)$; these forces are modeled by the elastic constraints with stiffness coefficients $\eta_q \gg \eta_s$.

Let us introduce the vector of variables $\mathbf{u}=\left(u,\frac{\partial u}{\partial t},\frac{\partial^2 u}{\partial x^2}\right)^T$, and the notation $\boldsymbol{\gamma}(t)=\mathbf{u}(t,x=-l)$, $\bar{\boldsymbol{\gamma}}(t)=\frac{\partial \mathbf{u}}{\partial x}(t,x=-l)$, $\boldsymbol{\beta}(t)=\mathbf{u}(t,x=l)$, $\bar{\boldsymbol{\beta}}(t)=\frac{\partial \mathbf{u}}{\partial x}(t,x=l)$. Then, following [9] and [6], the model of flexural-torsional vibrations of a thin-walled elastic rod can be written in the form of an initial-boundary value problem for the system of partial differential equations. This model is formulated for the distributed forces applied to an elastic thin-walled bar. To formulate this model in terms of the forces applied at a point, it is necessary to resort to the technique of the generalized functions [10].

As shown in [6], $\mathbf{u}(t,x)$, the considered generalized function from $\mathbf{S}'$, meets the condition of the partial differential equations system.

$$\frac{\partial \mathbf{u}}{\partial t}+\mathbf{A}\frac{\partial^2 \mathbf{u}}{\partial x^2}+\mathbf{B}\mathbf{u}=\mathbf{f}+\boldsymbol{\sigma}\delta(t)+\mathbf{b}$$

$$\mathbf{A}=\begin{pmatrix}0&0&0\\0&0&a^2\\0&-1&0\end{pmatrix};\quad \mathbf{B}=\begin{pmatrix}0&-1&0\\0&0&0\\0&0&0\end{pmatrix};\quad \mathbf{f}=\begin{pmatrix}0\\-g\left(\theta(x+l)-\theta(x-l)\right)+\frac{1}{\mu}\sum_f \mathbb{F}_f^*\,\delta\left(x-x_f\right)\\0\end{pmatrix}. \quad (1)$$

$$\mathbf{b}=\begin{pmatrix}0\\+a^2\bar{\gamma}_3(t)\delta(x+l)+a^2\gamma_3(t)\delta_x'(x+l)-a^2\bar{\beta}_3(t)\delta(x-l)-a^2\beta_3(t)\delta_x'(x-l)\\-\bar{\gamma}_2(t)\delta(x+l)-\gamma_2(t)\delta_x'(x+l)+\bar{\beta}_2(t)\delta(x-l)+\beta_2(t)\delta_x'(x-l)\end{pmatrix}$$

3 Computational algorithms

3.1 Algorithm based on global basis functions.

IOP Conf. Series: Materials Science and Engineering 1083 (2021) 012002 doi:10.1088/1757-899X/1083/1/012002

We define the system of linearly independent basis functions $H_p(x)$, for example, Legendre polynomials or Chebyshev polynomials on the segment $-l,l$. We will seek an approximate solution in the form $\mathbf{u}(t,x)=\sum_{p=1}^{P} H_p(x)\mathbf{u}^p(t)$. The functions of the $\boldsymbol{\varphi}^{p'}(t,x)\in\mathbf{S}'$ form, of the $\boldsymbol{\varphi}^{p'}(t,x)=H_{p'}(x)\boldsymbol{\chi}(t)$ kind, indexed $p'=1\div P$ will be taken as test functions, $\boldsymbol{\chi}(t)$ - is an arbitrary finite infinitely differentiable function. Following the stated in [6] to determine $\mathbf{u}^p(t)$, we obtain the Cauchy problem for the ordinary deferential equations system (ODE)

$$\begin{aligned}&\mathbf{D}^p_{p'}\frac{d\mathbf{u}^{p'}}{dt}+\mathbf{B}^p_{p'}\mathbf{u}^p=f^p\\&\mathbf{u}^p(t=0)=\boldsymbol{\sigma}^p\end{aligned} \tag{2}$$

Matrices $\mathbf{D}^p_{p'}$ are not unitary and densely filled, since the basic functions are not orthonormal and their support coincides with the entire segment $-l,l$. Having solved this Cauchy problem for the ODE system, we find an approximate solution to the original problem.

3.2 Algorithm Based on Finite Piecewise Linear Functions.

This algorithm differs from the one described above only in the choice of basic functions. Let us plot the mesh of nodes x_p, $p=1\div P$ on the segment $-l,l$ so, that $x_1=-l$, $x_P=l$. This grid splits the segment $-l,l$ at intervals. Let us connect the function $H_p(x)$ with each node x_p, which is equal at the point x_p, equal at all other grid points and linear at each interval. Matrices $\mathbf{D}^p_{p'}$ will also not be single, but unlike the previous algorithm, they will be strongly decompressed, due to the use of finite functions as a basis.

3.3 Algorithm based on the problem operator fundamental solution construction.

3.3.1 Fundamental solution.

Let us plot a fundamental solution to the operator of the problem (1). The fundamental solution to the operator of the problem (1) is called the generalized matrix function $\mathbf{G}(t,x)\in\mathbf{S}'$, which satisfies the equation

$$\frac{\partial\mathbf{G}(t,x)}{\partial t}+\mathbf{A}\frac{\partial^2\mathbf{G}(t,x)}{\partial x^2}+\mathbf{B}\mathbf{G}(t,x)=\mathbf{I}\delta(t,x) \tag{3}$$

Let us denote the Fourier transformation $\mathbf{G}(t,x)$ by the spatial variables by $\mathbf{V}(t,\xi)=F_x[\mathbf{G}](t,\xi)$. Let us perform the Fourier equations transformation (3) in spatial variables

$$\frac{\partial\mathbf{V}(t,\xi)}{\partial t}-\left(\xi^2\mathbf{A}-\mathbf{B}\right)\mathbf{V}(t,\xi)=\mathbf{I}\delta(t) \tag{4}$$

The solution of the problem (4) exists, is unique and has the form $\mathbf{V}(t,\xi)=\theta(t)e^{(\xi^2\mathbf{A}-\mathbf{B})t}$.

Matrix $\xi^2\mathbf{A}-\mathbf{B}$ is hyperbolic, i.e. has a complete set of linearly independent eigenvectors and therefore can be represented in the form $\xi^2\mathbf{A}-\mathbf{B}=\mathbf{\Omega}^{-1}\mathbf{\Lambda}\mathbf{\Omega}$. Here $\mathbf{\Lambda}$ - is a diagonal matrix of the matrix eigenvalues, ordered in non-decreasing order, $\mathbf{\Omega}$ - is the matrix which rows are the left eigenvectors of the matrix corresponding to the eigenvalues $\mathbf{\Lambda}$. Then $e^{\left(\xi^2\mathbf{A}-\mathbf{B}\right)t}=\mathbf{\Omega}^{-1}e^{\Lambda t}\mathbf{\Omega}$ and accurate to $O\quad at^{\;2}$

$$\mathbf{V}(t,\xi)=\theta(t)\begin{pmatrix} 1 & \dfrac{\sin a\xi t}{a\xi} & 0 \\ 0 & \dfrac{\partial}{\partial t}\dfrac{\sin a\xi t}{a\xi} & -a^2(i\xi)^2\dfrac{\sin a\xi t}{a\xi} \\ 0 & (i\xi)^2\dfrac{\sin a\xi t}{a\xi} & \dfrac{\partial}{\partial t}\dfrac{\sin a\xi t}{a\xi} \end{pmatrix} \tag{5}$$

Performing the inverse Fourier transformation, we obtain a fundamental solution to the operator of the problem (1) or Green's matrix function

$$\mathbf{G}(t,x)=\theta(t)\begin{pmatrix} \delta(x) & \dfrac{\theta(at-|x|)}{2a} & 0 \\ 0 & \dfrac{\partial}{\partial t}\dfrac{\theta(at-|x|)}{2a} & -a^2\dfrac{\partial^2}{\partial x^2}\dfrac{\theta(at-|x|)}{2a} \\ 0 & \dfrac{\partial^2}{\partial x^2}\dfrac{\theta(at-|x|)}{2a} & \dfrac{\partial}{\partial t}\dfrac{\theta(at-|x|)}{2a} \end{pmatrix} \tag{6}$$

The solution of the problem (1) is defined as the convolution of the fundamental solution with the right-hand side. Let us consider the point $(t+\Delta t,x_{\tilde{s}})$, corresponding to the location of one of the intermediate supports. Then

$$\begin{aligned} u_1(t+\Delta t,x_{\tilde{s}}) &= -g\frac{\Delta t^2}{2}+\frac{1}{2a\mu}\int_t^{t+\Delta t}\left(\frac{Mg}{N}-\eta_{\tilde{s}}u_1(\tau,x_{\tilde{s}})\right)d\tau\cdots \\ &\quad+\sum_r\frac{\theta\left(a\Delta t-|x_{\tilde{s}}-x_r|\right)}{2a\mu}\int_0^{t-|x_{\tilde{s}}-x_r|/a}\mathbb{R}_r^* d\tau+u_1(t,x_s)+\frac{1}{2a}\int_{x_{\tilde{s}}-a\Delta t}^{x_{\tilde{s}}+a\Delta t}u_2(t,\xi)d\xi \\ u_2(t+\Delta t,x_{\tilde{s}}) &= -g\Delta t+\frac{Mg}{2a\mu N}-\frac{\eta_{\tilde{s}}}{2a\mu}u_1(t+\Delta t,x_{\tilde{s}})\cdots \\ &\quad+\sum_r\left(1+\left(2\theta\left(x_{\tilde{s}}-x_r\right)-1\right)\frac{\partial x_r}{\partial t}\right)\frac{\theta\left(a\Delta t-|x_{\tilde{s}}-x_r|\right)}{2a\mu}\mathbb{R}_r^*\left(t+\Delta t-\frac{|x_{\tilde{s}}-x_r|}{a}\right)\cdots \\ &\quad+\frac{u_2(t,x_{\tilde{s}}-a\Delta t)+u_2(t,x_{\tilde{s}}+a\Delta t)}{2}+a\frac{\frac{\partial u_3}{\partial x}(t,x_{\tilde{s}}-a\Delta t)-\frac{\partial u_3}{\partial x}(t,x_{\tilde{s}}+a\Delta t)}{2} \\ u_3(t+\Delta t,x_{\tilde{s}}) &= -\frac{\eta_{\tilde{s}}}{2a^3\mu}u_2(t+\Delta t,x_{\tilde{s}})+\sum_r\frac{\theta\left(at-|x_{\tilde{s}}-x_r|\right)}{2a^3\mu}\frac{\partial\mathbb{R}_r^*}{\partial t}\left(t+\Delta t-\frac{|x_{\tilde{s}}-x_r|}{a}\right)\cdots \\ &\quad-\frac{\frac{\partial u_2}{\partial x}(t,x_{\tilde{s}}-a\Delta t)-\frac{\partial u_2}{\partial x}(t,x_{\tilde{s}}+a\Delta t)}{2a}+\frac{u_3(t,x_{\tilde{s}}-a\Delta t)+u_3(t,x_{\tilde{s}}+a\Delta t)}{2} \end{aligned} \tag{7}$$

At the points $(t+\Delta t, x_{\tilde{q}})$, corresponding to location of one of the limiting supports, an expression follows for the total force transmitted to the superstructure from the side of the s th intermediate support, $\mathbb{S}_s$ replaced by $\mathbb{Q}_q$ - total force transferred to the superstructure from the side of the q th limiting support. At the points $(t+\Delta t, x)$, that do not coincide with either intermediate or limiting supports, in all formulas the terms responsible for the total force transferred to the superstructure from the side of the intermediate support should be equated to zero.

Considering the points $(t+\Delta t, x=\mp l)$, coinciding with the left or right boundary, we obtain

$$\begin{cases} u_1(t+\Delta t,l) = -\dfrac{g\Delta t^2}{4} + u_1(t,l) + \dfrac{\int_{l-a\Delta t}^{l} u_2 d\xi}{2a} \\ u_2(t+\Delta t,l) = -g\dfrac{\Delta t}{2} + \dfrac{1}{2}u_2(t,l-a\Delta t) \\ u_3(t+\Delta t,l) = 0 \end{cases} ; \begin{cases} u_1(t+\Delta t,-l) = -\dfrac{g\Delta t^2}{4} + u_1(t,-l) + \dfrac{\int_{-l}^{-l+a\Delta t} u_2 d\xi}{2a} \\ u_2(t+\Delta t,-l) = -g\dfrac{\Delta t}{2} + \dfrac{1}{2}u_2(t,-l+a\Delta t) \\ u_3(t+\Delta t,-l) = 0 \end{cases} \quad (8)$$

3.3.2 Computational Algorithm and Numerical Experiments

Let us split the segment $-l,l$ with the nodes x_n, $n=1:N$ on $N-1$ intervals. In this case, we require that the position of each intermediate and each restraint support coincides with some of the nodes x_n. Let us denote the minimum length $h_n = x_{n+1} - x_n$, $n=1:N-1$ by h in these intervals. Also on the time interval $0,T$ we introduce a uniform grid t_m, $m=1:M$ with spacing $\Delta t = t_{m+1} - t_m$.

Let us connect the system of difference equations with each node x_n. At the internal nodes, we obtain these equations by approximating the equalities (7). The functions included in these equalities are approximated on each interval of the spatial grid by the linear functions plotted from the values at the nearest nodes of the spatial grid. We replace the integrals with finite-difference relations, for example, by the formula of rectangles or by the formula of trapeziums. At the nodes corresponding to the boundary points, the system of difference equations is obtained by approximating the equations (8) respectively. As a result, we obtain an explicit difference scheme for calculating the solution values at the next time layer.

A number of numerical experiments were carried out with the constructed computational model, including those with a moving active load at various parameters of this load. In these experiments, a computational scheme with the parameters described in detail in [11] and further used in [6], [8], [12], [13] was used. These experiments have shown, even in a single-processor implementation, high performance of a computational model based on this algorithm. The computation time is significantly less than the time required to calculate similar scenarios by other methods. For example, when compared with probably the most effective of the known methods - the modified Galerkin method, on piecewise-linear basis the functions developed by the authors in earlier works [8], the performance increase in the same scenarios reached 6-8 times.

Since the algorithm is explicit and the solution value at each node on the next time layer is calculated as a linear combination of the solution values at the nodes on the previous time layer and this linear combination involves the solution values only in several nodes of the previous time layer, then the solution at each node can be computed independently, which allows the algorithm to efficiently parallelize. And the potential for increasing the algorithm speed due to parallelization on multiprocessor or multicore systems is obviously great. This aspect of the proposed algorithm should and will be investigated in further works on its improvement.

Summary
In this work, using the example of an initial boundary value problem for an equation describing constructively nonlinear oscillations of the floating bridge span structure with intermediate supports and restrictive rigid supports, a review and comparative analysis of algorithms for finding an approximate solution is carried out. All considered algorithms are based on the original initial-boundary value problem's reformulation in the form of a partial differential equations system for generalized functions. This approach in a weak form includes boundary and initial conditions in the equations for generalized functions. This, in turn, leads to the fact that there is no need to subordinate the basic functions, according to which the solution is expanded to the boundary and initial conditions. This aspect is very important, especially in case of problems with many spatial variables. The use of the generalized functions technique makes it possible to pose and solve the problems with point and even instant sources of forces acting on the system.

The algorithms in which the solution was approximated by the global basic functions, i.e. the functions which support coincided with the entire domain of the solution search, the algorithms based on finite piecewise linear basic functions as well as the algorithms based on the construction of a fundamental solution to the problem or the Green's matrix function have been investigated.

The algorithms on the global basic functions, for example, on Legendre polynomials or Chebyshev polynomials reduce the original problem to the Cauchy problem for the ODE system dynamics in time of the coefficients of expansion of the sought solution in terms of basic functions. The resulting algorithm is implicit, since the matrix at the derivatives is not unit. Such algorithms make it possible to find only rather rough approximations of the initial-boundary value problems. The attempts to achieve higher accuracy by increasing the number of basic functions lead to a deterioration in the conditionality of the problem and, as a consequence, to the emergence of computational instability. They are also quite slow due to the implicit algorithm.

The algorithms in which the solution is approximated by finite piecewise linear basis functions also reduce the original problem to the Cauchy problem for the ODE system, they are also implicit, since the matrix at the derivatives is not unit, however, they make it possible to achieve a significantly higher accuracy of the approximate solution by increasing the number of basic functions. An increase in the number of basic functions does not lead to computational instability. However, these algorithms, due to the fact that they are implicitly not fast enough, especially in comparison with the explicit algorithms based on the problem operator fundamental solution plotting.

The algorithms based on the problem operator fundamental solution plotting reduce the finding of the solution at any point on the next time level to explicit formulas for recalculating the linear combination of the solution at the previous time level in a small number of nearby nodes of the spatial grid. Traditionally, numerical algorithms, including those described above, are based on one or another approximation of the differential operators of the problem. The algorithm based on the fundamental solution of the operator of the problem is based on the approximation of explicit formulas expressing the solution of the initial initial-boundary value problem for the system of partial differential equations on the next time layer through the solution on the previous layer. The algorithm is explicit, it gives an opportunity to achieve the required accuracy by increasing the number of nodes of the spatial grid and its performance on the identical scenarios is multiples of the performance of the other algorithms presented. Also, the algorithm has a high potential for increasing performance due to parallelization.

The study makes it possible to conclude that for calculating the dynamics of the elastically supported structural systems' behavior with intermediate and restraining supports, especially in computing systems for mass industrial calculations, it is most expedient to use the algorithms based on plotting a fundamental solution to the operator of the problem.

References

[1] Skalko Yu I, Mendel M A 2013 Application of the apparatus of generalized functions for constructing approximate solutions of the radiation transfer problem *MIPT Proceedings* **5 (4)** 133-144.

[2] Skalko Yu I, Tsybulin I V, Karasev R N, Akopyan A V and Mendel M A 2014 Marching algorithm for solving the problem of radiation transfer by the method of short characteristics *Computer research and modeling* **6 (2)** 203–215.
[3] Skalko Yu I Correct conditions on the boundary dividing subdomains 2014 *Computer research and modeling* **6 (3)** 347-356.
[4] Skalko Yu I 2016 The Riemann problem of discontinuity decay in the case of many spatial variables *MIPT Proceedings* **8 (4)** 169-182.
[5] Gridnev S Yu, Skalko Yu I 2017 Numerical analysis of nonlinear vibrations of an elastically supported deformable system with limiting supports at the ends *Marine Intelligent Technology St. Petersburg* **3 (4) (38)** 37-45.
[6] Gridnev S Yu, Skalko Yu I, Ravodin I V and Yanaeva V V 2018 Simulation of vibrations of a continuously elastic supported rod with varying boundary conditions under the action of a movable *MATEC Web of Conferences 196, 01053 (2018) XXVII R-S-P Seminar 2018, Theoretical Foundation of Civil Engineering* **196** 01053.
[7] Skalko Yu I, Gridnev S Yu and Yagfarova A 2019 Modeling the Focusing of the Energy of Elastic Waves in Block-Fractured Medium During the Long-Term Vibration Action *Proc. Int. Scient. Conf. Energy Management of Municipal Facilities and Sustain-able Energy Technologies EMMFT 2018. Advances in Intelligent Systems and Computing vol. 983 eds. Murgul V and Pasetti M (Springer Nature Switzerland AG)* **983** 534-546.
[8] Gridnev S Yu, Skalko Yu I, Minaeva N V and Yanaeva V V 2019 Comparative analysis of models of limiting supports in the study of structurally nonlinear oscillations of elastically supported bar from mobile load *IOP Conf. Series: Journal of Physics: Conf. Series 1203 (2019)* 012030. doi:10.1088/1742-6596/1203/1/012030.
[9] Vlasov V Z 1963 Thin-walled elastic rods. Principles of construction of the general technical theory of shells (Selected Works Vol. 2, publishing house of the Academy of Sciences of the USSR).
[10] Vladimirov V S 1979 Generalized functions in mathematical physics (Nauka, Moscow, Vol. 4).
[11] Laursen T A, Chawla V 1997. Design of energy conserving algorithms for frictionless dynamic contact problems *Int. J. for Num. Meth. Eng* **40** 863-886.
[12] Skalko Yu I, Gridnev S Yu and Minaeva N V 2020 Application of a fundamental solution to a problem operator for modeling vibrations of elastically supported load-bearing structures *IOP Conf. Series: Journal of Physics: Conf. Series 2020* 012111.
[13] Skalko Yu I, Gridnev S Yu, Minaeva N V 2020 Application of the fundamental solution of the problem operator for modeling the dynamics of elastically supported structural systems *In the proceedings of the International Scientific Conference. Voronezh State University: Actual Problems of Applied Mathematics, Informatics and Mechanics* 1077-1087.

Calculation of the triangular panels' orthotropic covers for the retained domes

S V Skuratov*, V N Samsonova

Don State Technical University, 1 Gagarin sq., Rostov-on-Don, 344002, Russia

E-mail: svskuratov@yandex.ru

Abstract. The experience of design and construction of single-mesh spherical dome coverings based on meshes with triangular cells was studied. Constructive solutions of wooden domes made of triangular panels with orthotropic plywood covers are considered. The method analysis for constructing a bar model of an orthotropic plate in the form of an oblique triangle has been carried out. It is proposed to reduce the calculation of the triangular dome panel's rod-sheet structure to the calculation of the hinge-bar system. The stiffness matrices of the contour and intra-contour rods of the super cell are obtained.

Introduction

Bar spatial structures, forming a mesh system on the surface are widely used in modern construction practice. Single-layer structures have a curved mesh surface and are called single mesh. Two-layer structures have two parallel mesh surfaces connected by rigid bar ties. Such structures are called two-mesh structures. One of the effective forms of spatial structures are single-mesh spherical dome coverings made of metal and wood. World experience in design and construction shows that currently single-mesh domes are designed primarily on the basis of meshes with triangular cells. [1, 2, 3, 4, 5, 7, 11, 13]. Domestic and foreign experience of dome construction confirms the rationality of using single-mesh spherical dome coverings made of wood and materials based on it.

When calculating single-mesh domes, many authors use an approximate technique based on the use of a continuous computational model in the form of an isotropic spherical shell [8, 9, 10, 12, 14]. In this case, the assumption that the structural network of the dome is rather dense and therefore each elementary cell of it is an equilateral triangle, is introduced. In addition, in these works, the elementary cell of the dome's structural network, located in such a way on the surface of the sphere that its height turns out to be lying in the meridional plane, is considered. The disadvantage of this approach is quite obvious, since the junctions between the panels in general do not coincide with the parallels and meridians, and the cells themselves have the shape of an arbitrary oblique triangle. It should also be noted that within the dome panel, the cladding can be made of an orthotropic material, and in this case a situation when the main directions of elasticity do not coincide with the directions of the plate sides is possible. To eliminate the noted drawbacks, below is a solution to the problem of constructing a bar model of an orthotropic plate in the form of an oblique triangle, reinforced along the contour by ribs.

Materials and methods

Based on the method proposed in [6] for constructing a bar model of an orthotropic plate in the form of an oblique triangle, it seems possible to reduce the calculation of a bar-and-sheet structure of a polyhedral dome to the calculation of a hinge-bar system consisting of a finite number of super cells interconnected at nodes. We will assume that the super cell is a triangular dome panel, which includes an orthotropic plate and the contour edges of the panel frame that reinforce it. As a design model of a triangular orthotropic plate, an element in the form of an oblique triangle, referred to the rectangular coordinate system *XOY*, is taken. One of its vertices 1 coincides with the origin, and the side 1-2 is aligned with the direction of the axis 0Y (Figure 1). Contour bars of the model 1-2, 2-3 and 3-1 are pivotally connected at the vertices of the triangle. The elastic modulus of these bars will be considered equal to *E*, and the cross-sectional area is assumed to be the same and equal to A. Let the interior angles at vertices 1, 2 and 3 be respectively equal to θ_1, θ_2 *and* θ_3, and the lengths of the contour bars - L_1, L_2 *and* L_3, moreover $L_1 + L_2 + L_3 = 2S$. Point D is connected to the vertices of the triangle using the elastic internal contour bars AD, BD and CD, which length is respectively equal to l_1, l_2 *and* l_3. The cross-sectional areas of these bars are assigned so that

$$A_{\iota 1} : A_{\iota 2} : A_{\iota 3} = cos(\theta_1 /2) : cos(\theta_2/2) : cos(\theta_3/2). \quad (1)$$

Let us introduce the designation of the bars' stiffness that are a part of the super cell. Let the stiffness of the contour bars of the design model of the orthotropic plate be EA, and the stiffness of the internal contour bars - E_1A_i. To determine the values EA and E_1A_i it is recommended to use the appropriate formulas given in the above-mentioned work:

$$EA = E_x th_2 /2B, \quad (2)$$

$$E_1A_i = EA(4sL_3/L_1L_2)[(B - C)/D]\cos(\theta_i/2), \quad (3)$$

where $B = l_{31}^2(l_{31}^2 - l_{23}^2) + \left(\frac{E_x}{G_{xy}}\right) l_{31}m_{31}(l_{31}m_{31} - l_{23}m_{23}) + \nu_{xy}(l_{23}^2 m_{31}^2 + l_{31}^2 m_{23}^2 - 2l_{31}^2 m_{31}^2) + (E_x/E_y)\, m_{31}^2(m_{31}^2 - m_{23}^2)$,

$$C = (h_2/h_2)\{l_{23}^4 + [(E_x/G_{xy}) - 2\nu_{xy}]l_{23}^2 m_{23}^2 + (E_x/E_y)m_{23}^4\},$$

$$D = l_{12}^2[l_{12}^2 + (h_3/h_1)l_{23}^2 + (h_3/h_2)l_{31}^2] + (E_x/G_{xy})l_{12}m_{12}[l_{12}m_{12} + (h_3/h_1)l_{23}m_{23} + (h_3/h_2)l_{31}m_{31}] - \nu_{xy}\left[2l_{12}^2 m_{12}^2 + \left(\frac{h_3}{h_1}\right)(l_{12}^2 m_{23}^2 + l_{23}^2 m_{12}^2) + (h_3/h_2) \times (l_{12}^2 m_{31}^2 + l_{31}^2 m_{12}^2)\right] + (E_x/E_y)m_{12}^2 \times [m_{12}^2 + (h_3/h_1)m_{23}^2 + (h_3/h_2)m_{31}^2].$$

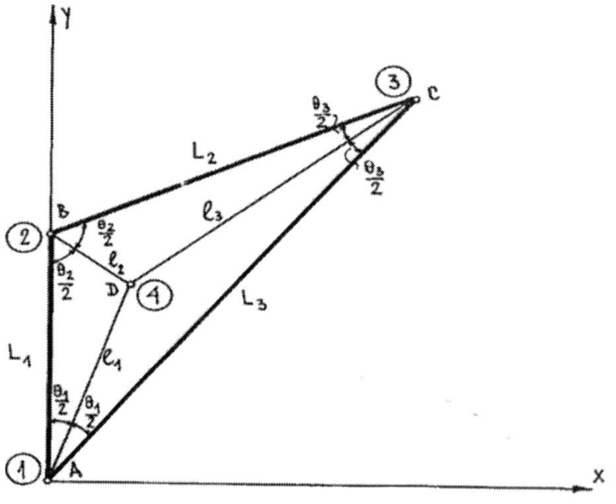

Figure 1. Design model of a triangular orthotropic plate.

Here, the following notation for the parameters of an orthotropic triangular plate is used:

E_x, E_y – are the elasticity moduli in the main directions X and Y;

G_{xy} – is the shear modulus in planes parallel to the median surface of the plate;
h_y – is the height dropped from the top i to the opposite side;
t – defines plate thickness;
l_{ij} and m_{ij} – are the direction cosines of the rib $i\ j$.
The stiffness of the contour ribs supporting the triangular covers of the dome panels is denoted by $E_s A_S$.

Results

When using the super cell approach for calculating domes, stiffness matrices of individual super cells are compiled, from which the stiffness matrix for the entire structure is formed. The expression for the stiffness matrix of the super cell contour bars, taking into account the stiffness of the contour ribs reinforcing the covering, is obtained in the form:

$$æ_{ks} = (1+k_1)\frac{E_sA_S}{L_1}\begin{bmatrix} a_{11} & a_{12} & a_{13} & a_{14} & a_{15} & a_{16} \\ & a_{22} & a_{23} & a_{24} & a_{25} & a_{26} \\ & & a_{33} & a_{34} & a_{35} & a_{36} \\ & & & a_{44} & a_{45} & a_{46} \\ & \text{symmetrically} & & & a_{55} & a_{56} \\ & & & & & a_{66} \end{bmatrix}, \tag{4}$$

$k_1 = EA/(E_sA_S)$. Matrix elements $æ_{ks}$ are calculated in the following way:

$a_{11} = -a_{15} = \frac{L_1}{L_3}\sin^2\theta_1$,

$a_{12} = -a_{16} = a_{25} = \frac{L_1}{L_3}\sin\theta_1\cos\theta_1$,

$a_{13} = a_{14} = a_{23} = 0$,

$a_{22} = 1 + \frac{L_1}{L_3}\cos^2\theta_1$,

$a_{24} = -1$,

$a_{26} = -\frac{L_1}{L_3}\cos^2\theta_1$,

$a_{33} = -a_{35} = \frac{L_1}{L_2}\sin^2\theta_2$,

$a_{34} = -a_{36} = -a_{45} = -\frac{L_1}{L_2}\sin\theta_2\cos\theta_2$,

$a_{44} = 1 + \frac{L_1}{L_2}\cos^2\theta_2$,

$a_{46} = -\frac{L_1}{L_2}\cos^2\theta_2$,

$a_{55} = a_{11} + a_{33}$,

$a_{56} = a_{12} + a_{34}$,

$a_{66} = -(a_{26} + a_{46})$.

Let us now proceed to the stiffness matrices construction of the model's internal contour bars (Figure 1). Let us express the displacement u_7 and u_8 points 4 through the displacements of the nodal points of the super cell 1, 2 and 3.

$$u_{7(8)} = au_{1(2)} + bu_{3(4)} + cu_{5(6)}, \tag{5}$$

where $$a = 1 - \frac{l_1}{L_1}\frac{\sin\theta_1/2}{\sin\theta_1}\left(1 + \frac{L_1}{L_3}\right); \tag{6}$$

$$b = \frac{l_1}{L_1}\frac{\sin\theta_1/2}{\sin\theta_1}; \tag{7}$$

$$c = \frac{l_1}{L_3}\frac{\sin\theta_1/2}{\sin\theta_1}, \tag{8}$$

The expression for the stiffness matrix of the internal contour bars has the form:

$$æ_i = k_2\,\frac{E_sA_s}{L_1}\begin{bmatrix} b_{11} & b_{12} & b_{13} & b_{14} & b_{15} & b_{16} \\ & b_{22} & b_{23} & b_{24} & b_{25} & b_{26} \\ & & b_{33} & b_{34} & b_{35} & b_{36} \\ & & & b_{44} & b_{45} & b_{46} \\ & symmetrically & & & b_{55} & b_{56} \\ & & & & & b_{66} \end{bmatrix}. \tag{9}$$

where $k_2 = E_1A_1L_1/(E_sA_sl_1)$.

The elements of the matrix expression included in (9) are as follows:

$b_{11} = (a-1)^2\sin^2\frac{\theta_1}{2} + a^2\left(d\sin^2\frac{\theta_2}{2} + f\cos^2\frac{\theta_2-\theta_1}{2}\right)$,

$b_{12} = \frac{1}{2}\{(a-1)^2\sin\theta_1 + a^2[f\sin(\theta_2-\theta_1) - d\sin\theta_2]\}$,

$b_{13} = ab\left(\sin^2\frac{\theta_1}{2} + d\sin^2\frac{\theta_2}{2} + f\cos^2\frac{\theta_2-\theta_1}{2}\right) - \left(b\sin^2\frac{\theta_1}{2} + ad\sin^2\frac{\theta_2}{2}\right)$,

$b_{14} = \frac{1}{2}\{ab[\sin\theta_1 - d\sin\theta_2 + f\sin(\theta_2-\theta_1)] + (ad\sin\theta_2 - b\sin\theta_1)\}$,

$b_{15} = ac\left(\sin^2\frac{\theta_1}{2} + d\sin^2\frac{\theta_2}{2} + f\cos^2\frac{\theta_2-\theta_1}{2}\right) - \left(c\sin^2\frac{\theta_1}{2} + af\cos^2\frac{\theta_2-\theta_1}{2}\right)$,

$b_{16} = b_{25} = \frac{1}{2}\{ac[\sin\theta_1 - d\sin\theta_2 + f\sin(\theta_2-\theta_1)] - [c\sin\theta_1 + a\sin(\theta_2-\theta_1)]\}$,

$b_{22} = (a-1)^2\cos^2\frac{\theta_1}{2} + a^2\left(d\cos^2\frac{\theta_2}{2} + f\sin^2\frac{\theta_2-\theta_1}{2}\right)$,

$b_{23} = \frac{1}{2}\{ab[\sin\theta_1 - d\sin\theta_2 + f\sin(\theta_2-\theta_1)] + (ad\sin\theta_2 - b\sin\theta_1)\}$,

$b_{24} = ab\left(\cos^2\frac{\theta_1}{2} + d\cos^2\frac{\theta_2}{2} + f\sin^2\frac{\theta_2-\theta_1}{2}\right) - \left(b\cos^2\frac{\theta_1}{2} + ad\cos^2\frac{\theta_2}{2}\right)$,

$b_{26} = ac\left(\cos^2\frac{\theta_1}{2} + d\cos^2\frac{\theta_2}{2} + f\sin^2\frac{\theta_2-\theta_1}{2}\right) - \left(c\cos^2\frac{\theta_1}{2} + a\sin^2\frac{\theta_2-\theta_1}{2}\right)$,

$b_{33} = d(1-b)^2\sin^2\frac{\theta_2}{2} + b^2\left(\sin^2\frac{\theta_1}{2} + f\cos^2\frac{\theta_2-\theta_1}{2}\right)$,

$b_{34} = \frac{1}{2}\{b^2[\sin\theta_1 + f\sin(\theta_2-\theta_1)] - d(1-b)^2\sin\theta_2\}$,

$b_{35} = bc\left(\sin^2\frac{\theta_1}{2} + d\sin^2\frac{\theta_2}{2} + f\cos^2\frac{\theta_2-\theta_1}{2}\right) - \left(cd\sin^2\frac{\theta_2}{2} + bf\cos^2\frac{\theta_2-\theta_1}{2}\right)$,

$b_{36} = b_{45} = \frac{1}{2}\{bc[\sin\theta_1 - d\sin\theta_2 + f\sin(\theta_2-\theta_1)] + [cd\sin\theta_2 - bf\sin(\theta_2-\theta_1)]\}$,

$b_{44} = d(1-b)^2\cos^2\frac{\theta_2}{2} + b^2\left(\cos^2\frac{\theta_1}{2} + f\sin^2\frac{\theta_2-\theta_1}{2}\right)$,

$b_{46} = bc\left(\cos^2\frac{\theta_1}{2} + d\cos^2\frac{\theta_2}{2} + f\sin^2\frac{\theta_2-\theta_1}{2}\right) - \left(cd\cos^2\frac{\theta_2}{2} + bf\sin^2\frac{\theta_2-\theta_1}{2}\right)$,

$b_{55} = f(1-c)^2\cos^2\frac{\theta_2-\theta_1}{2} + c^2\left(\sin^2\frac{\theta_1}{2} + d\sin^2\frac{\theta_2}{2}\right)$,

$b_{56} = \frac{1}{2}\{f(1-c)^2\sin(\theta_2-\theta_1) + c^2(\sin\theta_1 - d\sin\theta_2)\}$

$b_{66} = f(1-c)^2\sin^2\frac{\theta_2-\theta_1}{2} + c^2\left(\cos^2\frac{\theta_1}{2} + d\cos^2\frac{\theta_2}{2}\right)$.

The following notation is introduced here: $d = \frac{\sin\theta_2}{\sin\theta_1}$, $f = \frac{\sin\theta_3}{\sin\theta_1}$

Summary

The stiffness matrix of the super cell is written in its final form as the sum of two matrices (4) and (9):

$$æ_e = æ_{ks} + æ_i. \tag{10}$$

Having the stiffness matrices for triangular super cells in the local coordinate system XDY and using the congruent transformations, we further form the stiffness matrices of these elements in a common coordinate system. After constructing the stiffness matrix of the entire spatial configuration, the unknown displacements of nodes and stresses in the bar and plate elements of the multiple retained dome are determined. The validity of the proposed super cell approach is confirmed by the results obtained when solving the problem of calculating a dome for the vertical load action applied at its top [6].

References

[1] Ambratsumyan S A 1987 *Theory of Anisotropic Plates: Strength, Stability, Vibration* (Moscow, Nauka) 360.
[2] Evtushenko A I, Samsonova A N, Skuratov S V 2017 Formation of constructive networks of multiple retained non-sloping domes *Engineering Herald of Don* **1 (44)** 115.
[3] Ershov G F, Tavgen I A 1986 Stiffness matrix of the ribbed finite element of a spherical dome *Problems of construction and architecture* (Minsk, High School) 57-61.
[4] Zhuravlev A A, Nikolaenko L T, Rybina L A 1985 Investigation of the actual work of the triangular glued-veneer plate covering under transverse bending *Methods for calculating structures made of wood, plywood and plastics. SPSUACE,* 58-62.
[5] Zhuravlev A A, Skuratov S V 1988 On one way of representing an orthotropic triangular plate in the form of a bar model *Bulletin of the Universities. Construction and architecture* **5** 29-32.
[6] Zhuravlev A A 1989 Representation of an orthotropic triangular plate in the form of a hinge-bar system *Structural mechanics and calculation of structures* **1** 68-72.
[7] Zhuravlev A A, Muro G E, Kimsuore Long, Zhuravlev A A 2007 Bar structures of polyhedral domes (Rostov-on-Don, RSCEU) 318.
[8] Pshenichnov G I 1982 *The theory of exact elastic mesh shells and plates* (Moscow, Nauka, 1982) 352.
[9] Wright D T 1969 Large-span mesh shells *Large-span mesh shells* (Moscow) **1** 297-308.
[10] Savelyev V A 1966 Strength and stability of large-span metal mesh domes *Author's abstract. diss. Candidate of Engineering Sciences* (Moscow) 10.
[11] Skuratov S V, Skidan A A, Kevorkov A E 2019 Calculation of triangular covers of polyhedral mesh domes *Construction and industrial safety* **15 (67)** 21-24.
[12] Klöppel K, Schardt R Zur Berechnung von Netzkuppeln 1962 *Der Stahlbu* **5** 129-136.
[13] Skuratov S V, Shchutsky S V 2018 Coustructive Solutions of Bar Elements of a Poluhedral Wooden Dome *Materials Science Forum.* **931,** 300-303. ISSN: 1662-9752, doi: 10.4028.
[14] Wright D T 1965 Membran Forces and Bucklind Reticulated Shells *Journal of Structural Division* **91** 173-201.

IOP Conf. Series: Materials Science and Engineering 1083 (2021) 012004 doi:10.1088/1757-899X/1083/1/012004

Buckling of compressed rods taking into account their own weight under different boundary conditions

A S Chepurnenko*, S B Yazyev, D.A Vysokovskiy, G B Verzhbovskiy

Don State Technical University, 1 Gagarin sq., Rostov-on-Don, 344000

E-mail: anton_chepurnenk@mail.ru

Abstract. The article is devoted to the problem of calculating the stability of compressed rods taking into account their own weight. The fourth order resolving differential equation with respect to dimensionless deflection was obtained. A technique for the numerical solution of the resulting equation is proposed using finite differences method. As a result, the problem was reduced to a generalized eigenvalue problem. Comparison with the solution of other authors is given.

Introduction

Despite the fact that there is a significant amount of publications on the problem of stability of compressed rods, for example [1-5], the issues of taking into account their own weight are rarely considered.

We consider a homogeneous, straight, vertically located rod for various cases of fixing its ends under its own weight and compressive force (figure 1). In the case of small deviations from the rectilinear shape, the following relationship will take place between the bending moment and the deflection of the rod [6]:

$$EI\frac{d^2v}{dx^2} = M(x). \qquad (1)$$

where E – modulus of elasticity in tension and compression, I – axial moment of inertia, $M(x)$ – bending moment.

The bending moment for a section located at a distance x from the upper end of the rod will be written as:

$$M(x) = -Pv(x) - \int_0^x q\left[v(x) - v(t)\right]dt - Rx + M. \qquad (2)$$

Or

$$M(x) = -Pv(x) - qxv(x) + q\int_0^x v(t)dt - Rx + M. \qquad (3)$$

Substituting (3) into (1), we obtain

$$EI\frac{d^2v}{dx^2} + Pv(x) + qxv(x) - q\int_0^x v(t)dt = M - Rx. \qquad (4)$$

To obtain a solution that is valid for an arbitrary rod geometry and material, we turn to dimensionless quantities

$$\xi = x/l, \eta = t/l, y(\xi) = v(x)/l. \tag{5}$$

Then equation (4) will take the form:

$$\frac{d^2y}{d\xi^2} + \alpha^2 y(\xi) + \beta^3 \xi y(\xi) - \beta^3 \int_0^{\xi} y(\eta) d\eta = m - r\xi, \tag{6}$$

where $\alpha^2 = \frac{Pl^2}{EI}; \beta^3 = \frac{ql^3}{EI}, m = \frac{Ml}{EI}, r = \frac{Rl^2}{EI}$.

To eliminate the quantities m and r, we twice differentiate the equation (6) with respect to ξ :

$$\frac{d^4y}{d\xi^4} + \alpha^2 \frac{d^2y}{d\xi^2} + \beta^3 \left[\frac{dy}{d\xi} + \xi \frac{d^2y}{d\xi^2} \right] = 0. \tag{7}$$

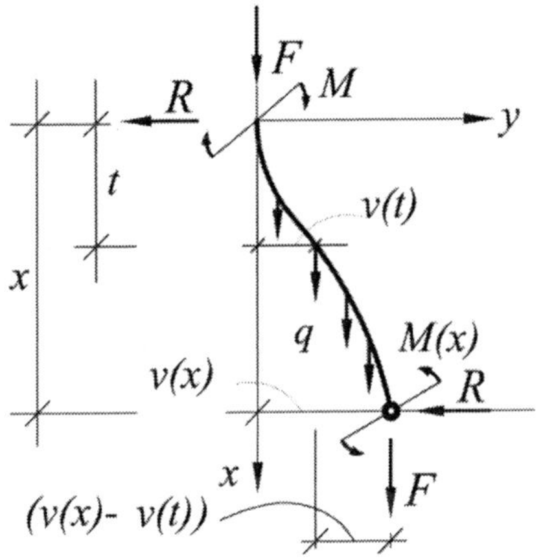

Figure 1. Calculation scheme

Methods

The solution of equation (7) is performed by the method of finite differences. On the interval $\xi \in [0;1]$, a uniform mesh with a step $\Delta\xi$ is introduced. For derivatives, the following approximation is used:

$$\left.\frac{d^4y}{d\xi^4}\right|_i = \frac{y_{i+2} - 4y_{i+1} + 6y_i - 4y_{i-1} + y_{i-2}}{\Delta\xi^4};$$

$$\left.\frac{d^2y}{d\xi^2}\right|_i = \frac{y_{i+1} - 2y_i + y_{i-1}}{\Delta\xi^2}; \tag{8}$$

$$\left.\frac{dy}{d\xi}\right|_i = \frac{y_{i+1} - y_{i-1}}{2\Delta\xi}.$$

The boundary conditions in the case of hinged fastening of both ends of the rod (figure 2) have the form:

$$y(0) = y(1) = 0; \quad y''(0) = y''(1) = 0. \tag{9}$$

When the upper end of the rod is hinged and the lower end is rigidly clamped (figure 3), the boundary conditions have the form:

$$y(0) = y(1) = 0; \quad y''(0) = 0; \quad y'(1) = 0. \tag{10}$$

IOP Conf. Series: Materials Science and Engineering 1083 (2021) 012004 doi:10.1088/1757-899X/1083/1/012004

If the rigid pinch is at the top and the hinge is at the bottom (figure 4), the boundary conditions take the form:

$$y(0) = y(1) = 0; \quad y'(0) = 0; \quad y''(1) = 0. \tag{11}$$

If both ends of the rod are rigidly clamped (figure 5), the boundary conditions have the form

$$y(0) = y(1) = 0; \quad y'(0) = y'(1) = 0. \tag{12}$$

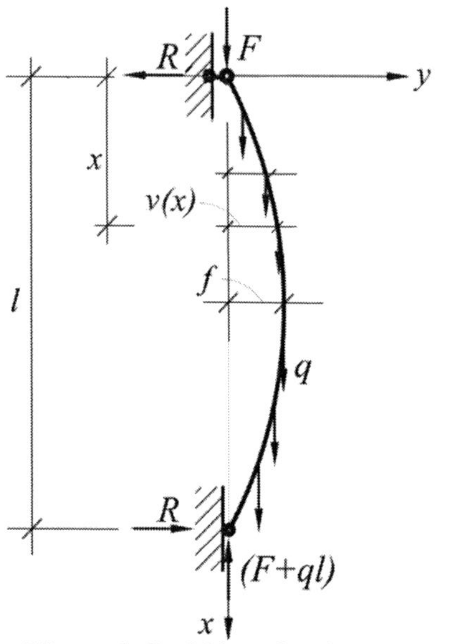

Figure 2. Rod hinged at the ends

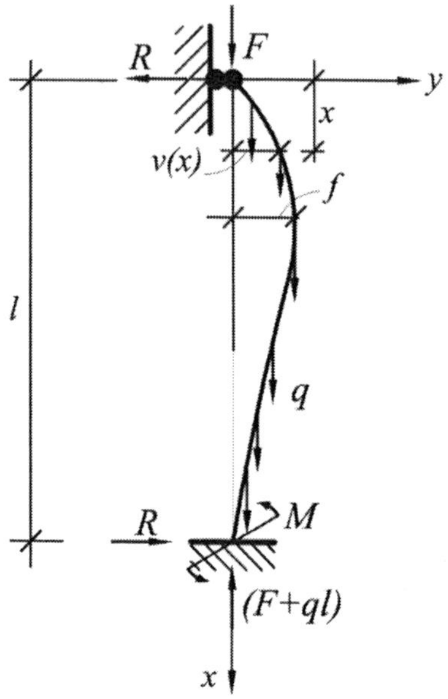

Figure 3. A rod, the upper end of which is hinged, and the lower end is rigidly clamped

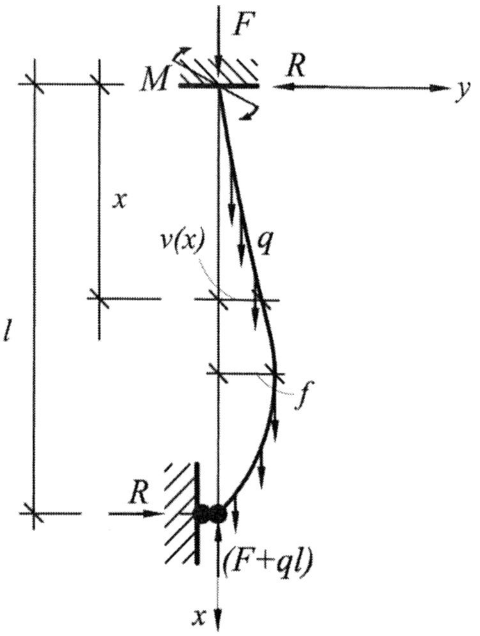

Figure 4. A rod, the upper end of which is rigidly clamped, and the lower end is hinged

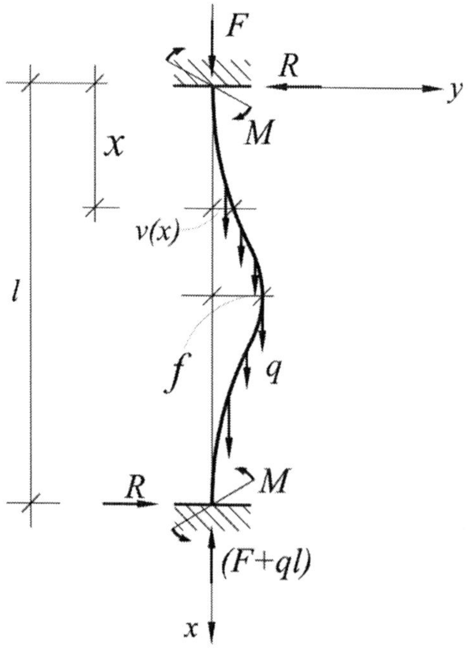

Figure 5. Rod with rigidly clamped ends

When the derivatives in (7) are replaced by approximate expressions according to formula (8) for each internal node of the mesh, the problem is reduced to a system of linear algebraic equations of the form

$$([A]+\alpha^2[B]+\beta^3[C])\{X\}=0, \tag{13}$$

where $\{X\}=\{y_2 \quad y_3 \quad \dots \quad y_{n-1} \quad y_n\}^T$ – vector of dimensionless displacements of nodes.
Matrices $[A]$, $[B]$, $[C]$ have the form

$$[A]=\frac{1}{\Delta\xi^4}\begin{bmatrix} a_{22} & -4 & 1 & 0 & 0 & 0 & \dots & 0 & 0 & 0 & 0 & 0 & 0 \\ -4 & 6 & -4 & 1 & 0 & 0 & \dots & 0 & 0 & 0 & 0 & 0 & 0 \\ 1 & -4 & 6 & -4 & 1 & 0 & \dots & 0 & 0 & 0 & 0 & 0 & 0 \\ 0 & 1 & -4 & 6 & -4 & 1 & \dots & 0 & 0 & 0 & 0 & 0 & 0 \\ \dots & \dots & \dots & \dots & \dots & \dots & \dots & \dots & \dots & \dots & \dots & \dots & \dots \\ 0 & 0 & 0 & 0 & 0 & 0 & \dots & 1 & -4 & 6 & -4 & 1 & 0 \\ 0 & 0 & 0 & 0 & 0 & 0 & \dots & 0 & 1 & -4 & 6 & -4 & 1 \\ 0 & 0 & 0 & 0 & 0 & 0 & \dots & 0 & 0 & 1 & -4 & 6 & -4 \\ 0 & 0 & 0 & 0 & 0 & 0 & \dots & 0 & 0 & 0 & 1 & -4 & a_{nn} \end{bmatrix},$$

$$[B]=\frac{1}{\Delta\xi^2}\begin{bmatrix} -2 & 1 & 0 & 0 & 0 & \dots & 0 & 0 & 0 & 0 \\ 1 & -2 & 1 & 0 & 0 & \dots & 0 & 0 & 0 & 0 \\ 0 & 1 & -2 & 1 & 0 & \dots & 0 & 0 & 0 & 0 \\ \dots & \dots & \dots & \dots & \dots & \dots & \dots & \dots & \dots & \dots \\ 0 & 0 & 0 & 0 & 0 & \dots & 1 & -2 & 1 & 0 \\ 0 & 0 & 0 & 0 & 0 & \dots & 0 & 1 & -2 & 1 \\ 0 & 0 & 0 & 0 & 0 & \dots & 0 & 0 & 1 & -2 \end{bmatrix},$$

$$[C]=\frac{1}{\Delta\xi^2}\begin{bmatrix} -2\xi_2 & \xi_2+\frac{\Delta\xi}{2} & 0 & 0 & 0 & \dots & 0 & 0 & 0 & 0 \\ & & & & & & & 0 & 0 & 0 \\ \xi_3-\frac{\Delta\xi}{2} & -2\xi_3 & \xi_3+\frac{\Delta\xi}{2} & 0 & 0 & \dots & 0 & 0 & 0 & 0 \\ & & & & & & & \dots & \dots & \dots \\ 0 & \xi_4-\frac{\Delta\xi}{2} & -2\xi_4 & \xi_4+\frac{\Delta\xi}{2} & 0 & \dots & 0 & & & \\ & & & & & & & -2\xi_{n-2} & \xi_{n-2}+\frac{\Delta\xi}{2} & 0 \\ \dots & \dots & \dots & \dots & \dots & \dots & \dots & & & \\ 0 & 0 & 0 & 0 & 0 & \dots & \xi_{n-2}-\frac{\Delta\xi}{2} & \xi_{n-1}-\frac{\Delta\xi}{2} & -2\xi_{n-1} & \xi_{n-1}+\frac{\Delta\xi}{2} \\ 0 & 0 & 0 & 0 & 0 & \dots & 0 & 0 & \xi_n-\frac{\Delta\xi}{2} & -2\xi_n \\ 0 & 0 & 0 & 0 & 0 & \dots & 0 & & & \end{bmatrix},$$

where $\xi_i=(i-1)\Delta\xi$.

The coefficient a_{22} is equal to 5 if the upper end of the rod is hinged and 7 if it is rigidly clamped. The a_{nn} coefficient for the lower end is determined similarly.

The system (13) is homogeneous and has a nonzero solution if its determinant is equal to zero:

$$\left|[A]+\alpha^2[B]+\beta^3[C]\right|=0. \tag{14}$$

For a known α, the problem of finding $\lambda=\beta^3$ is a generalized eigenvalue problem. Similarly, knowing β, it is not hard to find α.

Results and Discussion

Figure 6 shows the stability region as a dependence of the β parameter on α for a rod hinged at the ends. The dashed line in this graph shows the approximate formula proposed by V.P. Belousov [7]:

$$\alpha^2 + 0.532\beta^3 = \pi^2 . \tag{15}$$

It can be seen from the presented graph that the deviation of the results is insignificant. Figures 7-9 are the same for rods fixed according to fig 3-5. These graphs also show approximate formulas proposed in [7]. The results almost coincide.

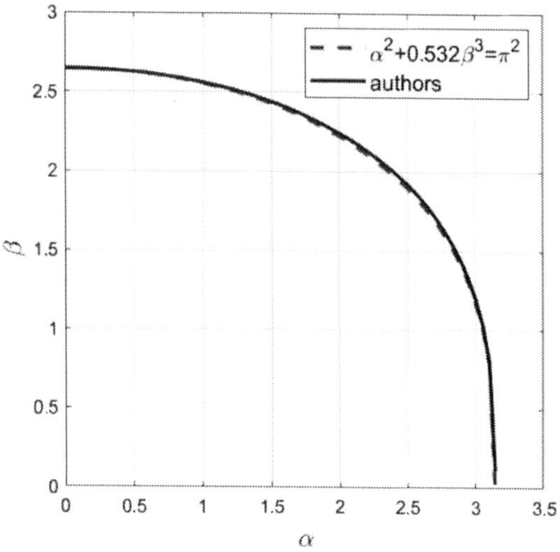

Figure 6. The stability region for a rod hinged at the ends

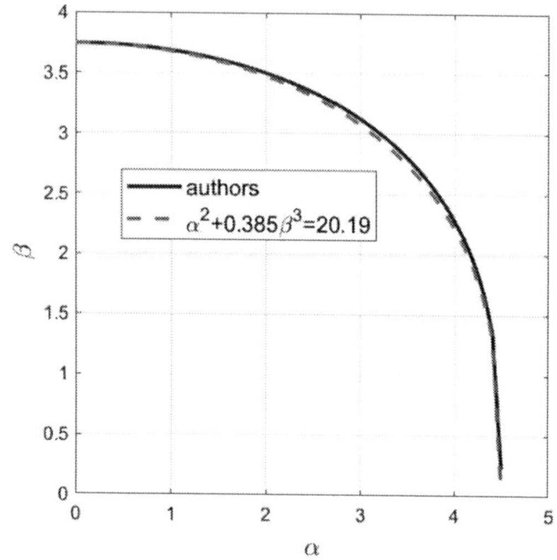

Figure 7. The stability region for a rod the upper end of which is hinged, and the lower end is rigidly clamped

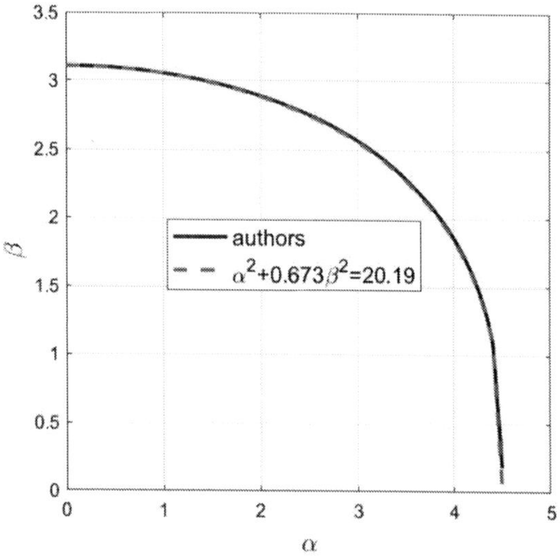

Figure 8. The stability region for a rod the upper end of which is rigidly clamped, and the lower end is hinged

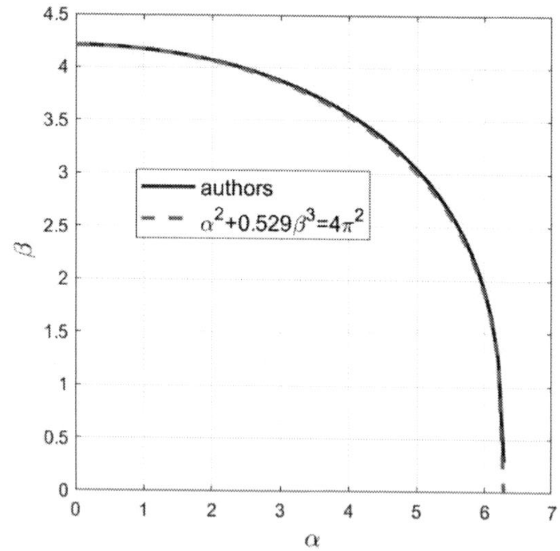

Figure 9. The stability region for a rod with rigidly clamped ends

Summary

The proposed technique, in contrast to analytical solutions, makes it possible to solve the problem with arbitrary fixing of the ends of the rod. It is also possible to take into account the length-varying stiffness and heterogeneity of the rod. The test problems showed good agreement with the literature data. In the future, it is planned to develop a calculation method taking into account creep

References

[1] Chepurnenko V, Yazyev B, Dubovitskaya L 2020 *E3S Web of Conferences* **164** 02003
[2] Litvinov S V, Klimenko E S, Kulinich I I, Yazyeva S B 2015 *International Polymer Science and Technology* **42** 23-25
[3] Andreev V I, Tsybin N Y 2015 *Procedia Engineering* **111** 42-48
[4] Andreev V I, Barmenkova E V 2014 *Procedia Engineering* **91** 20-25
[5] Deng J W, Xie C, Pandey M D 2014 *Journal of Sound and Vibration* **333** 6121-6139
[6] Yazyev S B, Ivanova P V, Konoplev Y G, Yazyev B M 2020 *IOP Conference Series: Materials Science and Engineering* **913 (2)** 022021
[7] Belousov V P 1983 *Investigation of the stability of rods under longitudinal and transverse loading: dissertation for the degree of candidate of technical sciences* (Dzhambul) 213.

IOP Conf. Series: Materials Science and Engineering 1083 (2021) 012005 doi:10.1088/1757-899X/1083/1/012005

Energy method for solving the stability problem of an elastic rod with variable stiffness taking into account its dead weight

S B Yazyev*, L Sh Akhtyamova, V S Chepurnenko, A A Reshetnikov, A D Ziganshin

Don State Technical University, 1, Gagarin Square, Rostov-on-Don, 344000, Russia

E-mail: serdariaziev@gmail.com

Abstract. The article is devoted to solving the stability problem of an elastic rod with variable stiffness taking into account its dead weight. The variable stiffness parameter is the rod section, in particular, the moment of inertia $I_z(x)$. The moment of the section inertia and the intensity of the distributed load (dead weight) vary according to the power law. As a result, the solution of the resolving integro-differential equation for this model (cantilever rod) by the energy method was obtained. It is shown that the obtained solution allows satisfying all boundary conditions and another approach to the energy method use is found. The energy method in this case made it possible to determine the exact values of the critical load with a particular change in the rod cross-section taking into account its dead weight.

1. Statement of the problem. Resolving Equations

The problems of rods stability in various formulations were considered in the works by the Russian and foreign authors [1-7]. But the methodology and problem setting are different.

Let us consider a rigidly clamped rectilinear rod located vertically (Figure 1). The cross-section of the rod will act as a parameter of variable stiffness, in particular, the moment of inertia $I_z(x)$. In the case of small deviations from the rectilinear form, the differential equation of the curved axis will have the following form:

$$\frac{d^2v}{dx^2} = \frac{M(x)}{EI_z(x)} \tag{1}$$

Let the moment of the section inertia change according to the power law

$$I_z(x) = a_1 x^n \tag{2}$$

The figure below shows a general view of the rod model with a uniformly distributed load $q = q(x)$.

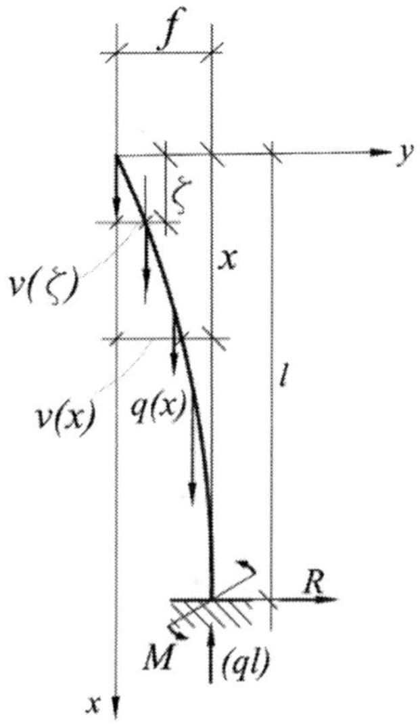

Figure 1. General view of the rod model with its dead weight

Distributed load intensity (dead weight)

$$q(\zeta) = a_2\zeta^n \tag{3}$$

here $a_1, a_2, m > 0, n > 0$ - are some constants.

Load element $qd\zeta$ causes bending moment $q \cdot [v(x) - v(\zeta)]d\zeta$ in the section with abscissa x. The total bending moment in this section is (Figure 1):

$$M(x) = -\int_0^x q(\zeta) \cdot [v(x) - v(\zeta)]d\zeta \tag{4}$$

Or, integrating the right-hand side by parts, the expression (4) taking into account (3) has the form:

$$M(x) = -\int_0^x \frac{a_2}{(n+1)E}\zeta^{n+1}\frac{dv(\zeta)}{d\zeta}d\zeta \tag{5}$$

Taking into account (5) and the expression for the moment of inertia (2), the ratio (1) takes the form:

$$x^m\frac{d^2v}{dx^2} + \int_0^x \frac{a_2}{(n+1)a_1E}\zeta^{n+1}\frac{dv(\zeta)}{d\zeta}d\zeta = 0 \tag{6}$$

or

$$\frac{d^2v}{dx^2} + \int_0^x \frac{a_2}{(n+1)a_1E}\frac{\zeta^{n+1}}{x^m}\frac{dv(\zeta)}{d\zeta}d\zeta = 0 \tag{7}$$

Further, it is convenient to consider the resolving equations in dimensionless quantities. Let us make the following substitution:

$$x = a\xi; \quad \zeta = a\psi; \quad v(x) = aw(x) \tag{8}$$

Then

$$\frac{d^2w}{d\xi^2}+\int_0^x \frac{a_2 a^{n+3-m}}{(n+1)a_1 E}\frac{\psi^{n+1}}{\xi^m}\frac{dw}{d\psi}d\psi=0 \tag{9}$$

We take the following notation:

$$a=\sqrt[\alpha]{\frac{(n+1)a_1 E}{a^2}} \tag{10}$$

where

$$\alpha=n+3-m, \text{ moreover } \alpha\geq 1 \tag{11}$$

since otherwise the ratios (10) and (8) lose their meaning.
Substituting the value of the ratio (10) into (9), we obtain the integro-differential equation:

$$\frac{d^2w}{d\xi^2}+\int_0^\xi \frac{\psi^{n+1}}{\xi^m}\frac{dw}{d\psi}d\psi=0 \tag{12}$$

The equation (12) solution in the form of a series does not give positive results, since for $\boldsymbol{x=0,\xi=0}$ (Figure 1) it turns to infinity. Below we propose a solution to this problem by the energy method.

2. Method of solution by energy method

The essence of this method is to study the change in the total energy of the system during the transition from the initial form of equilibrium to the adjacent form, i.e., in disturbed state. It is very important to choose the curved axis equation here. It can be multi-parameter.

For example, S.P. Timoshenko [1] suggests the following expression (Figure 1) for our case:

$$v=\delta_1\left(1-cos\frac{\pi x}{2l}\right)+\delta_2\left(1-cos\frac{3\pi x}{2l}\right) \tag{13}$$

Let us consider our case. The critical length and critical value of the external load is determined from the condition of the potential energy equality and the rod' dead weight work:

$$U=A \tag{14}$$

where the potential energy of the system

$$U=\int_0^l \frac{M^2(x)}{2EI_z(x)}dx \tag{15}$$

Or, taking into consideration (2) and (5)

$$U=\int_0^l \frac{1}{2a_1 E x^m}\left(\int_0^\zeta \frac{a_2}{(n+1)}\zeta^{n+1}\frac{dv(\zeta)}{d\zeta}d\zeta\right)^2 dx \tag{16}$$

The work with evenly distributed load (dead weight)

$$A=\int_0^l \frac{R}{2}\left(\frac{dv}{dx}\right)^2 dx \tag{17}$$

Here

$$R = \int_0^{\zeta} q(\zeta)d\zeta$$

According to the formula (3), the last expression has the form:

$$R = \frac{a_2}{(n+1)}x^{n+1}$$

Finally, the work from its dead weight is:

$$A = \int_0^{l} \frac{a_2}{2(n+1)}x^{n+1}\left(\frac{dv}{dx}\right)^2 dx \tag{18}$$

We take the following dependence ss an equation for the curved axis of the rod:

$$v(x) = \frac{k \cdot f}{(k-1)l^k}\left(l^{k-1}x - \frac{x^k}{k}\right), \tag{19}$$

where f is the maximum deflection value (Figure 1).
The equation (19) satisfies all boundary conditions at $k > 1$.
Let us introduce the designations

$$C = \frac{k \cdot f}{(k-1)l^k} \tag{20}$$

Ratio (19)

$$v(x) = C\left(l^{k-1}x - \frac{x^k}{k}\right), \tag{21}$$

After differentiating once (21):

$$\frac{dv(x)}{dx} = C\left(l^{k-1} - x^{k-1}\right), \tag{22}$$

Having substituted (22) into the ratios (16) and (18), taking into account (14), we obtain:

$$\frac{a_2 l^{n+3-m}}{(n+1)Ea_1}\left(\frac{1}{(n+k+1)^2(2n+2k+3-m)} - \frac{2}{(n+k+1)(n+2)(2n+k+4-m)} + \frac{1}{(n+2)^2(2n+5-m)}\right) = \left(\frac{1}{(n+2)} - \frac{2}{(n+k+1)} + \frac{1}{(n+2k)}\right) \tag{23}$$

Let us present the model problems corresponding to the condition $n + 3 - m = 0$

3. The problem solution

Task 1. $n = 3$ и $m = 6$

The cross-sectional dimensions correspond to the rod in Figure 2. Then the resolving equation (23) is simplified to the form [2, 7,8,9]:

$$\frac{a_2}{4Ea_1}\left(\frac{1}{(k+4)^2(2k+3)} - \frac{2}{(k+4)^2 \cdot 5} + \frac{1}{125}\right) = \left(\frac{1}{5} - \frac{2}{(k+4)} + \frac{1}{(2k+3)}\right) \tag{24}$$

Varying the values from $k = 3$ till $k \to 1$ in the equation (24) we obtain: (Table 1). That is, at $k \to 1$ the ratio a_2/a_1 tends to $40{,}0E$

Table 1.

No.	k	a_2/a_1
1	$k = 3$	$a_2/a_1 = 48.28E$
2	$k = 2$	$a_2/a_1 = 44.44E$

3	$k = 1.5$	$a_2/a_1 = 42.31E$
4	$k = 1.1$	$a_2/a_1 = 40.48E$
5	$k = 1.01$	$a_2/a_1 = 40.05E$

a) for a rod with a square cross section (Figure 2a).

$$b(x) = b\sqrt{\left(\frac{x}{l}\right)^3} \tag{25}$$

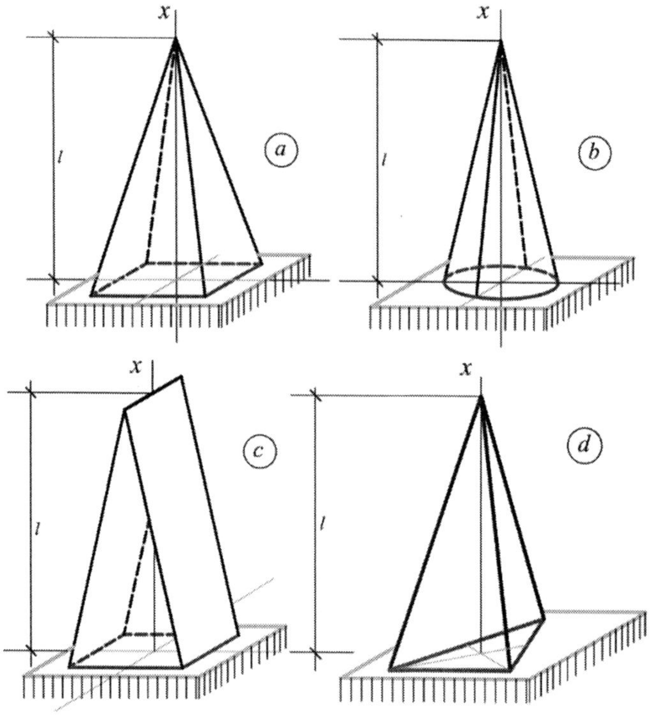

Figure 2. Variants of cross-sections of a rod with variable stiffness

Then, the moment of inertia $I_z(x)$ and the intensity of the distributed load $q(x)$:

$$I_z(x) = a_1x^n \rightarrow a_1 = \frac{b^4}{12l^6} \quad A_{\text{сеч}}\gamma = a_2x^n \rightarrow \quad a_2 = \frac{\gamma b^2}{l^3} \tag{26}$$

Then, substituting the coefficients (2*b*) into the ratio $a_2/a_1 = 40{,}0E$

$$l_{kp} = 1{,}49\sqrt[3]{\frac{b^2E}{\gamma}}\ ; \quad N_{kp} = 0{,}37b^2\sqrt[3]{b^2E\gamma^2}\ ; \quad \sigma_{kp} = 0{,}37\sqrt[3]{b^2E\gamma^2} \tag{27}$$

b) for a rod with round diameter d (Figure 2*b*).

$$I_z(x) = a_1x^n \rightarrow a_1 = \frac{\pi d^4}{64l^6} \quad A_{\text{сеч}}\gamma = a_2x^n \rightarrow \quad a_2 = \frac{\gamma\pi d^2}{4l^3} \tag{28}$$

Then, substituting the coefficients (28) into the ratio $a_2/a_1 = 40{,}0E$

$$l_{kp} = 1{,}36\sqrt[3]{\frac{d^2E}{\gamma}}\ ; \quad N_{kp} = 0{,}28d^2\sqrt[3]{d^2E\gamma^2}\ ; \quad \sigma_{kp} = 0{,}34\sqrt[3]{d^2E\gamma^2} \tag{29}$$

c) for a rod with a cross-section in the form of an equilateral triangle with a side b (Figure 2c).

$$I_z(x) = a_1 x^n \rightarrow a_1 = \frac{\sqrt{3}b^4}{96l^6} \quad A_{\text{сеч}}\gamma = a_2 x^n \rightarrow \quad a_2 = \frac{\gamma\sqrt{3}b^2}{4l^3} \tag{30}$$

Then, substituting the coefficients (30) into the ratio $a_2/a_1 = 40{,}0E$

$$l_{kp} = 1{,}18\sqrt[3]{\frac{b^2E}{\gamma}}\,; \quad N_{kp} = 0{,}13b^2\sqrt[3]{b^2E\gamma^2}\,; \quad \sigma_{kp} = 0{,}30\sqrt[3]{b^2E\gamma^2} \tag{31}$$

Task 2. $n = 1{,}5$ and $m = 4$,

With such values n and m, one side of a rectangular cross section is $b = const$, and the other side is determined by (25) and corresponds to the Figure 2*d*. The equation (23) has the following form:

$$\frac{a_2}{a_1}\frac{1}{2{,}5E}\left(\frac{1}{(k+2{,}5)^2(2k+1{,}5)} - \frac{2}{(k+2{,}5)^2 3{,}5} + \frac{1}{42{,}875}\right) = \\ = \left(\frac{1}{3{,}5} - \frac{2}{(k+2{,}5)} + \frac{1}{(2k+1{,}5)}\right) \tag{32}$$

So, with $k \rightarrow 1$ the ratio a_2/a_1 tends to $12{,}25E$

$$\frac{a_2}{a_1} = 12{,}25E \tag{33}$$

Wherein

$$I_z(x) = a_1 x^n \rightarrow a_1 = \frac{b^4}{12\sqrt{l^9}} \quad A_{\text{сеч}}\gamma = a_2 x^n \rightarrow \quad a_2 = \frac{\gamma b^2}{4\sqrt{l^3}} \tag{34}$$

Substituting (34) in (33)

$$l_{kp} = \sqrt[3]{\frac{b^2E}{\gamma}}\,; \quad N_{kp} = 0{,}4b^2\sqrt[3]{b^2E\gamma^2}\,; \quad \sigma_{kp} = 0{,}4\sqrt[3]{b^2E\gamma^2} \tag{35}$$

It is not difficult to prove that for $k \rightarrow 1$ the ratio a_2/a_1 tends to exact eigenvalue. Let us consider the rod in Figure 2 as an example. Here

$n = 2$ and $m = 4$. Then the ratios (23) take the form:

$$\frac{a_2}{a_1}\frac{l}{3E}\left(\frac{1}{(k+3)^2(2k+3)} - \frac{2}{(k+4)\cdot 3{,}0\cdot(k+3)} + \frac{1}{80}\right) = \left(\frac{1}{4} - \frac{2}{(k+3)} + \frac{1}{(2k+2)}\right) \tag{36}$$

We will tabulate the approximations into Table 2.

Table 2.

No.	k	a_2/a_1
1	$k = 3$	$a_2/a_1 = 33.95E/l$
2	$k = 2$	$a_2/a_1 = 32.31E/l$
3	$k = 1{,}1$	$a_2/a_1 = 31.03E/l$
4	$k \rightarrow 1$	$a_2/a_1 = 30.54E/l$

Having determined the coefficients (a_1), (a_2), (α) by the formulas (2), (3) and (11) we obtain the ratio

$$l_{kp} = x_1\sqrt[\alpha]{\frac{(n+1)a_1E}{a_2}}$$

From this we conclude that, it can be assumed, that the ratios (27), (29), (31), (35) should be considered exact.

Summary

Analyzing the solved problems, we can conclude that another approach to the use of the energy method has been found. Varying the coefficient k, it is possible to obtain the ratio (21), which will correlate to the real equation of the rod curved axis and then using the energy method without much effort, the exact eigenvalue of the critical force is determined with a special change in the cross section of the rod and taking into account its dead weight.

The developed methods are much simpler than the classical methods for solving the equations of this type and can be used in other areas of mechanics.

References

[1] Timoshenko S 1929 Stabilitätsprobleme der Elastizitat *Handb. der Phys. und techn. Mech.* 81–145.
[2] Belousov V P 1983 *Investigation of the stability of rods under longitudinal and transverse loading:dissertation for the degree of candidate of technical sciences* (Dzhambul) 213.
[3] Yazyev B, Chepurnenko A and Savchenko A 2018 *Materials Science Forum* **935** 144.
[4] Yazyev S B., Zotov I M, Lapina A P, Chepurnenko A S, Vysokovskiy D A 2020 The energy method for calculating the cantilever strip bending flat form stability taking into account its own weight development *IOP Conference Series: Materials Science and Engineering* **913** (2) 022023.
[5] Yazyev S B 2020 Energy method in the stability calculation of the cantilever strip plane bending taking into account its dead weight *Building materials and products* **3** **(1)** 76-82.
[6] Prantdle L 1899 Kipperscheinungen (Dissert, Nurnberg).
[7] Chepurnenko V, Yazyev B, Dubovitskaya L 2020 *E3S Web of Conferences* **164** 02003.
[9] Deng J W, Xie C, Pandey M D 2014 *Journal of Sound and Vibration* **333** 6121-6139.
[10] Yazyev S B, Ivanova P V, Konoplev Y G, Yazyev B M 2020 *IOP Conference Series: Materials Science and Engineering* **913** **(2)** 022021.

IOP Conf. Series: Materials Science and Engineering 1083 (2021) 012006 doi:10.1088/1757-899X/1083/1/012006

Comparative analysis of typical and reinforced support parts' design parameters for reinforced concrete roof beams

G V Degtyarev*, S K Saida, Ahmad Al Hajal

Kuban state agrarian University named after I. T. Trubilin, 13 Kalinina str., Krasnodar, 350044, Russia

E-mail: cst2007@mail.ru

Abstract. Reinforced concrete roof beams are the structural elements of industrial buildings that are most susceptible to destruction in emergency situations and abnormal operation are observed; the schemes for strengthening beams are proposed and the parameters' analysis in comparison with a typical solution is given.

Introduction

The covering beam outcome into the emergency category, occurs mainly due to the loss of the support part bearing capacity [1, 2, 3, 4].

Currently, the issue of strengthening the support parts of reinforced concrete roof beams is mainly solved by means of a device containing a tension device and unloading brackets installed on the support part of the beam on both sides and connected to each other; made of two rod elements forming an acute angle, when the end of one rod element is connected to the top of the beam, and the end of the other element is connected to the bottom [5, 6]. The opposite ends of the rods are joined by the support plates located on both sides of the beam and connected by the strips installed at the end of the beam. The connection of the bar elements with the upper and lower parts of the beam is made by means of strips forming a box filled with concrete. The tensioning device is made in the form of a screw pair of a coupling and a stop pin. The sleeve is attached to each support plate, and the pin-stop is installed on the supporting structure on which the beam leans [7, 8]. Shaft elements of relief brackets, support plates and strips are located with a gap from the beam [9, 10].

However, this device does not effectively use the capabilities of the unloading brackets due to the lack of connection between them and the impossibility of transferring forces to each other [11, 12]. Because of this drawback, it is necessary to assign the dimensions of the brackets according to the maximum load on each element, and this is the additional weight of the device and the load on the beam.

It is proposed to strengthen the support parts of reinforced concrete beams of the covering, on both sides of the beam to install the unloading brackets in the form of a rod elements' pair forming an acute angle, each end of rod elements located horizontally to connect to the bottom of the beam, and each end of rod elements located at an acute angle - with the upper part of the beam, the opposite ends of the bar elements are connected with the support plates located on both sides of the beam and connected by the strips [13]. Between the rod elements, mirrored on both sides additional tensioning devices connected to the rod elements, rigidly - with the upper element and flexibly, through a threaded connection, with the lower element, and the rod elements are rigidly connected to the upper and lower parts reinforced concrete girder, made in the form of anchors, are installed, as shown in Figure 1 for a duo-pitch roof beam. It is precisely due to the creation of adjustable forces in additional tensioning devices that an increase in the bearing capacity and reliability of the supporting parts of reinforced concrete beams is achieved, with the provision of the loads redistribution possibility not only between the rods of the unloading brackets, but also between the reinforcing structure and the reinforced concrete beam.

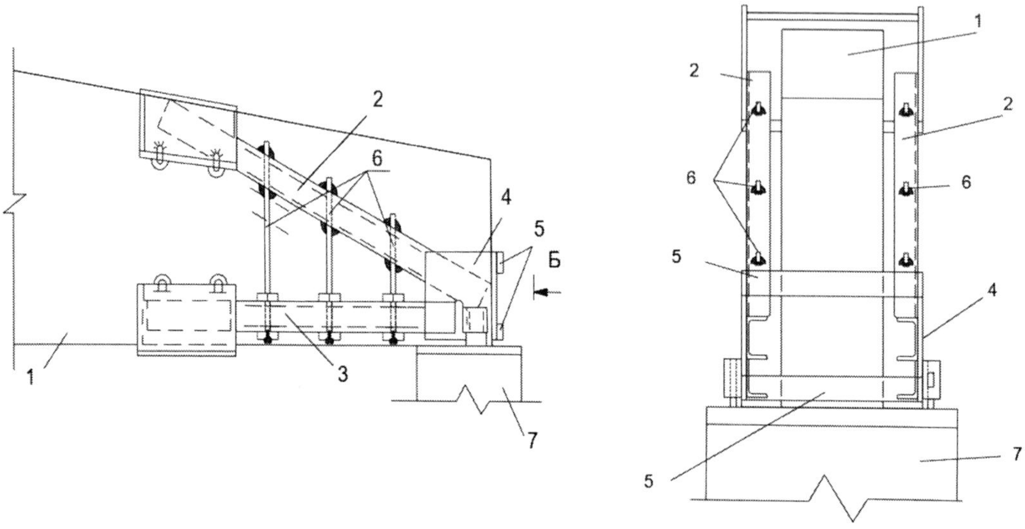

Side view View B

1 – reinforced concrete beam; 2, 3 – bar elements; 4 – base plates; 5 – battens; 6 – tensioners; 7 – column

Figure 1. Device for strengthening the supporting parts of reinforced concrete roof beams.

Materials and technique

In order to analyze the effect of unloading brackets, in the form of rod elements' pairs made of steel channels, installed on a beam with parallel chords of a square cross section, on its supporting parts operation, numerical modeling was carried out in the software package Etabs 2019 [14]. Figure 2 shows the sections of a reinforced concrete beam with a length L=6 m, and Figure 3 shows the dimensions of elements and parts.

As a result of numerical modeling, realized with a stepwise increase in concentrated loads F1 and F2, acting at a distance of a quarter span from each support, the values of bending moments were obtained, presented for the experiment No.10 on Figure 4.

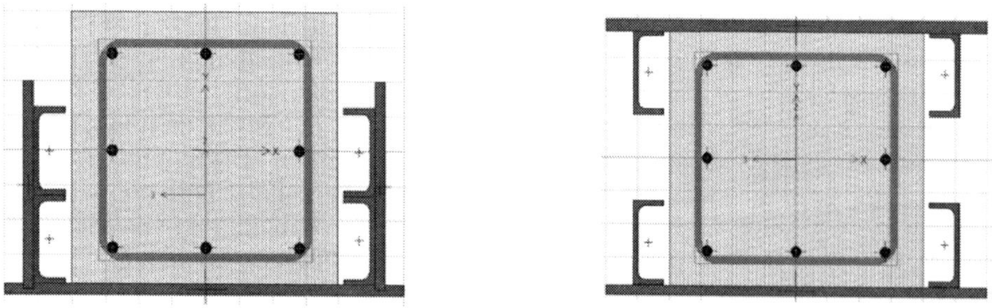

C – cross section at support D – cross section at a distance from the support

Figure 2. Cross sections taken into account for a reinforced concrete beam.

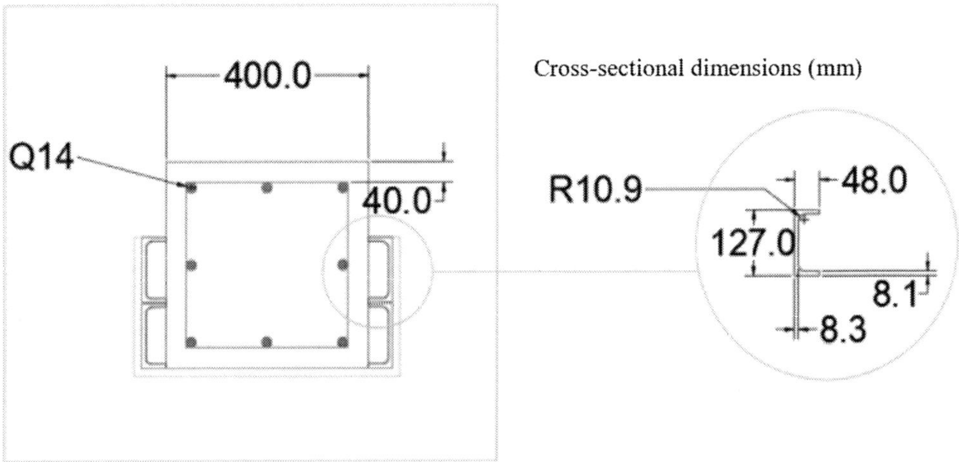

Figure 3. Dimensions of elements and parts in mm

Figure 5 shows a diagram of shear forces in a beam collapse situation. [15].

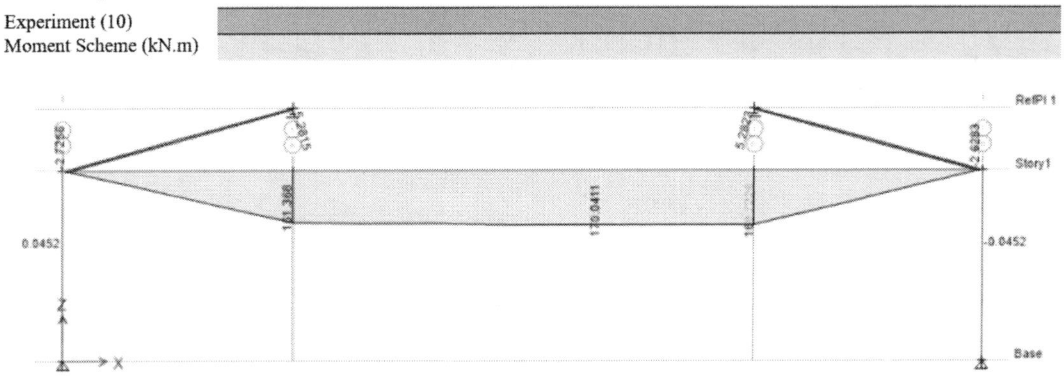

Figure 4. Bending moments plot in a beam

Figure 5. Plot of transverse forces during beam collapse

To illustrate the involvement of the reinforcing brackets in the work of the beam, Table 1 shows the values of the resulting transverse forces and moments, before and after reinforcing the beam with the brackets.

As it can be seen from Table 1, the concentrated loads on the beam are in some cases applied from both sides the same, and in other cases they are different. The influence degree of the reinforcing brackets on the maximum moments in the beam, expressed as a percentage, of the applied loads' values and the method of their application, is characterized as follows, presented in Table 2.

Table 1. Transverse forces and moments in beam sections from concentrated loads.

Experiment number	Applied loads (T), on distance L/4 (m)		Moment (max) in unpowered nom field (kNxm)	Moment (max) in enhanced field (kNxm)	Transverse force (max) in unreinforced field (kN)	Transverse power (max) in enhanced field (kN)	Beam collapse (yes / no)
	left F_1	right F_2					
1	6	6	105.01	100.85	64.5	73.89	no
2	6	7	112.98	104.3	66.94	81.25	no
3	7	7	119.36	115.12	74.3	83.7	no
4	7	8	127.2	118.57	76.75	91.1	no
5	8	8	133.6	129.38	84.1	93.51	no
6	8	9	141.5	132.84	86.55	100.86	no
7	9	9	147.89	143.65	93.9	103.3	no
8	9	10	155.77	147.1	96.36	110.67	no
9	10	10	162.15	157.9	103.7	113.1	no
10	10	11	170.04	161.14	106.17	120.5	yes

Table 2. Assessment of the influence degree of reinforcing brackets on the maximum moments in the beam.

Experiment number	1	2	3	4	5	6	7	8	9	10
The difference of moments in enhanced field and in non-enhanced field, %	3.96	7.7	3.55	6.78	3.15	6.12	2.86	5.56	2.62	5.23

Summary

Reinforcement of the supporting part of the reinforced concrete beam with unloading brackets, in the form of a bar elements' pair, increases the bearing capacity of the reinforced concrete beam, and with the same loads on both sides of the supporting parts of the beam, the bending moment is increased by 4%, at the minimum values and by 2.62% at the maximum load values. Whereas under unequal loads, the influence of the unloading brackets increases and reaches, at the minimum values, up to 7.7%, and at the maximum loads - up to 5,23 %.

References

[1] Sayda S K 2012 Strength of bent reinforced concrete sections without prestressing International *Journal of Experimental Education, Moscow* **11** 37-38.

[2] Degtyarev G V, Dats'o D A 2019 The seasonal regulation basin dam basis deformation forecast *IOP Conf. Series: Materials Science and Engineering* **698 022013**. doi:10.1088/1757-899X/698/2/022013

[3] Degtyareva O G, Degtyarev V G, Naydenov S Yu 2019 Seismic resistance of hydraulic structures, taking into account the work of structures beyond the elastic limit *Scientific journal of the Russian Research Institute of Melioration Problems, Novocherkassk* **1 (33)** 92-108.

[4] Degtyarev GV, Takhumova OV 2019 Designing an additional freshwater source infrastructure to ensure the environmental sustainability of coastal areas *IOP Conference Series: Earth and Environmental Science* **395** 012001. doi:10.1088/1755-1315/395/1/012001.

[5] Degtyarev V G, Sayda S K, Degtyarev G V 2015 Pat. 2 558 852 Russian Federation, MPK G01N 33/38. Device for measuring the characteristics of concrete samples prepared on the basis of expanding concrete; applicant and patent holder Kuban State Agrarian University (RU). - № 2014131510/15; appl. 29.07.2014; publ. 10.08.2015, Bul. № 22.

[6] Abdrazakov F K, Orlova S S, Pankova T A, Mirkina E N, Fedyunina T V 2018 The monitoring of condition of hydraulic structures *Journal of Advanced Research in Dynamical and Control Systems* **10 (13)** 1952-1958.

[7] Degtyarev G V, Degtyarev V G, Degtyareva O G, Kojchenko N V, Dast'o D A 2018 Modeling and calculation of reinforced concrete constructions in STARK ES software (Textbook, Under general editing of G. V. Degtyarev, Krasnodar).

[8] Degtyareva O G, Dats'o D A, Chebanova E F, Krylova N N, Khatkhokhu E I 2019 Numerical analysis of engineering geological elements while strengthening the base of a construction site *Research Journal of Pharmaceutical, Biological and Chemical Sciences* **10 (1)** 1554–1558.

[9] Sayda S K 1990 A truss of a composite structure with a pipe-concrete upper belt of a flexible lattice in the coatings of industrial buildings (author. dis. Cand. tech. sciences: 05.23.01, Sayda Salman Kamalovich, Leningrad).

[10] Degtyarev G V, Belokur K A, Sokolova I V 2018 Modeling of the Building by Numerical Methods at Assessment of the Technical Condition of Structures *International conference on Construction and Architecture: theory and practice of industry development (CATPID-2018). Trans Tech Publications, Switzerland* **931** 141-147.

[11] Degtyareva O G, Degtyarev G V, Lavrov N L, Aliev D U 2018 Constructive -technological decisions in regulating the flow of atmospheric precipitation *Magazine of Civil Engineering* **82 (6)** 32–48. doi: 10.18720/MCE.82.4.

[12] Olgarenko V I, Olgarenko I V, Olgarenko V I 2019 Technical condition diagnostics of the water supply facilities in the irrigation systems *IOP Conference Series: Materials Science and Engineering the International Scientific Conference "Construction and Architecture: Theory and Practice for the innovation Development" (CATPID-2019)* **022060.**

[13] Degtyareva O G, Dats'o D A, Degtyarev G V, Gumbarov A D 2017 Modeling in the CAE system of slab foundation settlement of a low-pressure dam *Proceedings of the Kuban State Agrarian University, Krasnodar* **1 (64)** 221–226.

[14] Degtyareva O G 2016 Investigation of the stiffness characteristics of gravity and buttress dams by means of the analysis of natural vibrations *Proceedings of the Kuban State Agrarian University* **6 (63)** 169–177.

[15] Sayda S K, Degtyarev V G 2014 Aspects of the theory of resistance of concrete reinforced with composite reinforcement *Proceedings of the Kuban State Agrarian University, Krasnodar* **48** 183-188.

The design parameters' analysis of typical support parts for the ferro-concrete roof beams reinforced with unweighted frame

G V Degtyarev*, M R Salloum, S G Abo Arraj

Kuban state agrarian University named after I. T. Trubilin, 13 Kalinina str., Krasnodar, 350044, Russia

E-mail: cst2007@mail.ru

Abstract. It is necessary to use ferro-concrete roof beams of considerable length, which, together with operational loads, often leads to an abnormal state of the latter, especially their support parts, in industrial buildings, especially in the presence of large spans, when they overlap. The option of strengthening the beam with an unweighted frame, in comparison with a typical solution is considered.

Introduction

The supporting part of ferro-concrete roof beams is the most vulnerable structural link of an industrial building, especially when it comes to large-span structures [1, 2, 3]. At the same time, the repair of ferro-concrete roof beams, today, basically boils down to their replacement. Such actions are associated, first of all, with large material and labor costs, but even more negative is the need, in some cases, to stop the main production process in the workshop for the period of the reconstruction and repair of load-bearing structures, namely the ferro-concrete roof beams. It is proposed to solve this issue by means of local strengthening the most vulnerable places of the ferro-concrete roof beams, that is, their support parts [4, 5]. At the same time, it was necessary to take into account the fact that if the spans are large, then the additional requirements arise for the reinforcing structures, namely, for their weight, when in fact they should be unweighted.

Today, the most widespread method is the strengthening the support parts of ferro-concrete roof beams by installing unloading brackets on both sides of the beam in the form of a pair of rod elements forming an acute angle [6, 7]. In turn, each end of a bar element running horizontally is connected to the bottom of the beam, and each end of a bar element running at an acute angle is connected to the top of the beam [8]. The opposite ends of the bar elements are united by the support plates located on both sides of the beam, which, in turn, are joined by strips [9, 10, 11]. Between the rod elements, mirrored on both sides, the additional tensioning devices, connected to the rod elements, rigidly, by welding, with the upper element and flexibly, through a threaded connection, with the lower element, are installed. At the same time, the bar elements themselves are rigidly connected to the upper and lower parts of the ferro-concrete roof beams by means of local concreting of special nodes for a gable roof beam, which is shown in Figure 1. Due to the creation of adjustable forces, in additional tensioning devices, an increase in the reliability of the supporting parts of ferro-concrete roof beams is achieved, as a result of an increase in the bearing capacity, since it is possible to redistribute the loads not only between the rods of the unloading brackets, but also between the ferro-concrete roof beams and the reinforcing structure [12, 13]. At the same time, pairs of rod elements forming an acute angle are currently supposed to be made of such rolled products as a channel, but this is a relatively heavy material, in comparison, for example, with angle beam. In our work, it is proposed to study the issue of the angle beam device made of an unweighted unloading frame through numerical modeling, which is shown in Figure 1. Otherwise, the technical solution for strengthening the supporting parts of reinforced concrete beams of the coating is similar to that when using a channel as a rod element.

Materials and technique

The technical solution for strengthening the support parts of ferro-concrete roof beams with an unweighted frame for a gable beam is shown in Figure 1.

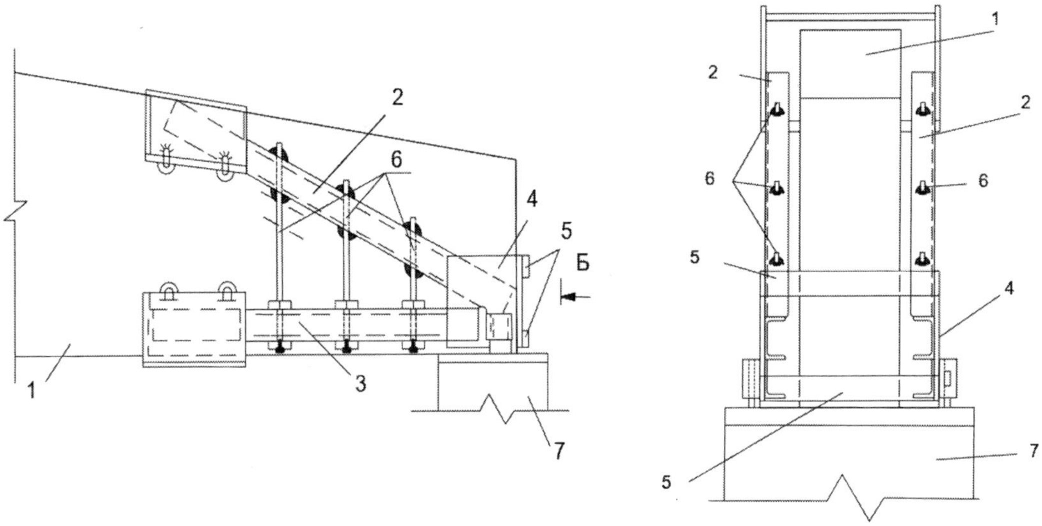

Side view View B

1- ferro-concrete roof beam; 2 and 3- the bar elements of unloading brackets; 4- supporting plates; 5- connecting strips; 6- additional tensioning devices; 7- support column

Figure 1. Device for strengthening the supporting parts of ferro-concrete roof beams with an unweighted frame

In order to analyze the effect of unloading brackets with an unweighted frame, in the form of a pair of rod elements made of steel angles, installed on a beam with parallel chords of a square section, on the operation of its supporting parts, numerical modeling was carried out in the Etabs 2019 software package. Figure 2 shows the sections accepted for the ferro-concrete roof beam calculation with the length L=6 m. and Figure 3 shows the dimensions of elements and parts.

As a result of numerical modeling, implemented with a stepwise increase in concentrated loads F1 and F2, acting at a distance of a quarter of the span from each support. In this case, in the numerical experiment, the concentrated loads F1 and F2 with odd attempts (1, 3, 5, etc.) were taken to be the same. For example, at the attempt 1 we have that F1 is equal to F2 and is equal to 6 tons, and at the attempt 3 - F1 is equal to F2 and is equal to 7 tons, and so on. Whereas for the even attempts (2, 4, 6, etc.), in the numerical experiment, the concentrated loads F1 and F2 were taken different, which is reflected in Table 1.

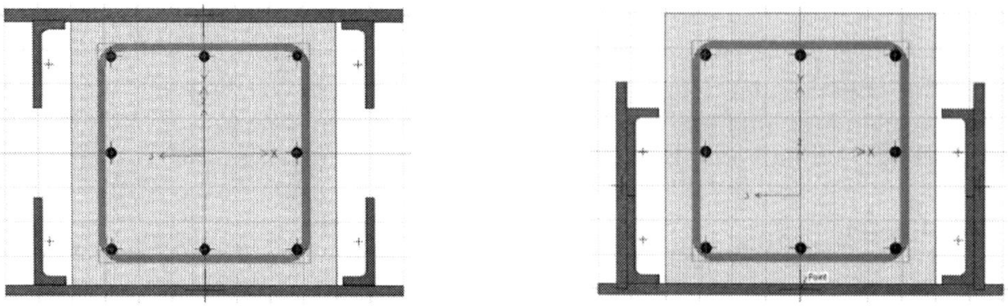

C- cross section at the support D- section at a distance from the support

Figure 2. The sections taken into account for a ferro-concrete beam

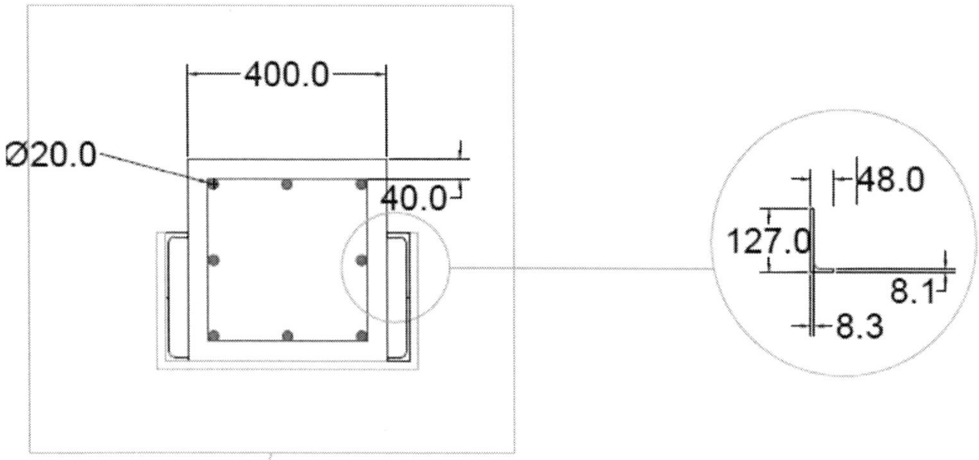

Figure 3. Dimensions of elements and parts in mm.

As a result of numerical modeling of the described structure to strengthen the supporting parts of a ferro-concrete beam by means of an unweighted frame made of angles, the diagrams of bending moments and shear forces were obtained for the attempts №1 and № 10 [14, 15].

Table 1. Transverse forces and moments in beam sections from concentrated loads with an unweighted reinforcement frame

Attempt No.	Applied loads (t), on distance L/4 [m]		Moment (max) in unreinforced field [kN*m]	Moment (max) in reinforced field [kN*m]	Shear force (max) in unreinforced field [kN]	Transverse force (max) in reinforced field [kN]	Beam collapse [yes / no]
	on the left F_1	on the right F_2					
1	6	6	105.5	101.1	64.43	73.79	no
2	6	7	113.13	104.65	66.89	81.19	no
3	7	7	119.74	115.76	74.24	83.66	no
4	7	8	127.69	119.01	76.71	90.9	no
5	8	8	134.14	129.85	84.05	93.46	no
6	8	9	141.95	133.12	86.4	100.79	no

7	9	9	148.3	144.1	93.82	103.24	no
8	9	10	156.1	148.2	96.25	110.55	no
9	10	10	162.5	158.3	103	112.9	no
10	10	11	170.46	161.8	111.6	119.6	yes

Figure 4 shows a diagram of bending moments in a beam for the attempt № 1 with the equivalent concentrated forces F1 and F2, and Figure 5 shows the diagram of shear forces for the same attempt.

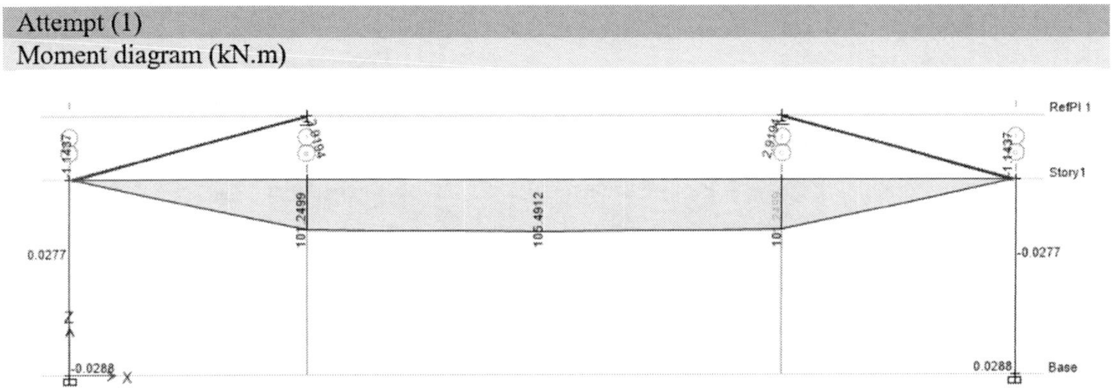

Figure 4. Plot of bending moments in a beam at sections, Attempt № 1

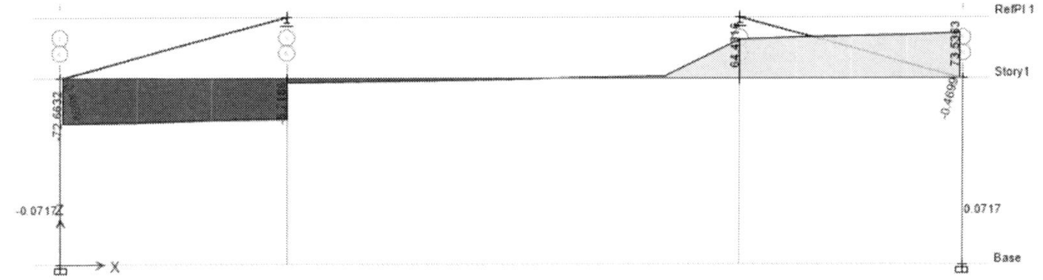

Figure 5. Plot of shear forces in a beam at sections, Attempt № 1

The following Figures 6 and 7 obtained by the attempt № 10, show the diagram of bending moments and the diagram of shear forces during the beam collapse, respectively.

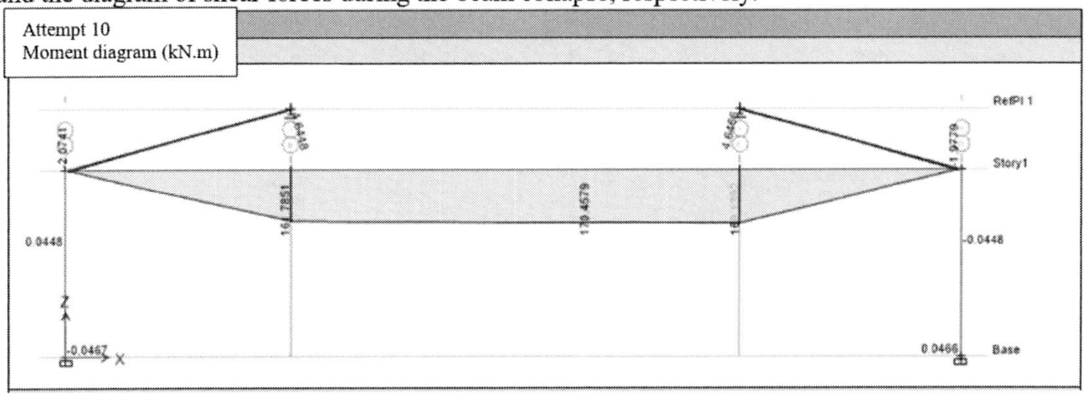

Figure 6. Bending moments plot in a beam at sections

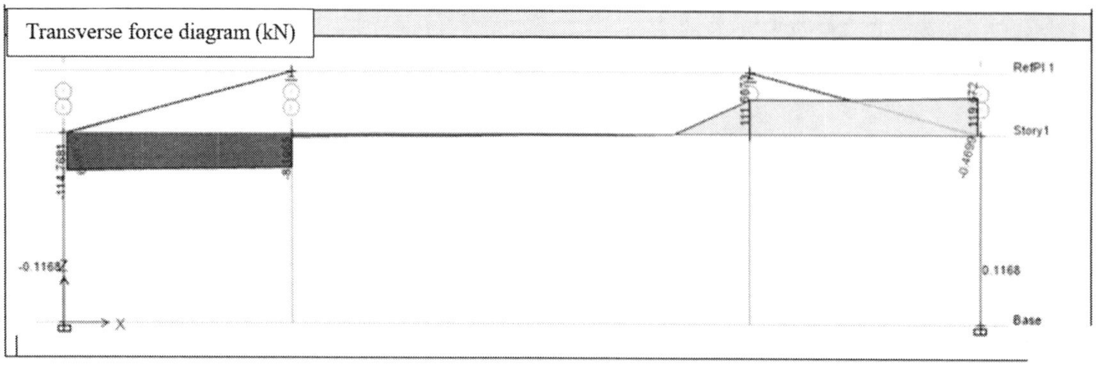

Figure 7. Plot of transverse forces at beam collapse

The analysis of Table 1 gives a possibility to state that the influence degree of equivalent and not equivalent concentrated forces applied to the beam on its state, assessed by the means of bending moments and shear forces in the beams, will be different. To assess this effect, let us consider the difference in moments in the experimental field of a beam reinforced with an unweighted frame and a beam without reinforcement, expressed as a percentage, which is reflected in Table 2.

Table 2. Influence degree assessment of the reinforcing brackets on the maximum moments in the beam

Attempt No.	**1**	**2**	**3**	**4**	**5**	**6**	**7**	**8**	**9**	**10**
The difference of moments in the reinforced field and without, [%]	4.17	7.7	3.32	6.79	3.2	6.22	2.83	5.06	2.58	5.08

Summary

Reinforcement of the ferro-concrete roof beam's supporting part with unloading brackets, in the form of a pair of bar elements, increases the bearing capacity of the ferro-concrete beam, and with the same loads on both sides of the supporting parts of the beam, the bending moment is increased by 4.17% at minimum load values and by 2, 58% at maximum load values. Whereas at unequal loads, the unloading brackets' influence is increased and reaches, at their minimum values, up to 7.7% and at maximum loads - up to 5.06%.

References

[1] Degtyareva O G, Dats'o D A, Degtyarev G V, Gumbarov A D 2017 Modeling the slab foundation settlement of a low-pressure dam in the CAS system *Proceedings of the Kuban State Agrarian University, Krasnodar* **1 (64)** 221–226.

[2] Sayda S K 2012 Strength of bent reinforced concrete sections without prestressing *Journal of Experimental Education, Moscow* **11** 37-38.

[3] Sayda S K, Degtyarev V G 2014 Aspects of the theory of resistance of concrete reinforced with composite reinforcement *Proceedings of the Kuban State Agrarian University, Krasnodar* **48** 183-188.

[4] Degtyarev G V, Belokur K A, Sokolova I V 2018 Modeling of the Building by Numerical Methods at Assessment of the Technical Condition of Structures *International conference on Construction and Architecture: theory and practice of industry development (CATPID-2018). Trans Tech Publications, Switzerland* **931** 141-147.

[5] Degtyarev G V, Degtyarev V G, Degtyareva O G, Kozheko N V, Datsio D A 2018 Modeling and calculation of reinforced concrete constructions in the software complex stark es (Edited by G.V. Degtyarev. Krasnodar).
[6] Sayda S K 1990 A truss of a composite structure with a pipe-concrete upper chord with a flexible lattice in the coatings of industrial buildings (Author. dis. Cand. tech. sciences: 05.23.01, Sayda Salman Kamalovich, Leningrad).
[7] Degtyarev G V, Dats'o D A 2019 The seasonal regulation basin dam basis deformation forecast. *IOP Conf. Series: Materials Science and Engineering* **698** 022013. doi:10.1088/1757-899X/698/2/022013
[8] Degtyarev V G, Sayda S K, Degtyarev G V 2015 Patent 2 558 852 Russian Federation, IPC G01N 33/38. Device for measuring the characteristics of concrete samples prepared on the basis of expanding concrete; applicant and patent holder Kuban State Agrarian University (RU). - № 2014131510/15; application 29.07.2014; publ. 10.08.2015, Bull. № 22.
[9] Degtyareva O G 2016 Study of the stiffness characteristics of gravity and buttress dams by analyzing natural vibrations *Proceedings of the Kuban State Agrarian University* **6 (63)** 169–177.
[10] Degtyareva O G, Degtyarev V G, Naydenov S Yu 2019 Seismic resistance of hydraulic structures, taking into account the work of structures beyond the elastic limit *Scientific journal of the Russian Research Institute of Melioration Problems, Novocherkassk* **1 (33)** 92-108.
[11] Degtyareva O G, Degtyarev G V, Lavrov N L, Aliev D U 2018 Constructive -technological decisions in regulating the flow of atmospheric precipitation *Magazine of Civil Engineering* **82 (6)** 32–48. doi: 10.18720/MCE.82.4.
[12] Degtyareva O G, Datso D A, Chebanova E F, Krylova N N, Khatkhokhu E I 2019 Numerical analysis of engineering geological elements while strengthening the base of a construction site *Research Journal of Pharmaceutical, Biological and Chemical Sciences* **10 (1)** 1554–1558.
[13] Abdrazakov F K, Orlova S S, Pankova T A, Mirkina E N, Fedyunina T V 2018 The monitoring of condition of hydraulic structures *Journal of Advanced Research in Dynamical and Control Systems* **10 (13)** 1952-1958.
[14] Degtyarev G V, Takhumova O V 2019 Designing an additional freshwater source infrastructure to ensure the environmental sustainability of coastal areas *IOP Conference Series: Earth and Environmental Science* **395** 012001. doi:10.1088/1755-1315/395/1/012001.
[15] Olgarenko V I, Olgarenko I V, Olgarenko V I 2019 Technical condition diagnostics of the water supply facilities in the irrigation systems *IOP Conference Series: Materials Science and Engineering the International Scientific Conference "Construction and Architecture: Theory and Practice for the innovation Development" (CATPID-2019)* 022060.

IOP Conf. Series: Materials Science and Engineering 1083 (2021) 012008 doi:10.1088/1757-899X/1083/1/012008

Technical and technological aspects of ensuring seismic stability during the construction of buildings and structures

A N Sekisov [1*], N A Shipilova[2], V N Bortsov[2], D M Dolbnya[2], O Kavkazi[2]

[1] Kuban State Agrarian University I.T. Trubilin, 13, Kalinina str., Krasnodar, 350044, Russia
[2] Kuban State Technological University, 2, Moskovskaya st., Krasnodar, 350072, Russia

E-mail: alnikkss@gmail.com

Abstract. The aim of the work is to consider the basic technical and technological aspects of ensuring the seismic resistance during the construction of buildings and structures, to study and analyze the existing technologies of earthquake-resistant construction, to identify their advantages and disadvantages, and the possibility of using them in mass construction. In addition, the analysis of the requirements of Building Codes of the Russian Federation for earthquake-resistant construction was carried out. This study presents the options for seismic isolation used in the world practice of earthquake-resistant construction. On the basis of the conducted studies, the conclusions about the effectiveness of seismic isolation methods, as well as the possibilities of achieving economic efficiency in the construction of earthquake-resistant buildings have been drawn. The paper gives recommendations on the earthquake-resistant construction technologies' dissemination in Russia and the development of a regulatory framework in this area of construction.

Introduction

About a quarter of the territory of Russia, which contains 27 constituent entities of the Federation with a population of about 20 million people, is exposed to earthquakes with an intensity of 7 points or more. The degree of seismic impact danger depends not so much on its strength, but on the social, economic, environmental damage which may arise as a result of human losses, destruction of buildings, structures, economic facilities, engineering infrastructure [2, 4, 6, 7, 8]. Against the background of devastating earthquakes and the results of their consequences occurring in Russia and neighboring countries, the efforts to implement the preparatory and organizational measures to improve the regulatory framework for the design of buildings' seismic protection active forms, as well as to ensure a coordinated approach to work on seismic strengthening of objects with a deficit of seismic resistance, have been made in our country [5, 10, 11, 14, 15, 16].

Main part

In the field of the calculating structures theory, the main one is currently the oscillatory doctrine, according to which the determining factor in the destruction of structures under seismic impact is bending deformations caused by the structures' oscillations from seismic movements [3]. Among both Russian and many foreign specialists, there is a point of view that shear deformations arising at the front of a wave traveling along the structures and causing large values of shear forces are the determining factor in fracture, which is not taken into account in the oscillatory concept.

An important factor, which is extremely difficult to predict by calculation, is that the properties of materials and structures change during relatively weak earthquakes, which occur much more often than the calculated ones. In this case, minor damage occurs in the form of cracks, often invisible, violations of adhesion between reinforcement and concrete, peeling of the protective layer of concrete with the reinforcement exposure, changes in the conditions of mating elements in the nodes. Over time, damage accumulates, increasing the vulnerability of the structure and making it unprepared to take the design load [1].

In modern design solutions, it is impossible to increase seismic resistance only by increasing the size of the sections, strength, and weight. The structure can be more stable, but not necessarily cost effective, because both weight and inertial seismic loading can increase even more. The new effective methods of seismic protection are required. Such solutions imply a change in mass and stiffness, damping of the system depending on its movements and speeds.

To date, more than 100 patented structures for seismic protection of buildings and structures are known. There are options that include additional energy sources and elements that regulate the operation of these sources. Examples of implementation of such systems exclude their active use for the construction projects. However, there are also special methods of passive seismic protection of the structures that do not use additional energy sources: seismic suppression and seismic isolation. In the first case, dampers and dynamic dampers are used. Thanks to them, the mechanical energy of the vibrating structure is transferred into other types of energy. As a result, oscillations are damped. Mechanical energy is redistributed from the protected structure to the absorber. The general classification of seismic protection systems, according to the authors [12,13, 17, 19], is as follows (Figure 1).

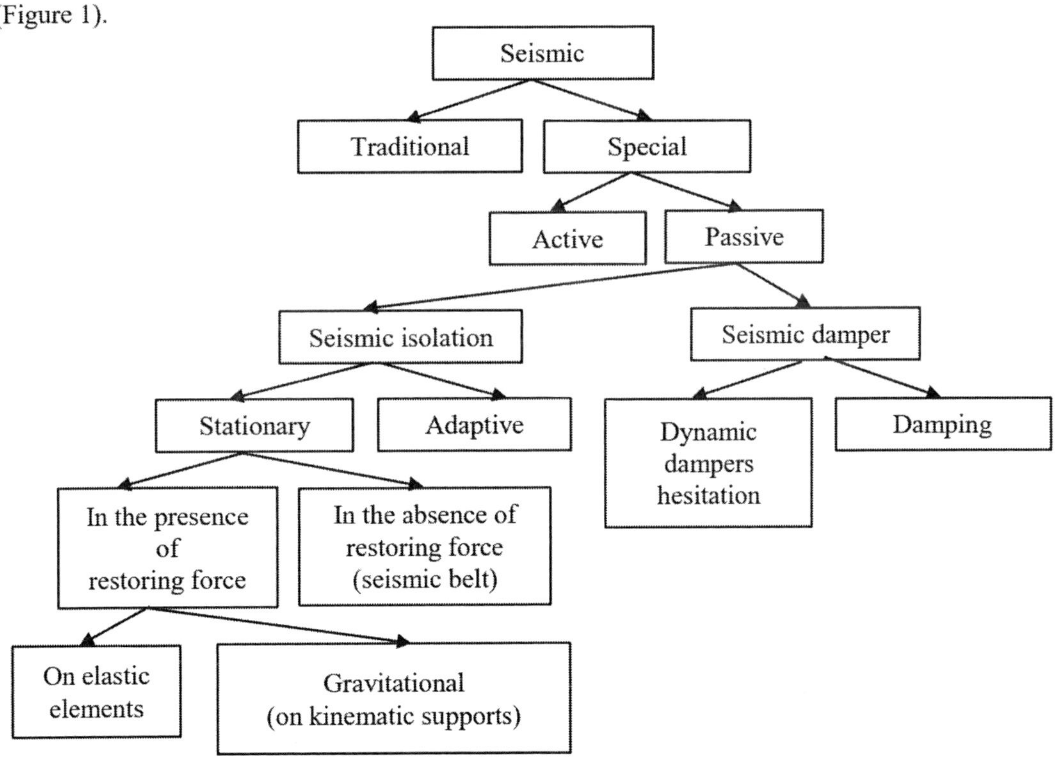

Figure 1. Classification of seismic protection systems according to the operation principle.

The very idea of seismic isolation is not new; even in ancient times, in some cases, builders in order to weaken the effect of earthquakes on structures tried to isolate the buildings from their foundations by installing soft gaskets at the top level of the foundations. In the III-VII centuries some of the monumental structures in Central Asia were built on sand blanket, in the X-XVII centuries - soft reed layers were laid on blankets of pure clay in the basement of the walls. The oldest and one of the most promising methods of active seismic protection is seismic isolation.

Seismic isolation – is a reduction of seismic loads on the structure due to the use of special structural elements:

- increasing the flexibility and periods of natural oscillations of the structure (flexible stands; swinging supports; rubber-metal supports, etc.);
- increasing absorption (dissipation) of seismic oscillation energy (dry friction dampers; sliding belts; hysteresis; viscous dampers);
- standby, shutdown elements [18].

In practice of the earthquake-resistant construction, adaptive systems have become widespread. In these systems, the dynamic characteristics of the structure change irreversibly during the earthquake, adapting to the seismic effect. The main idea is that to change the dynamic characteristics of the structure during an earthquake in such a way that as a result, the natural frequencies of the building "leave" the dangerous resonance range.

This can be achieved in two ways: by turning the links off and by turning the links on. The first type of connections is specially provided for the "weak" structural elements of the building, intended for destruction during an earthquake, which reduces the rigidity of the building and, accordingly, the natural frequencies of the structure's oscillation ("weak" mortar watering device at the border between the foundation top and the walls masonry). The second type of connections in a normal situation does not participate in the operation of the structure, but is turned on only during an earthquake at a certain amount of displacement: limiters installed with a gap; sagging extending braces, etc.

The nature of such links' connection ensures their certain mobility in an unloaded state, and under seismic action they are able to perceive a part of the load.

In the lower part of the building, between the bearing supporting frames of the lower floor, tie panels, which are turned off during intense seismic influences, are installed. After disconnecting the panels, the frequency of free oscillations decreases, the period of oscillations increases, and the seismic load decreases, since the period of natural oscillations of a building with tie panels is much lower than the values of the prevailing periods of ground oscillation, therefore resonance phenomena are weakly manifested and the tie panels are not destroyed.

The use of disconnecting links is most effective when the frequency composition of the expected seismic impact is confidently predicted.

Disadvantages: after the destruction of the disconnected connections during an earthquake, their restoration is necessary, which is not always practically feasible; in some cases, during an earthquake in its final stage, the prevailing frequency of impact decreases. Secondary resonance and loss of the building structures' bearing capacity are possible. In this case, the use of constructive measures is required, which leads to additional construction costs. Figure 2 shows seismic protection systems with links on and off [17].

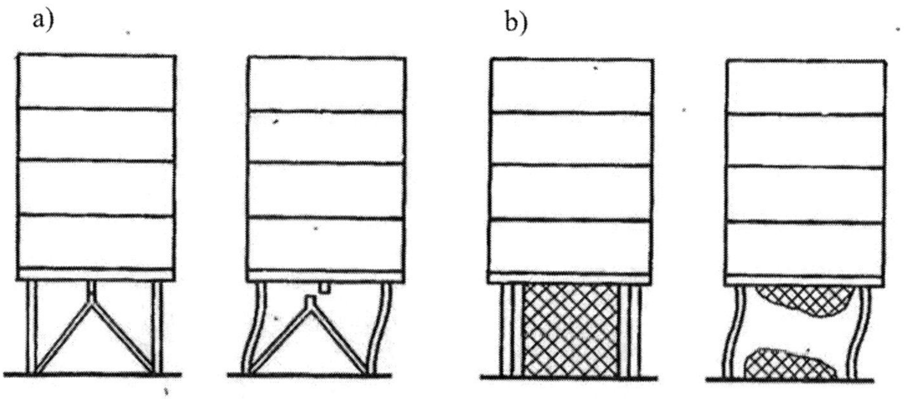

Figure 2. Seismic protection of a building using disconnecting links
a - special switching off elements; b - collapsing tie panels

Stationary seismic isolation system is a seismic protection device in the presence of a restoring force, a building with a flexible floor. The flexible floor can be made in the form of frame posts, elastic supports, piles, etc.

The most widespread are the buildings on flexible rubber-metal supports. Currently, several variants of rubber-metal elastic, compression supports are used: French, New Zealand, American and Italian versions of supports. To prevent excessive settlement of buildings under their own weight load, the supports are made rigid in the vertical plane and flexible in the horizontal plane. Due to the elastic properties of rubber, the rubber-metal supports have high compressive, tensile and torsional strength. However, the cost of the foundations themselves turns out to be significant and can reach up to 30% of the construction cost.

Seismic isolation, which does not provide a restoring force acting on seismically insulated parts of the structure, is implemented by means of a sliding belt. One of the most well-known technical solutions of this type is the seismic isolation foundations by Spie Batignolle and Electricite de France. This is an anti-seismic friction bearing structure. The support bearing the upper foundation slab consists of friction plates, a reinforced elastomer (neoprene) pad, a lower foundation slab, and a concrete polar resting on the lower foundation slab. The rigidity of the supports in the vertical direction is about 10 times higher than in the horizontal direction. Seismic isolation foundations by Spie Batignolle are a classic example of seismic isolation with a series of elastic and damping elements. With relatively weak influences, when the horizontal load on the support part does not exceed the friction forces, the system operates in a linear region; when the load increases, the friction force is overcome and the upper foundation slab slips relative to the lower one. At the same time, it is possible to reduce the load on equipment and the building several times.

It should be noted that kinematic supports are an example of stationary seismic isolation systems in the presence of restoring gravity. One of the examples of this seismic protection method is the swinging kinematic supports of Cherepinsky Yu.D. [17].

The task of seismic supports is to reduce horizontal dynamic impacts on a building during 7-, 8-, 9-point MSK-84 earthquakes up to the 6-point level. The task is solved by installing seismic-insulating pipe-concrete supports with special hinges in the upper and lower parts between the foundation (base) and the building itself (structure). These hinge joints allow the building (structure) to make significant displacements (up to 1000 mm) relative to the foundation without deformations (destruction) of structures. The essence of the structure lies in the fact that when the base is displaced by a certain calculated value, the building rises slightly, receiving some additional kinetic energy. In this case, a restoring moment arises, which brings the “foundation construction” system to its original state. A pipe-concrete seismic-isolating support is a round column of steel pipe with a diameter of 300 mm to 1500 mm and a height of 1500 mm to 10000 mm, filled with a reinforcing cage and heavy concrete of class B-25 and higher (Figure 3).

Figure 3. Pipe-concrete seismic support

Steel plates with a thickness of 10 to 40 mm are welded to the ends of the column from above and below. The plates are welded to working steel plates of the second level of inclusion with a thickness of 2 to 30 mm according to the calculation and working steel plates of the first level inclusion with a thickness of 2 mm to 30 mm. In the lower part, the column rests on a steel plate 10-40 mm thick, which is an embedded part in the base-foundation made of B-25 class concrete and higher. The floor slab of the seismically insulated part of the building (structure) is supported on the upper part of the column with a steel embedded part 10-40 mm thick. The steel plates are interconnected with the rolled steel oscillation dampers.

Summary

Conventional measures for seismic protection of buildings and structures are reduced mainly to increasing the bearing capacity of structures and increasing their rigidity, which leads to an increase in the consumption of steel and concrete during construction [9]. Such seismic protection is carried out in accordance with building codes BC 14.13330.2014. Moreover, the above-mentioned measures do not reduce the seismic loads impact on the building, but only take them into account.

Since the state does not define clear the requirements for the seismic isolation of buildings and structures, the operation of the seismic-isolated foundation structure is checked by the organization of field experiments, special equipment that allows creating the conditions that correspond to the calculated impact of an earthquake on a building under construction.

Unfortunately, it can be stated so far that the mass construction industry has switched to the extensive construction methods, shifting the costs associated with earthquake resistance of buildings to the end consumer, and the problem of earthquake-resistant construction, nevertheless, remains relevant for the residents in all the regions prone to earthquakes.

References

[1] Degtyarev G V, Molotkov G S, Sekisov A N, Datsjo D A 2018 Numerical modeling of condition of the bridge structure based on the results of national surveys *International Journal of*

Engineering and Technology (UAE) **7** **(2.13)** **(13)** 226-230. https://doi:10.14419/ijet.v7i2.13.11866

[2] Ovchinnikova S, Kalinichenko M, Markina N, Schneider E 2020 Energy modernization of housing stock *E3S Web of Conferences* **157** 06028. doi.org/10.1051/e3sconf/202015706028

[3] Nehaj R, Molotkov G, Rudchenko I, Grinev A, Sekisov A 2017 Algorithm of composing the schedule of construction and installation works *IOP Conference Series: Earth and Environmental Science* 1-8. doi:10.1088/1755 1315/90/1/012019.

[4] Kalinichenko M Y, Stojanov N I, Abornev D V, Ovchinnikova S V 2020 Optimization of the regeneration of the sodium-cation ion-exchange filter *IOP Conference Series: Materials Science and Engineering* **905 (1)** 012034. doi.org/ 10.1088/1757-899X/905/1/012034

[5] Sekisov A, Danilenko S, Klimenko V, Stepanov R 2020 Quick method of the grunt prisms gravity centers' exact compliance with vertical area planning *IOP Conference Series: Materials Science and Engineering* **913** 042036. doi:10.1088/1757-899X/913/4/042036

[6] Sekisov A, Ovchinnikova S, Grebneva V, Chernyshova M 2020 Modern directions of low-rise housing construction development in the world: economic and technological aspect *IOP Conference Series: Materials Science and Engineering* **913** 042035. doi:10.1088/1757-899X/913/4/042035.

[7] Sekisov A N 2019 Improving the efficiency of the organization of construction production based on the use of BIM-technologies *IOP Conference Series: Materials Science and Engineering* **698** 066005. doi:10.1088/1757-899X/698/6/066005.

[8] Sekisov A N 2019 Problems of organizing and conducting engineering surveys in construction *IOP Conference Series: Materials Science and Engineering* **698** 055016. doi:10.1088/1757-899X/698/5/055016.

[9] Sekisov A N, Degtyareva O G, Samsonova N V, Grigoryan M N 2018 Development of the Methods Improving the Production Costs Formation Process *International Conference on Construction and Architecture: theory and practice of industry development 2018* 1210-1213. doi.org/10.4028/www.scientific.net/MSF.931.1210.

[10] Sekisov A N, Varvarkina A V, Gura D A, Savenko A A, Shikhovtsov A A 2019 Pit design taking into account water level reduction using the software package MIDAS GTS NX. IIOAB *Journal.* **10 (S2)** 93-100. Information on http://www.iioab.org/IIOABJ_10.S2_93-100.pdf.

[11] Stepanov R R, Sekisov A N, Gura D A, Soloveva E V, Dreshpak V S 2019 «Quick Method» of Map Limit Exact Position Determination by Vertical Planning of Areas *Helix* **9(1)** 4744-4752. doi.org/10.29042/2019-4744-4752.

[12] Eisenberg Ya M 2007 Seismic isolation of tall buildings *Seismic construction. Safety of structures* **4** 41-43.

[13] Eisenberg Ya M 1976 Off-site structures for seismic areas (Stroyizdat, M.).

[14] Sekisov A N, Klimenko V V, Pismenniy V A, Bushneva I A, Matevosyan V S, Chernishova M V 2019 Database: Civil and Industrial Buildings; Database state registration certificate No. 2019621704 from 03.10.2019, No. 2019621105, application 28.06.2019, publ. 03.10.2019. Information on http: //www1.fips.ru/registers-docview/fips_servlet?DB=DB&DocNumber=2019621704&TypeFile=html.

[15] Mishulin G M, Sekisov A N, Bratishkina V E 2019 Database: Large-sized, heavy-weight and other machines for earth-moving operations; Certificate of state registration of the database Certificate of state registration of the database No. 2019621023 from 17.06.2019, No. 2019620940, application 04.06.2019, publ. 17.06.2019. Information on http:// www1.fips.ru/registers-doc-view/fips_servlet?DB=DB&DocNumber=2019621023&TypeFile=html.

[16] Sekisov A N, Shabanova I G 2019 Database: Concrete and Reinforced Concrete Machines; Database state registration certificate No. 2019621339 from 22.07.2019, № 2019621225, application 11.07.2019, publ. 22.07.2019. Information on http://www1.fips.ru/registers-doc-view/fips_servlet?DB=DB&DocNumber=2019621339&TypeFile=html.

[17] Polyakov V S, Kilimnik L Sh, Cherkashin A V 1989 Modern methods of seismic protection of buildings (Stroyizdat, M.).
[18] BC 14.13330.2014 – Construction in seismic areas SNiP II-7-81* (updated SNiP II-7-81* "Construction in seismic areas" (BC14.13330.2011)) (with Change N 1) article 3.44.
[19] Uzdin A M et.al. 1993 Fundamentals of the theory of seismic resistance and earthquake-resistant construction of buildings and structures (SPb).

Proposals on calculating the differential properties of concrete

L R Mailyan, S A Stel'makh, F M Shcherban'*, A P Korobkin, and E A Efimenko

Don State Technical University, 1, Gagarin Square, Rostov-on-Don, 344000, Russia

E-mail: au-geen@mail.ru

Abstract. Background: The paper dwells upon calculating the differential structural properties of centrifugated and vibration-centrifugated concrete as a function of the manufacturing parameters. Methods: One finding is that different layers of centrifugated and vibration-centrifugated concrete are exposed to forces of varying magnitude, resulting in these layers differing significantly in density, strength, strain, and elastic modulus. Another finding is that it is the centrifugal and centripetal forces that are key factors to be applied as arguments in calculational dependencies. Thus, the versatile calculational dependencies that make adjustments for changes in the structural properties of concrete should use these properties or their increments as functions, and the rotation inertia forces as arguments, which are in turn a function of distance from the center of rotation and the angular rotation speed. Results: This research has thus produced calculational dependencies for differential adjustment for change in all the concrete properties that have to be invoked in calculation. Conclusions: Strength- and strain-related properties of concrete that vary depthwise (i.e. across the section) are applied in the calculation procedures for more accurate and complete utilization of the available load-carrying capacity of concrete elements...

1. Introduction

Design of reinforced concrete elements based on nonlinear dependencies are listed in the standards of many countries [1-3], has set a question about their practical calculation for most of the engineers. This is because the usage of the formulas of these rules without a computer [1-3] is practically impossible. It is generally accepted that any calculation by using computers should be checked or evaluated through classical and practical techniques. Creating such a technique would allow making an assessment, verification and analysis conducted computer calculations. Simplified and approximate methods [4-7] not correspond to exact solution that is obtained by nonlinear dependencies in most of cases.

Calculation of resistibility performed by using the equation of equilibrium of external forces and internal forces, deformation diagrams of concrete and reinforcement and the function of changes of deformation by height section. As a function of diagrams of concretes deformation should be taken like function of stresses in the concrete which would correspond to the conditions of nonlinear deformation of concrete. The following dependences are proposed for calculations of nonlinear structures in the standards [1]: formula of rules Eurocode 2 (3.4), fifth degree polynomial (3.5), two- and three a linear dependence between the stresses and deformations. The dependence of stress-deformation for reinforcement is taken as two linear Prandtl diagrams. The distribution of deformation by height section at the moment of the destruction taking in a linear dependence like:

$$\varepsilon = \frac{1}{r} x \; or \; x = \frac{\varepsilon}{\frac{1}{r}}, or \frac{1}{r} = \frac{\varepsilon}{x}, \tag{1}$$

where ε – relative deformation of the material at a distance x from neutral line, 1/r – curvature of element in section.

In most cases, diagrams of concrete deformation are function of related strength and deformation characteristics: the calculated resistance of concrete to axial junction f_c, concrete deformation module E_c and limits of concrete deformation ε_{c1}, ε_{cu}. This can be expressed in the following functional dependence

$$\sigma_c = f(f_c, E_c, \varepsilon_{c1}, \varepsilon_{cu}). \tag{2}$$

There are many formulas that associate deformation and strength characteristics of concrete. The following expressions can be written by its summarizing

$$E_c = f(f_c), \varepsilon_{c1} = f(f_c), \varepsilon_{cu} = f(f_c). \tag{3}$$

With taking into account (3) the final dependence (2) will take the following form

$$\sigma_c = f(f_c). \tag{4}$$

The function laid down in the current norms [1,2] will be examined for further research. It is also called Eurocode function

$$\sigma_c = \frac{E_c\varepsilon_c - f_{cm}\left(\frac{\varepsilon_c}{\varepsilon_{c1}}\right)^2}{1 + \left(\frac{E_c\varepsilon_{c1}}{f_{cm}} - 2\right)\frac{\varepsilon_c}{\varepsilon_{c1}}} \; or \; \frac{\sigma_c}{f_c} = \frac{k\eta - \eta^2}{1 + (k-2)\eta}, \tag{5}$$

where $\eta = \varepsilon_c / \varepsilon_{c1}$, $k = 1.05\, E_c\varepsilon_{c1} / f_c$.
Here is the expression (4) for the function (5)

$$\sigma_c = f_c \times f(k, \eta) = f_c \times f(f_c). \tag{6}$$

The bending reinforced concrete elements at a single reinforcement are proposed to consider. After putting equilibrium equation and conducting simple transformation with considering the hypothesis of flat sections we will get:

- for non-overreinforced

$$\left(\frac{\int_0^{\varepsilon_c} \sigma_c\varepsilon_c d\varepsilon_c}{\left(\int_0^{\varepsilon_c} \sigma_c d\varepsilon_c\right)^2} - \frac{\varepsilon_c}{\int_0^{\varepsilon_c} \sigma_c d\varepsilon_c}\right)\rho_f^2 f_{yd}^2 + \rho_f f_{yd} = \frac{M_{Ed}}{bd^2}; \tag{7}$$

- for overreinforced

$$\left(\frac{\int_0^{\varepsilon_c} \sigma_c\varepsilon_c d\varepsilon_c}{\left(\int_0^{\varepsilon_c} \sigma_c d\varepsilon_c\right)^2} - \frac{\varepsilon_c}{\int_0^{\varepsilon_c} \sigma_c d\varepsilon_c}\right)\rho_f^2 E_s^2 \varepsilon_c^2 + \rho_f E_s \varepsilon_c = \frac{M_{Ed}}{bd^2}. \tag{8}$$

Both sides of equation (7) і (8) are divided into fc and taking into account (6), we will get:

- for non-overreinforced

$$\left(\frac{\int_0^{\varepsilon_c} f(f_c)\varepsilon_c d\varepsilon_c}{\left(\int_0^{\varepsilon_c} f(f_c) d\varepsilon_c\right)^2} - \frac{\varepsilon_c}{\int_0^{\varepsilon_c} f(f_c) d\varepsilon_c}\right)\frac{\rho_f^2 f_{yd}^2}{f_c^2} + \frac{\rho_f f_{yd}}{f_c} = \frac{M_{Ed}}{f_c bd^2}; \tag{9}$$

- for overreinforced

$$\left(\frac{\int_0^{\varepsilon_c} f(f_c)\varepsilon_c d\varepsilon_c}{\left(\int_0^{\varepsilon_c} f(f_c)d\varepsilon_c\right)^2}-\frac{\varepsilon_c}{\int_0^{\varepsilon_c} f(f_c)d\varepsilon_c}\right)\frac{\rho_f^2 f_{yd}^2 E_s^2 \varepsilon_c^2}{f_c^2 f_{yd}^2}+\frac{\rho_f f_{yd} E_s \varepsilon_c}{f_{yd} f_c}=\frac{M_{Ed}}{f_c b d^2}. \tag{10}$$

The following notations are introduced

$$\omega=\frac{\rho_f f_{yd}}{f_{cd}}, \eta_s=\frac{E_s \varepsilon_c}{f_{yd}}. \tag{11}$$

The name parameter ω is mechanical reinforcement coefficient [4, 5]. The formulas (9) and (10) with introduced notations (11) will take the following form:

- for non-overreinforced

$$\left(\frac{\int_0^{\varepsilon_c} f(f_c)\varepsilon_c d\varepsilon_c}{\left(\int_0^{\varepsilon_c} f(f_c)d\varepsilon_c\right)^2}-\frac{\varepsilon_c}{\int_0^{\varepsilon_c} f(f_c)d\varepsilon_c}\right)\omega^2+\omega=\frac{M_{Ed}}{f_c b d^2}; \tag{12}$$

- for overreinforced

$$\left(\frac{\int_0^{\varepsilon_c} f(f_c)\varepsilon_c d\varepsilon_c}{\left(\int_0^{\varepsilon_c} f(f_c)d\varepsilon_c\right)^2}-\frac{\varepsilon_c}{\int_0^{\varepsilon_c} f(f_c)d\varepsilon_c}\right)\omega^2\eta_s^2+\omega\eta_s=\frac{M_{Ed}}{f_c b d^2}. \tag{13}$$

The notation is introduced to both equations

$$k_z=f(f_c,\omega)=\frac{6M_{Ed}}{f_c b d^2}=\frac{M_{Ed}}{f_c W_c}. \tag{14}$$

Thereby

- for non-overreinforced

$$k_z=\frac{1}{6}\left(\left(\frac{\int_0^{\varepsilon_c} f(f_c)\varepsilon_c d\varepsilon_c}{\left(\int_0^{\varepsilon_c} f(f_c)d\varepsilon_c\right)^2}-\frac{\varepsilon_c}{\int_0^{\varepsilon_c} f(f_c)d\varepsilon_c}\right)\omega^2+\omega\right); \tag{15}$$

- for overreinforced

$$k_z=\frac{1}{6}\left(\left(\frac{\int_0^{\varepsilon_c} f(f_c)\varepsilon_c d\varepsilon_c}{\left(\int_0^{\varepsilon_c} f(f_c)d\varepsilon_c\right)^2}-\frac{\varepsilon_c}{\int_0^{\varepsilon_c} f(f_c)d\varepsilon_c}\right)\omega^2\eta^2+\omega\eta_s\right). \tag{16}$$

Connection between the calculated resistance of reinforced concrete and characteristics that derived above is following

$$f_{zM}=k_z f_c \tag{17}$$

Introduced parameter k_z generally depends on the mechanical reinforcement coefficient $\overline{\omega}$ and deformation characteristics of concrete. There are more details about the impact of deformation characteristics of concrete on bearing capacity of bending elements. Here is a submission of parameter dependence schedule k_z depending on the mechanical reinforcement coefficient $\overline{\omega}$ for different kinds of concrete. The impact of deformation characteristics of concrete (parameters $\eta = \varepsilon_c / \varepsilon_{c1}$, $k = 1.05\, E_c\varepsilon_{c1} / f_c$ on bearing capacity of bending reinforced concrete elements for considered kinds of concrete varies with differences within 10 %. This error is completely allowed by normative coefficients of variation of strength for such elements. This makes possible to offer slightly simplified force model of reinforced concrete elements calculation. The impact of deformation characteristics of strength sections is ignored and common functional dependence is taken $k_z = f(\overline{\omega})$ for all kinds of concrete. This dependence is true not only for different kinds of concrete and reinforcement and even for different duration of the load.

Formulas for non-central compression can be derived by similar arguments. Dependence of parameter k_z from the mechanical reinforcement coefficient π is presented in tables.

The strength conditions by a modified method of calculation resistances are formulated:

- bending elements

$$M_{Ed}=W_c f_{zM} \tag{18}$$

where f_{zM} – estimated resistance of reinforced concrete on bend, which depends on kinds of concrete and reinforcement, section shapes and percent of reinforcement section that defined by the expression (17). Parameter k_z in expression (7) is calculated by table depending on the mechanical reinforcement coefficient $\bar{\omega}$.

- non-central compressed

$$X_{Ed} = A_C f_{zN} \tag{19}$$

where f_{zN} – estimated resistance of reinforced concrete on bend, which depends on kinds of concrete and reinforcement, section shapes and percent of reinforcement section, relative initial eccentricity of the force application e_0 / d, that defined by the expression $f_{zN} = k_z f_c$, Parameter k_z is also calculated by table depending on the mechanical reinforcement coefficient $\bar{\boldsymbol{\omega}}$; A_c – working sectional area of concrete: for rectangular elements $A_c = bd$; for round ones $A_c = \pi d2 / 4$.

2. Materials and Methods

Concrete centrifugation is known [8-18] to utilize the ability of concrete mix to condense itself within a rotating mold due to the effects of centrifugal forces proportional to the mass of particles, the squared rotation speed and distance to the axis of rotation:

$$Fc.f. = \frac{4}{3g}\pi \cdot r_{c.a}^{3} \cdot \rho_{c.a.} \cdot l \cdot \omega^2 \tag{20}$$

where l is the center-of-grain to center-of-rotation distance, m; ω is the angular rotation speed, rad/s:

$$\omega = \frac{\pi \cdot n}{30} \tag{21}$$

Centrifugal force effects on a coarse aggregate grain (Figure 1) are proportional to the radius $r^3_{c.a.}$ and the density $\rho_{c.a.}$ of the particle. This causes larger grains to move towards the outer layer and the smaller ones towards the inner layer.

We herein propose focusing on a different, more generic formula for calculating the centrifugal force, where $F_{c.f.} = f(\omega, l)$:

$$F_{c.f.} = |m| \cdot \omega^2 \cdot l \tag{22}$$

where m is the mass of the rotating solid; ω is the angular rotation speed; l is the center-of-grain to center-of-rotation distance.

In this regard, centrifugated concrete differs from vibrated concrete in a sense that it has less uniform depthwise distribution of coarse aggregate grains, which results in different layers having different properties, i.e. the section becomes variatropic.

This applies even more to vibration-centrifugated concrete which has even stronger variatropy. In these elements, the outer layers have a greater concentration of coarse aggregate grains, thus greater strength.

All of this shows why the strength- and strain-related difference in variatropic-section concrete must be adjusted for.

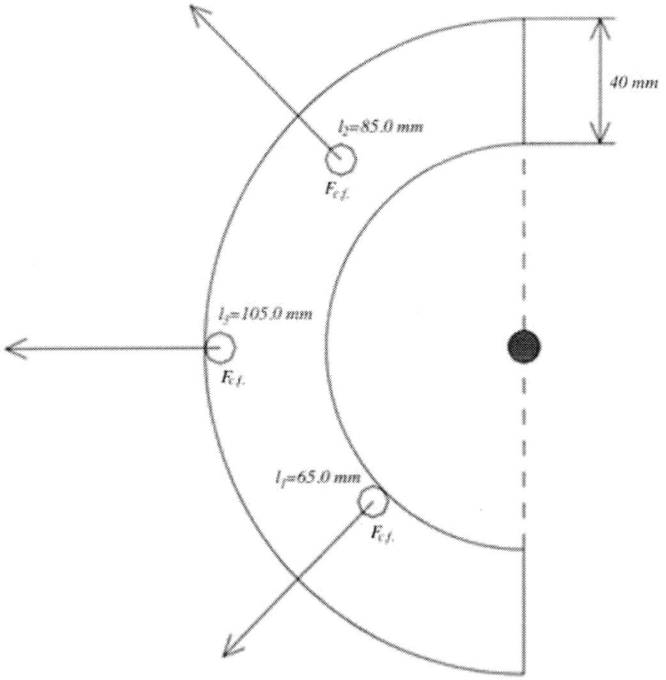

Figure 1. How centrifugal forces act on coarse aggregate grains across the wall of a centrifugated product.

To confirm this, calculate the effects of centrifugal forces, for instance, on the grains of a dense (granite) aggregate. Substitute the input data in (1) to find the centrifugal force acting on granite grains, see Table 1.

Table 1. Basic parameters of the acting centrifugal force.

Coarse aggregate grain radius, mm	Center-of-grain to center-of-rotation distance, m	Angular speed, rad/s	Mold rotation speed when compacting the concrete, rpm	Pressure on the concrete mix, p kgf/cm^2	Centrifugal force acting on the coarse aggregate grains, H
10.0	0.105	84	800	0.7	$316.5 \cdot 10^{-6}$
		157	1,500	2.16	$1105.6 \cdot 10^{-6}$

Using mathematical experiment design, assume a 2^k full factorial design FFD. Following our own recommendations [19-23], assume the concrete mix centrifugation time is 4.2 min. Table 2 presents the values of varying factors and their physical sense.

The following parameters are adopted as the response function:

- Y1 (X1, X2) is the axial compressive strength, MPa;
- Y2 (X1, X2) is the axial tensile strength, MPa;
- Y3 (X1, X2) is the ultimate axial compressive strain, mm/m;
- Y4 (X1, X2) is the ultimate axial tensile strain, mm/m;
- Y5 (X1, X2) is the elastic modulus, MPa;

Table 2. Varying factor values in FFD 2^k.

Factor code	Physical sense of the factor	Unit	Variation range	Levels of factors

				-1	0	+1
X_1	Center-of-grain to center-of-rotation distance.	mm	±50	100	150	200
X_2	Angular speed	rad/s	±36	84	120	156

3. Results and Discussion

The least squares method produced basic regression equations represented by 2nd degree polynomials:

$$Y(X_1, X_2) = B_0 + B_1 \cdot X_1 + B_2 \cdot X_2 + B_3 \cdot X_1 \cdot X_2 + B_4 \cdot X_1^2 + B_5 \cdot X_2^2 \qquad (23)$$

Statistical analysis of the obtained regression equations was tested for variance homogeneity, significance of the coefficient, and adequacy tested by the Fisher test.

The experiment design and the optimization parameters were based on the test results obtained in [27-31].

The significance of the equation coefficients was based on Student's t-test. Data was processed statistically in Mathcad, which ultimately produced 30 regression equations (2nd degree polynomials) for centrifugated and vibration-centrifugated concrete.

The equations were interpreted and transformed into calculational equations for finding the differential properties of variatropic section layers in centrifugated and vibration-centrifugated concrete.

The proposed calculational equations are presented in Table 3 and Table 4.

Table 3. Calculational equations for finding the differential properties of variatropic section layers in centrifugated concrete.

Concrete properties	Centrifugated concrete
	Written equation
	Inner layer
Axial compressive strength, MPa;	R_b= 23.25+0.877 l-1.446 ω-1.618 l ω-1.047 l^2+0.345 ω^2
Axial tensile strength, MPa	R_{bt}= 3.852+0.767 l -0.204 ω -0.892 l ω-0,311 1^2-0.189 ω^2
Ultimate axial compressive strain, mm/m	ε_{bR}=3.247-1.245 l+0.027 ω-0.417 l ω+0.006 l^2-0.032 ω^2
Ultimate axial tensile strain, mm/m	ε_{btR}= 2.679-0.788 l-0.215 ω-0,509 l ω-0.34 l^2-0.056 ω^2
Elastic modulus, MPa	E_b= E_{bt}= 21.32+4.86 l-0.639 ω+0.299 l ω+1.168 l^2-0.38 ω^2
	Middle layer
Axial compressive strength, MPa;	R_b= 27.54+1.147 l -1.226 ω-1.394 l ω-0.899 l^2+0.675 ω^2
Axial tensile strength, MPa	R_{bt}= 4.356+1.024 l -0.051 ω-0.567 l ω-0.181 l^2-0.137 ω^2
Ultimate axial compressive strain, mm/m	ε_{bR}=4.341-1.031 l+7.29·10^{-3} ω-0.203 l ω+0.038 l^2+0.01 ω^2
Ultimate axial tensile strain, mm/m	ε_{btR}= 3.021-0.532 l-0.015 ω-0.353 l ω-0.01 l^2+0.001 ω^2
Elastic modulus, MPa	E_b= E_{bt}= 26.507+5.67 l-0.442 ω+0.572 l ω +1.314 l^2-0.05 ω^2
	Outer layer
Axial compressive strength, MPa;	R_b= 35.89+3.166 l+1.287 ω+0.812 l ω-0.015 l^2+1.781 ω^2
Axial tensile strength, MPa	R_{bt}= 6.184+1.867 l-0.83 ω-0.336 l ω+0.452 l^2-0.06 ω^2
Ultimate axial compressive strain, mm/m	ε_{bR}= 5.31-0.543 l+1.456 ω +1.002 l ω+0.569 l^2+0.59 ω^2
	ε_{btR}=2.886+1.592 l-0.273 ω+0.987 l ω+0.256 l^2+0.043 ω^2

Ultimate axial tensile strain, mm/m Elastic modulus, MPa	E_b=E_{bt}= $29.111+6.49\,l+1.65\,\omega-0.399\,l\,\omega+1.678\,l^2+0.68\,\omega^2$

Table 4. Calculational equations for finding the differential properties of variatropic section layers in vibration-centrifugated concrete.

Concrete properties	Vibration-centrifugated concrete
	Written equation
	Inner layer
Axial compressive strength, MPa;	$R_b = 32.31+8.774\,l-0.450\,\omega+0.67\,l\,\omega+0.308\,l^2+0.809\,\omega^2$
Axial tensile strength, MPa	$R_{bt}= 3.882+0.459\,l-0.043\,\omega-0.49\,l\,\omega+1.282\,l^2+1.511\,\omega^2$
Ultimate axial compressive strain, mm/m	$\varepsilon_{bR}= 3.214-1.412\,l+0.037\,\omega-0.82\,l\,\omega-1.07\,l^2+0.086\,\omega^2$
Ultimate axial tensile strain, mm/m	$\varepsilon_{btR}= 1.95-0.825\,l+0.007\,\omega+1.16\,l\,\omega+0.778\,l^2+0.534\,\omega^2$
Elastic modulus, MPa	$E_b=E_{bt}= 31.11+7.626\,l-0.117\,\omega-1.12\,l\,\omega-5.134\,l^2-1.223\,\omega^2$
	Middle layer
Axial compressive strength, MPa;	$R_b = 38.467+10.506\,l-0.231\,\omega+0.7\,l\,\omega+0.558\,l^2-0.05\,\omega^2$
Axial tensile strength, MPa	$R_{bt}= 5.089+0.768\,l-0.02\,\omega-0.35\,l\,\omega-0.04\,l^2+0.046\,\omega^2$
Ultimate axial compressive strain, mm/m	$\varepsilon_{bR}= 3.983-1.202\,l+0.075\,\omega-0.14\,l\,\omega+0.041\,l^2+0.005\,\omega^2$
Ultimate axial tensile strain, mm/m	$\varepsilon_{btR}= 2.48-0.637\,l+0.017\,\omega-0.052\,l\,\omega-0.029\,l^2+0.083\,\omega^2$
Elastic modulus, MPa	$E_b=E_{bt}= 38.129+9.321\,l-0.51\,\omega+0.089\,l\,\omega-3.713\,l^2+0.638\,\omega^2$
	Outer layer
Axial compressive strength, MPa;	$R_b = 48.45+13.77\,l+0.65\,\omega+1.44\,l\,\omega+2.014\,l^2+0.45\,\omega^2$
Axial tensile strength, MPa	$R_{bt}= 8.132+1.56\,l+1.09\,\omega+0.321\,l\,\omega+0.563\,l^2+0.12\,\omega^2$
Ultimate axial compressive strain, mm/m	$\varepsilon_{bR}= 5.451+0.85\,l+1.345\,\omega+0.47\,l\,\omega+1.047\,l^2+0.06\,\omega^2$
Ultimate axial tensile strain, mm/m	$\varepsilon_{btR}= 5.98+1.12\,l-0.003\,\omega-0.008\,l\,\omega+0.844\,l^2+0.141\,\omega^2$
Elastic modulus, MPa	$E_b=E_{bt}= 43.18+14.23\,l+1.231\,\omega+1.157\,l\,\omega-0.09\,l^2+1.98\,\omega^2$

Summary

The findings of this study are as follows.

The paper experimentally and theoretically substantiates that different layers of centrifugated and vibration-centrifugated concrete are exposed to varying-in-magnitude inertia forces resulting in sectional (depthwise) variatropy and differentiation of strength- and strain-related properties.

It shows that the key affecting factors are centrifugal and centripetal forces, which are functions of the distance to the center of rotation and angular rotation speed.

To evaluate the differential properties of centrifugated and vibration-centrifugated properties, the paper proposes versatile calculational dependencies, where the concrete properties (strength, ultimate strain, and elastic modulus in compression and tension, or their absolute or relative increments) serve as the functions, while the distance to the center and the angular speed of rotation serve as the arguments.

The analytical dependencies for finding the differential strength- and strain-related properties of centrifugated and vibration-centrifugated concrete enable adjustment for the concrete properties that vary depthwise for a more complete utilization of the load-carrying capacity of structures.

References

[1] Filatov V B, Suvorov A A 2016 *Procedia Engineering* **153** 144–150.
[2] Panfilov D A, Pischulev A A, Romanchkov V V 2016 *Procedia Engineering* **153** 531-536.
[3] Pol'skoi P P, Mailyan D R, Georgiev S V 2014 *Engineering Herald of the Don* **4**.
[4] Radaikin O V 2019 *Bulletin of Belgorod State Technological University named after V. G. Shukhov* **10** 29–39.
[5] Kholodnyak M G, Stel'makh S A, Shcherban' E M, Tret'yakov D A, Dao V N, Zaikin V I 2018 *Bulletin of the Eurasian Science* **10** 6.
[6] Maruyama I, Lura P 2019 *Cement and Concrete Research* **123** 105770.
[7] Kim J-J, Yoo D-Y 2019 *Cement and Concrete Composites* **103** 213–223.
[8] Li K, Li L 2019 *Cement and Concrete Research* **124** 105811.
[9] Kirthika S K, Singh S K 2020 *Construction and Building Materials* **250** 118850.
[10] Hameed M, Maula B, Bahnam Q 2019 *International Review of Civil Engineering* **10 6** 17061.
[11] Alexander M, Beushausen H 2019 *Cement and Concrete Research* **122** 17–29.
[12] Geiker M R, Michel A, Stang H, Lepech M D 2019 *Cement and Concrete Research* **122** 189–195.
[13] Khalaf M A, Cheah C, Ramli M 2019 *Construction and Building Materials* **215** 73–89.
[14] Mailyan L R, Stel'makh S A, Shcherban' E M, Kholodnyak M G 2020 *Russian Journal of Building Construction and Architecture* **1 (45)** 6–14.
[15] Stel'makh S A, Shcherban' E M, Kholodnyak M G 2019 *IOP Conf. Ser.: Mater. Sci. Eng.* **687** 022008.
[16] Stel'makh S A, Shcherban' E M, Shuyskiy A I, Nazhuev M P 2018 *Materials Science Forum* **931** 502–507.
[17] Stel'makh S A, Shcherban E M, Zholobova O A 2018 *IOP Conf. Ser.: Mater. Sci. Eng.* **463** 022056.
[18] Shuyskiy A I, Stel'makh S A, Shcherban' E M, Kholodnyak M G 2018 *Materials Science Forum* **931** 508–514.

Investigation of the volumetric fractal structure shaping

G M Kravchenko*, L I Pudanova

Don State Technical University, 1 Gagarin sq., Rostov-on-Don, 344000, Russia

E-mail: galina.907@mail.ru

Abstract. On the basis of a specially developed algorithm, shaping of the fractal structures, the formation of which is regulated by the power index of the fractal, has been carried out. A sphere and an ellipsoid were used as the basic shapes. The formation analysis has made it possible to determine the characteristic principles of the fractal structure growth. The performed numerical experiment with respect to volumetric fractal structures of power 8 made it possible to determine the spectrum of frequencies and forms of natural oscillations. The offered objects of fractal architecture are analogs of unique high-rise and large-span buildings and structures. The application of the fractal shaping paradigm in combination with digital technologies and innovative materials will make it possible to create completely new objects of architecture.

Introduction

The variability in buildings and structures design expands significantly with the fractal shaping paradigm introduction. Fractal geometry is a comparative new theory, widely developed through the use of modern digital technologies. Volumetric fractal develops in space within the limits of mathematical equations and algorithms, is in constant motion and tends simultaneously to order and chaos [1-3].

Fractal structures can be used in the construction industry as a shell, a building frame, or as separate or groups of structures [4]. A feature of the fractal geometry use in the unique buildings and structures' design is the need to use innovative materials and technologies that allow the complex structures implementation [5]. According to the authors, the most rational technology for creating volumetric fractal structures is 3D printing technology. The development of 3D printing makes it possible to create the objects that differ in unique geometry, properties and material [6-7].

The object of the research is the Mandelbrot volumetric fractal.

The modeling of a three-dimensional Mandelbrot fractal and the study of the external fractal structure formation has been performed in this work.

Research methods and principles

3D Mandelbrot fractal is an analogue of the Mandelbrot set, created by Daniel White and Paul Nielander using hypercomplex algebra based on spherical coordinates. The study uses the concept of the power of a fractal or quasi-volume. The fractal power is taken as an analogue of the local dimension of the fractal set. For the power greater than three, the result is a three-dimensional Mandelbrot fractal [8].

Modeling a volumetric Mandelbrot fractal is carried out using a special program "External structure of a fractal set", developed by the authors of the study [9].

The purpose of the algorithm implemented in the program is to determine the points belonging to the outer surface of the Mandelbrot fractal with the subsequent construction of a finite element model in the PC SCAD.

Determination of the points coordinates is carried out by checking their belonging to the surface of the fractal structure after a given number of iterations or generations. Points membership is checked in spherical coordinates, changing cyclically the distance from the center, then the horizontal and vertical angles. The result of the calculations is the coordinates of the points and sets of finite elements that form the external structure of the fractal set. There is a rational distribution of the points density over the surface to create the most optimal mesh of finite elements.

To obtain a rational external fractal structure, it is necessary to correctly specify the initial data when calculating the shaping in the program "External structure of a fractal set". The initial data affect the quality of the finite element mesh and, accordingly, the accuracy of the stress-strain state calculations.

Modeling of fractal structures is implemented taking into account the introduction of 3D printing technology into the process. 3D printers create real objects based on a computer volumetric model using specialized methodologies. The advantages of 3D printers in construction are the low cost of the manufactured structure, efficiency and quality, reduced labor costs and reduced waste. A distinctive feature is the ability to erect the buildings with a unique frame in a short time. The development of robotics and the introduction of 3D printers that use innovative materials will contribute to the fractal architecture concepts' implementation.

Results

Complex modeling of the endostructure and exostructure of a fractal represents a new stage in the development of parametric design of fractal architecture objects. The proposed object of fractal architecture is not only based on a fractal form, but also repeats the nature algorithms.

Modeling of different forms of a fractal structure was carried out in the program "External structure of a fractal set" with the transfer of topological data to the PC SCAD. Models of the external fractal structure of the fourth, eighth and fourteenth powers with identical initial parameters have been developed (Figure 1). It should be noted that the fractal structure of the first power is a sphere that transforms into new types of structure with an increase in fractal power.

a) b) c)

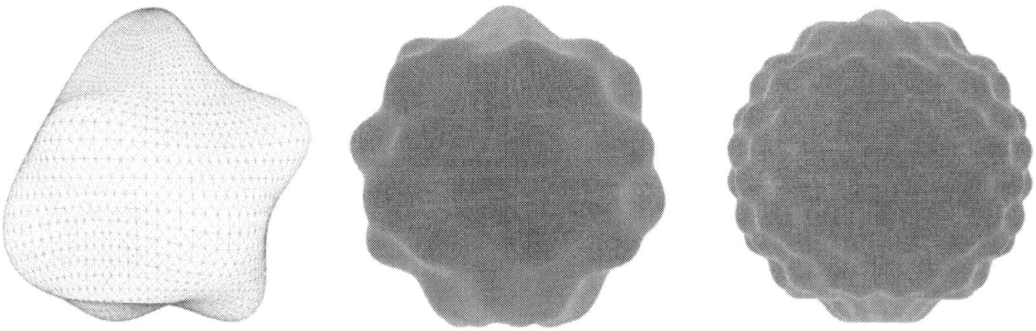

Figure 1. Volumetric Mandelbrot fractal: a) power 4; b) power 8; c) power 14.

The external structure of a bulk fractal includes an exobase, an f-quark surface, and a corona [10]. According to the authors, the fractal structure acquires its original and familiar form with a power value of 8. External structures of low power differ significantly from each other and have little in common with the classical form of the Mandelbrot volumetric fractal. Power 4 structure (Figure 1 (a)) has fundamental differences from power 8 and 14 structures (Figure 1 (b) and 1 (c), respectively). Low-power structures are characterized by shaping asymmetry. The higher powers of the fractal are distinguished by the symmetry of the forms.

During the study, it was found that the corona and exobase become more detailed with increasing power, but at the same time they occupy a smaller volume of the entire volume fractal [10]. The surface of f-quarks becomes pronounced, but with increasing power it occupies an ever-larger volume of the structure.

The program "External structure of a fractal set" makes it possible to change the structure of a three-dimensional Mandelbrot fractal by pulling it in different directions. For a more detailed study of the shaping, the developed program provides for a change in the initial parameters, which makes it possible to take the rotation ellipsoid as the first power. For the first time, the structures that are visually comparable with the concept of shaping of high-rise unique buildings, skyscrapers have been obtained.

Figure 2 shows the examples of fractal structures of 8 and 14 powers with development along the vertical axis. In the process of shaping, the relationship between the power value and the external structure components' detail remains.

a) b)

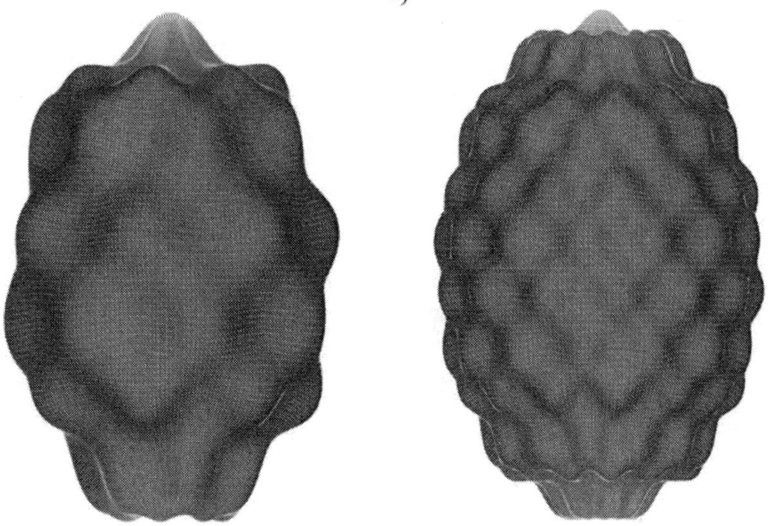

Figure 2. Mandelbrot ellipsoidal fractal: a) power 8; b) power 14.

As part of the fractal structure formation study, a modal analysis of the exostructures of the power 8 fractal was performed in the SCAD software package (Figure 1 (b), Figure 2 (a)). It should be noted that the free oscillations spectrum gives an idea of the possible ways of the structure deformation in case when the structure would be a shell of a building or structure. Modal analysis gives an opportunity to establish resonant frequencies corresponding to the most dangerous bending-torsional modes of natural oscillations.

In a numerical experiment, a frequency spectrum with twenty modes of natural oscillations was obtained. The results are summarized in Table 1.

Table 1. Forms and frequencies of natural oscillations.

Spherical Mandelbrot fractal, power 8		**Mandelbrot ellipsoidal fractal, power 8**	
The form, No.	**Frequency Hz**	**The form, No.**	**Frequency Hz**
1	0.0993663	1	0.120958
2	0.0993663	2	0.121018
3	0.215402	3	0.299469
4	0.215402	4	0.299624
5	0.224461	5	0.301273
6	0.224461	6	0.301327
7	0.290477	7	0.30596
8	0.290477	8	0.305993
9	0.362324	9	0.353494

10	0.48646	10	0.66468
11	0.498527	11	0.664875
12	0.498527	12	0.82499
13	0.547859	13	0.825148
14	0.547859	14	0.873489
15	0.762812	15	0.906053
16	0.762812	16	0.906104
17	0.769759	17	0.928005
18	0.781128	18	0.928083
19	0.78113	19	1.032162
20	0.874072	20	1.032211

It should be noted that for a volumetric fractal power 8 structure, the first two modes of oscillation are translational, the third is torsional (Figure 3, Figure 4).

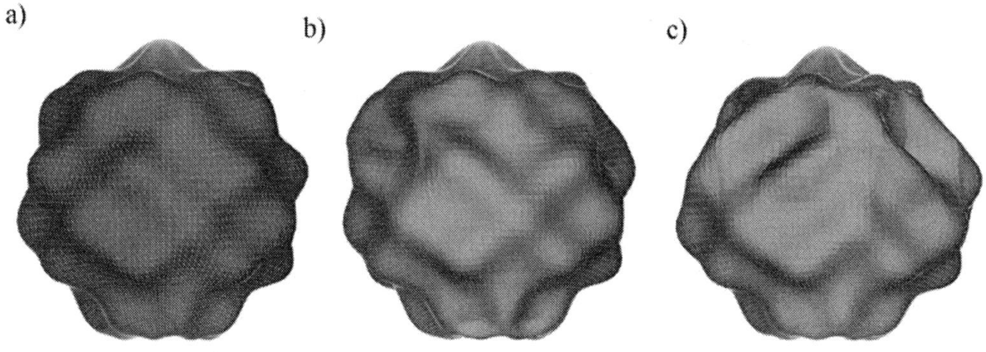

Figure 3. Mandelbrot spherical fractal a) initial structure; b) the first form of oscillation; c) the fourth mode of oscillation.

The frequencies of the Mandelbrot ellipsoidal fractal are higher than the frequencies of the spherical fractal of the corresponding oscillation modes. In this case, the modes of natural oscillations from 3 to 8 are characterized by significant displacements of the structure in the middle zone of the f-quark surface in different directions. In most cases, the structure corona movements are minimal or equal to zero.

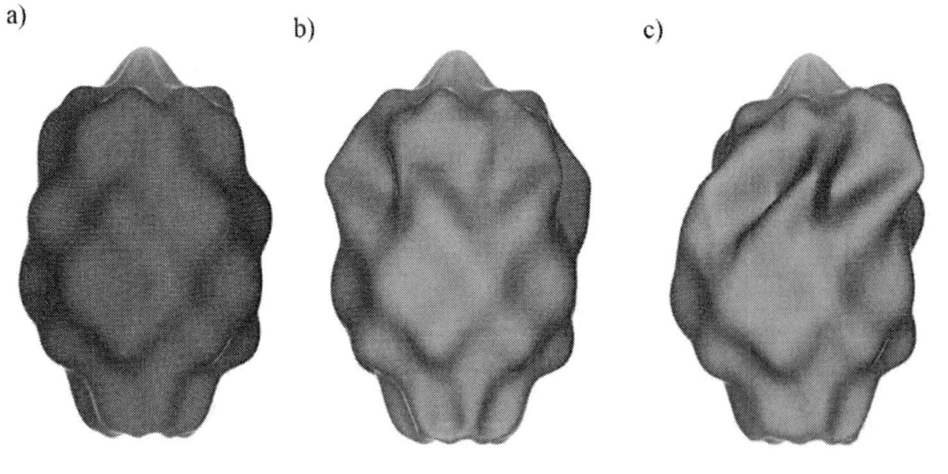

Figure 4. Mandelbrot ellipsoid fractal a) initial structure; b) the first form of oscillation; c) the fourth form of oscillation.

Discussion

The external fractal structure has features of existence and behavior, i.e., the fractals grow in a special way depending on the power and the iterative process. These features make it possible to form the structures of any type on the basis of a fractal structure, both long-span and high-rise. The unique buildings obtained as a result of shaping will have fractal properties, low weight, taking into account the use of innovative materials and create a comfortable interior space.

The design methodology for buildings and structures of fractal architecture can be significantly transformed with the help of mathematics, algorithms and digital technologies [11]. Application of the fractal shaping paradigm will make it possible to create completely new architectural objects that can be distinguished by their reliability and resistance to natural and man-made influences, aesthetics and a high degree of comfort of the endostructure.

Collaboration of fractal geometry, digital technologies and innovative materials will allow the construction industry to move to a completely new development level.

Summary

Models of volumetric fractal structures, which are the basis for further numerical experiments to study the stress-strain state of unique buildings, have been developed.

The external fractal structure formation study in the program "External structure of a fractal set" is carried out. For the first time, three-dimensional spherical and ellipsoidal fractal structures are obtained, which are analogous to high-rise and large-span buildings. External fractal structures were visualized using the SCAD software package and numerical experiments were carried out with the fractal power variation.

Dynamic calculation of the fractal architecture object made it possible to obtain a spectrum of twenty frequencies and modes of natural oscillations. The resonant frequencies corresponding to the most dangerous bending-torsional forms have been determined. An analysis of the exobase displacements, the surface of f-quarks and the corona is carried out for two models of the spherical and ellipsoidal Mandelbrot fractals. The results of the study allow us to conclude that fractal structures are resistant to natural and technogenic impacts. The structures are recommended to be implemented using innovative materials and 3D printing technology.

In conditions of dense urban development, it is necessary to use the principles of fractal shaping [12]. The architecture of modern large cities is a complex system with fractal properties that cannot be neglected when forming an urban network. Unique objects of fractal architecture will have cardinal differences from the existing buildings and structures, will be resistant to various types of external influences and will be able to harmoniously fit into the existing urban environment.

References

[1] Frame M L, & Mandelbrot B B 2002 Fractals, Graphics and Mathematical Education (Washington DC: Mathematical Association of America & Cambridge UK, The University Press).
[2] Mandelbrot B B 1982 The Fractal Geometry of Nature (San Francisco).
[3] Peitgen H-O, Richter P H 1986 The beauty of fractals (Springer-Verlag, Heidelberg).
[4] Kravchenko G M and Pudanova L I 2019 *IOP Conf. Ser.: Mater. Sci.* Eng. **698** 022017.
[5] Rayneau-Kirkhope D, Mao Y and Farr R 2012 Ultralight fractal structures from hollow tubes *Phys. Rev. Lett.* **109** 204301.
[6] Lipson H and Kurman M 2013 Fabricated: The new world of 3D printing (John Wiley & Sons).
[7] Das S, Bourell D L and Babu S S 2016 Metallic materials for 3D printing *Mrs Bulletin* **41** 729-741.
[8] Kravchenko G *2017* Modeling the External Structure of a Fractals *IOP Conference Series: Earth and Environmental Science* **90** 012100.
[9] Kravchenko G M, Vasil'ev S E, Pudanova L I 2017 *External structure of the fractal set* **2017615949,** 26.05.2017.

IOP Conf. Series: Materials Science and Engineering 1083 (2021) 012010 doi:10.1088/1757-899X/1083/1/012010

[10] Kravchenko G M, Pudanova L I 2018 The fractal Mandelbrot set and the shaping of the 3D fractal *Inženernyj vestnik Dona* **4**.
[11] Kravchenko G M, Vasil'ev S E, Pudanova L I 2017 Parametric architecture. The concept of sustainable development of science in modern conditions *Collection of papers of International scientific-practical conference Ufa* 264 – 267.
[12] Schumacher P 2009 Parametricism - A New Global Style for Architecture and Urban Design *AD Architectural Design - Digital Cities* **79 (4).**

Algorithmic method for modeling a parametric surface

G M Kravchenko*, E V Trufanova

Don State Technical University, 1, Gagarin Square Rostov-on-Don, 344000, Russia

E-mail: Galina.907@mail.ru

Abstract. The article discusses the peculiarity and relevance of the buildings and structures design using the latest computing technologies and digital modeling methods, which give a possibility to change the shape of a parametric architectural object. The formation and evolution of the surface of the crossed cap obtained from the Mobius strip is investigated. The key features of parametric modeling and the possibility of using this direction in the unique buildings and structures' design are considered. Recommendations for choosing a rational form of an object based on the frame dynamic characteristics' analysis and comparison are given. Visualization of the completed building of the sports and entertainment complex emphasizes the aesthetic component of parametric architecture.

1. Introduction

The use of computer technology has become an integral part in the parametric architecture objects' design. The latest digital modeling methods allow the variation of the surface equation parameters to create the new architectural forms of the object. The opportunity to embody complex geometric shapes, releasing from the images' linearity and symmetry [1,2] is relevant. A feature of parametric architecture is the consolidation of a complex spatial model and dynamics of natural forms [3].

In the parametric design environment, the following topical methods for obtaining object geometry can be distinguished: combinatorial modeling; algorithmic modeling method; adaptive systems [4]. Algorithmic modeling method is used to solve the design problems; based on changing the spaces parameters to form the new parametric shapes.

The aim of the study is to develop a unique building of unusual geometry by an algorithmic method using the analytical surface of a crosscap [5].

The crosscap is a non-orientable one-side surface obtained on the basis of the Moebius strip. The parametric form of specifying the surface of the crosscap has the form:

$$x = x(u,v) = (1+\cos\cdot 2u)\cdot\cos\cdot 2v;$$
$$y = y(u,v) = (1+\cos\cdot 2u)\cdot\sin\cdot 2v;$$
$$z = z(u,v) = \sin\cdot 2u\cdot\sin\cdot v,$$

where the parameters u, v vary within $-\pi/2 \le u \le \pi/2$; $0 \le v \le 2\cdot\pi$.

Figure 1 shows the parametric shapes of the investigated surface: a disk with a Mobius film, a dissected disk with a Mobius film, a self-intersecting Mobius disk.

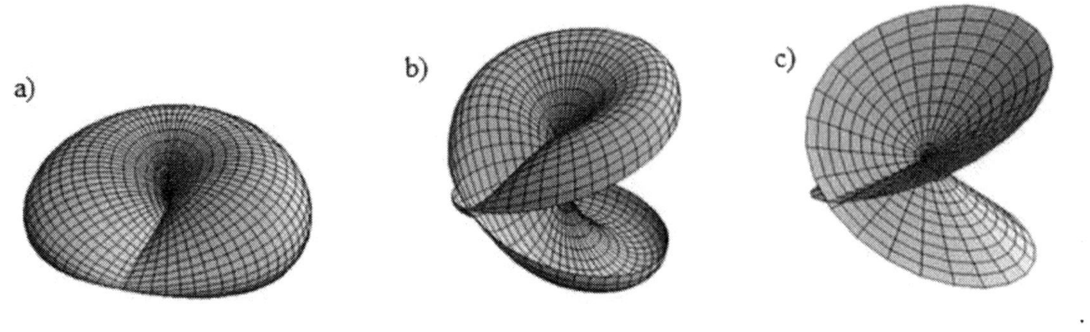

Figure 1. Parametric shapes: *a)* a disk with a Moebius film: *b)* a dissected disk with a Mobius film; *c)* self-intersecting disc

During the formation of a crossed cap, a Möbius surface with a circular boundary has two singular points, that is, self-intersection lines. Point with parameters (u,v)=(0;0) and u=π/2 are called Pinch points.

2. Materials and Methods

To study the primary forming evolution, the crosscap surface was constructed in the SAPFIR software package. The advantage of the SAPFIR software package is that it is possible to export the resulting file to the LIRA software package, which implements the finite element method.

In the process of studying the shape formation of the crosscap, the influence of the parameters u and v on the change in the Pinch point displacement, the shell bend radius was revealed.

The parameter u affects the displacement of the shell self-intersection singular point, respectively along the axes *x, y, z*. The parameter v affects the radius of shell bend during the surface formation along the *x, y, z* axes, respectively. The results of the algorithmic method for modeling a parametric surface are presented in Table 1.

Table 1. Results of the algorithmic method for modeling a parametric surface

Variation of the *u* parameter	Variation of the *v* parameter	Algorithmic modeling method
$x = (1+\mathbf{cos\cdot 1u})\cdot cos\cdot 2v$ 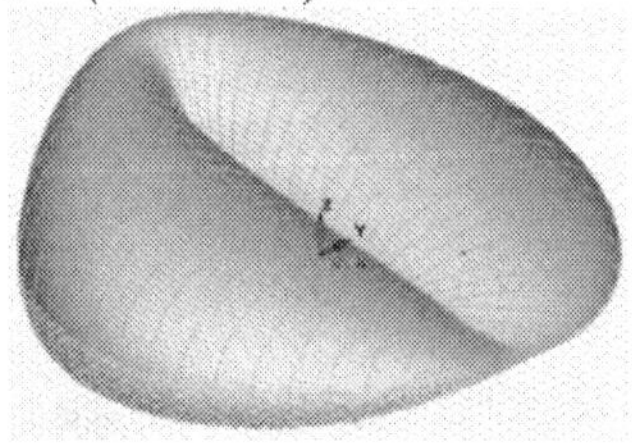	$x = (1+cos\cdot 2u)\cdot\mathbf{cos\cdot 1v}$ 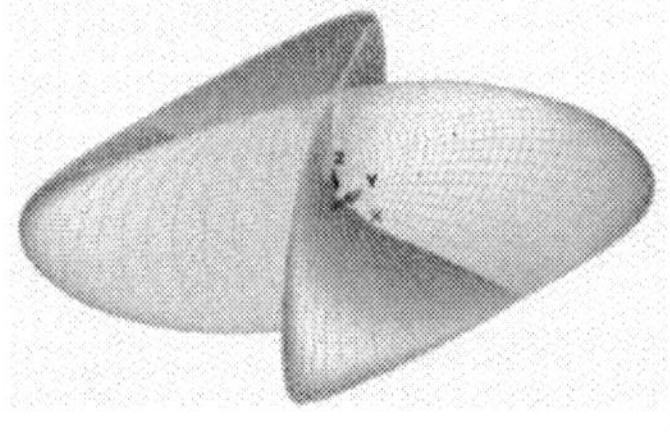	$x = (2+cos\cdot 2u)\cdot cos\cdot 2v$ 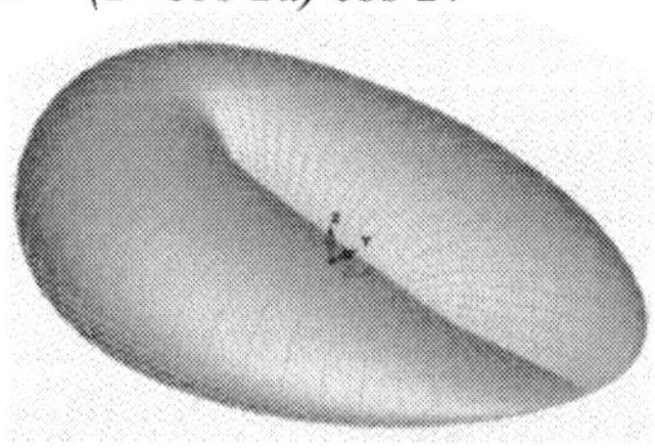
$y = (1+\mathbf{cos\cdot 1u})\cdot sin\cdot 2v$ 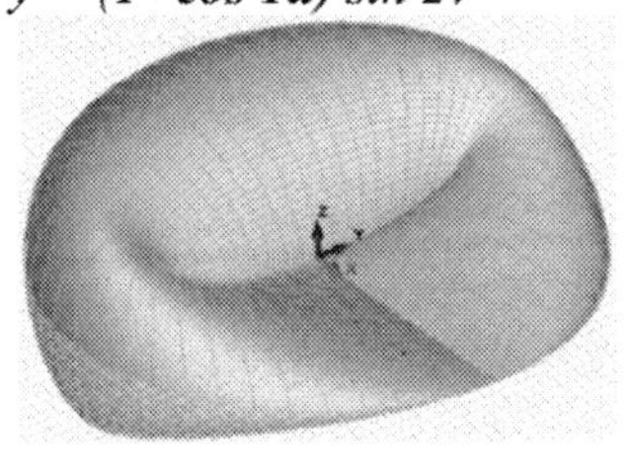	$y = (1+cos\cdot 2u)\cdot\mathbf{sin\cdot 3v}$ 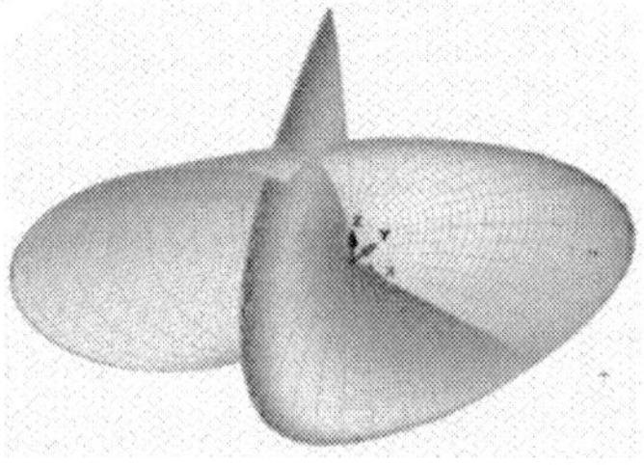	$y = (2+cos\cdot 2u)\cdot sin\cdot 2v$ 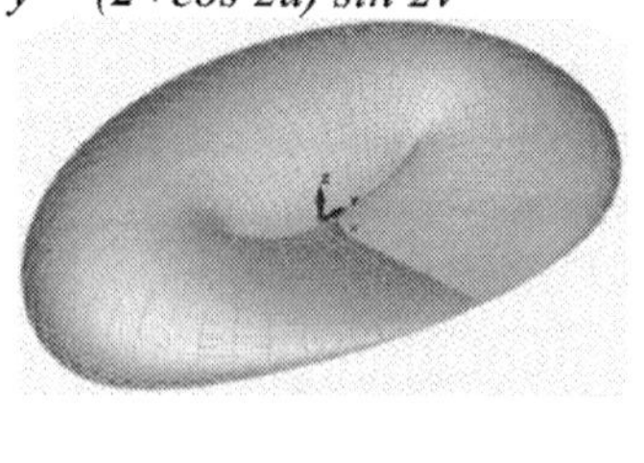

$z = \mathbf{sin \cdot 1u} \cdot sin \cdot v$ 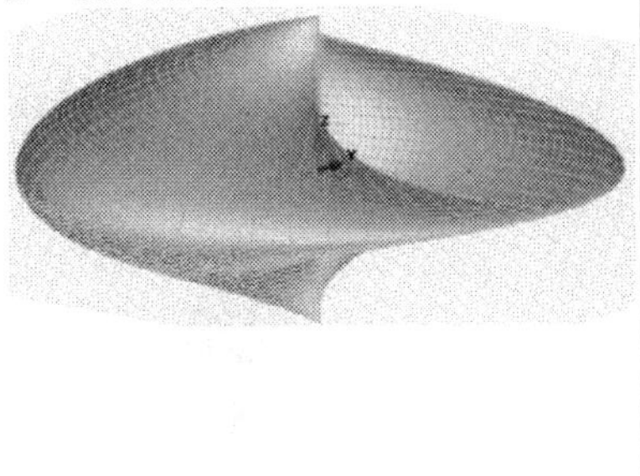	$z = sin \cdot 2u \cdot \mathbf{sin \cdot 3v}$ 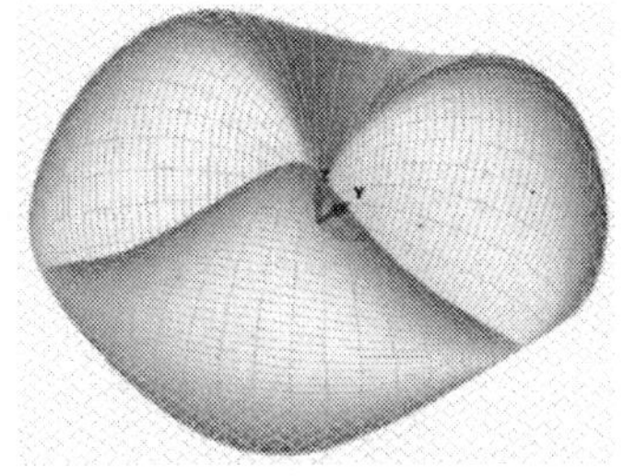	$z = 2 \cdot sin \cdot 2u \cdot sin \cdot v$ 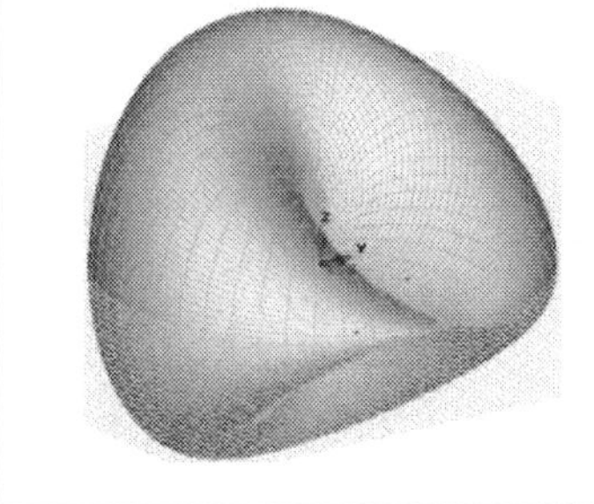

The result of studying the primary forming evolution is three types of algorithmic models obtained by varying the coefficients simultaneously in three surface equations. The types of algorithmic models are summarized in Table 2.

Table 2. Types of algorithmic models

Option 1	Option 2	Option 3
$x = (1+cos \cdot 1{,}5u) \cdot cos \cdot 2v$, $y = (1+cos \cdot 2u) \cdot sin \cdot 2v$, $z = sin \cdot 2u \cdot sin \cdot v$. 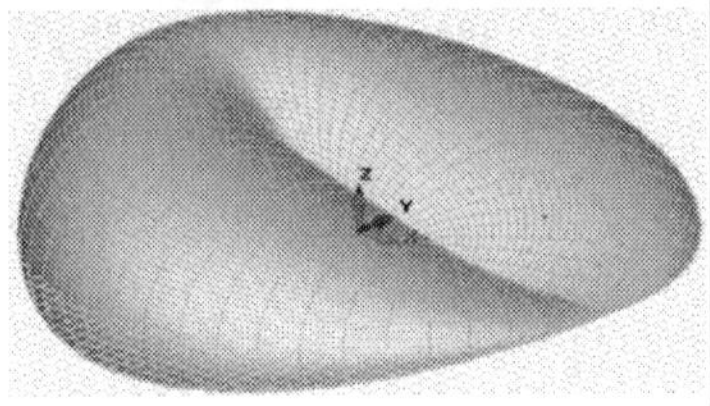	$x = (1+cos \cdot 1{,}1u) \cdot cos \cdot 2v$, $y = (1+cos \cdot 1{,}7u) \cdot sin \cdot 2v$, $z = sin \cdot 1{,}4u \cdot sin \cdot v$. 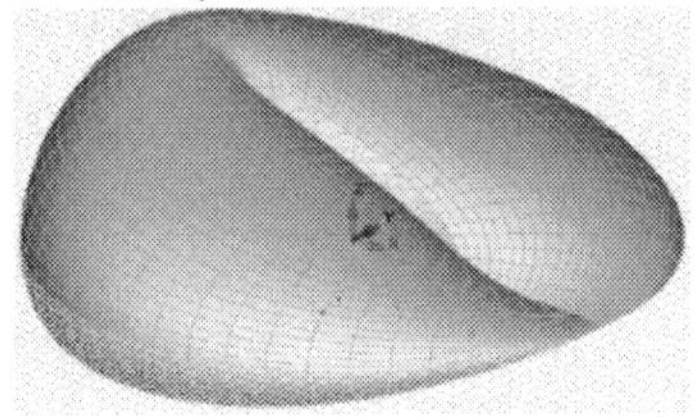	$x = (1{,}3+cos \cdot 5u) \cdot cos \cdot 2v$, $y = (1{,}2+cos \cdot 5u) \cdot sin \cdot 2v$, $z = sin \cdot 2u \cdot sin \cdot v$. 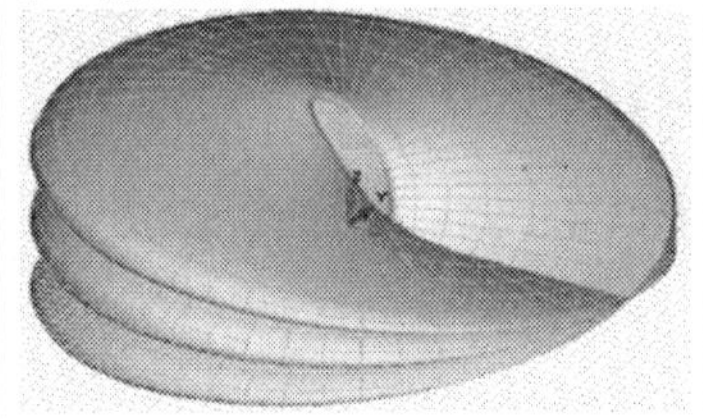

At the next survey stage, the selected variants of the crosscap surface were proposed for the development of the building design solutions for a sports and entertainment complex. A parametric architecture object modeling is carried out by the finite element method [6]. At the same time, the functional purpose of the object and the internal space zoning are taken into account.

Spatial rod finite element models have been developed in the SAPFIR and LIRA software complexes according to the selected primary forming options. The finite element model is approximated by the spatial bar finite elements with six degrees of freedom at the node. The boundary conditions of the computational models are set at the level of the foundation slab device in the form of constraints on six degrees of freedom.

The steel frame of the building is set by the rods of I-beam 25B1, the box section 80x80x6 mm and pipes TB60x7. Finite element diagrams of the building frame for three variants of algorithmic models are shown in Figure 2.

To select a rational shape of the investigated surface, it is necessary to obtain the results of the stress-strain state and determine the dynamic characteristics of the models under consideration [7].

3. Results

Three variants of the projected sports and entertainment complex building were obtained by the algorithmic method of modeling the crosscap surface. Without denying the aesthetic component of the architectural image, it is proposed to include a dynamic criterion corresponding to the characteristics of the building frame as the basis of the analysis. The spectrum of frequencies and modes of natural vibrations were obtained by the finite element method (Table 3).

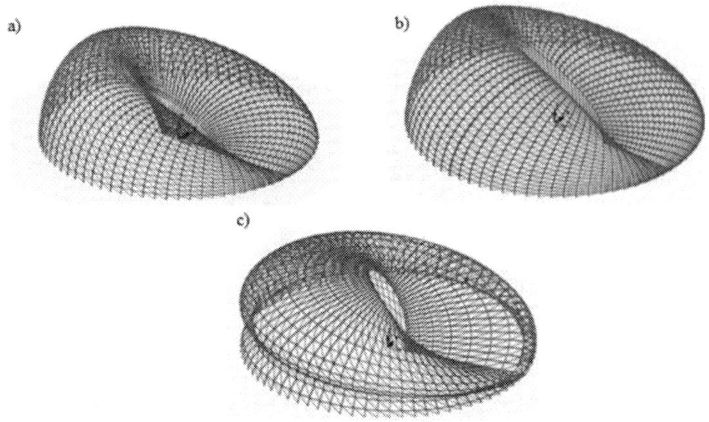

Figure 2. Finite element schemes: *a)* option 1; *b)* option 2; *c)* option 3

Table 3. Modal analysis of finite element schemes for the building frame

forming No.	Eigenvalues of frequencies, Hz		
	Model 1	Model 2	Model 3
1	2.84	2.74	2.89
2	3.00	2.84	3.06
3	3.23	2.97	3.13
4	3.26	3.02	3.20
5	3.40	3.06	3.27
6	3.47	3.17	3.42
7	3.53	3.28	3.45
8	3.61	3.29	3.48
9	3.78	3.44	3.59
10	3.86	3.50	3.63

Modal analysis makes it possible to clarify the design solutions and geometric characteristics of the frame.

The maximum vertical displacements of the finite element model nodes are 35.3; 38.0; 26.4 mm according to the model variant. One of the criteria for choosing a rational analytical surface is the structural stability coefficient. The stability loss of such structures occurs suddenly, with a flap, with the formation of deep dints or the shell snapping. This phenomenon is dangerous both for the frame as a whole and for its individual elements. The stability coefficient of the proposed analytical models in the Lira-CAD software package has been investigated; the stability coefficient varies from 8.7 to 12.4 [8.9]. The dynamic characteristics of options 2 and 3 do not meet the limit values of the design codes. The rational shape of the crosscap surface is a variant of the algorithmic model 1.

4. Discussion

A model for designing a unique building obtained by an algorithmic method is recommended. To clarify the dynamic movements of the frame, it is necessary to take into account the effect of the pulsating component of the wind load [10]. The preliminary results demonstrate the need to consider the first 10 eigenforms in the obtained spectrum.

Figure 3 shows the dynamic displacements of the finite element mesh nodes taking into account the wind pulsation.

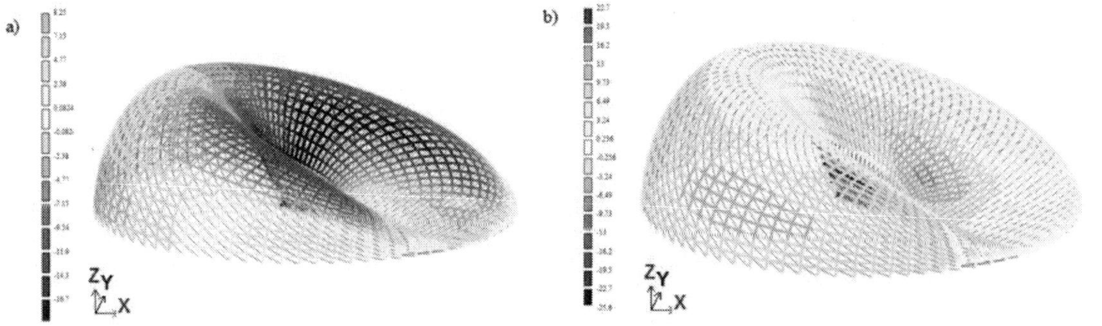

Figure 3. Dynamic displacements: *a)* along X-axis; *b)* Y-axis

It should be noted that the movements increased by an average of 30%. The design characteristics of a complex geometry object meet the current standards.

So, as part of the study:

1. The analysis of the given parametric surface formation in the form of a crosscap has been carried out. The algorithmic method for modeling the surface with varying the given parameters made it possible to choose a variant of the object primary forming for further research of the frame stress-strain state.
2. Using the finite element method in the LIRA software package, the characteristics of the object have been determined taking into account the pulsation component of the wind load.
3. The authors propose, when choosing an algorithmic model of a parametric architecture object, to regulate the stress-strain state, dynamic characteristics of the frame and the stability coefficient of a structure by changing the structural scheme and element stiffnesses.

5. Summary

Thanks to the development of innovative technologies and methods of information modeling, the relevance of creating parametric objects in the urban environment significantly increases [11, 12]. The finite element method allows performing the calculation of unique high-rise and large-span buildings and structures, the primary forming of which was developed with an algorithmic modeling method.

To select a rational shape of the crosscap that meets the requirements for bearing capacity, stiffness and stability, it is necessary to vary the coefficients simultaneously in all three equations and select several variants of the analytical surface for object design.

The stress-strain state study of a unique object of parametric architecture, obtained by the algorithmic method of modeling the crosscap surface, allows choosing the rational constructive solutions for the spatial frame of the building.

Improving the algorithmic method based on the analysis of the stress-strain state and dynamic characteristics will expand the boundaries of parametric architecture in the design of unique buildings and structures. The development of parametric architecture requires the improvement of software systems that implement not only analytical models of complex geometry objects, but also the numerical methods for calculating unique buildings and structures.

References

[1] Nadirshin N M 2013 Parametricism as a style in architectural design *Bulletin of the OSU* **1**.

[2] Kravchenko G M, Manoylenko A Yu, Litovka V V 2018 Parametric architecture *Engineering Bulletin of the Don* **2**. Information on ivdon.ru/ru/magazine/archive/N2y2018/5040

[3] Stessel S A 2015 Borrowing the natural principles of primary forming in parametric architecture *TSU science vector* **2** 52 - 57.

[4] Koporushkin P A, Partin A S 2004 Algorithm for calculating the parameterized geometric objects *Electronic journal "Investigated in Russia"* **7** 184-197. Information on http://zhurnal.ape.relarn.ru/articles/2004/018.pdf.

[5] Krivoshapko S N, Ivanov V N, Halabi S M 2006 *Analytical surfaces: materials on the geometry of 500 surfaces and information for calculating the strength of thin shells* (Moscow, Nauka) 544.

[6] Kravchenko G M, Trufanova E V, Kochura V G 2019 Information modeling of a parametric architecture object *Proceedings of higher educational institutions. North Caucasian region. Technical science* **1 (201)** 17-22.

[7] Kravchenko G M, Trufanova E V, Ladnaya E V 2017 Rational design of the building spatial frame elements *Engineering Bulletin of the Don* **1.** Information on ivdon.ru/ru/magazine/archive/n1y2017/3989

[8] Belostoskiy A M, Akimov P A, Petryashov N O, Petryashov S O, Negrozov O A 2015 Computational studies of the stress-strain state, strength and stability of the supporting structures of a high-rise building, taking into account the actual position of reinforced concrete structures *Bulletin of MSCEU* **4** 50—68.

[9] Kravchenko G M, Trufanova E V, Boyko A G, Adleyba T S 2019 Investigation of the overall stability of the building frame by the finite element method *Construction and architecture* **7 (1)** 45-48.

[10] Agakhanov E K, Kravchenko G M, Osadchiy E V, Trufanova E V 2017 Calculation of buildings of complex geometric shapes for wind effects *Bulletin of the Dagestan State Technical University. Technical science* **2** 8.

[11] Talapov V V 2011 *BIM Fundamentals: An Introduction to Building Information Modeling* (Moscow, DMK-Press) 392.

[12] Astrakhantsev V D, Zolotarev I I 2019 Topical issues of BIM technologies implementation *Interexpo Geo-Siberia* **1.**

Trihedral lattice towers with optimal cross-sectional shape

A S Chepurnenko[1*], L S Sabitov[2], B M Yazyev[1], G R Galimyanova[2], L Sh Akhtyamova[1]

[1]Don State Technical University, 1 Gagarin sq., Rostov-on-Don, 344000, Russia
[2]Kazan Federal University, 18 Kremlevskaya st., Kazan, 420008, Russia

E-mail: anton_chepurnenk@mail.ru

Abstract. The article discusses a new design of triangular lattice supports and suggests a technique for optimizing their cross-sectional shape. Optimization is carried out from the condition of the cross-section being equally stable in two planes. Such relationships are found between the geometric parameters of the section, at which the moments of inertia take the maximum value.

1. Introduction

This article will consider a new rational type of trihedral lattice structures used in the construction of tower structures for various purposes. The structure under study (patent for invention RU2584337 [1]) (Figure 1) contains belts 1 of a polyhedral closed section, lattice bars 2, attached to sheet gussets 3.

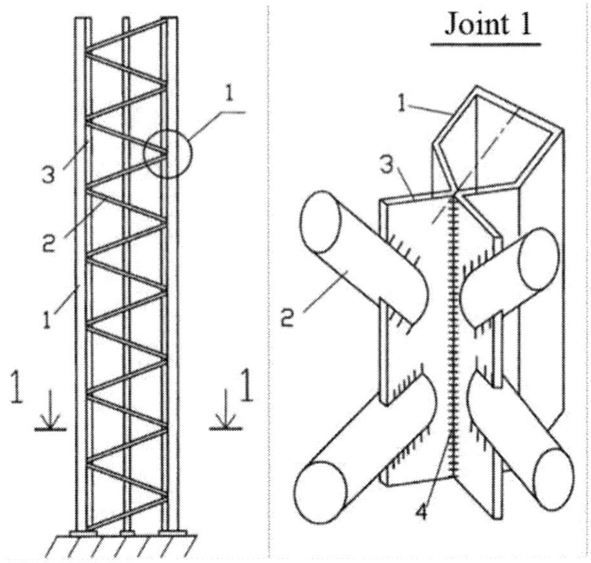

Figure 1. Trihedral lattice tower

Polyhedral belts of a closed section are made of sheet steel by bending it, while for the formation of gussets, the edges of the sheet steel are bent symmetrically in the opposite direction at an angle of 60 °, and the walls of the lattice 2 of each face are attached to the gussets by welding. The formation of the closed polyhedral section 1 is carried out by performing the longitudinal weld 4 at the point of bending and the contact of the bent edges.

The purpose of this work is to determine the optimal parameters b, h, l of the pentagonal section of the belts (Figure 2). The l_1 size is assigned for design reasons to ensure the required length of the weld, and we will consider it as given. An axial moment of inertia relative to the main central axis y is selected as the objective function. A limitation is introduced in order to ensure the equal stability of the belt in two planes. Thus, optimization is performed from the stability condition. In the process of optimization, the perimeter of the section without taking into account the gussets is taken to be constant, which corresponds to the constancy of material consumption.

2. Methods

For simplicity, we will assume that the section is thin-walled and $\delta \ll L$. We represent the cross-section shown at fig. 2 as the set of seven rectangles (Figure 3).

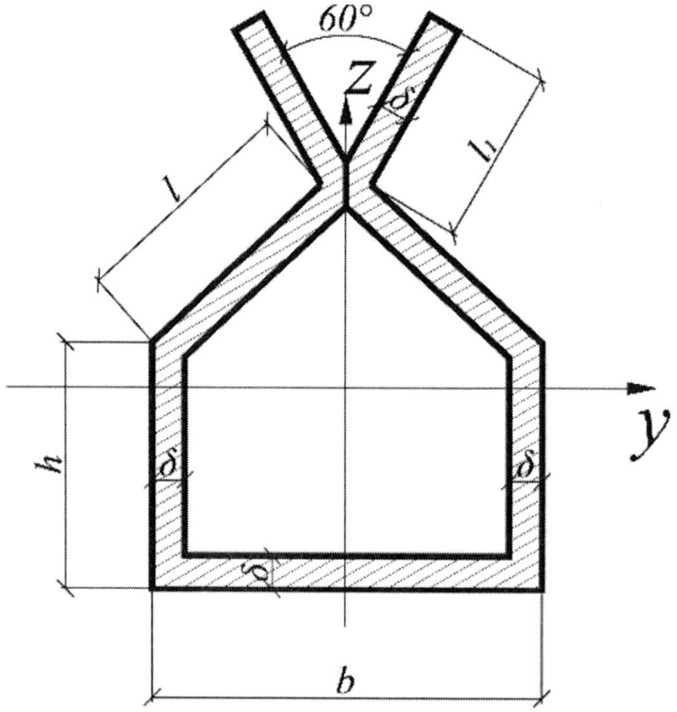

Figure 2. Optimized cross section

The areas of each of the rectangles are defined as:

$$A_1 = b\delta; A_2 = A_3 = h\delta; A_4 = A_5 = l\delta;\ A_6 = A_7 = l_1\delta. \tag{1}$$

Let us choose the coordinate system zC_1y_1 as the auxiliary one. The coordinates of the figures 1-7 centers of gravity are determined by the formulas:

$$z_{c1} = 0; z_{c2} = z_{c3} = \frac{h}{2}; z_{c4} = z_{c5} = h + \frac{l}{2}\sin\alpha = h + \frac{l}{2}\sqrt{1 - \frac{b^2}{4l^2}}.$$
$$z_{c6} = z_{c7} = h + l\sin\alpha + \frac{l_1}{2}\cos 30° = h + l\sqrt{1 - \frac{b^2}{4l^2}} + \frac{l_1\sqrt{3}}{4}. \tag{2}$$

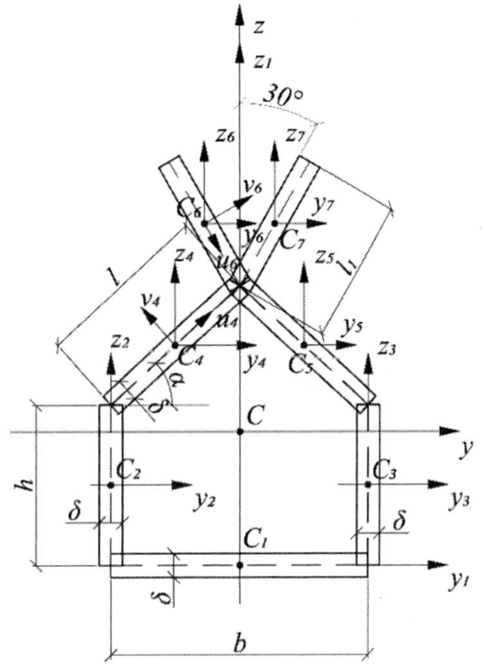

Figure 3. Simplified representation of the optimized section

The position of the entire section center of gravity is determined by the formula:

$$z_c = \frac{2\left(A_2 z_{c2} + A_4 z_{c4} + A_6 z_{c6}\right)}{A}, \tag{3}$$

where A is the total cross-sectional area.

The moments of inertia of figures 1-7 about their central axes y_i, z_i are calculated as follows:

$$\begin{aligned}
&J_{y1}^{I} = \frac{b\delta^3}{12}; J_{z1}^{I} = \frac{b^3\delta}{12}; J_{y2}^{II} = J_{y3}^{III} = \frac{\delta h^3}{12}; J_{z2}^{II} = J_{z3}^{III} = \frac{h\delta^3}{12};\\
&J_{y4}^{IV} = J_{y5}^{V} = J_{u4}^{IV}\cos^2\alpha + J_{v4}^{IV}\sin^2\alpha = \frac{l\delta^3}{12}\frac{b^2}{4l^2} + \frac{l^3\delta}{12}\left(1-\frac{b^2}{4l^2}\right);\\
&J_{z4}^{IV} = J_{z5}^{V} = J_{u4}^{IV}\sin^2\alpha + J_{v4}^{IV}\cos^2\alpha = \frac{l\delta^3}{12}\left(1-\frac{b^2}{4l^2}\right) + \frac{l^3\delta}{12}\frac{b^2}{4l^2};\\
&J_{y6}^{VI} = J_{y7}^{VII} = J_{u6}^{VI}\sin^2 30° + J_{v6}^{VI}\cos^2 30° = \frac{3}{4}\frac{l_1\delta^3}{12} + \frac{1}{4}\frac{l_1^3\delta}{12};\\
&J_{z6}^{VI} = J_{z7}^{VII} = J_{u6}^{VI}\cos^2 30° + J_{v6}^{VI}\sin^2 30° = \frac{1}{4}\frac{l_1\delta^3}{12} + \frac{3}{4}\frac{l_1^3\delta}{12}.
\end{aligned} \tag{4}$$

The moments of inertia of the entire section relative to the main central axes y and z are determined by the formulas:

$$J_y = J_{y1}^{I} + A_1\left(z_{c1}-z_c\right)^2 + 2\left(J_{y2}^{I} + A_2\left(z_{c2}-z_c\right)^2\right) + \\ +2\left(J_{y4}^{IV} + A_4\left(z_{c4}-z_c\right)^2\right) + 2\left(J_{y6}^{VI} + A_6\left(z_{c6}-z_c\right)^2\right); \tag{5}$$

$$J_z = J_{z1}^{I} + 2\left(J_{z2}^{II} + A_2\frac{b^2}{4}\right) + 2\left(J_{z4}^{IV} + A_4\frac{b^2}{16}\right) + 2\left(J_{z6}^{VI} + A_6\frac{l_1^2}{16}\right).$$

Finally, the expression for the moment of inertia J_z takes the form:

$$J_z = \frac{\delta\left(2b^3 + 12hb^2 + 3\delta^2 l_1 + 4h\delta^2 + 4l_1^3\right)}{24} + \frac{\delta l\left(b^2+\delta^2\right)}{6} - \frac{b^2\delta^3}{24l}. \tag{6}$$

The expression for the moment of inertia J_y is not presented here due to its cumbersomeness. Formula (6) can be presented in a simplified form if we neglect the terms that include the quantities δ^3 and δ^2.

In the problem under consideration, the objective function and constraints are nonlinear, therefore, to solve it, it is necessary to apply nonlinear optimization methods.

We carried out the solution in the Matlab environment using the Optimization Toolbox package. The function *fmincon* was used, which determines the minimum of the nonlinear objective function J_y^{-1} with nonlinear constraints. The interior point method is selected as the nonlinear optimization method.

3. Results and Discussion

Table 1 shows the optimal values of the ratios b / L, h / L and l / L depending on the ratio l_1 / L.
No solution was found for $l_1 / L > 0.125$. It probably does not exist for such relations between l_1 and L.

Table 1. Optimal parameters of the pentagonal section depending on the ratio l_1 / L

l_1/L	b/L	h/L	l/L	α, °
0	0.2677	0.2093	0.1569	31.5
0.025	0.2851	0.1826	0.1749	35.4
0.05	0.3022	0.1604	0.1885	36.7
0.075	0.3188	0.142	0.1986	36.6
0.1	0.3345	0.1265	0.2063	35.8
0.125	0.3495	0.1131	0.2122	34.6

Note that the pentagonal section without gussets at $l_1 = 0$ can act as a replacement for square tubes in truss chords of the Molodechno type [2-5]. Compared to a square tube, a pentagonal tube with optimal parameters b, h, l for the same cross-sectional area has 5.2% higher moments of inertia. The optimum angle α is 31.46°. When designing real structures, for convenience, you can take $\alpha = 30°$.

The equality of the moments of inertia J_y and J_z is necessary to ensure equi-stability in the case of the same fixation of the rod in the xOz and xOy planes. At different reduced lengths in two planes, it becomes necessary to find the optimal section parameters for which the ratio J_y / J_z is specified. The technique proposed by the authors and the developed program in the Matlab environment allows to do this. Table 2 shows the optimal values of b / L, h / L and l / L for various ratios J_y / J_z at $l_1 = 0$.

Table 2. Optimal parameters of the pentagonal section for various ratios J_y / J_z

J_y/J_z	b/L	h/L	l/L
0.6	0.3135	0.1683	0.175
0.7	0.3	0.1803	0.1698
0.8	0.288	0.191	0.165
0.9	0.2773	0.2006	0.1608
1	0.2677	0.2093	0.1569

1.1	0.259	0.2171	0.1534
1.2	0.251	0.2244	0.1501
1.3	0.2437	0.2311	0.1471
1.4	0.2369	0.2373	0.1442
1.5	0.2307	0.243	0.1417
1.6	0.2249	0.2483	0.1393
1.7	0.2195	0.2533	0.137
1.8	0.2144	0.258	0.1348
1.9	0.2097	0.2624	0.1328
2	0.2052	0.2667	0.1307

In a similar way, you can optimize the cross-section of the trihedral lattice tower with additional bends of the edges (Fig. 4). The values b / L, h / L and l / L obtained from the condition of equi-stability and the maximum moment of inertia, depending on the ratio l_1 / L, are presented in table. 3.

Table 3. Optimum parameters of the trihedral lattice tower section with additional bends of the edges, depending on the ratio l_1 / L

l_1/L	b/L	h/L	l/L	α, °
0	0.2677	0.2093	0.1569	31.5
0.025	0.2933	0.1714	0.1819	36.3
0.05	0.3175	0.1432	0.1981	36.7
0.075	0.3398	0.1214	0.2087	35.5
0.1	0.3604	0.1039	0.2159	33.4

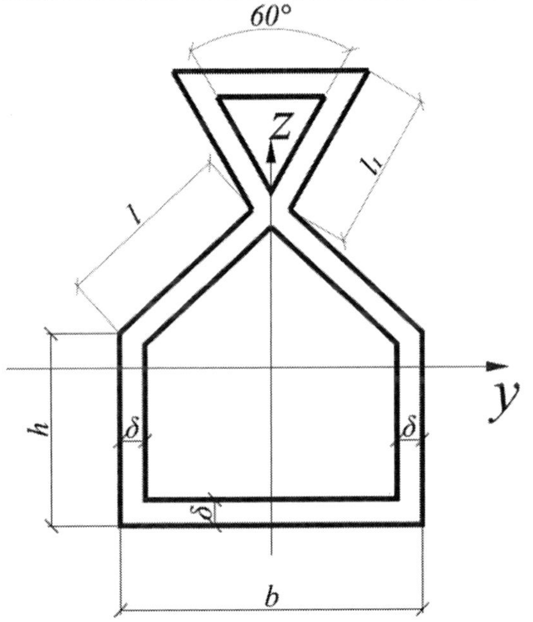

Figure 4. Triangular lattice tower cross-section with additional bends of the edges

4. Summary

The optimal ratios between the pentagonal section sides sizes of the trihedral lattice supports are determined, providing their maximum rigidity and uniform stability at a given material consumption. The optimal angle at the top was close to 120 degrees, and this value can be used in the design for simplification. In the future, it is advisable to consider the issues of local stability of the support elements with the proposed cross section.

References

[1] Sabitov L, Kuznecov I, Badertdinov I 2016 RU Patent No. 2584337.
[2] Marutjan A 2016 *Stroitel'naja mehanika i raschet sooruzhenij [Structural Mechanics and Analysis of Constructions]* **3** 25-35.
[3] Kopytov M, Erohin K, Matveev A, Kosincev A, Jashin S 2001 RU Patent No. 2174576.
[4] Marutjan A 2017 *Stroitel'naja mehanika i raschet sooruzhenij [Structural Mechanics and Analysis of Constructions]* **275** 35-41.
[5] Marutjan A 2017 *Zbirnik naukovih prac' Ukraïns'kogo institutu stalevih konstrukcij imeni VM Shimanovs'kogo [Collection of scientific works of the Ukrainian Institute of Steel Structures named after V.M. Shimanovskii]* **19** 43-53

IOP Conf. Series: Materials Science and Engineering 1083 (2021) 012013 doi:10.1088/1757-899X/1083/1/012013

The use of composite glass-fiber reinforcement (CGFR) in construction

N I Zakieva*, A P Lapina

Don state technical university, 1 Gagarin sq., Rostov-on-Don, 344002, Russia

E-mail: nikorani@mail.ru

Abstract. The article considers the use of composite glass-fiber reinforcement in construction. Its main advantages and disadvantages, installation technology, scope, comparison and some technical characteristics of the material are observed. The following conclusion is made: CGFR, in comparison with metal, wins in almost all points of comparative characteristics.

Introduction

Today, composite reinforcement is becoming stronger and stronger in the construction market. The concept of "composite reinforcement" (or as it is also called - "polymer reinforcement") is quite comprehensive, in fact, composite reinforcement can be of different types, depending on the main supporting fiber type. For example, there are "glass-fiber reinforcement", "basalt reinforcement", "carbon fiber reinforcement" - all of them are the subspecies of the same "composite reinforcement". The article considers the history of this type of reinforcement, the advantages and disadvantages of composite reinforcement, the news of its current use.

Main part

Glass-fiber reinforcement is a building material created on the basis of fibers connected with a complex composition. It is made on the basis of basalt, glass and carbon fiber, and they can be combined. However, the most popular are basalt-plastic rebar and glass-fiber.

It consists of two parts. The first one is the trunk, due to which the high strength of the material is achieved. The fibers are bonded together by the polyester resin of the composite. The outer layer serves for reliable adhesion to concrete: it is a fibrous body that is wound around the trunk in a spiral. It is due to this composition that the plastic reinforcement received positive reviews as a reliable material for construction.

There are various variations of the reinforcement model, some of which are quite unusual. Glass-fiber is used for the production of these building reinforcement. Its peculiarity is that there are practically no analogues in the world, and its positive qualities significantly expand the scope of application.

Figure 1. View of glass-fiber reinforcement

In addition, this material is modern and effective, and therefore meets the requirements of the construction process in the best possible way. Any glass-fiber reinforcement is based on two components. The first is directly a reinforcing material, the second is a binder (a mixture based on epoxy resin). The ratio of these components is 75 to 25. In composite reinforcement, all mechanical loads fall on the reinforcing component, while the bonding materials are a kind of matrix that evenly distributes the load over the entire length of the bar and protects it from the external influences. The following most common recipe can be considered: glass roving or basalt fiber act as a reinforcing link, epoxy resin is used for bonding, in addition, a hardener and an accelerator will be included in the material. However, there is not a universal composition, since each manufacturer creates its own technological process.

The essence of the technology for the plastic materials' production lies in the following algorithm:

1. The fibers are unwound and formed into a bundle.
2. Then the bundles are saturated with a specific binder and a cross-section is formed.
3. A glass-fiber cord is wound on the surface of the bundle, forming a spiral and sent to the oven for polymerization.
4. Finally, rods of a certain length are cut or wound into bights.

Installation of composite reinforcement

Reinforcement technology with composite rebar is almost the same as with steel rods. It does not require much effort or knowledge, and it is quite possible to do. After all, working with this material is much easier and faster. It is possible to cut glass-fiber like steel with a grinder. In just a couple of seconds, several rods with a diameter of 12 mm are cut through. It is also possible to use a conventional metal saw. The plastic rods are knitted in several ways: with soft metal wire. This process is identical to knitting steel rods. It is carried out manually or using a special crochet hook. It can be tightened with plastic ties, which greatly reduces the time spent on the knitting process. Special fasteners that simply snap into place on the reinforcing bars.

Like any building material, glass-fiber reinforcement has its own advantages and disadvantages compared to similar metal ones, which can become a serious help or hindrance in its application in various fields of construction.

Advantages:

1. Low specific gravity. This advantage allows it to be used in lightweight structures, such as, for example, aerated concrete, etc. This property of glass-fiber reinforcement allows to reduce the weight of the entire structure.

It is worth noting that the use of glass-fiber reinforcement in ordinary concrete will not significantly affect the mass of the structure, taking into consideration that the bulk of the weight will be given by concrete itself.

2. Low thermal conductivity. As it is known, glass-fiber conducts heat through itself much worse than metal.

This advantage of glass-fiber reinforcement allows it to be used where it is necessary to reduce cold bridges, which are so remarkably created by steel reinforcement.

3. Packing in bights. For the private houses' construction, this is a very significant advantage of glass-fiber reinforcement, because it is possible to reduce its delivery cost to the site. In addition, the use of glass-fiber reinforcement in bights reduces its consumption, since there will be practically no overlaps in the reinforcement cage, and this will also slightly reduce financial costs.

4. Durability. Manufacturers rely on the fact that glass-fiber, in comparison with metal, is much more durable (The dubious advantage of glass-fiber reinforcement, given that the metal inside the concrete is practically not subject to corrosion and inside the reinforced concrete structure will also last a very long time).

5. Dielectric. This property, most likely, in private construction does not give any advantages of glass-fiber reinforcement over metal, but it should be taken into account.

6. Resistance to chemical attack. This means that in acidic and other aggressive chemical environments, glass-fiber reinforcement is much more comfortable than steel.

In low-rise private construction, this advantage of glass-fiber, like the previous one, practically does not play any role, except for construction in winter, when various salts are added to the solution or concrete, which adversely affect the metal.

7. Radio transparency. This means that the glass-fiber reinforcement does not generate any radio interference, unlike the metal circuits created by the steel reinforcement.

Such an advantage of glass-fiber reinforcement as radio transparency will play a significant role only if there are a lot of reinforcement in the walls of the building. Then the use of glass-fiber reinforcement will reduce radio interference inside it.

Disadvantages:

1. Glass-fiber reinforcement is more expensive than conventional steel when comparing reinforcement of the same diameter.
2. Not thermally stable. Glass-fiber reinforcement does not withstand high temperatures.
3. Does not bend. Thus, if it is necessary, for example, to bend the reinforcement at an angle of 90 degrees, this will not be possible. Although on the other hand, we can make all the bends from ordinary steel and build them up with glass-fiber.
4. Low modulus of elasticity for fracture. This means that glass-fiber reinforcement cannot withstand the same fracture loads as metal reinforcement.
5. Difficulty in the construction of a rigid reinforcement cage. In other words, the frame made of glass-fiber reinforcement is not as rigid as that of metal, and, accordingly, is less resistant to vibration and loads that will be present when pouring concrete from a car mixer.

Comparison of glass-fiber reinforcement with steel

The main competitor of glass-fiber composite reinforcement (CGFR) is steel reinforcement. Their characteristics are largely similar, but in some respects, glass-fiber is clearly superior to the usual type of metal equipment.

Let us compare glass-fiber with steel according to some parameters:

1. Deformability. Steel reinforcement - elastic-plastic, CGFR - perfectly elastic.
2. Ultimate strength: steel - 390 MPa, glass-fiber - 1300 MPa.
3. Coefficient of thermal conductivity. In the first case, it is equal to 46 W / m^{O}C, in the second - 0.35.
4. Density. Steel reinforcement has a value of 7850 kg / m^3, CGFR - 1900 kg / m^3.
5. Thermal conductivity. Glass-fiber is not thermally conductive, unlike steel.
6. Corrosion resistance. CGFR is a stainless metal, steel corrodes relatively quickly.
7. Ability to conduct electricity. Dielectric is glass-fiber reinforcement. The disadvantages of steel rods are that they are 100% current conductors.

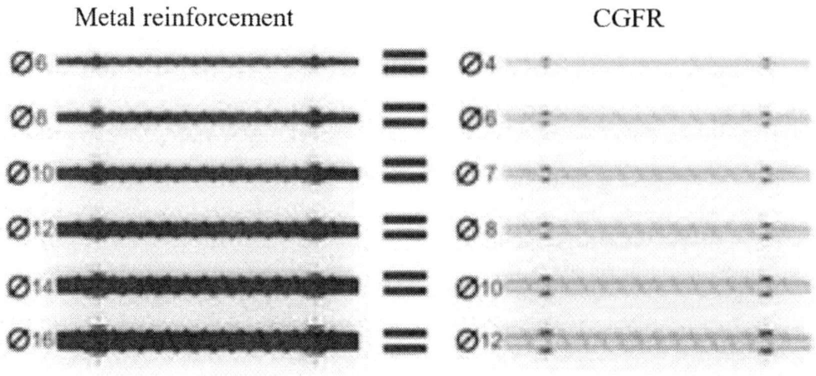

Figure 2. Comparison table

Almost all the main advantages and disadvantages of glass-fiber reinforcement are listed. Judging by them, it is impossible to say with great certainty that it is much better or worse than metal reinforcement, therefore, in which building structures and frames the use of glass-fiber reinforcement will be justified and expedient.

The use of glass-fiber reinforcement is justified in some cases both in industrial construction and in private low-rise construction.

Scope of glass-fiber reinforcement in private low-rise construction

1. Glass-fiber reinforcement is used in some types of foundations, such as strip - buried below the freezing depth, slab foundation.

Figure 3. An example of the glass-fiber reinforcement use

It should be noted that this applies only for the low-rise private buildings, on good soil. On floating soils, there will be increased fracture loads, which glass-fiber reinforcement may not withstand.

2. It is advisable to use glass-fiber reinforcement in the reinforcement of brick walls, walls from blocks, very often it is possible to find the reinforcement of walls from gas silicate blocks with glass-fiber reinforcement.

The use of glass-fiber reinforcement in wall reinforcement is very popular among the developers. Moreover, such reinforcement is used as a reinforcement element of the walls themselves, and as a bundle of the facing wall with the load-bearing one.

3. In multi-layer panels as links. Since there is usually a dense insulation inside the panels, glass-fiber reinforcement is used to bond the concrete parts together.

4. Justified the use of glass-fiber reinforcement in the bearing parts of elements subject to increased corrosion, swimming pools, for example.

Metal reinforcement will corrode when concrete is in water, and glass-fiber reinforcement is devoid of such a disadvantage, based on one of its advantages.

5. Also, glass-fiber reinforcement is widely used in the reinforcement of glued wooden beams, increasing their rigidity.

6. Reinforcement of asphalt, in places of increased loads, though can be rarely seen.

As it is seen, the scope of glass-fiber reinforcement application in construction is quite wide, although there are some limitations.

Examples of the composite glass-fiber reinforcement use in the construction of public buildings

CGFR is slowly gaining market share as government agencies begin to incorporate it into building codes and design guides, and as the test methods are developed. As a result, an increasing number of such agencies as the Manitoba Floodway Authority have chosen CGFR as the most efficient and durable product for public projects. An example is the Floodway Bridge over the Red River in Winnipeg, Manitoba, Canada, which was completed in 2006. The bridge contains 16 spans between supports, each approximately 50 to 143 feet (15.3 to 43.5 meters). All concrete elements above the main beams are reinforced with Aslan 100 CGFR rods manufactured by Hughes Brothers Inc. (Seward, Nebraska). Consuming 310,000 lb / 140.614 kg of CGFR rebar, the project is the largest concrete bridge in the world that is not reinforced with metal.

Figure 4. Floodway Bridge

Glass-fiber composite rebar was selected for the Miami-Dade MetroRail project, a 2.5-mile (4 km) overhead “heavy” rail moving walkway that will carry passengers (and luggage) from Miami International Airport (Miami, Florida) to downtown. The reinforcement shown in the first photo is wired and ready for a team of construction workers to pour concrete on the guide track. In the second photo, the glass-fiber composite rebar in the Miami-Dade MetroRail track, after pouring the concrete, forms a stable platform for the steel rails. The use of a corrosion-resistant composite reinforcement guarantees the guide track durability and thus will increase the useful life cycle of the rail system and reduce maintenance and repair costs.

Figure 5. Miami-Dade MetroRail Project

Summary
Many professionals and ordinary consumers in the construction equipment choice choose such a material as composite glass-fiber reinforcement. Reviews about it in most cases are positive. Ease of operation and ease of use are noted. Glass-fiber rods are practically waste-free, since one of the main advantages of glass-fiber is that the reinforcement is produced in bights, which means that it is possible to work with the required length. CGFR, in comparison with metal, wins in almost all positions.

References

[1] Taran V V, Yankov A V 2013 Features of the composite reinforcement use in the construction of building structures and frames *Bulletin of the Donbass National Academy of Civil Engineering and Architecture* **6 (104)** 35-39.

[2] Klimov Yu A 2010 Modern composite basalt rebar for reinforcing concrete structures *Concrete technologies* **11/12** 56–57.

[3] Klimov Yu A, Soldatchenko O S, Oreshkin DA 2010 Experimental studies of adhesion of composite non-metallic reinforcement to concrete *Composite reinforcement.* Information on http://www.frprebar.com/frprebar_test_adhesion_concrete.html

[4] Khozin V G, Piskunov A A, Gizdatullin A R, Kuklin A N 2013 Adhesion of polymer composite reinforcement to cement concrete *KSACEU Herald* **1 (23)** 214–220.

[5] Okolnikova G E, Gerasimov S V 2015 Prospects for the use of composite reinforcement in construction *Ecology and construction* **3.** Information on https://cyberleninka.ru/article/n/perspektivy-ispolzovaniya-kompozitnoy-armatury-v-stroitelstve (date of access: 11/13/2020).

[6] Maximov S P, Bashkova Yu B, Vshivkov E P 2015 Experimental studies of the operation of glass-fiber reinforcement in the reinforcement of concrete structures. *Universum: Technical sciences* **6 (18).** Information on https://cyberleninka.ru/article/n/eksperimentalnye-issledovaniya-raboty-stekloplastikovoy-armatury-pri-armirovanii-betonnyh-konstruktsiy (date accessed: 11/13/2020).

Research of the additional building influence on the foundation in a stress superposition existing along the radius zone

Al-Bukhaiti Ali Yahya Nasser[1*], Yaya Keita[1], V V Ledenyov[1], Ya V Savinov[2], O V Umnova[1]

[1]Tambov State Technical University, 106, Sovetskaya Str., Tambov, 392000, Russia
[2] LLC Mostostroy, 7 D, Yuzhnaya St., Periksa village, Tambov district, Tambov Region, 392551, Russia

E-mail: al_bukhaiti@mail.ru

Abstract. The comprehensive studies' results of the soil environment stress-strain state from the loaded and unloaded foundation model action are considered and compared:

1. Laboratory experiments with foundation models were carried out, on the basis of which a numerical SSS soil foundation analysis was carried out, empirical relationships between the influencing parameters and the foundations' additional settlement amount due to their mutual influence on each other were obtained.
2. Numerical modeling of foundations located in the mutual influence zone by the finite element method of numerical research using the software package Plaxis 3D was performed.
3. A numerical analysis of the influence of the distances between the stamps' axes on the additional stresses' values from the loaded foundation is given using the method of angular points, as a method recommended by the current regulatory documentation [13].

These results can be used to refine the existing calculation methods. The difference between the results of numerical and experimental modeling is noted.

Introduction

In a tight urban environment, high-rise buildings are often constructed in close proximity to the existing buildings. Mutual influence zone size clarification of the foundations for different schemes of their loading is a very urgent topic. The soil located in the building load area is a part of the building-foundation system, it is an integral structural element of this system. The influence zone size depends on numerous factors [1, 3 ... 9, 12], for example: the size and shape of objects, their location, the depth of the underground part, climatic engineering-geological and hydraulic conditions, the magnitude of loads and impacts, the presence of unfavorable processes (for example, landslide), structural, planning and technological (construction and operation) features of the facility under construction [9, 10]. In addition, taking into account the influence of one or a number of factors is rather complicated due to the calculation schemes and models' conventionality, numerous mistakes at all stages of the building construction life cycle, low quality of work and materials. In some cases, scientific support for design and construction is used [2, 6, 11].

Unilateral axial loading can cause adverse consequences: inclination, additional settlement, horizontal displacement, curvature and distortion of the building, provoke rheological processes (creep). BC 22.13330.2010 [13] in such cases requires a geotechnical forecast (assessment) of the construction impact on the change in the stress-strain state of the surrounding soil mass. Influence zone radius r_{iz} starts from the boundaries of the projected pit and is allowed to be limited by the distance at which the calculated value of the additional soil mass settlement or the surrounding building existing structure base does not exceed 1 mm.

The problem formulation

The purpose of the study was an experimental-theoretical study of the mutual influence of foundations, the determination of functional dependencies between the movements of closely located foundations, the distances between them, loading schemes, the study of the foundations' stress-strain state.

Numerical modeling and calculations soil deformations were performed using the program ***Plaxis 3d*** finite element method. Fine sand with physical and mechanical characteristics typical to Neogene deposits in the central part of Tambov was taken as the base soil (Table 1).

Table 1. Physical and mechanical characteristics of the sandy soil model fine sand with characteristics:

c_s, kgf /cm^2	ϕ, °	*E*, MPa	ρ, g/cm^3	Porosity coefficient *e*
0.02	32	28	1.5-1.55	0.65

After constructing the above-mentioned models of the "basement-foundation" system, a number of iterations were made in the calculation program with varying parameters of influencing factors, which are reflected in the Table 2.

Table 2. Variability of influencing parameters

No.	*D*, mm	$\bar{L}=L/D$	ρ, g/cm^3	
			1	2
1	200	2.5	1.5	1.55
2		3		
3		3.5		

The results of the foundation soil displacements calculations are reflected in the diagrams (Figure 1). In addition, the movements of the foundation models' control points were observed, which made it possible to compare the calculated data with the results of field observations.

Based on the calculated data, the dependences of displacements on varied influencing factors are plotted, reflected in the summary Diagram 7.

In order to verify the calculated data, the authors carried out an experiment, in which the factors reflected in the above-mentioned calculations were varied in steps. The experiment was carried out in a steel tray with the dimensions: 180 - 100 - 85. As models of shallow foundations, rigid steel round dies with a diameter D_{st}, equal to 200 mm (Figures 1, 3). In the experiments, the parameters were investigated according to Table 2. To ensure the sand-sand friction coefficient, the lower contact surfaces of the models are pasted over with sand.

a)

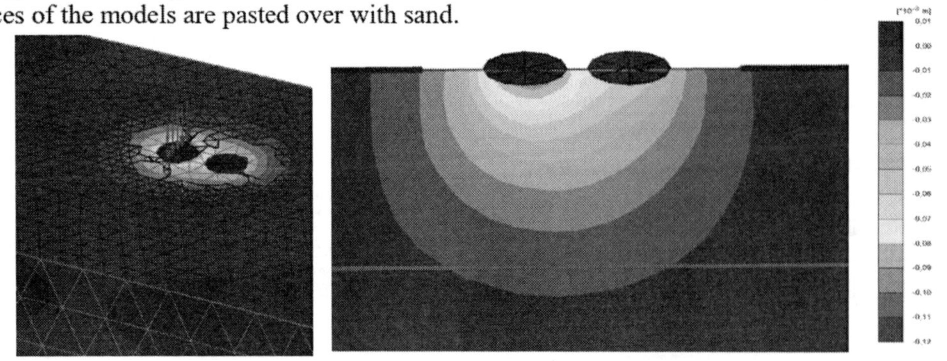

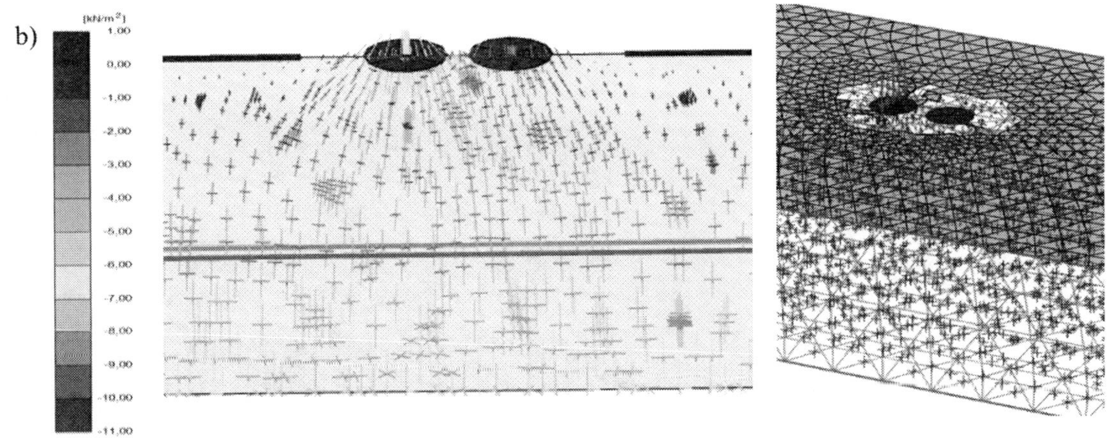

Figure 1. The stress-strain state of the foundation under the mutual influence of a number of foundations located: a mesh of finite elements (a), complete displacements (b), at ρ=1.5 g/cm³, δ=0°, D= 200 mm, e_0= 0, L=250 mm, F=4.8 kN (static)

On the models' top, rigid steel plates with grooves are welded to fix the position of the cylindrical hinges. The load on the model was transferred in steps of approximately 1/10 ... 1/15 of the breaking load. Each stage of loading was maintained until the conditional stabilization of deformation equal to 0.1 mm for 1 hour of observation. In some experiments, the load was increased after 1 hour of observation. The experiment replication was 2-3.

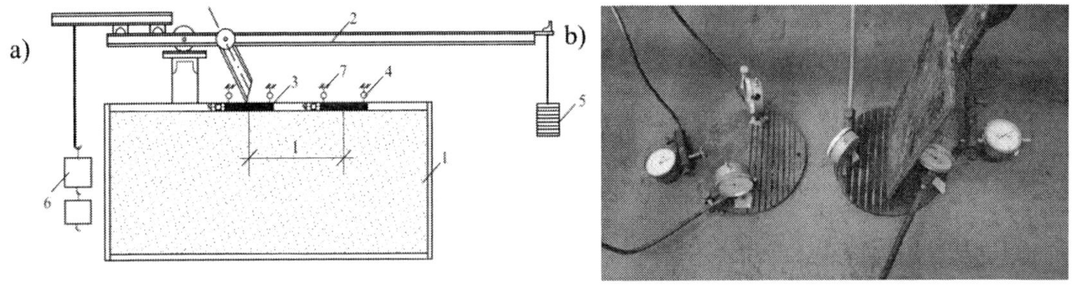

Figure 2. a) Schematic laboratory installation: 1- subsoil; 2- lever with 10x magnification; 3- loaded dies; 4- unloaded die; 5- cargo; 6- counterweight; 7- dial indicators; 8-steel patch size 180×100×85 (*h*) cm; 9- lever knife; b) Dies - models of shallow foundations: a) location of ICh-10 displacement indicators.

The experimental observations' results of the settlement and foundation models roll are shown in Figure 3.

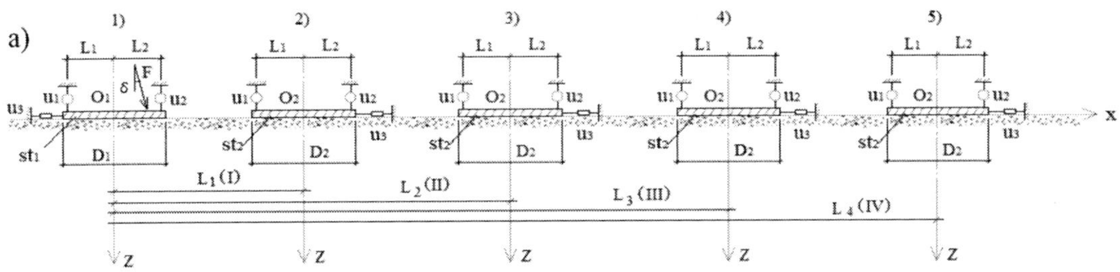

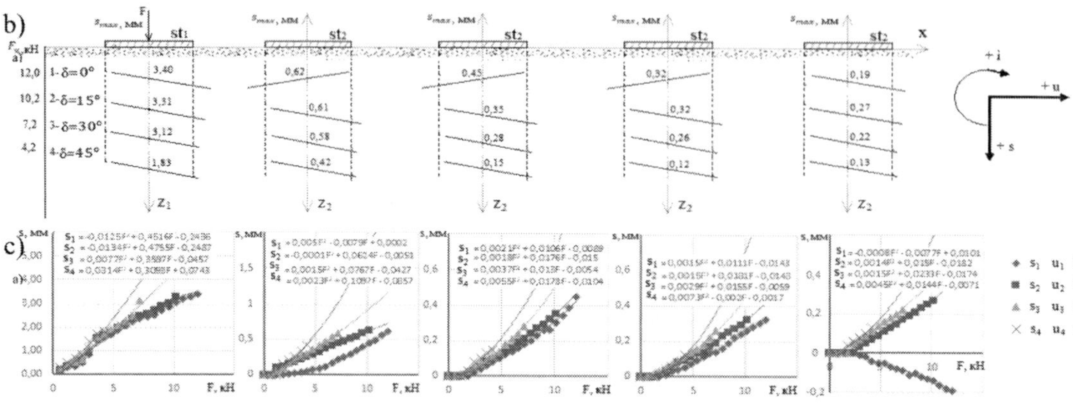

Figure 3. Results of the foundation models' displacements observations when changing load patterns and distances L from each other: a) loading scheme and changing the distance between foundation models; b) the foundation models' settlement value when the distance is changed; c) the corresponding empirical functions of settlement of the models obtained; horizontal displacements of the centers of gravity of the dies base for loads F, (kN): 4.2; 7.2; 10.2; 12.

The foundations mutual influence zones' determination using the method of corner points by BC 22.13330.2010 [13]

The calculation of the bases for structures' deformations of the surrounding buildings located in the influence zone of new construction or reconstruction is performed from the condition

$$s_{ad} \leq s_{ad,u}$$

where s_{ad} - is an additional settlement of the foundation base (joint additional deformation of the foundation and structure), determined taking into account the totality of effects associated with new construction or reconstruction;

$s_{ad,u}$ - limit value of additional foundations settlement [13].

The settlement of the foundation base s, cm, using a design scheme in the form of a linearly deformable half-space is determined by the method of layer-by-layer summation according to the formula:

$$s = \beta \sum_{i=1}^{n} \frac{\sigma_{zp,i \times h_i}}{E_i}$$

where β - is a dimensionless coefficient equal to 0,8;

$\sigma_{zp,i}$ - is average value for vertical normal stress.

The total load on the base from the maximum loading step $N = 12$ kN (1.224 tf). Depth of laying $d = d_n = 0$ m. The dimensions of circular foundation models in plan Ø200 mm the size of an alternative square foundation of the same area is 17.72×17.72 mm =314 cm² = 0.0314 m². In our case, the total pressure under the foundation base from an external load for 4 stages of loading is:

$$p_1 = N_1 / S = \frac{4,2}{0.0314} = 133 \text{ kPa} = 1.33 \text{ kgf/cm}^2,$$

And loading p_2 = 2.29; p_3=3.24; p_4=3.82 kgf /cm², respectively.

The values $\sigma_{zp,nf}$ along the loaded foundation F-1 axis, we obtain by summing the stresses $\sigma_{zp,1}$ from pressure p_0 under the foundation itself and additional stress $\sigma_{zp,2}$ from the foundation F-2 influence. The latter is determined by the method of corner points as the algebraic sum of stresses at the considered depth at the corner point *M* four loaded areas (fictitious foundations): *MLAT and MNDL* with positive pressure p_0 and *MKBIu MNCK* - with negative (see Figure 5). The aspect ratios of the indicated rectangular foundations are equal:

Table 3.

Foundation spacing options	Aspect ratio values: η		
	η_3 *MKBJ and MNCK*	η_2 *MLAJ and MNDL*	η_1 *ABCD*
Option I	1.8	3.8	1
Option II	2.4	4.4	1
Option III	2.95	4.95	1
Option IV	3.5	5.5	1

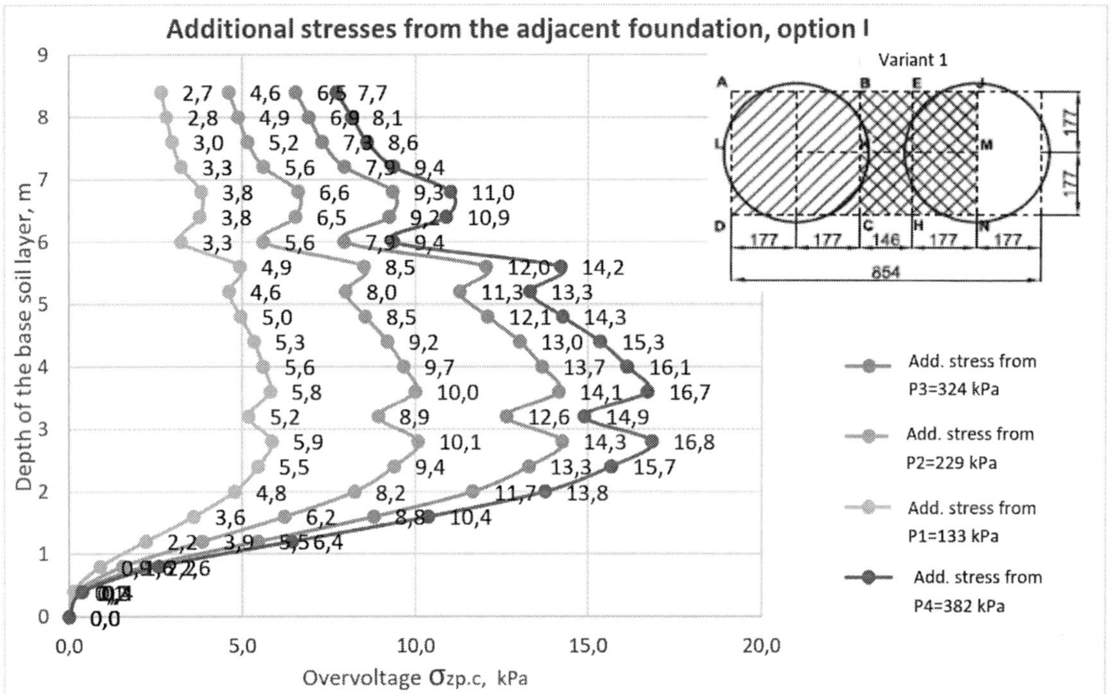

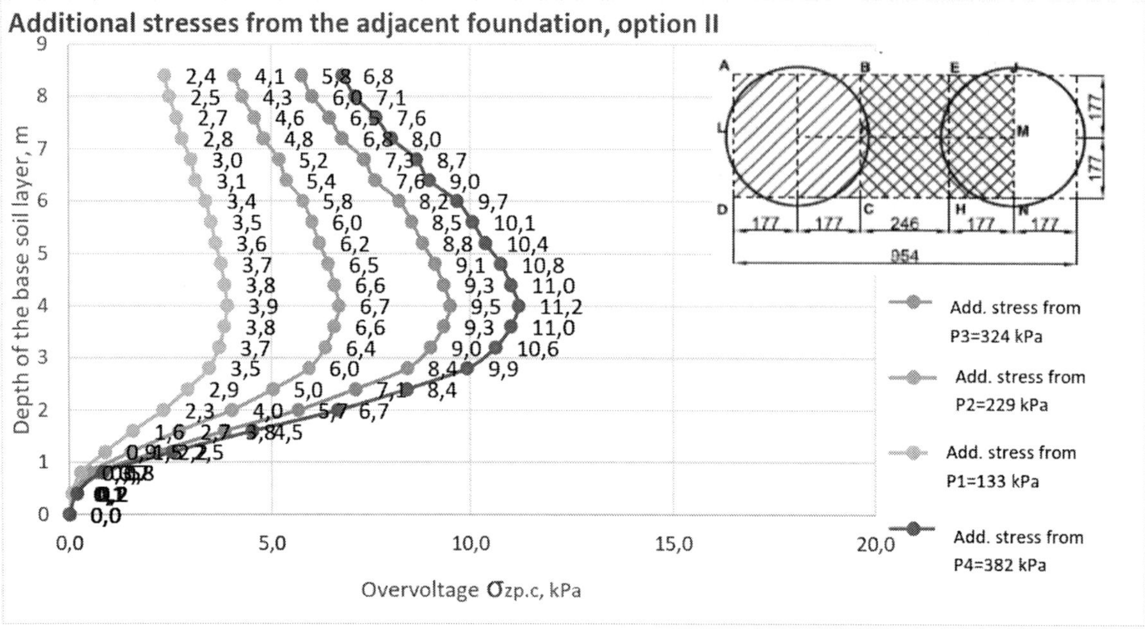

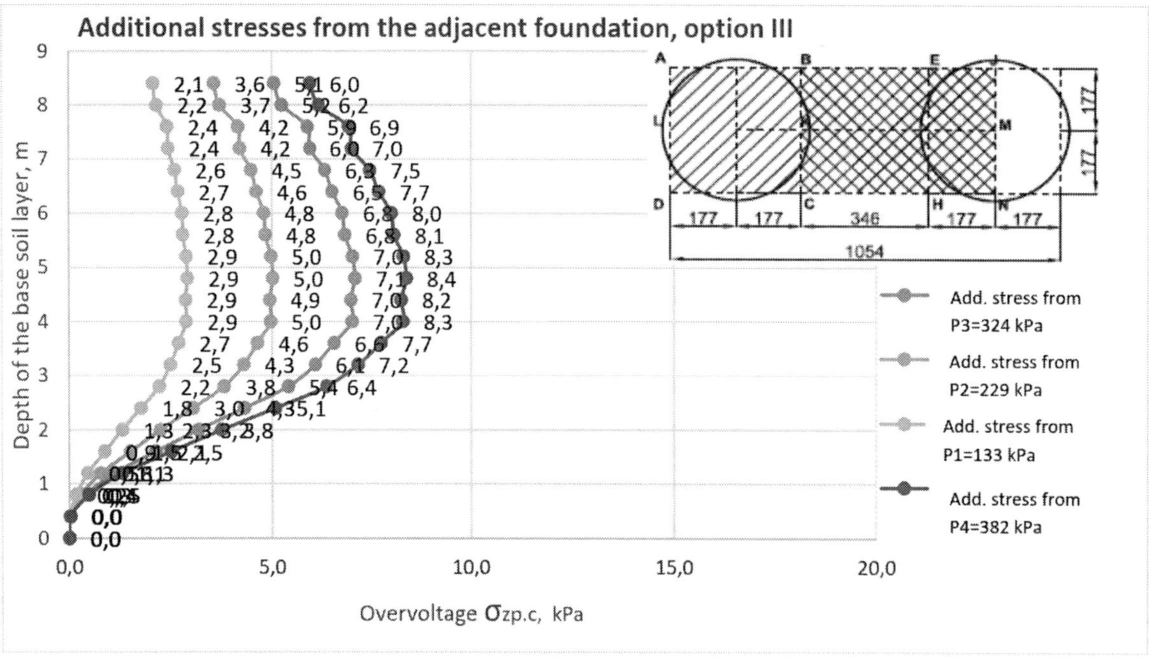

Figure 4. Calculation by the method of corner points of additional stresses in the soil from the adjacent foundation pressure influence. Variants with different spacing between the foundation models: I (L*=100m), II (L*=200mm), III (L*=300mm).

* After bringing round foundations to square foundations, the L values changed: I (L*=146 mm), II (L*=246 mm), III (L*=346 mm).

The lower boundary of the compressible base stratum is taken at the depth $z = H_c$, where the condition is satisfied $\sigma_{zp} = 0{,}2\sigma_{zg} = 8{,}4$ m (here σ_{zp} - is additional vertical stress).

It takes into account the load effect from the F-2 foundation. Coefficient α: $\alpha = \alpha_1 + \alpha_4$

External load voltages $\sigma_{zp,1}, \sigma_{zp,2}, \sigma_{zp,nf}$ calculated by the formulas:

$$\sigma_{zp} = \alpha p_0; \qquad \sigma_{zp,c} = \frac{\alpha p_0}{4},$$

The result of our research is a summary diagram in Figure 6, where it is possible to evaluate the results of the applied methods for the numerical assessment of the loaded die effect on an adjacent foundation located in the mutual influence zone.

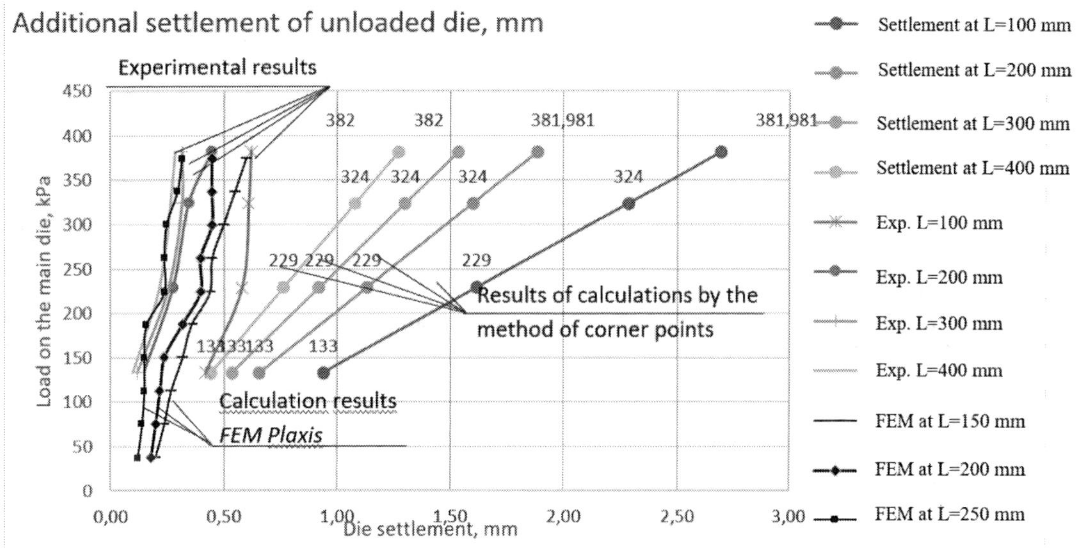

Figure 5. Summary graph of the indicator die additional displacements from the load on the main die according to the results of experiments, computer modeling and calculations by the method of corner points.

Summary

1. Based on the analysis of the summary graph (figure 5), it can be seen that the experimental modeling results are significantly lower than the calculated data obtained by the method of corner points. This can be explained by an error caused by the experimental tray geometry. The thickness of the soil being compressed in it is 1000 mm, and according to the calculations, the depth of the compressed soil at the maximum loading step is 8400 mm.
2. The additional displacements' calculation of the foundations from mutual influence by the method in corner points used in the current regulatory documents is based on the outdated method of calculating the foundations settlement by the method of layer-by-layer summation.
3. The functions of various parameters' influence (distance between the foundations, eccentricities of force application, the soil foundation density).
4. Computer simulation by the FEM method gives the results that are very close to the experimental values, since the edge effect of the tray is taken into account.

References

[1] Al-Bukheiti A Ya, Ledenev V V 2014 Displacement of an unloaded foundation from the influence of a loaded one *Structural mechanics and construction* **1 (8)** 125-139.

[2] Al-Bukheiti A Ya, Ledenev V V 2019 Experimental and theoretical studies of displacements of closely spaced foundations under the action of a plane system of forces *Structural mechanics and construction* **4 (23)** 93-107.

[3] Danisch A B, Ya Al-Bukheiti A A, Ledenev V V, Savinov Ya V 2019 Damage to the walls of closely located buildings due to the mutual influence of their foundations *Structural mechanics and structures, scientific journal* **2 (21)** 109-117.

[4] Emelyanov V V, Komarov I A, Ledenev V V, Savinov Ya V 2004 Mutual influence of adjacent foundations, buildings and structures. Defect of buildings and structures. Strengthening building structures *Materials of the VIII Scientific and Methodological Conference of VETU (March 18, 2004)* (SPb. VSTU) 27-31.

[5] Karaulov A M 2009 Bearing capacity of bases of axisymmetric foundations of buildings and structures *abstract of Ph.D. dis. ... doct. tech. sciences* (SPb.) 46.

[6] Ledenyev V V 2015 Strength and Crack Resistance of Walls *Monograph* (Tambov, Publishing house of FSBEI HPE "TSTU") 224.

[7] Ledenev V V, Skrilyev V I 2017 Accidents, destruction and damage. Causes, consequences and warnings *monograph* (Tambov, Publishing house of FSBEI HE "TSTU") 440.

[8] Savinov Ya V 2008 The main causes of damage to load-bearing wall structures and recommendations for their prevention, for example, the city of Tambov *dis. ... Cand. Tech. sciences. 05.23.01* (Voronezh, VGASU) 235.

[9] Simagin V G 2010 Design and construction of foundations near existing structures in dense buildings *a tutorial* (Moscow, ASV) 128.

[10] Ghosh P, Rajesh S, Chand J Sai 2017 Linear and nonlinear elastic analysis of closely spaced strip foundations using Pasternak model *Front. Struct. Civ. Eng.* 15.

[11] Das B M, Said Larbi-Cher If 1986 Ultimate Bearing Capacity of Closely Spaced Strip Foundations *Department of Civil-Engineering, Indian Institute of Science* Bangalore 560012: India.

[12] Singh A, Punmia B C, On M L 1973 Intellizence between adjacent gmaze footings on cohesionless soil *Indian geotechnical journal* **13 (4)** 275-284.

[13] *BC 22. 13330.2011 Foundations of buildings and structures. Updated edition of SNiP 2.02.01-83. - Instead of SP 22.13330.2010; access date 05/20/2011* (Moscow, LTD CPP) 162.

Stress-strain state of the soil mass under the uniformly distributed load action adjacent to a vertical excavation

Z G Ter-Martirosyan, I N Luzin*, Yu V Vanina, A Z Ter-Martirosyan

National Research Moscow State University of Civil Engineering (NRU MGSU), 26, Yaroslavskoye highway, 129337, Moscow, Russia

E-mail: luzinin@mgsu.ru

Abstract. This article provides a solution to the problem of the stress-strain state of a soil mass adjacent to a vertical excavation when applying a uniformly distributed load on its surface. The solution of the problem was obtained by the method of trigonometric Ribere-Fileon series. The final expressions for the SSS components are provided. Contours of stresses and displacements are also shown.

1. Introduction

When excavating soil from excavations with vertical slopes under the protection of enclosing structures, a complex non-uniform stress-strain state is created in the adjacent soil mass. In the case of application of a distributed load on the soil surface in the soil mass adjacent to the working, additional normal and shear stresses, which have additional effects on the enclosing working structures, are created. This article provides an example of an analytical solution to the stress-strain state problem of a base of limited width and thickness adjacent to a vertical working under the influence of a uniformly distributed load.

It should be noted that the issue of stress distribution in the soil layer resting upon a rigid foundation was studied by many scientists - Fileon (1903), Melan (1919) [12], Marger (1931), Burmister (1956), Sovints, Davis, Taylor, Shekhter O. Ya. (1937), Gorbunov - Posadov M.I. (1946-1953) [2], Egorov K.E. (1961) [3], Tsytovich N.A. (1943) [11].

In these works, the authors note that the SSS formation in a subsoil under the local load action has a certain specificity. Under the foundation of finite width, a certain closed area is formed, within which the stresses exceed the structural strength, and in this area significant deformation of the soil occurs (Fig. 1a). Outside this area, these deformations can be neglected. The shape and size of such an area depend on the mechanical properties and on the structural strength of the soil, as well as on the load area (b = 2a). In this work, a rectangular area with the dimensions hx2l is considered as a computational one (Fig. 1b). To simulate a situation when a uniformly distributed load of width (a-b) is applied to the soil surface at a distance b from the vertical excavation, a rectangular area is considered in the range from 0 to h in height and from –l to 0 in width.

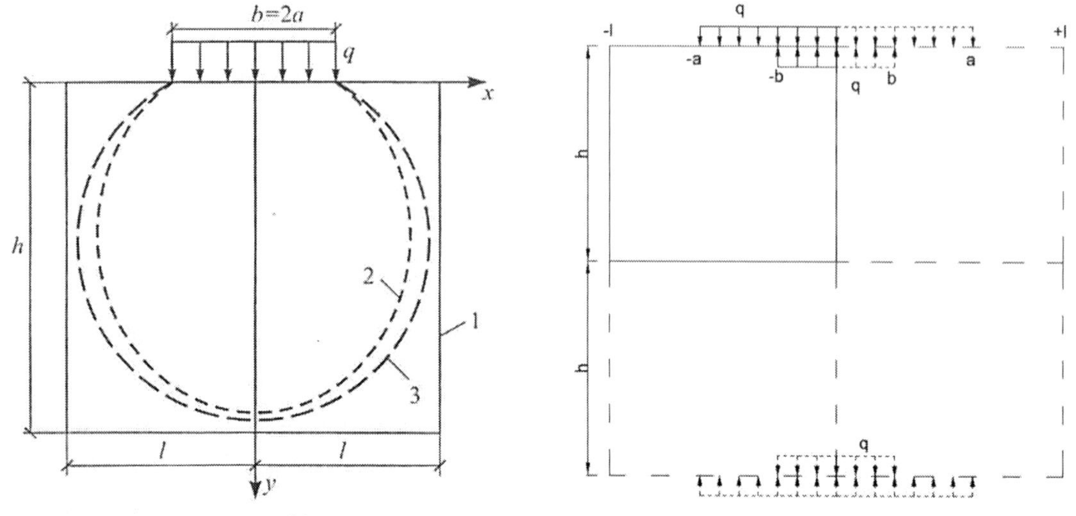

(a) (b)

Figure 1. Geo-mechanical model of the foundation of finite thickness and width (a): 1-boundaries of the model, 2,3- contours of an area with a broken structure; massif of soil with limited thickness (h) and width ($2l$) (b).

2. Problem setting

It is known that the plane problem solution of the theory of elasticity is reduced to the integration of two equilibrium equations and the equation of compatibility (continuity of deformations) with the obligatory satisfaction of the boundary conditions. In addition, this solution assumes a linear relationship between stresses and strains, i.e., the application validity of the generalized Hooke's law [9].

It is also known that when the only volumetric force is the weight of the body, then the plane problem solution can be reduced to finding some function φ(x,y) (Airy function), which is related to stress components by the following dependencies [6-10]:

$$\sigma_x = \frac{\partial^2 \varphi}{\partial y^2} - \rho g y;\ \sigma_y = \frac{\partial^2 \varphi}{\partial x^2} - \rho g y;\ \tau_{xy} = -\frac{\partial^2 \varphi}{\partial x \partial y} \tag{1}$$

and satisfies the biharmonic equation:

$$\frac{\partial^4 \varphi}{\partial x^4} + 2\frac{\partial^4 \varphi}{\partial x^2 \partial y^2} + \frac{\partial^4 \varphi}{\partial y^4} = 0 \tag{2}$$

In the presented work, the solution to the problem formulated is obtained by the of Ribière-Fileon trigonometric series method. The general solution of the biharmonic equation (2) can be represented as an infinite series [9-10]:

$$\begin{aligned}\varphi(x,y) = \sum_{m=1}^{\infty} [\cos\alpha x\left(A_m ch\alpha y + B_m y ch\alpha y + C_m sh\alpha y + D_m y sh\alpha y\right) + \\ + \sin\alpha x\left(A'_m ch\alpha y + B'_m y ch\alpha y + C'_m sh\alpha y + D'_m y sh\alpha y\right)]\end{aligned} \tag{3}$$

The constants $A_m, B_m, \ldots, C'_m, D'_m$ are determined from the conditions on the computational domain contour. Using the stress function (2), adding, if necessary, power polynomials, it is possible to obtain the solutions to the plane problem of the theory of elasticity.

3. The formulated problem solution

To create a uniformly distributed load at a distance b from the vertical excavation, we present the design scheme as a result of the uniformly distributed loads of intensity q addition over a strip with width 2a and intensity – q over a strip with width 2b.

For the problem under consideration, we will accept the following boundary conditions at the upper and lower boundaries of the array area.

$$\begin{gathered}\text{at } y=0 \text{ and } y=2h:\ \sigma_y(x,0)=\sigma_y(x,2h)=q \\ (-a\le x\le a);\sigma_y(x,0)=\sigma_y(x,2h)=0 \quad (-l\le x\le -a) \text{ and } (a\le x\le l); \\ \text{at } y=0 \text{ and } y=2h:\ \tau_{xy}(x,0)=0;\ \tau_{xy}(x,2h)=0\end{gathered} \tag{4}$$

If horizontal displacements at $x=\pm l$ are absent, then we have one more boundary condition in the form:

$$u(\pm l)=0 \quad v(\pm l)\neq 0 \tag{5}$$

Taking into account the equations (4) - (5), we finally obtain the expressions for the stress components, $\sigma_y(x,y)$, $\sigma_x(x,y)$ and $\tau_{xy}(x,y)$ in the soil:

$$\sigma_x(x,y)=\frac{qa}{l}+$$

$$+\frac{4q}{\pi}\sum_{m=1}^{\infty}\frac{\sin\frac{m\pi a}{l}}{m}\frac{\left[\left(\frac{m\pi h}{l}ch\frac{m\pi h}{l}+sh\frac{m\pi h}{l}\right)ch\frac{m\pi(y-h)}{l}-\frac{m\pi(y-h)}{l}sh\frac{m\pi(y-h)}{l}sh\frac{m\pi h}{l}\right]}{sh\frac{2m\pi h}{l}+\frac{2m\pi h}{l}}\cos\frac{m\pi x}{l} \tag{6}$$

$$\sigma_x(x,y)=\frac{qa}{l}\frac{v}{1-v}-$$

$$-\frac{4q}{\pi}\sum_{m=1}^{\infty}\frac{\sin\frac{m\pi a}{l}}{m}\frac{\left[\left(\frac{m\pi h}{l}ch\frac{m\pi h}{l}-sh\frac{m\pi h}{l}\right)ch\frac{m\pi(y-h)}{l}-\frac{m\pi(y-h)}{l}sh\frac{m\pi(y-h)}{l}sh\frac{m\pi h}{l}\right]}{sh\frac{2m\pi h}{l}+\frac{2m\pi h}{l}}\cos\frac{m\pi x}{l} \tag{7}$$

$$\tau_{xy}(x,y)=-\frac{4q}{\pi}\sum_{m=1}^{\infty}\frac{\sin\frac{m\pi a}{l}}{m}\frac{\left[\frac{m\pi h}{l}ch\frac{m\pi h}{l}sh\frac{m\pi(y-h)}{l}-\frac{m\pi(y-h)}{l}ch\frac{m\pi(y-h)}{l}sh\frac{m\pi h}{l}\right]}{sh\frac{2m\pi h}{l}+\frac{2m\pi h}{l}}\sin\frac{m\pi x}{l} \tag{8}$$

For the problem under consideration, the expressions (6) - (8) take the form:

$$\sigma_x(x,y)=\frac{qa}{l}-\frac{qb}{l}+$$

$$+\frac{4q}{\pi}\sum_{m=1}^{\infty}\frac{\left(\sin\frac{m\pi a}{l}-\sin\frac{m\pi b}{l}\right)}{m}\frac{\left[\left(\frac{m\pi h}{l}ch\frac{m\pi h}{l}+sh\frac{m\pi h}{l}\right)ch\frac{m\pi(y-h)}{l}-\frac{m\pi(y-h)}{l}sh\frac{m\pi(y-h)}{l}sh\frac{m\pi h}{l}\right]}{sh\frac{2m\pi h}{l}+\frac{2m\pi h}{l}}\cos\frac{m\pi x}{l} \tag{9}$$

$$\sigma_x(x,y)=\left(\frac{qa}{l}-\frac{qb}{l}\right)\frac{v}{1-v}-$$

$$-\frac{4q}{\pi}\sum_{m=1}^{\infty}\frac{\left(\sin\frac{m\pi a}{l}-\sin\frac{m\pi b}{l}\right)}{m}\frac{\left[\left(\frac{m\pi h}{l}ch\frac{m\pi h}{l}-sh\frac{m\pi h}{l}\right)ch\frac{m\pi(y-h)}{l}-\frac{m\pi(y-h)}{l}sh\frac{m\pi(y-h)}{l}sh\frac{m\pi h}{l}\right]}{sh\frac{2m\pi h}{l}+\frac{2m\pi h}{l}}\cos\frac{m\pi x}{l} \tag{10}$$

$$\tau_{xy}(x,y)=-\frac{4q}{\pi}\sum_{m=1}^{\infty}\frac{\left(\sin\frac{m\pi a}{l}-\sin\frac{m\pi b}{l}\right)}{m}\frac{\left[\frac{m\pi h}{l}ch\frac{m\pi h}{l}sh\frac{m\pi(y-h)}{l}-\frac{m\pi(y-h)}{l}ch\frac{m\pi(y-h)}{l}sh\frac{m\pi h}{l}\right]}{sh\frac{2m\pi h}{l}+\frac{2m\pi h}{l}}\sin\frac{m\pi x}{l} \tag{11}$$

IOP Conf. Series: Materials Science and Engineering 1083 (2021) 012015 doi:10.1088/1757-899X/1083/1/012015

These equations can be solved in the Mathcad software package and the isolines of normal σ_x, σ_y and tangent τ_{xy} stresses in the soil mass. Figures 2-5 show the results of solving the equations (9) - (11) in the Mathcad software package.

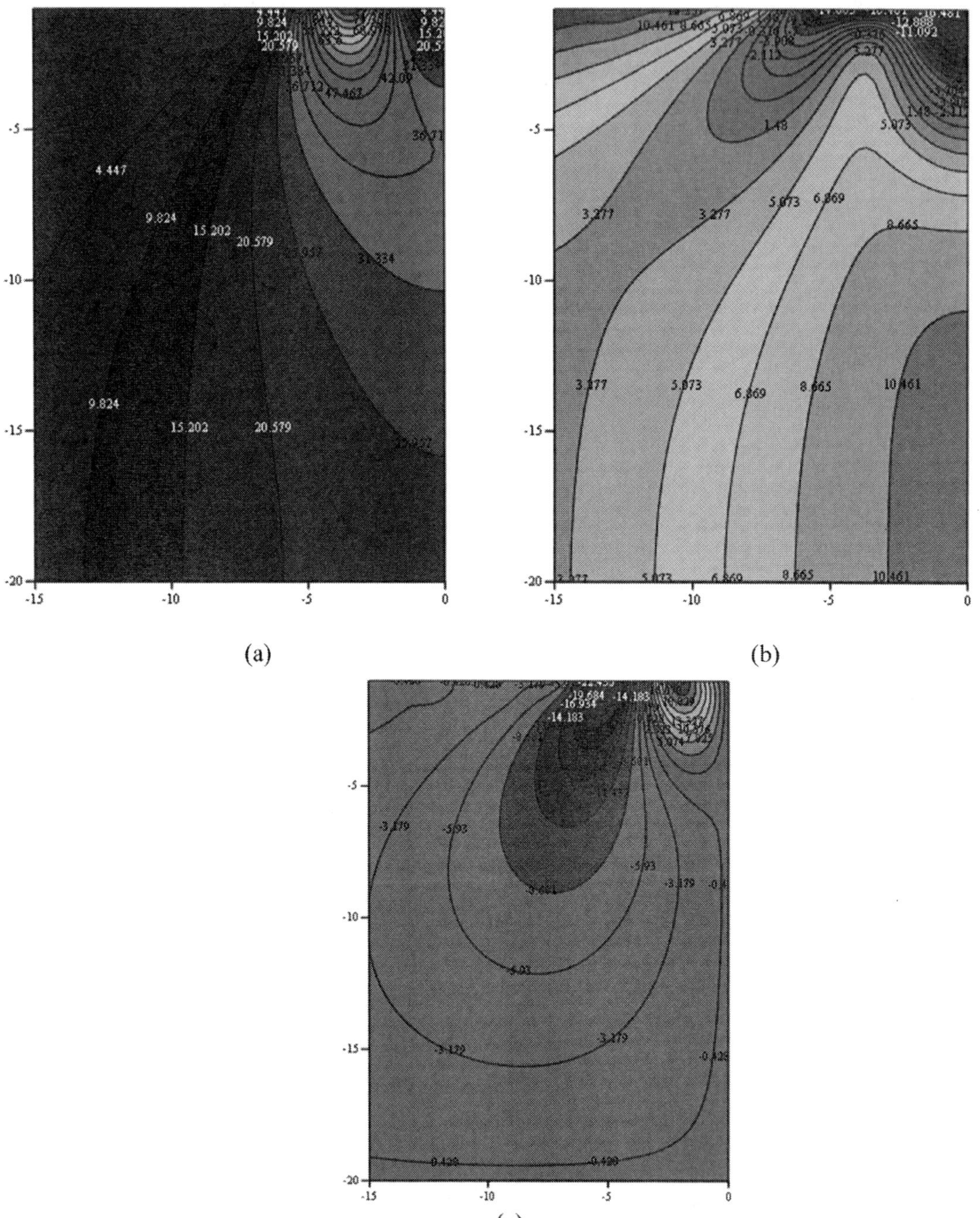

(c)

Figure 2. Isolines of stresses when applying a distributed load at a distance b from the vertical excavation, obtained by the equation (9): a) vertical stresses σ_y ; b) horizontal stresses σ_x ; c) shear stress contours τ_{xy}

Figure 2 shows that when a distributed load is applied near a vertical excavation, the normal and shear stresses are somewhat different and tend to increase when approaching the excavation boundaries. Thus, an analytical solution to the problem of the stress-strain state of a soil mass near a vertical excavation is obtained when a uniformly distributed load is applied.

4. Summary

As a result of the deformation problem analytical solution of a layer with limited width and thickness adjacent to a vertical excavation, the formulas for determining the components of normal and shear stresses when a uniformly distributed load is applied nearby a vertical excavation, are obtained. The analysis of the solution results shows that the load application nearby the excavation significantly affects the stress-strain state of the entire massif, and the stresses tend to increase when approaching the boundary of the vertical excavation.

The results of the solution can be used to determine the active pressure on the enclosing structures of the excavation to take into account additional stresses when applying a uniformly distributed load on the excavation edge from the equipment and stored building materials.

References

[1] Vyalov S S 1978 Rheological foundations of soil mechanics (High school, Moscow).
[2] Gorbunov-Posadov M I 1979 Problems of nonlinear soil mechanics 1961 *Experimental and theoretical studies of nonlinear problems in the field of foundations and foundations. NPI interuniversity collection. Novocherkassk* 3-8.
[3] Egorov K E On the deformation of the base of finite thickness *Foundations, foundations and soil mechanics* **1**.
[4] Zaretsky Yu K, Kabantsev M I 2006 Influence of the sequence of erection of nearby high-rise buildings on precipitation and roll of foundation slabs *MGSU Herald* **1** 51-56.
[5] Mangushev R A, Nikiforova N S, Konyushkov V V, Osokin A I, Sapin D A 2013 Design and construction of underground structures in open pits. Tutorial (ASV Publishing House, Moscow, St. Petersburg).
[6] Nadai A 1961, 1969 (2) Plasticity and Fracture of Solids, ed. (Mir, Moscow).
[7] Ter-Martirosyan Z G 1990 Rheological parameters of soils and calculations of the foundations of structures (M .: Stroyizdat).
[8] Ter-Martirosyan Z G 2009 Soil mechanics (ASV, Moscow).
[9] Timoshenko S P and Goodyer J 1975 Theory of elasticity (Nauka, Moscow).
[10] Tikhonov A N, Samorskiy A A 1966 Equations of Mathematical Physics (Nauka, Moscow).
[11] Tsytovich N A, Ter-Martirosyan Z G 1981 Fundamentals of applied geomechanics in construction: textbook (Moscow, Higher school).
[12] Melan E 1932 Der Spannungzustand der duch eine Einzelkraft im innern beanspruchten Halbschiebe *Zeitschrift fur augewandte Mathematik und Mechanik* **B. 12, H. 6.**

The technical state experimental studies of construction with free support in the non-stationary oscillations' analysis

M N Grigoryan[1*], T V Polyakova[1], Ya A Kokareva[1], I A Parinov[2], A V Cherpakov[2]

[1]Don State Technical University, 1, Gagarin sq., Rostov-on-Don, 344000, Russia
[2]Southern Federal University, 8a, Milchakov str., Rostov-on-Don, 344090, Russia

E-mail: manigri@inbox.ru

Abstract. An algorithm for the experimental determination of the defective zones of the construction with free support is presented. A reinforced concrete floor plate freely supported on both sides is studied as an example. The oscillation method is used as a diagnostic method. Impact loading is used as the excitation of non-stationary oscillations. The studies were carried out using an oscillation measuring setup. The research is based on the analysis of oscillation modes under non-stationary dynamic loading of an object. The analysis of oscillation modes makes it possible to detect the conditionally defective zones of floor plate.

1. Introduction

It is well-known, any actual constructions lose its performance over time. This is due to the loss of the operational properties of materials under the external negative factors' influence, the creation of emergency situations and further construction destruction. Technical diagnostics require for preventing the critical state of the object and for its technical conditions' analysis for further operation.

Floor plates are the research objects in the study. Restoration of the integrity and reliability of floor plates is carried out during the repair and restoration works. Carrying out these works in practice should be accompanied by planned technical monitoring. With this aim, at the first stage, it is required to conduct construction and technical examination, one of the algorithms of which is proposed in [1]. The paper presents a draft of schedule for the comprehensive technical monitoring of industrial building constructions. The features of construction and technical monitoring of industrial building constructions are described. The algorithm scheme is presented. During the analysis, the object is divided into zones, where various research methods and specialized equipment can be used.

In the research process, different procedures, the examples of which have been presented in various publications, could be applied:

(i) visual and technical detection of defects in the structure [2];

(ii) description and establishment of the object's technical conditions category [3];

(iii) development of repair technology procedures (the creation of a map of technological preparation of the construction surface [4]; the selection of repair, the connection and restoration of structural integrity [5, 6]);

(iv) quality control of repair works (the establishment of a criterion for assessing the quality of work performed; verification tests and calculations of structural elements [7]; development of methods for testing and preliminary calculations [8]).

For instrumental examinations of floor plates, specific examination methods can be used, for example, oscillation diagnostics. Some approaches to oscillation diagnostics have been developed in [9-11].

Some approaches of the authors of that use algorithms for solving problems based on assessing the influence of damage parameters on the change in natural oscillation frequencies and modal characteristics of signals are known [12-14]. In these studies, both analytical and numerical methods of calculation can be used.

Application of specialized calculation methods, evolutionary algorithms and neural networks can improve the constructions analysis accuracy [16-18]. In these studies, an experimental approach can be used to determine the criterion-based assessment of the applied theory and to minimize the time of mathematical operations by calculating the object's characteristics.

2. Formulation of the problem

The basic aim of the research is to draw up a technique for oscillation diagnostics of floor plates and its experimental testing.

3. Description of experimental research

The research was carried out in the Don State Technical University, Rostov-on-Don by using the hollow floor plates (with dimensions of $h \times b \times l = 220 \times 1190 \times 6000$ mm^3) of one of the university buildings. The oscillation analysis of floor plates is based on the natural frequencies and oscillation modes' analysis. The use of modeling and comparison of results makes it possible to restore the missed parameters of the construction. The calculation of the stress state can be performed on the base of the previously obtained data on the structure oscillations by comparing the numerical analysis of the floor plates' deflection and experimental data.

At the first stage, the oscillation diagnostics system was assembled, tested and the sensors were placed on the floor plate. The frequency of the analog-to-digital converter (ADC) was chosen equal to 100 kHz. The number of accelerometer sensors simultaneously interrogated by the ADC was 16. The distance between the sensors was fixed and equal to 38 cm. Thus, the sensors covered an area of 5.7 m. *Figure 1* shows an example of the measuring setup arrangement.

Oscillations were grouped at a fixed impact loading for a group of floor slabs. The fixed impact loading was applied by using a weight on a tripod, dropping from a certain height at a point located at the edge of the slab near to the sensor No. 1. A significant signal level was recorded in the far registration zone. Thus, the most optimal effect on the plate in the force terms was selected. In this case, a series of floor plates oscillations' indications were recorded on a computer. A sampling of a representative signal was made. Acceleration diagrams were built (*Figure 2a*) for a group of points and the comparative analysis production. The natural frequencies of plate oscillations were calculated based on the fast Fourier transform of the signal. Using Fourier transform and accelerations $a(t)$ oscillations integration, the velocity $v(t)$ and displacement $u(t)$ were calculated for the measured data.

Figure 2b shows an example of accelerations $a(t)$ oscillations diagrams and displacements $u(t)$ for 16 points on the floor plate.

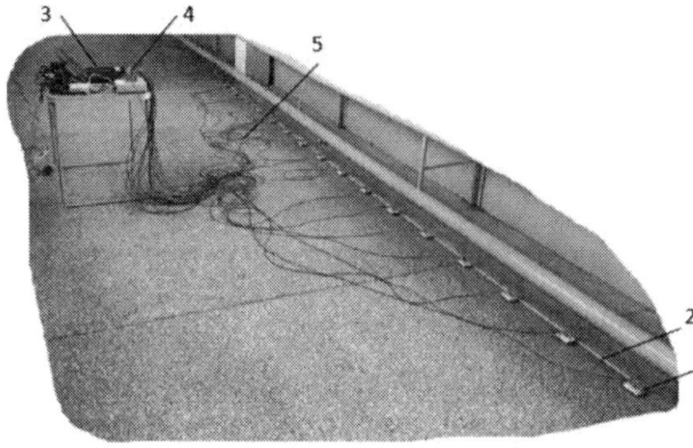

(a)

(b)

Figure 1. (a) Measuring equipment of the diagnostic setup on the full-scale section: 1 - sensors, 2 - tape measure, 3 - PC; 4 - matching device; 5 - transmitting path; (b) impact mechanism.

$$\bar{a}(\omega) = \int a(t)e^{-i\omega t}dt \quad (1)$$

$$\bar{v}(\omega) = \int v(t)e^{-i\omega t}dt;\ \bar{v}(\omega) = \frac{\bar{a}(\omega)}{i\omega} \quad (2)$$

$$\bar{u}(\omega) = \int u(t)e^{-i\omega t}dt;\ \bar{u}(\omega) = \frac{\bar{v}(\omega)}{i\omega} \quad (3)$$

where ω is the circular oscillation frequency, i is the imaginary unit, $\bar{a}(\omega)$, $\bar{v}(\omega)$ and $\bar{u}(\omega)$ are the Fourier vectors for acceleration, velocity and displacement, respectively, t is the time parameter.

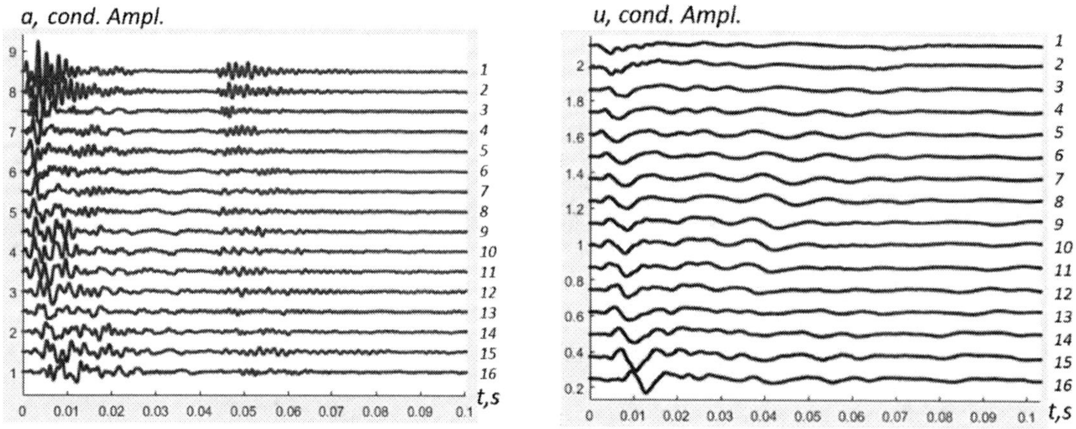

(a) (b)

Figure 2. Diagrams of accelerations $a(t)$ (a) and displacements $u(t)$ (b) experimentally measured for 16 points on the floor plate

The oscillations analysis shows that the accelerations $a(t)$ are represented as a series of oscillations within a period of $t = 0.03$ s. For a range equal to $t = 0.1$ s, the velocity $v(t)$ and displacement $u(t)$ are calculated by integration.

In a comprehensive comparative analysis of natural frequencies and the largest oscillation amplitudes for the studied floor plates, a conclusion was made about the weakest floor slabs. *Table 1* shows an example of comparing the first natural frequencies of some floor slabs at the same row. The maximum resonance frequencies obtained by Fourier analysis of the acceleration signal $a(t)$ are investigated. Analysis shows that the lowest frequency is the first frequency for floor plate 3, which may be related to its most defective state.

Table 1. Natural frequencies of floor slabs

Floor slab No.	Floor slab Natural frequency, Hz		
	w_1	w_2	w_3
1	122	134	207
2	122	146	195
3	97	146	219

At the next stage, a comparative analysis of damped oscillation modes was carried out for all floor plates. The criterion was a modal sign about the maximum angles of the construction deflection during damping. The deflection angles were calculated by the formula:

$$\alpha = \arccos\left(\frac{(\overline{AB})(\overline{BC})}{\|\overline{AB}\|\|\overline{BC}\|}\right) \quad (4)$$

where $\overline{AB}$ and $\overline{BC}$ are the vector representations of two segments between the corresponding points of the normalized oscillation mode for points $[i-1, i]$ and $[i, i+1]$.

Figure 3 shows an example of oscillation modes during impact loading of a floor plate at different times. Fractures of oscillation modes, expressed by the greatest curvature of oscillations, were revealed at the points 2, 7, 10, 15. Thus, the zones for which it is necessary to monitor the defect state of the floor slab in more detail, were identified.

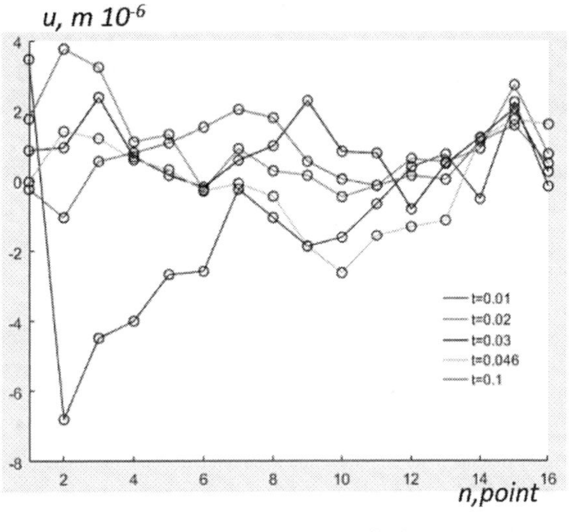

Figure 3. Forms of the floor slab structure oscillations at some points in time t.

Summary

The problem of experimental analysis of oscillations and identification of defective zones in a reinforced concrete floor plate is considered. An analysis of damped oscillations was carried out under non-stationary action on the plate in the form of impact loading. The natural oscillation frequencies of a set of floor plate constructions were considered. The construction with the lowest first natural frequency was chosen as the most damping one. By analyzing the forms of damped oscillations, the criterion of the largest bends of the oscillation modes was applied, which may correspond to a loss of structural rigidity or the presence of defects in the form of cracks in concrete in the vicinity of this zone.

In accordance with the normative and technical base, the state of the existing monolithic plate could be classified as limited-workable. Reinforcement of the construction is required to restore the bearing capacity. Before reinforcement, it is necessary to limit the loads on the floor slab. A more detailed analysis requires calculations and additional studies of hazardous areas of the facility.

Acknowledgments
The authors acknowledge the support by the Southern Federal University, grant No. VnGr-07/2020-04-IM (Ministry of Science and Higher Education of Russia).

References
[1] Serebryanaya I A, Egorochkina I O, Shlyakhova E A, Matrosov A A, Serebryanaya D S 2020 Integrated construction and technical structural analysis of the industrial buildings *IOP Conference Series: Materials Science and Engineering* **913(2)** 022071. doi: 10.1088/1757-899X/913/2/022071.
[2] Cherpakov A, Egorochkina I, Shlyakhova E, Kharitonov A, Zarovny A, Dobrohodskaya S 2017 Using technique oscillation diagnostics for assessing the quality of power transmission line supports repairs *Matec Web Conf.* **106** 04009. doi:10.1051/matecconf/201710604009.
[3] Sheina S G, Girya L V, Burakova A A, Gorskaya A L 2019 Methodology for a comprehensive assessment of the state of architectural monuments Construction and industrial safety. **15 (67)** 7 - 13.
[4] Bolotskikh O N 2001 Surface preparation of concrete and reinforced concrete structures for repair (Waterpas, Moscow).
[5] Matrosov A A, Serebryanay I A, Lukinova N A, Nizhnik D A, Terekhina Yu V 2018 The Calculation of the Stress-Strain State in Masonry with Various Types of Bricks *2018 International Conference on "Physics and Mechanics of New Materials and Their Applications" (PHENMA 2018), Busan, South Korea, August 9-11, 2018* 238. Information on https://phenma2018.wordpress.com/
[6] Serebryanaya I A, Matrosov A A, Poryadina N A, Soloviev A N 2020 Analysis of the stress-strain state of ceramic brick when tested for compressive strength *Materials Science Forum* **974** MSF 510-514. doi: 10.4028/www.scientific.net/MSF.974.510.
[7] Nguyen H T N, Tan K-H 2021 Shear response of deep precast/prestressed concrete hollow core slabs subjected to fire *Engineering Structures* **227** 111398. doi: 10.1016/j.engstruct.2020.111398.
[8] Brovko T N, Silver I A, Volokhova A V 2015 On methods of measuring strength Construction and architecture - 2015 *Mater. conf. - Rostov-on-Don.: Publishing house of FSBEI HPE RSSU* 412-413.
[9] Worden K, Dervilis N, Farrar C 2019 Applying the concept of complexity to structural health monitoring *Structural Health Monitoring 2019: Enabling Intelligent Life-Cycle Health Management for Industry Internet of Things (IIOT) - Proceedings of the 12th International Workshop on Structural Health Monitoring* **2** 3226-3234. doi: 10.12783/shm2019/32481.
[10] Dimarogonas A D 1996 Oscillation of cracked structures: astate of the art review *Eng. Fract. Mech.* **55(5)** 831–857. doi:10.1016/0013-7944(94)00175-8.
[11] Friswell M I 2007 Damage identification using inverse methods. Phil. Trans. R. Soc. A *Mathematical, Physical and Engineering Sciences* **365(1851)** 393–410. doi:10.1098/rsta.2006.1930
[12] Cherpakov A V, Soloviev A N, Gricenko V V, Goncharov O U 2016 Damages identification in the cantilever-based on the parameters of the natural oscillations *Def. Sci. J.* **66** 44–50. doi:14429/dsj.66.8182.
[13] Shevtsov S N, Soloviev A N, Parinov I A, Cherpakov A V, Valery A C 2018 Piezoelectric Actuators and Generators for Energy Harvesting. *Springer IP AG.* ISBN 978-3-319-75629-5.
[14] Akopyan V A, Solov'Ev A N, Cherpakov A V, Shevtsov S N 2013 On a deformation sign for identifying defects on the basis of the analysis of the forms of the natural oscillations of a cantilever with a notch *Russian Journal of Nondestructive Testing* **49(10)** 579-583. doi: 10.1134/S1061830911060088.
[15] Cherpakov A V , Shilyaeva O V, Grigoryan M N, Polyakova T V 2019 Simulation of wave processes in the multilayer structure surface layer properties identification by the finite element method *IOP Conference Series: Materials Science and Engineering* **698(6)** 066021. doi:

10.1088/1757-899X/698/6/066021.
[16] Tran V-L, Thai D-K, Nguyen D-D 2020 Practical artificial neural network tool for predicting the axial compression capacity of circular concrete-filled steel tube columns with ultra-high-strength concrete *Thin-Walled Structures* **151** 106720. doi: 10.1016/j.tws.2020.106720.
[17] Krasnoshchekov A A, Sobol B V, Solov'ev A N, Cherpakov A V 2011 Identification of crack-like defects in elastic structural elements on the basis of evolution algorithms *Russian Journal of Nondestructive Testing* **47(6)** 412-419. doi: 10.1134/S1061830911060088.
[18] Abambres M, Lantsoght E O L 2019 ANN-based fatigue strength of concrete under compression *Materials* **12(22)** 3787. doi: 10.3390/ma12223787.

Determination of stress-strain state of short eccentrically loaded concrete-filled steel tubular (CFST) columns using finite element method with reducing the problem from three-dimensional to two-dimensional

V S Chepurnenko, B M Yazyev*, P M Urvachev, A A Avakov

Don State Technical University, 1 Gagarin sq., Rostov-on-Don, 344002, Russia

E-mail: ps62@yandex.ru

Abstract. The article presents the derivation of the FEM equations, which make possible to calculate short concrete-filled steel tubular columns in a plane elastic formulation. Steel tube is modeled by 1D bar finite elements. The obtained results are compared with the results of calculations using 3D solid and shell finite elements.

Introduction

In modern high-rise construction, steel-reinforced concrete structures are widely used, including structures consisting of a concrete core and a steel tube. More acute is the question of improving regulatory documentation, increasing the economic efficiency and reliability of such structures based on new theoretical and experimental data. Numerical modeling of the CFST columns deformations in existing finite element software requires significant computing power and time. Correct accounting of the tube affect on the concrete core stress-strain state is possible only when using volumetric finite elements for concrete and shell ones for steel [1-2].

In this work, a technique has been developed for converting the calculation of a short eccentrically compressed CFST column in an elastic formulation to a plane problem, which significantly reduces the dimension of the stiffness matrix used in the calculation, the required machine time and memory.

Derivation of resolving equations

We consider a short concrete-filled steel tubular column of unit length shown in Figure 1, loaded with force F and moment M. Cross-section characteristics are shown in Figure 2.

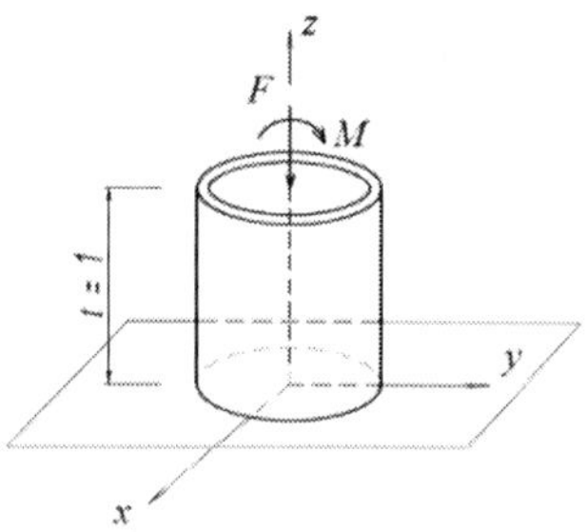

Figure 1. Column loading scheme

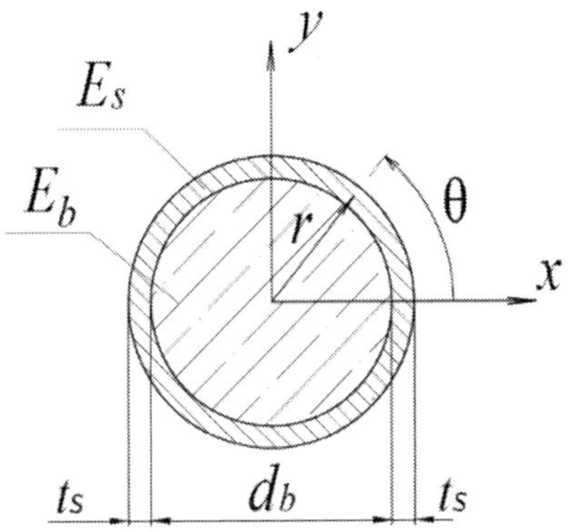

Figure 2. Cross section of the column

According to the variational principle of the minimum total potential energy [3]:

$$\delta\Pi + \delta V = 0, \tag{1}$$

Π is the elastic strain energy, V is the potential of applied forces.

In the further derivation of the equations, it is assumed that the Navier's hypothesis is satisfied. Steel and concrete work together without sliding. Normal strain along z axis:

$$\varepsilon_z = \varepsilon_0 + y \cdot \chi_0, \tag{2}$$

ε_0 is the normal strain of the column's central longitudinal axis and

$$\chi_0 = -\frac{d^2 f_y}{dz^2}$$

is the curvature of the longitudinal central axis of the column, which are constant along the length.

For the case shown in Figure 1 with the fixed support of the lower end of a unit length column, taking into account the direction of action of the load F:

$$\delta V = F\delta\varepsilon_0 + M\delta\chi_0 = \delta\begin{Bmatrix}\varepsilon_0\\ \chi_0\end{Bmatrix}^T \begin{Bmatrix}F\\ M\end{Bmatrix}. \tag{3}$$

The variation of the potential strain energy of the column is the sum of the strain energies variations of the concrete core and the steel tube:

$$\delta\Pi = \delta\Pi_b + \delta\Pi_s . \tag{4}$$

$$\delta\Pi_b = \int_{A_b} (\sigma_z\delta\varepsilon_z + \sigma_x\delta\varepsilon_x + \sigma_y\delta\varepsilon_y + \tau_{xy}\delta\gamma_{xy})dA, \tag{5}$$

A_b is the area of the concrete cross section part. Using the generalized Hooke's law, we get:

$$\int_{A_b}\sigma_z\delta\varepsilon_z dA = \int_{A_b}(\lambda_b\left(\varepsilon_z+\varepsilon_x+\varepsilon_y\right)+2\mu_b\varepsilon_z)\delta\varepsilon_z dA = \int_{A_b}(\lambda_b+2\mu_b)\varepsilon_z\delta\varepsilon_z dA + \int_{A_b}\lambda_b(\varepsilon_x+\varepsilon_y)\delta\varepsilon_z dA = \delta\Pi_{b1} + \delta\Pi_{b2}, \tag{6}$$

here λ_b, μ_b are Lamé parameters [4].

$$\int_{A_b}(\sigma_x\delta\varepsilon_x+\sigma_y\delta\varepsilon_y+\tau_{xy}\delta\gamma_{xy})dA=$$

$$=\int_{A_b}\left(\begin{bmatrix}\lambda_b+2\mu_b & \lambda_b & 0\\ \lambda_b & \lambda_b+2\mu_b & 0\\ 0 & 0 & \mu_b\end{bmatrix}\cdot\begin{Bmatrix}\varepsilon_x\\ \varepsilon_y\\ \gamma_{xy}\end{Bmatrix}\right)^T\cdot\delta\begin{Bmatrix}\varepsilon_x\\ \varepsilon_y\\ \gamma_{xy}\end{Bmatrix}dA+ \tag{7}$$

$$+\int_{A_b}\lambda_b\begin{Bmatrix}\varepsilon_0+y\chi_0\\ \varepsilon_0+y\chi_0\\ 0\end{Bmatrix}^T\cdot\delta\begin{Bmatrix}\varepsilon_x\\ \varepsilon_y\\ \gamma_{xy}\end{Bmatrix}dA=\delta\Pi_{b3}+\delta\Pi_{b4}.$$

$$\delta\Pi_{b1}=\int_{A_b}(\lambda_b+2\mu_b)[1\quad y]\begin{Bmatrix}\varepsilon_0\\ \chi_0\end{Bmatrix}[1\quad y]\delta\begin{Bmatrix}\varepsilon_0\\ \chi_0\end{Bmatrix}dA=$$

$$=\int_{A_b}(\lambda_b+2\mu_b)\begin{Bmatrix}\varepsilon_0\\ \chi_0\end{Bmatrix}^T\begin{Bmatrix}1\\ y\end{Bmatrix}[1\quad y]\delta\begin{Bmatrix}\varepsilon_0\\ \chi_0\end{Bmatrix}dA= \tag{8}$$

$$=\int_{A_b}(\lambda_b+2\mu_b)\begin{Bmatrix}\varepsilon_0\\ \chi_0\end{Bmatrix}^T\begin{bmatrix}1 & y\\ y & y^2\end{bmatrix}\delta\begin{Bmatrix}\varepsilon_0\\ \chi_0\end{Bmatrix}dA.$$

Transposing the right-hand side of expression (8), and taking into account that the values of ε_0, χ_0 and their variations are constant over the cross-section, we get:

$$\delta\Pi_{b1}=\delta\begin{Bmatrix}\varepsilon_0\\ \chi_0\end{Bmatrix}^T\int_{A_b}(\lambda_b+2\mu_b)\begin{bmatrix}1 & y\\ y & y^2\end{bmatrix}dA\begin{Bmatrix}\varepsilon_0\\ \chi_0\end{Bmatrix}. \tag{9}$$

When modeling the concrete core, we will use triangular finite elements. The contribution of a single finite element to the total part of the potential energy variation $\delta\Pi_{b1}$ will be:

$$\delta\Pi_{b1}^e=\delta\begin{Bmatrix}\varepsilon_0\\ \chi_0\end{Bmatrix}^T(\lambda_b+2\mu_b)\begin{bmatrix}1 & y_m\\ y_m & y_m^2\end{bmatrix}\cdot\Delta^e\cdot\begin{Bmatrix}\varepsilon_0\\ \chi_0\end{Bmatrix}, \tag{10}$$

Δ^e is the area of the finite element, y_m is the center mass coordinate of the triangle.

We transform the second term from equation (6):

$$\delta\Pi_{b2}=\int_{A_b}\lambda_b(\varepsilon_x+\varepsilon_y)[1\quad y]dA\delta\begin{Bmatrix}\varepsilon_0\\ \chi_0\end{Bmatrix}, \tag{11}$$

for a single finite element:

$$\delta\Pi_{b2}^{e}=\lambda_b\{b\}^T\{U_{pl}^{e}\}[1\quad y_m]\cdot\Delta^e\cdot\delta\begin{Bmatrix}\varepsilon_0\\ \chi_0\end{Bmatrix}+$$
$$+\lambda_b\{c\}^T\{U_{pl}^{e}\}[1\quad y_m]\cdot\Delta^e\cdot\delta\begin{Bmatrix}\varepsilon_0\\ \chi_0\end{Bmatrix}=\lambda_b\{g\}^T\{U_{pl}^{e}\}[1\quad y_m]\cdot\Delta^e\cdot\delta\begin{Bmatrix}\varepsilon_0\\ \chi_0\end{Bmatrix}. \tag{12}$$

Here $\{U_{pl}^{e}\}$ is the nodal displacements vector of the finite element in the section plane, $\{U_{pl}^{e}\}=\begin{bmatrix}u_i & v_i & u_j & v_j & u_k & v_k\end{bmatrix}^T$, where *i, j, k* are node's numbers,

$\{b\}^T=\frac{1}{2\Delta^e}\begin{bmatrix}b_i & 0 & b_j & 0 & b_k & 0\end{bmatrix}$, $\{c\}^T=\frac{1}{2\Delta^e}\begin{bmatrix}0 & c_i & 0 & c_j & 0 & c_k\end{bmatrix}$,

$\{g\}=\{b\}+\{c\}$, the vector's $\{b\}$ and $\{c\}$ elements values *b* and *c* depend on the coordinates of the element nodes and are calculated similarly to the case of solving the plane problem of the theory of elasticity using the FEM [5].

Transforming expression (12) and transposing with multiple use of the property for the product of two matrices $([A][B])^T=[B]^T[A]^T$, we get:

$$\delta\Pi_{b2}^{e}=\delta\begin{Bmatrix}\varepsilon_0\\ \chi_0\end{Bmatrix}^T\lambda_b\begin{Bmatrix}1\\ y_m\end{Bmatrix}\{g\}^T\cdot\Delta^e\cdot\{U_{pl}^{e}\}. \tag{13}$$

Denote as:

$[D_b]=\begin{bmatrix}\lambda_b+2\mu_b & \lambda_b & 0\\ \lambda_b & \lambda_b+2\mu_b & 0\\ 0 & 0 & \mu_b\end{bmatrix}$ is stress-strain elastic matrix, $\{\varepsilon_{pl}\}=[\varepsilon_x\quad \varepsilon_y\quad \gamma_{xy}]^T$ is elastic strain vector in the section plane. Then:

$$\delta\Pi_{b3}=\int_{A_b}\{\varepsilon_{pl}\}^T[D_b]\cdot\delta\{\varepsilon_{pl}\}dA \tag{14}$$

or after discretization:

$$\delta\Pi_{b3}^{e}=([B_{pl}]\{U_{pl}^{e}\})^T[D_b]\cdot[B_{pl}]\delta\{U_{pl}^{e}\}\cdot\Delta^e=$$
$$=\left(([B_{pl}]\{U_{pl}^{e}\})^T[D_b]\cdot[B_{pl}]\delta\{U_{pl}^{e}\}\right)^T\cdot\Delta^e=\left([D_b][B_{pl}]\delta\{U_{pl}^{e}\}\right)^T[B_{pl}]\{U_{pl}^{e}\}\cdot\Delta^e=$$
$$=\delta\{U_{pl}^{e}\}^T[B_{pl}]^T[D_b]^T[B_{pl}]\{U_{pl}^{e}\}\cdot\Delta^e=\delta\{U_{pl}^{e}\}^T[B_{pl}]^T[D_b][B_{pl}]\cdot\Delta^e\cdot\{U_{pl}^{e}\}, \tag{15}$$

$[D_b]^T=[D_b]$ due to the symmetry of the stress-strain elastic matrix, $[B_{pl}]$ is strain-displacement matrix:

$$[B_{pl}]=\frac{1}{2\Delta^e}\begin{bmatrix} b_i & 0 & b_j & 0 & b_k & 0 \\ 0 & c_i & 0 & c_j & 0 & c_k \\ c_i & b_i & c_j & b_j & c_k & b_k \end{bmatrix}.$$

$$\delta\Pi_{b4}=\int_{A_b}\lambda_b\begin{Bmatrix}\varepsilon_0+y\chi_0 \\ \varepsilon_0+y\chi_0 \\ 0\end{Bmatrix}^T\cdot\delta\{\varepsilon_{pl}\}dA=\int_{A_b}\begin{Bmatrix}\varepsilon_0 \\ \chi_0\end{Bmatrix}^T\begin{bmatrix}1 & y \\ 1 & y \\ 0 & 0\end{bmatrix}^T\lambda_b\delta\{\varepsilon_{pl}\}dA. \quad (16)$$

Discretizing equation (16), we obtain:

$$\delta\Pi_{b4}^e=\begin{Bmatrix}\varepsilon_0 \\ \chi_0\end{Bmatrix}^T\begin{bmatrix}1 & y_m \\ 1 & y_m \\ 0 & 0\end{bmatrix}^T\lambda_b[B_{pl}]\cdot\Delta^e\cdot\delta\{U_{pl}^e\}=$$

$$=\begin{Bmatrix}\varepsilon_0 \\ \chi_0\end{Bmatrix}^T\begin{bmatrix}1 & y_m \\ 1 & y_m \\ 0 & 0\end{bmatrix}^T\lambda_b[B_{pl}]\cdot\Delta^e\cdot\delta\{U_{pl}^e\}=\delta\{U_{pl}^e\}^T[B_{pl}]^T\begin{bmatrix}1 & y_m \\ 1 & y_m \\ 0 & 0\end{bmatrix}\lambda_b\cdot\Delta^e\cdot\begin{Bmatrix}\varepsilon_0 \\ \chi_0\end{Bmatrix}. \quad (17)$$

Thus:

$$\delta\Pi_b^e=\delta\Pi_{b1}^e+\delta\Pi_{b2}^e+\delta\Pi_{b3}^e+\delta\Pi_{b4}^e, \quad (18)$$

the terms $\delta\Pi_{bi}^e$ are determined for each plane finite element using expressions (10), (13), (15), (17). The concrete core total potential energy variation is calculated by summing over the elements:

$$\delta\Pi_b=\sum_e\delta\Pi_b^e. \quad (19)$$

Let $[K]$ be the CFST column stiffness matrix of dimension $(2n+2)\times(2n+2)$, n is the number of nodes in the section, in each of which there are 2 degrees of freedom, 2 is the number of additional degrees of freedom (ε_0 and χ_0), that determine the column behavior along z axis direction. Then:

$$[K]=[K_b]+[K_s]=\sum_e[K_b^e]+\sum_e[K_s^e], \quad (20)$$

$[K_b]$, $[K_s]$ are matrices that determine the contribution of concrete and steel to the total stiffness matrix. Due to the fact that equality (1) must be valid for arbitrary values of variations, for a triangular concrete finite element we get:

$$[K_b^e]=\begin{bmatrix}[K_{b11}^e] & [K_{b12}^e] \\ [K_{b21}^e] & [K_{b22}^e]\end{bmatrix}, \quad (21)$$

$[K_b^e]$ is the 8×8 finite element stiffness matrix. A vector that determines the displacements of the FE with nodes *i, j, k*: $\{U^e\}=[\varepsilon_0 \;\; \chi_0 \;\; u_i \;\; v_i \;\; u_j \;\; v_j \;\; u_k \;\; v_k \;\;]^T$,

the values ε_0 and χ_0 are common to all elements. The submatrices included in the composition $[K_b^e]$ in accordance with (10), (13), (15), (17) will be equal:

$$[K_{b11}^e]=(\lambda_b+2\mu_b)\begin{bmatrix}1 & y_m\\ y_m & y_m^2\end{bmatrix}\cdot\Delta^e,$$

$$[K_{b12}^e]=\lambda_b\begin{Bmatrix}1\\ y_m\end{Bmatrix}\{g\}^T\cdot\Delta^e,$$

$$[K_{b21}^e]=[B_{pl}]^T\begin{bmatrix}1 & y_m\\ 1 & y_m\\ 0 & 0\end{bmatrix}\lambda_b\cdot\Delta^e=[K_{b12}^e]^T,$$

$$[K_{b22}^e]=[B_{pl}]^T[D_b][B_{pl}]\cdot\Delta^e.$$

Total nodal force vector: $\{P\}=[-F \;\; -M \;\; 0 \;\; 0 \;\; \cdots \;\; 0]^T$, with dimensions $(2n+2)\times 1$.

We model the steel tube with 1D bar finite elements with two translational degrees of freedom in the section plane at each of the nodes (Figure 2).

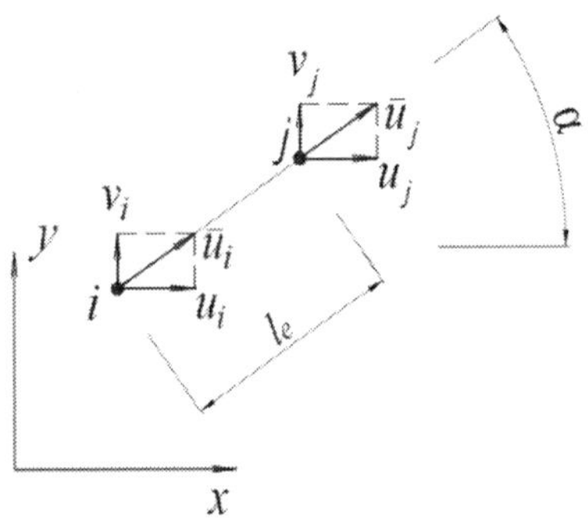

Figure 3. 1D bar finite element

The tube has a smooth shape. It is loaded with pressure from the side of the concrete core, continuously and smoothly changing along the circumference of the inner face, therefore, outside the fixed points there is a momentless biaxial stress state of the shell, the rotational degrees of freedom in the nodes of the steel bar FE are therefore not taken into account.

Strain potential energy of steel tube variation:

$$\delta\Pi_s=\int_{A_s}(\sigma_z\delta\varepsilon_z+\sigma_\theta\delta\varepsilon_\theta)dA, \tag{22}$$

A_s is area of the steel part of cross-section. Radial stress in the tube $\sigma_r\approx 0$ due to the small steel thickness $t_s \ll D_b$.

Using the generalized Hooke's law and taking into account the equality of radial stresses to zero, we obtain:

$$\sigma_z = E_s\varepsilon_z + \nu_s(E_s\varepsilon_\theta + \nu_s\sigma_z) \Rightarrow \sigma_z = \frac{E_s}{1-\nu_s^2}\varepsilon_z + \frac{\nu_s E_s}{1-\nu_s^2}\varepsilon_\theta. \tag{23}$$

We denote $\tilde{E}_s = \dfrac{E_s}{1-\nu_s^2}$, then the expressions for the relationship between normal stresses and strains will take the form:

$$\sigma_z = \tilde{E}_s\varepsilon_z + \nu_s\tilde{E}_s\varepsilon_\theta, \tag{24}$$

the equation for the circumferential stress is obtained similarly by the permutation of the indices:

$$\sigma_\theta = \tilde{E}_s\varepsilon_\theta + \nu_s\tilde{E}_s\varepsilon_z. \tag{25}$$

Substituting (24), (25) in (22):

$$\begin{aligned}\delta\Pi_s &= \int_{A_s}\tilde{E}_s\varepsilon_z\delta\varepsilon_z dA + \int_{A_s}\nu_s\tilde{E}_s\varepsilon_\theta\delta\varepsilon_z dA + \int_{A_s}\tilde{E}_s\varepsilon_\theta\delta\varepsilon_\theta dA + \\ &+ \int_{A_s}\nu_s\tilde{E}_s\varepsilon_z\delta\varepsilon_\theta dA = \delta\Pi_{s1} + \delta\Pi_{s2} + \delta\Pi_{s3} + \delta\Pi_{s4}.\end{aligned} \tag{26}$$

We transform the term $\delta\Pi_{s1}$ similarly to the term $\delta\Pi_{b1}$:

$$\delta\Pi_{s1} = \delta\begin{Bmatrix}\varepsilon_0\\ \chi_0\end{Bmatrix}^T \int_{A_s}\tilde{E}_s\begin{bmatrix}1 & y\\ y & y^2\end{bmatrix}dA\begin{Bmatrix}\varepsilon_0\\ \chi_0\end{Bmatrix}, \tag{27}$$

For a single bar finite element with length l_e:

$$\delta\Pi_{s1}^e = \delta\begin{Bmatrix}\varepsilon_0\\ \chi_0\end{Bmatrix}^T \tilde{E}_s\begin{bmatrix}1 & y_m\\ y_m & y_m^2\end{bmatrix}l_e\cdot t_s\begin{Bmatrix}\varepsilon_0\\ \chi_0\end{Bmatrix}. \tag{28}$$

$$\delta\Pi_{s2} = \delta\begin{Bmatrix}\varepsilon_0\\ \chi_0\end{Bmatrix}^T \int_{A_s}\begin{Bmatrix}1\\ y\end{Bmatrix}\nu_s\tilde{E}_s\varepsilon_\theta dA. \tag{29}$$

Let $\{\bar{U}_{tr}^e\} = \begin{Bmatrix}\bar{u}_i\\ \bar{u}_j\end{Bmatrix}$ be the nodal displacements of the bar finite element in the section plane in the local coordinate system, $\{U_{tr}^e\} = [u_i \quad v_i \quad u_j \quad v_j]^T$ are in the global coordinate system, then

$$\{\bar{U}_{tr}^e\} = [L]\{U_{tr}^e\}, \tag{30}$$

where $[L]$ is directional cosine matrix, $[L] = \begin{bmatrix}r & s & 0 & 0\\ 0 & 0 & r & s\end{bmatrix}$, $r = \cos\alpha$, $s = \sin\alpha$. Then:

$$\varepsilon_\theta^e = [-\frac{1}{l^e} \quad \frac{1}{l^e}]\{\bar{U}_{tr}^e\} = [B_{tr}]\{\bar{U}_{tr}^e\} = [B_{tr}][L]\{U_{tr}^e\}. \tag{31}$$

Equation (29) for a single finite element takes the form:

$$\delta \Pi_{s2}^e = \delta \begin{Bmatrix} \varepsilon_0 \\ \chi_0 \end{Bmatrix}^T v_s \tilde{E}_s \begin{Bmatrix} 1 \\ y_m \end{Bmatrix} [B_{tr}][L] l_e t_s \{U_{tr}^e\}. \tag{32}$$

$$\delta \Pi_{s3}^e = \tilde{E}_s ([B_{tr}][L]\{U_{tr}^e\})([B_{tr}][L]\delta\{U_{tr}^e\}) l_e t_s. \tag{33}$$

Transposing (33), we get:

$$\delta \Pi_{s3}^e = \delta\{U_{tr}^e\}^T [L]^T [B_{tr}]^T [B_{tr}][L] \cdot l_e t_s \{U_{tr}^e\}. \tag{34}$$

$$\delta \Pi_{s4}^e = \delta \varepsilon_\theta^e \cdot \tilde{E}_s v_s [1 \quad y_m] \begin{Bmatrix} \varepsilon_0 \\ \chi_0 \end{Bmatrix} l_e t_s = $$

$$= \delta\{U_{tr}^e\}^T \tilde{E}_s v_s [L]^T [B_{tr}]^T [1 \quad y_m] l_e t_s \begin{Bmatrix} \varepsilon_0 \\ \chi_0 \end{Bmatrix}. \tag{35}$$

The total strain potential energy variation of the steel tube calculated by summing over the elements:

$$\delta \Pi_s = \sum_e \delta \Pi_s^e = \sum_e (\delta \Pi_{s1}^e + \delta \Pi_{s2}^e + \delta \Pi_{s3}^e + \delta \Pi_{s4}^e). \tag{36}$$

Basing on (28), (32), (34), (35) and (36) for bar finite elements we obtain:

$$[K_s^e] = \begin{bmatrix} [K_{s11}^e] & [K_{s12}^e] \\ [K_{s21}^e] & [K_{s22}^e] \end{bmatrix}, \tag{37}$$

$[K_s^e]$ is FE stiffness matrix of dimension 6×6. Vector defining the FE displacements with nodes *i*, *j*: $\{U^e\} = [\varepsilon_0 \quad \chi_0 \quad u_i \quad v_i \quad u_j \quad v_j]^T$,

$$[K_{s11}^e] = \tilde{E}_s \begin{bmatrix} 1 & y_m \\ y_m & y_m^2 \end{bmatrix} l_e \cdot t_s,$$

$$[K_{s12}^e] = \tilde{E}_s v_s \begin{Bmatrix} 1 \\ y_m \end{Bmatrix} [B_{tr}][L] l_e t_s,$$

$$[K_{s21}^e] = \tilde{E}_s v_s [L]^T [B_{tr}]^T [1 \quad y_m] l_e t_s = [K_{s12}^e]^T,$$

$$[K_{s22}^e] = [L]^T [B_{tr}]^T [B_{tr}][L] \cdot l_e t_s.$$

Using equation (20), we find the stiffness matrix $[K]$. The problem is reduced to solving a system of linear equations:

$$[K]\{U\} = \{P\}, \tag{38}$$

where $\{U\} = [\varepsilon_0 \quad \chi_0 \quad u_1 \quad v_1 \quad u_2 \quad \cdots \quad u_n \quad v_n]^T$ is nodal displacements vector, n is the number of nodes in the section.

Method of calculation

System (38) will be composed and solved using the MATLAB software package. An example of meshing using the «initmesh()» function is shown in Figure 4. The concrete core diameter is 0.2 m here, node numbers are shown in black.

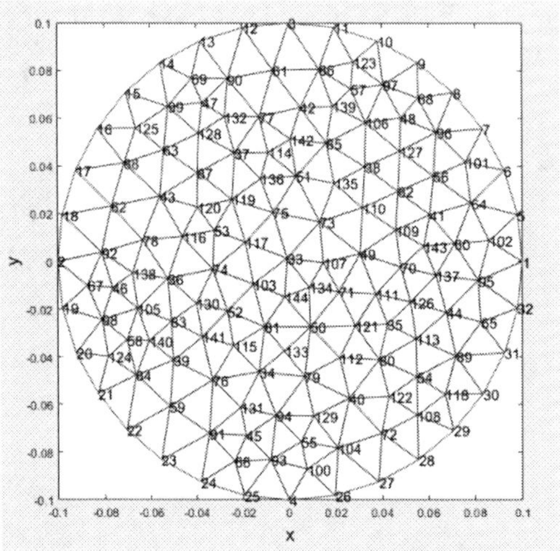

Figure 4. Finite element mesh made with «initmesh» function

The stiffness matrix and the load vector composing is performed using the formulas specified in section 2 of the article.

The stresses are calculated after determining the nodal displacements according to the formulas, obtained from the generalized Hooke's law for concrete and steel elements:

$$\begin{Bmatrix} \sigma_x \\ \sigma_y \\ \tau_{xy} \end{Bmatrix}_b^e = [D_b][B_{pl}]\{U_{pl}^e\} + \lambda_b \begin{Bmatrix} \varepsilon_0 + y_m \chi_0 \\ \varepsilon_0 + y_m \chi_0 \\ 0 \end{Bmatrix},$$

$$\sigma_{zb}^e = E_b(\varepsilon_0 + y_m \chi_0) + \nu_b(\sigma_{xb}^e + \sigma_{yb}^e),$$

$$\sigma_{\theta s}^e = \tilde{E}_s[B_{tr}][L]\{U_{tr}^e\} + \nu_s \tilde{E}_s(\varepsilon_0 + y_m \chi_0).$$

The obtained results will be compared with the results of calculations in the ANSYS software using volumetric finite elements for concrete and shell ones for a steel tube.

Results and discussion

When solving the test problem, we will use the following initial data: $d_b = 0.2\ m$, $E_b = 3 \cdot 10^4\ MPa$, $E_s = 2 \cdot 10^5\ MPa$, $\nu_b = 0.2$, $\nu_s = 0.3$, $t_s = 0.008\ m$, $F = 1\ MN$, $M = 0.1\ MN \cdot m$, the column length is taken equal 1 m. The loaded model of the problem when solving in the ANSYS is shown in Fig. 5. The obtained results are shown in Figures 6-7.

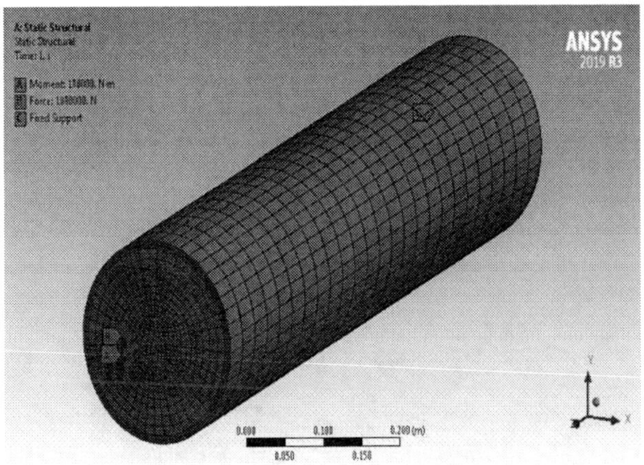

Figure 5. The finite element mesh used in the simulation in ANSYS.

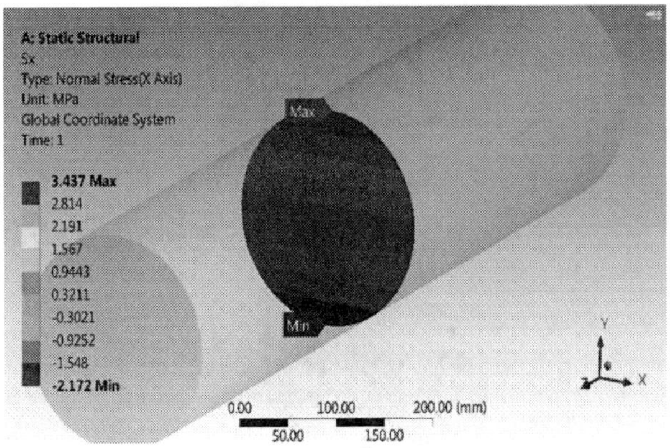

Figure 6. Concrete stress σ_x distribution in the middle section of the CFST column (ANSYS).

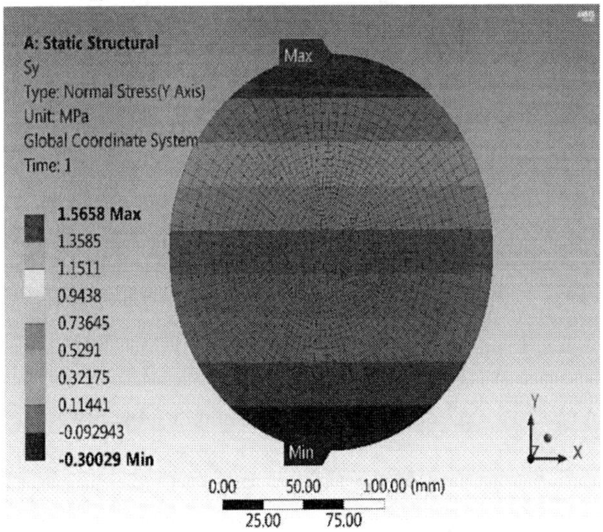

Figure 7. Concrete stress σ_y distribution in the middle section of the CFST column (ANSYS).

The results of calculations in the MATLAB software package according to the technique presented in the article by reducing the problem to two-dimensional are shown in Figures 8-9.

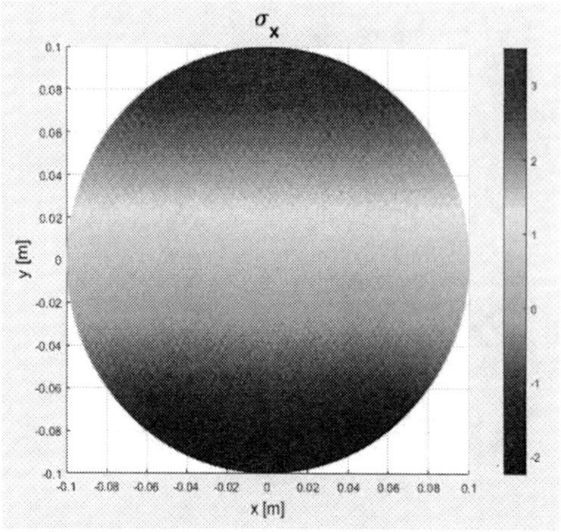

Figure 8. Concrete stress σ_x distribution in the section of the CFST column (MATLAB).

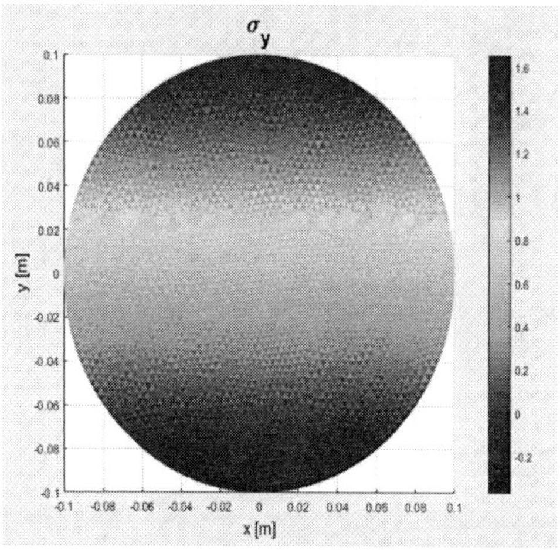

Figure 9. Concrete stress σ_y distribution in the section of the CFST column (MATLAB).

The maximum values of σ_x in concrete as calculated in ANSYS and MATLAB were 3.44 MPa and 3.50 MPa, respectively. The error is about 1.7%. The maximum normal stresses in concrete were 1.57 MPa and 1.65 MPa, respectively.

It is important to note that a simplified elastic model of concrete was used in the calculations, the material physical nonlinearity, as well as the dilatation effect, are not taken into account. As a result, the negative stresses of compression from the side of the tube are insignificant and are located in a smaller part of the concrete section.

Summary

In the presented article, a method has been developed for reducing the problem of determining the stress-strain state of short CFST columns to two-dimensional. Expressions for the stiffness matrices elements of the column are obtained. To implement the algorithm described in the work, a program was written in MATLAB. The results obtained are compared with the results of calculations in the modern finite element software ANSYS by modeling the deformation of concrete with volumetric finite elements, and the steel tube with shell ones. The resulting error is within acceptable limits, while the use of plane finite elements together with bar ones according to the presented method significantly reduces the order of the FEM system of equations and requires less resources.

In the calculations, the concrete was modeled as linearly elastic. Using the proposed approach, it is possible to solve the problem in a physically nonlinear formulation, including taking into account creep. When solving creep problems, the approaches presented in [6-10] can be used.

References

[1] Grigoryan M N, Urvachev P M, Chepurnenko A S, Polyakova T V 2020 Determination of the ultimate load for centrally compressed concrete filled steel tubular columns based on the deformation theory of plasticity *IOP Conference Series: Materials Science and Engineering* **913/** DOI 10.1088/1757-899X/913/2/022069.
[2] Krishan A L, Melnichyuk A S 2012 Strength of square-section tubular concrete columns under axial compression *Vestnik Magnitogorskogo gosudarstvennogo tekhnicheskogo universiteta im. G.I. Nosova* **3** 51-54 (in Russian).
[3] Reddy J N 2002 *Energy principles and variational methods in applied mechanics. 2nd Edition* (New York, John Wiley) 608.
[4] Samul V I 1982 *Fundamentals of the theory of elasticity and plasticity* (Moscow, Vysshaya shkola) 264.
[5] Segerlind L J 1976 *Applied finite element analysis* (New York, John Wiley) 422.
[6] Mailyan L R, Chepurnenko A S, Ivanov A 2016 Calculation of prestressed concrete cylinder considering creep of concrete *Procedia Engineering* **165** 1853-1857. DOI 10.1016/j.proeng.2016.11.933.
[7] Yazyev B M, Chepurnenko A S, Litvinov S V, Kozel'skaya M Yu 2014 Stress-strain state of a prestressed reinforced concrete cylinder with the consideration of concrete creep *Nauchnoe obozrenie* **11(3)** 759–763 (in Russian).
[8] Yukhnov I V, Yazyev B M, Chepurnenko A S, Litvinov S V 2014 Stress-strain state of a short eccentrically compressed reinforced concrete rod with nonlinear creep *Nauchnoe obozrenie* **8** 929–934 (in Russian).
[9] Chepurnenko A S, Chepurnenko V S, Savchenko A A 2017 Finite element modeling of a three-layer plate creep *Molodoy issledovatel' Dona* **3** (in Russian). Information on http://mid-journal.ru/upload/iblock/508/95_102.pdf.
[10] Andreev V I, Yazyev B M, Chepurnenko A S, Litvinov S V 2015 Calculation of a three-layer shallow shell taking into account the creep of the middle layer *Vestnik MGSU* **7** 17–24 (in Russian).

Eco-sustainable architecture and comfortable living environment

S B Yazyeva*, I A Mayatskaya

Don State Technical University, 162, Sotcialisticheskaya str., Rostov-on-Don, 344022, Russia

E-mail: iskra1917@bk.ru

Abstract. Eco-sustainable architecture strives to achieve the integration of the human habitat with natural ecosystem. It gives a rise for the harmonious development of all the components in this direction, allows to apply the methods of structures' entire complex optimal design, taking into account the natural features of the area and the comfort of people living in this urban space. The problems associated with the peculiarities of the eco-houses' construction are considered. The article focuses on the relationship of architectural, engineering, economic, socio-cultural and environmental aspects of modern structures' design.

1. Introduction

The sustainable architecture's goal is to achieve a balance of the surrounding urban environment and humans from the ecological point of view [1]. These issues have been considered by the UN Commission on the Environment since 1987. The directions of sustainable architecture are manifested in the concept of "green" construction, energy saving, the use of alternative energy sources, for example, wind turbines and solar panels (Figure 1,2). Six wind farms will be built in the Rostov region, some have already been built.

Figure 1. Wind power plants.

Figure 2. Solar power plant.

Attention should be paid to the development of innovative methods of using resources, for example, energy [1-3]. Alternative energy sources are wind, sun, ebb and flow, biomass, geothermal sources. Recently, much attention has been paid to the first two components. Geothermal energy in Russia is developing in Kamchatka, at a very slow pace, though. For example, a house in Hamburg, Germany is covered with transparent panels with water which is inhabited by algae. These algae convert the energy from the sun into biogas. Due to this, energy and heat energy is saved. Such panels have thermal insulation properties. But being carried away by energy saving, the architects did not take into account the aesthetic component of this project. It is unlikely that the residents of this building are comfortable in it. But in this respect the secondary use of materials should not be ignored, but getting too carried away with this direction might cause some problems as well.

Some architects offer domed structures with translucent cover, which cover vast territories, entire cities. The development of new technologies in the future will make it possible to do this, but is it advisable? Of course not. This or that project should be assessed from the point of view of the construction possibility and from the point of view of its necessity. And it is quite possible to use shell, cable-stayed and other large-span coatings, for example, for the construction of sports facilities, for shopping, entertainment and exhibition complexes. Such structures were used in the construction of stadiums in Russia in preparation for the football championship [2].

It is necessary to use the methods of optimal design of buildings and structures, it is possible to apply the principles of architectural bionics when creating building structures, as well as to introduce and use new materials and technologies [3-5]. It is these issues that should be considered from the point of view of a person's comfortable living in harmony with the surrounding nature. Modern designers can consider a variety of options using mathematical modeling methods using BIM technologies.

2. Ways of developing a modern urban environment in harmony with nature and human comfort

The modern urban environment is determined by the peculiarities of architecture, heritage of the past and the natural environment. When an architect takes on a project, he should take into account all these features, not forgetting about the architectural value of adjacent buildings and their strength. And sometimes after the new modern structures' construction, it is necessary to deal with the reconstruction and strengthening the old buildings. It is very important to build not a stand-alone facility, but an architectural ensemble that harmoniously fit into the environment. At present, much attention is paid to the combination of urban objects and nature, which contributes to the comfortable coexistence of humans, flora and fauna. Of course, speaking about the urban environment, we focus on plants [1, 6]. Since the space for construction is limited, plants can be located both near the object and on the structure itself. This makes the building unique and human-friendly.

Figure 3. View of a modern city, London, UK.

Nature, urban environment, people - all this makes up our world, and it should be harmonious and comfortable for humans [1]. The beauty of nature, architecture has a beneficial effect on a person, giving the opportunity to work, live, create. "Green" architecture concept is not only structures, but also progressive technologies that make it possible to create a structure that meets modern requirements; these are the structures that make it possible to combine human needs and vivid natural beauty without destroying it.

Figure 4. Combination of modern building architecture and plant objects

Architects realize that the relief effect of buildings can be completely arbitrary, but nature prompts them for optimal solutions [6-8]. To do this, it is necessary to study the structural laws of natural objects and use the design ideas of both structural elements and the entire structure. Often, in the construction of buildings, the properties of symmetry, asymmetry and similarity are used in architectural elements, for example, the arrangement of columns, windows, domes. Since rod and glass structures are very often used in modern architecture, the properties of symmetry, asymmetry and similarity can be successfully used in the structures design. [9].

Figure 5. The interior space of uniquely shaped buildings.

In architectural bionics, it is possible to search for new, functionally justified architectural forms and create new rational structures in harmony with nature and architecture. Many works of the architect G. Boehm also use the architectural style, where the structural elements are the block structures of the most diverse forms and spatial composition.

To enhance the impact of architectural structures on the emotional state of a person, illumination is currently actively used. Designers create interesting designs for city streets and facades using light. This is especially beautiful on New Year and Christmas holidays. But luminous fluxes can also be used to create unique structures, using the structural features of a building, glass structures and reflective coatings on the buildings' surface. The direction of light flux can be changed with the help of awnings, cornices, balconies, making it possible to enjoy the beauty of nature and the building itself.

Figure 6. Glass structures of awnings and galleries of modern buildings.

During the construction of the quarter of mixed development in the city of Lyon (France) "International City" (architect R. Piano), the natural environment was taken into account: the park and the Rhone river. The solution to the arrangement of the streets of this complex was interesting. This is the glazing of the inner streets of the complex. There are also other options for street decoration using a wide variety of glass structures. (Figure 7) [9].

Figure 7. Glass structures in the streets' improvement of the city blocks with unique shape.

Glass structures are widely used in the construction of skyscrapers (Figures 8.9). R. Piano was one of the first architects to use these trends. His most famous skyscraper is the Shard in London (UK). The skyscraper got its name "The Shard" because of the architectural elements on the top of the building in the form of fragments.

Figure 8. Modern skyscrapers.

Figure 9. Skyscraper "The Shard", London, UK.

For the further use of various materials during construction, it is necessary to take into account the variety of architectural solutions [10, 11]. Bionics makes it possible to find the most unusual in shape and structure optimal for the design of a wide variety of structures. The development of architectural bionics has led to the emergence of buildings in architecture using a wide variety of lines and surfaces. This makes the building unique in its shape [1,3-5]. But modern architects are not limited to this. They use a combination of the most interesting in terms of structure and layer properties as structural elements when decorating the building both inside and outside. In this case, a wide variety of materials are used: polymer composites, metals, glass, wood.

Buildings built using the bionic principles are resilient, able to withstand the adverse natural events and natural disasters. The Coral Reef project by architect Vincent Callebaut is an example of such structures' design (Figure 10). He tried to solve the problem of resistance of structures to earthquakes. Two thousand modular houses are combined into a wave-like structure built on anti-seismic piles. Houses work on the energy of waves, and inside there are all the amenities. This architect also works on projects where people not only live in buildings, but also grow plants.

Figure 10. Coral Reef Project (architect Vincent Callebaut).

Currently, there are many natural manifestations in the world. Architects design buildings to withstand this. Using the achievements of architectural bionics and using modern materials, they create structures that can withstand the elements. They study the laws of nature, use them to model the structural elements of building structures and to model the very impact of natural manifestations. For example, they build earthquake resistant houses using anti-seismic piles and undulating building structures. For the last years some architects have been designing structures resistant to wind loads due to their spherical dome shape.

A very interesting hurricane-resistant dome building in Florida (Figure 11). The idea for its construction belongs to Mark and Valerie Sigler. They started building safe houses after the destruction of the old house by the hurricane. The home owners used new technologies and improved cement and steel structures during the construction, which will have to withstand the hurricane.

Figure 12 shows a house in Singapore. It is an example of a sustainable architecture. This structure is adapted to the effects of volcanoes and hurricanes.

Figure 11. Dome of the House Building, Pensacola Beach, Florida.

Figure 12. House in Singapore.

A disaster can happen in any part of the world, and architects are concerned about the building structures resistant to such phenomena. As a result of the negative impact, buildings do not remain intact and people are at risk. Architects try to design the structures that could protect against this.

Eco-sustainable architecture is a set of design solutions of architects and engineers that allow balancing the quality of a comfortable human life and maintaining the environmental sustainability of the surrounding nature.

Summary

"Green" architecture is one of the sustainable architecture directions, which, when designing structures, tries to solve the problems of harmonious development of the territory with the rational use of material resources, energy, taking into account the ecology, with the reduction of waste, with the improvement of people's lives with the aim of minimizing negative impact on nature [1, 10, 11]. But it is necessary to take into account both the physical and psychological comfort of a person living in an urban space, not forgetting about the climate situation on the planet and the aesthetic component of architecture.

Sustainable architecture tends to take into account such conditions as the use of sustainable materials; reduction of waste during construction and operation, their processing; environmental pollution; energy efficiency and water saving; healthy living environment and comfortable area around the complex of structures. It would be worth noting the social, cultural and economic components in the buildings design. This factor should be taken into account. Currently, the interest in sustainable architecture is increasing in the world. When designing, it is necessary to take into account two sides of harmonious development: a comfortable living environment for a person and a minimum impact on the nature around him.

To develop a sustainable architecture, it is necessary to apply optimal design methods; be able to change the project if not previously taken into account negative conditions arose; apply modern methods of mathematical modeling and conduct sociological research to improve the functional, microclimatic and aesthetic parameters of the human environment. It is necessary to look for the new ways, developing architectural bionics, which this direction includes.

References

[1] Yazyeva S B, Seferyan L A, Oparina L A, Golubeva A Y 2018 Greening technology organization of multi-storey buildings, in the reconstruction of architectural and planning solutions with the use of modern building materials *Materials Science Forum* **931** 883-888.

[2] Yazyeva S B, Mayatskaya I A, Kashina I V, Nesterova A N 2019 The manifestation of fractality in the architecture of buildings and structures *IOP Conference Series: Materials Science and Engineering* **698 (3)** 033046.

[3] Mayatskaya I A, Eremin V D, Strelnyikov G P 2019 Modern trends in architectural bionics *Integration of world science in a crisis: theoretical approaches and practical results, Materials of the All-Russian Scientific and Practical Conference* 30-32.

[4] Mayatskaya I A, Eremin V D 2019 Bionics and the choice of rational structural form *E3S Web of Conferences* **110** 01042.

[5] Mayatskaya I A, Eremin V D, Yazyeva S B 2019 Green architecture: the unity of the beauty of nature, environmental friendliness and architectural forms *Construction and architecture* **7 (1)** 91-96.

[6] Yazyev B M, Mayatskaya I A, Yazyeva S B, Yazyev S B 2019 Fractality in architectural forms and in organization of space in buildings *IOP Conference Series: Materials Science and Engineering* **698 (2)** 022087.

[7] Mayatskaya I A 2016 On the possibility of improving building structures based on bionic principles *Vestnik VolgGASU Series Building & Architecture, Proceedings of Volgograd State University of Architecture and Civil Engineering* **45(64))** 27-34.

[8] Mayatskaya I A, Yazyev B M, Yazyeva, S B, Kulinich P B 2017 Building Constructions: architecture and nature *MATEC Web of Conferences* 106.

[9] Mayatskaya I A, Yazyeva S B, Zakieva N I, Lapina A P 2018 Modern glass constructions and comfortable urban environment *Materials Science Forum* **931**754-758).
[10] Mayatskaya I A, Yazyev B M, Yazyeva S B 2017 *Optimal design in terms of architectural bionics Construction and industrial safety* **9 (61)** 7-11.
[11] Kaschina I V, Nesterova A N, Yazyeva S B, Danilova-Volkovskaya G M 2019 Material and function: on the specifics of formation in modern architecture *IOP Conference Series: Materials Science and Engineering* **698 (3)** 033031.

Bamboo in modern construction and architecture

A P Lapina*, N I Zakieva

Don state technical university, 1 Gagarin sq., Rostov-on-Don, 344002, Russia

E-mail: anastasialapina7@gmail.com

Abstract. The use of bamboo in construction and architecture is considered in the article. The main advantages of bamboo in comparison with modern building materials are considered, some technical characteristics of the material are given. The conclusion about bamboo as a renewable natural material, which in the future may replace the steel and reinforced concrete structures people are used to, was made. The design characteristics of bamboo make it possible to use it in complex architectural forms' creation.

Introduction

In the modern world of innovative materials and technologies, it is difficult to imagine that somewhere such a plant as bamboo is used as a building material. But in vain. China, Oceania, Latin America, South America, Central and South Africa widely use bamboo as a building material.

Bamboo grows mainly in the humid tropics and subtropics. Not only the frame of the dwelling (walls, columns) is made of it, but also the roof, scaffolding, it can also be used as a finishing material.

Main part

The population in Asian countries is growing every year. More and more people want to live in metropolitan areas, such as Hong Kong, which has a population of over 7.5 million, and population density is over 6500 people / km^2. A super-modern metropolis, where every seventh inhabitant is a dollar millionaire, has to grow not in breadth, but in height. And bamboo is helpful in this situation.

There are more than 1,500 buildings over 100 meters in height in Hong Kong. The pace of construction is amazing. Builders construct 1 floor per week. The buildings can be said to grow as fast as bamboo. And it will be both literally and figuratively. Because the main building technology in Hong Kong is bamboo scaffolding (Figure 1).

Bamboo woods are used for the construction and reconstruction of any buildings and structures. All nodes are connected with plastic clamps (Figure 2) without a single nail. It is amazing how harmoniously modern skyscrapers and the oldest building technologies in China are combined. So why do engineers in Hong Kong and other countries prefer bamboo?

Figure 1. Bamboo scaffolding on a high-rise building

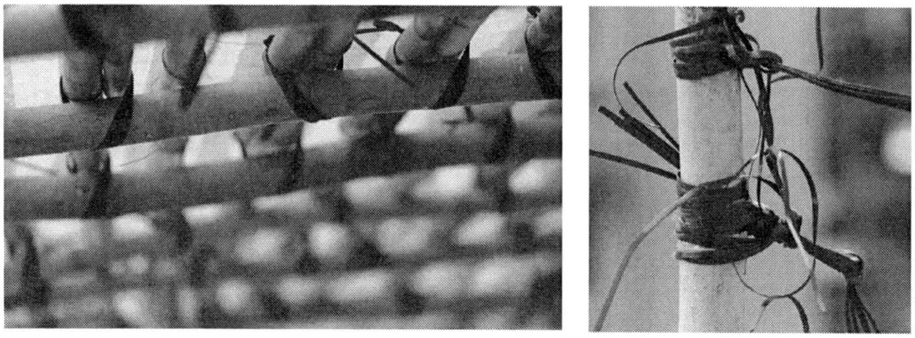

Figure 2. Nodes' connection

The beneficial peculiarities of bamboo are as follows:

- Environmentally friendly and renewable. Currently, the destruction of the wood cover of our planet is in progress. The scale of deforestation is amazing. Each year, the planet irrevocably loses 95 million trees. The use of bamboo as a building material can stop this process at least a little. After all, the maximum growth rate of bamboo is 1200 mm per day. In 3-5 years, bamboo becomes fully suitable for construction purposes.
- Low cost.
- Durability. With proper care, bamboo structures can last over 100 years. For example, in the Chinese town of An-Lan, a suspension bamboo bridge across the Ming River was built in the third century AD and is still in operation (Figure 3).

Figure 3. Suspended bamboo bridge over the Ming river

- Seismic resistance. Bamboo has high strength and low weight, which increases the building's resistance to seismic impact. With proper anti-corrosion treatment, the building will not collapse even in the event of a strong earthquake. For example, in 1991, Costa Rica was hit by a 7.7 magnitude earthquake. The buildings in the epicenter of the usual concrete and brick were completely destroyed, and the buildings made of bamboo had only minor damage.
- High temperature resistance and low heat transfer coefficient. Thermal conductivity coefficient of bamboo is 0,30 – 0,40 W \(m*°C) (depending on the variety). In order to preserve the thermal insulation effect, the thickness of bamboo should be 1/15 of concrete and 1/400 of steel.
- Durability.

The design characteristics of bamboo include:

- tensile strength. Longitudinal fibers of bamboo make it quite strong. For this ability it received the name "vegetable steel".
- tubular section that resists torsional forces.
- Under central compression, no significant deformation occurs. Bamboo retains and restores geometric dimensions even if stability is disturbed.
- Due to the large number of external fibers, bamboo has low shear resistance.

Ways of connecting nodes

The main ways of connecting structural nodes are shown in Figure 4.

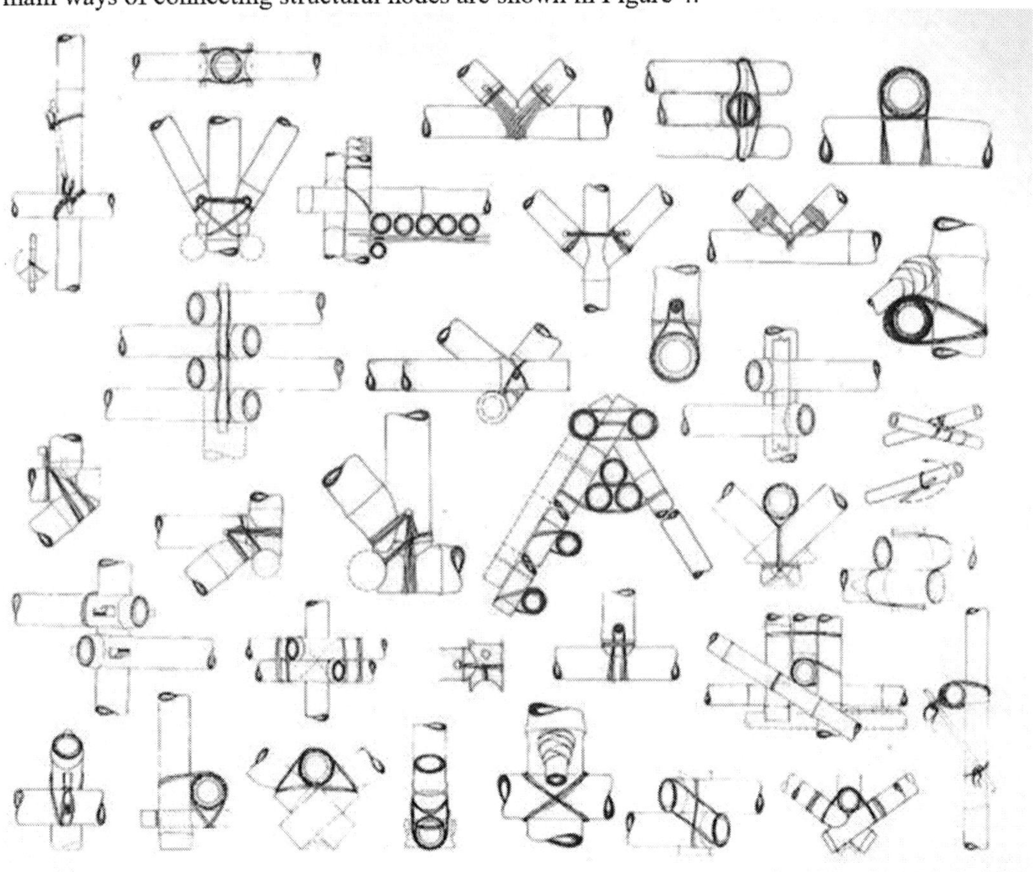

Figure 4. Types of nodes' connection

Contemporary architects

In the modern world, there are engineers and architects who specialize in the design and construction of bamboo structures, for example:

- Simon Velez. On the website of his design office the following slogan "Vegetarian architecture" can be seen. In his work, the architect mainly uses the laguadua bamboo class, as it is most common in Colombia and Ecuador, and is also highly durable. Also, Simon Velez developed a new type of nodes connection - using bolts and steel plates, which is used for corner joints and overhangs. The assembly consists of axial or longitudinal screws, depending on the work and the plates connecting the bamboo units (Figure 5).

Figure 5. Node designed by Simon Velez.

Figure 6 shows some examples of Simon Vele'ss work. For example, Crosswaters Ecolodge (Figure 6, b), a bridge created in the woods of the Nankun Shan mountain reserve in Guangdong province. This project was the first project of this kind in Asia to use bamboo as a structural element in a house. Simon Velez received the 2006 American Society of Landscape Architects' Honorary Award for Analysis and Planning.

a) Contemplation Bamboo Pavilion, Arles, France (2018)

b) Crosswaters Ecolodge, Guangdong (2006).

Figure 6. Examples of works performed by architect Simon Veles.

– Markus Heinsdorff, who is one of the activists in this field. In 2008, under his leadership, the company erected 21 unique pavilions (Guangzhou). In these pavilions, a bamboo frame structure was used, and a combination of steel elements, leaf springs and grout was used in the nodal connections (Figure 7).

Figure 7. Pavilion designed by Markus Heinsdorff (2010).

– Xiao Yang, a professor at Hunan University (China), developed a technology for the production of bamboo plywood, from which a lightweight low-rise residential building was constructed. Structural nodes were carried out using bolted connections, which ensured the speed and convenience of construction.

An important event in the history of bamboo was the holding of the first Biennial of Bamboo Architecture in 2017. Its purpose was as follows: to create such buildings and structures that could serve people, would be integrated into the urban environment precisely for a practical purpose.

The task of the Biennale was completed successfully. The result of the work was such structures as: a bridge, three hotels, cafes and restaurants, museums and workshops for working with wood and ceramics (a strong tradition of work with which is still manifested in the Baoxi settlement).

The interest of the participants in this Biennale gave the organizer inspiration to create more events of this type. Workshops, conferences, practical seminars enable people to work with professionals, join their communities, discuss current issues and create new projects.

Of course, we are talking about the construction of construction projects only in the region where bamboo grows, since its transportation increases the cost and reduces the strength characteristics. In addition, it cannot be used in the countries with cold climates, while in other countries with hot climates there is a "local" alternative to bamboo.

Summary

Taking the above-said into consideration, we can say that bamboo has a future in construction and architecture. Modern computer technology helps to design bamboo nodal connections, based on the classic principles of building reinforced concrete and steel structures. It is possible to make rather cumbersome mathematical calculations to determine the operation of a particular node, integrate it into the project.

Taking into account the scale of deforestation, environmental pollution from the production of traditional materials, as well as the calculated characteristics of bamboo, its cost, we can conclude that it will soon completely replace the usual metal and reinforced concrete structures.

References

[1] Yao Wei 2015 Development and calculation of nodal connections of bearing spatial bamboo bar structures *Dissertation for the degree of Candidate of Technical Sciences, St. Petersburg State University of Architecture and Civil Engineering*, St. Petersburg, 163.

[2] Shuvalov V. M., Solanilla Y. M. 2016 Using bamboo in modern architecture in Colombia *RUDN Bulletin, Engineering Research Series* **2** 116-125.

[3] Solanilla Medina Y M 2019 Features of traditional architecture and bamboo construction in countries with hot humid climates *Architecture and Modern Information Technologies* **3 (48)** 175-184. Access mode: https://marhi.ru/AMIT/2019/3kvart19/PDF/13_solanilja.pdf

[4] Solanilla Medina Y M 2018 Architectural Design of Bamboo as an Ecological Alternative Material of the XXI Century *Architecture and Modern Information Technologies* **1 (42)** 201-211. Information on http://marhi.ru/eng/AMIT/2018/1kvart18/15_solanilla_medina/index.php

[5] Information on http://www.berlogos.ru/article/bambukovaya-arhitektura/

Synthesis of virtual reality technologies and design

E V Vishnevskaya*, N B Strekalova

Togliatti Academy of Management (TAoM), building 5, health-improving complex Alye Parusa, Stavropol district, Samara region, 445144, Russia

E-mail: lady.vishnja@mail.ru

Abstract. The article examines the phenomenon of synthesis of virtual reality and the design of a digital environment and possible areas of its application. The main terms and definitions are given. The historical background of the development and formation of virtual reality technology is given. The main technologies providing design means and virtual reality technologies' interaction are described. The theoretical basis for the interaction of virtual reality technologies and design tools, such as visualization and image creation techniques, is considered. A design solution for using virtual reality technologies in design as a method for solving design and artistic problems is presented.

Introduction

Virtual reality technology has started its active development in recent years, becoming an integral part in many areas of human life. So, thanks to the high-tech developments of Sony, HTC, Microsoft, created for games and entertainment, this technology has been actively used in medicine to create simulators for the rehabilitation of people who have suffered a stroke, in driver training, in online stores, in construction and game design.

In the study of the virtual reality technologies' possibilities and their application in the field of design, first of all, it is necessary to emphasize that this area is promising and multifaceted. In parallel with VR technology, which creates the effect of complete immersion, affecting the human brain with the further perception of virtual space as natural, augmented reality (AR) technology, in turn, expands the range of applications and virtuality by superimposing visual images on the real world that a person sees in front of himself. Augmented reality (AR) technology is actively used in quest games.

The study of the latest technologies, novelties and technical capabilities of virtual reality made it possible to conclude that it is necessary and relevant to integrate VR technology, augmented reality (AR) technology and design, which will solve the problem with the static space and move to a new digital level in design - engineering.

The practical orientation of the research demonstrates the mechanisms of such interaction on the example of creating a digital environment of various virtual spaces designed by a designer with a person immersed in augmented reality glasses into the virtual world of travel.

The phenomenon of virtual reality synthesis and digital environment design and possible areas of its application

The objectives of the study were to study the virtual reality technologies' development, their application in the field of design with solving the problems of substantiating the relationship between design and virtual reality technologies and a practical design solution for creating an interactive, self-changing space (non-residential premises) using augmented reality glasses.

Considering the possibilities of virtual reality technologies in terms of adaptation in modern design needs to highlight the specifics of the application, tools and their scope.

The subject of the research is digital technologies for solving the interior design problems based on virtual reality technologies.

The object of research is the phenomenon of virtual reality and possible areas of its application, features of the digital environment design.

Study boundaries. The temporal boundaries of the study cover the turn of the XX - XXI centuries. Territorial boundaries cover the development of virtual reality technology that began in foreign countries. Foreign experience and technologies are involved in the analysis of modern and experimental trends.

The main research methods are determined by the goals and objectives. During its implementation, a systematic approach was applied, which made it possible to consider the main characteristics of the object and the subject of research in their relationship; a comprehensive analysis of the features of technology and design of the digital environment, historical, cultural and methodological positions; analytical modeling method; experimental design method.

The scientific novelty of the research consists in proposing a new method of using virtual reality technologies in design, for solving a number of problems and searching for the technological solutions for the non-standard creative ideas' implementation.

The theoretical foundations of the virtual reality and design technologies integration are considered as a complex of knowledge in the field of technology, psychology of perception, philosophy, innovative design in design activity. The key works that influenced the formation of an innovative design holistic view are the works of S. I. Orekhov, N. L. Sokolova, in which the concept of virtual reality, its role in modern culture, its philosophical and psychological component, types and modern trends are considered as an innovative trend. The opinion of such experts as K. Rashid, V. Ryzhonkov, as well as the founders of such major companies as Sony, HTC, Microsoft, was of great importance for the research part. The main provisions on innovative trends and the problems associated with the entertainment centers' organization were studied in connection with the works of I. S. Ivyanskaya, M. V. Monastyrskaya, K. V. Auer. In addition, it was possible to learn about the digital environment design and game design from the English-language works of such authors as R. Koster, J. Shell, T. Sylvester. The position of foreign scientists and the priority of considering the phenomenon of virtual reality as a unique technology deserves close attention.

In the process of research, the virtual reality technologies' development chronology has been established, from the creation of a stereoscope in 1837 by Chalz Whiston, in 1950 the Sensorma attraction and in 1957 a telescopic mask, as the first prototype of VR glasses, by Marton Hayling, the creation of the Atari virtual reality laboratory in 1982 and the appearance the term "virtual reality" itself, the development of VPL Research cyber gloves in 1983 by Tom Zimmerman, Scott Fischer and in 1991 VR machines by the Virteality group until the technology turning point in 2014. With the invention of the virtual reality helmet HZT-1 in 2011 by Sony and the Oculus Rift DK glasses in 2012, a technological breakthrough occurred in the possibilities of using virtual reality in various spheres of human life.

Virtual reality as a part of technical informational progress is described by the representatives of reality comprehensive philosophy [1-3]. According to Segal A.P. the term "virtuality" also carries a mathematical understanding [4].

In the course of the study, an optimal solution - chroma key technology was found. The problem of using AR technologies in space was that the final image is not of high enough quality, interference is created in illuminated places, and a sharp color change. Based on the virtual reality glasses analysis and the search for the model and technology most adaptive to design, it was concluded that the implementation is compatible with augmented reality glasses and chroma key technology. The key components of the chromakey application have been identified. Light is primarily needed for the screen to be distributed as evenly as possible. Cool green is preferred, but when shooting green subjects, they will dissolve in space. The study of the chroma key technology possibilities made it possible to solve the technical drawbacks for this idea implementation. In the course of studying innovations in the field of virtual reality, a model of glasses, functionality and technical capabilities that will solve design problems were identified. This is the Microsoft Hololens model. Using virtual reality and chroma key technologies together makes it possible to maximize the image quality of the digital space created specifically for augmented reality glasses. The combination of chromakey technology with the necessary interior lighting will allow the visitor to fully experience the digital interior. It should be noted that the analysis of consumers allowed us to conclude that 40% of people aged 17 to 20 and 35% of people aged 31 to 45 are interested in this modern technology.

Creation of virtual reality scenes is based on a design tool - empathy. Empathy works for the image. G.V. Ioivleva writes about psycho-emotional impressions of image perception [5].

Being in the virtual space, the visitor, as a user of a unique technology, is a part of another world and is able to interact with his environment in meaningful relationships. This is the so-called telepresence effect: a combination of immersion and interactivity.

It is necessary to note the words of V.V. Kulenenok about the possibilities of three-dimensional graphics [6].

From the point of view of the symiotic, the immersion method is described by the researchers Yu.M. Shaev and Morozov M.G. [7-8]

Effective immersion requires the user to be able to explore the life-size visuals in a virtual environment and to take in the entire space. The user should be able to view the space.

The process is similar to the process described by D.V. Agadzhanyan, M.I. Chukarin [9].

It was possible to implement the idea of integrating the design ideas and the technological capabilities of virtual reality using the example of designing public interiors. Within the framework of the project, the object of the GorkiPark shopping and entertainment center in Kazan was selected. In the initial situation, the total area of the premises is 323 m^2. Load-bearing walls divide the room into two parts: the main room and the corridor. After the redevelopment, it was decided to divide the premises into three spacious zones, two utility rooms, a bathroom and a corridor. One of the zones is intended for a virtual reality cafe (Figure 1).

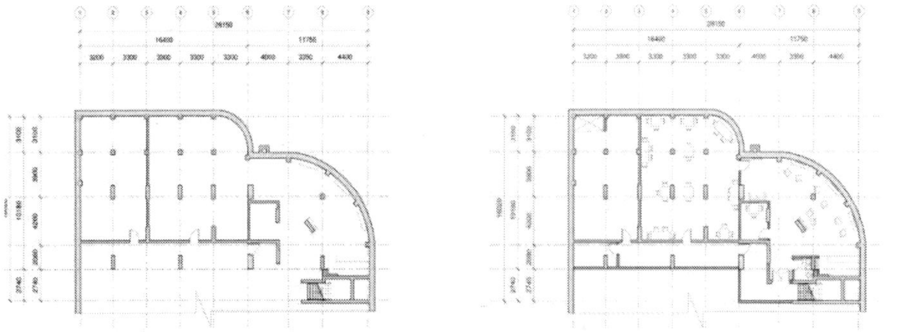

Figure 1. Floor plan. Initial situation and redevelopment

It was revealed that thanks to the integration of AR (MR) technology and chroma key technology, the cafe visitor will be able to completely immerse himself in the virtual world, while still remaining in reality, that is, being able to see his interlocutor. It should be noted that the lighting in the room should be uniform and well illuminate the entire surface of the chroma key. And also, all surfaces should be perfectly flat, not distorted in places of poor illumination and surface deformation. All coverings of walls, ceilings, floors, furniture in the cafe hall are painted in matte blue. This color is the chroma key coating for the glasses to work correctly. All corners are smoothed and rounded with blue fabric to avoid dark areas. The styling of the interior is due to the chroma key technology requirements (Figure 2).

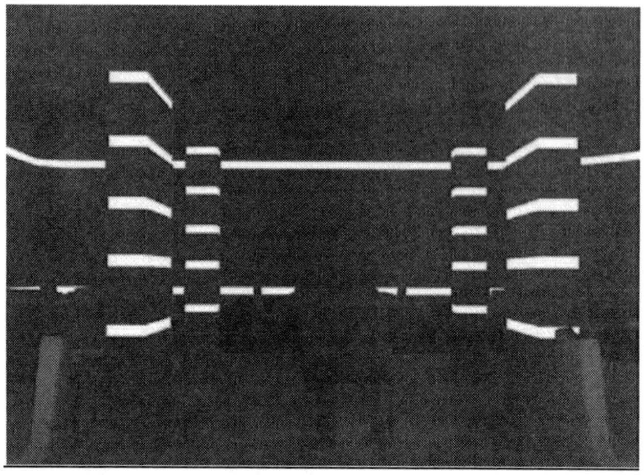

Figure 2. Virtual reality cafe interior.

A key part of the research-based project is to design a digital environment for augmented reality glasses. The number of digital scenes can be unlimited. The search for the actual approach to solving problems is presented in three scenes. The filling of each scene is a strict adherence to the requirements of technology and at the same time the expression of the design idea in the subject shaping of images and a color scheme that amazes the visitors' imagination.

By choosing Microsoft Hololens glasses with a specific theme of the scene, visitors get the impression of aesthetic unity and the interpenetration reception of artistic and technical means.
The virtual reality interaction is mentioned in their works by Averbukh, N.V. A. A. Shcherbinin [10].

The scene "Breakfast at the tavern" developed by 3D-design imitates the pristine medieval nature, overlooking the mountains, in the courtyard of a rare tavern. The scene will allow the visitor to be transported from the noisy city, causing the visitor to be pacified and calm. The boundaries of the room are highlighted to indicate the area covered by the space in the initial situation before immersion in an unreal virtual world (Figure 3).

Figure 3. Scene "Breakfast at the tavern".

The "Lunch Inside a Picture" scene is designed for the fans of fantasy and thrills. In this digital scene, the visitor moves to lunch inside a real painting, where the entire space is composed of artists' brushstrokes. This scene should induce a contemplative emotional state in the viewer (Figure 4).

Figure 4. Scene "Lunch inside a painting"

The Dinner at the Space Station scene is designed for dreamers. Restrained monochrome colors reflect the vast space. Looking at the glass floor, the visitor observes the illusion of a green forest. The artistic image of the scene, thanks to its color combinations, creates a dreamy state in the viewer (Figure 5).

Figure 5. "Dinner at the Space Station" scene. Overall plan.

The rendered virtual reality scenes become a kind of theatrical stage. These are the new 3D visualization tools. The synthesis of virtual reality and design technologies opens up vast possibilities for creativity.

Summary

Key findings and results of the study.

- The design problem was identified due to the lack of innovative methods of using virtual reality technologies in the field of design, due to the small amount of scientific research and technology novelty
- The technological capabilities of new products in the field of virtual reality were considered.
- The drawbacks of modern virtual reality glasses are revealed, the main of which are the visual image insufficient elaboration and the imperfection of the methods of transmitting the necessary information. The solution to the problem is chromakey technology and the correct light setting.
- The synthesis of virtual reality and design technologies will solve the problem of a static and closed space illusion, provide a breakthrough in design - engineering, that is, the transition from a "real" interior to a "digital".
- The ways to solve the highlighted problem, based on the example of developing a virtual reality cafe space using design tools and chroma key technology with 3D visualization of virtual travel scenes were proposed and visualized.

References

[1] V I Gnatyuk 2014 Philosophical foundations of the technological approach (Monograph Moscow).
[2] V N Pervushina, S N 2016 Hutornoy Virtual reality: methodological approaches to the definition of the concept *VSU Herald. Series. Philosophy* **4** 52.
[3] Sayapin V O 2005 From "Possible" worlds to the nanosphere *Questions of modern science and practice. Vernadsky University* **2**.
[4] Segal A P 2017 "Image of the Future" or Virtual Reality? On the space of personality formation *Professional education in the modern world* **7(4)** 1380–1387. DOI: 10.15372/PEMW20170407
[5] Ioivleva G V 2014 2012 Virtual reality: structural and functional features and the specificity of its influence on consciousness *Humanities and social sciences* **3** 88.
[6] Kulenenok V V Virtual environment and multimedia technologies in the design of the subject-spatial environment *Art and culture* **3(7)** 94-102.
[7] Shayev Yu M 2015 To the question about the semiotic structure of virtual reality *Fundamental Research* **2(9)** 2036-2040.
[8] Morozov M G 2014 Typology of approaches to the analysis of social reality and the problem of illusion *SUCCESSES OF MODERN NATURAL SCIENCE* **11** 125-127.
[9] Agadzhanyan D V, Chukarin M I 2016 Description of a prototype face recognition application for mobile development using augmented reality. Virtual and Augmented Reality-2016: State and Prospects *Collection of scientific and methodological materials, abstracts and articles of the conference. Under the general editorship of Doctor of Technical Sciences, prof. DI. Popov. - M.: Publishing house SBPEI MSEC* 12-23.
[10] Averbukh N V, Shcherbinin A A 2011 The phenomenon of presence and its influence on the efficiency of solving intellectual problems in virtual reality environments *Psychology: Journal of the Higher School of Economics / Ed. T.N. Ushakova* **8 (4)** 102-119.

IOP Conf. Series: Materials Science and Engineering 1083 (2021) 012021 doi:10.1088/1757-899X/1083/1/012021

Transformation of the regional settlement system and the factors affecting the spatial development in the Ivanovo region

D V Chebotarev

Don state technical university, 162, Socialisticheskaya street, 344022, Russia

E-mail: chebotarev5@yandex.ru

Abstract. The article deals with the settlement system transformation of the Ivanovo region in modern socio-economic, demographic, spatial and other conditions. The main factors causing impact on the spatial development transformation processes in the conditions of "shrinking" territories are identified on the analysis basis.

Introduction

One of the important issues of urban planning and design, as well as state policy in the field of urban planning and strategic planning, is the determination of the main directions of spatial development of the country and its regions. Determination of the main trends in the processes of spatial transformation of territories requires the collection and analysis of a significant amount of information on various, but interrelated processes in the field of social, demographic, economic, infrastructural and other spheres of the general complex within the urban planning system.

Like most regions of Russia, the Ivanovo region faced a number of processes and problems caused by them, which are a characteristic at the present stage both for the region itself and the central federal district, and for the country development as a whole. The main problems include the rapid processes of depopulation of the territory and spatial "compression" of the territory.

Factors influencing the transformation and development of the territorial spatial system

One of the most important factors in this matter is the socio-demographic situation in the region. The official source for the social and demographic processes' analysis in the Ivanovo region is the data posted on the public resource of the Territorial body of the Federal State Statistics Service in the Ivanovo region. One of the key indicators are the demographic processes indicators in a given territory, which include - population dynamics, birth and mortality rates, gender and age structure of the population, migration processes and the potential of the territory's labor resources. The population of the Ivanovo region on the date 01.01.2020, according to the Federal State Statistics Service, was 997,135 people [1]. The maximum size of the region at 1.32 million (Rosstat) was recorded in 1970 and since then the population has been declining almost continuously. Table 1 and Figure 1.present the data on population changes in the region for different time periods [2].

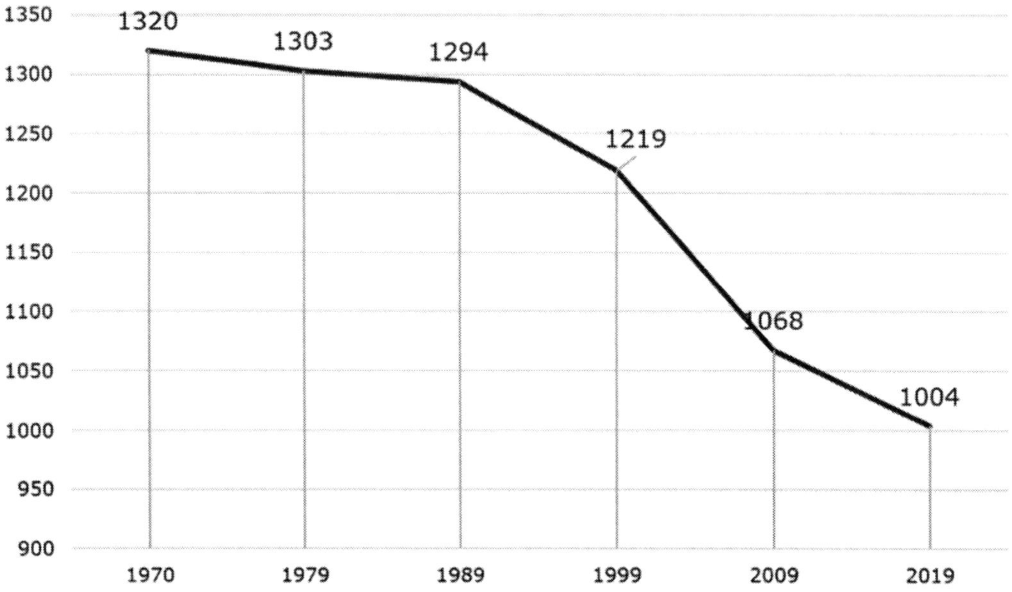

Figure 1. Change in the population of the Ivanovo region during 1970-2019

Table 1 shows these processes in more detail in the context of the last six years.

Table 1. Dynamics of the population in the Ivanovo region for 2015-2020.

Population / year	2015	2016	2017	2018	2019	2020
Population (thousand people) total	1036.91	1029.84	1023.17	1014.65	1004.18	997.14
Urban population	842.36	837.24	832.55	826.73	819.46	814.78
Rural population	194.55	192.60	190.62	187.92	184.72	182.36

Thus, the total population decreased by 3.8% over six years, while the urban population share increased from 81% to almost 82%, which indicates a relatively faster decline in the rural population. If to look at demographic processes in the context of municipalities, it is possible to see the different nature and speed of their course. Figure 2 shows a cartogram of changes in the population size by municipalities in the period from 2010 to 2019, which clearly shows the processes not only of changes (%) of the population, but also the processes of population redistribution, and most importantly the processes associated with the so called "compression" of territory, concentration of the population around the main centers and planning nodes [2].

The determining factors include the actual spatial characteristics of the territory that have been developed in the course of historical development: the existing settlement system, the planning structure of the territory, the position of the region in the general federal structure of the settlement.

Figure 2 shows the position of the Ivanovo region in the central federal district [2].

An analysis of the Ivanovo region location in the Central Federal District shows that the region is part of the Central macroregion and is in the zone of strong influence of the largest agglomeration system of the country (the Moscow region, on the territory of which 13% of the country's population is concentrated), which largely determines labor and migration processes in the region [3]. Negative migration indicators are due, inter alia, to this situation.

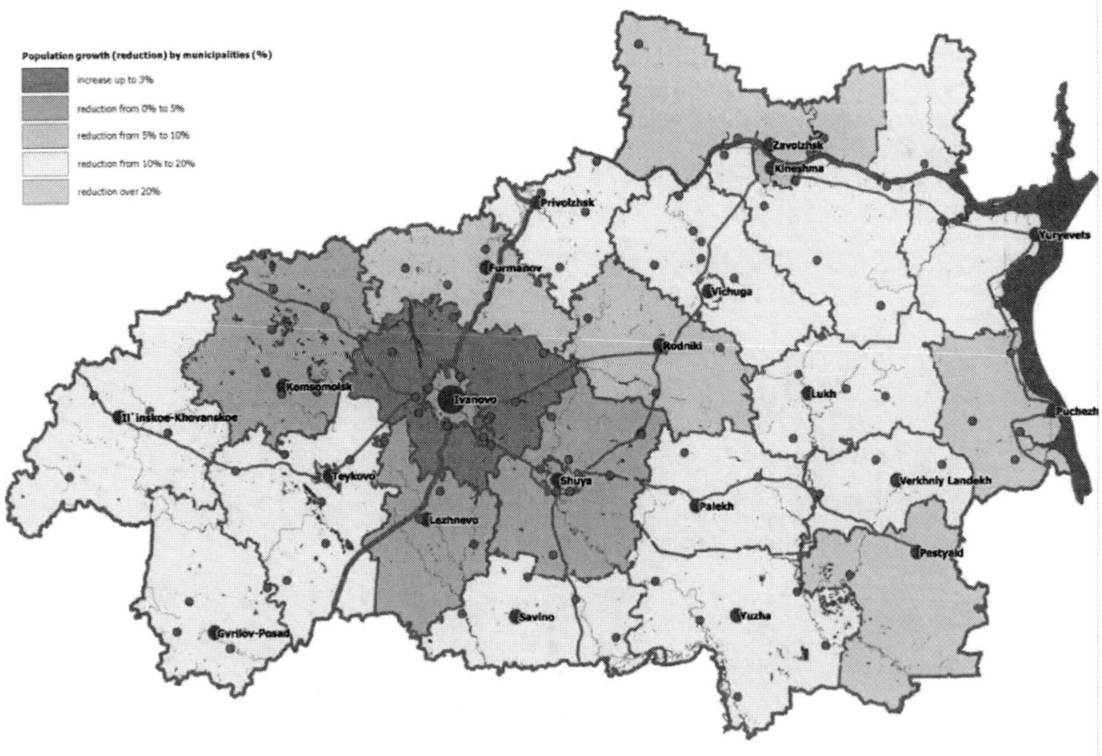

Figure 2. Dynamics of change in the population (population growth and decline in%) in the period from 2010 to 2019.

To determine the trends in changes in the structure of spatial development and the system of settlement in the region, it is necessary to analyze the existing population density. Fig. 3. shows a cartogram of population density in the context of municipalities, which, to a certain extent, also illustrates the emerging picture of population distribution and the tendency of "territorial compression" around the regional (Ivanovo) and the local (city) settlement centers [2].

Historically, the settlement system of a significant part of the region's territory was predominantly dispersed (distributed), which also left its mark on the transport network. A large number of small settlements, relatively evenly distributed over the territory, determined the subsequent process of mass migration (in combination with negative population growth) and the "dying" of hundreds of villages and rural settlements.

According to the 2010 census, there were 3,018 rural settlements in the region, in 634 of which no one lived for the census date. In 1040 settlements there were less than 10 people, and in other 781 rural settlements - from 11 to 50 people. For this period, the situation has only worsened. There is a process of further contraction of the population from rural settlements to urban settlements, which in the 2000s caused a wave of the administrative-territorial division optimization (unification and merger of dozens of rural settlements) [4].

The factor of transport infrastructure development is directly related to the spatial characteristics of the territory.

The planning framework of the Ivanovo region is formed by a combination of planning axes of different order, passing in the meridional, latitudinal and intermediate directions. Historically, the formation of the planning frame was primarily influenced by the geographical location of the region, the combination of the large natural planning axes' location (the Volga River) and the existing settlement system. So, for the most part, the main axes of settlement were the river valleys and the location of transport communications. The basis of the modern planning frame of the territory in the region is the existing transport frame, represented by the network of highways of federal and regional significance, the railway network, to a lesser extent, the formation of the planning frame was influenced by natural planning axes and elements of engineering infrastructure. In the places of their intersection, as a rule, planning centers are formed - national economic complexes, the largest cities and other organizational and economic centers.

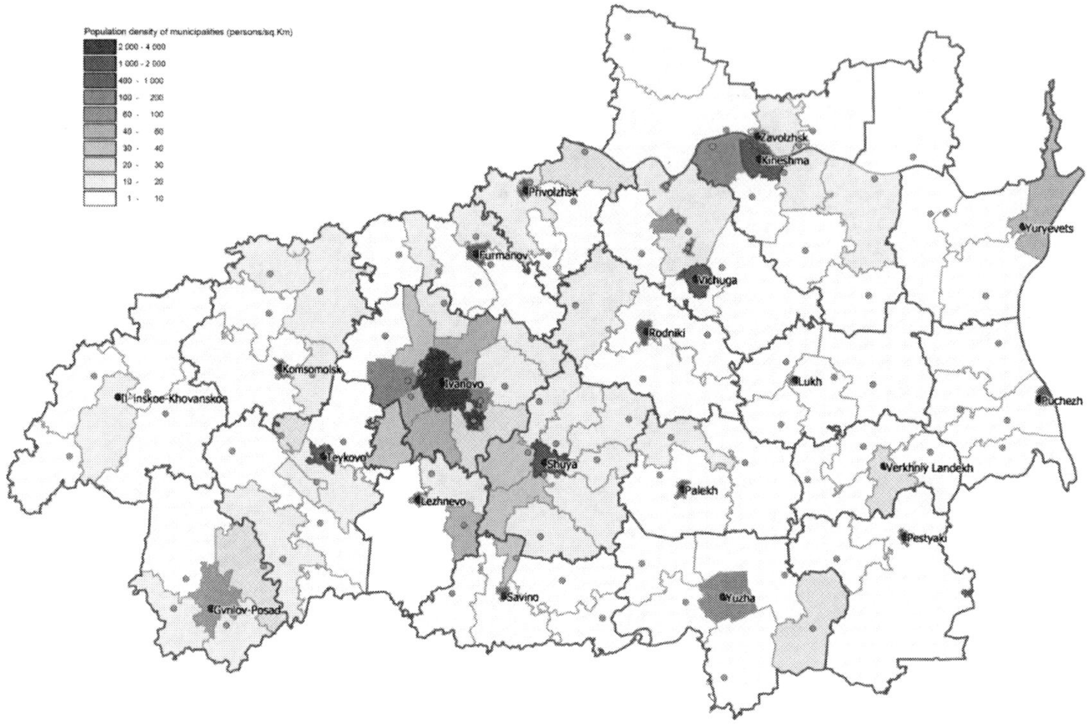

Figure 3. The population density of the Ivanovo region in the context of municipalities [2].

The existing transport network in the region is distinguished by good branching, and generally satisfies the internal needs of the region, however, if we take into account the condition and quality of the road surface, especially local and part of inter-municipal roads (the proportion of hard-surface roads in total length of roads does not exceed 63%), then it is possible to see the problems and internal connectivity of the territories. One of the important indicators of the transport road network development is the density of public roads with hard surface (km of tracks per 1000 km^2 of territory), in the Ivanovo region this figure is 334, which is an average value in comparison with other regions of the Central Federal District [2].

At the same time, the road density in the Ivanovo region itself is not uniform. Figure 4 shows a cartogram showing the distribution of the highways' density in the context of municipalities.

The main problem is the interregional transport position of the region. A significant part of the territories that make up the Ivanovo region are at a disadvantage, located at the junction of regions, but remaining away from federal highways, large transport corridors, differ in minimal communications with neighboring regions.

Thus, despite the proximity to such centers as Moscow and Nizhny - Novgorod, the region's transit potential has not been fully explored. The solution to this problem may become one of the growth drivers in the future (especially the gravitation zones of future transport corridors).

The level of engineering support in the territory (gas supply, electricity supply, water supply, etc.) has a significant impact on the economic and social development processes, which can become an additional negative factor or a significant resource of this territory.

IOP Conf. Series: Materials Science and Engineering 1083 (2021) 012021 doi:10.1088/1757-899X/1083/1/012021

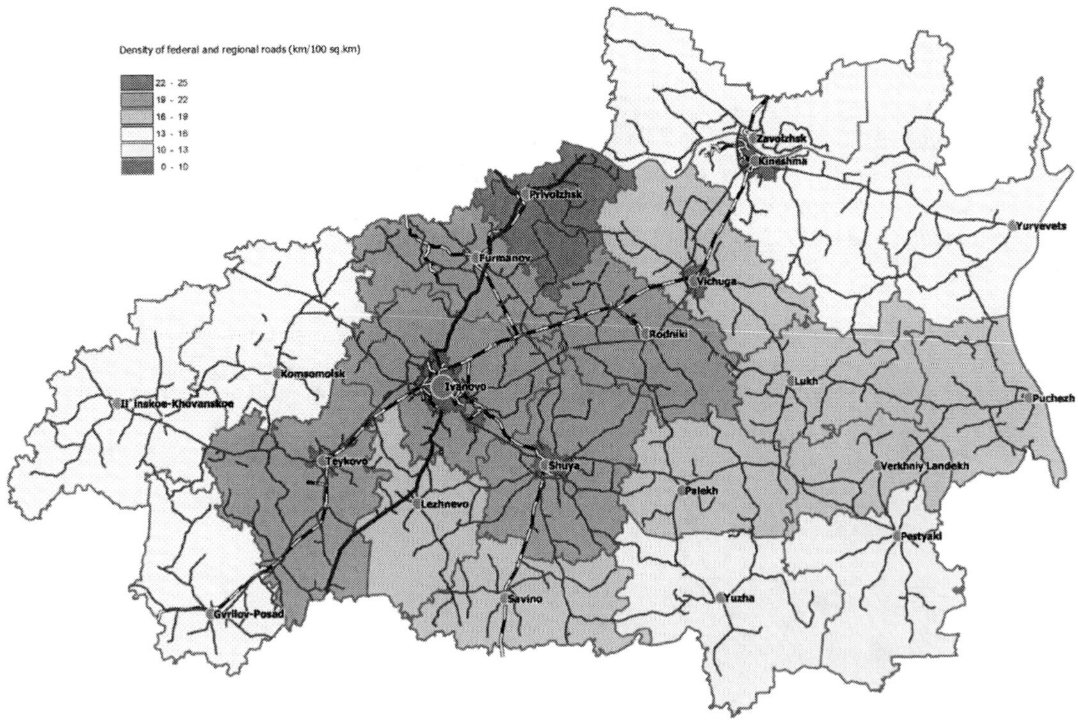

Figure 4. Cartogram of the density of the highways of federal and regional significance in the context of municipal districts. Federal and regional road network.

Summary

The transformation of the regional settlement system in the Ivanovo region is currently associated with the predominance of a number of main trends: a decrease in the population of almost all municipalities in the region, the redistribution of the population from the smallest settlements to larger ones (internal migration), the departure of part of the population outside the region (external migration), which in turn reflects the process of territorial "compression". This transformation leads to a change in the settlement system from a distributed state to a more "focal" and point distribution of the population, there is an increase in poorly urbanized territories.

The main factors influencing the transformation of the settlement system in the Ivanovo region are:

- Social and economic processes;
- Demographic and migration processes;
- Position of the territory in the macro-region;
- The existing spatial structure of the region;
- The existing system of settlement;
- Transport frame of the territory and its condition;
- Engineering provision of the territory;
- General trends of urbanization around the main planning units.

The identification of these factors, as well as the determination of their role and level of impact on the spatial development processes in the Ivanovo region will allow finding approaches and developing recommendations for the most optimal and at the same time realistic way of transforming the urban planning system at the regional level. All these factors significantly affect the territorial development, in turn, by influencing them, it is possible to influence these processes. This analysis can become one of the stages in the development of a strategy for spatial development at the regional and federal levels [5].

References

[1] *Territorial body of the Federal State Statistics Service in the Ivanovo region: official website.* – Information on http//Ivanovo.gks.ru/folder/27479 (date of the application 10.10.2020).

[2] Prokhorov A Yu, Trukhachev S Yu, Trukhachev Yu N, Batunova E Yu, Kryukova V V, Hokhlachev R V, Chebotarev D V 2019 The concept of the project for making changes to the territorial planning scheme of the Ivanovo region *Recommendations and suggestions for amending the territorial planning scheme of the Ivanovo region "[Text]: report on research (concluding): 22-2019/ LLC SPO YURGTs; supervised by [et.al.]* **147** 951.

[3] Trukhachev Yu N 2020 *General theory of settlement systems* (Moscow, LLC "Printing House") 288.

[4] Gun'ko M S, Glezer O B 2015 Small regional centers and their surroundings in Central Russia in 1970-2010: dynamics and distribution of the population *RSA News. Geographic series* **1** 64-76.

[5] *Concept of the Spatial Development Strategy of the Russian Federation for the Period up to 2030* (Moscow, Ministry of Economic Development of the Russian Federation) 2019.

Cultural landscape in the sociospatial development strategy of a small town in Bashkortostan

K V Ishmukhametova[1*], S H Baymuratova[1]

[1] Ufa State Petroleum Technological University, 30, Kosmonavtov Street, Ufa, 450064, Russian Federation

E-mail: ikv23@mail.ru

Abstract. This article examines the cultural landscape as a prerequisite for the socio-spatial strategy formation for the development of a small town in Bashkortostan. The authors of the article analyze the social, cultural contexts of small towns in the republic as the aspects of the city cultural landscape formation. Revealing the continuity of cultural, national and regional patterns of land use and management among the small towns' residents, the authors of the article propose to consider the cultural landscape as an identity element of a small town. The article draws the conclusions about the need to use the patterns of the cultural landscape of a small town as the fundamental design frameworks for the new city spaces' development. Following the general algorithms of urban strategizing, namely the identification of the city's self-identity, the article proposes to use the characteristics of the cultural landscape as a semantic "core" of the small towns' self-identification in Bashkortostan.

1. The cultural landscape as an aspect of the development strategy formation of a small town in Bashkortostan

1.1. Approaches to understanding and defining the term "cultural landscape".

The term "cultural landscape" covers the variety of manifestations of human interaction with his natural environment [8]. As an impact product of social, economic and cultural forces on nature, a cultural landscape can become a representation of a particular geo-cultural region. In turn, the urban cultural landscape can illustrate the identical cultural codes of a particular city.

According to T.A. Smolitskaya, the cultural landscape of a city ("townscape" or "urban landscape") is a process of forming an urban environment generated by the society culture, including the tradition of land use inherited by the city from the territory on which it arose and developed [1].

Depending on the approaches to the definition of the term "cultural landscape" and "urban cultural landscape", the emphasis of the influence different aspects' priority in the formation of these elements is shifted.

In the Russian geographical science in the classification of cultural landscapes, there are three main approaches to the definition and understanding of the cultural landscape [3].

With the classical landscape-geographical approach, the cultural landscape is considered as a special case of the anthropogenic landscape. And the priority object of research, as a rule, is agricultural cultural landscapes [3].

The ethnological-geographical approach examines the influence and transformation of the natural component by the ethnos that lives and develops this territory. With this approach, the type of settlement, management, communication, language (especially toponymy) [4] is distinguished.

With the information-axiological approach, the cultural landscape becomes the result of joint co-creation of man and nature. The cultural landscape is a complex system of material and spiritual values with a high degree of ecological, historical and cultural information content [4, 5].

The most acceptable in urban planning science are the second and third approaches in the cultural landscapes' classification. It is the ethnological-geographical and information-axiological approaches that give rise to the consideration of the cultural landscape of a small town as the basis for the urban socio-spatial strategy formation.

According to the typology adopted in the Guidelines for the World Heritage Convention (UNESCO) Implementation, all cultural landscapes are divided into three main categories: landscape designed and created intentionally by man, including garden and park landscapes; "Naturally formed" (organically evolved landscapes), usual rural or urban historical landscapes, among which there are subcategories - "relict" (or fossil) landscapes and "continuing progressive development" (continuing landscapes); and the third category is "associative cultural landscape", the examples of which are the memorial landscapes, places of creativity or residence of great people, sacred places [8, 9].

The cultural landscapes of small towns in Bashkortostan are difficult to attribute strictly to one of the listed categories, but they still possess the hidden features of each category as a whole. Thus, they are more of a hybrid categorical nature.

The genesis of the city's cultural landscape includes a reciprocal process of anthropogenic and natural structures' development. There is a natural continuity in the formation process of the anthropogenic landscape under the influence of natural landscapes, which, in turn, are also changeable in the course of human activity. The development of both structures occurs at each stage of settlement and affects the change of historical eras. Thus, a humanized cultural landscape becomes a reflection of the chrono-tope of city spaces, the result of event-time changes in the socio-cultural field of the city, material evidence of the change and transformation of material needs and needs of urban communities [5]. In this sense, the cultural landscapes of small towns in Bashkortostan, despite their relative youth, have the features of urban historical landscapes with the "associative landscapes" elements [2].

In UNESCO documents, the following main criteria of the cultural landscape were identified in order to meet the status of World Heritage: outstanding universal value, authenticity and integrity [8, 9].

Considering the small towns of Bashkortostan as integral authentic formations, an approach in their further development strategy is formed. The city appears as a cultural material evidence of the unique ethnic groups in the republic, and from this point of view, it becomes a universal value. Defining a small town in the republic as an integral cultural landscape, the basis for the concept of its public urban spaces' development, for its territorial eco-revitalization strategy is formed. Understanding the genesis of the city cultural landscape provides a direction for choosing the course of the city's economic strategy based on the hybrid rural-urban mentality of its inhabitants [7]. Such concepts and strategies will become the socio-spatial development drivers of a small town.

1.2. Research aspects of the cultural landscape of a small town.

When defining the concept of "cultural landscape", the geographer Vladimir Kagansky uses the metaphor of "carpet" [2]. In the simplest view, a cultural landscape is a natural combination of natural and cultural components, spatially organized in a certain way. The cultural landscape, made up of diverse elements, has its own complex patterns, which are built like a pattern on a carpet.

It is necessary to clearly structure the laws of the cultural landscape of a small town, to analyze the laws of the "patterning" of its landscape "carpet", as V. Kagansky defined it. The analysis of the cultural landscapes of small towns can cover the following aspects:

• location on the ground and interaction with the urban environment - spatial differentiation of cultural landscapes in the city structure, identifying the patterns of urban landscapes;

- identification of the initial impulse of development - determination of the reasons, fundamental conditions, development in the existing key of the landscape pattern; studies of soils and biodiversity of the area, criteria for the conservation and selection of the existing biodiversity;
- contrast or similarity to the original landscape - defining the degree of impact and "cultivation" of the original landscape;
- planning structure and history of planning development - typological features of the cultural landscapes' various patterns of the city, in particular, the study of private land plots of the city, their sizes, proportions, the system of parceling within the quarter, structural features (location of the main house entrance, backyard, greenhouses, structure of economic ridges, etc.); placement in the city structure of gardens, flower beds and gardens, mastered by the local residents; the nature and level of water and geo-landscape natural components' development in the structure of the city (banks of rivers, lakes and other bodies of water, hills, precipices, mountains);
- main dominants and main planning axes - defining the cultural landscape role as a town-planning framework;
- the main architectural landscapes and natural boundaries - defining parts of the cultural landscape as the cores of the town-planning framework;
- basic visual, planning, functional links - the cultural landscape as a condition for urban planning links;
- basic natural structures and processes as geographic determinants;
- sensory perception, phenomenology of urban space - colors, sounds, smells, climatic features of the city;
- toponymy;
- semiotics - symbolistic components of the landscape;
- semantics - historical associativity;
- geoglyphs [1].

The role of an architect in changing the cultural landscape is currently reduced, on the one hand, to the restoration of historical landscapes, on the other, to the creation of new city spaces filled with cultural meaning and content [1]. By identifying the patterns of cultural landscapes in the structure of the city, the architect forms a project framework for the development of the city's environmental spaces. Evaluating the historically formed cultural relationship between man and the nature of a particular city, the architect has the conditions and reasons for the concept and strategy formation for the development of the city as a whole.

The recipes of each successful city are unique, but in one way or another, the general algorithms for the city development strategies boil down to the formation of a unique city brand based on local identity as a new urban economy [6].

Analyzing the experience of successful urban strategies for small towns, the curators of the Creative Environment and Urban Studies section of the VIII St. Petersburg International Cultural Forum identified several areas around which a successful brand for a small town can be formed: heritage (historical monuments and historical and cultural complexes of the city), cultural and educational initiatives, sports infrastructure, site-specific tourism, transport infrastructure (convenient transport logistics and the possibility of comfortable and quick communication with other cities) and special unique public spaces ("Zaryadye effect") [13].

Considering all of the above-mentioned areas in the context of their application as the strategies for small towns of Bashkortostan, the authors of this article see their possible application only under the condition of a detailed study of the contextual features of these cities, namely their cultural landscapes as the initial spatial patterns of the city's territory development and use by its inhabitants. This approach gives a correct idea about the social portrait of the city, about its social and communicative processes, about the imagery and cultural significance of urban spaces for its inhabitants. The level of ecological ethics of the townspeople and the manifestation degree of the genetic cultural code, welded with the nature of the Bashkir land - all this will become the correlation aspects of the strategies presented above.

2. Small town as an integral cultural landscape

2.1. Conditionality of the cultural landscape characteristics by the social context of a small town.
More than two-thirds of the cities in the Republic of Bashkortostan are classified as small towns with a population of no more than 50 thousand people. Thus, a small town is numerically the preferred type of cities in the republic.

Today, small towns do not have positive dynamics of socio-economic development. Such spheres of small towns as ecology, quality of public spaces, cultural and leisure spheres, and the sphere of education and medicine are also problematic. Becoming less and less attractive for the life and self-realization of their citizens, small towns of Bashkortostan become cities-donors of human capital for the larger cities of the republic [7].

Despite the internal problems, the small towns have a high potential for the "humane" urban environment formation. A "humane" urban environment is an environment where high quality indicators of urban infrastructure and mental environmental characteristics, expressed through the cultural identity of the city, popular mentality, and social cohesion, remain in a balanced ratio. Against the background of the global urbanization process and the numerical hypergrowth of cities, the percentage of problems, especially ethical, cultural, environmental and social, of a big city to the advantages of living in it is increasing. The "humane" urban environment is becoming an unattainable allusion to the big city. Unlike big cities, small towns of Bashkortostan preserve the patterns of cultural landscapes to form a "humane" urban environment on their basis.

Ferdinand Tennis, describing the "city dweller" of a metropolis, emphasized its ability to "be in a society" where "mutually beneficial exchange" prevails. Communication of a "city dweller" is built on rationality and selective connection with urban individuals. Such communication acts as a reflexive social connection, where everyone achieves his own benefit. The urban community of a megalopolis is essentially selfish, it is formed according to the principle of rational collective participation in certain activities [11].

According to George Simmel, human fluidity and crowding in big cities do not allow a person to focus on the perception sensibility, personal assessment of everything and everyone, so abstract and typical prevail over the individual. The big city was originally formed as a center of monetary and commodity relations, in which the personality of the seller and the buyer is as abstract as possible, in such relations rationality prevails, and not the warmth of personal ties. The tradition of a business relationship not only to things, but also to people is continued by a modern large city, and in the megalopolis it reaches its apogee [10].

In contrast to the big city, the mentality of a villager, rural settlement and small town resides in the field of communal communication, not the urban community. Community refers to the property of "having something in common". This is an organic, essential, primordial, non-reflective connection between people, which is possible in a family or a peasant community [11,12].

Big cities are multicultural centers of globalization, while small towns are more homogeneous in their composition. The ethnic composition of the small towns' population, in contrast to large cities, is less variegated. The main nationalities of both large and small towns of Bashkortostan are Russians, Tatars and Bashkirs. But the share of other nationalities in a city in a small town is only 5%, and in a large city it reaches up to 56% [7].

Most of the small towns' inhabitants have ties with the countryside. The small town is a transit point in the process of the migration mainstream from rural areas to large cities [7]. Most of the contingent of small towns remains the bearers of the cultural traditions of their people, the archaic ties of the village community, the folk language and the agrarian traditional imagery of life. The inhabitants of a small town retain their ancestral attachment to the land, to nature. In the context of world globalization, the blurring of the lines between cultures, peoples, with the loss of points of support for human self-identification, all these traits of a resident of a small town are becoming rare and valuable. K. Jaspersa wrote that a person becomes himself precisely thanks to the traditions that "allow him to look into the dark depths of his origin and live, feeling responsibility for the future of his own and his loved ones; immersed for a long time in the substance of his historicity, he is really present in the world he creates from the inheritance he has received" [15].

Thus, the small towns preserve the cultural codes of the folk traditional way of life. The residents of small towns maintain a vital need for a private household with gardens and plantings, vegetable gardens and even mini arable land. All this is manifested in a large percentage in the structure of small towns, the prevalence of the private sector over the apartment buildings' development [7].

Partially retaining the features of the agrarian community mentality, the inhabitants of a small town are no longer villagers. The life cycles of a resident of a small town are subject to daily labor migrations not related to agriculture (most of the small towns of Bashkortostan are the industrial cities, and they have a high percentage of factory workers) [7]. The desire to cross the lines of behavior of a "person tied to his land" and a city dweller who uses the benefits of the city's communal areas with their infrastructure - all this forms a hybrid mentality in a small town dweller. On the one hand, he has not lost cultural codes for self-identification as a representative of "his" people, as a part of the integral people of Bashkortostan, on the other hand, he is adaptive to the adoption of the new, to the common achievements of the global world civilization. Without losing the roots, a resident of a small town is capable of crossing the traditional and the universal, the generic and the universal. He is receptive and mentally flexible, he is able to develop and connect his folk model of world perception to the world universe [10,12].

The formation of a hybrid mentality and the nature of the social portrait of a small-town resident has a connection with the genesis of small towns in Bashkortostan.

2.2. The genesis of small towns in Bashkortostan.

There are 9 small towns in the Republic of Bashkortostan - Agidel, Baimak, Birsk, Blagoveshchensk, Davlekanovo, Dyurtyuli, Mezhgore, Uchaly, Yanaul - all of them have a unique fate that has formed a special unique "face" of each city. The genesis of their typology is associated with a certain stage of economic and state development in the region [7].

Formed as a fortress on the way to the capital of the Republic - Ufa in 1667, the city of Birsk is one of the oldest small towns in the republic. Having a "city" historical nucleus with a square and the Holy Trinity Cathedral in the center, the structure of the city is dominated by such a form of land use as private housing construction with a land plot. In 1939, the village of Kalmazy was included in the structure of Birsk. Thus, in the city, the percentage of people whose lifestyle and land use tends to be closer to the "rural", increased in relation to people who do not have a "relationship to the land" and lead a predominantly "urban" way of life.

Unlike Birsk, which back in the 17th – 19th centuries, was called a city, all other small towns of the republic grew out of rural settlements.

Blagoveshchensk was formed as a mining settlement in 1756 at the Blagoveshchensk copper-smelting plant, passed the stage of town formation as a workers' settlement, and only in 1941 acquired the status of a city. Baymak is associated with the discovery of gold and copper deposits at the end of the 19th century and the construction of the Baymak copper smelter in 1913. Already in Soviet times, in 1938, it received the status of a city. The city of Davlekanovo grew out of a village located at the intersection of transport communications, in 1928 it became an urban-type settlement, and in 1942 - a city. Uchaly as a city was formed in 1963 by the merger of two villages: Malye Uchaly (1943) and Novye Uchaly (1960), which arose during the development of the Uchalinsky copper-pyrite deposit. The city of Dyurtyuli grew out of a village due to the discovery of oil fields, in 1964 it received the status of an urban-type settlement, and in 1989 - the status of a city. Thus, apart from Birsk, all other small towns on the map of Bashkortostan appeared in Soviet times due to the industrialization policy and the needs of the region's industrial development.

It is worth highlighting the cities that arose in the post-Soviet era - Yanaul, Agidel, Mezhgore.

Since 1938 Yanaul has been a workers' settlement and only in 1991 acquired the status of a city of republican significance. But the first written mention of the Yanaul village, where mainly patrimonial Bashkirs lived, dates back to 1750. In the village they were engaged in agriculture, cattle breeding and beekeeping, they leased land to alien peasants. The construction in 1909-1914 of the Moskava-

Yekaterinburg railway, and the opening of the Yanaul station near the village radically influenced the further development of the settlement.

The city of Agidel appeared in 1980 as a settlement in connection with the construction of a nuclear power plant; in 1990, under pressure from the "green" movement, the station construction was stopped. The village of nuclear scientists was virtually left without a livelihood. In 1991 the village with a large number of highly qualified specialists was elevated to the rank of a city. And in 1993, the decree of the Presidium of the Supreme Soviet of Bashkortostan No. 6-2 / 408 "On the measures to stimulate jobs in the city of Agidel" was signed, which made it possible to largely ensure the viability of the city [7]. But despite the absolute industrial history of the city emergence, it did not arise from scratch (before the construction of typical housing, the village of Starokabanovo was located on the site of the city) and in its planning structure the city demonstrates the “fusion” of the village of Novokabanovo and the young industrial city with the modernist building structure of the industrialization times.

Mezhgore occupies a special place in this group of small towns. The city of Mezhgore was founded in 1995 by the Decree of the President of the Russian Federation of July 8, 1995 and has the status of a closed administrative-territorial entity (CATE). It is located at the foothills of the highest mountain in the Southern Urals, Yamantau, where the Bashkir villages of Tatly and Kuz’elga were previously located [11].

One way or another, “village roots” can be traced both in the modern structure of small towns and in the “pedigree” of urban education.

The mechanical increase in the small towns’ population in Bashkortostan was formed due to the intensive migration of people from the villages that surround them. According to some estimates, the number of migrants from rural areas in small towns is ¼ of the city’s population [16]. Most of the small towns in the Republic of Bashkortostan in the post-Soviet period did not have the opportunity to develop their production functions. Also, the socio-cultural potential inherent in them is not fully realized. As a result of the influx of migrants from the rural areas who did not have time to join the urban way of life, to assimilate and master the city values, a “semi-urban” way of life has been formed in small towns. Lacking sufficient means of subsistence, the arrived former villagers continued to grow vegetables and fruits on their local plots of land for their own consumption.

According to the Federal State Statistics Service for the Republic of Bashkortostan in 2019, agricultural organizations account for only 39.4% of the total agricultural production. In farms, this figure did not exceed 13%. The largest percentage remains with the population of Bashkortostan, that is, almost half (47.7%) of all agricultural products are grown by people on their own plots. For certain types, such as potatoes and other vegetables, this figure reaches 95% and 65% respectively [17].

When analyzing residential development in the structure of small towns in Bashkortostan, the type of private housing with a land plot prevails in percentage terms over the development of multi-apartment residential buildings. Also, with a detailed analysis of land plots in the structure of the city, it is possible to trace the nature of land use: on a personal plot, residents of a small town set up gardens, plant fruit trees around the perimeter, organize greenhouses, and cultivate land for a "potato field". It can be assumed that the land use patterns of a small-town dweller are similar to those of a villager. The intertwining of the urban lifestyle with the rural lifestyle remains a characteristic feature of small towns in Bashkortostan.

2.3. Agrarian patterns of urban dweller mentality as a guarantor of economic stability of small towns.
By the order of the Government of the Russian Federation of July 29, 2014 No. 1398-r, the lists of Russian cities were compiled according to the degree of complexity of their socio-economic situation in connection with the crisis situation of the city-forming enterprises of the city. Among 9 small towns of the republic, only Blagoveshchensk and Uchaly were included in these lists and only in the category “Single-industry municipalities of the Russian Federation (monotowns) with a stable socio-economic situation” [14]. This gives rise to the formation of a hypothesis about the sustainability of small towns in the context of economic instability in the era of globalization, due to their close connection with the rural district. The relative economic stability of small towns, in contrast to medium-sized and large

single-industry towns, was made possible by the diversification of their economies through the opening of different-level enterprises for processing agricultural products - Yanaul Dairy Plant, Belebeevsky Dairy Plant, Dyurtyulinsky Butter Factory, Bashkir Delicatessen Meat Processing Plant in Baymak, the Turbaslinsky Broiler poultry farm in Blagoveshchensk, etc.

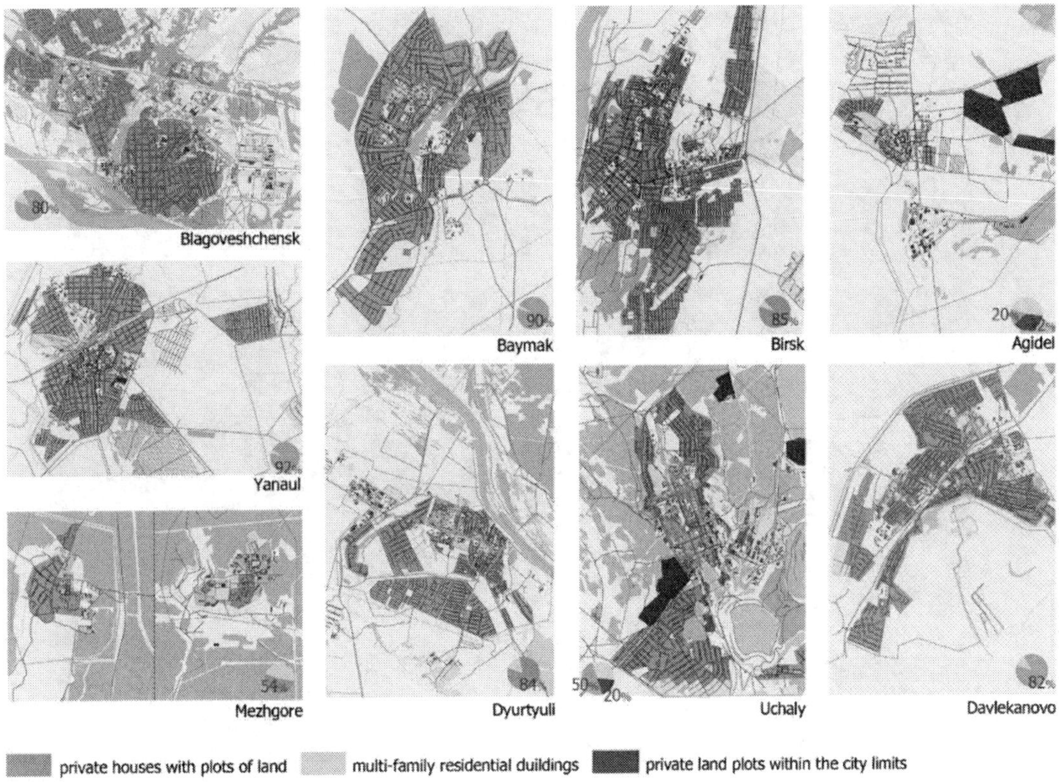

Figure 1. Percentage of the private development area with land plots to the area of apartment buildings' development in the structure of small towns in Bashkortostan.

When developing a strategy for the development of small towns, an essential aspect is precisely the connection of the local population with the traditional rural way of life. The ability and desire of the small towns' residents to cultivate their private household plots, a large percentage of private land plots in the structure of the city, the close proximity and contact of the city and the countryside should be taken into account when choosing a city development strategy.

2.4. Natural artifacts in the structure of small towns.

In addition to the residents' household style close to the rural way, the presence of natural artifacts both in the structure of the city and in its surroundings serves as a special context of small towns. Natural contextual artifacts are presented in the city both as the developed cultural landscapes and as natural inclusions or city components. The arboretums of Birsk, the geoglyphs of Blagoveshchensk, made by planting trees, and many other are the examples of the developed cultural landscapes.

Nature is closely related to the toponymy of small towns, with the semantics of urban landscapes. It is the organic interweaving of natural components, assimilated and cultivated, that each time forms a special structure and imagery of urban spaces and planning decisions of the city [5].

The traditional settlement of the peoples of Bashkortostan is associated with the presence of a water component in the area as a drinking resource, for economic and industrial needs, and also as a water transport artery. Thus, a village or a rural settlement was often located near a river, lake, or network of water bodies. Many small towns in Bashkiria have a unique natural water frame that influenced their development and settlement patterns. The etymology of the very name of Agidel, one of the small towns, reflects the importance of the urban water component as a "city on the river". Also, from the hydronym Bir came the name of the city of Birsk.

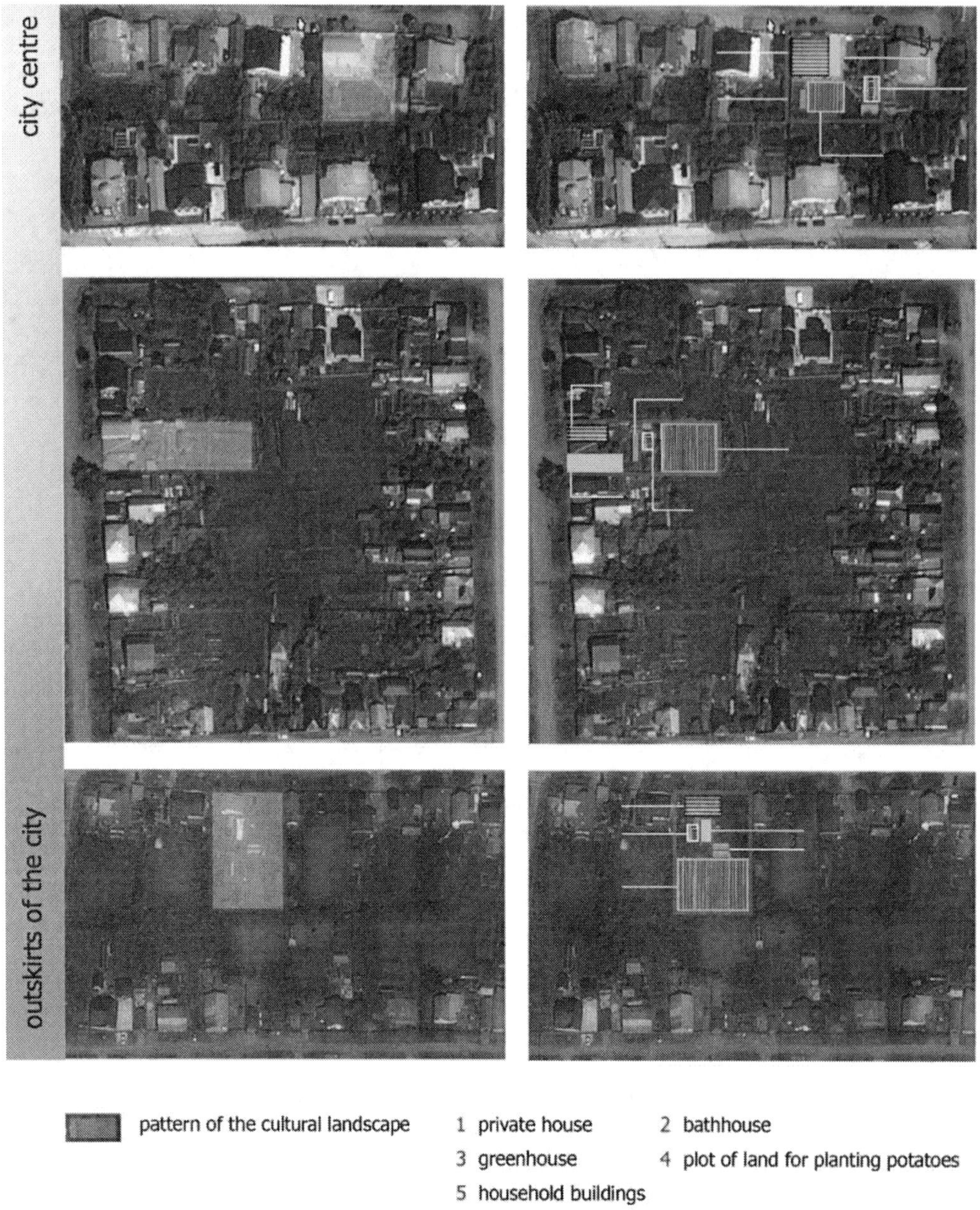

Figure 2. Analysis of the cultural landscape pattern structure on the example of the Blagoveshchensk city.

2.5. Environmental ethics of small towns in Bashkortostan.

Considering the natural component of small towns in Bashkortostan, it is necessary to touch upon the environmental friendliness topic, especially against the background of the fact that it is this parameter that affects the life quality index in the city as a whole.

For small towns with a rich industrial history, which include most of the small towns of Bashkiria [7], environmental issues are acute. And this is not only about the absolutely necessary modernization of enterprises in accordance with modern high standards of environmental protection. The concept of sustainable development changes the very paradigm of the human relationship to nature: from the source of an "almost inexhaustible" resource to minimizing the exploitation of natural resources and a responsible, careful attitude to natural objects. Environmental friendliness is becoming the most important factor in the territory attractiveness. Thus, we are talking about creating an ecological environment not only in the city itself, but also outside it.

Despite the industrial complexes of city-forming enterprises, most of the small towns of Bashkortostan have a strong potential for creating their own brands as "green" cities with unique natural components both within their territory and near it. As an example, the small town of Mezhgore is located in the most beautiful state nature reserve, the habitat of Siberian stags and the growth of rare useful plants on Mount Yamantau. The Blagoveshchensk inland lakes network is a unique natural framework on the basis of which an eco-friendly brand of the entire city can be formed [7]. At the same time, the emphasis can be placed both on the high aesthetic, ecological qualities of natural objects in the structure of the city, and on their cultural significance as the objects of self-identity of the city and its inhabitants.

Considering ecological ethics in the cultural landscape paradigm, it is necessary to highlight two main directions - biocentrism and anthropocentrism [4]. From the standpoint of anthropocentrism, a person does not take into account the value characteristics of nature as a spiritual and cultural component for the formation of an anthropogenic landscape. At the same time, the ethics of biocentrism strives in the maximum way to direct a person into a nature-saving channel as the sources of urban form-creation.

Self-identification and ecological ethics as values of a cultural landscape follow from such a characteristic of human existence as creativity - the creation of something new that did not exist in nature before - and further, integration around the results of this creativity [5].

The fusion of nature and the cultural environment (material and spiritual) of a person is expressed in folk tales, myths and legends of the people, as well as in songs, customs, rituals and sayings. At the same time, in a binary pair, culture and nature, culture is a part of the archaic form of understanding the phenomenon of unity and integrity of oneself as a person with a natural environment [4].

"Culture, reflecting or practically transforming the natural landscape, endows it with images, meanings, new meanings, relief and rhythm of life. Along with this, the natural landscape, entering the mythological, religious, historical consciousness, artistic creation, influences the image of the world, mentality, serves the national self-identification of the people, giving rise to a branched landscape topic and metaphor" [1, p. 55].

Thus, the cultural landscape is fully one of the key forms of cultural reflection. Interaction with him in the historical development process contributed to the formation of the traditional axiological foundations of society: creativity, integration, self-identification, environmental ethics - which indicate, first of all, the spiritual, conscious component of a cultural landscape phenomenon [5].

The city as an integral cultural landscape becomes an expression of the material and spiritual culture of the peoples of Bashkortostan. The urban landscape elements, their structure and spatial relationship represent a cultural value layer that requires special treatment.

Summary

The territory where the natural landscape has been transformed and a full-fledged bodily, spiritual and mental life and therefore social and state of a person is possible, becomes a full-fledged cultural landscape. A paradoxical consequence is derived from this definition - there are territories that have been greatly transformed by man, but are not the cultural landscapes.

In this sense, the landscapes of large cities are becoming more and more "anthropogenic", as much as possible divorced from their spiritualizing natural component, in contrast to small towns, where cultural landscapes predominate.

The unique cultural urban landscapes of small towns in Bashkortostan are a part of the cultural landscape framework diversity of Russia as a whole [2]. The unique natural component of the urban environment of small towns in Bashkortostan has the potential to form the core of branding for each of the individual cities. The formed brand of the city, which is based on the specificity and uniqueness of the cultural landscape of each of the small towns, can become a driver of the economic growth positive dynamics for the small towns in the republic.

A small town should not strive to be like a large dynamic metropolis. The structure of small towns reflects the spatial codes of their cultural landscapes that are very different from the urban landscapes of large cities. In a small town, it is possible to trace "lyricism", "poetics" of urban and natural connections, "soulfulness", "regularity and philosophically meaningful simplicity", a distinct embeddedness of human existence in the world order reflected in the laws and cycles of nature - all this can be an attractor for people, looking for an alternative to the fast-paced lifestyle of megacities, cut off from the natural basis, from the genetic cultural codes [1,2,4].

References

[1] Smolitskaia T A Korol T O Golubeva E I 2018 *Urban cultural landscape: Traditions and modern development trends* (Moscow, LIBRIKOM) 272.
[2] Kaganskiy V L 2011 Research of the Russian cultural landscape as a whole and some of its results. *International journal of cultural studies*. **4 (5)** 26-40
[3] Galkov O V 2008 Cultural landscape as part of the world heritage (according to UNESCO documents) *Bulletin of the Saratov state socio-economic University* **4** 156-159
[4] Leonova V A 2015 Cultural landscape: concepts, history and issues *Forestry bulletin* **19 (5)** 83-87
[5] Shishkina A A 2016 "Second nature" and "Cultural landscape" as philosophical categories *Manuscript* **1 (63)** 203-206
[6] Makashev A O 2017 Branding as a tool to increase the investment attractiveness of the territory. *Bulletin of the State University of management* **2** 146–150
[7] Bukanova R G 2019 Small towns of Bashkortostan: historical and legal status and modern development trends *Problems of Oriental studies* **2 (84)** 58-63. DOI:10.24411/2223-0564-2019-10209
[8] *Operational Guidelines for the Implementation of the World Heritage Convention, UNESCO, WHC-99/2* (March) 1999 38.
[9] *Operational Guidelines for the Implementation of the World Heritage Convention, UNESCO WHC.19/01 10* July 2019 177.
[10] Simmel G 2018 *Big cities and spiritual life* (Moscow, Strelka Press) 112.
[11] Tonnies F 2002 *Community and society. Basic concepts of pure sociology* (Saint Petersburg, Vladimir Dal) 450.
[12] Webber M 2017 *City* (Moscow, Strelka Press) 252.
[13] *VIII St.Petersburg International Cultural forum «Creative Enviroment and Urban Planning» section 2019 Regional development strategies* (Moscow, United projects LLC) 47.
[14] *List of single-industry municipalities Russian Federation (single-industry towns) No. 1398-R dated July 29* 2014. Information on http://static.government.ru/media/files/41d4f68fb74d798eae71.pdf
[15] Jaspers K 2013 *The spiritual situation of the time* (Moscow, AST) 288.
[16] Khilazheva G F, Shamsutdinova N K 2018 *The Republic of Bashkortostan. Demographic* (Ufa, Bask. Encycl) 132.
[17] Agricultural activity of farms of the population of the Republic of Bashkortostan: statistical collection 2020 (Ufa, Bashkortostanstad) 67.

Housing construction design taking into account dust content index of climate in the Russian Federation

N I Zakieva*, A P Lapina, A P Aleksandrov, A S Kulikov, V A Zheleznyakov

Don state technical university, 1 Gagarin sq., Rostov-on-Don, 344002, Russia

E-mail: nikorani@mail.ru

Abstract. The article deals with the residential buildings' design taking into account the dust content of dry climate. At the same time, such certain principles of designing buildings and structures in special conditions should be taken into account as: orientation to the cardinal points, wind direction, the amount of solar energy, the number and size of windows, wall thickness, roof slope, etc.

Introduction

The object of research is the design of residential buildings taking into account the dust content of the dry climate of the Russian Federation.

The subject of the research is the principle of designing residential buildings in a climatic zone of a dry climate. When designing buildings, it is important to pay attention to the climatic zone in which the building will be located. The main goal of this approach is to provide the residents with the most comfortable conditions throughout the year. Dry, hot climate has a number of special properties that should be considered when designing buildings. Cold, dry climate, in turn, has some troubles, depending, however, almost exclusively on the low temperature, against which a person can successfully fight and therefore can arrange his life tolerably and without harm to health.

1. Natural and climatic factors affecting housing construction

The quality level of individual residential buildings is determined by their functional, architectural, constructive, hygienic, aesthetic, economic and many other indicators. In this regard, the main goal of the design and construction of individual housing is to achieve only comfort. And in order for the house to be comfortable, during its construction it is required to fulfill a number of complex requirements and take into account a whole range of specific factors. Before getting to grips with the project, it is necessary to study what features the air, water, relief, vegetation in the area of the land plot have. The greatest influence on a private house design is exerted by atmospheric conditions. This is due to the fact that state of the atmosphere determines how much the ground on the site will warm up, whether there is enough moisture for the plants planted near the house, and finally, whether the wind that will rip off the roof or not. Of course, you can build a universal house suitable for any climatic region, but it is much easier and more economical to focus on local conditions and take them into account as much as possible in the design.

Building climatology is a section of industry codes that defines the requirements for the design of buildings and housing construction, taking into account climatic conditions. Climatic elements are: air temperature and humidity, wind, solar radiation, precipitation. Temperature, relative humidity and wind make it possible to compile the climatic characteristics of the territory, city. The critical values of the main climatic factors help to determine the climate characteristics and the critical values of various climatic factors are functionally related to each other. Knowing the criterion values of the elements, it is possible to identify the climate specifics, to establish the elements' deviation degree from convenient conditions and to formulate a set of requirements to be taken into account when designing buildings, adjoining territory and development. The design is carried out in such a way as to compensate for sudden temperature fluctuations, to exclude the possibility of overheating in the south and hypothermia in the north.

Wind mode

The strength and direction of the wind significantly affect the dwelling, so the designers develop a special diagram - the "wind rose". This graph in the form of a multi-pointed star (rose) clearly shows the frequency of winds in summer, winter or any other period of the year in one direction or another. Figure 1 shows the "wind rose" for the city of Rostov-on-Don.

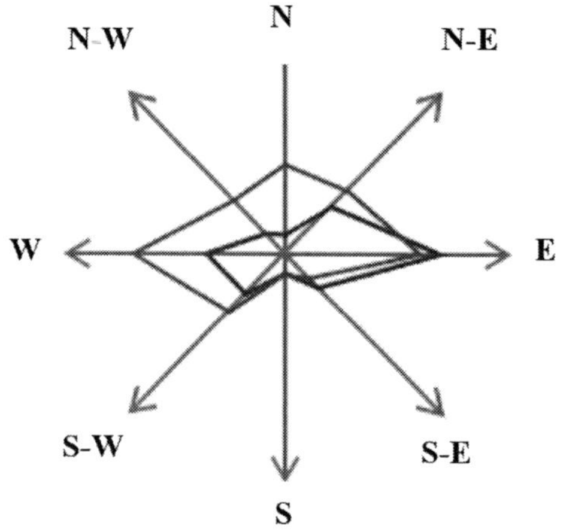

Picture 1. "Rose of the Winds" of Rostov-on-Don.

——— July ——— January

On the windward side, where the wind pressure creates a zone of increased pressure, windows and vents can be smaller than the vents on the opposite side of the house. Taking into account the wind regime is very important when developing a master plan for a complex estate development of a residential building together with outbuildings, a garage, a land plot and neighboring buildings. The architect should choose the right objects' layout, the configuration of buildings and the spacing between them. For example, in hot, humid climates, significant gaps should be left between buildings to increase air circulation.

In harsh climatic conditions with strong winds in the 1st climatic region and other regions, the opposite uncomfortable situation arises, which should also be taken into account when designing housing. So, in the Far North, the combination of strong winds with low temperatures leads to increased heat loss in the building and ventilation of enclosing structures. To prevent these phenomena, special architectural solutions are used to reduce the effect of cold wind on a residential building. In addition, in the 3rd and 4th climatic regions with hot and dry climates, strong winds often carry a significant amount of dust and dirt. To protect against this, special shielding properties of buildings, green spaces, closed courtyards, houses with gaps and voids to the entire depth of the body, and other measures can be used.

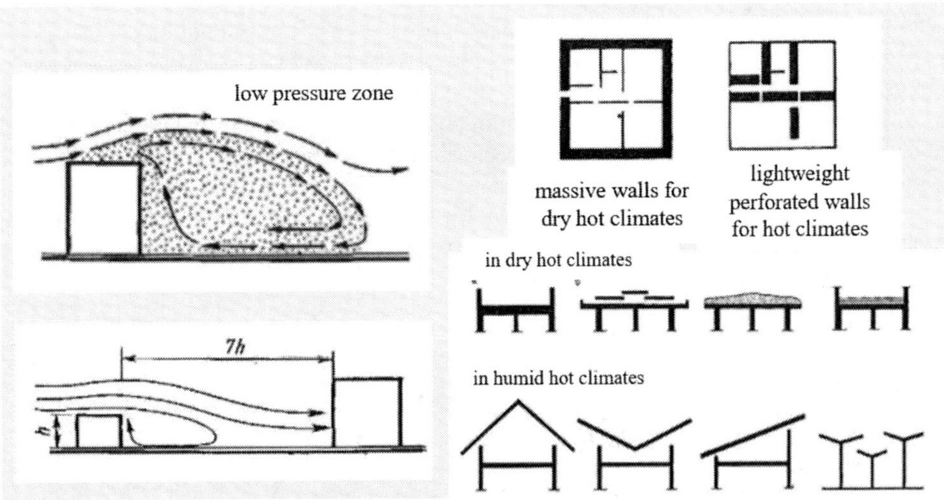

Figure 2. Features of building design in dry or humid climates

Snow and rain regimes

The entire territory of the Russian Federation is divided into three zones of humidity: wet, normal and dry zones. For example, St. Petersburg is in the wet zone, Moscow is in the normal zone, and Orenburg is in the dry zone. In general, snow and rain are the unfavorable factors for the housing operation, since they, like wind, carry harmful substances contained in the atmosphere. To protect against snow and rain, the following techniques are provided:

- waterproofing of structures and reliable drainage from the roof;
- external drainage from the house into ditches and other structures;
- special snow-protective building of the territory;
- installation of a drainage system and slopes of the territory;
- snow protection of external surfaces for the buildings with cladding and painting with special resistant compounds;
- intermediate cornices, pediments and ventilated air layers in the walls, as well as other urban planning and constructive solutions.

Characteristics of dust content in a dry climate and its influence on the building design

The high dust content of the urban environment is the cause of serious ecological and hygienic problems in cities. The negative impact of dust is manifested in various spheres of human activity, especially in urban construction and economy. Dust content leads to an increase in economic losses due to an increase in the cost of operating and repairing buildings and structures and a harmful effect on vegetation. Dust content of the urban environment, associated with an increase in the amount of man-made dust in residential buildings, significantly exceeds the volume of natural sources. The drier the air and soil and the higher the wind speed, the higher the dust content in the air. The sizes of dust particles range from the particles visible without any special gadgets to the particles that are barely visible under a microscope. The smaller the dust particles, the longer they do not settle. Reducing the dust content of the urban environment is effective with an integrated approach, in which urban planning and construction techniques are used. The measures carried out with the aim of improving the environment can prove the ecological reconstruction of residential buildings or its reorganization, where it is necessary to conduct a comprehensive assessment of the dust content of the environment and include three main stages:

The first stage of the ecological reconstruction pre-design analysis is to determine the background assessment of the dust content of the environment and the qualitative characteristics of the dusty air flow entering the residential group.

The aeration regime and the soil type of a given territory is determined for:

- local sources of dust formation and their zones of influence on the residential group;
- urban planning factors that increase the dust content of the urban environment.

Zones of potential dusting within the boundaries of an unfavorable wind sector are marked on the general plan of the area on a scale of 1: 10000, 1: 5000. The result of the first stage of the pre-project analysis is to determine the sequence of ecological reconstruction of the urban living environment.

The second stage of the pre-project analysis consists in building a map of dust-wind activity of the territory of a residential group, which is carried out on a scale of 1: 1000. The zones are determined according to the groups of wind speed coefficients, where the following are distinguished:

- dust-wind shade zone;
- possible dust transfer zone ;
- dust-wind activity zone;
- dust-wind hyperactivity zone.

The third stage of the ecological reconstruction pre-design study solves two problems:

- determines the sources of dust formation on the territory of the residential group;
- determines the dust resistance of building facade materials.

Determination of dust formation sources on the territory of a residential group is carried out on the basis of coatings' differentiation according to the principle of dusty and non-dusty coatings, of artificial or natural origin. Combining the map of dust-wind activity of the residential group territory with the coverage scheme inside the courtyard territory makes it possible to assess the dust content of the sites and to zonate the territory according to the degree of dust content:

- dust deposit zone
- low dust content zone;
- medium dust content zone;
- high dust content zone.

Reducing the dust content of the environment by the construction methods is carried out by differentiating the surface of the building facade into the tiers that satisfy the level of the basement, first, second floors and above. According to the tiers of the building facade, it is recommended to use the facade building materials with certain properties of the surface texture. The result of the pre-design analysis is the development of comprehensive measures for the ecological reconstruction of the urban living environment in order to reduce its dust content. The ecological reconstruction of the city dwelling consists in carrying out the activities that have urban planning, construction directions and activities related to improving the microclimate in the territory.

Summary

When designing residential buildings, it is important to target the climate zone in which it will be located. After all, the goal of a successful design depends on many factors. For example, dry, hot climate has a range of special properties that should be considered when designing buildings. In a hot, dry climate, air temperature during the day in summer reaches 45 degrees Celsius, and the humidity does not exceed 20 percent. This type of climate is most noticeable in the areas located away from the sea coasts and in which there is rather low rainfall. Consequently, during construction, first of all, the impact of solar energy, its maximum use is taken into account:

1. Location of the building. The longer walls of the building should face the North and the South to minimize solar exposure. The kitchen should be located on the leeward side of the building. This will help avoid hot air circulation and odors from the kitchen.
2. Windows and openings in the walls. If it can be permitted in the project, windows and openings should be larger and more in the North and West of the building. It is preferable to build awnings and verandas, which in turn act as a barrier to radiation and reduce the level of hot air entering the house.
3. Thickness and painting of walls. This factor plays an important role in the thermal insulation of a building. In a dry, hot climatic zone, thick walls are most preferable, because they will serve as an insulating barrier against the excessive heat penetration. Painting the walls is also essential. Light

colored walls as well as reflective paints do not absorb heat, so they are a good option in hot climates. Walls with a smooth finish will help achieve the best effect. The presence of voids in the walls will also enhance the insulation effect.

4. Roof. Good insulating material, upwind slope are the main requirements for an insulated roof. A suspended ceiling can also be used to improve insulation from external heat.
5. Vegetation. It is necessary to include the planting of large shady trees near the walls in the project of the territory adjacent to the building. This will provide shade and cool air around the house.

For example, in a cold climate zone, compact volumes with asymmetric gable roofs are common. The steep slope of the roof is oriented to the South, longer, gently sloping - to the North, because in this case it is better able to withstand snow and wind loads. Basement and basement floors, attics are heavily insulated, the entrance to the house passes through the vestibule. City houses are blocked to reduce heat loss. In cold climates, compact, "assembled" volumes with a minimum of protruding parts and niches protruding into the total volume are appropriate. A gable roof is advisable, it can be asymmetric. The steep roof slope should be oriented to the South. If the slope faces North longer and gentler, then it resists snow and wind loads better. Basement and basement floors, attics should be heavily insulated, the entrance to the house should be done with a single or double vestibule. Urban and rural houses can be blocked to reduce heat loss. The north facade has minimal glazing. Colder areas are characterized by a compact layout of structures without kinks in the outer walls, insulated floors of the first floor, increased areas of entrance vestibules, and reduced window openings.

In addition to these factors, the terrain and surrounding buildings can have a significant impact on houses. The relief of the area with a slight slope of the site, as a rule, does not affect the architecture of the house and permits almost any layout. With an increase in the slope to 10-15%, it is necessary to make already special decisions on the first floor, and with slopes of more than 15%, it is advisable to move to special types of residential buildings - terraced and others. At the same time, additional preparation of the territory, special calculations for the rooms lighting, the selection of appropriate types of houses and the methods of their construction are needed. It is inappropriate to use slopes facing North for housing construction, because in this case the premises and the site itself will not be sufficiently illuminated by the Sun. The surrounding area can also influence the architecture of the house. Urban cottages in the historical area are characterized by such a layout of new buildings, which supports and repeats the existing architectural styles, geometry and proportions.

On the contrary, it may be preferable not to assimilate, but to harmonious contrast of the designed private housing, achieved by complicating the architecture of facades, plans and details. All over the territory of Russia in the settlements there are special protection zones, established by the state to preserve the architectural and historical heritage. They define the maximum permissible height of newly erected buildings. The number of storeys of the projected building can also be limited to preserve the visibility of landscape and architectural landmarks - rivers, hills, palaces, bridges and other objects.

References

[1] Mirzai Reza 2009 Medium-rise residential buildings erected by industrial methods for a hot-dry climate zone: on the example of large cities of Iran *Dissertation for the degree of candidate of architecture* (Moscow) 341.

[2] Maximova M V, Melnikova S O 2017 Space-planning solutions for low-rise residential buildings in hot climates *Bulletin of SibADI* **3 (55).** Information on https://cyberleninka.ru/article/n/obemno-planirovochnye-resheniya-maloetazhnyh-zhilyh-zdaniy-v-usloviyah-zharkogo-klimata (access date: 13.11.2020).

[3] Maximova M V, Nemtseva O G 2017 Design features of building design in hot climates *Bulletin of SibADI* **4-5** 56-57. Information on https://cyberleninka.ru/article/n/konstruktivnye-osobennosti-proektirovaniya-zdaniy-v-usloviyah-zharkogo-klimata (access date: 13.11.2020).

[4] Fam T Kh T, Solovyev A K 2018 Influence of light and climatic factors on the architectural design of buildings in tropical countries (on the example of Vietnam) *Engineering Herald of the Don* **4 (51).** Information on https://cyberleninka.ru/article/n/vliyanie-sveto-klimaticheskih-faktorov-na-

arhitekturnoe-proektirovanie-zdaniy-v-usloviyah-tropicheskih-stran-na-primere-vietnama (access date: 13.11.2020).

Features of art and design schools on the example of the southern Federal and Riga technical universities

A Y Mokina [1]*, A Yu Mokina [1], Ulme Andra[2]

[1]The Southern Federal University, 105\42, B. Sadovaya Street, 344006, Rostov-on-Don, Russia
[2]The Riga technical University, 1, Kalku Street, Latvia, LV-1658, Riga

E-mail: aymokina@list.ru

Abstract. The article examines the issues of modern pedagogical science in the field of art and design activities on the example of universities in Russia and Latvia. A comparative analysis of approaches to the formation of creative skills for students while working on creative projects in the field of design, decorative and applied arts is performed. The main criteria for evaluating the final students' results when working on art and design projects are also identified, as far as these criteria are similar or different depending on the region, province and countries where educational activities are conducted to prepare the designers and artists of decorative and applied arts. The novelty of this research is the choice of art and design areas for training specialists and a comparative analysis of two schools design methodology: in Latvia and Russia. The results are the analysis tasks of the research problem, namely, how different the students creative process is depending on the region or not stand out at all, if the teachers' criteria in grading art projects of future specialists are different, whether the quality of the students' professional training in the field of artistic and design activities differs in Latvia and Russia on the example of two Universities: Riga Technical University and Southern Federal University.

1.Introduction

Today, in modern conditions, the education system is the subject to modernization and various kinds of changes, the main goal of which is to create, improve and change the educational environment filled with an atmosphere of personal growth, but also does not exclude the development of spirit and humanity in future professionals. The creative atmosphere forms a humanistic project thinking, which is able to direct its professional activities to create a comfortable human environment, both at the level of architecture, interior, subject environment, and humanity itself as a whole. The task of the educational process is to create a system of specialized professional education that takes into account the humanistic ideals at the cross-cultural level, as well as at the level of social society, personal choice of everyone and at the level of consideration of human dignity. Within the framework of the institutional approach, education is «nothing but interaction of social groups, communities, organized in a certain way for the achievement of personal enhancement, socialization, professional training» [1].

The system of art and design education is a synthesis of a number of activities that form the development of the future professional's personality, the fullness of his spirit, cultural and moral perception of various cultures and societies' values. Building a comfortable professional educational environment in the field of architecture and art requires a special approach from the teaching staff of the University in the educational programs development that take into account cross-cultural interaction, the dialogue of human activities, social issues that can be addressed in further communication in society, regardless of the region or country where the future specialist will work. Purposeful pedagogical work that contributes to the professional self-awareness formation occurs to a greater extent within the framework of professional education. [2]. The level of preparation and readiness for the creative activity depends on the educational process content and learning path of the future architect, artist and designer. Harmony and dialogue between the teacher and the student, professional orientation, clearly defined goals and objectives of each project, the result that the future specialist strives for are the factors that affect the effectiveness and future specialist level of training, especially in the field of architecture, art and design.

Today, when the process of improving the education system in all areas, especially art and design, the search for new forms of conducting creative disciplines, the problem of education in the field of architecture, art and design was interesting to a number of researchers: Ya. K. Chernikhov, N. A. Ladovsky, V. V. Baburov, M. ya. Ginzburg, I. V. Zholtovsky, L. p. Kholodova, M. S. Tupolev and others. Design pedagogy is revealed in the works of V. R. Aronov, A.V. Efimov, V. T. Shimko, V. L. Glazychev, M. M. Mikheev, D. L. Melodinsky, E. F. Runge, V. V. Senkovsky, and others. We are interested in modern approaches and problems of design education, principles and methods of teaching design disciplines that develop the future specialists thinking who have linked their professional activities with art and design. Teachers-practitioners also actively participate in the development of theories, techniques and methods of design management in the training process of future artists and designers. The basis for the study of the modern architectural and artistic practice problem also became the latest architectural objects of some trends: modernism of the new wave of the end of the XX century - A.Isozaki, F. D. Israel, B. Tsumi, P. Pran, And K. Zapata, R. Kulhas, D. Libeskind, K. L. Malamaki, E. Matenell, Asymptote, and A. Arribas. Rationalistic and irrationalist trends, neo-expressionism - Blrintz (Prince), P. Floppo (Porro) and Renaud de La Hoye (Renaud de la Noue), S. Calatrava (Calatrava), J. Hubbell (Hubbell), M. Muermig (Muermig), R. Pitela (Pictila). Also, the activities of such schools as the BAUHAUS, VKHUTEMAS, and VKHUTEIN have made a great contribution to the main aspects' formation of world-class design and art education. The education is organized as a process of developing creative abilities and creative potential of citizens, as the creative individuals' education who have the ability to create and implement their individual professional paths [3]. In modern conditions of the social order of training creative activity specialists, the upbringing of a subject, capable of self-development and independence even at the stage of training is mobilized.

Today, the specialists in the field of design and art are trained in many specialized institutes and universities around the world, including Latvia and Russia. The lack of systematic analysis related to the study of modern trends in art and design creativity and the education system in these countries give reason to assert the relevance of the study.

The experimental basis for this study was the teaching project activities methods to students of the Riga Technological University (Riga) in the field of study «Design» and future artists of the Southern Federal University (Rostov-on-Don), in the field of study «Decorative and applied arts». The purpose of this study is to compare the methods and stages of conducting the discipline «Design», evaluation criteria and students results on the execution of course projects.

2. Methods and techniques

The empirical aspect of this research required conducting a survey, observation, and photo-fixation to identify the specifics of the art education organization at the international level and identify the unity of its components.

The hypothesis of this research is that the methods of conducting the discipline «Design» work equally effectively regardless of the region, country and language of this discipline teaching, as well as the fields of training in art.

According to Adolf V. A. and Savchuk A. N., professional training is the process of students' mastering fundamental knowledge, skills and abilities that ensure their readiness to effectively perform certain production functions [4]. Especially, this can be attributed to the fundamental discipline «Design», during which the student receives and develops his creative abilities, forming hisown author's method and way of working in the field of design. To model the professional activity of future specialists in the context of their future profession, the students form the qualities and skills they need in the future, as well as fill the learning process with a personal-oriented meaning. According to A. A. Veritsky, «context» reflects the system of internal and external conditions of behavior and activity in a particular situation [5]. Currently, the main aspects of design training include complex thinking, interdisciplinarity, personal-oriented orientation in work, intuition and self-development. The humanization of learning, the formation of young people value orientations based on familiarization with domestic artistic traditions together with the priority direction of the modern education system are an integral part of world culture [6]. Another positive factor in the high-quality educational environment formation in higher art and design universities is the authority of a successful «teacher-mentor», a teacher-practitioner who inspires confidence in the student, through active design and exhibition activities.

Design is effective under the following conditions: design thinking is formed correctly and in a professional direction, and the specialist has the following qualities: - the ability to set the task and the result of the project; - to give a number of versatile creative original ideas and solutions; - the ability to make a selection in the process of solving such tasks as «Planning is important» [7]. Planning the learning process for a specific project, as well as for the entire discipline, is one of the important principles of effective learning. Three main methods of teaching in higher education improve the learning process, especially in the field of art and design: - visualization; - wisely managed classroom technology; - Active learning: peer instruction, discussion groups, and collaborative problem solving [8]. All of them are actively used by teachers, both in Riga and Rostov-on-Don. The current challenges of creating and modernizing educational technologies require improving the quality of motivation and interest in training for future specialists, but this situation involves rethinking established and traditional teaching models, especially in the high school system. In modern conditions of the social order of training specialists in creative activity, it is mobilized the education of a subject capable of self-development and independence even at the stage of training. The priority direction of the modern education system is the humanization of learning, the formation of value orientations of young people based on familiarization with domestic artistic traditions, which are an integral part of world culture [6]. As well as in the process of professional maturation, students begin to understand why they need to learn and what they need to learn [9, p.43]. In addition, as adults manage other aspects of their lives, they are able to guide or at least assist in planning and implementing their own learning [10]

During the research, the authors found that the stages and methods of teaching students in the field of «Design» in Riga and in the field of «Decorative and applied arts» in Rostov-on-Don are identical and consist of several stages:

- issue and define tasks for the upcoming project;
- analytical part;
- generating ideas;
- project implementation and execution.

During the creative process, there is a search for harmony between the external form of expression and the content of the underlying idea. For the effectiveness of such creative search in design, it is necessary to develop the artistic thinking of a designer and artist. The application of practical methods makes it possible to activate the creative initiative of future specialists in the field of design and decorative arts in the learning process, learn to use individual creative abilities and talents in the profession, develop logical thinking in a professional field. Thus, the results of the project tasks for the students of the applied arts, the aim is to craft an applied art work, for example, a panel for public interior, takingthe purpose of this public interior, scale spaces, colour content and technical execution in the material into consideration.

At the first analytical stage, the student needs to conduct a research, study analogues in art and architecture, analyze these cultural heritage sources and objects, determine how the task will be solved, in a realistic manner or associative-figurative, collect the samples for an associative and visual series that corresponds to the chosen theme of the panel.

Figure 1. Podgornova E. Visual series for the project panel in mixed technique «Silent song»

During the research work on the project, the student presents a portfolio of selected analogues, compares and justifies why this choice was made, and determines the methodology for further activities in achieving goals.

For comparison, the project «Secret dreams of an architect» in the discipline «Artistic composition and forms» of the Riga technical University students can be demonstrated, where the task is to figuratively solve the topic through compositional exercises on rhythm, contrast and analogues, the main forms for transformation: circle, square, triangle.

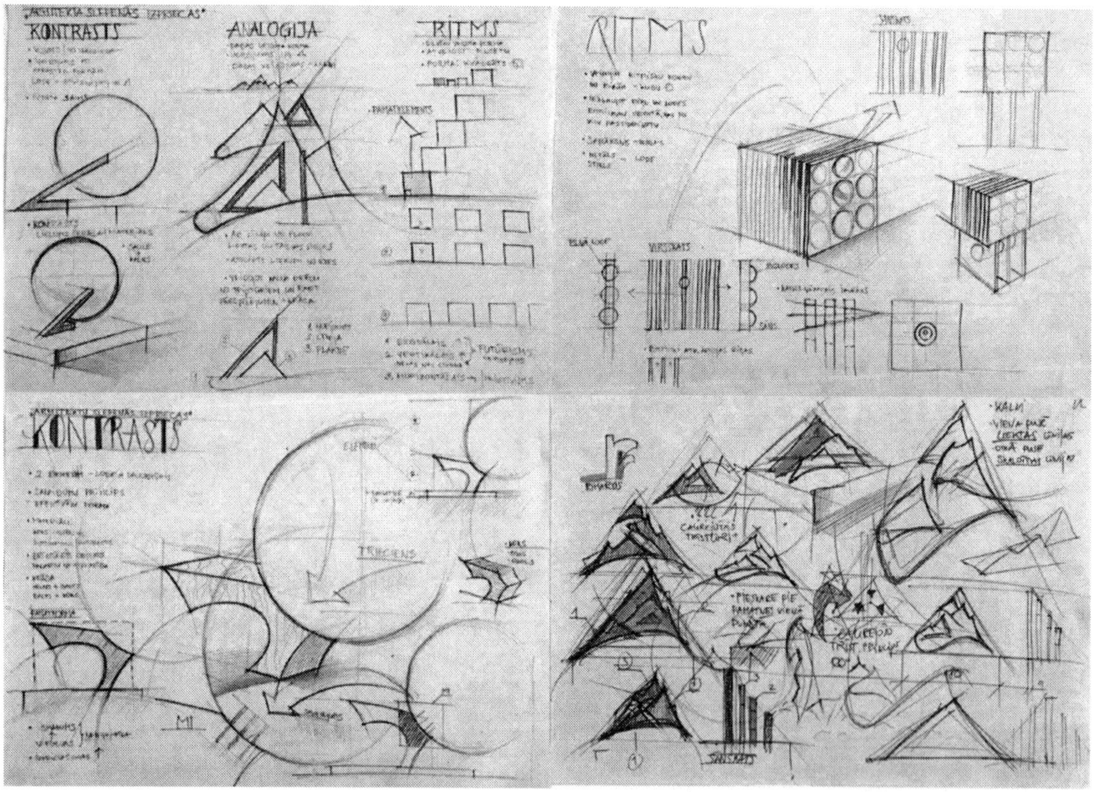

Figure 2. Anna Katrina Kupse and Richard Runkovskis. Search thumbnails: circle, square, triangle.

At the second stage, the student is faced with the task of generating ideas, issuing a number of clausures (diagrams, sketches) for a short period of time on the selected topic of the panel.

Associative-figurative project management is one of the main forms of visualization of ideas for the applied artist. This method makes it possible to most effectively transfer the creative imagination of the student-artist, makes it possible to develop the thinking apparatus of a creative person ready for professional activity. This way of thinking makes it possible to visualize abstract, graphic, psychological, emotional, and subject associations at a high level, transmitting the chosen topic in the project. Associative and imaginative representation of subjects, objects and phenomena of the environment, often transmitted from memory – this is the basis of artist, designer and architect professionalism, The sources for creative ideas and images in projects are events, objects, phenomena, nature, songs, and many other things that make up the human world.

Figure 3. Podgornova E. Sketches of the panel series «Silent song»

The second phase of generating ideas involves a number of stages: sketching, selection of main paths of the project, color the search implementation plan, working with scale and linkage to interior design, working drawings, and one of the most important steps of this period of work on the project, it is possible to highlight the selection technique and execution of samples in the material. At the end of the design phase, students should submit a portfolio of graphic and colour sketches, series of drawings with reference of the object to the interior (exterior) as well as a series of samples in the material (ceramics, textiles, wood or material synthesis) the rationale for choice of technique.

Students in Riga perform layouts, demonstrating an understanding of space, transformation in the environment, scale and ability to work in the material, namely, with paper and plastic in the process of layout.

Figure 4. Anna Katrina Kupse and Richard Runkovskis.

Layouts in the material on the topic «Secret dreams of an architect»

Figure 5. Podgornova E. Testers to a series of panels «Silent song»

At the end of this stage of work on the project, the student together with the teacher makes the final choice of the main line of design, execution technique, the final version of the scale, and the concept. At the final stage of design management, the future specialist is faced with the task of completing the project, both in the graphic version and in the material, meeting the established deadlines, achieving all the initially set goals.

The finished project for students of the field of study «Decorative and applied art» includes: A0 tablet, a portfolio of search material and sketches, a series of samples in the material, a product or its fragment, depending on the scale of the design object.

Figure 6. Podgornova E. Project of the panel series «Silent song».

A compulsory part of the tablet is the presence of sketches, visuals, floor plan for which is designed the decorative-applied art product, the walls covering plans of this room with the product, summary of the project and, of course, a photo of the product, made in the material, in this case, it's a mixed technique: hand weaving and felting.

To sum up the design and evaluation of the work done are imposed certain requirements and criteria:

- quality of «Graphics demos» (tablet project), their fullness of content, correspondence to the topic, the urban planning quality, functional planning, structural, compositional and artistic architectural solutions, quality graphic for project materials submission;
- the decorative and applied art products quality made in the material (textiles, ceramics, wood, etc.);
- the student's portfolio volume and quality as an artist;
- the project concept novelty and relevance;
- the public speech quality, the student's project validity and defense, his ability to justify the choice of his own project solution and execution technique in a reasoned, professional manner.

For effective work, the students should be able to adequately evaluate their own efforts and work. The question of self-esteem is an important characteristic of a professional specialist. «For formative assessment to be productive, students should be trained in self-assessment so that they can understand the main goals of their learning and thus understand what they need to do to achieve this goal» [11]. However, there is a possibility that self-assessment and peer assessment can become a superficial process, since much depends on whether students understand the design goal and objectives correctly and what is required of them. To avoid such casual situations in the course of conducting classes in the discipline «Design», an individually-oriented dialogue is conducted, both with each author of the project, and a collegial one, in order to identify the positive and insufficiently professional approaches during the set tasks of the project. Also, one of the participants in productive work can be called a competitive environment around the student, which is created by students themselves in a group, participating in the discussion of each other's projects. According to Gravells [12 p. 119], peer assessment includes a student's assessment of another student's progress. Self-assessment involves the student's assessment of their own progress. Both methods encourage students to make decisions about what has been learned so far and reflect on aspects for further development. With this construction of the design results evaluating process, students can better understand the evaluation criteria, and be more fair and objective in their further professional activities. Peer assessment is useful and effective for further self-development and motivation in achieving higher goals as artists and designers. Group discussions are one of the most effective measures that help achieve positive results both for the student in the learning process and for the teacher in his/her teaching activities.

3. Summary

According to the results of the comparative analysis of methods of training and assessment criteria for the discipline «Design» the students of the specialty «Design» of Riga technical University (Riga) and the direction of «Decorative arts», southern Federal University (Rostov-on-Don) to conclude that equally effective these methods «work» and produce results regardless of region, country and language teaching this discipline, as well as areas of training in the field of art.

Projects based on feelings and emotions, thoughts and impressions, abstractions or realities of the surrounding world, or quite the opposite, on a conceptual approach, generally accepted or subjective concepts, regardless of the geographical location of the place where they are created and developed, are based on similar principles of the designer or artist thinking. If the teacher and the student mutually set themselves an increased task at each stage of the design process: choosing a theme, sketching, performing, then each such project is successful, relevant and filled with novelty. The aesthetics of the «correct» educational artistic and design environment creates a professional education world, where the emphasis is the construction of humanistic values based on a personal-oriented approach to the educational environment formation, an interdisciplinary dialogue that develops the student creative independence. Today the focus on personality, on identity are of particular importance in the methodological support of educational process. They become dominant in the formation of learning paths, the focus of activity and ways of personality development of the future professional architectural and artistic sphere.

Creative professions are quite diverse, related to various types of human activity, in the course of work should be solved a variety of environmental, social, cultural tasks related to the human life organization and society as a whole. It is impossible to create a high level professional environment of modern architecture, art or design without «laying» a number of cultural-value and aesthetic-moral principles in the education system of such specialists. Moral, humanistic, and aesthetic values combine artistic and humanitarian training into an inseparable, complementary system of artistic and creative education. [1].

The educational process humanization and the formation of quality and professional conditions is more based on the relationship and mutual responsibility in the «same boat» of teacher-student. The readiness of participants in the educational environment for interaction, mutual influence and mutual enrichment determines the level of humanistic content of educational activities in higher education. This principle defines the professional level of personality, ready to solutions of various kinds of circumstances in the process of the profession implementation, generating their paths, as well as self-development and self-improvement that makes it possible to perform work adequately and efficiently regardless of region, country, language, media, carriers of beauty in the objective world.

The main thing, we can conclude, is that cultural and art history categories serve as a reference point and «tags» for interpreting and understanding modern art currents and trends, reading symbolically complex images and the associations that fill the fine art world. Thus, the study of architectural traditional forms and canons, art and design play a huge role in the future specialists' aesthetic education in the creative sphere. The orientation of the training path to the relationship in society, the tolerant consideration of all aspects of artistic and aesthetic principles of the society development, culture and art activates the spiritual and moral and creative and aesthetic potential of the specialist.

In conclusion we can say that, interdisciplinary approach to building a learning path is quite dominant and it makes possible to count on success in the educational process in the architecture and art education field today. This principle is manifested in the relationship of theoretical and humanitarian disciplines with practical specialized courses. Especially important in these conditions is the designed life and professional attitudes of the pedagogical process subject, the lifestyle organization, improving project-creative thinking behavior in work situations. The creative personality of the future specialist has the ability to think globally, creatively, involving ordinary people in the process of creativity, opening them new moral and aesthetic aspects of the society formation and development.

References

[1] Poidina T V 2014 Formation of students' professional competence culture in the design

education system *Cultural heritage of Siberia. Barnaul* **15** 126-132.1.

[2] Anfimova E B 2012 Experimental study of the formation of a professional and creative position of the future architect and designer in Higher Education *TSPU Bulletin* **5 (120)** 32-36.

[3] Nefedov V A 2011 How to enter Russian architecture in the international format *Accreditation in education* **6 (50)**. JSC: Accreditation in education, 2011. Information on http://www.akvobr.ru/rossiiskoe_zodchestvo_v_mezhdunarodnom_formate.html, free. - Blank from the screen. [accessed 29.10.20]

[4] Adolf V A, Savchuk A N 2014 Forecasting the formation of professional competence of a University graduate: monograph (Krasnoyarsky gosudarstvenny, PED UN-t named after V. p. Astafiev, Krasnoyarsk).

[5] Verbitsky A A 1991 Active learning in higher school: contextual approach [Text]: Method. manual (Higher school, M.).

[6] Rashidov O K, Damadanova S R 2009 Traditions and modernity in Dagestan decorative and applied arts *International journal of experimental education* **6** 53.

[7] Information on https://thevirtualinstructor.com/blog/the-art-of-teaching-art [accessed 29.10.20]

[8] Information on https://education.cu-portland.edu/blog/classroom-resources/3-innovative-methods-of-teaching-for-high-school-educators/ [accessed 29.10.20]

[9] Knowles M 1980 The Modern Practice of Adult Education (Cambridge Adult Education, Cambridge).

[10] Information on http://teachinglearningresources.pbworks.com/w/page/30310516/Andragogy-Adult%20Learning%20Theory [дата обращения 29.10.20]

[11] Black P J & Wiliam D 1998 Inside the Black Box: Raising standards through classroom assessment (King's College, London).

[12] Gravells A 2012 Preparing to Teach in the Lifelong Learning Sector: The New Award (Fifth edition. Learning Matters, London) pp.116-124.

Features of the Kuban life environment historical formation

O S Subbotin

Kuban State Agrarian University named after I.T. Trubilin, 13, Kalinin st., Krasnodar, 350044, Russia

E-mail: subbos@yandex.ru

Abstract. The specificity of the Kuban settlements' historical formation is considered in the context of regional peculiarities, as well as the folk traditions and customs of the peoples inhabiting the indicated territory. The main socio-economic prerequisites for the living environment formation in the settlements are indicated. Much attention is paid to the historical settlements of the Kuban, the objects of cultural heritage, their preservation and restoration. The specific planning structure of the settlements and regional identity in the constituent territories' diversity of these settlements are noted. Attention is focused on the settlement structure, on its formation over a long historical period of architectural and urban development. The following features have been proven: historical continuity, the transfer of cultural values, folk traditions from generation to generation.

Introduction

Kuban is the most ancient fertile region of Russia, where cultures of various nationalities have interacted for several centuries, while maintaining their artistic uniqueness. The excellent natural conditions of the three climatic zones of the Kuban: sea, steppe and mountain, left a significant imprint on the living environment of various settlements, which largely determined the way of life of the population living in this region. Especially significant is the fact that it was possible to maintain unity in diversity: since a real opportunity was established for each multinational people to preserve their traditions and historical and cultural heritage, which supported their sustainable development, promoted a peaceful neighborly existence and strong strategic cooperation, despite the existing multi-religious composition of the Russian Empire population.

It should be emphasized that the basis of traditions is the historical and cultural heritage, which includes the monuments of architecture and urban planning [1]. Particularly noteworthy is the traditional living environment of the Kuban - material culture, the basis of which is the architectural heritage, as well as the way of life of the previous generation. Therefore, the study is focused on identifying the patterns of formation and development of the specified environment, in order to solve the geostrategic tasks of the region under study.

Historical settlements, which have a unique historical and architectural value, play a special role in national architecture. The specificity of these settlements proceeded not only from the geographical location, but also from the social, historical and cultural heritage. These are the unique excavations of the ancient Greek policies' antique era in the Black Sea coast - Gorgippia (Anapa), the Meotskoe settlement on the territory of the future Yekaterinodar (Krasnodar), the settlement of the Tmutarakan Russian principality (modern - Taman), exhibits of the Sarmatian culture of the Yeisk peninsula (Yeisk), the heyday of the early Sochi culture period, a settlement of the Circassians near the river Urup (Armavir).

Purpose, objectives, research methods

The study is aimed at an objective assessment of the conditions for the formation of settlements in the Kuban, in the context of the socio-cultural environment of life, in order to determine the regional resource potentials, including historical and cultural, based on their specific characteristics and the addition of the settlement system heritage for the strategic territorial and spatial development of populated areas.

Accordingly, the goal sets the following research objectives are:

- to analyze the history of the Kuban settlement from the point of view of the historical and urban regional development;
- to identify the patterns of the economic development processes in the territories, their structural and functional elements;
- to determine the influence of climatic conditions and geographical location on the formation of settlements, including the entire settlement;
- to analyze the urban planning situations of the Kuban settlements on the basis of cartographic plans characterizing the silhouette of the building in a certain period of the studied region development.

The object of the research is the historical planning structure of the settlements' development in the Kuban, its volumetric spatial solutions, in the context of the living environment formation.

The subject of the research is the main regularities of the historical and urban planning development and formation of the Kuban territory.

The methodological basis of the study is the historical-compositional, town-planning and architectural-artistic analysis of the planning structure of settlements, the determination of the living environment value characteristics from the point of view of temporal evolution.

Analysis of the settlements' historical development in the context of the living environment spatial evolution process

From the very beginning of the development and settlement of the Kuban territory, the Cossacks, united by the common goal of protecting the southern borders of the Russian state and at the same time acting in the vanguard of the settlements' formation for a long period, are included in all spheres of human life. The military class, based on the principles of ethnic and historical community, as well as the way of life, has always been the backbone of the state. The Cossacks were significantly different from other peoples of the Russian Empire; therefore, they had their own specific features of the living environment, namely the patriarchal approach aimed at improving life in the settlements, this was especially clearly manifested in the initial period of the region's development. The main role in the settlements was played by the historical buildings, namely the center, which at different times of the settlement's development has always been an attractive place and had its own unique architectural appearance.

The historical buildings' value is seen, on the one hand, in its diversity, in compositional merits, on the other, in addressing the culture of the past, in the opportunity to feel the living presence of this culture, to associate modern culture with it [2].

It is especially important to note that in the central part of the settlements, the historical environment is compositionally interconnected with the surrounding areas of the natural landscape, forms a regular system of building blocks, which are an integral urban ensemble, reflecting the architectural and stylistic unity of objects for various purposes, while organizing visual-spatial connections.

Analyzing the initial organization of the territories in a number of settlements of the Kuban, it should be especially noted that the leading role of the historical center, regulating the development of the areas adjacent to the center, was directly reflected in the living environment formation.

For the most part, the historical center, which includes a significant number of architectural and urban planning heritage monuments, as well as historical buildings and structures, is geographically included in the city center. Therefore, it serves as a planning basis for the city development. The architectural and artistic interconnections of the historical center are a vivid example of an integral system that forms the environment with active urban planning components [3].

A striking example of the Kuban settlements' initial development can be the formation of the appearance in the central part of Armavir at the intersection of the main arteries of the settlement - Nikolaevsky avenue and Pochtovaya str. (Figure 1).

Figure 1. Armavir. Nikolaevsky avenue, 1911.

Here in 1903 in memory of the last Russian emperor coronation, local artisans erected a chapel in the name of Saint Nicholas, which became a visible accent of the village [4]. Not being a city, Armavir nevertheless followed such a broad path of social economy and improvement development, which is difficult to achieve for many ancient cities [5].

Thus, an example of the use of urban planning techniques to create a favorable living environment is the compositional solutions of the Kuban historical settlements' centers. In most cases, the well-thought-out development of ensembles, as well as the individual architectural structures in the central part, harmoniously combined with the natural environment. The architectural and planning organization of the historical environment carefully preserves the traditions due to which this or that settlement was born. The qualitative level of the historical environment, its aesthetic and artistic components, largely predetermined the corresponding level of the living environment.

The specifics of the historically established settlement and continuity in the Kuban architecture

Urban and large-village forms of settlement are noted, a kind of agglomeration, which significantly affect the settlement of nearby territories. They become attractive for living, have social and engineering infrastructure, investment and cultural potential. As a result of these actions, the ratio of urban and rural population is gradually changing. At the same time, the specific structure and peculiar silhouette of the Kuban rural settlements also have valuable architectural and artistic buildings and an organic planning structure and monuments of architectural heritage characteristic in this territory. Great importance in the formation of an appropriate environment for the life of rural settlements belongs to the specifics of the architecture development in these settlements. For example, for the coastal settlements, such a specificity is made up of dachas, villas, sanatorium-resort facilities, for steppe settlements - manor houses, traditional dwellings with the peculiarities of the Kuban way of life, for the mountain settlements - a variety of architectural objects, successfully inscribed in the complex mountainous terrain.

At the same time, "water spaces: seas, rivers, lakes are an important factor that has a positive effect not only on the emotional state and energy of a person, but also on the architectural and artistic expressiveness of the urban and rural settlements territories, in the context of creating the architectural and artistic ensembles subordinated to a single spatial solution and placed taking into account certain points of viewer's perception [6].

At the same time, the main and most important supporting town-planning nodes and dominants in the planning structure of urban and rural settlements were the religious buildings (Figure 2).

Figure 2. Ekaterinodar. Alexander Nevsky Army Cathedral, 1853-1872 I.D. and E.D. Chernik brothers-architects.

So, the Alexander Nevsky ArmyCathedral in Yekaterinodar compositionally had very advantageous viewpoints, as well as a dominant location over the nearby architectural environment.

The domes of the cathedral, covered with solid Demidov's iron, made by the famous Demidov's factories, were painted with green paint, and the crosses were gilded. In the western domes there were bells, the largest of them weighed 506 pounds, the ringing of which in a fine summer hour could be heard from tens of miles away, and the cathedral itself, sparkling with crosses, was visible both from the northern steppe expanse and from the distant southern Trans-Kuban plain [7].

Simultaneously with the resettlement process, the measures to improve the territories and transform the already established territories were taken. The proper condition of the streets started to be monitored.

In particular, the improvement of Yekaterinodar started with paving streets and digging ditches to drain the central part of the city. The streets were paved with bricks - iron ore, which turned out to be a more practical material than stone [8].

In the future, in order to optimize the settlement system of the Kuban, the centralization of nearby settlements was carried out, between which there are socio-economic, cultural, economic and ethnic ties in the largest settlement. An example would be Krasnodar, Armavir, Kropotkin, Labinsk and other cities. If initially neighboring settlements were at an insignificant distance from each other, then for the purpose of sustainable development, they were combined into a single urban-planning formation, which forms the living environment, while creating a historically formed image of an urban settlement. At the same time, it should be noted that this integration is observed not only at the level of urban settlement systems, but also at the level of rural settlements and contributes to the organization of a single supporting frame of the settlements.

At the same time, the natural landscape, the spatial organization of territories, architectural objects, the historical core and monuments of historical and cultural heritage were the most important elements of the urban environment. Each of these elements performed its own function, which together with others formed the planning structure of the city [9].

It has been proven that the continuity in the architecture of the Kuban traditional dwelling was clearly reflected in the formation of an appropriate environment for life, based on the material and spiritual needs of a person and, above all, on the Orthodox faith. Household and cultural traditions and customs of the Cossacks were sacredly observed. Along with this, the existing methods and methods of building housing buildings were directly related to the level of economic well-being, as well as the proper quality of the living environment. Most of the Kuban dwellings were mainly covered with thatch or reed, later with iron or tiles, special importance was attached to local building materials.

In conditions when the construction of stone buildings was beyond the means of most developers, with the help of "exemplary" projects, the possibility of using the local building materials to replace scarce bricks was clearly shown [10]. Particularly indicative in this respect are the rural settlements of the Kuban, which have now retained their original, distinctive appearance. An accent role in the development of which belongs to the objects of unique folk architecture, made by outstanding folk masters of that time.

Summary

In the course of the study, it was proved that an example of the urban planning techniques' use to create a favorable living environment is the compositional solutions of the Kuban historical settlements centers. In most cases, the well-thought-out development of ensembles, as well as the individual architectural structures in the central part, harmoniously combined with the natural environment. The architectural and planning organization of the historical environment carefully preserves the traditions due to which this or that settlement was born. The qualitative level of the historical environment, its aesthetic and artistic components, largely predetermined the corresponding living environment level.

At the same time, among the many problems of regional urban planning, the leading place is taken by the problem of preserving the historical appearance of settlements, namely, to prevent its distortion due to ill-considered new development. Monuments of architectural heritage, unique folk architecture, as well as their complexes, are harmoniously combined with building blocks, following the principle of building settlements according to the regular system of the 18th – 19th centuries. At the same time, they create a wonderful composition and a unique original image of the Kuban populated areas, which has been formed over two centuries. The indicated development forms a clear, historical spatial organization of the settlement.

This is clearly confirmed by a retrospective analysis, a comparison of the planning structure in different historical periods, since in each subsequent era the plan was transformed, but the historical center, its expressive silhouette, remained unchanged. Therefore, an important strategic task in the field of urban planning policy in the region is not only creating the conditions for sustainable settlements' development, but also maintaining the continuity of the planning structure, adopting the experience of the previous generation in order to create a favorable living environment. The historical environment of the Kuban settlements should be carefully preserved, and at the same time be viable and comfortable.

Therefore, when developing the projects for the urban and rural settlements quarters' regeneration, it is necessary to strictly know and take into account that the architectural and urban planning heritage, which carries valuable information about the historical formation of the Kuban living environment, the historical planning structure of settlements, traditions, artistic tastes of previous generations, is a unique material basis preserving the culture of the multinational people inhabiting this region, and the memory of generations as well as the symbolic "spirit of the place" is the guarantor of the settlements' historical and cultural identity preservation.

Refrences

[1] Subbotin O S 2019 Features of the building materials use in architectural and urban heritage restoration *IOP Conference Series: Materials Science and Engineering* **698(3)** 033045.

[2] Schenkov A S 2011 Reconstruction of historical buildings in Europe in the second half of the XX century: Historical and cultural problems (LENAND, M.).
[3] Subbotin O S 2018 Problems of reconstruction of historical center of the city *Materials Science Forum MSF* **931** 745-749.
[4] Ktitorov S N 2002 The history of Armavir (pre-Soviet period: 1839-1918) (Armavir).
[5] Scherbina F A 2010 History of Armavir and the Circassians (Krasnodar: Tradition).
[6] Subbotin O S 2018 Architectural and planning principles of the organization of coastal areas *Materials Science Forum MSF* **931** 750-753.
[7] Bardadim V P 2009 The architecture of Yekaterinodar (Krasnodar: Lebedev Yu.Yu).
[8] Apostolov L Ya 2010 Geographical outline of the Kuban region (Krasnodar: Tradition).
[9] Subbotin O S 2020 Cultural and historical potential of the urban environment (regional aspect). *IOP Conference Series: Materials Science and Engineering* **775 (1)** 012036.
[10] Kriger L V, Chesnokov G A 2009 The architecture of the historical cities of the Voronezh region (Center for Spiritual Revival of the Black Earth Region, Voronezh).

Updating the architectural heritage of the Krasnodar Territory, in the context of its preservation and restoration

O S Subbotin

Kuban State Agrarian University named after I.T. Trubilin, 13, Kalinin st., Krasnodar, 350044, Russia

E-mail: subbos@yandex.ru

Abstract. The priority for the preservation and restoration of objects of architectural heritage, which are an important historical and cultural potential of the Kuban is indicated. It was noted that this potential has actual effectiveness and allows it to be realized in the strategic development of the region. The expediency of applying the preservation mechanism for the objects of architectural heritage for the investors and non-budgetary organizations wishing to rent premises in these objects is recommended. An assessment of the economic significance of this mechanism implementation is given. The necessary measures are identified, in the context of "smart space" to carry out appropriate work to create the historical environment comfort, in terms of the innovative technologies' application.

Introduction

Krasnodar Territory possesses a unique architectural heritage, which is by right an incomparable and priceless heritage of not only the multinational people living in the territory of this region, but the Russian Federation as a whole. The architectural heritage has been created over a long historical period, during which its formation as a cultural and historical potential took place.

This potential is 'the most important resource for the architectural formation of the material-spatial environment of the city, which should be considered in the relationship between the value characteristics of old and new cultural samples, traditional and modern forms of architectural objects and structures. Therefore, primarily architectural, urban planning, construction, and other activities should be aimed at ensuring the preservation, development and use of this potential [1]'.

Separately, the high importance of monuments of the medieval period with the influence of Byzantine culture should be noted, as well as the combination of various styles in the architectural heritage of the Kuban in the late 19th - early 20th centuries: neoclassicism, modernism, early constructivism, eclecticism. The peculiar original traditions of regional architecture also deserve special attention. At the same time, it quite 'can be stated that they were based on the humanistic principle [2]'.

Studying the history of any region and even a small settlement on the basis of existing monuments of architectural heritage spiritually enriches and significantly raises the cultural level of the modern generation and is of great importance in the spiritual and moral formation of the future. The preservation of this heritage is impossible without due respect, love and care for it.

It should be emphasized that over a long period of society development, the objects of historical and cultural heritage, in particular, some of the types of this heritage – the monuments and ensembles are the most vulnerable and poorly protected in comparison with the objects from other spheres, and the measures taken to maintain them did not always lead to the desired result, and sometimes they were wrong and further aggravated their situation. Failure to take prompt measures to preserve individual objects of architectural heritage led to the fact that we lost them irretrievably or they gradually deteriorated over time and were no longer suitable for use.

A specific manifestation of the heritage ideological significance in all periods was a special attitude towards "living" monuments (used for their original purpose) [3]. As a result, it is necessary, first of all, to carefully preserve and strengthen the appropriate measures for updating the architectural heritage, improving the operating conditions and ensuring its accessibility (Figure 1).

Figure 1. Krasnodar. Architectural monument - Yekaterinodar office of the State Bank. Architect I.K. Malberg (1902, reconstruction in 1954). Modern state. An extensive range of restoration activities have been completed.

It is also necessary to create and further improve the comfortable conditions for the implementation of the listed activities, commensurate with this strengthening of the material and technical base.

Relevance, purpose, objectives, research methods

The relevance of the research topic is dictated by the insufficient degree of study of preserving and restoring the architectural heritage problems, in the context of the proposed constructive measures. Some of the measures currently being taken are not always effective, and sometimes even erroneous. Attention is focused on the fact that this area is the most vulnerable and unprotected in comparison with other territories.

The aim of the study is a systematic approach to studying the problems of preserving and restoring the architectural heritage, as well as developing a conceptual proposal for attracting investments in this area.

The object of research is the monuments of architectural heritage in the Krasnodar Territory. The subject of the research is based on an integrated approach to the current state study of the monuments of architectural heritage.

Research methods are based on the systematization and analysis of the existing situation in the field of the specified heritage preservation.

The scientific significance lies in the recommendations' application of the research carried out in the process of implementing the architectural and urban planning activities by legal entities and individuals, as well as in developing a concept for the future development of the tourism sector.

The practical significance lies in the use of materials from scientific work in the preparation of a strategy for updating the preservation and restoration of the architectural heritage, which makes it possible to increase the efficiency of its use.

Conceptual proposals for regulation in the field of conservation of monuments of architectural heritage

When strengthening and restoring the monuments of architectural heritage, great importance is attached to the restoration of these objects. At the same time, attention should also be paid to the objects of historical and architectural value, in spite of the fact that the objects located in the historical environment do not have the status of a monument [4].

At the same time, the main problem that is most often encountered in the preservation and restoration of the lost parts or the entire object of architectural heritage is the lack of sufficient financial resources. In most cases, it is required to attract the additional investments from the individuals or legal entities in order to reimburse the necessary costs for carrying out these activities.

Therefore, to solve this problem, it is advisable to develop an appropriate mechanism to stimulate investors, to determine the directions and methods of their support, in particular the tax instruments. In this case, a proper agreement should be legally formalized, which specifies the rights and obligations of the parties, as well as their responsibility for their obligations' fulfillment.

Assessing the economic significance of the presented scheme of actions, namely, the expediency of using the funds from individuals and legal entities, makes it possible to speed up the process of preserving and restoring monuments of architectural heritage without attracting public funds.

A clear example for solving this problem in the Krasnodar Territory is the preferential rental program "1 ruble per 1 square meter", aimed at restoring the decaying architectural monuments, which has been implemented in Moscow since 2012. The key point is that, provided that tenants fulfill their security obligations, for the purpose of effective functioning and targeted use of these monuments, the object is preserved as an important strategic basis for the society development, being at the same time a solid foundation of our historical memory.

So, according to the calculation performed by the author of the study, this program promotes the restoration and repair work on architectural monuments of the federal and regional categories of historical and cultural significance, which are in an emergency condition in the amount of approximately 1.2 billion rubles, while the rent for the year will be make up to 360 thousand rubles.

It should be emphasized that in the approximate calculation (as an example) only 1/5 of the objects of the architectural heritage in the Krasnodar Territory from their total number, which presumably require restoration or appropriate repair, were taken. The total number of architectural and urban planning monuments located in the Krasnodar Territory with newly identified objects, according to the state register as of July 15, 2020, is over 1,300 units.

The economic significance of the results obtained, by attracting the funds from private investors and non-budgetary organizations in order to possibly adapt the individual monuments of architectural heritage that have specific utility, which is limited to a certain type of use, consists in obtaining the actual total economic effect from the action plan proposed by the author. In general, this contributes not only to the individual objects' preservation, but also the valuable historical environment of the Krasnodar Territory.

Consequently, granting investors the right to preferential lease of the monuments of architectural heritage, after carrying out repair and restoration work, has a decisive role in preserving the historical and cultural potential.

At the same time, one of the measures for the preservation of objects of historical and cultural heritage is 'the creation of an open information system on objects of protection and restrictions on possible transformations for the territory with historical buildings [5]'.

Criteria for the "smart space" of the settlement in order to create the proper comfort of the historical environment

Also, in most cases, when solving the problem of preserving and restoring the architectural heritage, we often face the problem of engineering networks deterioration, in particular, the life-support systems that do not meet the normal conditions of their operation, which undoubtedly affects the state of this heritage.

For example, in the case of the historical center of Krasnodar, we are talking about the utilities' deterioration - energy, heat and water supply facilities, sewerage, as well as the modernization of public transport. Also, one of the most problematic issues in the coastal cities: Sochi, Novorossiysk, Gelendzhik and the settlements of the Crimean region is unsatisfactory water supply, high water consumption in the summer season, all this negatively affects the architectural heritage, in the context of the tourist space in the Krasnodar Territory.

At the same time, the scale of unjustified individual "invasions" into the historical center created a great threat to the historical and cultural heritage, namely to the monuments of architecture and urban planning in terms of preserving the unique works of outstanding masters of their time [6]. Therefore, a comprehensive historical and cultural study of the city allows solving a number of issues related to the assessment, preservation and use of the architectural and urban planning heritage [7].

Along with this, the analysis of the engineering networks state of the past years has revealed that even the construction of new boiler houses to replace the worn-out ones gives a significant economic effect. The payback period for these structures is also reduced by attracting private investment in the modernization and further operation of heat supply systems. In some cases, it is necessary to use autonomous engineering equipment, while preserving the historical appearance of an architectural heritage monument.

An important role should be given to the investment targeted programs for the residents' resettlement from architectural heritage sites that are in disrepair.

At the same time, the process of preservation and restoration should take place without destroying the integrity of the historical building structure. It is necessary to develop a comprehensive model for the historical environment regeneration of the settlements with backbone pillars in the form of objects of architectural heritage using various elements of urban planning tools for further sustainable territorial development.

The preservation and restoration of the architectural heritage is inextricably linked with the preservation and regeneration of the historical background environmental buildings, as well as with the creation of public spaces, taking into account the cultural context, existing restrictions and existing urban planning regulations.

The formation of the so-called "smart space" of the settlement with innovative infrastructure, including transport, digitalization of restoration and repair of architectural heritage objects in conjunction with functional transformation and development within the boundaries of historical territories is of fundamental importance.

The identified criteria for the "smart space" of a settlement based on assessing the historical environment comfort quality, in order to preserve and restore the architectural heritage, include the following activities:

- the use of energy-saving technologies in the restoration or restoration of both whole objects and their individual parts;
- complex integration of all management and control systems for the energy and water resources consumption;
- the use of solar panels for converting solar energy into electrical energy, solar systems for converting solar rays into heat;
- use of devices for autonomous lighting, in particular innovative laser lighting;
- the use of innovative solar lighting technology, such as a world-famous brand Solatube®;
- introduction of green technologies, as well as the proper formation of green environmental infrastructure;

- introduction of current modern environmental safety systems for the territorial environmentally balanced development;
- development of a unified tourist map with the main routes for the objects of architectural heritage and unique attractions.

At the same time, the complexity of the internal organization of objects for various purposes is a general pattern not only of architectural monuments, but also of the usual buildings of past eras [8].

At the same time, the developed strategy of "smart space" in conjunction with the concept of preserving and maintaining in proper form monuments of architectural heritage, based on domestic foreign experience in the restoration of these objects, makes it possible to solve the complex problems of historical quarters' regeneration.

The specified concept, which is a structural and logical model, should reflect the sequential nature of the actions performed, while determining the primary goals and priorities when carrying out the relevant activities.

Summary

The architectural heritage is an integral part of the social, cultural and economic life of any settlement.

At the same time, these settlements are also a kind of scientific center - a "genetic" data bank on the formation and preservation of the historical architectural and spatial environment during certain development periods [9].

The importance of its preservation in the modern world and its indispensable maintenance in good condition is the guarantor of urban and rural areas' sustainable development, in the context of unique historical development. The inclusion of monuments of architectural heritage in conjunction with the surrounding natural landscape and ordinary historical buildings makes it possible to create a favorable and comfortable living environment.

Not only architectural objects that are the monuments of cultural heritage are significant, but also the historical town-forming buildings and structures that are of architectural and artistic value and are an element of the town-planning context that forms the spatial historical environment of the settlement and the compositional silhouette of the building. These buildings and structures, united by scale, structure and other elements, constitute an integral part of the historical and urban planning environment.

At the same time, the solution of the problems related to the preservation and restoration of the architectural heritage in some cases is possible only in the presence of public-private partnership, namely through the joint efforts of the state, private and legal entities. With a lack of appropriate budgetary funds necessary for the preservation and restoration of monuments of architectural heritage, the joint use of material resources of economic entities is the most urgent.

To a greater extent, this partnership is of great importance when it comes to socially significant projects, as well as the objects of particular historical value. Thus, fundamental in defining the actualization of the architectural heritage is the immutability of its preservation and restoration, effective use and integration into modern life. For the full-fledged work of this cooperation, an appropriate legislative base, which determines the procedure and nature of actions in this area, a special program must be developed to identify and eliminate problems associated with the preservation and restoration of the architectural heritage should be adopted.

Therefore, one of the most important tasks of our time is to preserve, restore and protect the historically important architectural objects and pass them on to the younger generation. 'At all times and all over the world the monuments of the past tell us about their time [10]. This is our memory, spiritual and material values, and we are obliged to preserve our unique heritage.

References

[1] Subbotin O S 2020 Cultural and historical potential of the urban environment (regional aspect) *IOP Conference Series: Materials Science and Engineering* **775(1)** 012036.
[2] Savarenskaya T F 1987 Western European urban planning XVII - XIX. Aesthetic and theoretical premises (Stroyizdat, Moscow).

[3] edited by Shchenkova A S 2002 Architectural Monuments in Pre-Revolutionary Russia: Essays on the History of Architectural Restoration (TERRA-Book Club, Moscow).
[4] Subbotin O S 2019 Features of the building materials use in architectural and urban heritage restoration *IOP Conference Series: Materials Science and Engineering* **698(3)** 033045.
[5] Kurbatov Yu I 2008 Petrograd - Leningrad - St. Petersburg: Architectural and urban planning lessons (Art-SPb, St. Petersburg).
[6] Subbotin O S 2020 Actual issues of the settlements' historical preservation center of the Kuban *IOP Conference Series: Materials Science and Engineering* **913(3)** 032020.
[7] Monuments of Russian architecture and monumental art: Style, attribution, dates 1983 (Science, M.).
[8] Mityagin S D 2016 Urban planning. An era of change (Architect, St. Petersburg).
[9] Subbotin O S 2020 Architectural and spatial environment in the historical settlements of regional significance *IOP Conference Series: Materials Science and Engineering* **913(3)** 032021.
[10] Novikov F A 2006 Architects and architecture (Editorial URSS, Moscow).

Three-layer composite panel with discrete layer and bionic principles

I A Mayatskaya*, B M Yazyev, V D Eremin

Don State Technical University, Socialisticheskaya str., 162, Rostov-on-Don, 344022, Russia

E-mail: irina.mayatskaya@mail.ru

Abstract. A very interesting direction in bionics, which uses the properties of natural, complex in design structures. The purpose of these studies is to increase strength, thermal insulation and sound insulation by changing the shape and thickness of the intermediate layer and using composite materials in a lamellar structure.

1. Introduction

Three-layer structures are widely used in aircraft and helicopter construction. The use of thin ones, for example, plates and shells, in conditions of bending deformation are the most rational from the point of view of strength and rigidity. The use of lamellar structures in construction and mechanical engineering is also relevant [1-5]. Lamellar structures, which are three-layer panels with discrete cores of low density, make these objects lighter [6-8].

When creating modern structures, the issues of increasing strength, thermal insulation and sound insulation by changing the shape and thickness of the intermediate layer and using composite materials in a layered structure are relevant. The paper considers a three-layer structure with two outer layers, which are unidirectional polymers with mutually perpendicular directions and an intermediate spongy core layer.

2. Bionics and Building Structures Using Lamellar Plates and Shells

Currently, when designing, the latest composite materials and intermediate layer models are preferably used. Lamellar plates and shells are widely used in aircraft and helicopter construction in the manufacture of wings, empennage and fuselage skin, one of the variants of which is a three-layer shell with a honeycomb core made of a composite material (Figure 1).

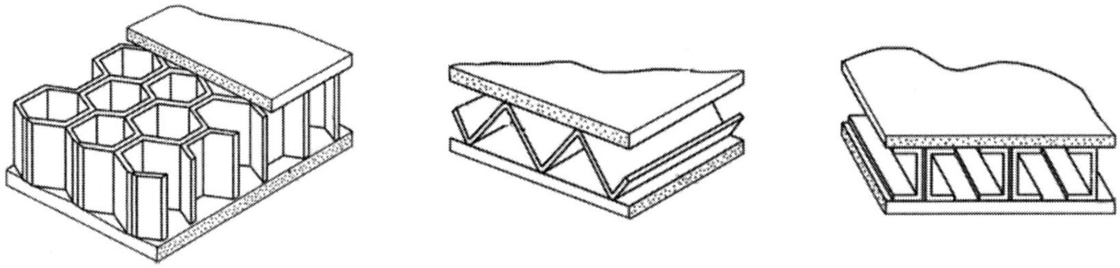

Figure 1. Discrete-filled three-layer plates

The use of thin three-layer skins with a lightweight filler with a low density or in the form of honeycombs makes these structures lighter. High strength and resistance to vibration and acoustic loads determine the growth in the use of lamellar structures as load-bearing elements of the fuselage. Three-layer plates with a ribbed intermediate layer (Fig. 1) are also used in construction as thermal insulation cladding of buildings and for sound insulation inside the structure.

The disadvantage of three-layer plates and cladding shells is the constancy of the intermediate layer thickness and the impossibility of using more complex geometric shapes and anisotropic polymeric materials for the outer and inner layers.

The closest technical solution to a bionic structure in the form of a three-layer structure with a discrete intermediate layer is a three-layer shell with a spongy structure of the intermediate layer. At present, much attention is paid to bionic principles in structures design [5, 9, 10]. Let us consider the bones of a human skull. It is a three-layer shell with different thickness, surface shape and material properties. The outer and inner thin layers of the shell element represent an anisotropic structure, and the intermediate one has a spongy structure of two types: porous and rod with variable thickness. It is this structural feature of the bones that determines its increased strength, heat and sound insulation of the inner part of the skull.

The paper investigates the problem of increasing the strength of a three-layer structure with a discrete intermediate layer and proposes to use a discrete intermediate layer in the form of cylindrical blocks of variable thickness. This discrete layer of a three-layer structure has a structure for which each block is a cylindrical shell or a rod with a circular cross-section of variable thickness (Figure 2).

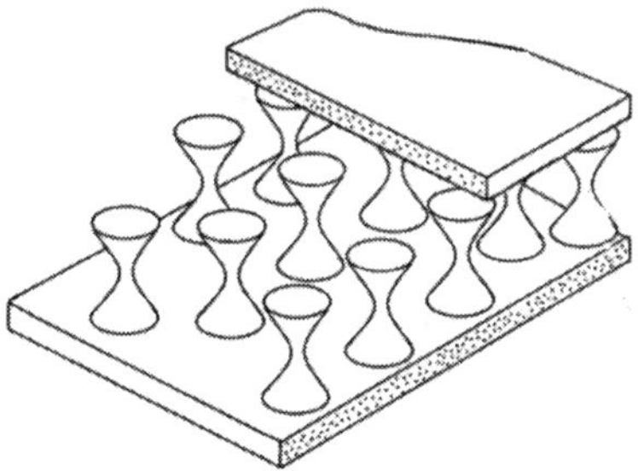

Figure 2. Three-layer plate with discrete aggregate in separate blocks

A three-layer structure with two outer layers, which are unidirectional polymers with mutually perpendicular directions, and an intermediate spongy core layer works as a structure with isotropic properties, which is more rational [5].

The specified technical result is achieved by the fact that the intermediate layer has a block structure. Each block is a cylindrical shell or bar with a circular cross-section of variable thickness with a curved surface. This surface can be in the form of a rotating hyperboloid, which has a continuous surface or the rectilinear agents of which are the intermediate layer's fibers. A catenoid can be used as a block. This is a body which surface is formed by the rotation of a chain line$y = ach(x/b)$. This surface belongs to the minimum surfaces. One of the examples is a soap film obtained by tension on two rings, the planes of which are perpendicular to the rotation axis. This block may resemble an hourglass shape, which is quite common in engineering structures, or two cones in contact, it is possible to see both a truncated cone and a hexagonal cross section. These blocks can have different sizes, and a similar shape, only its height should be constant for panels with constant thickness. If the thickness of the structure changes, then for a given structure of the intermediate layer, the change in height does not cause difficulties.

In spar structures of helicopters, the bending moment is perceived mainly by the longitudinal elements – spars, and the skin (lamellar panels) perceives local loads, shearing force and torque. And in a monoblock structure, the skin, together with the frame elements, also perceives normal forces from bending moments. The load on the panel is quite varied and it is desirable that the properties of the plate react less to the material properties in different directions.

As blocks in the intermediate layer, cylinders or thin-walled cylindrical shells with a constant radius, a rotating hyperboloid or a thin-walled shell of rotation formed by an algebraic curve, for example, of the second order, circular cones or thin-walled conical shells, as well as the bodies of variable thickness with a cross-section in the polygons form.

The device works as follows. The blocks included in the intermediate layer of a three-layer structure are loaded differently and the block system starts operating as a structure in which the influence of material properties in different directions in the outer and inner layers is reduced.

Such three-layer shells, as, for example, the structure of the human cranium or the structure of a turtle can be met in nature. The tortoise shell can withstand very significant loads and has a block structure (Figures 3, 4).

Figure 3. Turtle species

The carapace consists of shields in the form of curved polygons: quadrangles, pentagons, and hexagons with variable thickness and a slightly convex surface. This natural construction is very durable. But most often in the shell structure is the hexagonal shape of the blocks. This structure is very strong and reliable. Octagons can also be used as blocks.

Figure 4. Types of block structure of turtle shells

In the three-layer structures' design, a block structure with three-layer hexagonal blocks can be used, each layer of which has different mechanical properties. The outer layer should be strong, and its strength is increased due to variable thickness and convex shape, which is widely used in natural objects of rational shape. The intermediate layer should have shock absorption damping properties. The inner layer is made of an elastoplastic material with shape memory or a material with improved thermal insulation properties.

For the intermediate layer of a three-layer structure, the elements that are encountered when constructing semiregular polyhedrons should be considered, for example, a truncated cuboctahedron (Figures 5, 6).

Figure 5. Types of semiregular polyhedron

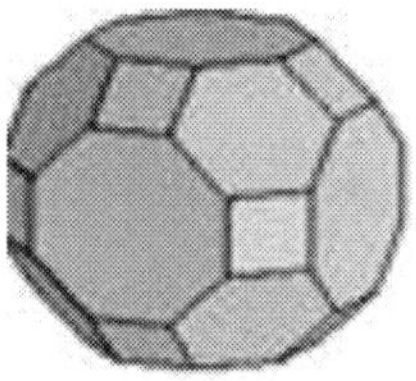

Figure 6. Truncated cuboctahedron view

Three-layer panels are used in the construction of residential buildings. This is a three-layer "sandwich" of two layers of reinforced concrete or metal sheet and a modern light intermediate layer between them, as well as a decorative layer that can take on any texture (Figure 7). During construction, the slabs are joined together with minimal seams at the joints.

Figure 7. Types of three-layer panels in building structures

The multi-layer structures using spherical elements or thin-walled spheres in the intermediate layer can be applied in construction, although this direction in the multilayer structures design is not sufficiently developed. Development of new types of three-layer structures with an intermediate discrete filler, the possibility of using them in structures with complex geometry is possible. There is an opportunity to construct a three-layer panel with a discrete filler in the form of rods of both rectilinear and curved shapes.

Currently, one of the important tasks is the development of software for performing such calculations and automation of the design process, creation and assembly of a three-layer structure with a discrete filler of complex shape. Such structures could be used much more widely than the structures with honeycomb and corrugated or lightweight cores.

3. Summary

The proposed three-layer structure with two outer layers, which are unidirectional polymers with mutually perpendicular directions, and an intermediate spongy core layer operates as a structure with isotropic properties. It is also possible to use a panel with an intermediate discrete layer in the form of separate blocks in the form of rotating hyperboloids or two touching frustums of a cone.

When designing thin-walled structures, a problem arises, which consists in determining the parameters of a rational and optimal structure for a given load. To solve the problems of optimal design, it is necessary not only to analyze the optimality conditions for thin-walled shell structures and an algorithm for determining the optimal parameters for different types of shells and loading schemes, but also to pay attention to the structural structure.

Determination of the best design solution is an extremely complex process, consisting of work to ensure the best operating conditions for loading, the choice of rational schemes, shapes of parts and effective materials that contribute to obtaining the minimum mass of the structure, taking into account manufacturability and cost. To find the correct design solution that ensures the optimal design, you can use the methods of bionic research.

Further research is needed to improve structures' design using the bionic principles. Using these principles, it is possible to create the objects of amazing shape using the layered shells and to design the structures with a rational form [5, 9, 10].

References

[1] Mayatskaya I A, Demchenko D B, Demchenko B M 2018 The theory calculation of plates (Publishing house Don State Technical University).
[2] Mayatskaya I A, Krasnobaev I A 2011 Fundamentals of calculation for the output of thin rigid plates (Publishing house Rostov State University of Civil Engineering).
[3] Mayatskaya I A, Krasnobaev I A, Smirnov I I, Yazyev B M 2012 Plate and shell theory (Publishing house Rostov State University of Civil Engineering).
[4] Mayatskaya I A, Demchenko D B 2020 The fundamentals of the theory and calculation of thin-walled open profile rods (Publishing house Don State Technical University).
[5] Mayatskaya I A, Eremin V D 2019 Bionics and the choice of rational structural form *E3S Web of Conferences* **110** 01042.
[6] Mayatskaya I A, Krasnobaev I A, Vitols O E 2017 On the issue of calculating three-layer structures with light filler *Scientific works of the National University of Architecture and Construction of Armenia* **3 (66)** 67-73.
[7] Mayatskaya I A, Krasnobaev I A 2016 Calculation of rounded three-layer plates under the influence of a uniformly distributed load *Scientific works of the National University of Architecture and Construction of Armenia* **2 (61)** 94-99.
[8] Mayatskaya I A, Krasnobaev I A 2016 Strength calculation of three-layer cylindrical under the action of axial force *Scientific works of the National University of Architecture and Construction of Armenia* **3 (62)** 35-40.
[9] Mayatskaya I A, Demchenko B M, Shvetsov P A 2016 On the possibility of improving the meshy plates and shells based on bionic principles *Regional architecture and construction* **2 (27)** 137-145.

[10] Mayatskaya I A 2016 On the possibility of improving building structures based on bionic principles *Bulletin of the Volgograd State University of Architecture and Civil Engineering. Series: construction and architecture* **45 (64)** 27-36.

Fractality and urban environment

I A Mayatskaya*, B M Yazyev, S B Yazyeva

Don State Technical University, Socialisticheskaya str., 162, Rostov-on-Don, 344022, Russia

E-mail: irina.mayatskaya@mail.ru

Abstract. The article considers a very interesting trend in architecture, which uses fractal geometry in the complex structures' design. Architectural bionics gives a possibility to find the most unusual in shape and structurally optimal building structures. If earlier the architects got their inspiration from natural objects, then modern designers also use mathematical methods, for example, fractal geometry.

1. Introduction

A city is a dynamic system that has a random and fractal character. The urban environment includes the most diverse spheres of human life, for example, social or architectural space. The work examines the architectural appearance of the city from the point of view of its fractal structure [1-3]. In the world, architects often construct buildings using the principles of fractal geometry (Figures 1, 2).

Figure 1. An example of a building with fractal elements, Barcelona, Spain.

Figure 2. View of modern urban space, London, UK

Modern architects and designers working in this direction use methods of mathematical modeling with the use of modern software systems. Using these computational tools, it is possible to create a complex of structures of an amazing shape that will organically fit into the previously established urban environment [4].

2. Architectural style in a modern urban environment

Back in the twentieth century, architects used methods of architectural shaping based on fractal geometry, surface theory and nonlinear dynamics. We can mention such architects as Le Corbusier, F.L. Wright, Zaha Hadid and P. Frankhauser.

Fractality is a property of an object that has a fractional dimension, the property of self-similarity, and has a complex structure with a recursive organization. Typically, architecture uses fractal algorithms for constructing the shape of a structure using self-similar elements with a limited number of repetitions [5-7]. But fractal geometry has other properties as well. These are dissymmetry and scaling. Dissymmetry is the unity of symmetry and asymmetry. Scaling is a changing likeness or likeness with sliding.

In the objects built by the great modern architect Zaha Hadid, it is possible to see an amazing combination of lines, fractal elements, it seems that the structure moves in space (Figure 3). It is this perception of unusual and unique structures that creates a unique urban style.

Figure 3. Unique structures by architect Zaha Hadid.

Le Corbusier introduced asymmetry into architecture, fractal elements with sliding, creating unusual facades of raw concrete. He created the architectural objects, combining industrial and natural materials (House of Culture, Firminy, France), often used support pillars, while the buildings looked as raised above the ground, built roof-terraces on which a recreation area could be arranged (Figure 4). Le Corbusier preferred a free layout of the interior space, created buildings with tape windows and a free facade, making the object unique in shape.

Figure 4. Unique structures by architect Le Corbusier.

One of the most famous houses of F.L. Wright is the house above the waterfall (Figure 5). This building fits organically into the environment. Frederick Roby's house is also very famous. The structure of the building is built on fractal elements in the form of rectangular volumes with varying fractality degrees. It combines both properties of fractality: dissymmetry and scaling. The laconic design of the building and the surrounding nature harmoniously interact, creating a unique architecture. The architect often uses fractal elements to organize the interior space. For the construction, not only appearance is important, but also the internal space. The internal space makes the environment comfortable for humans.

Figure 5. Unique structures by architect F.L. Wright.

In the building complex in Moscow City, the architects also used fractal principles of construction. This is a unique architectural complex that includes skyscrapers of amazing shape (Fig. 6): One Tower, Federation Tower, Eye, Neva Tower, Mercury, Eurasia, City of Capitals, Grand Tower, Capital Tower, Waterfront Tower, iCity, Evolution and Empire, which have a height of 404 m to 239 m.

The complex of two high-rise towers "City of Capitals" represents the objects of rectangular blocks, slightly rotated relative to each other. These blocks are the self-similar objects with a slight slip around the axis (Figure 7).

The structure of the Evolution tower, which is made in the form of a 55-storey skyscraper twisted in a spiral around its own axis, is unusual (Fig. 8). This is a very impressive building and is the decoration of the city. The spiral shape is repeated only by four corner supports, creating a unique openwork and lightness of the structure.

Figure 6. Unique complex "Moscow-City".

Figure 7. Tower complex"City of Capitals"

Figure 8. “Evolution” Tower.

The city is a fractal space consisting of the structures' buildings and complexes. When designing buildings in an urban space, it is necessary to take into account its fractal properties. For a harmonious and comfortable living environment in a city, it is necessary to study the architectural appearance development patterns [8, 9].

The city is a spatial multifractal of the stochastic type [5-7]. The use of repetitive self-similar shapes is widespread in structural design. Architects of our time are looking for the architectural forms of urban space, involve the methods of fractal geometry and mathematical modeling in their research.

To set a fractal structure means to set the principle of constructing form and structure, taking into account change and growth. Fractal objects are capable of creating all new, most diverse forms, which may have stochastic properties from a predetermined initial structure on the basis of some algorithm. These are the advantages of the fractal method. Fractals have two sides - they are dynamic systems capable of development and change, and they contribute to the creation of a whole from fragments, where the next element is based on the development of the previous one, but on a different level.

Fractal algorithms have found application in urban environments as well. Fractal geometry and the development of the urban environment is a revolution in changing the view of the city architectural form, in creating the environment for a person, taking into account his comfortable living.

Fractal architecture can be divided into two types: artificially created, which is intuitive and conscious, and naturally formed. The principles of fractal-like shaping are used in the construction of both individual objects with fragments and details, and in the space of the city as a whole. Architects of the past made extensive use of the fractal principle. Building elements were at times self-similar objects. The examples of intuitive fractal architecture are the buildings of A. Gaudí, ancient temples of India and the East (Figures 9-12).

Figure 9. The complex of Indian temples in Khajuraho, India.

Figure 10. Mosque of Omar ibn al-Aasa, Egypt

Figure 11. Umayyad Mosque, Syria. Exterior view

Figure 12. Umayyad Mosque, Syria. Interior view

Voluntary fractal architecture is represented in the architecture of Moscow, where it is possible to find modern skyscrapers and amazing historical buildings. The city is a fractal space consisting of buildings, complexes of structures, a transport network. This space includes both historical buildings and modern structures.

For the harmonious development of the urban environment, it is necessary to carry out the comprehensive studies, taking into account the fractality of the city space itself.

3. Summary

It is necessary to conduct further research related to fractal analysis. In fractal architecture, the following principles are applied: self-similarity; dynamism, ability to develop; irregularity; recursiveness; fractionality. It is necessary to look for the new forms and technologies, developing this direction. With the help of fractal principles, it is possible to create amazingly shaped objects of architecture, study the architectural composition of buildings, design structures with a rational organization of space and taking into account the ecological requirements of the environment [10].

The beauty of fractals revealed the patterns of harmonious development in the world, contributed to the creation of unique structures by modern architects. This was made possible by the talent of architects, their ability to creative solutions, as well as the methods of mathematical modeling and digital technologies.

References

[1] Mudrova A Yu 2014 *The great masterpieces of architecture. 100 buildings that are admired the world* (Publishing Tsentrpoligraf).
[2] Gnedich P P 2012 *World architecture* (Publishing house Eksmo-Press).
[3] Kouchner M 2016 *Future architecture. 100 of the most unusual buildings* (Publishing house ACT).
[4] Mayatskaya I A, Yazyev B M, Yazyeva S B, Kulinich P B 2017 Building Constructions: architecture and nature *MATEC Web of Conferences* **106** 01031.
[5] Yazyeva S B, Mayatskaya I A, Kashina I V, Nesterova A N 2019 The manifestation of fractality in the architecture of buildings and structures *Materials Science and Engineering* **698 (3)** 033046.
[6] Yazyev B M, Mayatskaya I A, Yazyeva S B, Yazyev S B 2019 Fractality in architectural forms and in organization of space in buildings *Materials Science and Engineering* **698 (2)** 022087.
[7] Mayatskaya I A, Yazyev B M, Yazyeva S B 2018 Fractality and symmetry in architectural structural elements *Construction and technogenic safety* **12 (64)** 29-32.
[8] Mayatskaya I A, Eremin V D 2019 Bionics and the choice of rational structural form *E3S Web of Conferences* **110** 01042.
[9] Mayatskaya I A, Yazyeva S B, Zakieva N I, Lapina A P 2018 Modern glass constructions and comfortable urban environment *Materials Science Forum* **931** 754-758.
[10] Mayatskaya I A, Yazyev B M, Yazyeva S B 2017 Optimal design in terms of architectural bionics *Construction and technogenic safety* **9 (61)** 7-11.

The concept of creating an ethnographic miniature park "Armenia: historical capitals and masterpieces of architecture" in Armavir, Russian Federation

A Yu Safaryan[1*], O S Subbotin [2], Yu A Safaryan [1], R G Sargsyan [1]

[1]National University of Architecture and Construction of Armenia,105, Teryan Street, Yerevan, Republic of Armenia
[2] Kuban State Agrarian University,13, Kalinin Street, Krasnodar, 350004, Russian Federation

E-mail: ayusafaryan@gmail.com

Abstract. Today, the problem of preserving the huge material and intangible cultural heritage is very relevant in general, and for the Armenian people in particular. One of the problem solutions is the creation of the national ethnographic park "Armenia of all times" with the "Pan-Armenian Center for Armenian studies" in its structure, a network of regional ethnographic miniature parks in the territory of the Republic of Armenia, as well as abroad - in the countries with a significant Armenian Diaspora. The exposition of these parks can present the most significant objects of architectural and cultural heritage in miniature, located both on the territory of the Republic of Armenia as well abroad and created by the Armenian architects.

Based on the research and design works carried out, the territory for the ethnographic miniature park "Armenia: historical capitals and masterpieces of architecture" creation in the city of Armavir in the Russian Federation was selected as an example. The conceptual proposals have been submitted, and a draft architectural layout solution was designed. 13 historical Armenian capitals, as well as the models of Armenian architectural monuments of great cultural and historical value, can be placed and displayed in miniature on the selected territory.

1. Introduction

The current stage of society development is characterized by a sharply increased role of information processes, which, along with an effective positive impact on all spheres of human life, at the same time had its influence on blurring the national identity boundaries, in particular, in the context of national culture. People slowly moved away from their national, historical origins, and the transnational tendencies started prevailing in culture and architecture. For decades, in part due to insufficient attention to the preservation of cultural heritage, many historical and architectural monuments have been destroyed. The settlements where traditional lifestyles, crafts, handicrafts and national customs had been preserved, eventually disappeared. Globalization has many pros but at the same time, globalization has become a serious challenge to national identity, a factor that affects it, due to such phenomena as the standardization of culture, economization, and universalization of values, while the only condition for preserving national identity is the paradigm of maintaining and developing national identity and traditions (with architecture being a major part of cultural heritage)

[1]. We consider the creation of ethnographic parks as one of the most effective ways to solve this problem [2, 3].

Society's vision of the cultural heritage values started changing, which resulted in an increase in the level of demand for the preservation of objects of the historical and ethnographic environment. At the same time, the growing migration of the population and the interpenetration of different nations' cultures contribute to the increased interest in architectural monuments of different nations and periods, which has become one of the main incentives for creating the new types of ethnographic parks. The basis for the exposition of such parks were the objects-copies of the immovable cultural heritage of various nations and countries, presented in the form of models and layouts of individual religious and public buildings, complexes, streets and entire cities. These educational and entertainment parks have become one of the most popular and profitable: "My Russia", "Taltsy", "Angarsk village" in the Russian Federation (RF), "Mini Europe" in Belgium, "Italy in miniature" in Rimini, "Spanish village", "Switzerland in miniature" in Lugano, and many others. It is the ethnographic parks of miniatures that are an effective form of folk architecture monuments' preservation and the historical and cultural heritage demonstration of ethnic communities [3, 4, 5, 6, 7].

The Armenian civilization, being one of the most ancient ones, has made a significant contribution to the global culture. It should be noted, that being the first and for two decades the only Christian state in the world, Armenia in the 4th century began to independently develop the types of church buildings, without having the examples for borrowing, which found their spread and development throughout the Christian world [8, 9, 10, 11].

The problem of preserving the huge material and intangible cultural heritage is very relevant for the Armenian people. This problem can be solved by creating a network of ethnographic miniature parks [1, 7, 11, 12]. The creation of ethnographic parks is of primary importance for demonstrating the Armenian nation's achievements in various fields, including science, culture and art. At the same time, these parks will host various performances, festivals, exhibitions, sales of home-made goods, souvenirs, etc., which will ensure economic profitability.

Earlier in a number of works, the idea of creating a miniature ethnographic park "Armenia of all times" in Republic of Armenia and a complex of "Pan-Armenian Center for Armenian studies" involved in it, where the cultural heritage of the Armenian people can be presented centrally, was proposed and developed in the form of a draft scientific and project proposal [11, 12, 13].

It was proposed to create five national parks in Armenia: "Ararat Ashkhar" (in Ararat, Aragatsotn, Kotayk and Armavir marzes), "Gugarkh Ashkhar" (in Lori and Tavush marzes)," Sevan Ashkhar "(in Gegharkunik marz),"Syunik Ashkhar"(in Syunik and Vayots Dzor marzes), "Shirak Ashkhar" (in Shirak marz) [12, 13, 14, 15]. It was also suggested to select the territories for ethnographic zones in the capital of Armenia - Yerevan (as well in the Kond, Sari Tah districts), in the city center (Abovyan, Nalbandyan, Mashtots streets, etc.), where historical buildings are of particular interest [12, 14, 16, 17, 18]. We have made the proposals to provide the ethnographic zones and their pedestrian sections in Yerevan related to the problem of total modernization of the city's residential formations [17]. Proposals for zoning the territories of miniature parks have been developed [11, 12, 14]. The idea of creating an ethnographic miniature parks abroad in the territories of the embassies of Armenia, monastic complexes, churches has also been proposed.

2. Proposals on ethnographic miniature park "Historical capitals of Armenia" and "Masterpieces of Armenian architecture" creation in Armavir (RF)

Taking into account the peculiarities of the Armenian people settlement (the Armenian Diaspora is functioning in many countries, as well as the fragmentation of a huge cultural fund), the problem of ethnographic parks creation in different countries on the territory of which the Armenian Diaspora was established, also becomes urgent.

In the countries where the Armenian Diaspora lives, there are cities which names are associated with historical Armenia, which is explained by the role and contribution of Armenians in the creation of specific settlements. These are, in particular, Armavir in Kuban, Armyansk in Crimea, a number of cities in the South of Russia and elsewhere.

The formation of the Armavir city is inseparably linked with the historical and cultural heritage of the Armenian people. At the end of 1836, the mountain Armenians (circassianshai) appealed to the chief of the Kuban line, major General Baron G. H. von Zass, with a request to "take them under the protection of Russia and give them the means to settle near the Russians". In 1839, the Armenian settlement moved closer to the mouth of the Urup river, and just this year is considered to be the official date of Armavir foundation. Although the ancestors of the Armenian settlers left their historical homeland - Armenia, they named the new village Armavir (1848) dedicated to the ancient Armenian city of Armavir, the capital of ancient Armenia founded by Aramais, grandson of Haik, who is believed to be the ancestor of the Armenian people in the 4-2 centuries BC [19, 20].

Figure 1. Armenian Apostolic Church of the Blessed Virgin Mary Assumption in Armavir.

A conceptual solution for creating an ethnographic park of miniatures in Armavir is proposed. A research and design search with a comprehensive assessment of the territory made it possible to choose a very favorable territory in the city for the ethnographic park (Figures 1, 2) - the city district between marshal Zhukov, Dzerzhinskiy, Kirov and Polina Osipenko streets, where the current Armenian Apostolic Church of the assumption of the Blessed Virgin Mary (5/2 Kirov street) is located. The ethnographic park is supposed to be located at the corner of Dzerzhinskiy street and marshal Zhukov streets (along the street); Marshal Zhukov, on the left side – the courtyard of the Church of St. Astvatsatur, on the right - which is expected to be the place for the exposition of miniatures "Historical capitals of Armenia" and "Masterpieces of Armenian architecture") (Figure 2). The lower part of the block along Polina Osipenko street, between Kirov and Dzerzhinskiy streets is also planned to be allocated for the exhibition.

The park will be of a mixed type, that is, it will be formed on the basis of the "in situ" monument (Blessed Virgin Assumption of the Armenian Apostolic acting church) and will be supplemented with miniature models of significant architectural monuments.

13 historical Armenian capitals, as well as a number of architectural monuments that are of great cultural and historical value, will be placed and displayed in miniature on the selected territory (Table 1).

Figure 2. Model of the ethnographic miniature park "Armenia: Historical capitals and masterpieces of architecture" in Armavir.

The layouts will include 13 Armenian capitals, according to their chronology (Table 1).

Table 1. Armenian Histoical Capitals

No.	City	Foundation Date	Period	State
1.	Tushpa-Van	825 BC	825-585 BC	• Kingdom of Van (9th- 5 th BC) • Kingdom of Orontid Dynasty (570-201 BC)
2.	Armavir	4000 BC	331-220 BC	• Kingdom of Orontid Dynasty
3.	Ervandashat	220 BC	220-201 BC	• Kingdom of Orontid Dynasty • Kingdom of Artashesid Dynasty (189- 1 BC)
4.	Artashat	187 BC	160-80 BC 67 BC -336 BC	• Kingdom of Artashesid Dynasty • Kingdom of Arshakid Dynasty (52/66- 428)
5.	Tigranakert	87 BC	80-67 BC	• Kingdom of Artashesid Dynasty
6.	Vaghasharat	2th century BC	117-140	• Kingdom of Arshakid Dynasty
7.	Dvin	336	336-885	• Kingdom of Arshakid Dynasty • Marzpan Armenia • Vicegerency Armenia
8.	Bagaran	3th century BC	885-890	• Kingdom of Bagratuni Dynasty
9.	Shirakavan (Yerazgavors)		892-828	• Kingdom of Bagratuni Dynasty
10.	Kars		928-961	• Kingdom of Bagratuni Dynasty

11.	Ani	6th century BC	961-1045	• Kingdom of Bagratuni Dynasty
12.	Sis		1197-1375	• Kingdom of Cilician Armenia
13.	Yerevan	782 BC	1918	• First Republic of Armenia (1918-1920) • Armenian SSR (1920-1991) • Republic of Armenia (since 1991)

Models of the following masterpieces of Armenian architectural monuments were also selected to be presented:

1. Teishebaini Citadel (founded between 650 and 600 BC - abandoned at 585 BC)
2. Cathedral of Ani (groundbreaking at 989 - completed at 1001)
3. Gandzasar Monastery complex (groundbreaking at1216 - completed at 1238)
4. Zvartnots Cathedral (groundbreaking at 643 completed at 652)
5. Shengavit Settlement (Early Bronze Age)
6. Erebuni (groundbreaking at 780-destroyed at 595 BC)
7. Armenian cross-stones "khachkars", etc.

The park itself will become not only a business card, but also an information platform where the historical past and present of the Armenian people will be presented by means of using modern technologies, including information technologies [21]. It is planned to create an information center, digital libraries and a database (including 3D images of monuments and surrounding territories), and the possibility of organizing multimedia excursions using virtual, interactive expositions.
The space-planning solution is based on the idea of a complex ring-shaped paths formation, which allows the visitors to see the exhibits presented in the park from all sides in an easy manner.

3. Summary

For the first time, the idea of the Armenian miniature ethnographic park organization abroad - on the territory of Armavir in the Russian Federation has been suggested and a specific site has been selected for its development.

The draft proposals for the organization of the ethnographic miniature park "Historical capitals of Armenia" and "Masterpieces of Armenian architecture" in Armavir on the territory of the quarter between marshal Zhukov, Dzerzhinsky, Kirov, Polina Osipenko streets have been developed, where the current Armenian Apostolic Church of the Blessed Virgin Mary Assumption (5/2 Kirov street) is located.

References

[1] Gasparyan M A, Safaryan Yu A 2014 The Problem of Preservation of National Traditions in Modern Armenian Architecture *Advanced Materials Research***1020** 702–706.
[2] Gnedovsky B V, Dobrovolskaya E D 1987 Open-air museums. Development of principles of formation and structure (Moscow, SLL).
[3] Fotii L A et al 1985 Creation of architectural and ethnographic complexes in open-air museums: Guidelines (Leningrad).
[4] Marcus A S and Levine H T 2011Knight at the museum: learning history with museums *The Social Studies - University of Connecticut* **102** 104-109.
[5] Massing Katharina 2018 Safeguarding intangible cultural heritage in an ethnic theme park setting The case of Binglanggu in Hainan Province *China International Journal of Heritage Studies* **24** 66-82.

[6] DuanXialei, Chan Chung-shing and Marafa Lawal M 2018 Does authenticity exist in cultural theme parks? A case study of millennium city park in Henan *China Journal of Tourism and Cultural Change* 1-18.
[7] Safaryan A Yu 2017 Concepts of perspective types of ethnographic parks *Proceedings of National University of Architecture and Construction of Armenia* **II (65)** 96-101.
[8] FrancisChing B K, Mark J M and Prakash Vikramaditya 2017 *A Global History of Architecture* (Hoboken: John Wiley & Sons) 268-270.
[9] Leng David 2004 *Armenians. The people-creator* (Moscow, Centrpoligraf).
[10] Khan-Magomedov S 2007 Reflections on the topic of "The Phenomenon of Armenian Architecture" *Architecture, construction* ***10(20)*** 21-22.
[11] Safaryan A Yu 2019 The concept of creating a network of ethnographic parks of the Armenian nation *IOP Conf. Ser.: Mater. Sci. Eng.* **698** 033053
[12] Safaryan A Yu 2020 Proposals on nomenclature, functional orientation and territorial zoning of the Armenian people's ethnographic parks *IOP Conf. Ser.: Mater. Sci. Eng.* **913** 032030.
[13] Safaryan A Yu, Tovmasyan S A, Muradyan J N 2015 Selection and comprehensive assessment of the territory of the national ethnographic Park in the Republic of Armenia*Bulletin of National University of Architecture and Construction of Armenia* **3** 28-33.
[14] Safaryan A Yu 2015 Prerequisites and prospects of formation of ethnographic parks in the Republic of Armenia Proceedings of 7th International Conference Contemporary *Problems of Architecture and Construction, Florence-Italy* 727-732.
[15] Aloyan A A, Safaryan A Yu 2014 Peculiarities of Urban Strategy Formation of the Republic of Armenia *Advanced Materials Research (Switzerland, Trans Tech Publications)* **1020** 875–878 4.
[16] Safaryan Yu A, Gasparyan M A, Aloyan A A 2011 The master plan of Yerevan *IET Conference Publications* **1-30** 357.
[17] Safaryan A Yu, Safaryan Yu A 2020 Housing architecture of Yerevan. Typology, Forecast *IOP Conf. Ser.: Mater. Sci. Eng.***913** 032015.
[18] Gasparyan M A 2018 Architecture of Yerevan of the 19th and early 20th century City in space and time (Riga, Palmarium Academic Pubishing) 468.
[19] Subbotin O S 2012 Architectural and urban development of historical settlements in the South of Russia (Moscow, ACB Publishing house).
[20] Subbotin O S 2020 Architectural and spatial environment in the historical settlements of regional significance *IOP Conf. Ser.: Mater. Sci. Eng.***913** 032021.
[21] Safaryan A Yu 2019 The new patents usage in the Republic of Armenia during the ethnographic parks creation *IOP Conf. Ser.: Mater. Sci. Eng.***698** 033047.

Acknowledgements
This work has been carried out in the frame of "Architectural issues of the creation of ethnographic park-museums of the Armenian people" theme, financed by Science Committee of Republic of Armenia.

IOP Conf. Series: Materials Science and Engineering 1083 (2021) 012030 doi:10.1088/1757-899X/1083/1/012030

New type of hybrid housing structure for the city of Yerevan

A Yu Safaryan*, Yu A Safaryan

National University of Architecture and Construction of Armenia, 105, Teryan Street, Yerevan, Republic of Armenia

E-mail: ayusafaryan@gmail.com

Abstract. The accumulated development experience in Yerevan over the past century (1920-2020) is analyzed taking into account the practice of housing designing and building of the past stage. Based on the analysis of the existing development experience and the studied material, a number of significant shortcomings in the space-planning solutions available for the housing currently in exploitation are identified.

Using the principles of sustainable architecture and the foreign experience, we offer the suggestions for the new type hybrid house structure with the possibility of variations in the number of stories and planning solutions. The expected effectiveness of these proposals' implementation in the development execution in Yerevan is assessed.

1. Introduction. Analysis of contemporary foreign experience

One of the main directions in modern housing architecture formation remains the problem of finding the optimal solutions to create a comfortable living environment for the main category of the inhabitants.

Today, among the main principles of shaping the architecture of residential structures are: the principles of sustainable architecture, the principle of autonomy provision, the principle of land resources' optimal use, the principle of dynamic forms development, the principle of residential structures' spatial composition uniqueness [1].

Based on the accumulated experience and trends in creating the sustainable architectural solutions, the following principles of creating a sustainable architecture can be emphasized [2]:

- Maintaining the ecological balance between natural and artificial components,
- Application of waste-free or low-waste industrial and construction technologies,
- Reducing resource consumption, improving the urban planning solutions through the energy-efficient technologies' use for the renewable energy sources application,
- Application of structural and spatial solutions that fit into the context of the natural environment,
- Erection of cost-effective buildings and structures,
- Increase the comfort of people by improving the functional, microclimatic and aesthetic parameters of the environment, as well by using smart technologies,
- Introduction of nature components into the building structure,
- Focus on local natural, landscape and cultural conditions,

- Adaptability to natural, climatic and anthropogenic challenges and risks [2].

We considered the influence of some of these principles in the architectural form.

The modern stage in the formation of residential complex architecture can be characterized by an appeal to natural forms. The inclusion of nature elements into the structure of residential buildings can be implemented by various architectural means: in the apartment interior, by means of the landscaping terraces use, winter gardens, etc.; in the structure of a residential building, by means of vertical and horizontal landscaping of building facades, roofs, open and glazed atrium spaces, maximum disclosure of internal spaces to the external environment, glazed openings of a large area, extensive terraces, balconies of various configurations and sizes; in the structure of urban development, by means of horizontal and vertical buildings with open green floors, multi-level atriums. One remarkable example of an organic combination of natural and artificial forms is the residential complex "Vertical forest" in Milan (architects Stefano Boeri, Gianandrea Barreca and Giovanni La Varra). Two towers of 80 and 112 floors of 110 and 76 *m* are covered with green spaces, which determined the original design of the building facades. A number of Ken Young's works, who also used the principles of continuous landscaping in the structure of the building: hanging gardens form the facades of a 26-story tower (EDITT tower, Singapore, 2010), which is designed so that its functional purpose can vary from office to residential. All modular elements of the structure are assembled mechanically meaning that the movable walls make it possible to change the facade surfaces according to changing needs [3].

An example of using the natural environment elements in the space of a residential building is the Hundertasser house "Hundertwasserhaus" (architects Friedensreich Hundertwasser and Joseph Krawina), where green surfaces on roofs, facades, balconies and terraces, two-light glazed spaces, winter gardens, winter courtyards have been used. All these planning elements of a residential building help to increase the comfort of living and improve the environmental background of a residential building [3].

The principle of ensuring the autonomy of residential structures is based on reducing energy consumption in the residential buildings operation, minimizing the use of traditional energy systems and actively used autonomous systems with alternative renewable energy sources (solar collectors, solar cells, wind devices on facades or roofs, geothermal and other natural systems). The use of autonomous energy generation systems (both passive and active) together with energy-efficient solutions for the space-planning structure of houses increases the economy and comfort level. The operation of the solar systems' specialized equipment, in turn, affects the formation of modern housing both during the existing housing stock reconstruction and the newly built houses [4].

2. Analysis of the housing current state in Yerevan and the proposals for a new hybrid type of housing

In a number of works [4, 5, 6, 7, 8, 9], the authors analyzed the accumulated experience of development in Yerevan over the past century (1920-2020), considered the practice of designing and building housing of the past stage, as well as a number of works by other authors [10, 11, 12, 13, 14, 15, 16, 17] dedicated to the same problem.

The analysis of the considered material shows, that there are a number of significant drawbacks in the operated housing in Yerevan, namely:

1. The absence of an elevator in residential buildings above 3 floors (at a time when, according to building codes, an elevator is required to be installed above the planning mark of 1000 *m* in the climatic conditions of Yerevan),
2. An obliged organization decision of the entrance to the staircase-elevator unit, usually, after lifting ≈ half of the flight of stairs, leads to an uneconomical increase in the planning level of the first floor by at least ≈ 90 *cm*, and also causes some discomfort to the residents,
3. The absence of complete maintenance of the roof, even if the roof is flat,
4. A single passenger elevator in the stairwell in residential buildings above 5 floors (usually 9-story residential buildings). If the elevator is stopped for repairs, there are difficulties in unusual situations – in cases of patients' transportation on stretchers, transferring oversized cargo, etc.,

5. The presence, as usually, of 1-2-room apartments that do not have through, corner ventilation,
6. The solution of summer facilities in the form of attached verandas, mainly in 5-story buildings, which leads to their mass glazing, and negatively affects the architectural appearance of the building, as well as leads to natural lighting and darkening disturbance of the residential facilities adjacent to the veranda,
7. Dominance of uniform monotonous solutions of the entire residential development due to the use of a single type of sectional house, set by the house-building plant products.

A new type of housing for the city of Yerevan was also developed taking into account the sustainable architecture principles. The proposed structure of a new type of hybrid house with the possibility of variations in number of floors and planning solutions is considered by the authors not as the only direction for solving the housing problem as it was in the past, but as an addition to a diverse range of proposals [4].

The authors proposed:

- In the structure of the building on the 1st and 2nd floors, the place apartments of the "maisonette" type ("duplex") with an entrance-exit from the street, without connection to the staircase/staircase-elevator unit. In front of each apartment, a green area is provided on both sides (front garden, garden with a depth of max $\approx 9m$, taking into account the fire protection requirements provision). There can be basically several solutions (Figure 1.a).
- Organization of entrance and exit to the apartments through the open galleries. The proposed solution will contribute to creating a more comfortable and eco-friendly home and living environment in general [4]. Residential buildings will have two stairwells instead of the previously traditionally used four ones. In residential buildings, an elevator unit will be provided with 2 elevators (passenger and cargo-passenger) instead of one, which will significantly increase comfort - when repairing one of the elevators, the other will serve residents. The presence of a cargo and passenger elevator will also significantly increase the comfort level for the residents. Finally, instead of 4 elevators, the building will have only 2. If galleries are used, the hygienic quality of the air environment on the way in and out of the apartment will increase. Entrance and exit to the stairwell from the main platform will allow more variable placement of the house on the site. The entrance and exit to the autonomous elevator unit should increase the comfort for residents. Since the combination of the staircase and elevator unit will lead to the mandatory overcoming of half a flight of stairs to enter the elevator (Figure 1.b, c).
- The presence of a flat operated roof can be used both for the installation of photovoltaic systems (energy-saving roof), and for the recreational area organization.
- Landscaping of courtyard areas around the perimeter of $\approx 9m$ will also provide protection from noise. The presence of a green zone will help to improve the climate of the residential environment.

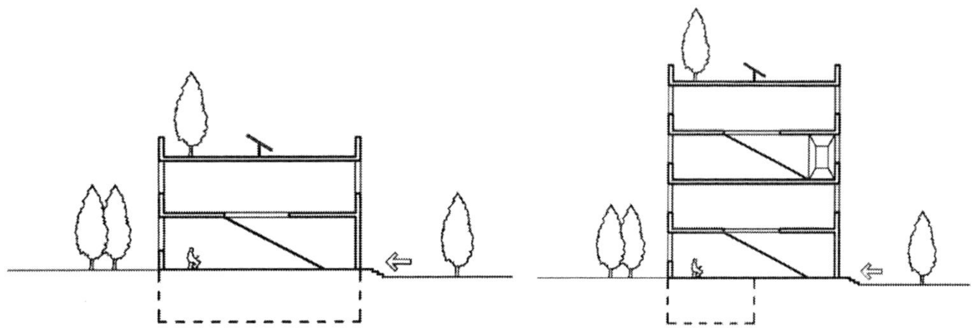

(a) (b)

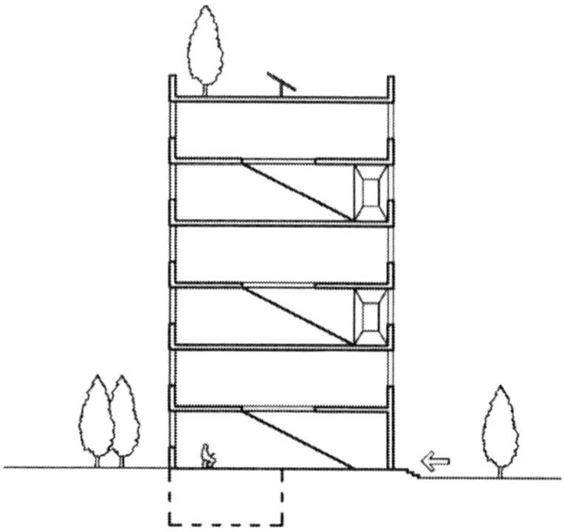

(c)

Figure 1. Structure of hybrid houses, sections.

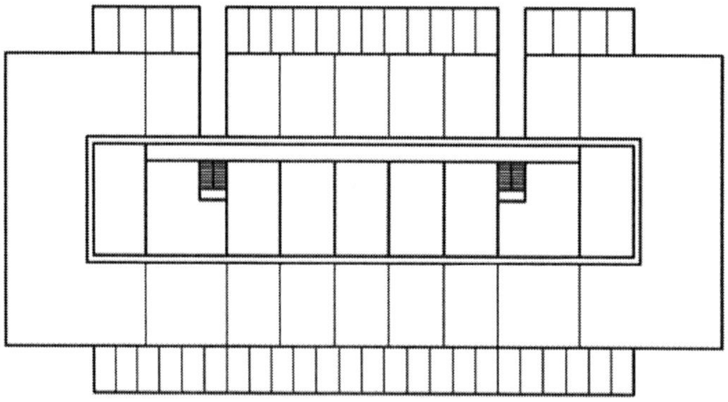

Figure 2. The structure of a hybrid house, a master plan example.

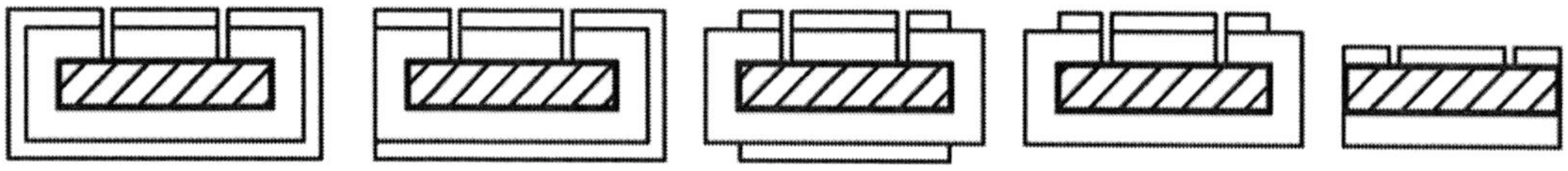

Figure 3. Hybrid home master plan options.

Summary

Analysis of the housing current state in Yerevan has been carried out, with several issues being identified.

The proposals for a hybrid structure are as follows:

1. It is recommended for a residential house to be consisted of two autonomous combinations of apartments, differing in the type of "entry and exit" arrangement: the entrance to the apartment directly from the street and entrance from the gallery (Figure 1.b, c).
2. All apartments are usually offered to place "maisonette" ("duplex") two-level-type (Figure 1.a, b, c; Figure 2).

3. It is recommended to use an earthquake-proof compartment apartment with a length of about ≈60 *m* as a basis.
4. The residential building is proposed to be surrounded from two (optionally from three or four) sides by a zone of green spaces (≈9 *m* deep) and along the front by the sections which width corresponds to the apartment facade size. Each apartment will have two sections (with an entrance from the street and from the courtyard). This principle will also ensure proper care for the entire area surrounding the residential building (Figure 2, 3).
5. Around the green zone, it is proposed to organize parking for personal use by the residents (Figure 2, 3).
6. A residential building will usually have a flat roof that is used to get solar energy, provide a recreational zone to the residents, etc. (Figure 1.a, b, c).
7. The gallery is proposed to be located usually through the floor (Figure 1.b, c).
8. The residential building will have 2 staircases (instead of the traditional 4 in the case of the apartment entrance from the staircases) (Figure 2).
9. Elevator-free 4- and 6-story houses are proposed, respectively, with the residents' lifting to the 3rd and 5th floors (Figure 1.b).
10. Residential buildings in 8, 10, 12, 14, 16 floors will be equipped with 2 elevators: passenger and cargo-passenger at one of the stairwells (staircase-elevator unit) or elevator unit, solved independently (figure 1.c).

The proposed solutions for organizing the structure of a hybrid residential building will allow to save the cost of two staircases (only 2 instead of 4) for each residential building, as well as save on the number of elevators (2 instead of 4) with a significant increase in comfort. In practice, each of the residential buildings above 6 stories in the traditional sectional structure should have one passenger elevator per section.

It should also be noted, that entering the apartments through the air zone will help solve many hygiene and sanitary issues.

In the context of the residential formations' modernization and development problem in the city of Yerevan, it is also proposed to allocate the territories for ethnographic zones and their pedestrian sections [4, 18, 19, 20].

References.

[1] Esaulov G V 2012 Sustainable architecture as a design paradigm (on the issue of definition) *"Sustainable architecture: present and future". Proceedings of the intern. Symp. November 17-18, 2011 Scientific works of the Moscow architectural Institute (state Academy) and the KNAUF CIS group* 76-79

[2] Salmina O E, Bystrova T Yu 2015 The principles of creating sustainable architecture *Academic Bulletin of the Ural SRI Project Russian Academy of Architecture and Construction Sciences RAACS* **4** 36-37

[3] Belogolovsky V 2010 "Green style" by Ken Yang [Electronic resource] *SPEECH*-**5.** Information on http: / / www.arhi.ru./ press/ world/ 25523 / zelenyi-stil-kena-yanga

[4] Safaryan A Yu, Safaryan Yu A 2020 Housing architecture of Yerevan. Typology, Forecast *IOP Conf. Ser.: Mater. Sci. Eng.***913** 032015

[5] Safaryan Yu A, Gasparyan M A, Aloyan A A 2011 The master plan of Yerevan *IET Conference Publications*pp 1-30 -1-35

[6] Aloyan A A, Safaryan A Yu 2014 Peculiarities of Urban Strategy Formation of the Republic of Armenia *Advanced Materials Research* **1020** 875–878

[7] Gasparyan M A, Safaryan Yu A 2014 The Problem of Preservation of National Traditions in Modern Armenian Architecture *Advanced Materials Research* **1020** 702–706

[8] Safaryan Yu A 1988 *Architecture of buildings and structures erected by lifting* (Yerevan, Hayastan) 240.

[9] Safaryan A Yu, Avakyan R E 2010 Some issues of modernization of residential areas in the center of Yerevan *Proceedings of YSUAC* **2 (38)** 100-103
[10] Zakaryan E, Petrosyan H 2009 *Yerevan. Historical and geographical Atlas* (Yerevan, Center of geodesy and cartography) 95.
[11] Harutyunyan V M, Hasratyan M M, Melikyan A A 1968 *Yerevan* (Moscow, Stroyizdat) 302
[12] Grigoryan A G, Tovmasyan M L 1986 *Architecture of Soviet Armenia* (Moscow, Stroyizdat) 318
[13] Gasparyan M A 2018 *Architecture of Yerevan of the 19th and early 20th century. City in space and time* (Riga, Palmarium Academic Pubishing) 468.
[14] Tigranyan E A 2000 *Construction of Armenian architects in Transcaucasus near the end of 19th and at the beginning of 20th century* (Yerevan, VoskanYerevantci) 263.
[15] Dolukhanyan L K 1980 *Architecture of Soviet Armenia, 20th years* (Yerevan, Sovetakangrogh) 83
[16] Harutyunyan E P 2009 *The permanent traditions of Armenian architecture* (Yerevan, Mugni) 199
[17] Rashidyan G H 1999 *Development of architecture of mass housing in Armenia and tasks of the modern stage (*Yerevan, David Publishing house) 87
[18] Safaryan A Yu 2019 The concept of creating a network of ethnographic parks of the Armenian nation *IOP Conf. Ser.: Mater. Sci. Eng.* **698** 033053
[19] Safaryan A Yu 2019 The new patents usage in the Republic of Armenia during the ethnographic parks creation *IOP Conf. Ser.: Mater. Sci. Eng.* **698** 033047
[20] Safaryan A Yu 2020 Proposals on nomenclature, functional orientation and territorial zoning of the Armenian people's ethnographic parks *IOP Conf. Ser.: Mater. Sci. Eng.***913** 032030

Acknowledgements
This work has been carried out in the frame of the "Architectural issues of the creation of ethnographic park-museums of the Armenian people" theme and in the frame of the "Creating the ways for sustainable urban, architectural and construction complexes development in Republic of Armenia and elaboration of directions with use of permanent monitoring system" programme, financed by Science Committee of Republic of Armenia.

IOP Conf. Series: Materials Science and Engineering 1083 (2021) 012031 doi:10.1088/1757-899X/1083/1/012031

Analysis of the welded pipe joints state and the methods for increasing the bearing capacity of gas transmission systems' pipelines

N V Pirumyan[1*], M G Stakyan[1], H A Khazaryan [2]

[1] National University of Architecture and Construction of Armenia, 105 Teryan Street, Yerevan, Republic of Armenia

[2] "Gazprom Armenia" CJSC, Tbilisyan Highway, 43, Yerevan, Republic of Armenia

E-mail: pirumyannarine@gmail.com

Abstract. The changes in the structural state of the weld material of gas pipes at different seam cross-sections are considered. The classification of external and internal weld defects caused by different types of factors carried out, the stress concentration in the transition zone of the weld to the pipe metal is the most significant factor. The analysis of atmospheric effects on the welded seams of pipelines, that initiate adsorption and electrochemical processes of the metal microstructure, which cause corrosion on the external surface of the weld, is given. Taking into account the non-stationary loading mode of pipes due to fluctuations in the internal gas pressure, corrosion and fatigue damages are also added to these defects, which cause the development of fatigue cracks, leading to a violation of the system's tightness and even to the welds destruction. The method for increasing the bearing capacity of pipelines is proposed by applying a technological operation of the external surface plastic deformation of the weld and of the transition zone to the pipe metal, which creates residual compressive stresses that relieve working tensile stresses in the stress concentration zone.

Introduction

The development of the main spheres of industries, agriculture of the countries requires consecutive expansion of the gas transmission system (GTS), the expansion of which over the past decades has been interstate and regional, and requires overcoming of difficult relief, climatic zones and water (sea) areas. At the same time, an important technological problem is the main pipelines network creation by using the large-diameter steel pipes connected by welds, and this process is constantly being improved in order to accelerate the laying time, ensure the necessary operational reliability and extend the service life of pipeline structural elements [1, 2, 3].

According to statistical data and the results of technical survey and expertise carried out by the relevant divisions of gas transmission organizations, it was fixed that about 70...75% of failures and damage to GTS elements are fatigue-related, and 20...25% of failures are caused by corrosion. Taking into account the continuous growth of production and transport services over the past decades, emissions of industrial waste and transport gases into the atmosphere are increasing annually, which leads to increased corrosion effects on various purposes metal structures, including GTS. Corrosion losses are estimated at up to 30% of the weight of structural steels and alloys in the form of unacceptably deformed or destroyed structural elements which need repair or replacement, of which about 5...7% permanently disappear in the form of corrosive dust or various chemical compounds. This dictates the necessity for a detailed study of corrosion processes in GTS structures and the development of effective anti-corrosion procedures in order to minimize the losses volume of structural steels and alloys, the cost of which tends to increase due to the interstate GTS network expansion [4, 5, 6, 7, 8].

Corrosion phenomena in GTS structural elements

Corrosion of structural steels and alloys is usually manifested in two types - chemical and electrochemical. In the first form, it occurs and develops as a result of contact with various gases and liquids which do not have electrical or electromagnetic properties [4, 5, 6]. More common type is electrochemical corrosion, when contact occurs with moist air, water, salt and acid solutions, etc., in which anodic and cathodic interaction processes occur in metal microstructures and in liquids' electrolytes, herewith the intensity and of the corrosion process speed depends on the *pH* hydrogen index of the liquid (Figure1.a). The most significant type of GTS welded joints destruction in a corrosive environment is the O_2 and H_2 adsorption in ionic, atomic or molecular states in the metal surface layer and the creation of oxide and hydride phases caused by embrittlement and delamination of this layer and the appearance of surface micro cracks (Figure 1.b). Eventually, significant decrease in the physical and mechanical characteristics of the alloy under static and cyclic loading occurs [6, 8]. Since the corrosion process is continuous, physical limits of strength and endurance are not observed, which compel, to determine their values, to fix the service life of the structure in the interval of durability of $N = 5 \cdot 10^4 \cdots 8 \cdot 10^8$ loading cycles and obtain the corresponding experimental functions. But as fatigue tests show, the gradient of these functions varies in the N interval, due to the different nature of the corrosion failure mechanism [3, 9, 10, 11].

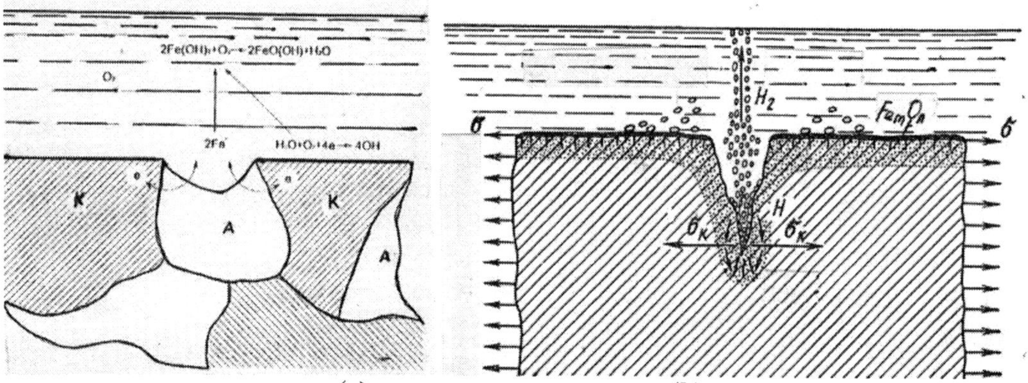

(a) (b)

Figure 1. Scheme of electrochemical corrosion in an aquatic environment. *a* - oxygen depolarization of the steel microstructure, *b* - development of a corrosion crack that causes the metal surface layer cracking.

In many studies, the hypothesis of adsorption-electrochemical corrosion failure has been put forward, which more precisely represents the specified changes in the corrosion process. Under more common loading modes (re-static and cyclic), local micro-fractures and the first fatigue micro-cracks also appear in the areas where dislocations reach the metal surface. Anodic processes, occurring inside micro cracks, stimulate to their growth, and at high cyclic overstraintions, the adsorption process increases and prevails. At low overstraintions, cyclic micro-damages mainly occur due to electrochemical processes, when overstraintions destroy the oxide micro-layers of the surface and stimulate to the penetration of the corrosive medium into new metal layers [5, 6, 12].

Structural state of the pipelines weld material

The operating conditions of the GTS network are characterized by internal gas pressure, which creates a static or variable stress state in the system, and by variable temperature conditions (- 45... +50°C), as well as natural and climatic conditions of the region (landscape, solar radiation, atmospheric pressure, wind, humidity, rainfall, etc.), which form a natural corrosive environment that can cause irreversible physical and chemical changes in the welded joint and the pipe's metal of the pipeline due to corrosion, sorption, erosion and other processes that eventually lead to micro-damage, loss of strength and structural failure [1,10, 11, 12, 13, 17].

The pipeline network, usually, formed using circular butt welds, which are the main type of welded joints in GTS. In this case, the weld and the pipe's metal in the cross section have a layered microstructure consisting of several sections (Figure 2) [4, 14].

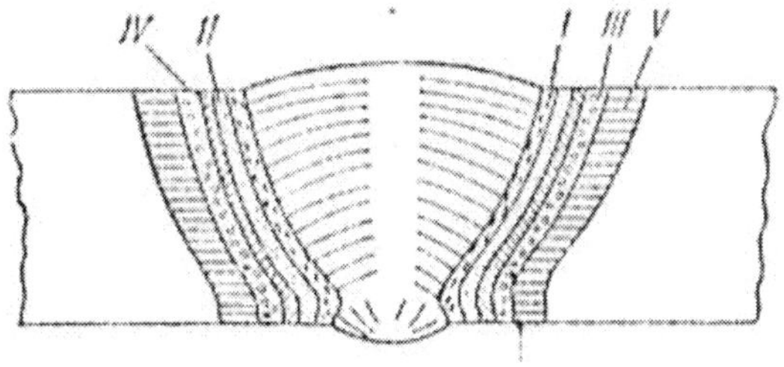

Figure 2. Micro structural layers in the contact zone of the weld and the pipe's metal.

Section I, immediately adjacent to the weld, is a zone of thermal influence, where the welding temperature reaches up to $t_1 = 1000 \cdots 1100^0 C$, has a thickness of $\delta_1 \approx 2\,mm$ and a characteristic coarse-grained microstructure. Section II $(\delta_2 = 0.5 \cdots 2.0mm)$ undergoes recrystallization and gets a micro grain structure, and the temperature regime is close to the thermal normalization process $(t_2 = 800 \cdots 900^0 C)$. Section III of incomplete recrystallization is located in the $\delta_3 = 0.5 \cdots 1.0\,mm$ zone $(t_3 = 720 \cdots 900^0 C)$, where a partial change of grains in the micro structure of the pipe metal occurs.

On sections IV $(\delta_4 = 2 \cdots 5mm,\ t_4 = 500 \cdots 550^0 C)$ and V $(\delta_5 = 4 \cdots 5mm,\ t_5 = 200 \cdots 350^0 C)$ there are no visible microstructural changes and thermal aging processes occur.

The hardness of these sections' layers is also variable, it increases close to the seam. Since the stress concentrators are located in the weld fusion zone, sections I and II are of the main interest.

Defects and damages of GTS welded joints

Butt joints have the greatest endurance of all types of welded joints, but even in this case, the stress concentration and residual tensile stresses that occurred during the formation of the joint in this zone significantly affect the fatigue joint strength. Even in the case of proper implementation of welding technologies, absence of defects, cracks, and nonmetallic inclusions the weakest link in determining the bearing capacity of the connection is the transition area of the weld to the pipe's metal (Figure 2), and only in the presence of significant defects and damages in another section of the weld residual tensile stresses can be moved to this section [2, 3, 12, 13, 18].

Technological process of welding has its own features and quality of its implementation depends on the following factors:

- poor pipe's material quality,
- inaccurate pre-processing of the edges of the seams,
- presence of delamination and corrosion layer on the surfaces of the seam edges,
- stress concentration in the transition zone of the weld to the pipe`s metal,
- violation of the welding mode - fluctuations in the electric current voltage, changes in the movement and location speed, as well as the number of the electrode passes,

- poor qualified service personnel.

These factors predetermine the appearance of various types of seam defects which can be classified as follows:
1. *By time-* during the seam edges preparation before welding and occurred during welding,
2. *According to the location in the cross-section of the seam*-internal and external.

- The most characteristic defects of this type are: improper processing of the seam edges, which causes a non-constant gap between the edges and mismatch of their joined areas along the entire circumference of the seam, delamination of the surface layer of the seam edges and unacceptable surface irregularities after the mechanical treatments, that will lead to the formation of an inhomogeneous structure in the cross-section of the seam.
- After welding, the seams may have irregularities and deviations from the profile due to unstable welding mode, speed and incorrect electrode supply. Herewith, incisions may appear on the surface layer of the seam, which reduce the cross-section of the seam and cause stress concentration. Internal seam defects include cracks that are sharp stress concentrators and most often appear during welding of high-carbon and alloy steels as a result of sharp cooling of the seam after its installation. Pores and slag inclusions are formed due to the presence of scale and rust in the welding zone, and non-welding and non-melting–at a gap between the edges or a small angle of their bevel, as well as at incorrect electrode supply and high speed of the welding process.

These defects, if they are located unfavorably in the seam section, can also violate the tightness of the pipeline, so after performing welding operations, additional technological operations are mandatory: mechanical cleaning and grinding of the weld contour, strengthening the joint and applying an anti-corrosion coating in order to increase the tightness, corrosion resistance and strength of the welded joint.

The environmental impact on GTS elements is enhanced in the case of the above-mentioned defects and technological omissions presence in the welded joints of pipelines, where corrosion processes are initiated, which would eventually lead to the destruction of welds [7, 12].

The main types of corrosion failure of welded joints are (Table):
1. Entire or common corrosion - can be evenly distributed over the entire surface of the welded joint and the role of the stress state is secondary. It becomes more dangerous, if it is concentrated in a certain welding zone: on the surface of the seam, in the zone of thermal influence, or on the base of the pipe's metal,
2. Local corrosion is more dangerous than the first type and depends on the stress state, which can lead to a sudden failure of the pipeline. Intergranular corrosion significantly reduces the bearing capacity of pipes made of chromium and aluminum alloys, in which the pitting corrosion can also be developed [3, 6, 18].
3. Static and fatigue corrosion are the most dangerous types of failure when the combined action of the corrosive environment and the stress state caused by extreme static or re-static and cyclic loads. In the first case, static cracking occurs, and in the second, inter-crystalline or trans-crystalline corrosion-fatigue failure of the weld occurs.

Table. Main types, forms and pattern of corrosion damage of welded joints.

N	Type of destruction	Forms of destruction	Pattern of destruction
1	General corrosion	Equal	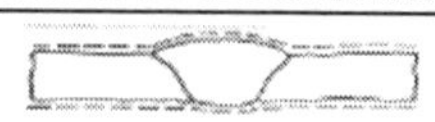
2		On the weld	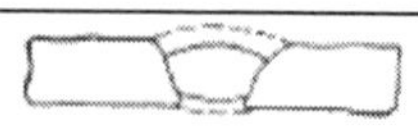
3		In the zone of thermal influence	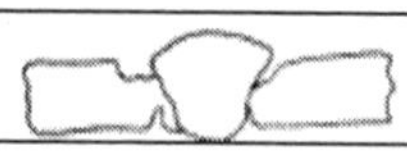

4		On the pipe`s metal	
5	Local corrosion	Inter-crystalline, in the zone of thermal influence	
6		In the fusion zone	
7		In welds	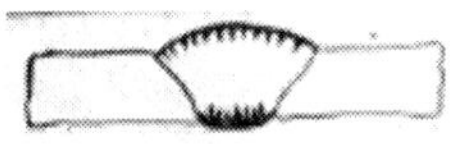
8		Pitting	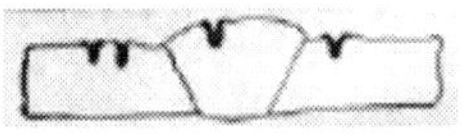
9	Statical and fatigue corrosion	Statical	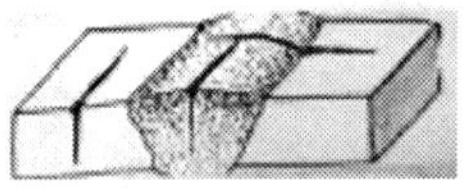
10		Corrosion-fatigue	

A significant factor affecting the bearing capacity of GTS elements also is the stress concentration in the welded joint caused by the structural form, assembly technology and operation of the system [18]. The stress concentration zone is a section of the weld where structural and technological defects and various types of damage can be accumulated.

A similar effect is also affected by the material science factor - the heterogeneity of the mechanical properties of the metal in the weld zone, where other types of defects may occur, which will lead to a decrease in the joint strength. During the application of high-strength and alloy steels which are more sensitive to these defects and stress concentrations, the reduction will become more noticeable [1, 2, 3]. For this reason, for the GTS structures, preference is given to low-carbon and low-alloy structural steels, which have good weldability and low sensitivity to these factors' effects [15].

Methods for improving the performance of GTS welded joints

Increasing the bearing capacity, corrosion resistance and fatigue strength of GTS welded joints is mainly implemented by materials science and technological methods, which include the following procedures [1, 2, 3, 16, 18]:

- selection of the type and grade of the steel pipe's metal, as well as the hardening method that ensures the strength of the welded joint with minimal expenses of the hardening process and relatively low cost of pipe material,
- reduction of stress concentration by means of mechanical processing of the seam surface and by creating a smooth transition zone to the metal of pipe,
- complete formation of the weld volume and elimination of non-welding in its lower part in order to prevent a new zone of stress concentration and microcracks,
- cleaning the surface layers of the weld and the transition zone from non-metallic coatings and inclusions which occurred during welding,

- additional heat treatment of the weld by means of using an electric arc operation in a CO_2 or inert gas environment to regulate the microstructural state and to reduce internal tension in the seam area,
- creation of residual compressive stresses in the weld zone by the methods of plastic deformation of the seam surface layer and the transition zone to the pipe metal – shot blasting, stamping, rolling by rollers,
- application of new protective coatings made of polymer composite materials which increase the corrosion and crack resistance of welds.

The structural strength of GTS pipelines in the area of welds depends on the mechanical characteristics of the metal to plastic deformation and the performance of the weld hardening process, as well as the type of heat treatment. For these steels, plastic deformation of the surface layers of welded joints can increase their endurance limits up to 1.5...2.0 times. In this case, the weld and the metal of pipe act as a joint force system with mutual penetration of their microstructural layers in the contact zone. Strengthening the contour of the welded joint in the surface layer creates residual compressive stresses, which interact with the residual tensile stresses resulting from the welding process, as well as with the operating stresses from the gas pressure [17], reduce the effect of stress concentration and ensure the normal loading mode of the GTS elements [10,11].

The most technologically advanced and easily implemented of these hardening methods is the method of plastic deformation by rolling rollers, which create a uniform layer of deformation on the welded joint surface. The resulting residual compressive stresses C (Figure 3, curve 2) compensate for the influence of residual tensile stresses C (Figure 3, curve 1), creating a favorable initial stress state (Figure 3, curve 3) for receiving operating stresses.

Surface plastic deformation has also a positive effect on the bearing capacity of welded joints under the negative temperatures influence $\left(t_{з} = 0 \cdots -50^{0} C\right)$ and corrosive environments, actually, being a universal technology for increasing the corrosion-fatigue strength of these joints. Another important advantage of this technique is its implementation by mechanisms and devices that have a simple, reliable and portable design that makes it possible to perform weld strengthening in the field conditions, directly during welding operations and laying the GTS pipeline [1, 2, 11, 17,18].

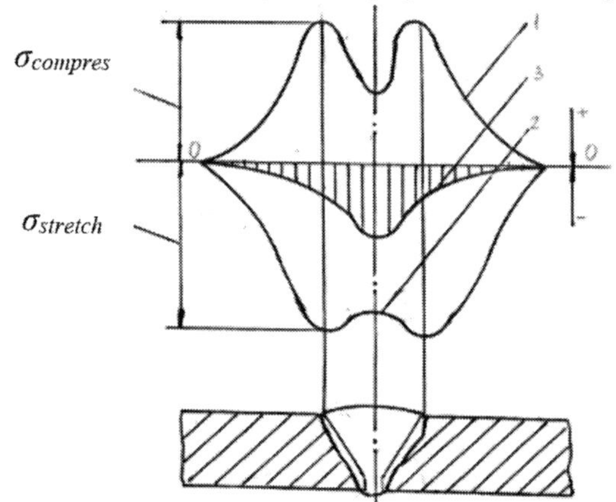

Figure 3. Stress distribution in the area of the welded joint:
1, 2- residual tensile and compressive stresses, accordingly,
3-initial total stresses before the start of the GTS operating mode.

For medium-, high-carbon and alloy structural steels that are not sensitive to plastic deformation, and the stress concentration degree after welding is significant, heat treatment is preferable (quenching with high tempering), which reduces the stress concentration and provides favorable microstructural changes in the weld material.

The application of new protective coatings made of polymer composite materials, due to their high adhesive properties, not only increases corrosion resistance and tightness, but also due to filling micropores and cracks on the surface of the weld, reduces the effect of stress concentration and increase the fatigue strength of welded joints.

Summary

The increased volume of engineering work on the design and laying of main pipelines of the GTS brings new requirements for ensuring the reliable operation of the mentioned big engineering structures. To implement this research, it is necessary to perform the following works:

1. Study, analysis and classification of all possible defects and damages of critical structural elements of the existing GTS, which include welded pipe joints;
2. Development of new design and technological measures to improve their performance over of the pipeline service life [1, 2, 3];
3. On the basis of the activities performed according to the paragraphs 1 and 2, creation of a mathematical model of the specified design and technological procedures for its further application in the designed main pipelines of the GTS [3, 10, 11, 17].

References

[1] Pirumyan N, Stakyan M 2020 Measures to Increase the Building Steel Structures' Bearing Capacity, *IOP Conf. Ser.: Mater. Sci. Eng.***913** 022006

[2] Pirumyan N, Stakyan M 2020 The strengthening technologies' application in building steel structural elements, *IOP Conf. Ser.: Mater. Sci. Eng.***913** 022036

[3] Pirumyan N, Stakyan M 2020 The building structures' bearing capacity assessment with the nomogram method, *IOP Conf. Ser.: Mater. Sci. Eng.***913** 022035

[4] Trufyakov V I 1973 *Fatigue of welded joints* (Kiev, Sciences.Dumka) 216

[5] Pokhmursky V I 1985 *Corrosion fatigue of metals* (Moscow, Metallurgy) 207

[6] Pirumyan N, Stakyan M 2019 Asessment of corrosion fatigue strength of gas-transport system constructions under atmospheric forcing *IOP Conf. Se.: Materials Sci.Eng,* **698** 022077.

[7] Kurkin S A 1986 *Strength of welded thin-walled vessels working under pressure* (Moscow, Mashinistroyenie) 184.

[8] Oleynik N V, Volchev A V, Bershak S V et al. 1990 *Calculation of machine parts for corrosion fatigue* (Kiev, Technique) 150

[9] Pirumyan N V, Khazaryan H A, Stukyan M G 2018 Study of work reliability of gas subsurface storage by the method of finite elements *Proceedings of the 10th Int.Conf. on Contemporary Problems of Architecture and Construction (Beijing)* 76-78

[10] Pirumyan N, Stakyan M, Galstyan G 2020 Mathematical Modeling of the Test Process of Construction Materials *Key Engineering Materials* **828** 115–120.

[11] Pirumyan N, Stakyan M, Galstyan G 2020 Software Processing of the Test Results of Building Materials Key *Engineering Materials* **828** 121–128.

[12] Steklov O Yu 1976 *Strength of welded structures in aggressive environ*ments (Moscow, Mashinistroyenie) 200

[13] Kudryavtsev I V, Naumchenov N E 1976 *Fatigue of welded structures* (Moscow, Mashinistroyenie) 270

[14] Dashchenko A F, Kravchuk V S, Iorgachev V D 2004 *Bearing capacity of reinforced machine part* (Odessa, Astroprint) 160

[15] Khazaryan H A, Chibukhchyan S S, Stakyan M G 2016 Improving the reliability of the gas transmission system (GTS) according to the criteria of structural strength and vibration resistance *Proceedings of 8th International Conference on Contemporary Problems of Architecture and Construction, October 26th -28th, 2016, Yerevan-Armenia* 168-170

[16] Torosyan M S, Pirumyan N V, Stakyan M G 2020 Complex changes in the fatigue resistance indices of constructions at applying of stenghening technology *Bulletin of NUACA* **1 (66)** 58-66

[17] Pirumyan N, Stakyan M 2019 Mathematical model of pressure distribution in the main gas-transport system *IOP Conf. Se.: Materials Sci. Eng* **698** 066022

[18] Pirumyan N, Stakyan M 2019 Bearing capacity of elements of a gas transportation system, *E3S Web of Conferences* **97** **04027** 1-9

Acknowledgements
This work has been carried out in the frame of "Creating the ways for sustainable urban, architectural and construction complexes development in Republic of Armenia and elaboration of directions with use of permanent monitoring system" programme, financed by Science Committee of Republic of Armenia.

The individual components' influence on aging of bitumen mineral materials

M G Salikhov[1], E V Veyukov[1*], V M Vainshtein[1], L I Malianova[2]

[1] Volga State University of Technology, 3 Lenin sq., Yoshkar-Ola, 424000, Russia
[2] I.N. Ulyanov Chuvash State University, 45 Moskovsky Ave., Cheboksary, 428017, Russia

E-mail: VeukovEV@volgatech.net

Abstract. When heating bitumen-mineral mixtures at high temperatures under the air access conditions, their accelerated aging is observed, i.e. the values and quality indicators' decrease of the compacted samples from them. Due to the different structure of bitumen films on the mineral components' surfaces of different compositions and sizes, aging processes may occur in different ways. In this regard, a need for a differentiated study of the aging process on the surfaces of large fine aggregates and the crushed part of the bitumen-mineral (asphalt concrete) mixture arose. The experiments in laboratory conditions are carried out according to the Volga State Technological University developed at the Department of Construction Technologies and Automobiles (according to RF patent No. 2654954). The aging processes analysis of the samples from bitumen-mineral mixtures is carried out using the dimensionless aging coefficient in terms of the ultimate compressive strength at +50 °C. The nature and dynamics of the samples' aging over time are assessed using special indicators - the aging coefficient and the aging intensity. As a result of the experimental study, the optimal ratios of bitumen with mineral components, taking into account their susceptibility to aging over time, were established. The changes' dynamics in the bitumen aging intensity in time in the bitumen-mineral mixture as a whole and separately on the mineral components' surfaces separately has been studied. The results of this study can be taken into account when revising the standard service life values of the roads with structural layers of bitumen-mineral materials.

Introduction

The purpose of this study is to establish the aging process dynamics of bitumen in bitumen-mineral mixtures (asphalt-concrete mixtures) in time in laboratory conditions, and separately on the components' surfaces.

It is known that the physical and mechanical properties' values of bitumen-mineral materials significantly decrease in the aging process, as a result of which their performance indicators can reach the maximum permissible values. The need to take into account the aging processes on the bitumen-mineral materials' durability has been pointed out by many domestic and foreign authors [1-6]. At the same time, the various factors' influence on the oil bitumen aging process at the stages of its production, storage and use was revealed [6-13]. At the same time, the proposals to take into account the aging processes at all the above-mentioned stages have been put forward.

Despite the existing experience in studying the aging processes of bitumen and bitumen-mineral mixtures, there is little information about bitumen aging on the surfaces of their individual mineral components to date.

Materials and research methods

The hot-laid fine bituminous concrete of the B type according to GOST 9128-2013 and the mixtures of its individual components with construction bitumen CB 90/130 were taken for the experiments:

1) asphalt concrete type B composition, % by weight: granite crushed stone M 1200 fr. 5-20 mm–47,0; crushed sand (screenings of crushing of hard rocks) - 43.0; limestone mineral powder - 10.0; bitumen CB 90/130 – 5,5 (in excess of 100,0 %);

2) mixture of limestone mineral powder with bitumen CB 90/130 in the ratio (by weight) 0,85:0,15;

3) crushed sand mixture fr. 0,14-5,00 mm with bitumen in the ratio 10:1;

4) mixture of crushed granite M 1200 fr. 5-20 mm with bitumen in the ratio by weight 25:1.

The ratio of the components corresponds to the optimum and were assigned based on a preliminary study of their average density and compressive strength at +20 °C and +50 °C degrees, including after artificial aging at +150 °C.

Since there is currently no standard technique, the study of the bitumen-mineral mixtures' aging was carried out according to the new technique proposed by the authors (according to the RF patent № 2654954) [14]. The essence of the technique is as follows:

1. Sampling of mineral materials under study, drying and heating them to the operating temperatures (150-170 °C);

2. Preparation of a petroleum bitumen sample by dehydration and heating to the operating temperature: for heavy asphalt CB 90/130 – 140-150 °C;

3. Mixing the components in specified proportions;

4. Placing the prepared mixture samples on trays in even layers and holding in an electric furnace (furnace RTFOT) at high temperature +150 °C and in the conditions of free air access during the estimated time (0, 1, 3, 5, 7, etc. hours).

After the intended period of time expiration, the mixtures are removed from the furnace, the standard cylindrical samples are formed from them, according to the standard method of GOST 12801-98, they are tested and the values of their physical and mechanical properties are set.

Results and discussion

The results of the experimental data are shown in Table. 1.

Table 1. Some experimental results

No.	Warm-up time at +150 °C	Compression strength values at +50 °C ($R_{comp}^{+50\ °C}$) in MPa and aging coefficient							
		Asphalt concrete type B		Bitumen mixture with mineral powder		Bitumen mixture with crushed sand		Mixture of crushed granite with bitumen	
		$R_{comp}^{+50\ °C}$, MPa	$K_{ag}^{R_{comp}^{+50\ °C}}$	$R_{comp}^{+50\ °C}$	$K_{ag}^{R_{comp}^{+50\ °C}}$	$R_{comp}^{+50\ °C}$	$K_{ag}^{R_{comp}^{+50\ °C}}$	$R_{comp}^{+50\ °C}$	$K_{ag}^{R_{comp}^{+50\ °C}}$
1	0	2.10	1.00	0.80	1.00	1.37	1.00	0.73	1.00
2	1	1.40	0.67	0.80	1.00	1.40	1.02	0.53	0.32
3	3	1.50	0.71	0.90	1.13	1.97	1.43	0.11	0.15
4	5	1.40	0.67	1.30	1.63	2.76	2.00	0.07	0.01
5	7	1.40	0.67	1.10	1.38	1.47	1.37	0.08	0.01
6	9	-	-	1.05	1.31	-	-	-	-

Table 2 shows the results of calculating aging intensity of the studied mixtures according to the aging coefficient values.

Table 2. Aging intensity values

No.	Warming-up time t_w at +150 °C	Aging intensity values of bitumen-mineral mixtures $I_{ag}^{K_{ag}^{R_{comp}^{+50°C}}}$

		Asphalt concrete type B	Bitumen mixture with mineral powder	Bitumen mixture with crushed sand	Mixture of crushed granite with bitumen
1	0	0.00	0.00	0.00	0.00
2	1	-0.33	0.00	0.02	-0.68
3	3	0.02	0.07	0.21	-0.08
4	5	-0.02	0.25	0.29	-0.07
5	7	0.00	-0.13	-0.32	0.00
6	9	-	-0.04	-	-

For visual presentation, the obtained experimental results are shown in Fig. 1 and 2 as graphs.

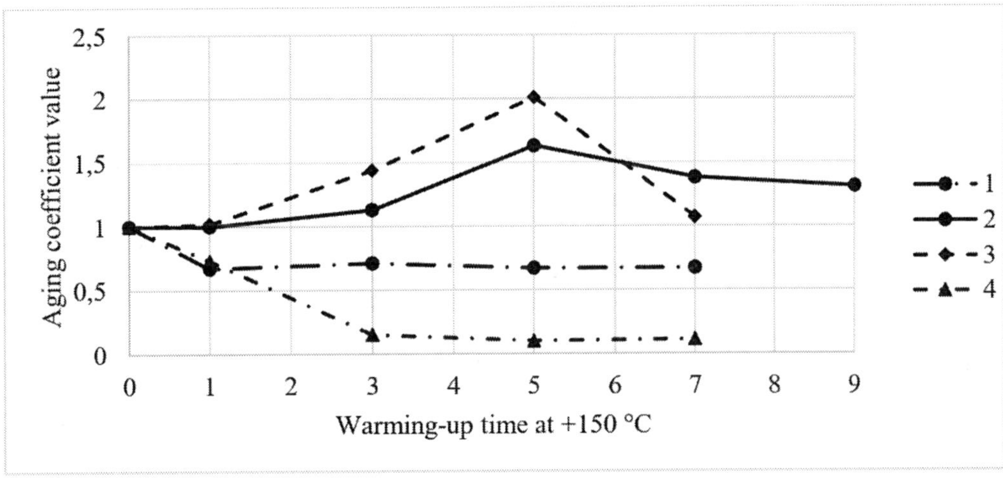

Figure 1. Graphs of the aging coefficient values' dependencies in terms of ultimate strength in compression at +50 ^{0}C from the time of warming up at a temperature +150 ^{0}C: 1 – asphalt concrete mix of type B (bitumen content 5.6% of the mass of the mineral part, over 100%); 2 - mixtures of limestone mineral powder with bitumen in the ratio by weight B:MP=1:5,67; 3 – a mixture of crushed sand (granite screenings) with bitumen in the ratio 1:10; 4 – mixture of crushed granite fr. 5-10 mm with bitumen in the ratio 1:25

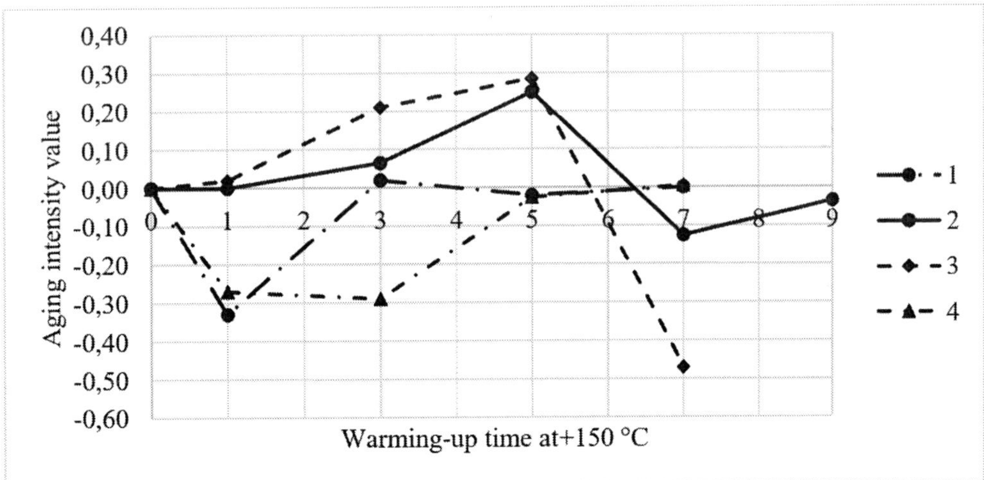

Figure 2. Graphs of the aging intensity values' dependences according to the aging coefficient on the heating time at temperature +150 ^{0}C: 1 – asphalt concrete mix of type B (bitumen content 5.6% of the mineral part mass, over 100%); 2 - mixtures of limestone mineral powder with bitumen in the ratio

by weight B:MP=1:5,67; 3 – crushed sand mixture fr. 0-5 mm with bitumen in the ratio B:CS=1:10; 4 – mixture of crushed granite fr. 5-20 mm with bitumen in the ratio B:CG=1:25

As it can be seen from the graphs in Fig. 1, change in the aging coefficient values in terms of $R_{comp}^{+50\,°C}$ for asphalt concrete and a mixture of crushed granite with bitumen occur according to the law of the following classes loss, as the heating time increases, respectively:

$$y = e^{(-A\cdot t_w)^B}, \quad (1)$$
$$y = C \cdot e^{-D\cdot t_w}, \quad (2)$$

where A, B, C, D – are the constant coefficients, the values of which depend on the type of bitumen-mineral mixture.

For the mixtures of fine-grained mineral materials (limestone mineral powder and crushed sand) with bitumen, the aging coefficient values in terms of $R_{comp}^{+50\,°C}$ can be extrapolated by the dependency:

$$y = a \cdot e^{\frac{(t_w - b)^2}{2\cdot c^2}}, \quad (3)$$

where a, b, c – are the constant coefficients.

Selection accuracy (approximation rank) by the dependencies (1 ... 3) is within 0,9796…0,9909.

It can be seen from the graphs in Figure 2 that coarse-grained bitumen-mineral mixtures age most intensively during 1 ... 3 hours of heating and with further heating there is a sharp attenuation of this process. The intensity of the aging process in fine-grained mixtures is relatively weak up to 3 hours of warming up, then it increases within 3 ... 5 hours.

The analysis of the experimental results given in Table 1 and Table 2 and in Figure 1 and Figure 2, show:

1) Bitumen mixtures with crushed components of bitumen-mineral materials (mineral powder and crushed sand) for 1 ... 3 hours of heating lead to some increase in the samples' strength, and with further heating, the process of its decrease starts;

2) Bitumen mixtures with crushed stone age to the greatest extent during the first hour of heating at a high temperature, and then this process stabilizes and the aging coefficient of the samples tends to zero;

3) The greatest contribution to the mechanical strength of type B asphalt concrete is provided by asphalt binder (a bitumen mixture with mineral powder) and a bitumen mixture with crushed sand. Moreover, this conclusion turns out to be true for all the mixtures' aging stages at high temperatures;

4) By the highest values of aging intensity, the studied mixtures are arranged in the following row: bitumen mixture with crushed stone; asphalt concrete mix type B; bitumen mixture with crushed sand; bitumen mixture with mineral powder.

The change in the mechanical strength of bitumen mineral materials during aging can be explained from the previous studies [15]:

1) When bitumen is combined with powdery materials in comparison with coarse-grained, firstly, on the particles' surfaces, especially of carbonate origin, relatively thin bituminous films are formed and the binder is more structured in them. As a result, the bitumen-mineral system becomes more resistant to external influences, including the action of high temperatures;

2) Due to an increase in the asphalt-resinous component in bituminous films during heating in the initial period, a certain increase occurs in the adhesion strength of the bitumen film on the surfaces of mineral particles, which leads to a slight increase in the mechanical strength of the entire system. In the future, this process goes into a zone of weakening due to bituminous films' embrittlement and a decrease in their adhesive ability;

3) In coarse-grained systems, due to the less structured bitumen and the greater thickness of the bitumen films, the adhesion strength of the bitumen films on the minerals' surfaces is relatively weak. Despite a slight increase in the cohesive strength of bituminous films in the initial period of heating, the embrittlement processes' intensity prevails in them and cannot prevent the general weakening of the

entire system as a whole from the beginning of the process.

The results of the work performed in this study are fully consistent and explained by the previously known provisions on the bitumen binders' interaction with the stone materials' surfaces and the change in the bitumen properties when heated.

Summary

1. The aging process of bitumen-mineral materials and its intensity are influenced by both the composition and the mineral components' size. At the same time, bitumen compositions on the large particles' surfaces and with inactive and acidic surfaces at high temperatures age faster than on the fine-grained components;

2. The greatest contribution to the mechanical strength in compression of type B asphalt concrete at all stages of aging is made by a bitumen mixture with fine-grained aggregates and crushed mineral components. In general, it should be expected that the aging coefficient values' change of bitumen-mineral mixtures during heating will closely depend on the comparative content, size and mineralogical composition of the mineral components relative to each other.

References

[1] Bakhrakh G S, Pechenyi B G, Borisenko O A 1973 Consideration of the "aging" process when designing the compositions of bitumen-mineral mixtures *Automobile roads* **9** 8-9.

[2] Iliopolov S K, Mardirisova B V, Uglova E V 1994 Development of the aging process of bitumen in asphalt concrete pavements of highways *HEI News. "Construction" Series* **3** 48-52.

[3] Stukalov A A 2015 Aging of asphalt concrete mixtures, asphalt concrete and methods of increasing thermal oxidative resistance (dis. ... Cand. tech. sciences. - Makeevka (Donetsk region): Don. NASA).

[4] Gulyak D V 2010 Technological and operational aging of tar-concrete mixtures and tar-concrete and methods of slowing down: Author's abstract (dis. ... Cand. tech. sciences. - Makeevka (Donetsk region): Don. NASA).

[5] Chan Nyat Tan 2010 Regulation of thermo-oxidative stability of road bitumen and bituminous materials (dis. ... Cand. tech. sciences, M.).

[6] Siddiqui Mohammad Mahid 2002 Use of X-ray diffraction in assessing the aging pattern of asphalt fractions *Fuel* **81 (1)** 51-58.

[7] Zhao Zhaohui 2003 A study on aging kinetics of asphalt based on softening point *Petroleum science and technology* **21 (9-10)** 1575-1582.

[8] Skripkin A D, Starkov G B, Kolesnik D A 2007 Evaluation of Bitumen Aging in Thin Films Using a Thin Chromatography Analyzer "Iatroscan Mk-5" (CJSC "NOMBUS", M.).

[9] Anosuke M 1971 Methods of preventing embrittlement and destruction of asphalt concrete pavements. Japan patent № 51-41766; E 01 07/18.

[10] Saenko S S 2008 Methods for minimizing the aging of bitumen in a working boiler in the preparation of hot asphalt concrete mixtures (Abstract of thesis ... Ph.D. Rostov-on-Don: RSCEU).

[11] Salnikova A I 2017 Biological and climatic resistance of modified bituminous binders and composites (Author's abstract. dis. ... Cand. tech. sciences. - Penza: PSUAC).

[12] Ivanski M, Leonovich I I 2002 The influence of the aging process on the physical and mechanical properties of asphalt concrete *Bulletin of BSTU* **5** 17-22.

[13] Salikhov M G, Veyukov E V, Malyanova L I, Gaifullina A Z 2019 Study of temperature aging of bitumen-powder mixtures *Scientific journal "Vestnik MarSTU". Series "Materials. Constructions. Technology". - Yoshkar-Ola: PSTU* **4 (12)** 25-33.

[14] Salikhov M G, Veyukov E V, Sabirov L R, Malyanova L I 2018 Method for determining the rate and intensity of aging of asphalt concrete. Invention patent № 2654954 from 13.02.2017. SC. G01N 17/00 (2017/08); G01N33/42 (2017/08). – Publ. 23.05.2018. Bul. № 15.

[15] Korolev I V 1986 Ways to save bitumen in road construction (Transport, M.).

Structural features of a cement matrix modified with additives of sedimentary origin

I V Korchunov*, E A Dmitrieva and E N Potapova

Mendeleev University of Chemical Technology of Russia, 9 Miusskaya square, Moscow, 125047, Russia

E-mail: korchunov94@bk.ru

Abstract. The possibility of improving the properties of cement materials through the use of limestone and thermoactivated clays additives is considered. It has been established that the optimal content of limestone and clays in the composition should not exceed 10% and 15% by cement weight, respectively. Structural transformations in the phase composition of the cement matrix at various stages of hardening have been established by the methods of X-ray phase analysis and electron microscopy. It has been proven that the additives used affect the structure formation process over the entire hardening interval, due to which such properties as strength, water resistance and corrosion resistance of the composite material can be improved by 30-50%.

1. Introduction

In cement technology, processes for obtaining various types of this material have long been developed. Each of the types is designed for use in specific conditions. The narrow field of materials use based on special cements undoubtedly creates a need for unique properties of the binder, achieved through strict adherence to certain technological features of the production process, as well as the increased requirements for the initial components' purity and the mineralogical composition of the product [1-4]. All of these factors affect the clinker cost. This pushes modern science to research aimed at finding more economical ways to obtain building materials with target characteristics. One of these methods is the development of cement compositions with a reduced clinker factor, obtained through the rational use of mineral additives in cement composition.

The experience of using cements with mineral additives throughout the world is very extensive. However, the clinker factor of cements in the Russian market, as a rule, ranges from 0.83 to 0.92, which indicates a rather high clinker content in the cements produced on the market (in Europe, this indicator is 0.6–0.65). The state standard for general construction cements GOST (state standard) 31108-2016 is in force on the territory of the Russian Federation. In accordance with this document, all cements are divided into five types according to the amount and type of additive used in the composition. In this study, when developing the composition of a unique cement, preference was given to the additional cements of the type CEM II with a total content of additives up to 36%. Preliminary research in the field of additive and composite cements, conducted at the Department of the Mendeleev Chemical Technology University of Russia showed the positive effect of additives of sedimentary origin on the stability of the phases of the cement stone in different conditions. Therefore, the purpose of this research work was to study the effect of finely ground fractions of thermally activated clays and limestone on the phase composition and structure of cement stone during the first 28 days of hardening.

2. Materials and research methods

Portland cement CEM I 42,5H GOST 31108-2016 produced by LLC HeidelbergCement (Rus), Novogurovsky settlement was tested. Limestone screening of the Skripkinsky crushed stone deposit was chosen as one of the constituent components of the composition. This material is a by-product formed during crushing and is a finely ground pulverized fraction of limestone with a specific surface area (S_{sp}) equal to 390 m^2/kg. The prepared compositions are assigned with the following symbols: Cem – no additive cement, CemL10, CemL20 and CemL30 – cement with the addition of 10%, 20% and 30% limestone screenings, respectively. Aluminosilicate components were also considered as additives. The properties of aluminosilicates with different chemical compositions were investigated (Table 1). The clay deposit has a significant effect on the basic clay oxides content (silicon oxide and alumina) in the composition of aluminosilicate. Therefore, the clays produced by PJSC "Gornozavodskcement" in the Western Urals (Clay 1), and LLC "HeidelbergCement (Rus)" in Novogurovsky (Clay 2) were selected for the study. Cement compositions with the addition of thermoactivated clays are designated as CemCl-1 and CemCl-2, respectively.

Table 1. Chemical composition of aluminosilicate components.

Clays	Content, %								
	SiO_2	Al_2O_3	Fe_2O_3	CaO	MgO	SO_3	R_2O	P_2O_5	TiO_2
Clay 1	42.28	13.25	7.40	12.72	3.22	1.00	2.27	-	0.66
Clay 2	64.29	14.62	6.18	2.49	1.12	0.11	2.77	-	-

Determination of normal density, setting time of cement paste and strength was carried out according to the standard methods, real density (ρ_{real}) samples were measured using an automatic helium pycnometer AccuRus 1340 (Micromeritics Instrument Corp., USA) accurate to ± 0,0001 g/cm^3 at the Center for Collective Use of the Mendeleev Chemical Technical University of Russia.

Aluminosilicates were burnt in a muffle kiln in the temperature range from 650 to 800 °C at different isothermal exposures (from 30 to 120 min). Further, the clays activity was determined by the method of absorption by an active mineral additive of lime from a lime solution.

3. Influence of limestone additives on the cement stone properties

Limestone or $CaCO_3$ itself is quite inert to chemical influences. With regard to the possibility of the calcium bicarbonates formation together with the hydrated cement phases, the ability of calcite to be incorporated into the hydro silicates structure is noted. Hydro carbonates discovered using the known analytical methods such as electron microscopic analysis and X-ray analysis can arise through the layer-by-layer distribution of water groups in the original calcium carbonate [5-11].

In addition, the indirect interaction of calcium carbonate with the products of cement hydration is due to a variety of calcite crystal habits (up to several tens of thousands), which allows it to act as a substrate for growing on crystals $CaCO_3$ of not only inorganic hydrate compounds, but also for the organic substances adsorption (for example, plasticizing chemical additives). Thus, the incorporation of calcite into the structure of the hydrated composition occurs.

According to the experimental data obtained, it was found that with an increase in the limestone content in cement, there is a decrease in water demand of the mixture from 29 to 26%. Reducing water demand has a positive effect on the cement stone structure, the strength gain is more intensive and more uniform than that of Cem (Figure 1). However, the effect depends on the amount of limestone contained in the composition (Figure 1).

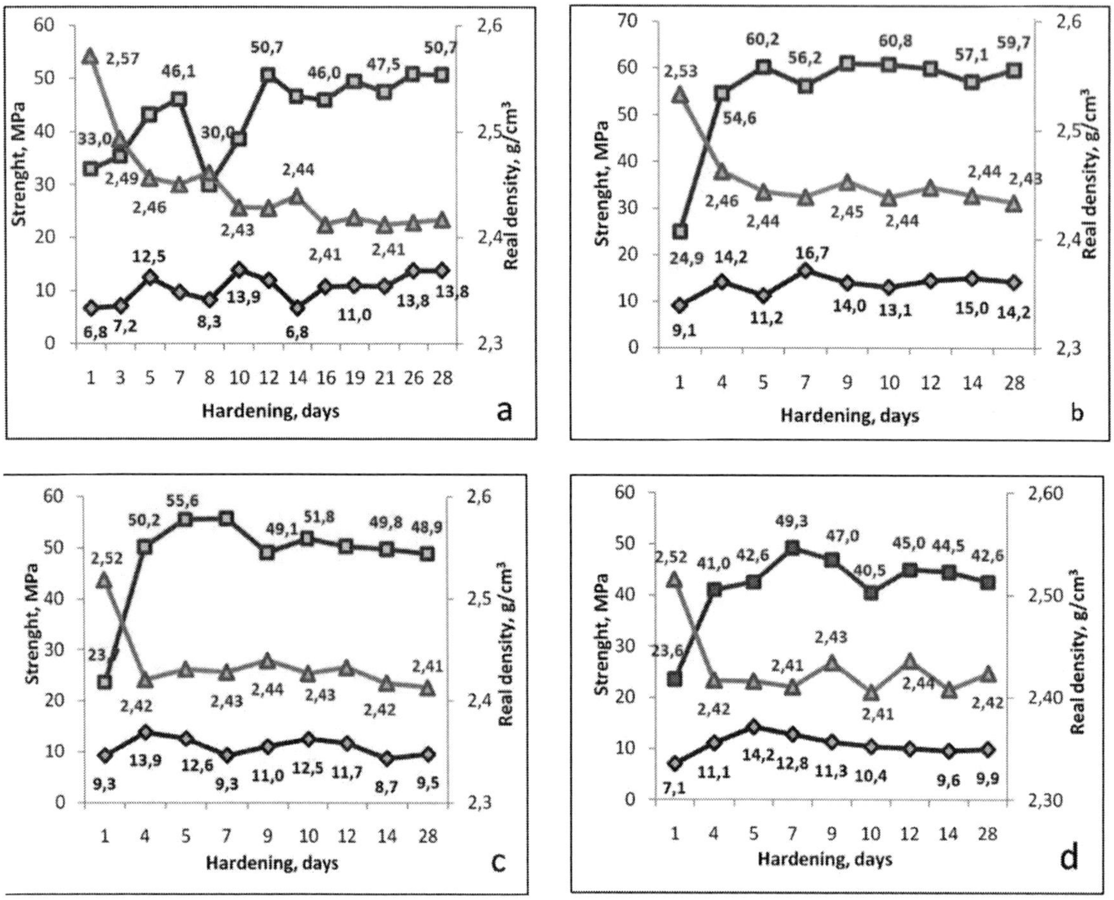

Figure 1. Bending strength (◇), compressive strength (□) and density of cement stone (Δ) dependence on the cements hardening duration: a – Cem; b – for CemL10; c – CemL20; d – CemL30.

The analysis of the experimental data obtained shows that when the interval for taking cement strength readings is narrowed (every 2-3 days instead of 7), a more plausible cement hydration curve can be seen. Hardening process for Cem proceeds unevenly, by the 8th day there is a significant inflection with a sharp drop in strength and a gradual re-strength gain by 12 days. Moreover, changes in strength (first, a drop from 46.1 MPa to 30.0 MPa, then an increase to 50.7 MPa) correlates with the data on changes in the real density (change from 2,4503 g/cm³ to 2,4613 g/cm³ and then to 2,4278 g/cm³) for 7, 8 and 12 days, respectively. Based on the differences in strength and density, it is possible to draw the conclusions about the occurrence of phase transformations, and the fluctuations in strength for no additive cement are more significant than for the compositions with limestone. The phase composition of the cement stone at this time was investigated by the X-ray phase analysis (X-rPhA) and electron microscopy (EM). It was found that these fluctuations are caused by the fact that the less dense etringite-like hydrates formed at the initial moment of hydration (AF_t) rebuild into AF_m mono hydro sulfa aluminates of higher density (Figure 2.b). As a result of this process, the total density of the cement stone increases by 8 days. By the tenth day of the composition Cem hardening SO_4^{2-} excess ions accumulate occurs, which leads to the reverse reaction of the transformation AF_t in AF_m (Figure 2.c and 2.d). The whole cycle of transformations is described by the reactions 1 - 3.

$$C_3A{\cdot}3CaSO_4{\cdot}H_{32}{=}C_3A{\cdot}CaSO_4{\cdot}H_{12} + 2SO_4^{2-} + 20H_2O \quad (1)$$

$$C_3A{\cdot}CaSO_4{\cdot}H_{12} + 2SO_4^{2-} + 20H_2O = C_3A{\cdot}3CaSO_4{\cdot}H_{32} \quad (2)$$

$$C_4AH_{(13\text{-}19)} = C_3AH_6 + (7\text{-}13)\ H_2O \qquad (3)$$

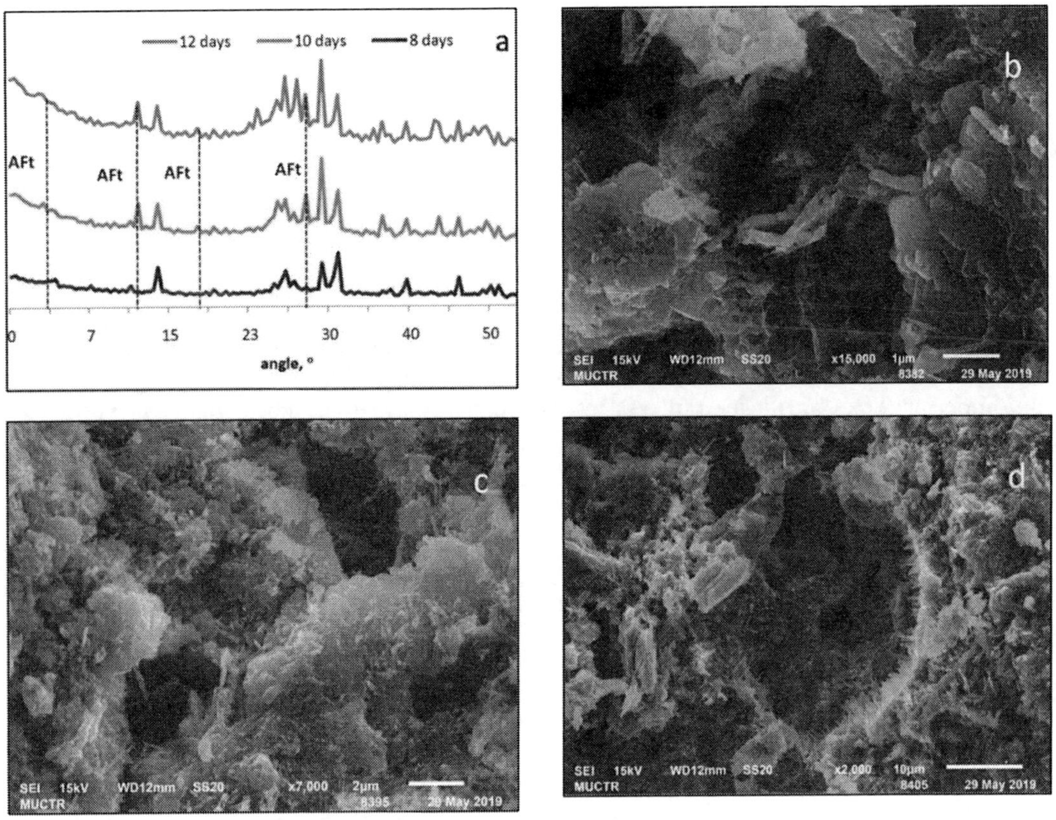

Figure 2. Cement stone structure: *a* – X-rPhA compositions Cem for 8, 10 and 12 day; EM Cem for 8 (*b*), 10 (*c*) and 12 (*d*) day of hardening.

Density of the initial phases $C_3A{\cdot}3CaSO_4{\cdot}H_{32}$ (diffraction reflections: d = 10,0047; 5,6357; 4,9587; 3,8899; 2,7671; 2,5097; 2,2121 Å) and $C_4AH_{(13\text{-}19)}$ (7,3877; 2,9008; 2,5685; 2,3249 Å) less than that of the main products of reactions (1) and (3) and are, respectively, 1,75 and 2,05 g/cm^3 vers 1,99 and 2,55 g/cm^3 for $C_3A{\cdot}CaSO_4{\cdot}H_{12}$ (4,2956; 3,8869; 2,9008; 1,6300 Å) and C_3AH_6 (4,2956; 2,7827; 2,3249; 2,1930 Å), respectively. By the tenth day, the density and strength indices level out, moreover, the diffraction reflections that AF_t phase disappeared by the 8th day, are formed again. This transformation occurs with a significant change in volume (the volume of the phase AF_t 2,5 times more than phases AF_m), which causes temporary structural transformations and is reflected in the strength characteristics of the composition Cem. Generally, AFt negatively affects the cement properties, however, due to the elongated prismatic shape of the crystal, this phase has a positive effect on the flexural strength.

With the introduction of limestone, the transformations of the sulfa aluminate phases weaken, the changes in strength and density in the period of 8-12 days are not so significant. This can be explained by the fact that the carbonate phases formed by the mechanism described at the beginning of the section AF_t-CO_2 ($C_3A{\cdot}3CaCO_3{\cdot}H_{30}$) and AF_m-CO_2 ($C_3A{\cdot}CaCO_3{\cdot}H_{11}$) are inert to the action of sulphate ions of the pore solution. SO_4^{2-} ions are located much to the right from CO_3^{2-} according to their substituting ability in the series of anionic activity (4), therefore, they also have a lower adsorption capacity. However, this substitution is unlikely, but possible from the point of view of the process thermodynamics ΔG^0_{298}= -18,1 kJ / mol (5):

$$OH^- > CO_3^{2-} > PO_4^{3-} > CNS^- > Cl^- > F^- > SO_4^{2-} \qquad (4)$$

$$3CaO{\cdot}Al_2O_3{\cdot}CaCO_3{\cdot}11H2O + SO_4^{2-} + H_2O = 3CaO{\cdot}Al_2O_3{\cdot}CaSO4{\cdot}12H_2O + CO_2^{2-} \qquad (5)$$

The phases AF_t-CO_2 and AF_m-CO_2 are very problematic to identify by the methods known in modern science, however, the probability of their presence in the composition can be calculated by the analytical methods using the available thermodynamic databases [12, 13]. From the point of view of the process thermodynamics, the chemical interaction of hydrate phases AF_m with limestone is impossible (ΔG^0_{298} >> 0). However, at the initial moment of hydration, before the hydrates' formation, this reaction is possible in the absence of sulfate ions (ΔG^0_{298}= -183,7 kJ / mol):

$$3CaO\cdot Al_2O_3 + CaCO_3 + 11H_2O = 3CaO\cdot Al_2O_3\cdot CaCO_3\cdot 11H_2O \quad (6)$$

Thus, limestone, subject to its dosage, is an effective additive to Portland cement, and contributes to its strength characteristics' improvement. We tend to consider the changes in the phase composition of cement to be the main reason for this effect.

4. Addition effect of thermally activated aluminosilicates on the cement stone properties

It is known that aluminosilicates in their initial state are unsuitable for use in the cement mixture composition, since they are the inert fillers. But under certain conditions of heat treatment, clay materials can be decomposed to amorphous aluminum and silicon oxides, which affects their pozzolanic properties [14-16].

The original clays contain a mixture of various clay minerals such as kaolinite, montmorillonite, illite, halloysite and most of the impurity minerals such as quartz, sulfides, calcite, mica, etc. These impurities are difficult to remove and require enrichment followed by drying of the feedstock. Therefore, special attention should be paid to the selection of raw materials in order to obtain reactive aluminosilicate pozzolans [14].

As a result of the clays' heat treatment, the resulting amorphous silicon and aluminum oxides are able to react with calcium hydroxide already at room temperature in the presence of water, forming C-S-H-gel and other hydrated phases containing aluminum oxide. This contributes to the cement stone structure compaction, an increase in strength characteristics, a decrease in porosity and has a positive effect on corrosion resistance.

After thermal activation of aluminosilicates, their pozzolanic activity was investigated by the method of lime absorption from a lime solution. This method for determining the activity of mineral additives is traditional in Russia. Activity is determined by the $Ca(OH)_2$ amount, absorbed 1 g of supplement within 30 days.

The results obtained showed (Table 2) that the pozzolanic activity of the heat-treated components directly depends on the burning temperature and time parameters. Activity of Clay 1, heat-treated in the temperature range 650-800 °C within 60 min, varied within 320-410 mg CaO/g additives. The heat treated *A* at calcination temperatures 650-800 °C, Clay 2 absorbs 340-450 mg CaO/g additives.

Table 2. Pozzolanic Activity of Clays Thermally Activated at Different Burning Regimes.

Clays	T, °C	Activity, mg CaO/g	$T_{optimal}$, °C	τ, min	Activity, mg CaO/g
Clay 1	650	310	750	30	340
	700	340		60	360
	750	360		90	330
	800	330		120	310
Clay 2	650	340	700	30	400
	700	450		60	450
	750	390		90	395
	800	330		120	360

In this regard, for clays of different origins, it is necessary to select the optimal mode of burning thermal activation. Thus, for Clay 1, the optimum burning temperature is 750°C. At this temperature, complete decomposition of aluminosilicates has already taken place, which led to the amorphous phase formation, reflecting the amount of absorbed calcium oxide from the solution. A similar conclusion can be made for Clay 2. With an isothermal holding of 60 minutes, the optimal burning temperature is 700 °C

Since the heat treatment time affects the recrystallization processes, aluminosilicates were additionally calcined at the optimum temperature for each of them for 30, 90 and 120 minutes.

The data obtained (see Table 2) confirm the earlier studies [17-19], in which it was found that the calcination time significantly affects the pozzolanic clays activity. With a short isothermal clearing time (30 min), the minerals do not have time to completely decompose to the amorphous phase, and as a result, the activity does not reach its maximum. The studies have shown that for all the studied clays, the maximum activity is achieved by calcining for 60 minutes.

Thus, as a result of the experiment, it was found that the parameters (temperature and time) of thermal treatment of polymineral clays should be selected individually for each composition, depending on their chemical and mineralogical composition, which has a significant effect on their activity.

Using X-ray diffraction analysis, it was confirmed that when replacing Portland cement clinker with thermoactivated clays in an amount of 15%, the content of calcium hydroxide, which is an undesirable phase of cement stone in the mortar mixture, is significantly decreased (Figure 3), which causes an increase in strength (Figure 4).

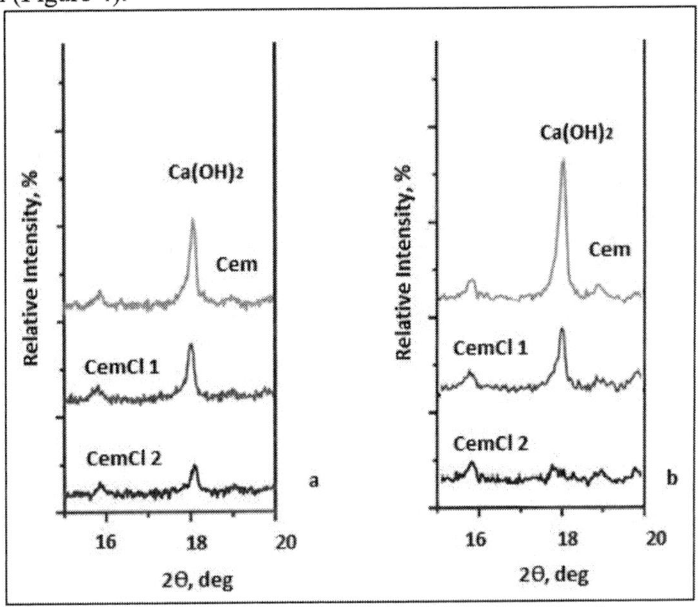

Figure 3. Radiographs of cement stone in the presence of thermally activated clays on 7 (*a*) and 28 (*b*) days of hardening.

For additive-free cement (Cem) the content of calcium hydroxide is the highest, which is characterized by the most intense peak, both at 7 and 28 days of hardening. Under the same conditions, the intensity of the cements' peaks with thermoactivated clays is significantly lower, which indicates their pozzolanic properties. The most pronounced effect has a sample CemCl-2 (the peak completely disappears on day 28), and for the sample CemCl-1 the peak practically does not differ in intensity on the days 7 and 28. This reflects the sample's reactivity, which also affects the strength indicators (Figure 4).

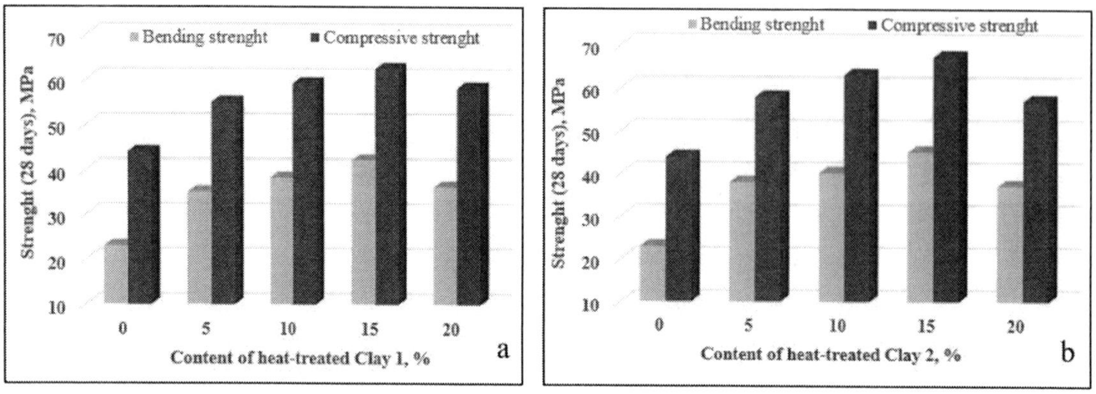

Figure 4. Strength of Portland cement in the presence of thermally activated clays: Clay 1 – a; and Clay 2 – b (28 days of hardening).

For the additive-free composition, the flexural strength was 24.6 MPa, and the compressive strength was 45.4 MPa. The introduction of 5% heat-treated clay increases the strength by 1.5 times. Cement stone reaches its maximum strength with a pozzolan content of 15%. With the introduction of the 20% additive, the strength starts declining. However, at the same time, the strength indicators are still significantly higher than for conventional cement. So, the composition with thermally activated Clay 1 in the amount of 15% (750 °C, 60 min) is characterized by the highest strength - for compression within 28 days - 62.4 MPa, for bending - 42.1 MPa. With the introduction of 15% Clay 2 (700 °C, 60 min) – compression resistance is 67.4 MPa, bending resistance – 45,1 MPa.

5. Summary

If the limestone dosage in cement composition is not more than 10%, the strength characteristics increase. Non-additive cement is gaining its strength unevenly, the critical hardening period should be considered the interval from 8 to 12 days from the mixing moment. However, in the presence of limestone, it is possible to achieve the structural transformations' stabilization that take place in the case of no addition cement and lead to a structure softening. The changes in the phase composition of the cement and the transition of some of the unstable (during the hardening period from 8 to 12 days) sulfa aluminate phases to carbo aluminate ones should be considered as the main reason for this effect. These conclusions are confirmed by the obtained results of electron microscopy and X-ray analysis.

The study of the thermally activated clays' effect on the Portland cement properties shows that with an increase in the pozzolanic additives' concentration up to 15% of the cement mass, the cement stone strength increases by 1.5-2 times compared to the non-additive composition. This is confirmed by the data of X-ray phase analysis, as well as the determination of the thermally activated aluminosilicates activity - the maximum activity (450 mg / g of the additive) is reached by the pozzolanic additive with a kaolinite content of about 90%.

The studied additives of sedimentary origin generally have a positive effect on the cement stone properties. However, a positive effect can be achieved only with strict adherence to the original components' ratio. When the ratio is violated, the properties of the composition deteriorate rapidly. The results obtained will be used in further research in the composition development of the composite cements containing both thermoactivated clays and a limestone component in the composition.

References

[1] Nikolaev N I and Leusheva E 2019 *Journal of Mining Institute* **236** 194-200.
[2] Chernyshov S E, Galkin V I, Ulyanova Z V and Macdonald D I M 2020 *Journal of Mining Institute* **242** 179-190.
[3] Taylor H F W 1990 England, London.

[4] Brykov A S J 2012 *Cement and its application* **4** 36-40.
[5] Yu C, Yuan P, Yu X and Liu J September 16-20 2019 *15th International Congress on the Chemistry of Cement Prague.*
[6] Ingram K, Poslusny M, Daugherty K and et al 1990 *Carboaluminate reactions as influenced by limestone additions.*
[7] Voglis N, Kakali G, Chaniotakis E and et al 2005 *Cement & Concrete Composites* **27 (2)** 191–196.
[8] Potapova E N and Korchunov I V 2020 *Technique and technology of silicates* **27 (2)** 45–52.
[9] Vance K, Aguayo M, Oey T et al 2013 *Cement & Concrete Composites* **39 (5)** 93–103.
[10] Potapova E, Sivkov S and Korchunov I 2018 *Durability and Sustainability of Concrete Structures* **326** 82.1–82.9.
[11] Stark J and Wicht B 2004 (Kiev, Oranta).
[12] Lottenbach B, Kulik D A, Matschei T et al 2018 *Cement and Concrete Research* **115** 472-506.
[13] Babushkin V I, Matveev G M, Mcheldanov-Petrosyan O P 1986 (Moscow, Stroyizdat).
[14] Potapova E, Zyryanov M and Dmitrieva E 2018 *20.International Baustofftagung. Weimar. Bundesrepublik Deutschland. Tagungsbericht* **2** 2-519/2-526.
[15] Almenares R S, Vizcaíno L M, Damas S, Alujas A and Martirena F 2017 *Case Studies in Construction Materials* **6** 225-232.
[16] Heisig A, Nigg K, Urbonas L and Heinz D 2018 *20.International Baustofftagung. Weimar. Bundesrepublik Deutschland. Tagungsbericht.* **1** 1-779/1-786.
[17] Potapova E and Dmitrieva E 2019 *J.Materials Science Today* **974** 319-324.
[18] Stanek T and Bohac M 2016 *J. Cement and its application* **4** 103-107.
[19] Scrivener K L, John V M and Garther E M 2016 *United National Environment Programme: Paris* **50** 85-93.

IOP Conf. Series: Materials Science and Engineering 1083 (2021) 012034 doi:10.1088/1757-899X/1083/1/012034

Strength and deformation of bending normal sections made of concrete frame structure

D R Mailyan, G V Nesvetaev, Yu I Koryanova* and E V Lesniak

Don State Technical University, 1, Gagarin sq., Rostov-on-Don, 344010, Russia

E-mail: Koryanova.yi@mail.ru

Abstract. Concrete of the frame structure, obtained by the separate concreting technology by immersing a coarse aggregate in a low-viscosity mortar component due to increased concentration of coarse aggregate, have a number of advantages in comparison with the traditional concretes of vibration compaction.

1. Introduction

In order to effectively use the frame structure in reinforced concrete bent concrete structures, it is necessary to study the physical and mechanical properties of such concretes: cubic and prismatic strength, elastic modulus, longitudinal and transverse deformations. In addition, to study the operation of structures under load, it is necessary to have experimental data obtained on the experimental basis.

2. Methodology

Portland cement CEM I 42,5; quartz sand with a fineness module of 1.3 as fine aggregate; sandstone crushed stone M 1200 with factions 5 – 20 mm as coarse aggregate; coarse aggregate void 0,408 with density in a piece 2703 kg/m^3; superplasticizer Vinavil Flux 3 were used for the research.

The composition of the mortar component: C/S = 1/1,11 by mass. Crushed stone from quartz sandstone with elastic modulus 40300-41800 N / mm^2 was used as a coarse aggregate.

Compressive Strength Samples - Cubes 150x150x150 mm. Samples for determining the elastic modulus - prisms 150x150x600 mm.

The following indicators have been determined: limit of cubic and prismatic compressive strength, modulus of deformations, dependence of "stress-strain", moment of beams cracking, dependence "moment-deflection", limiting value of bending moment, relative deformations at the level of compressed and stretched reinforcement (Figure 1).

The rectangular beams 125x250 mm with a span of 2400 mm, tested by the concentrated forces in the span thirds were used as experimental samples.

The beams were made of concrete with a design strength of 46 MPa. The beams were reinforced 2∅12 A500 in the extended area and 2∅6 A400 in the compressed zone.

The test procedure included stepwise loading with a short-term static load with the beams being brought to failure. The tests were carried out on a special stand equipped with a mechanical jack. The magnitude of the loading steps was 5% and 10% of the expected maximum load, respectively, at the initial stages and after the cracks' appearance.

Figure 1. Procedure for testing the reinforced concrete beam.

3. Results

The research results are presented in Table 1 and in Figures 2 - 6.

Table 1. Research results

Index	Value
Prismatic strength, N/mm^2	46.7
Elastic modulus, N/mm^2	37420
Flexural moment maximum, T·M	3.19
Deflection maximum, mm	18.1

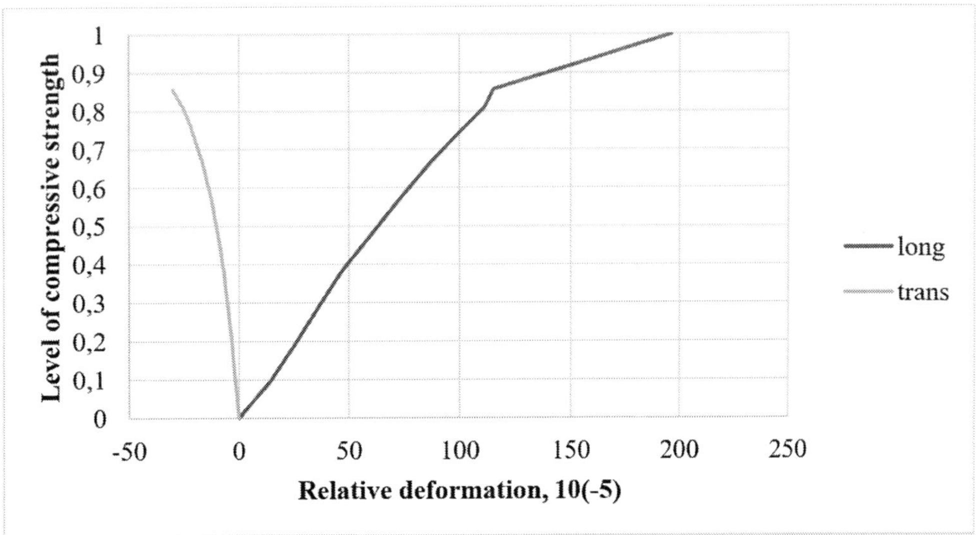

Figure 2. Relationship between level of compressive strength and relative deformation of concrete. long – longitudinal deformation; trans – transverse deformation.

IOP Conf. Series: Materials Science and Engineering 1083 (2021) 012034 doi:10.1088/1757-899X/1083/1/012034

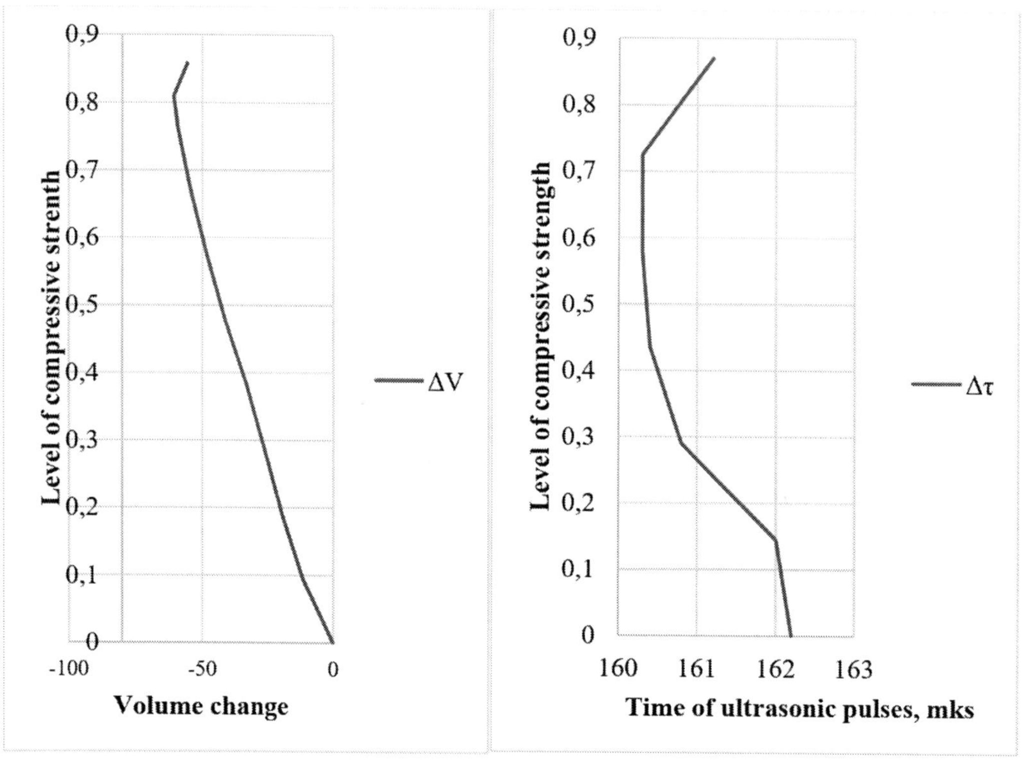

Figure 3. Relationship between level of compressive strength and volume change of concrete (left) and time of ultrasonic pulses (right).

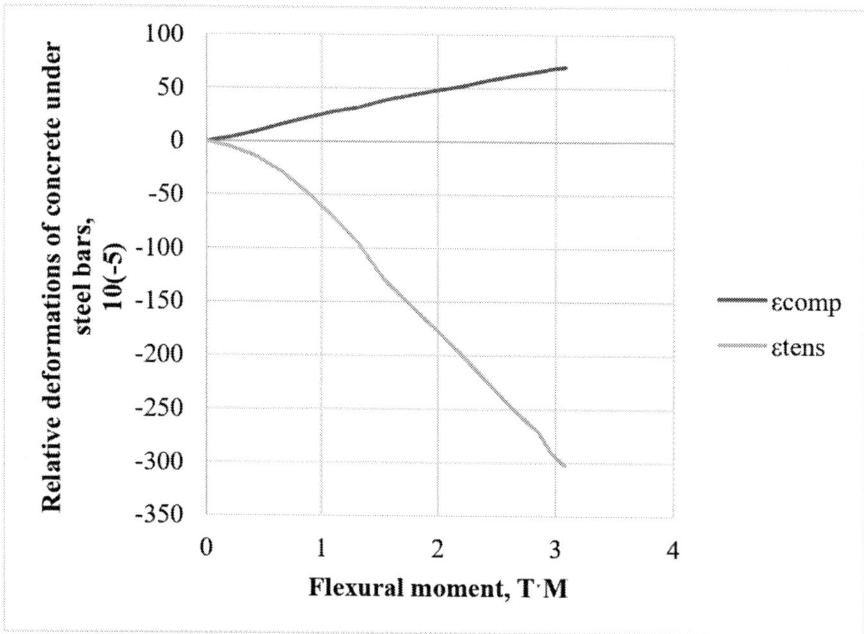

Figure 4. Relationship between flexural moment and relative deformations of concrete under steel bars.

εcomp – concrete under compressive strength; εtens - concrete under tensile strength.

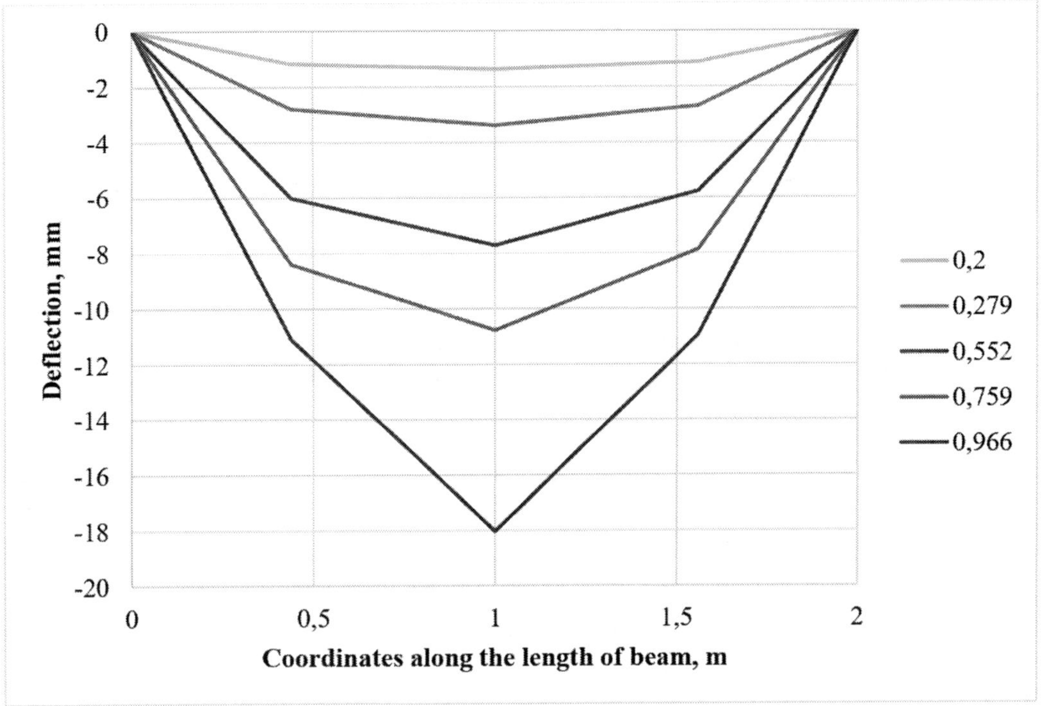

Figure 5. Relationship between deflection and level of flexural moment.
0,2 … 0,966 – level of flexural moment.

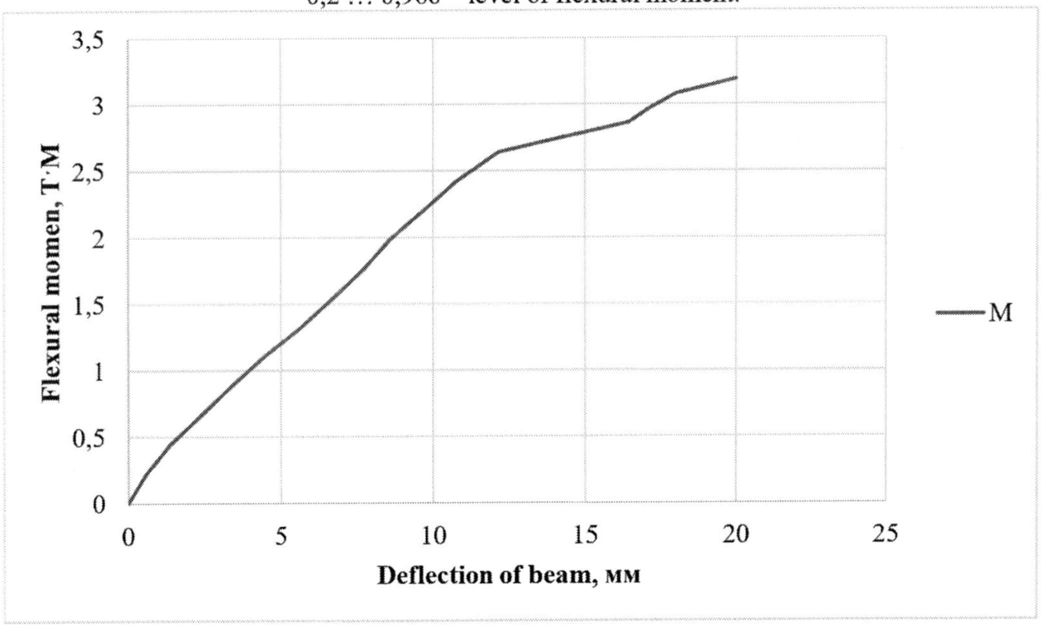

Figure 6. Relationship between deflection of beam and level of flexural moment.

4. Discussion

The beams failure started with reinforcement in the tensile zone and ended with concrete in the compression zone. The beams failure was accompanied by the formation of a large number of cracks and their opening, an increase in deflections and deformations and a reduction in the height of the compressed zone. Compared with the beams made of ordinary concrete [1-3], the number of cracks was greater, but the width of their opening was less.

With the end of the stable deformation zone, the formation of concrete spalling in the compressed zone started. The stability loss of the compressed reinforcement occurred simultaneously with the delamination of concrete spalls. Moreover, the area of the spalls was less than in ordinary concrete [4, 5], but their number is greater.

The strength of the normal sections of the test beams was characterized by the value of the greatest moment in the pure bending zone. The stresses in the concrete of the extreme compressed fiber and tensile reinforcement were determined using the experimental diagrams "$\delta - \varepsilon$" by measured deformations.

Reinforcement failures that could not be measured were calculated using the hypothesis of flat sections through the deformation of the extreme compressed fiber and the height of the compressed zone.

Analysis of the critical values of the experimental parameters of the stress-strain state of normal sections shows that their change obeys, in general, the laws characteristic of bending elements made of ordinary concrete [5-8]. The beams showed a stable decrease in the height of the compressed zone with increasing load.

All beams' extreme compressed fiber experimental deformations showed that their values, when the maximum load is reached, are on the descending branch of the compression diagram of the experimental concrete prisms. In this case, the stresses in the extreme compressed concrete fibers are significantly lower than the prismatic strength and reach $(0{,}6 - 0{,}8)\ R_b$.

Framing concrete has a positive effect on the compressive concrete strength.

This is fully manifested in flexible elements, the bearing capacity of which is higher than that of the elements made of ordinary concrete [7-9].

Reinforcement failure at the maximum moment turned out to be higher than the deformations when the yield point was reached.

In order to test the possibility of applying the methodology of norms when calculating the strength of the reinforced concrete elements made of concrete of the frame structure, the theoretical values of the breaking moments for the author's experimental beams were calculated. Comparing them with the experimental ones shows the excess of the experimental moments over the theoretical ones (up to 25%).

In general, such results are typical for the beams made of ordinary concrete [8-12].

At the same time, the resistance of the frame structure reinforced concrete beams made of concrete has a number of features. For the purpose of their determination, the experimental values of the compressed zone diagram completeness ξ_o were revealed as well as the boundary value ξ_R.

The experiments show that the coefficient of completeness of the stress diagram ξ_o of the frame structure for the elements made of concrete is 10-15% lower than for equal strength heavy concrete [13-16].

Reduced values ξ_o and ξ_R testify to the less practical work nature of the bending elements made of concrete of the frame structure at the limiting stage. These data correspond to the diagrams "$\delta - \varepsilon$" concrete prisms of the frame structure, which, in comparison with the diagrams for heavy concrete [15-19], are more straightened and have flatter ascending and steep descending branches, which reflects their less plastic deformation.

Comparison of the experimental deflections in the range of operational loads with the calculated ones according to the norms showed their satisfactory convergence.

The crack resistance of the beams was estimated from the moments of formation and the width of normal cracks. The process of cracking and crack opening in general does not differ from the beams made of ordinary concrete [1-12]. At the same time, the characteristic drop in the load at the moment of the first crack formation and a sharp increase in deformations are less pronounced.

The relative value of the cracking moment was 0.4-0.5, which is significantly higher than in the ordinary structures [20-25].

5. Summary

It is possible to assume, that the frame structure bending reinforced concrete elements made of concrete have a higher resistance to external loading. Their strength and crack resistance are higher, and the width of crack opening and deflections is lower than the similar bending elements made of ordinary concrete.

Therefore, according to the experimental reinforced concrete beams studies results, it can be concluded that the concrete of the frame structure has a positive effect on the operation of bending structures under load.

References

[1] Nesvetaev G V, Lesniak E V, Kolleganov A V and Kolleganov N A 2020 *Materials science forum* **1011** 66-71.

[2] Kaprielov S S, Sheinfeld A V, Kardumyan G S and Dondukov V G 2006 *Concrete and reinforced concrete* **2** 2-7.

[3] Kaprielov S S, Sheinfeld A V and Kardumyan G S 2010 New modified concretes (Printing house "Paradise", Moscow).

[4] Kaprielov S S, Sheinfeld A V, Karpenko V I and Kuznetsov E N 2003 *Concrete and reinforced concrete* **6** 8-12.

[5] Komohov P G 2007 *Cement, concrete, mortars and dry mixes: Creep. Part 1* (NPO "Professional", S.-Pb.).

[6] Solomatov V I and Potapov Y B 1986 *Math. universities: Construction and architecture* **7** 57-8.

[7] Solomatov V I and Tahir Shah Md Taher M K 1989 *Intensive technology of concrete* (Stroyizdat, Moskow).

[8] Bazhenov Y, Salamanova M, Murtazaev S-A and Saidumov M 2016 *Int. j. of environmental & sci. education* **11 (18)** 12779 – 86.

[9] Bleschik N P, Cancer A N and Kotov D S 2009 *Problems of modern concrete and reinforced concrete* **2** 132 –58.

[10] Yen T, Pann K S and Lin S K 2005 *Concrete and reinforced concrete - the path of development: sci. works conf.* **3** 81 –7.

[11] Okamura H and Ouchi M 2003 *J. of advanced concrete technol.* **1** 5-15.

[12] Domone P L 2007 *Cement and Concrete Composition* **1** 1 – 12.

[13] Nesvetaev G V and Kardumyan G S 2013 *Concrete and reinforced concrete* **5** 6-8.

[14] Nesvetaev G V, Mailyan D R and Davidyuk A N 2011 *Concrete technology* **1-2 (54-55)** 57-9.

[15] Mkrtchyan A M, Aksenov V N, Mailyan D R, Blyagoz A M and Smorgunova M V 2013 *New technologies* **3** 135-143.

[16] Nesvetaev G V 2013 *Concretes* (Feniks, Rostov-on-Don).

[17] Kaprielov S S, Sheinfeld A V, Kardumyan G S and Dondukov V G 2006 *Eight CANMET/ACI Int. Conf. on superplasticizers and other chemical admixtures in concrete* 87-102.

[18] Nesvetaev G V, Koryanova Y I and Zhilnikova T N 2018 *Matec web of conferences* **196.**

[19] Nesvetaev G V, Koryanova Y I and Kolleganov A V 2018 *Solid State Phenomena* **284** 963-9.

[20] Yuvaraj L Bhirud and Keshav K Sangle 2017 *OJCE* **1(7)** 130-40.

[21] Demyanova V S, Kalashnikov V I and Ilyina I E 2002 *Construction materials* **9** 4-6.

[22] Zotkin A G 2014 *Concrete technol.* **9** 22–6.

[23] Yuvaraj L Bhirud and Keshav K Sangle 2017 *OJCE* **7 (1)** 26-30.

[24] Udodov S A and Bychkov M V 2013 *Sci. studies* **4 (17)** 41.

[25] Nazhuev M P, Yanovskaya A V, Kholodnyak M G, Khalyushev A K, Scherban E M and Stelmach S A 2017 *Eng. J. of Don* **3 (46)** 99.

IOP Conf. Series: Materials Science and Engineering 1083 (2021) 012035 doi:10.1088/1757-899X/1083/1/012035

Modeling the dependence of the foam concrete structural quality coefficient on the speed and duration of mixing the foam concrete mixture

A N Pavlov, L R Mailyan, S A Stel'makh, E M Shcherban' *, A K Sysoev

Don State Technical University, 1, Gagarin Square, Rostov-on-Don, 344000, Russia

E-mail: au-geen@mail.ru

Abstract. Experimental studies of the technological factors' influence on the physical and mechanical properties of foam concrete were carried out. Additive Portland cement was used as a binder. Finely ground quartz sand of fractions 0.08–0.16 and 0.16–0.315 mm was used as a filler. Protein foaming agent Arecom-4 was used as a foaming agent. The porization time of the aerated concrete mixture was controlled using a stopwatch, and the rotation speed of the vertical rotor mixer shaft was controlled with a digital photo-tachometer. In the experiments, the following factors were varied: mixture porization time, mixing speed. The main characteristics of foam concrete were determined by testing the prototypes or their fragments according to the methods of the relevant regulatory documents using verified equipment. In the research laboratory of the DSTU, the studies on the influence of the speed and duration of mixing the foam concrete mixture on the density, strength and coefficient of the resulting foam concrete structural quality were carried out. The article presents the experimental dependences of the investigated characteristics on the speed and time of mixing the foam concrete mixture. The parameters of density and strength show a monotonic decline during pore formation, while the structural quality coefficient exhibits its maximum. The behavior of density, strength, coefficient of structural quality can be described in a single model using the parameters that reflect bonds loosening, a general decrease in the number of bonds in the material and the bonds' change spatial distribution.

1. Introduction

The most generalized definition of foam concrete, compiled from the versions of various authors, is as follows: foam concrete is an artificial stone material obtained as a result of hardening a mixture of a hydraulic binder, sealing compound, fine aggregate and building foam, characterized by the required porosity degree [1].

Foam concrete porosity is achieved mechanically by mixing the concrete mixture with the foam mass and is uniformly distributed open, communicating or closed pores of three types: cellular, capillary and gel. The optimal volume of all pores (for heat-insulating foam concrete) varies within 83 – 86 % [2]. Porosity is also classified as true, open and closed. The volume of true porosity depends on the solid phase content (interpore partitions) in the material, which, in general, determines the mechanical and operational properties of the products. Therefore, it is necessary to adhere to rational values of the total porosity volume in order to avoid a sharp decrease in the strength and increase in the deformative indicators of foam concrete.

Open porosity deteriorates the operational properties of the material due to the intensive migration of moisture and gases into the product, leading to acceleration of the solid phase corrosion processes. Therefore, in foam concrete they tend to increase the proportion of closed pores [3].

Closed porosity is a closed non-interconnected bubble that provide the increased operational resistance of the thermal insulation material. An increase in the proportion of closed porosity in the total porosity volume is controlled during the production of heat-insulating materials by increasing the density of interpore partitions, regulating the rheological properties and the formation of air cells [4].

Also, when assessing the foam concrete porosity, it is necessary to take into account the nature of the air cells' structure, that is, the pores size, their shape, and distribution heterogeneity. Ideally, the pores should be as small as possible and evenly distributed throughout the material. Reducing the pore size is achieved by increasing the dispersion of the binder and filler, as well as by regulating the rheological properties and the mixture porosity kinetics [5].

Taking into account the fact that the foam concrete production cost is 1.4 - 1.8 times lower than its counterpart aerated concrete [6], the issue of improving the characteristics of foam concrete (especially heat-insulating) becomes even more urgent and requires a deeper and more thorough application of recipe-technological variations [7-16].

2. Materials and Methods

To accomplish the assigned tasks, the experimental studies on the influence of technological factors on the physical and mechanical properties of foam concrete were carried out.

2.1. Materials

For the foam concrete production, the binders, aggregates and foaming agents are used, in accordance with the requirements of GOST 25485-2019 and BC 277-80.

No additive Portland cement CEM I 42.5N of the industrial holding "Eurocement group" was used as a binder in accordance with GOST 31108-2016, corresponding to PC 400-D0 in accordance with GOST 10178-85.

The properties of this cement are given in Table 1; the mineralogical composition - in Table 2.

Table 1. Properties of Portland cement used in research as a binder

Material name	Properties
Portland cement without additives CEM I 42,5H "Eurocement group"	Activity R_c = 43.0 MPa Setting Start– 110 min Setting End – 390 min The uniformity of the volume change withstands Normal density of cement paste (NDCP) – 26 % Flexural strength at 28 days – 6.0 MPa True density – 3.16 g/cm^3 Bulk density – 1182 kg/m^3 Grinding fineness (sieve residue № 008) – 5.9 %

Table 2. Mineralogical composition of the used cement clinker.

Manufacturer	C_3S, %	C_2S, %	C_3A, %	C_4AF, %
"Eurocement group"	64.0	14	7.4	14

Fine quartz sand with fractions 0.08–0.16 and 0.16–0.315 mm was used as a filler in the study of the technological factors' influence on the aerated concrete properties. The properties of sand were determined according to the methods of GOST 8735-88 and compared with the requirements of GOST 8736-2014.

The main properties of sand are presented in Table 3.

Table 3. Applicable Filler's Properties.

Material name	Properties
Quartz sand	Size module (M_s) – 1.2 True density – 2.65 g/cm^3 Bulk density – 1410 kg/m^3

Protein foaming agent Arecom-4 was used as a foaming agent. Its main properties are shown in Table 4.

Table 4. Properties of the applied foaming agents.

Name	Density, g/cm^3	Foam ratio	Hydrogen exponent (pH)
Arecom -4	1.12	10	TU 31-10

2.2. Methods

The production of foam concrete mixtures was carried out in laboratory conditions.

Prototype manufacturing. Before adding to the mixture, the sand was sieved through a standard set of sieves to obtain the required fractions. The pre-dosed components - water, Portland cement and sand - were sequentially loaded into a special container for subsequent mixing using a WiseStir HS-120A vertical rotor mixer with analogue mixing speed control. At the same time, the aerated concrete mixture porization duration also changed [17-22].

To study the technological factors' influence on the foam concrete properties, the following ratio of the mixture components was taken:

- ratio of sand and cement S/C = 0.4;
- water-solid ratio W/S = 0.47;
- the amount of injected foaming agent – 0.3 % from the cement mass.

Slip viscosity remained approximately constant (270-280 mm spread).

After loading the materials, their mixing started for a set time until a homogeneous mixture was obtained. Then the viscosity of the slip was determined on the Suttard's device and, with the mixer activator stopped, the dosed foaming agent was poured into it. Then the mixture was porous.

The porosity time of the aerated concrete mixture was controlled using a stopwatch, and the rotational speed of the vertical rotor mixer shaft was controlled with a digital photo-tachometer (model DT2234B), shown in Figure 1.

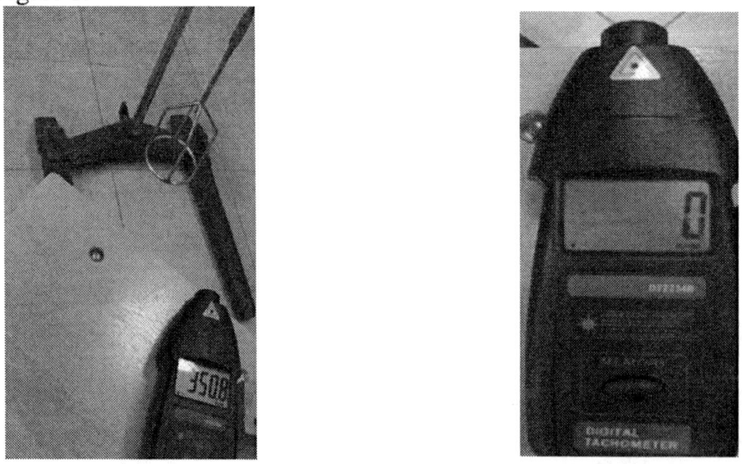

(a) (b)

Figure 1. Measurement of the stirring shaft rotation speed: *(a)* digital photo-tachometer; *(b)* photo-tachometer measurement of the rotation speed of the stirring body.

The following factors were varied in the experiments:

- time of mixture porization (from 2 to 6 with an interval of 1 min);
- stirring speed (from 500 to 1500 with an interval of 250 rev/min).

Slip viscosity was determined using a Suttard's viscometer according to the method of paragraph 6 of GOST 23789-2018.

The average density of the foam concrete mixture was determined by the method of paragraph 5 of GOST 10181-2014.

The prepared foam concrete mixture was poured by gravity flow into the prepared (pre-lubricated with mineral oil) four metal molds, 3 cells each with a cell side of 7.07 cm. The molds were lightly compacted by tapping on a metal table and smoothed with a ruler.

After that, the samples were hardened in natural conditions. After 3 days after the mixtures' preparation, the samples were removed from the molds and they continued to harden in natural conditions until they reached 28 days of age.

Testing prototypes. In each experiment of the conducted research, 12 sample cubes of size 7.07×7.07×7.07 cm. The main characteristics of foam concrete were determined by testing the prototypes or their fragments according to the methods of the relevant regulatory documents on the verified equipment.

The compressive strength of the samples was determined according to the method of GOST 10180-2012.

The breaking load on the samples was recorded on a breaking machine P-10. Press characteristics: range of indications 0…20 kN, measuring range 4…20 kN.

The average density in a dry state was determined according to the GOST 12730.1-78 method according to the results of measuring the dimensions of each sample (length, width and height) and weighing with subsequent clarification of its moisture content by drying to a constant mass of foam concrete fragments taken from the samples. The average concrete density of a samples series was calculated as the arithmetic mean of the all samples' test results [17-22].

3. Results and Discussion

Research laboratory of DSTU conducted the research on the effect of speed (n_{rev}) and mixing time (*t*) foam concrete mixture for density (ρ), strength *(R)* and the constructive quality factor (*CQF*) the resulting foam concrete. Figure 2-4 and Figure 5-7 show the experimental dependences (+ signs) of the investigated characteristics on, respectively (n_{rev}) (at $t = 4$ min) and *t* (at ($n_{rev}=10^3$ min^{-1}). The density and strength parameters show a monotonic decline during pore formation, while *CQF* detects a maximum that correlates with the performance. The authors of this work also note the non-monotonic behavior *CQF* depending on the mixing speed and duration. The task of this work was to comprehensively simulate the effect of mixing conditions on ρ, *R* and *CQF*.

Let us describe the density by the expression (1), where *m* and *V* – are the weight and volume of foam concrete:

$$\rho = \frac{m}{V} \tag{1}$$

The mass of foam concrete does not change during pore formation, while the density decreases due to an increase in the mixture volume (ΔV).

$$V = V_0 + \Delta V. \tag{2}$$

Here V_0 – is the volume of a given mass without pore formation. Let us substitute the expression (2) into (1)

$$\rho = \frac{m}{V_0\left(1+\frac{\Delta V}{V_0}\right)} = \frac{\rho_0}{1+\frac{\Delta V}{V_0}}, \tag{3}$$

$$\rho_0 = \frac{m}{V_0} \tag{4}$$

Here ρ_0 - is the mixture density without pore formation.

ΔV– an increase in the volume of this mass occurs due to the involvement of atmospheric air in the foam concrete mixture during the mixer operation, which depends on the speed and time of mixing.

$$\frac{\Delta V}{V_0} = K_V n_{rev} t \tag{5}$$

Here K_V – is the empirical coefficient of proportionality. Substituting the expression (6) into expression (3), we obtain a model expression for describing density:

$$\rho = \frac{\rho_0}{1+K_V n_{rev} t} \tag{6}$$

The calculation results using the expression (6) at K_v=15·10^{-5}, ρ_0 =990 kg·m^{-3} are shown in Figure 2 and 5 (solid curves) and describe the dependence of the density as a function of the velocity (n_{rev}) (Figure 2) in a unified model; and on the duration of mixing (*t*) (figure 3).

In view of the foam concrete heterogeneity, we estimate its strength as the bond energy (W) per volume unit ($V_c < V$) , determining strength:

$$R = \frac{W}{V_c}, \tag{7}$$

$$W = N_b \varepsilon. \tag{8}$$

Here N_b is a number of bonds in volume; ε– one bond energy.

$$N_b = N_{b0} - N_{pf}. \tag{9}$$

Here V_p is an average pore size, N_{pf} is the number of bonds not realized due to pore formation during the formation of one pore.

Let us take into account the expressions (9), (10), (5) in the expression (8).

$$W = \varepsilon N_{b0}\left(1\text{-}\frac{N_{pf}}{N_{b0}}\right) = \varepsilon N_{b0}\left(1\text{-}\frac{\Delta V}{V_{\Pi}}\frac{N_{pf}}{N_{b0}}\right) = \varepsilon N_{b0}(1\text{-}K_R n_{rev} t). \tag{10}$$

Here K_R – is an empirical proportionality factor.

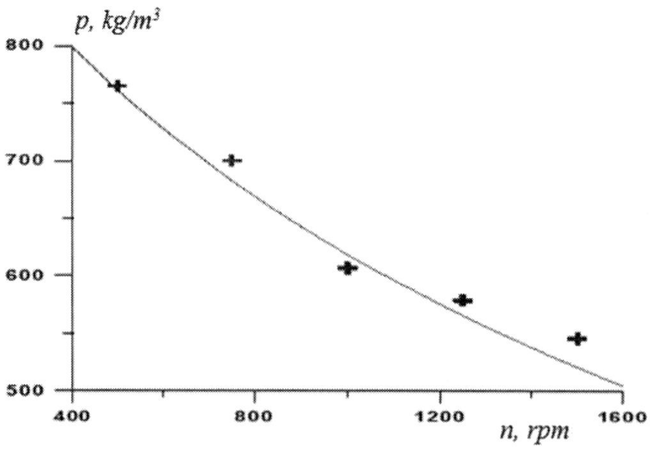

Figure 2. Experimental (+ signs) and theoretical (solid line).

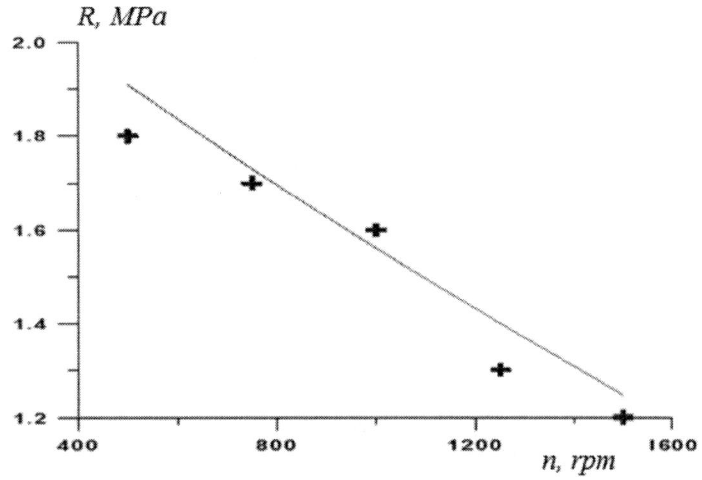

Figure 3. Experimental (+ signs) and theoretical (solid line) dependences R from n_{rev}

By the analogy with the expression (2), we write down the expression for V_C:

$$V_c = V_{0c} + \Delta V_c = V_{0c}\left(1 + \frac{\Delta V_c}{V_{0c}}\right) = V_{0c}\left(1 + K_{V_c} n_{rev} t\right) \quad (11)$$

Here ΔV_C– is an increase due to pore formation of the volume determining strength; K_{VC} is an empirical proportionality factor.

Taking into account the relations (11) and (12) in the expression (7), we obtain a model expression for describing strength.

$$R = R_0 \frac{1 - K_R n_{o6} t}{1 + K_{V_c} n_{rev} t}. \quad (12)$$

The calculation results using the expression (12) at $K_R = 6 \cdot 10^{-5}$, $K_{VC} = 3 \cdot 10^{-5}$ and $R_0 = 2.3$ MPa are shown in Figures 3 and 6 (solid curves) and describe the dependence of strength as a function of speed n_{rev} (Figure 3) in a unified model, and on the mixing duration (t) (Figure 6).

The quantity CQF is described by the expression (14) where we substitute the relations (5) and (13).

$$CQF = \frac{R}{\rho} \quad (13)$$

Taking into account the relations (5) and (13) in the expression (14), we obtain a model expression for describing CQF.

$$CQF = \frac{R_0}{\rho_0} \frac{(1 - K_R n_{rev} t)(1 + K_V n_{rev} t)}{(1 + K_{V_c} n_{rev} t)} \quad (14)$$

The calculation results using the expression (15) at $K_V = 15 \cdot 10^{-5}$, $K_R = 6 \cdot 10^{-5}$, $K_{VC} = 3 \cdot 10^{-5}$ and $R_0 = 2.3$ MPa, $\rho_0 = 990$ kg·m^{-3} are shown in Figures 4 and 7 (solid curves) and reflect the dependence CQF like speed (n_{rev}) (Figure 4) in a unified model, and on the mixing duration (t) (Figure 7).

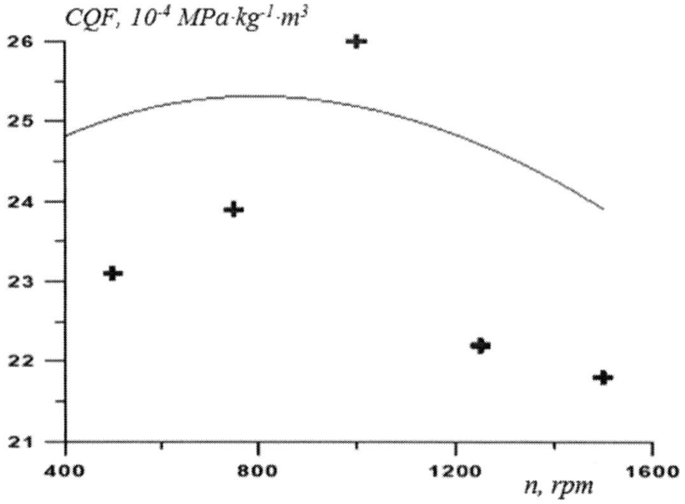

Figure 4. Experimental (+ signs) and theoretical (solid line) CQF dependences from n_{rev}

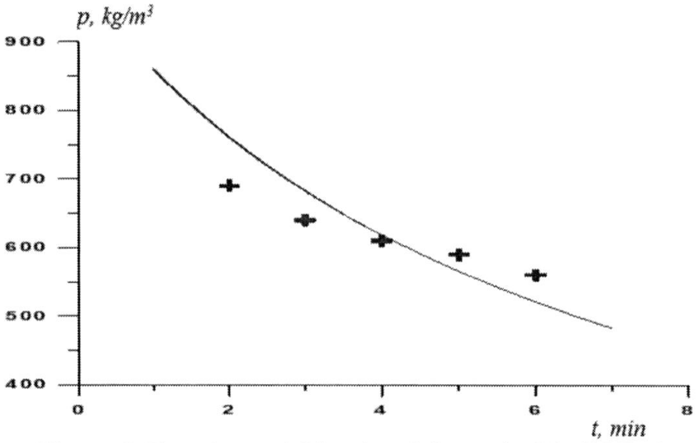

Figure 5. Experimental (signs) and theoretical (solid line) dependences ρ from t

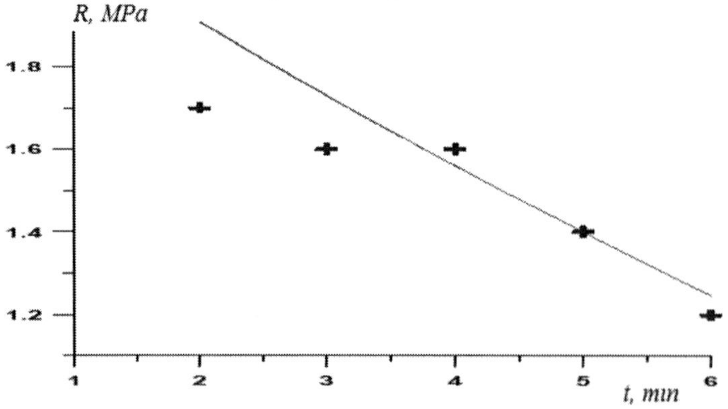

Figure 6. Experimental (+ signs) and theoretical (solid line) dependences R from t.

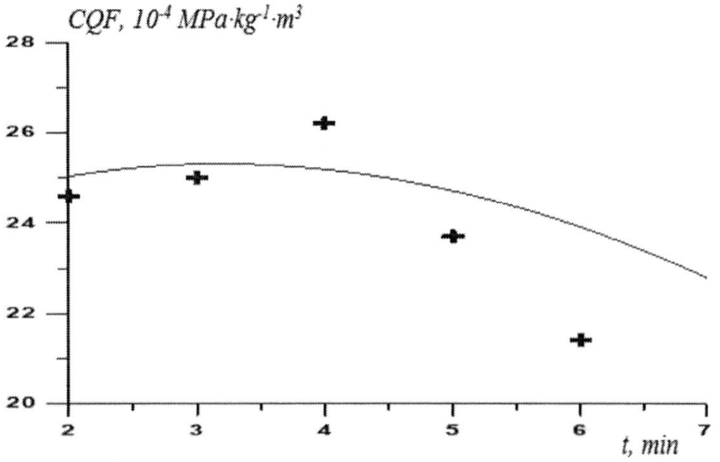

Figure 7. Experimental (+ signs) and theoretical (solid line) *CQF* dependences from *t*.

4. Summary

Density, strength, CQF behavior can be described in a single model using the parameters K_v, K_R, K_{VC} which reflect the loosening of bonds, a general decrease in the number of bonds in the material and a change in the spatial bonds distribution.

References

[1] Zakorchemnyi Yu O 2004 *Managing the structure and properties of foam concrete* (Odessa, OGASA) 162.
[2] Gorlov Yu P, Merkin A P and Ustenko A A 1980 *Technology of thermal insulation materials*. (Moscow, Building Publishing) 397.
[3] Silaenkov E S 1986 *Cellular Concrete Products Durability* (Moscow, Building Publishing) 176.
[4] Geniev G A, Kissyuk V N, Levin N I and Nikonova G A 1978 *Strength of lightweight and cellular concrete under difficult stress conditions* (Moscow, Building Publishing) 166.
[5] Creuse M 1997 *Composants* (Paris, Pelagrave) 176.
[6] Konovalov V M 2003 *Building materials* **6** 6–7.
[7] Kim J-S, Chung S-Y, Han T-S, Stephan D and Elrahman M A 2020 *Construction and Building Materials* **248** 118637.
[8] Asadzadeh S and Khoshbayan S 2018 *Construction and Building Materials* **170** 101-110.
[9] Yaseen Z M, Deo R C, Hilal A, Abd A M, Bueno L C, Salcedo-Sanz S and Nehdi M L 2018 *Advances in Engineering Software* **115** 112-125.
[10] Mugahed Amran Y H, Farzadnia N and Abang Ali A A 2015 *Construction and Building Materials* **101 (1)** 990-1005.
[11] Vinith Kumar N, Arunkumar C and Srinivasa Senthil S 2018 *Materials Today: Proceedings* **5 (2) 3** 8753-8760.
[12] Hilal A A, Thom N H and Dawson A R 2015 *Construction and Building Materials* **75** 227-233.
[13] Hilal A A, Thom N H and Dawson A R 2015 *Construction and Building Materials* **85** 157-164.
[14] Ramamurthy K, Kunhanandan Nambiar E K and Indu Siva Ranjani G 2009 *Cement and Concrete Composites* **31 (6)** 388-396.
[15] Schumacher K, Saßmannshausen N, Pritzel C and Trettin R 2020 *Construction and Building Materials* **264** 120167.
[16] Kunhanandan Nambiar E K and Ramamurthy K 2006 *Cement and Concrete Composite* **28 (9)** 752-760.

[17] Shcherban' E M, Gol'tsov Yu I, Tkachenko G A and Stel'makh S A 2012 *Engineering journal of Don* **3.**

[18] Shcherban' E M, Tkachenko G A, Gol'tsov Yu I and Stel'makh S A 2012 *Modern problems of science and education* **1.**

[19] Pavlov A N, Gol'tsov Yu I, Stel'makh S A and Shcherban' E M 2015 *Science Review* **10-1** 147-150.

[20] Shcherban' E M 2014 *Regulation of structure formation and properties of insulating foam and fibroconcrete, activated by low-energy variable electrical and physical impact, technological and compounding factors* (Rostov-on-Don, RGSU) 216.

[21] Stel'makh S A 2014 *Influence of the parameters of low-energy variable electric fields on the properties of activated thermal insulating foam and fibropen concrete* (Rostov-on-Don, RGSU) 185.

[22] Mashkin N A and Barten'eva E A 2016 *Herald. Technical and physical-mathematical sciences* **3** 21-30.

Investigation of the technological factors' influence on the formation of the repair compositions' properties

I A Serebryanaya*, E A Shlyakhova, I O Egorochkina, A V Serdyukova, R G Sarkisyan

Don State Technical University, 1, Gagarin Square, Rostov-on-Don, 344000, Russia

E-mail: silveririna@mail.ru

Abstract. Power transmission lines (PTL) belong to strategic facilities, on the reliability of which the life support of the country depends. Most of the currently operated PTLs were put into operation in the twentieth century, and now there is a problem of assessing their technical condition, repairing, if necessary, and maintaining them in working order. The defects appeared during the operation of PTL reinforced concrete elements have different localization and manifestation degree, but destruction of the concrete surface layer with reinforcement exposure is a characteristic, more often at the base of the PTL frame. The aim of the work is to select the optimal composition of the repair mortar based on non-shrinking and expanding cements, ensuring reliable adhesion of the repair mortar to the concrete of the structure, which will allow restoring the bearing and operational characteristics in the repaired PTL reinforced concrete elements at the design level.

Introduction

The considerable experience accumulated at present in the operation of power transmission lines (hereinafter PTL) both in our country and around the world gives an opportunity to judge the practical operational reliability of all PTL elements [1].

Most of the currently operated PTLs were commissioned back in the twentieth century, and over the past few decades, the deterioration of high voltage power grids, according to most experts, has increased by 40%. Taking the high degree of deterioration of existing facilities into consideration it is necessary to periodically carry out diagnostic measures in order to maintain them in working condition, [2], as well as repair work.

The search for the new solutions for the PTL technical condition effective restoration and the problem of timely diagnosis of damage are the urgent tasks of the modern electric power industry [2, 3].

The measures known to date to restore the operational reliability of PTL reinforced concrete elements in most cases, they require significant expenditures of materials and special repair compositions [4]. These activities are very time consuming. Their implementation requires special equipment and appropriately trained personnel. In addition, the complexity is caused by the diversity of the raw material base for the repair compounds' manufacture in different climatic regions of the country, and often the low quality of the raw materials used [5].

The relevance of this work is due to the fact that PTL, representing a complex technological and constructive complex containing a significant number of components, refer to strategically important facilities in any country and the country's life support depends on their reliability.

We have carried out the experimental studies on the effective repair compositions' development for the restoration of the operated reinforced concrete PTL frames. These structural elements perform the main load-bearing function in the PTL composition and the long-term and trouble-free operation of the entire PTL. Destruction of PTL support frames can lead to early failure of the entire line, which will require its partial or complete dismantling.

Defects appearing during the PTL reinforced concrete elements' operation can have different localization and manifestation degree [5]. Outwardly, this can manifest itself in the form of various destructions (Figure 1).

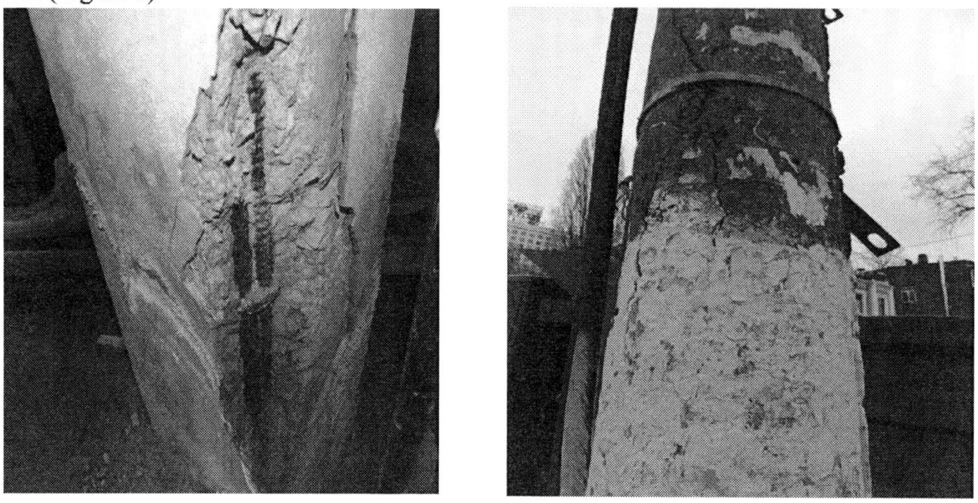

a) solid PTL frames *b)* centrifuged frames PTL

Figure 1. Defects that appeared at the stage of operation of PTL reinforced concrete elements

An alternative to replacing PTL individual structural elements or even the entire sections, the restoration of reinforced concrete structures as a result of repair work can serve. Bearing and performance characteristics in restored PTL reinforced concrete elements can be maintained at the level of design indicators, provided that the repair restoring compounds adhere to the restored structures.

The experience accumulated to date in the use of conventional mortars or sand concretes based on Portland cement for grouting precast reinforced concrete structures indicates that such compositions do not sufficiently provide the required structures' solidity. This is mainly due to the fact that such compositions undergo shrinkage during the setting and hardening process. In this regard, the problem of creating expandable and non-shrinking cements for the designated purposes is urgent.

In the 70s of the last century [6], a method for the preparation of expandable waterproof cements, which are characterized by the fact that they are subject to shrinkage during the hydraulic hardening process, was developed. The cement used in this case has a high setting rate and the achievement of the required strength degree. The used binder is obtained from a mixture of aluminate cement, two-water gypsum stone ($CaSO_4 2H_2O$) and tetra calcium aluminate, which are finely ground in a ball mill. The tetra calcium aluminate is obtained by autoclaving a mixture of alumina cement and lime in a mass ratio of 1: 1 with a water-solid ratio W / T of 0.30. Next, the resulting product is dried and milled. The advantage of the developed expanding cement is that its expansion degree can be regulated within a fairly wide range by changing the ratio between the components.

It should be noted that the formation of a hydro sulfa aluminate framework when cement is mixed with water occurs during the induction period of the cement stone structure formation, when coagulation processes still prevail in it. Therefore, expansion occurs without breaking the continuity of the structure, which ensures the achievement of the required strength and density for the resulting material.

The disadvantage of this cement is the complexity of storage due to its high hygroscopicity, as well as technological difficulties due to its fast setting. In addition, the negative properties of this cement include the fact that the expansion degree depends on the temperature and humidity conditions of hardening, as well as the reduced ability, in comparison with Portland cement, to protect the steel reinforcement of reinforced concrete.

The widespread use of expanding composites based on aluminate cement is constrained by its high cost, as well as technological difficulties during the construction work. These difficulties are caused by the shortened setting time and the high sensitivity of the cement in relation to changes in the temperature and humidity conditions of hardening.

The characteristic disadvantages of expanding cements based on alumina high-aluminate components are largely overcome when using expanding composites based on Portland cement [7].

When developing the expanding compositions based on Portland cement, the principle of compensated expansion was put forward and substantiated [6, 7]. According to this principle, in such formulations, expansion occurs mainly due to the inclusion of a small amount of aluminum powder additive in the mixture. The latter enters into chemical interaction with the resulting hydration products of Portland cement, as a result of which many small bubbles of hydrogen gas are formed in the mixture. At the same time, the external volume of the system increases, ensuring its expansion. To prevent a decrease in the strength characteristics and water resistance of the resulting expanding composite, in addition to aluminum powder, chemical additives are additionally introduced into the mixture [8-11]. These additives include calcium chloride, aluminate sulphate and a surfactant - sulfite-alcohol mash SAM (currently not produced by the pulp and paper industry).

Thus, expanding Portland cement contains a four-component complex additive with the following ratio of components in percentage: by weight of cement consumption - 0.01 aluminum powder; 2.0 aluminum sulfate; 2.0 calcium chloride; 0.15 sulfite-alcohol mash SAM. These components are introduced into the concrete mix together with the mixing water. The presence of additives and SAM provides uniform distribution of finely dispersed aluminum powder in a concrete or mortar mixture without first removing the paraffin film from its particles. The positive effect of the SAM additive lies in the fact that the calcium lignosulfonate included in its composition is deposited on the surface of the aluminum powder particles, which contributes to their uniform distribution in the form of a suspension in the liquid phase. That is why, if usually a system of interconnected capillaries and pores is formed in the cement stone, then in this case, the smallest closed gas pores are formed in the cement stone, evenly distributed throughout the cement stone volume.

Aluminum sulfate and calcium chloride are included in the expanding additives, they enter into chemical interaction with the hydrating cement, contributing to the partial filling of the above-described gas pores with crystalline neoplasms. These components of the expanding additive provide clogging (filling) of the pores of the cement stone by such crystallizing complex salts as hydro sulfa aluminates and calcium hydro chloro aluminates. This results in a very dense cement stone in which the smallest partially sealed pores are evenly distributed. As a result, the strength indicators, water resistance and frost resistance of the resulting material are significantly increased, which is confirmed by the comprehensive studies [12-14].

At the same time, along with all the advantages of the considered expanding binders based on Portland cement, it is also necessary to note their certain disadvantage, which consists in the technological complexity of dosing and the practical use of a four-component expanding complex additive. In this case, the SAM plasticizer is already out of production.

Targets and goals

The purpose of this work is to study the possibility of creating effective expanding compositions intended for carrying out repair and restoration measures on the PTL reinforced concrete supports in operation, based on Portland cement, local fine sand contaminated with impurities of dusty and clay particles, and the components of a complex expanding additive available in regional conditions.

When setting the problem, we proceeded from the following considerations. In all known expanding cements, both based on alumina cement and based on Portland cement, such neoplasms as calcium hydro sulfa aluminate play a decisive role. Calcium hydro sulfa aluminate can influence the structure formation processes of cement stone in two ways, depending on the time factor of its formation in the system. One of the first to come to this conclusion was Lafuma, who formulated the fundamental proposition that in the process of direct interaction of substances in the solid phase, dangerous deformations can appear in the hardening cement stone. This is due to the fact that when crystallization of neoplasms on the interconnected structural particles' surfaces, the structure of the cement stone as a whole is damaged. The reaction of calcium hydro sulfa aluminate formation in the liquid phase proceeds in a different way. In this case, neoplasms crystallize in the free gaps of the structure, leading to its compaction and strengthening [15].

When hardening ordinary Portland cement, there are always the above-mentioned conditions for the formation of calcium hydro sulfa aluminate. In particular, in the hydration process of the prevailing mass content in the composition of the clinker mineral C_3S (tricalcium silicate) lime is inevitably formed in the form of portlandite $Ca(OH)_2$. In the production of cement, gypsum is compulsorily added, which is a regulator of the cement setting time.

To achieve the expansion effect of the formed Portland cement stone, the necessary increase in the amount of the formed hydro sulfa aluminate calcium should be ensured in the earliest period of structure formation.

It is possible to directionally regulate the composition of Portland cement clinker, increasing the content of clinker minerals such as tricalcium aluminate involved in the formation of calcium hydro sulfa aluminate C_3A, or tricalcium silicate C_3S, only when cement is manufactured in the factory. Naturally, it is not possible to control the kinetics of calcium hydro sulfa aluminate formation in this way when making mixtures for carrying out repair and restoration work in relation to the PTL supports.

The content of gypsum in cement can be increased by adding it while mixing the repair mixture, however, it is practically difficult to distribute it in cement in the field without having additional appropriate mixing equipment, which is equipped with conventional stationary concrete mixing units.

It is known that it is possible to increase the gypsum content with its uniform distribution in the hardening system under normal conditions by adding easily dissolving sulfate-containing substances to the mixing water, for example, Glauber's salt Na_2SO_4 (sodium sulfate CH) [14-15].

Experimental part

The use of aluminum powder as an expanding component in the complex additive composition is associated with the difficulties of its dosage and uniform distribution in the mixture. This is explained by the fact that finely ground particles of metallic aluminum in the process of its production are covered with a paraffin protective film in order to avoid its oxidation and spontaneous combustion. This paraffin coating prevents wetting of the aluminum powder particles, which complicates its uniform distribution in the mixing water volume when preparing mixtures. In order to eliminate this phenomenon, the aluminum powder is freed from the paraffin film by introducing surfactants into the mixing water.

In this work, we studied the possibility of using for this purpose the additive of the superplasticizer ST 2.1, which is introduced into the mixture in a ratio with aluminum powder by weight 1: 1.

It was decided to use sodium sulfate SS in the expanding additive composition as a sulfate-containing component. To obtain an expanding composition based on Portland cement, the content of the gas-forming additive of aluminum powder in the composition of the complex developed additive was taken equal to 0.01% of the Portland cement mass.

In order to establish a rational amount of sodium sulfate SS as a part of the expansion additive developed for the repair and restoration work of PTL frames, a special calculation was made. The calculated amount of the addition of sodium sulfate in the composition of the expanding additive is determined from the condition that the contraction shrinkage of the hardening Portland cement stone is compensated due to the formed calcium hydro sulfa aluminate. At the same time, clogging of the so-called "open" capillaries and pores in its structure should be ensured. The quantitative ratios of both the reagents and the products of the formation reaction in the solidifying system of calcium hydro sulfa aluminate after the corresponding chemical reaction completion were determined from the process conditions.

Calculations have shown that the sodium sulfate content in the composition of the developed expanding additive should be in the range of 1.5-2.5% of the cement mass. As a result of the analysis, for further experimental studies of expanding compositions based on local substandard raw materials, it seemed expedient to use the proposed expanding additive in the composition of repair mixtures, including 0.01% by weight of cement additive ST 2.1, aluminum powder and 1.5-2, 5% sodium sulfate.

When studying the formulation effect on the construction and technical properties of sandy concrete used as a repair material for the restoration of PTL frames, a set of indicators was adopted as the optimized response functions:

- from the condition of ensuring the uniform strength of concrete materials for PTL frames and their restored sections, the requirement was formulated to achieve a branded strength of at least 30 MPa (class B22.5 in strength);
- as a criterion for operational reliability under typical operating conditions of supports PTL cyclic aggressive environmental influences, such as alternating moistening and drying, freezing and thawing, etc., in experimental studies, strength characteristics of concrete were taken - compressive strength at the age of 7, 28, 250 days, MPa; tensile strength, MPa; elastic modulus.

To obtain the experimental data, we used Samara sand of the Rostov region and Portland cement of Sebryakovcement PC500 D0. To prepare repair compositions, cement-sand mixtures of the composition cement: sand were used in a ratio of 1:2 at a W / C ratio in the range of 0.4-0.6, in combination with the introduction of the proposed complex expanding additive into the mixture. From the mixtures obtained, beams of standard sizes 4x4x16 cm were molded, tested at the age of 28 and 250 days of normal hardening. Features of the manufacturing full-scale samples' technology are presented in the work [12].

Age and conditions for their hardening (water, air-dry, etc.) have a significant impact on the formation of the structure and properties of concrete. It is well known that expanding cements have the potential to gain design strength in a shorter time frame when providing moisture care. We assume that it is not possible to provide a special temperature and humidity regime of hardening to restore the operated PTL frames in natural conditions. In this regard, the dynamics of strength gain by the investigated expanding composition over time under averaged conditions for temperature and humidity is of interest; therefore, we included a factor that takes into account the age of the concrete of the repair composition into the test plan.

Investigated factors:

$X1$ – W/C ratio – 0,4-0,6;

$X2$ – sodium sulfate additive content, % by cement weight – 1,5-2,5;

$X3$ – aggregate grain spread ratio – 1,1-1,6;

$X4$ – dosage of aluminum powder, % by cement weight – 0,005-0,015;

$X5$ – concrete age, day – 7-250.

The main mechanical and deformative properties of concrete were selected as responses:

Feedback:

$Y1$ – $Y3$ – compressive strength at the age of 7, 28, 250 days, MPa;

$Y4$ – tensile strength, MPa;

$Y5$ – elastic modulus.

In the work, a factorial experiment was implemented according to the Hartley-5 (Na-5) design. Number of plan points - 27. Number of points in the center of the plan -1. The number of coefficients in a second-order polynomial with five factors is 21. The analysis of the Ha-5 design was carried out using the Matcad statistical package. The reproducibility variance was determined from the results of parallel experiments at each point of the design.

To describe the results, a second-degree polynomial of the form was used:

$$Y = B_0+\sum B_iX_i + \sum B_{ii}X_i^2 + \sum B_{ij}X_{ij}$$

As a result of mathematical processing of experimental data, adequate models of the studied response functions were obtained. Experimental Statistical Model Coefficients (ESM) the response functions of the main mechanical characteristics of concrete are presented in Table 1.

Table 1. The main mechanical characteristics response functions experimental-statistical models' regression equations coefficients of form (*Y1-Y4*) concrete

$$Y = B0+\sum BiXi + \sum BiiXi2 + \sum BIjXiXj$$

Polynomial coefficients	Response function			
	Y_1	Y_2	Y_3	Y_4
B_0	25.97	40.3	48.2	42.2
B_1	-6.76	-6.8	-5.94	-6.12
B_2	0.004	-0.01	-1.92	-2.6
B_3	-2.71	-2.8	-0.73	-1.6
B_4	-0.11	1.12	2.12	2.5
B_5	11	1.6	2.85	0.8
B_{11}	-0.95	0.48	-0.93	0.97
B_{22}	2.65	-1.12	0.56	2.01
B_{33}	2.4	1.33	-0.34	-0.13
B_{44}	-3.56	-0.17	-3.1	-3.3
B_{56}	2.38	-2.39	2.16	-1.3
B_{12}	0.91	0.93	2.19	1.3
B_{13}	-0.3	-0.83	-0.74	-1.4
B_{14}	1	1.1	1.1	0.7
B_{15}	-1.53	-0.7	-0.9	0.9
B_{23}	0.76	1.46	0.3	0.96
B_{24}	1.11	-0.61	-1.6	-0.7
B_{25}	-0.21	-0.35	-1.5	-0.6
B_{34}	0.93	0.95	1.3	-0.3
B_{35}	0.999	-1.32	-0.44	-1.4
B_{45}	0.163	-0.99	0.54	-0.3

Geometric images of the studied response functions were constructed using the regression equations using the linear algebra methods. The geometric images' analysis of the investigated response functions made it possible to establish the optimal limits of variation of the main input parameters for obtaining concrete for repair compositions with an expanding additive. The analysis of the factor space subregions allowed us to recommend the sodium sulfate expanding additive component content concentration - 2.4%, aluminum powder - 0.010%. The ages of hardening of repair compounds for concrete optimal for various production purposes have been determined.

Summary

The tests carried out showed that the proposed expanding additive of compounds for repair compounds for restoration of PTL frames provides an increase in the strength of the resulting material, increases the restored structure's durability and reliability degree (Table 1).

Experimentally, using the methods of mathematical planning, it was found that the developed expanding complex additive significantly increases the frame resistance and strength of sandy concrete and improves its structure. This may be the basis for using the proposed solutions when performing PTL repair work.

Evaluation of the repair and restoration works quality, carried out in accordance with the methodology described in [13], confirmed the high rates of adhesive adhesion of the developed four-component repair composition with concrete of the PTL frame structure.

References

[1] Bazhenov Yu M 2012 Ways of building materials science development: New concrete *Concrete technology* **3-4 (68-69)** 39-42
[2] Mailyan L, Yazyev S, Sabitov L, Konoplev Y, Radaykin O 2020 Stress-strain state of the combined tower-reinforced concrete foundation-foundation soil system for high-rise structures *E3S Web of Conferences* **164 02035**.
[3] Bazhenov Y M, Erofeev V T, Rimshin V I, Markov S V, Kurbatov V L 2016 Changes in the topology of a concrete porous space in interactions with the external medium *Engineering Solid Mechanics* **4** 219-225. DOI: 10.5267/j.esm.2016.5.001
[4] Shlyakhova E A, Egorochkina I O, Cherpakov A V 2019 Expanding additive to repair compositions for restoration of power line supports *Proceedings of the 2018 international conference on "Physics, mechanics of new materials and their applications"* 227-232.
[5] Cherpakov A V, Shlyakhova E A, Egorochkina I O, Kokareva Ya A 2018 Identification of concrete properties in beam-type structures with defects based on dynamic methods *Materials Science Forum* **931** 373-378.
[6] Mikhailov V V 1974 *Expanding and stressing cements and self-stressed reinforced concrete structures* (Stroyizdat, Moscow) 312 (In Russian).
[7] Mcheldov-Petrosyan O P, Borovskaya I V 1983 New complex additives to cement from metallurgical waste *Cement* **6** 6-8.
[8] Kharchenko I Ya, Pustovgar A P, Pashkevich S A, Bazhenov Yu M, Kharchenko A I 2017 Expanding cements hardening within the limited deformations conditions *Magazine of Civil Engineering* **75 (7)** 161-170.
[9] Alimov L, Kharchenko I, Voronin V 2017 Self-compacting fine-grained concretes with compensated shrinkage *MATEC Web of Conferences* **117 00006.**
[10] Nesvetaev G V, Grigoryan M N, Yazyev B M, Dolgova A V, Postoj L V 2020 Effect of dosage of redispersible powders on the properties of fine concrete *Materials Science Forum* **974** 413-418.
[11] Shlyakhova E A, Egorochkina I O, Serebryanaya I A, Matrosov A A 2018 Methods of Investigating and Quality Assessing Structure of Concretes with Combined Aggregates *Materials Science Forum* **931**.
[12] Shcherbinina T A, Serebryanaya I A 2015 Concrete shrinking *Izvestia of the Rostov State University of Civil Engineering* **2 (20)** (In Russian).
[13] Titov M Yu 2012 Concretes with increased strength based on expanding additives *Building materials* **2**.
[14] Efremova I A 1997 Concrete with Combined Aggregate Based on Portland Cement with Expansion Additives *Rostov Construction Institute, Rostov-on-Don*, 24 (In Russian).
[15] Egorochkina I O, Khalyushev A K, Scherban E M, Stelmakh S A, Sysoev A K, Zholobova E A 2020 Force modeling of interparticle pair interaction of spheroid in composite cement *IOP Conference Series: Materials Science and Engineering* **905 (1) 012061.**

IOP Conf. Series: Materials Science and Engineering 1083 (2021) 012037 doi:10.1088/1757-899X/1083/1/012037

Analytical optimization of the dispersion-reinforced fine-grained concrete composition

A D Zhukov[1,2], I V Bessonov[2], A D Bekele[1], E A Zinoveva[1*]

[1] Moscow state university of civil engineering, 26, Yaroslavl highway, Moscow, 129337, Russia

[2]Research Institute of Building Physics of the Russian Academy of Architecture and Building Sciences, 21, Lokomotivny driveway, Moscow, 127238, Russia

E-mail: zinkate@mail.ru

Abstract. The article presents the results of studying the problem of the concrete scrap accumulation and disposal. It is noted that after the implementation of comprehensive programs, including the renovation and modernization of the housing stock, a significant amount of concrete scrap has already been accumulated, and its amount will only increase. The results of scientific research on the compositions development of composite binder and fine-grained dispersed reinforced concrete based on this binder are presented. The purpose of the research described in the article was to develop the elements of a methodology for selecting the dispersed reinforced composite material composition based on finely ground waste of concrete scrap, mineral binder, thin basalt fiber and superplasticizers. The goal was solved using the methodology of technological processes system analysis and, in particular, mathematical planning, processing the experimental results and applying the analytical optimization method. It is noted that fine grinding of mineral components and the use of fibers with a diameter of 0.1 ... 6 microns allows not only to modify the concrete structure with an increase in strength characteristics, but also changes the chemical and physicochemical processes' nature on the contact surfaces of the mineral matrix and fiber, as well as the same binder and finely ground reactive mineral particles with the formation of microsystems qualified as nanostructures.

Introduction

Technology development in modern society presupposes the emergence of many challenges and the formation of new problems associated with an excessive load on technology. Since the 1950s, the need to recycle industrial waste, including construction industries has become one of the serious problems [1-3]. For Russia, this problem is exacerbated by the renovation of the housing stock and the demolition of dilapidated housing. As a result of the comprehensive housing programs implementation, a huge amount of construction (including concrete) scrap has already been accumulated at the relevant facilities. The need for its use arises both from the fact of alienation of territories for the storage of these wastes, and from the additional burden on the environment. Disposal of this waste is necessary and will become mandatory in the near future.

The most technologically advanced way of recycling waste generated both as a result of the old buildings and structures' demolition and waste from the building materials' production is the use of these by-products in building products' manufacturing [4–6]. It should be noted here that any by-products of the production of mineral binders, Portland cement, concrete, or their derivatives, classified as waste, contain some percentage of clinker minerals. As a result of fine grinding or mechanical or mechanochemical activation, this component ceases to be an inert filler and starts exhibiting astringent properties [7–9].

The introduction of fibers or thread lines into the composition of raw mixtures also makes possible to significantly modify the properties of the finished product. In domestic and foreign practice, sufficient experience has been accumulated in the use of fibers of various natures in the composition of concretes, building mixtures or their analogues: steel, fibers based on alkali-resistant glasses or basalts, polymer, cellulose, nanotubes, etc. [10–12].

Fine grinding of mineral components and the use of fibers with a diameter of 0.1 ... 6 microns allows not only to modify the concrete structure, with an increase in strength characteristics, and, first of all, bending strength, but also changes the nature of chemical and physicochemical processes on the contact surfaces of the mineral matrices and fibers with the formation of ordered microsystems classified as nanostructures [13–15]. This suggests the emergence of a new subclass of building composites.

The purpose of the research described in the article was to develop elements of a methodology for selecting the composition of a dispersed reinforced composite material based on finely ground waste of concrete scrap, mineral binder, thin basalt fiber and superplasticizers.

Materials and Methods

System analysis of technological processes on the provisions of which the developed technique is based is prdicated on considering technology as a cybernetic system. This system is divided into blocks adequate to technological advancements, for each of which the "black box" model is implemented. In the classical system chain, there are blocks of dosage (consumption) of components, their preparation and pretreatment, grinding, mixture preparation, molding, heat treatment [16–19].

The article presents the results of studying the first block. It is associated with the concrete mixture main components' consumption optimization.

As variable factors, the costs of cement, fine waste, screening, plasticizer and reinforcing component are taken. The experimental conditions are shown in Table 1. The sand consumption is taken equal to 300 kg/m^3. Water consumption is set in accordance with W / C and is not an independent factor. The response functions are the compressive strength of concrete (Y_1) and its average density (Y_2).

As an optimization parameter at the third stage of the experiment, the coefficient of constructive concrete quality (CCQ), equal to the ratio of the compressive strength of concrete (Y_1) to its average density (Y_2):

$$CCQ = (Y_1)/ (Y_2)$$

The obtained regression equations are checked for all statistical hypotheses and the models' adequacy is checked by the Fisher's criterion. As a result of statistical tests as a result of comparison with confidence intervals (Δbj) only significant factors are left, and as a result of the Fisher test, a conclusion about the adequacy (or inadequacy) of the obtained models is made.

Evaluation of each factor's influence on the result is carried out according to the value and sign of the coefficient facing the factor (its linear value or quadratic function) or their paired interaction. It should be noted that the experiment is carried out in the coded values of the factors (reduced to the interval [–1, +1].

The interpolation of the results consists in determining the strength and average density of the material by calculation, depending on the variable factors' values and is carried out by implementing the programs for electronic computer. Such a program includes data input blocks (values of factors in natural terms), a factor coding block, a calculation block, and a block for outputting results to the display. Testing of statistical hypotheses, modeling, processing and optimization of the results was carried out in accordance with the methods of the software package G-BAT-2011.

The equations are optimized using the analytical method, which is based on the following provisions: the obtained regression equations adequately describe the studied technological process; each equation is an algebraic function of several variables (according to the number of significant variable factors) and methods of mathematical analysis are applicable to study this function [18–20].

Table 1. Factor variation intervals.

Factor name	Symbol X_i	Average factor, $\overline{X}_i$	Variation interval, ΔX_i	Factor values at the levels	
				-1	+1
Portland cement consumption, [kg/m^3]	X_1	450	50	400	500
Plasticizer consumption, [kg/m^3]	X_2	3	0.5	2.5	3.5
Consumption of finely ground waste, [kg/m^3]	X_3	320	80	240	400
Reinforcing component consumption, [%]	X_4	1	0.5	0.5	1.5

Sand consumption 200 [kg/m^3]

Results

Mathematical processing of the experimental results made it possible to obtain the regression equations for the compressive strength (Y_1) and average density (Y_2). The following mathematical models (polynomials) are obtained:

- for compressive strength

$$Y_1 = 41.7 + 5.2X_1 + 2.7X_2 + 2.3X_3 + 3.0X_4 + 1.7X_1X_3 - 1.6X_2^2 - 1.7X_4^2$$

- for average density:

$$Y_2 = 1920 + 52X_1 + 34X_2 + 43X_3 + 13X_4 + 11X_1X_3 - 6X_2^2 - 4X_4^2$$

The significance of the coefficient was checked by confidence intervals, respectively, the confidence interval for strength was $\Delta b_1 = 0.8$ MPa, and for average density $\Delta b_1 = 3$ kg/m^3.

The resulting models were tested for adequacy according to the Fisher's criterion. The calculated values of the Fisher criteria are equal for the average density model F_2=15,1 and for the compressive strength model F_1=16,7. The table values of the criteria are 19.2 and 19.3, respectively. The calculated values of the F-criterion do not exceed the tabular one, and with the corresponding confidence level (98%) the model can be considered adequate.

Analysis of the equation coefficients $Y_1 = f_1(X_1, X_2, X_3, X_4)$ shows that the strength increases with an increase in the Portland cement and finely ground filler consumption in the intervals adopted in the experiment (positive coefficients at X_1, X_3. With an increase in the consumption of the plasticizer, first there is an increase in strength, and then, at high consumption, a decrease is observed (coefficients at X_2 and X_2^2). A similar picture takes place with an increase in the consumption of the reinforcing component (coefficients at X_4 and X_4^2). This suggests that the function $Y_1 = f_1(X_1, X_2, X_3, X_4)$ has local extrema in X_2 and by X_4, and it is possible to use analytical optimization.

Analysis of the equation coefficients $Y_2 = f_2(X_1, X_2, X_3, X_4)$ shows that an increase in the consumption of Portland cement and finely ground filler (coefficients at X_1 and X_3). An increase in the plasticizer and the reinforcing component consumption contributes to an increase in density: at first intensive, and at high costs - insignificant.

Analytical optimization is based on the fact that the functions for strength and density $Y_1 = f_1(X_1, X_2, X_3, X_4)$ and $Y_2 = f_2(X_1, X_2, X_3, X_4)$ – are the mathematical methods and it is permissible to apply the methods of mathematical analysis to them, provided that the condition of adequacy is not violated. In the case under consideration, the following scheme is adopted:

- the equation $Y_1 = f_1(X_1, X_2, X_3, X_4)$ is differentiated by X_2 and equate to zero, determining the extremum of the function Y_1 by X_2;
- the equation $Y_1 = f_1(X_1, X_2, X_3, X_4)$ is differentiated by X_4 and equate to zero, determining the extremum of the function Y_1 by X_4;

IOP Conf. Series: Materials Science and Engineering 1083 (2021) 012037 doi:10.1088/1757-899X/1083/1/012037

- the functions $Y_1 = f_1(X_1, X_2, X_3, X_4)$ and $Y_2 = f_2(X_1, X_2, X_3, X_4)$ are solved at the optimized values X_2 and X_4 ($X_2 = opt_2$ and $X_4 = opt_4$), then local optimization is carried out.

Discussions

Analysis of the polynomial describing the relationship between compressive strength and variable factors shows that this function (which is essentially a function of several variables) in two of these variables, namely, the consumption of the plasticizer (X_2) and the consumption of the reinforcing component (X_2), has local extrema. Therefore, we can use the mathematical apparatus of analytical local optimization.

The methodology of analytical local optimization is based on the fact that the nonlinear dependences obtained as a result of mathematical processing of the statistical analysis results are, on the one hand, mathematical models that adequately describe the phenomenon under study, and this adequacy is confirmed by testing the statistical hypotheses.

On the other hand, these dependences are the second-order polynomials and mathematical functions of several variables, to which the methods of algebraic analysis of such functions in general and, in particular, the determination of their local extrema can be applied. The local extremum of such functions is determined by differentiating the function with respect to the parameter of interest to us (in this case, X_2) and setting the result to zero. The subsequent solution of the linear dependence gives a possibility to find the factor value corresponding to the extremum in the domain of this factor.

In the optimization process, we perform the following sequence of actions:

1) Determine the local extrema of the function Y_1 by X_2 and X_4:

$$\partial Y_1 / \partial X_2 = 2.7 - 3.2X_2 = 0 \rightarrow X_2 = \frac{2.7}{3.2} = 0.84$$
$$\partial Y_1 / \partial X_4 = 3.0 - 3.4X_2 = 0 \rightarrow X_2 = \frac{3.0}{3.4} = 0.88$$

2) Solve the functions $Y_1 = f_1(X_1, X_2, X_3, X_4)$ and $Y_2 = f_2(X_1, X_2, X_3, X_4)$ at the optimized values X_2 and X_4 ($X_2 = 0.84$ and $X_4 = 0.88$)

- for compressive strength

$$Y_1 = 46.7 + 4.9X_1 + 2.3X_3 + 1.5X_1X_3$$

- for average density:

$$Y_2 = 1953 + 52X_1 + 43X_3 + 11X_1X_3$$

3) Determine the natural values of the optimized factors, taking into account the data in Table. 1:
Plasticizer consumption: 3 + 0.5x0.84 = 3,42 ± 0,2 kg/m^3
Reinforcing component consumption: 1 + 0.5x0.88 = 1,44 ± 0,1%
4) Form the analytical expressions for the coefficient of concrete constructive quality (CCQ):

$$CCQ = \frac{Y_1}{Y_2} = \frac{46.7 + 4.9X_1 + 2.3X_3 + 1.5X_1X_3}{1953 + 52X_1 + 43X_3 + 11X_1X_3}, \text{ [MPa·m}^3\text{/kg]}$$

The graphical interpretation of the obtained equation is shown in Figure 1.

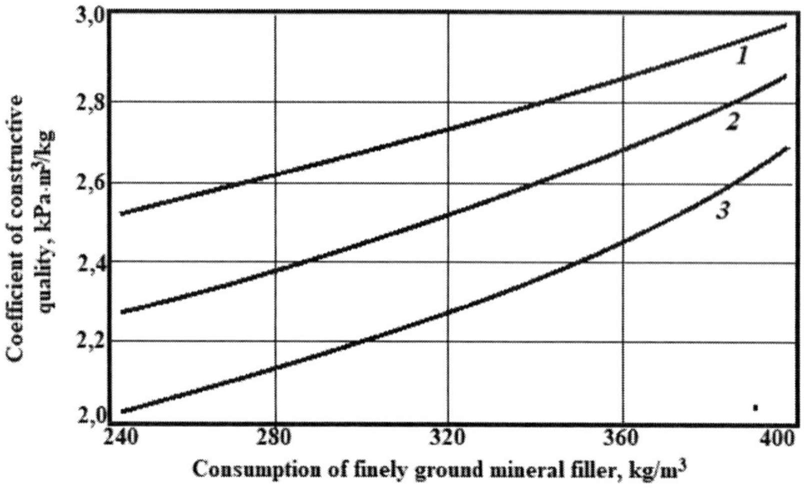

Figure 1. Coefficient of constructive quality dependence on the consumption of finely ground mineral filler and the consumption of Portland cement, kg/m^3: 1 – 500; 2 – 459; 3 – 400

Analyzing the graph of the coefficient of constructive quality (CCQ) dependence from Portland cement and the consumption of mineral filler obtained by fine crushing of concrete scrap, the following can be stated. First, the obvious fact that with the analytically obtained optimal values of the plasticizer consumption (3,42 ± 0,2 kg/m^3) and consumption of reinforcing fiber (1,44 ± 0,1%) with an increase in the Portland cement consumption CCQ increases. Second, it was found that in the intervals of factor variation provided by the experimental conditions (Table 1), there is a tendency for additional CCQ growth with an increase in filler costs. This may well be explained by its hydraulic activity, but requires additional research.

Summary

In connection with the increasing volumes of construction waste accumulation in the process of housing renovation, demolition of dilapidated ones, replacement of bridge structures and road surfaces, the problems of processing elements of the destroyed structures and buildings in order to obtain secondary nonmetallic building materials become extremely urgent. The research results presented in the article show the possibility of using a fine-grained waste concrete scrap as an active filler for composite binder and structural concrete based on it.

The experimental technique based on the system analysis of technological processes allows obtaining adequate mathematical models, the engineering interpretation of which allows solving technological problems, including the selection of the composition of fine-grained concrete with dispersed reinforcement, as well as obtaining the optimization solutions in the process of processing the experimental data.

References

[1] Telichenko V I, Oreshkin D V 2015 Material science aspects of geoecological and ecological safety in construction *Ecology of urbanized areas* **2** 31–33.

[2] Ivanov L A 2019 Establishment of technological startups based on research and development *(interview with Polad Malkin, researcher and developer Ph.D., professor, serial technology entrepreneur, CEO of «StartUpLab»). Nanotechnologies in Construction* **11 (2)** 207–216. DOI: 10.15828/2075-8545-2019-11-2-207-216.

[3] Almusaed A, Almassad A, Alasadi A 2019 Analytical interpretation of energy efficiency concepts in the housing design process from hot climate *Journal of Building Engineering* **21** 254-266. DOI: 10.1016/j.jobe.2018.10.026

[4] Lipunova I N, Yupatov A A, Alikin V I 2009 The use of solid industrial waste in the production of materials for construction purposes *Ecology and Industry of Russia* **3** 19–23.
[5] Efimenko A Z, Pilipenko A S, Kolyadin A A 2009 Facade basement tiles on construction waste filler *Concrete technology* **7(8)** 31
[6] Artamonova O V, Slavcheva G S, Chernyshov E M 2017 Effectiveness of Combined Nanoadditives for Cement Systems *Inorganic Materials* **53** 1080–1085. DOI: 10.1134/S0020168517100028.
[7] Ivchatov A L, Glyadenov S N 2007 The use of phosphorus production waste as a mineral additive in building materials *Ecology and Industry of Russia* **9** 10–14.
[8] Murtazaev S-A Yu, Dena K-S Batayev, Magomed S Saydumov 2016 Cement concrete composites on the basis of bypassed stone and stone milling wastes *Acta Technica CSAV (Ceskoslovensk Akademie Ved)* **61(4B/2016)** 327–336.
[9] Bullard J, Livingston J, et al. 2011 Mechanisms of cement hydration *Cement and Concrete Research* **41** 1208–1223. DOI: 10.1016/j.cemconres.2010.09.011.
[10] Falliano D, De Domenico D, Ricciardi G, Gugliandolo E 2019 Compressive and flexural strength of fiber-reinforced foamed concrete *Construction and Building Materials* **198** 479-493. DOI: 10.1016/j.conbuildmat.2018.11.197
[11] Efimov Boris , Isachenko Sergey, Kodzoev Mukhamad-Basir, Dosanova Gulzar and Ekaterina Bobrova 2019 Dispersed reinforcement in concrete technology 01032 *E3S Web Conf.* **110**. Published online: 09 August 2019. DOI: https://doi.org/10.1051/e3sconf/201911001032
[12] Xu S, Lyu Y, Li Q 2019 Enhancing the initial cracking fracture toughness of steel polyvinyl alcohol hybrid fibers ultra high togness cementitious composites by incorporating multy-walled carbon nanotube *Construction and Building Materials* **195** 269-282. DOI: 10.1016/j.conbuildmat.2018.10.133
[13] Pisarenko Zh V, Ivanov L A, Wang Q 2020 Nanotechnology in Construction: State of the Art and Future Trends *Nanotechnologies in Construction* **12(4)** 223–231. doi: 10.15828/2075-8545-2020-12-4-223-231.
[14] Sanchez F, Sobolev K 2010 Nanotechnology in concrete – A review *Construction and Building Materials* **24** 2060–2071. DOI: 10.1016/j.conbuildmat.2010.03.014.
[15] Paul S C, Rooyen A S, Gideon P A G. van Zijl et al. 2018 Properties of cement – based composites using nanoparticles: A comprehensive review *Construction and Building Materials* **189** 1019–1034. DOI: 10.1016/j. conbuildmat.2018.09.062.
[16] Zhukov A, Shokodko E, Bobrova E, Bessonov I, Dosanova G, Ushakov N 2019 Interior Acoustic Materials and Systems *In: Murgul V., Pasetti M. (eds) International Scientific Conference Energy Management of Municipal Facilities and Sustainable Energy Technologies EMMFT 2018. EMMFT-2018. Advances in Intelligent Systems and Computing* **983**. Springer, Cham. 740-747. DOI: 10.1007/978-3-030-19868-8_72
[17] Gnip I, Vaitkus S, Kersulis V, Vejelis S 2012 Long-term prediction of creep strains of mineral wool slabs under constant compressive stress *Mech Time Depend Mater* **16** 31–46.
[18] Zhukov A, Shokodko E 2020 Mathematical Methods for Optimizing the Technologies of Building Materials *In: Popovic Z., Manakov A., Breskich V. (eds) VIII International Scientific Siberian Transport Forum. TransSiberia 2019. Advances in Intelligent Systems and Computing,* **1116** 413-421. Springer, Cham. First Online 31 January 2020. DOI: https://doi.org/10.1007/978-3-030-37919-3_40
[19] Bessonov I, Zhukov A, Shokodko E, Chernov A 2020 Optimization of the technology for the production of foam glass aggregate *TPACEE 2019, E3S Web of Conferences* **164** 14016. Published online: 05 May 2020. DOI: https://doi.org/10.1051/e3sconf/202016414016
[20] Zhukov A D, Chugunkov A B 2011 Local analytical optimization of technological processes *Vestnik MGSU* **1(2)** 273–278.

Lightweight concrete based on crushed foam glass aggregate

I V Bessonov[1], B I Bulgakov[2], A D Zhukov[2], V A Gradov[1], N A Ivanova[1], M-B Kh Kodzoev[2*]

[1]Research Institute of Building Physics of the Russian Academy of Architecture and Building Sciences, 21, Lokomotivniy Tr., Moscow, 127238, Russia
[2]National Research Moscow State University of Civil Engineering, 26, Yaroslavskoye highway, Moscow, 129337, Russia

E-mail: basir731@yandex.ru

Abstract. Increased requirements for energy saving in the operation of buildings and structures, as well as the need to adjust the raw material base, presupposes the development and improvement of lightweight concrete technologies, including the use of effective lightweight aggregates, which include granulated foam glass. The goal of the study was to develop a method for selecting the composition of lightweight concrete with foam glass gravel as a large aggregate and foam glass powder as a fine aggregate. To achieve this goal, an experiment was carried out to optimize the composition of lightweight concrete and determine the optimal proportions of the used fractions of foam glass gravel. The raw mix for producing lightweight concrete included the following components: Portland cement, foam glass gravel (mixture of fractions), glass powder (crushed foam glass gravel of 0.315 mm in size), soda glass, super plasticizing admixture, air-entraining agent and water. As a result of the implementation of the experiment and the use of the apparatus for mathematical planning, processing and optimization of the data obtained, the optimal costs of main components were established, as well as the proportions of the content of coarse aggregate fractions.

1. Introduction

For decades, lightweight concrete has been positioned as a material that allows the manufacture of products that have the necessary bearing capacity and a relatively low density and thermal conductivity. This makes it possible to erect load-bearing or self-supporting building envelopes without the use of additional thermal insulation, which has a positive effect both on the durability of such structures and the parameters of energy saving and thermal comfort, and on reducing labor costs during construction and heat losses during operation [1-3].

Lightweight concrete is used in monolithic construction or as masonry blocks in two main groups: cellular concrete (foam and aerated concrete), lightweight aggregate concrete, as well as expanded polystyrene concretes [4–6]. Traditionally, expanded clay gravel and, in much smaller volumes, aggloporite and perlite gravel are used as lightweight aggregates. As a fine aggregate, crushed expanded clay or aggloporite is used, and expanded perlite sand or mineral granules, as well as expanded polystyrene granules [7–9].

The increased requirements for energy saving during the operation of buildings and structures, as well as the need to expand the raw material base, makes it expedient to use foam glass as a light aggregate for concrete [10–12]. Foam glass is a rigid cellular material that is obtained by high-temperature foaming of softened glass. The composition of the glass mixture is determined by the requirements for the final product. Mainly, modified aluminum silicate compositions are used; it is also possible to use cullet of packing or sheet glass [13–15].

According to the purpose, materials and products made of foam glass are divided into heat-insulating, sound-insulating and sound-absorbing (to protect against noise and create conditions for acoustic comfort in buildings), as well as into tailored materials (protection against electromagnetic waves, road construction, waterproofing systems, for small architectural forms, etc.). According to their shape, they are subdivided into blocks (Figure 1b), slabs, shaped articles, crushed stone (gravel) (Fig. 1a), as well as granular and microgranular foam glass. Foam glass is characterized not only by high strength and excellent thermal insulation properties, but also by complete environmental and hygienic safety.

a *b*

Figure. 1. Foam glass gravel dispersed into fractions (a), and a block of foam glass (b).

Foam glass gravel is a loose foam glass in the form of particles and pieces of irregular shape (Figure 1a). Features of the technology for the production of gravel on specialized production lines are in the correct selection of the mixture composition and determination of the optimal firing temperature. The mixture for producing foam glass gravel consists of finely ground glass (glass powder), a foaming agent and water. Unlike foam glass slabs, where soot or coke is used as a foaming agent, food glycerin is used in the technology of foam glass gravel. The fundamental difference in the production of slabs in comparison with gravel is that after production the slab requires additional annealing in a furnace with a gradual decrease in temperature for 10-12 hours, as well as trimming to give the slab the correct shape. The foam glass gravel leaves the smelting furnace already in the form of a finished product.

Foam glass gravel is used as backfill for winterizing the foundations of buildings, brick walls with well masonry, blind area and flat roofs. This article considers a new area of its application — as an aggregate for lightweight concrete.

2. Experimental conditions

The goal of the study, the results of which are presented in the paper, was to develop a method for selecting the composition of lightweight concrete with foam glass gravel as a large aggregate and foam glass powder as a fine aggregate. To achieve this goal, an experiment was carried out to optimize the composition of lightweight concrete and determine the optimal pro-portions of the used fractions of foam glass gravel.

Other The raw mix for producing lightweight concrete included the following components: Portland cement, foam glass gravel (mixture of fractions), glass powder (crushed foam glass gravel of 0.315 mm in size), soda glass, superplasticizing admixture, air-entraining agent and water.

During the experiment, the methods of mathematical planning and processing of results were used [16, 17]. The following parameters were taken as variable factors in the experiment: consumption of Portland cement (X_1), consumption of foam glass gravel (X_2) and consumption of glass powder (X_3).

The compressive strength of the samples (Y_1) was taken as a response function and optimization parameter. In parallel, the thermal conductivity of the resulting products and their average density were monitored. See table 1 for experimental conditions.

Table 1. Experimental conditions.

Factor name	**Symbol X_i**	**Average factor value, $\overline{X}_i$**	**Variation interval, ΔX_i**	**Factor values at levels**	
				-1	+1
Portland cement consumption, [%]	X_1	28	4	24	32
Foam glass gravel consumption,	X_2	26	3	23	29
Glass powder consumption, [%]	X_3	20	4	16	24

a *b* *c*

Figure 2. Test of samples: a, b – general view of samples; c – determination of the thermal conductivity coefficient.

Concrete production was carried out in the following sequence: First, the dry components were dosed according to the composition of the mixture. Then foam glass gravel of three fractions was poured into a paddle mixer of periodic action and stirred for 3 minutes; after that Port-land cement was added, stirred for 1-2 minutes; then glass powder was added; stirred for 1-2 minutes; soda glass was added and stirred for 1-2 minutes; water and superplasticizing admixture were added and stirred for 1-2 minutes, and last of all air-entraining agent was added; then the whole mixture was stirred for 5 minutes until a uniform consistency of the raw mixture was achieved.

The molding of standard samples with dimensions of 100×100×100 mm and 250×250×50 mm was carried out with vibration compaction of the mixture for 1-2 min. The manufactured samples were placed in a humidity chamber for 28 days. For compression strength test, samples of 100×100×100 mm were used (Figure 2a); For thermal conductivity test, samples of 250×250×50 mm were used (Figure 2b).

The study of the obtained polynomials was carried out by the method of analytical optimization [18, 19]. This method is based on two principles:

- obtained mathematical models based on the results of the adequacy check, describe the process with an acceptable degree of probability;
- equations (polynomials) are functions of several variables to which the rules of mathematical analysis are applicable, including methods for studying functions for extremum in partial derivatives.

3. Results

As a result of carrying out a three-factor experiment, the matrix of which (in coded values of factors) was built according to the D-optimal design for a full quadratic polynomial, processing experimental data and testing statistical hypotheses, the following regression equation was obtained:

$$Y_1 = 8.9 + 0.6X_1 + 0.4X_2 + 0.3X_3 + 0.2X_1X_3 - 0.3X_2^2 \quad (1)$$

The equation includes significant coefficients only; coefficients smaller than the confidence interval Δb equal to 0.14 MPa were omitted (equal to zero), and the F-criterion turned out to be equal to 0.96, which is quite consistent with the result forecasting accuracy adopted in solving tasks of building materials technology.

$$\partial Y_1/\partial X_2 = 0.4 - 0.6X_2 = 0 \rightarrow X_2 = 4/6 = 0{,}67$$

Substitute $X_2 = 0.67$ into equations (1):
The result is an optimized equation:

$$Y_1 = 9.1 + 0.6X_1 - 0.3X_3 + 0.2X_1X_3 \quad (2)$$

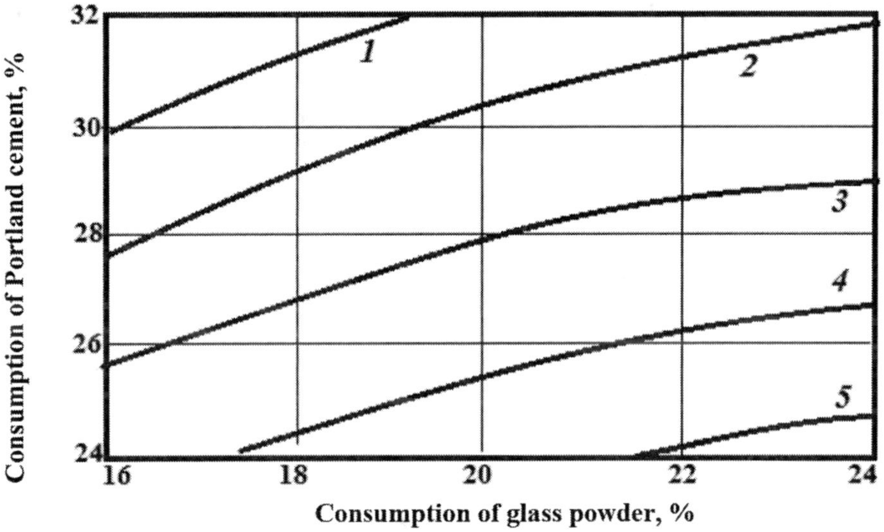

Figure 3. Graphical interpretation of the relationship $Y_1 = 9.1 + 0.6X_1 - 0.3X_3 + 0.2X_1X_3$ at $X_2 = 0.67$ (or in natural terms with a foam glass gravel content of 27±1%). Compressive strength, MPa: 1 – 9.6; 2 – 9.4; 3 – 9.2; 4 – 9.0; 5 – 8.8.

According to the graphical interpretation of equation (2), a nomogram was obtained, which makes it possible to select the preliminary values of the Portland cement and glass powder consumption to manufacture products of a given strength (Fig. 3). To clarify and verify the calculated data, a full-scale experiment is required.

4. Discussion

By the magnitude and sign of the coefficients for variable factors, one can judge the strength and influence of each factor on the response function, that is, on the result of the experiment.

An increase in the consumption of all variable factors in the intervals stipulated by the experimental conditions leads to an increase in the compression strength of lightweight concrete samples. The greatest influence on the strength is exerted by the Portland cement consumption (coefficient at X_1, equal to 0.6).

Analysis of the influence of foam glass gravel consumption shows that at average values of consumption, the strength of the samples increases, but at high consumption, a tendency to decrease in strength appears (coefficients at X_2 and X_{22} equal to 0.4 and –0.3, respectively). The carried out analytical optimization for this factor made it possible to establish the calculated in-tervals of the optimal consumption of coarse aggregate in the range of 26-27 %.

When implementing and analyzing the results of the experiment, the expediency of fractionation of foam glass gravel was established. To obtain the maximum strength of lightweight concrete, it is rational to use a mixture of fractions of 5-10 mm, 10-20 mm and 20-40 mm of foam glass gravel with a density of 150-180 kg/m^3 as a coarse aggregate in a ratio of 30:30:40 %.

Attention should also be paid to the increase in strength as a result of the combined effect of the Portland cement and finely ground glass consumption (the coefficient for pair interaction X_1X_3 equal to 0.2). This can be explained by the activation of silicates and aluminates contained in finely ground glass when interacting with minerals of Portland cement clinker. Therefore, glass powder obtained by crushing foam glass gravel with a density of 700 kg/m^3 is recommended to be ground to a particle size of 0.315 mm or less.

The gel formed as a result of the alkali-silicate reaction is distributed in the pores of the foam glass gravel and therefore does not lead to the destruction of concrete. Moreover, the amount and rate of gel formation depend on a number of factors: temperature, humidity and permeability of concrete, its operating conditions, etc. For this reason, the conditions for the formation of an expanding gel in foam glass concrete as a result of the interaction of silica aggregate with free calcium hydroxide and cement alkalis require separate studies.

In the course of the experiment, it was found that the thermal conductivity of lightweight concrete of the average density grade D700 depends, first of all, on the correct selection of the fractional composition of the coarse aggregate.

As a result of the implementation of the methodology for selecting the main components and carrying out corrective mixes, the composition of lightweight concrete was obtained with the following percentage: Portland cement — 24-30%; foam glass gravel — 26-27%; glass powder — 16-23%; liquid glass — 0.5-0.6%; superplasticizer — 0.05%; water — 21-26%.

The average density of the samples of the obtained foam glass concrete was 700 kg/m3; dry heat conductivity — 0.13-0.15 W (m·K); compressive strength — 8.1-9.6 MPa. The analogue of comparison, cellular concrete with a similar average density, has a compressive strength of 3-4 MPa and a dry thermal conductivity of 0.18 W (m·K).

5. Summary

Lightweight concrete based on foam glass gravel using finely ground foamed glass in its properties fully meets the requirements for heat-insulating and structural lightweight concrete. To obtain the maximum strength and minimum thermal conductivity of concrete with an average density of D700, it is rational to use a mixture of foam glass gravel fractions 5-10 mm, 10-20 mm and 20-40 mm as a coarse aggregate in a ratio of 30:30:40 %.

The synergistic effect of the combined action of the Portland cement and finely ground glass consumption on an increase in the strength of foam glass concrete can be explained by the activation of silicates and aluminates contained in finely ground glass when interacting with minerals of Portland cement clinker.

In the course of the experiment, the method was tested for designing the composition of lightweight concrete containing the following components: Portland cement, foam glass gravel, glass powder, liquid glass, superplasticizing admixture, air-entraining agent, and water. Strength and thermal characteristics of the obtained lightweight concrete are 1.2–2.7 times higher than those of equivalent - cellular concrete of the same average density.

References

[1] Bobrova E, Pilipenko A and Zhukov A 2019 Insulating sheath system and energy efficiency of buildings *TPACEE 2018* 02019. DOI: https://doi.org/10.1051/e3sconf/20199102019.

[2] Rubtsov O I, Bobrova E Yu, Zhukov A D, Zinovieva E A 2019 Ceramic bricks, stones and full-brick walls *Building materials 2019* **9** 8-13. DOI: https://doi.org/10.31659/0585-430X-2019-774-9-8-13.

[3] Zhukov A, Dovydenko T, Kozlov Se, Ter-Zakaryan K and Bobrova E 2019 Innovative technologies for low-rise construction *TPACEE 2018* 02032. DOI: https://doi.org/10.1051/e3sconf/20199102032.

[4] Umnyakova N, Bessonov I, Zhukov A, Zinoveva E 2019 The effectiveness of facade systems *MATEC Web of Conferences (ICMTMTE 2019)* 00013. DOI: https://doi.org/10.1051/matecconf/201929800013.

[5] Lesovik V S, Bessonov I V, Bulgakov B I, Larsen O A, Puchka O V, Vaysera S S 2018 Approach on Improving the Performance of Thermal Insulating and Acoustic Glass Composites. International Multi-Conference on Industrial Engineering and Modern technologies *Materials Science and Engineering* **463** 042030. DOI:10.1088/1757-899X/463/4/042030(2018).

[6] Bessonov I V, Larsen O A, Alimov L A, Voronin V V, Sapelin A N. and Komorova N A 2018 Effect of alumina-silica microsphere additives on properties of ceramic wall products. Proceedings of XXI International Scientific Conference in Advanced in Civil Engineering "Con-struction – The Formation of Living Environment" *Materials Science and En-gineering (FORM 2018)* **365** 032011. Doi:10.1088/1757-899X/365/3/032011.

[7] Tang L V, Dinh Th V, Kim D V, Bulgakov B, Aleksandrova O, Bazhenova S 2018 Combined Effect of Bottom Ash and Expanded Polystyrene on Light-weight Concrete Properties *VI International Scientifis Conference «Integration, Partnership and Innovation in Construction Science and Education» (IPICSE-2018). Moscow, Russia, November 14-16, 2018. MATEC Web of Conference* **251** 01007. DOI: https://doi.org/10.1051/matecconf/201825101007.

[8] Tang V L, Dien V K, Nguyen T Ch, Bulgakov B I 2019 Predicting light-weight concrete properties containing fly ash and expanded polystyrene *Vietnam-Japan Science and Technology Symposium 2019 (VJST 2019).*

[9] Tho V D, Lam T V, Korol E A, Bulgakov B I, Aleksandrova O V, Larsen O A 2020 Thermal insulation properties of effective lightweight concrete for three-layer building envelope *Industrial and civil construction* **5** 36-42.

[10] Bessonov I V, Shubin I L, Umnyakova N P, Spiridonov A V 2017 Prospects for the use of materials and products of foam glass in thermal insulation systems *Bulletin of construction technology (BCT)* **6** 12-14.

[11] Bessonov I V, Lesovik V S, Alekseev S V, Vaysera S S 2018 Control of the structure and properties of acoustic materials based on glass-foam composites *Construction Materials* **6** 41-45. DOI: https://doi.org/10.31659/0585-430X-2018-760-6-41-44.

[12] Bessonov I V, Starostin A V, Ivanova N A and Bulgakov B I 2020 The composition of the raw mix for the preparation of lightweight concrete with increased strength and low thermal conductivity *Application for invention* **2020113197** C04B 38/00, C04B 28/04, C04B 38/10.

[13] Zhukov A, Medvedev A, Poserenin A and Efimov B 2019 Ecological and energy efficiency of insulating systems *E3S Web of Conferences* **135** 03070. DOI: https://doi.org/10.1051/e3sconf/201913503070.

[14] Tang L V, Kim D V, Xuan H N, Tho V D, Bulgakov B I, Alexandrova O V 2020 Modelling of effect of water-cement ratios on properties foam concrete *International Scientific Conference «Modeling and Methods of Structural Analysis» (MMSA 2019) November 13-15, 2019, Moscow, Russian Federation Journal of Physics* **1425** 012189. DOI: 10.1088/1742-

6596/1425/1/012189.
[15] T V L, Kim D V, Bulgakov B I 2019 Study of the effect of gyp-sum and alumina cement on the propertiesof binder on the basis of Portland cement used for the preparation of non-autoclaved foam concrete *Construction Publishing House Conference. Vi-etnam Institute for Building Science and Construction* **XV(11)** 13-24.
[16] Bessonov I, Zhukov A, Shokodko E, Chernov A 2020 Optimization of the technology for the production of foam glass aggregate *TPACEE 2019 E3S Web of Conferences* **164** 14016. DOI: https://doi.org/10.1051/e3sconf/202016414016.
[17] Tang L V, Kim D V, Bulgakov B I, Bazhenova S I 2019 Mathematical Simulation of Bottom Ash Effect and Expanded Polystyrene on the Polystyrene Concrete Properties *International Scientific Conference «Construction and Architecture: Theory and Practice for the Innovation Development» (CATPID-2019) October 1-5, 2019, Kislovodsk. Materials Science Forum «Materials and Technologies in Construction and Architecture II»* **974** 312-318.
[18] Krashchenko N, Tretyakov A, Chernov I, Shaykhalov A, Zhukov A 2020 Modeling and thermal calculation of a pipeline insulation system *TPACEE 2019 E3S Web of Conferences* **164** 14021. DOI: https://doi.org/10.1051/e3sconf/202016414021.
[19] Zhukov A, Shokodko E 2020 Mathematical Methods for Optimizing the Technol-ogies of Building Materials *VIII International Scientific Siberian Transport Forum TransSiberia 2019 Advances in Intelligent Systems and Computing* **1116** 413-421. DOI: https://doi.org/10.1007/978-3-030-37919-3_40.

IOP Conf. Series: Materials Science and Engineering 1083 (2021) 012039 doi:10.1088/1757-899X/1083/1/012039

Technologies for creating directional orientation of fiber reinforcement in concrete

L R Mailyan, P A Shilov, A A Shilov*

Don State Technical University, 1 Gagarin sq., Rostov-on-Don, 344000, Russia

E-mail: ale381420082008@yandex.ru

Abstract. Information on the analysis of methods for creating directional orientation of steel fiber rodding of reinforced concrete structures is presented. The design of a mobile experimental setup for the manufacture of fiber-concrete and fiber-reinforced concrete linear elements with a directional orientation of steel fibers by mechanical means has been developed. The test results of the pilot plant are presented. The required dimensions of the bunker for the fiber-reinforced concrete mixture, the type of the leveling element and the method of increasing the fiber-reinforced concrete mobility during casing using a pilot unit have been determined. The creation of a directed orientation of fiber threads in fiber-reinforced concrete makes it possible to make the most of this material's advantages. Despite a sufficient number of theoretically possible ways to create a directional orientation of fibers in a concrete mixture, not all of them can be applied in real construction conditions. Also, a number of methods introduce additional difficulties into the concreting process, which can adversely affect the construction processes duration. The use of a mechanical method for creating a directional orientation of steel fiber reinforcement in concrete made it possible to obtain the desired result on a mixture already containing fiber threads, without the need for their separate laying. The optimal design of the mechanical element leveling and the features of the process of leveling steel fibers in concrete have been established.

Introduction

Obtaining a fiber-reinforced concrete mixture with the specified parameters is possible only with the correct selection of a combination of raw materials and their manufacturing technology.

At the moment, the steel fiber reinforced concrete compositions selection issues, depending on its purpose, have been well studied.

The main problem that restrains the massive use of steel fiber reinforced concrete in construction is a more complex technology for its production and laying in molds compared to ordinary concrete, which leads to an uneven distribution of fiber in the concrete body, the appearance of fibers' fisheye, so-called "balls", which, as a result, do not allow to get the specified design parameters.

As stated in previously published works of the authors [1], even the usual distribution of fibers in concrete leads to a significant increase in its compressive and tensile strength characteristics (more than 25 and 50%, respectively).

Even more (up to 40 and 100%, respectively) the compressive and tensile strengths of concrete grow with the aggregated distribution of fibers [2], which makes it possible to get rid of fisheye and achieve a more uniform distribution of fibers over the concrete volume.

The next jump in the strength properties of concrete, both in compression and tension, occurs with the directional orientation of the fibers, in which the fibers are directed along the tensile stresses acting in the structure. At the same time, the creation of an oriented structure of the fibers' arrangement in concrete in practice is a complex and difficult technical task.

At the moment, we can conditionally distinguish two main directions in the steel fiber reinforced concrete manufacturing technologies:

- separate packing technology;
- technology of joint mixing of concrete and fibers.

The technology of separate packing consists of preliminary layer-by-layer fiber reinforcement placing into the formwork with the formation of a fiber cage, followed by the introduction of a concrete mixture into it and vibration shaping. The same technology includes the method of spraying fibers into the mixture at high speed and the vibro-extrusion method.

This technology can significantly reduce the labor intensity of work, but practically excludes the use of coarse aggregate and requires special equipment - a technological line with a vibro-extruder, which provides an oriented arrangement of fibers in a cement-sand matrix.

The technology of joint mixing of a concrete mixture with fiber at the preparation stage is easier to use, does not require any additional equipment or a special technological line, allows to obtain structures of any shape, to use a large aggregate, but leads to a chaotic arrangement of fibers in concrete, as a result of which some of the fibers are simple not included in work.

Most of the earlier studies [3-6] were aimed at improving the fibers' uniform distribution technology in a concrete matrix and, only recently, experimental work to create a method for the oriented arrangement of fibers after they are mixed with concrete in a concrete mixer and fed into the formwork has started.

Currently, the following methods for creating an oriented structure of the fibers' arrangement in the concrete body using the joint mixing technology are known:

- constraint of fibers by formwork;
- roller pressing;
- mechanical orientation of fibers, achieved by passing the mixture through special meshes;
- intense vibration of the mixture, in which the fibers occupy a horizontal position in one plane;
- various kinds of electromagnetic influences at the stage of molding products.

The above-mentioned methods, tested in laboratory conditions, turn out to be rather complicated, difficult to implement in practice and, as a rule, only partially provide directional orientation of fibers in full-scale structures.

At the same time, the methods of constraining fibers by formwork, roller molding of products, intense vibration of the mixture, passing fibers through the meshes do not initially focus on the fibers' orientation along the stresses acting in the finished product and do not allow achieving the maximum effect from the use of reinforcing steel fibers [7].

The most promising in this direction is the method of electromagnetic alignment of fibers at the stage of forming products, which is aimed at a clear unidirectional orientation of fibers in a concrete matrix. At the same time, this technology is still at the scientific development stage and is not applied in practice, it provides for the use of a concrete mixture of increased mobility without the use of a large aggregate and requires specific equipment to create magnetic fields. The application of this method involves the use of formwork made of non-magnetic materials, the need to create constant or variable electromagnetic fields of certain parameters, taking into account their interaction with linear steel reinforcement and the possible effect of "interference" from steel working reinforcement on the fibers' orientation [8-11].

The method of the fibers' directional orientation aggregated creation in concrete with separate placement of components in the formwork, at the moment, is also at the stage of theoretical development and has not been introduced into construction production practice, since there are still no mobile installations that allow using this method on a construction site.

Moreover, the use of the above-mentioned technologies does not allow obtaining directional orientation of fibers in two or more directions for use, for example, in slab elements.

Materials and methods

An additional factor preventing the implementation of previously developed methods of fibers orientation in a concrete body in a given direction into construction practice is that the proposed methods initially exclude the use of large aggregate in the form of crushed stone.

In connection with the above-mentioned, in order to obtain steel-fiber concretes with an oriented arrangement of fibers with increased characteristics, it is necessary to create a technology that allows solving the following problems:

- to ensure uniform distribution of fibers in the concrete body, completely eliminating the appearance of fisheye from it in the "balls" form;
- to ensure the use of this technology not only in fine-grained heavy concrete, but also in concretes containing a large aggregate in the form of crushed stone;
- to include a larger number of fibers in the work, due to their orientation along the tensile stresses acting in the structure;
- to ensure the simplicity and manufacturability of the applied method of orientation of fibers, which makes it possible to use it not only in laboratory conditions, but also in construction practice;
- to ensure the mobility of the used installation, allowing it to be used directly on the construction site.

Taking into account the tasks set in this work, a significantly simpler, mechanical, technology for producing fiber-concrete and fiber-reinforced concrete elements with a directional orientation of fibers is proposed, which is much easier to implement both in the factory and on the construction site.

The proposed technology is based on the following basic principles:

- assembly and mobility of the installation, allowing it to be used for concreting linear elements of any length, width and height;
- preliminary orientation of the fibers when feeding the fiber-concrete mixture into the formwork from the conical hopper, which is carried out due to the combination of vibration action and the use of the fibers restraining method with a cone at the place where the mixture exits the formwork;
- concreting placing, allowing to obtain uniform distribution of a given thickness fibers in each layer;
- the use of a manifold-shaped working body, operating in a vibration-translational mode and allowing to create a given orientation of fibers in each layer and in the element as a whole.

These principles are the distinctive features of the proposed technology, to which its basing not on an electromagnetic, but on a mechanical basis is added.

A schematic structural diagram of the proposed installation for creating fiber-concrete and fiber-reinforced concrete elements with directional orientation of fibers according to the proposed mechanical technology is shown in Figure1.

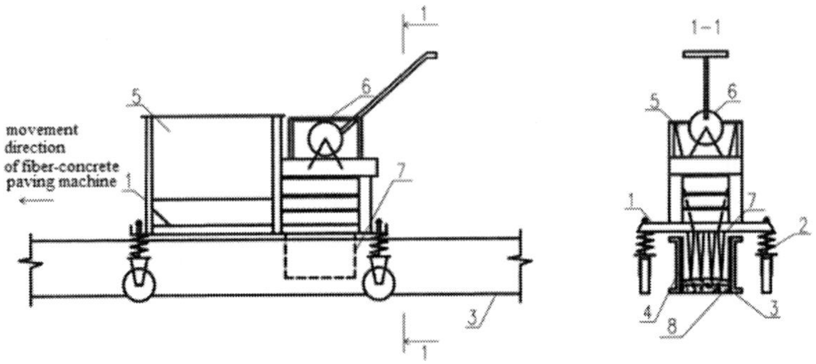

Figure 1. Diagram of a mobile installation for concreting linear elements with directional fiber orientation.

1 – steel frame installation; 2 – spring chassis; 3 – a concrete element formwork; 4 – reinforcement cage of a concrete element; 5 – fiber-concrete mix feed bunker; 6 – vibrating element; 7 – manifold with teeth rows; 8 – first layer of fiber-reinforced concrete mix.

Steel frame 1 on a spring chassis 2 continuously and evenly moves at a given speed along form 3 with a reinforcing cage 4 installed in it (or without it), into which a movable steel-fiber-concrete mixture is fed from a bunker 5 to a given thickness of the first (lower) layer 8.

An electric motor with an eccentric 6 is installed on the frame, which sets the vibrational vertical reciprocating movements to the working body of the installation in the form of a manifold with teeth 7, which can move in height in the range of 25-50 mm - depending on the aggregate size and the concrete layer thickness.

The preliminary orientation of the fibers is achieved when the fiber-concrete mixture is fed from the conical bunker. At the same time, the bunker walls converge on a cone with a rectangular finish, oriented along the formwork and, at the place where the mixture exits, narrow to a size that allows the large aggregate to pass freely, but forces the fibers to unfold when vibrated in the longitudinal direction of the installation movement.

The final orientation of the fibers takes place directly in the formwork at the contact of the fiber-reinforced concrete mixture with longitudinally arranged elements (knives). At the same time, the distance between the leveling elements of the manifold is greater than the size of the coarse aggregate, but much less than the fibers length. This allows the coarse aggregate to pass freely between the manifold knives, but forces the fibers, upon contact with them, to align along the movement of the machine.

After the first heading, vibration compaction and creation of a directed orientation of the mixture first layer fibers, the mold is filled with the second layer of the fiber-concrete mixture and the cycle is repeated again. This happens until the form is completely filled, the proposed mobile installation is completed and the element is concreted.

The proposed mechanical technology for creating directional orientation of fibers can be successfully applied to linear fiber-reinforced concrete and fiber-reinforced concrete elements.

However, the application area of the proposed technology is far from being exhausted by the linear elements.

If we provide a mutually perpendicular direction in the orientation of the fibers in different layers of the structure, this will allow the proposed technology to be used for the manufacture of not only linear, but also plate elements with different support conditions and working in two directions.

To create the first layer of a slab element with a longitudinal arrangement of fibers, the first installation is made. The formwork of the slab is filled from the installation bunker to a given layer thickness with preliminary orientation of the fibers in a given direction. The layer thickness is regulated by the movement speed of the bunker and the mobility of the fiber-reinforced concrete mixture. After that, the final orientation of the fiber to the width of the manifold is carried out with simultaneous vibration and the formation of the first layer.

To create a second layer with a transverse arrangement of fibers, the installation is turned at an angle of 900 and the form is filled with a second layer of concrete mixture, in which the fibers are located strictly perpendicular to the first layer.

This happens until the form is filled in layers and in different directions to the entire height of its section with a vibrated fiber-reinforced concrete mixture.

The use of this technology makes it possible to set any layer-by-layer orientation of the fibers in the structure for the tensile stresses perception and get the maximum effect from the fiber-reinforced concrete use.

The difference in the mobile installation design for concreting slab elements with a directed orientation of fibers from the installation for linear elements lies only in the design of the feed bunker and a wider manifold.

Discussions and results

To work out the optimal parameters of the proposed technology for creating a directed orientation of fibers, a prototype of a mobile installation for concreting linear elements, shown in Fig. 2, has been performed, the composition of fiber-reinforced concrete has benn selected, which was subsequently used for the manufacture of prototypes.

During the experimental studies, heavy concrete of design class B30 and anchor fiber made of high-carbon steel wire from spring steel 70-85 with a tensile strength of 2200 MPa, with a brass coating, nominal diameter 1.0 mm, length 50 mm were used (FSR-A-0,7/50), produced by JSC "Tanis" TC RB 400518274.004-2009.

The used fiber has curved anchors at the ends, preventing it from being pulled in the concrete composition, and the ratio of length to diameter is 50 (l_f/d_f=50). The brass coating of the fiber increases its resistance to corrosion and, at the same time, improves adhesion to the cement stone at a microscopic level.

For the manufacture of prototypes, the following was used: Portland cement of the OJSC "Novoroscement" brand PC 500-D20 according to GOST10178-85; crushed stone from the Beslan quarry, fraction 10 mm, strength 140 MPa, with a content of silt and clay particles - 0,3…0,8%; quartz river sand with a size modulus was used as a fine aggregate ($M_к$) within 1.6 ... 1.8, bulk density 1510 kg/m^3.

In order to increase the concrete mixture mobility and improve the concrete adhesion with fiber reinforcement elements, the plasticizing additive "Optimist superplasticizer C 409" was used.

The composition of heavy concrete of B30class was selected by the calculation and experimental method, concrete and fiber-reinforced concrete mixtures were made with the same amount of all main components.

Table 1. Composition of B30 class heavy concrete with natural hardening without fiber reinforcement

Concrete class	Material consumption for 1m^3 concrete, kg				
	C	S	CS	W	Plasticizer
B30	459	456	1267	177	8.5

Table 2. Composition of B30 class heavy concrete with natural hardening with fiber reinforcement

Concrete class	Material consumption for 1м^3 concrete, kg					
	C	S	CS	W	Plasticizer	Fiber
B30	459	456	1267	177	8.5	152.7

a) b)

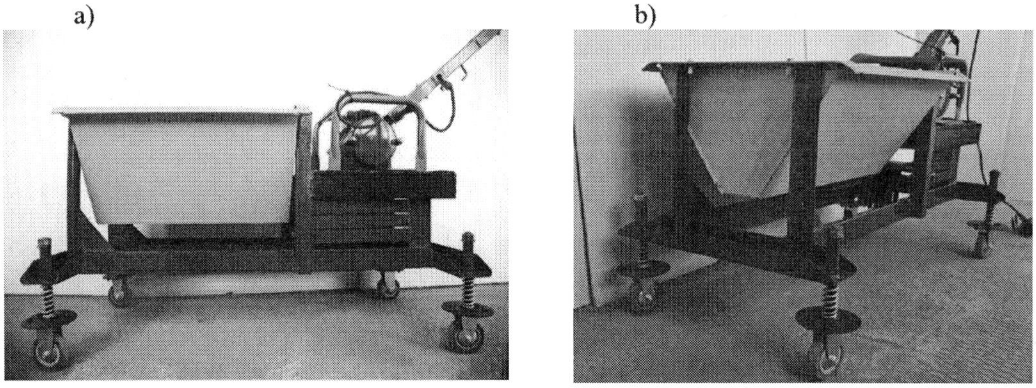

Figure 2. General view of a prototype of an experimental mobile installation for concreting linear elements with directional orientation of fibers.

Based on the previous studies, when designing a mobile unit, a choice was made in favor of a platform vibrator, with a vibration frequency of the working body 90 Hz, and the design of the feeding bunker is based on the fibers restraint principle in combination with intense vibration.

When designing a bunker for feeding a fiber-concrete mixture into the form, we used a conical cross-sectional shape, which, under vibration action, allows the mixture to move without "hanging" on the walls and, at the point of its exit into the formwork, using the constraint method, force the fiber to turn in the direction we need. At the same time, the rotation of the fibers, located initially across the walls of the bunker, occurs under the gravity action in combination with vibration action, and the specified layer thickness is controlled by the installation speed.

The greatest effect on the preliminary orientation of the fibers is achieved with the maximum possible narrowing of the walls of the bunker finish feeding opening. At the same time, its width depends on the length of the fiber used and should not lead to sticking of the fiber-reinforced concrete mixture at the exit to the formwork.

Taking into account the use of steel fiber with a diameter of 1 mm, a length of 50 mm and a combination of compositions of different plasticity with a maximum size of a coarse aggregate of 10 mm in our study, we tested various options for the bunker design with a change in the walls inclination angle and the feeding opening width as well as several types of manifolds that differ from each other by the arrangement of the alignment elements.

During experimental concreting, the optimal movement speed of the installation along the mold was selected, regardless of the type of manifold, which was about 2 m / min, which ensured the orientation of the fibers, the orientation of the layer and its high-quality vibration. In this case, the thickness of the formed layer at a given movement speed was 30 .... 35mm.

A general view of a fiber-concrete mixture with a chaotic arrangement of fibers is shown in Figure 3 (a), and the view of the mixture with the oriented fiber arrangement is shown in Figure 3 (b).

a) b)

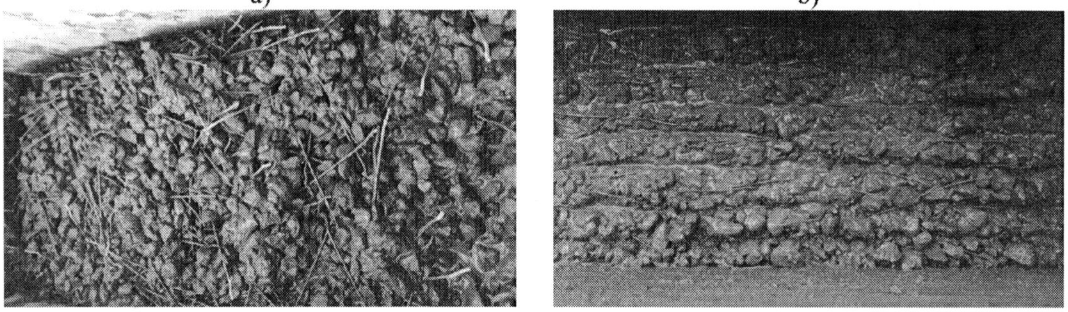

Figure 3. General view of the chaotic (a) and directional (b) arrangement of fibers in concrete when using the proposed mechanical technology.

Summary

Analyzing the above-said, we can draw the following conclusions:

1. The mechanical method of steel fibers orientation in concrete is the most promising for use in real construction conditions, since it provides the required result with minimal complication of the steel fiber reinforced concrete placing process;

2. The optimal inclination of the feeding bunker walls from the vertical axis for the steel fiber concrete mixture is in the range from 35° till 45°, and the optimal finish width is in the area 20÷25mm ($0,4 \div 0,5 l_f$), which allows the coarse aggregate to pass freely under vibration and, at the same time, ensures the fiber rotation in the installation movement direction;

3. The final orientation manifold must have a distance between the knives greater than 0,3÷0,5 the length of the fibers, while the manifold knives should be of the sufficient width, close to the length of the fibers used, allowing them to act on the fiber for a time sufficient to flatten it;

4. The optimal speed of the experimental setup for mechanical orientation of steel fibers in concrete mixture was 2 m / min, which provided the necessary orientation of fibers and high-quality concrete vibration.

References

[1] Mailyan L R, Shilov P A, Shilov A A 2020 Fiber reinforcement as an alternative to the compressed zone linear reinforcement and the flexible concrete elements stretched zone prestressing *IOP Conference Series: Materials Science and Engineering* **913** 022038.

[2] Bezborodov V G, Matus E P 1999 Influence of the magnetic orientation of fibers on the tensile strength of fiber-reinforced concrete *Proceedings of the Novosibirsk State University of Architecture and Civil Engineering (SIBSTRINE)* **3** 24-28.

[3] Matus E P, Bezborodov V G 1999 Application of a magnetic field to increase the tensile strength of fiber-reinforced concrete *Bulletin of higher educational institutions. Construction* **11** 139-142.

[4] Suhas S Joshi, Nikesh Thammishetti, Suriya S Prakash 2018 Efficiency of steel and macro-synthetic structural fibers on the flexure-shear behaviour of prestressed concrete beams *Engineering Structures* **171** 47–55.

[5] Mailyan L R, Mailyan R L, Shilov A V 1997 Strength calculation of bent fiber-concrete elements with high-strength reinforcement *Bulletin of higher educational institutions. Construction and architecture* **4** 4–7.

[6] Mailyan R L, Mailyan L R, Shilov A V, Abdallah M T 1995 Bending elements made of expanded clay-fiber-reinforced concrete with high-strength reinforcement without preliminary stress and with partial stress *Bulletin of higher educational institutions. Construction* **12** 19–23.

[7] BC 52-104-2006* *Steel fiber concrete structures* (Moscow, OJSC "SRC Construction).

[8] Furlan Junior S, J. Bento de Hanai 1999 Prestressed fiber reinforced concrete beams with reduced ratios of shear reinforcement *Cement & Concrete Composites* **21** 213–221.

[9] Trambovetskiy V P 1974 Concrete reinforced with dispersed reinforcement, concrete and reinforced concrete **2** 40–42.

[10] Rad'kova I N, Gritsayenko V I, Koval I V 2012 Effective reinforcing component of reinforced concrete structures - steel fiber *Casting and metallurgy* **1** 38–42.

[11] Pukharenko Yu V, Panteleev D A, Morozov V I, Magdeev U Kh 2016 Strength and deformability of poly-reinforced fiber-reinforced concrete using amorphous metal fiber (Academia. Architecture and construction) **1** 107–111.

IOP Conf. Series: Materials Science and Engineering 1083 (2021) 012040 doi:10.1088/1757-899X/1083/1/012040

Destruction of polymers under the action of laser radiation

B I Kunizhev[1], Z S Torshkhoeva[2*], R N Zhelikhazhev[1], A V Starov[3]

[1]Kabardino-Balkarian State University, 173, Chernyshevsky St., Nalchik, 360004, Russia
[2]Ingush State University, 25 Moskovskaya st., Nazran, 386001, Russia
[3]Volgograd State Technical University, 28, Lenin avenue, Volgograd, 400005, Russia

E-mail: torshhoeva.zina@yandex.ru

Abstract. Laser ablation is one of the main mechanisms of polymers destruction under the action of laser radiation. In this paper, the basic information on laser ablation of polymers is presented. Under the action of laser radiation on the polymer, processes of thermal, thermo-oxidative, and mechanical degradation occur simultaneously, resulting in the evaporation of macromolecules fragments (up to oligomers), and in some cases, separation of polymer particles and filler by a gas or plasma jet is observed. In addition, a substance ablation is accompanied by a large number of concomitant effects: vapor condensation and liquid phase dispersion.

Introduction

The unique properties of polymers have ensured their widespread use in construction. For example, polymethyl methacrylate is used to produce sheets that replace ordinary silicate glass. Polymethyl methacrylate is used for making door panels and finishing windows in gates.

Due to its low specific gravity, resistance to sunlight and ease of processing, polymethyl methacrylate is used as the main material in the construction of visors and transparent balustrades. Sound barriers from PMMA are used to create reflective acoustic baffles along the highways and busy roads.

Obtaining the experimental data on the polymers' dynamic strength is important for solving many problems in various areas of the national economy. We have investigated the polymers destruction by laser exposure.

The process of a solid destruction, similar to evaporation or sublimation, characterized by the presence of a condensed phase in the destruction products, is called laser ablation.

The energy efficiency of polymeric materials laser ablation depends on the amount of absorbed energy at a certain density. In the infrared, visible and near ultraviolet frequency ranges, radiation absorption is described by the Bouguer-Lambert law:

$$I(z) = I_0 \eta exp(-\chi z) \tag{1}$$

where I_0 — is the intensity of the light beam on the surface; χ — is an absorption coefficient; z — is the coordinate measured into the depth of the material; η — is the coefficient determining the absorbed energy share.

Energy absorbed in the sample layer during *t:*

$$E = S\int_0^t I(z)dt = S\eta I_0 e^{-\chi z}t, \tag{2}$$

where S — is the cross-sectional area of the laser beam on the sample surface. The intensity lost in a layer of thickness z is

$$\Delta I = I_0 - I(z) = I_0(1-\eta)e^{-\chi z}. \quad (3)$$

Table 1 shows the values of the absorption coefficients χ, obtained by irradiating polymer films with radiation CO_2- laser.

Table 1 gives the data indicating that the value of the coefficient χ significantly depends on the method of obtaining a polymer film.

Table 1. The absorption coefficients χ for laser radiation with a wavelength of 10.6 μm.

Polymer	Method of obtaining	χ_T, $[cm^{-1}]$			χ_T, $[cm^{-1}]$
		Frequency			
		943	927	104	
Polyethylene	Extrusion	37	31	39	35
Polypropylene	Hot pressing	118	255	131	169
Polystyrene	From solution in benzene	112	131	347	176
Polyvinyl chloride	Hot pressing	147	176	104	147

All polymers, depending on their behavior when irradiated with a laser beam, can be divided into three groups, as it was done in the work [1]:

polymers that melt and splash (group A);

polymers that form a coke layer on the surface (group B);

polymers that turn into a gaseous state without residue (group C).

In some cases, it is necessary to take into account the ignition of polymers, which has an additional destructive effect.

Group B behavior is only polymers characteristic and group A and B behavior is also observed in metals and semiconductors.

Group A includes thermoplastic polymers (polyethylene, polypropylene, polyethylene succinate, nylon, kapron, etc.). Some polymers, such as polystyrene, polymethyl methacrylate, belong to group A - at low radiation power densities, and to group B - at high radiation power densities.

For polystyrene and polymethyl methacrylate, it was found that in the surface layers inside the sample under study, destruction occurs along the supramolecular structures' boundaries, as a result of which a cloud appears above the surface, shielding the sample by absorbing a part of the radiation and containing solid, liquid and gaseous products, and the composition of the cloud depends on laser beam intensity [1].

Figure 1 shows the destruction zones of the sample along the cross-section perpendicular to the surface: **zone**: ***I*** — solid polymer with exponential temperature distribution;

II — the polymer is in a highly elastic state;

III — liquid-like molten polymer (pyrolysis occurs in this area with increasing temperature);

IV — the smallest droplets resulting from the molten polymer crushing;

V — gaseous state of decomposition products.

Zones ***IV and V*** can be divided only by analyzing the products' composition in these zones.

Here T_{gt}, T_s, T_{dc} and T_d — glass transition temperature, softening, decomposition, destruction front (for thermoplastics it does not exceed 773K) respectively.

IOP Conf. Series: Materials Science and Engineering 1083 (2021) 012040 doi:10.1088/1757-899X/1083/1/012040

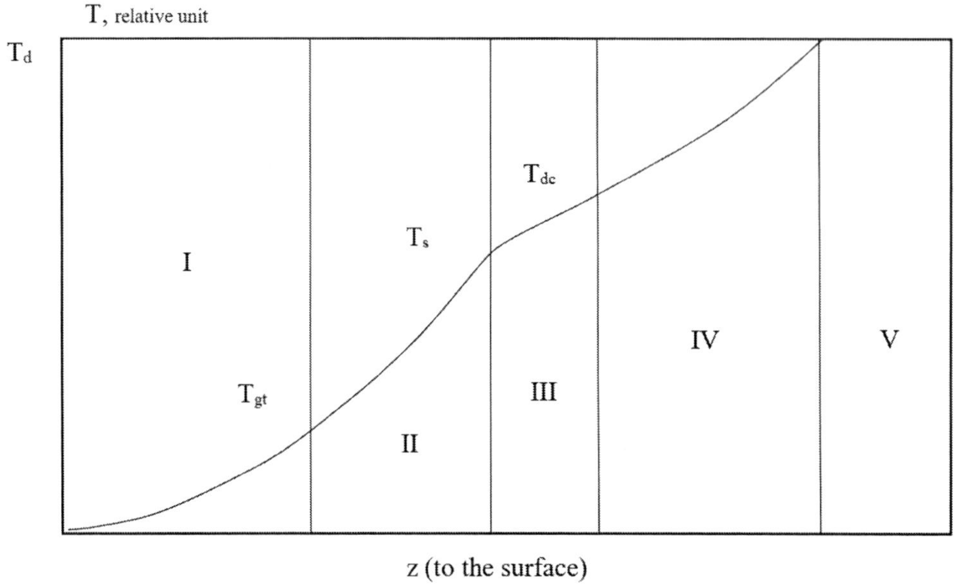

Figure 1. Schematic representation of the destruction zones of thermoplastics under the influence of infrared laser radiation with a wavelength 10.6 μm.

Figure 1 shows that for the group *A* polymers it is possible to subdivide the destruction zones and even determine their width. The appearance of rare cross-links in polymers of group A, for example, in radiation-crosslinked polyethylene [2], does not lead to significant deviations in their behavior in a laser beam. The ablation energies of polymers in this group lie within the bounds of the bond strengths GT-GT and are equal approximately 2—3 kJ / g (Table 2).

When exposed to radiation I_1 or I_2 the first-order phase transitions are possible.

Table 2. Threshold energy characteristics of the laser radiation effect on some polymers of group A.

Polymer	Power density, [W / cm^2]			
	I_1	I_2	I_3	I_4
Nitron	3	6	18.5	25
Lavsan	7	11	17	22
Lavsan	3	5	17	23

I_1-I_4 — are the threshold power densities at which respectively, the rearrangements in the supramolecular structure, softening, destruction process, decay, weight loss due to molecular fragments' evaporation, etc. start in polymer.

Group B includes aromatic thermosetting polymers (polybenzimidazole, polycyanurates, polyphenylenes, etc.), as well as some thermoplastics capable of crosslinking when heated (polyphenylene oxide, polyphenylquinoxaline, polyarylate, etc.), after which they behave like thermosetting plastics and form a coke layer when heated in a laser beam. After the formation of the secondary structure, such a polymer is no longer able to soften. As the irradiation dose increases, the network structure first becomes denser to the limiting value, and then condensation of aromatic cycles, or dehydrogenation with the formation of multiple bonds, and then aromatic cycles according to the Diels – Alder reaction start, followed by the cycles' condensation. The result is parquet and graphite-like structures [1, 4].

For the group B polymers when exposed to laser radiation, the zones *II, III, IV* are either absent, or significantly narrowed and the transition through them temperature occurs in a short time. These

IOP Conf. Series: Materials Science and Engineering 1083 (2021) 012040 doi:10.1088/1757-899X/1083/1/012040

polymers usually do not melt under irradiation (their crosslinking temperature is less than or close to the softening temperature), and along with the formation of a coke layer simultaneously with the processes of destruction and crosslinking, the process of ablation from the surface of oligomeric products is also possible.

The growth rate of the coke layer and its radiation-insulating properties strongly depend on the thermophysical characteristics associated with the morphology and the nature of the coke porosity (pore size, their closedness or openness, etc.), which should be taken into account when measuring ablation resistance. When exposed to laser radiation with an intensity less than 50 W / cm^2 on polycarbonate and polychlorotrifluoroethylene, a coke layer is formed, but in the second polymer the thickness is stabilized at a value of 0.1 mm, while in the first it grows to 3 mm or more. When the radiation intensity is increased to 200 W / cm^2 and more, the presence of a coke layer on the surface is not noted, since its formation rate is equal to or less than the ablation rate by sublimation [3].

Table 3. Laser ablation energies for a number of polymers capable of forming secondary polymer structures.

Polymer	I, [W/cm^2]	E_{abl}
Polycarbonate	10.5	49.0
	21.1	10.3
	31.7	10.6
	42.3	13.6
	55	13.3
	58.8	12.6
Polyimide	55	13.3
Polyphthalocyanine	55	37.0

The third group B includes polymers that ablate without residue. The mechanism of this process can be depolymerization, complete destruction to gaseous products, or the polymer entrainment in the form of individual chain fragments (oligomers) and their settling on cold surfaces near the site of exposure. Table 4 gives the values of the laser ablation energy E_{abl} for a number of polymers in this group.

Table 4. Laser ablation energies for a number of group C polymers.

Polymer	I, [W/cm^2]	E_{abl}
Polymethyl methacrylate	10.5	17.1
	21.1	4.8
	31.7	4.7
	42.3	4.2
	52.8	3.3
	55.0	3.5
Polystyrene	55.0	3.9
Poly-α-methylstyrene	55.0	2.4
Polybutadiene acrylate	55.0	3.4
Polytetraftoethylene	55.0	3.8

Table 4 data confirm that polymethyl methacrylate and polystyrene belong, apparently, to both group A and group C.

In polymethyl methacrylate, the radiation intensity I less 5-7 W/cm^2 destruction is not observed at all; at I>7-10 W/cm^2 the ablation process starts with E_{abl}>10 kJ / g (the value is overestimated due to additional heat losses for convection and thermal conductivity at long exposure times); further, the ablation process intensification occurs with a simultaneous decrease in ablation energy to a constant value in the range of 3.3–3.5 kJ / g (see Table 4). Consequently, the mechanism of decomposition of this polymer changes at elevated surface temperatures. In the case of laser ablation at low power densities, decomposition process of polymethyl methacrylate with an activation energy of 92 kJ / mol (decomposition of allyl bonds and depolymerization of radicals) is most likely realized, and with increasing values I till 50-60 W/cm^2 in addition, the processes of destruction according to the law of chance and depolymerization of the main polymer chain are additionally initiated, and the dynamic equilibrium is shifted towards the process with a higher activation energy (252 kJ / mol) [4].

Summary

The ablation energy of group C polymers is lower than that of group B polymers. The lowest E_{abl} value is observed in Poly-α-methylstyrene, which is due to the occurrence of uncomplicated depolymerization at any values I. Close values E_{abl} for other polymers allow us to conclude that the ablation mechanism itself is similar [4]:

- at low and medium power densities, the ablation mechanism of such polymers can include the processes of depolymerization, destruction with the elimination of side substituents, decomposition of the formed fragments, etc.;
- at high power density (I > 50—60 W/cm^2) the main chain breaks according to the law of chance, followed by partial depolymerization.

References

[1] Said-Galiev E D, Nikitin L N 1992 *Mechanics of Composite Materials* **2**152-157.
[2] Krasovskiy A M, Tolstopyatov E M 1989 Obtaining thin films by sputtering polymers in a vacuum (Minsk).
[3] Kononenko T V 1996 Study of the mechanism of pulsed laser micromachining (Author's abstract. Ph.D. thesis in physics and mathematics, M.).
[4] Vinogradov B A, Pereplkin K E, Mesherakova G P 2006 Effect of laser radiation on polymer materials (Science, St. Petersburg).

IOP Conf. Series: Materials Science and Engineering 1083 (2021) 012041 doi:10.1088/1757-899X/1083/1/012041

On the problems of controllability and uncontrollability for some mechanical systems described by the equations of vibrations of plates and beams with integral memory

A V Romanova[1*] and I V Romanov[2]

[1] Ishlinsky Institute for Problems in Mechanics RAS, 101-1 Vernadskogo ave., Moscow, Russia

[2] National Research University Higher School of Economics, 20 Myasnitskaya st., Moscow 101000, Russia

E-mail: avromm1@gmail.com

Abstract. Controllability problems for some models of plates and beams with integral memory are considered. The vibrational equation of the plate contains an Abelian kernel in the integral term, and the vibrational equation of the beam contains a continuous kernel consisting of a finite sum of decreasing exponential functions. It is proved that by controlling the whole domain, the first system cannot be driven to a state of rest, and for the second system, controllability to rest is possible.

1. Introduction

Integro-differential equations with integral terms in the convolution form often arise in various applications in the mechanics of homogeneous media, the theory of viscoelasticity, thermal physics and the kinetic theory of gases. In some cases, the convolution kernel is a sum (finite or infinite) of exponentially decreasing functions. For example, in the theory of heterogeneous media, it has been proven that the model describing a two-phase medium (elastic medium and viscous fluid) is represented using a system of integro-differential equations with convolution kernels in the form of a finite or infinite sum of decreasing exponentials. The laws of heat conduction with integral memory have been studied in various works (see, for example, [1]). Much attention is paid to controllability issues for models of plates and beams in some sections of engineering.

A large number of works are devoted to controllability problems for similar systems, but not considering integral memory (see, for example, [2-6]). For the systems with memory, the articles [7-12] should be noted.

This research is devoted to the problems of the vibrations' distributed control of a plate and a beam with integral memory. The question about the possibility of driving such systems to a state of rest is raised. It should be noted that, generally speaking, this concept for the systems with memory is not equivalent to driving the system to the zero state. As it will further be clear, controllability to rest is not always possible for such models. In this paper, we will not dwell in detail on the questions of the solvability of initial-boundary value problems, but we will consider the qualitative aspects of this theory.

Two cases will be considered. In the first, the integral term of the equation contains the Abel kernel, that is, a kernel with a singularity; in the second, a kernel, consisting of the sum of a finite number of decreasing exponential functions, is considered. Both types of kernels are used in viscoelasticity models. For example, a beam model with memory and a kernel consisting of the sum of exponential functions describes the process of viscoelastic vibrations.

2. The problem of irreducibility to a rest state of a system described by the equation of a plate with integral memory and Abel kernel

This section is of an overview nature and does not contain any fundamentally new results. Let us consider the initial-boundary value problem of the following form:

$$\theta_{tt}(t,x)+\alpha^2\Delta^2\theta(s,x)+\int_0^t K(t-s)\Delta^2\theta(s,x)ds=v(t,x),\ t>0,\ x\in\Omega, \tag{1}$$

$$\theta|_{t=0}=\xi_1(x),\ \theta_t|_{t=0}=\xi_2(x), \tag{2}$$

$$\theta(t,x)=\Delta\theta(t,x)=0,\ x\in\partial\Omega, \tag{3}$$

where $K(t)$ – is a kernel and $v(t,x)$ defines control. Here Ω – is the bounded domain in R^2 with a boundary consisting of a finite number of disjoint closed infinitely smooth curves, Δ^2 – is the Laplace operator squared with domain

$$D(\Delta^2)=\{y\in H^4(\Omega):y(x)=\Delta y(x)=0,x\in\partial\Omega\}.$$

In mechanics models, kernels of various types are used to describe the process of viscoelastic vibrations. For example, the kernel of the following form is quite widespread:

$$K_1(t)=\sum_{j=1}^N d_j e^{-\gamma_j t},\quad d_j,\gamma_j>0, j=1,...,N. \tag{4}$$

For the equations "similar" to (1) and in a number of special cases (see, for example, [10] and [11]), it is possible to prove that the system oscillations can be completely stopped in finite time if the control is applied to the entire domain Ω. In [9], it is proved that the solution of the Gurtin-Pipkin equation for two-dimensional domains and a wide class of continuous kernels cannot be driven to rest by a control applied only to a subdomain which closure is contained in Ω. It should be noticed, that $K_1(t)$ at $N>1$ is an example of such a kernel. In this regard, it would be interesting to investigate the question of controllability for the system (1) - (3) if control is applied only to a part of the domain.

Now let the kernel have the form:

$$K_2(t)=\frac{1}{\Gamma(1-\sigma)}t^{-\sigma},\ \sigma\in(0,1), \tag{5}$$

where $\Gamma(\lambda)$ – is Gamma function. This is Abel kernel. It is proved in [8] that this system is uncontrollable to rest if control is applied to one end of the segment and the other end is fixed; for a one-dimensional heat equation with integral memory and kernel (5).

Without pretending to be a fundamental novelty of the result and using the methods of [8], we will show that in the problem of controlling vibrations of a plate with memory for a kernel $K_2(t)$ it is impossible to bring the solution to rest. Namely, let us consider the control $v(t)\in L_2((0,T);L_2(\Omega))$, continued by zero at $t>T$. Let us suppose the first initial condition (bias) is identically zero. Next, we will show that there is such an initial velocity $\xi_2(x)$, at which the vibrations of the plate cannot be stopped. More precisely, there is an initial condition $\xi_2(x)$, stating that for each control $v(t,x)$, which vanishes on the set $\{t:t>T\}$ with some $T>0$, the corresponding solution cannot be identically zero outside the bounded segment (with respect to the variable t).

For the proof, let us consider the orthonormal system of eigenfunctions $\{\psi_n\}$ and eigenvalues a_n^2 ($n=1,2,..$) of the operator Δ^2 with respect to the boundary conditions (3). Let us suppose, that

$$\xi_2(x)=\xi_{2k}\psi_k(x),$$

where k – is some natural number and $\xi_{2k}\neq 0$. Let us expand the solution $\theta(t,x)$ and the control action $v(t,x)$ into the Fourier series with respect to the aforementioned system of eigenfunctions (this system is a basis in $L_2(\Omega)$). Then we obtain a countable system of integro-differential equations:

$$\theta_n^{(2)}(t)+\alpha^2a_n^2\theta_n(t)+a_n^2\int_0^t K_2(t-s)\theta_n(s)ds=v_n(t),\ t>0,\ n=1,2,.... \tag{6}$$

It should be noted that only the k -th equation from the system (6) has a nonzero initial condition.

We apply the Laplace transform to both sides in (6) for $n=k$ and express $\hat{\theta}_k(\lambda)$ in terms of:

$$\hat{\theta}_k(\lambda)=\frac{\hat{v}_k(\lambda)+\xi_{2k}}{\lambda^2+a_k^2\left(\alpha^2+\hat{K}_2(\lambda)\right)}. \tag{7}$$

Let PW_+ – be the linear space of the Laplace transforms of elements of $L_2(0,+\infty)$, so that they are equal to zero on the set $\{t:t>T\}$ for some $T>0$. It is known that $\varphi(\lambda)\in PW_+$ if and only if it is an entire function such that:

1) there are real such numbers C and T that $|\varphi(\lambda)|\leqslant Ce^{T|\lambda|}$. It should be noted, that C and T depend on $\varphi(\lambda)$.

2) $\sup\limits_{x\geqslant 0}\int_{-\infty}^{+\infty}|\varphi(x+iy)|^2\,dy<+\infty$.

Next, we will apply the method associated with the so-called "branch point" and described in [8] for solving another problem. Let us suppose that the solution to problem (1) - (3) is controllable to rest, then the functions $\hat{\theta}_k(\lambda)$ and $\hat{v}_k(\lambda)$ are the elements of space PW_+. So these are the entire functions. Since the functions are holomorphic $\hat{\theta}_k(\lambda)$ and $\hat{v}_k(\lambda)$ it follows that the equalities:

$$\lim_{\varphi\to 0}\hat{\theta}_k(\rho e^{i\varphi})=\lim_{\varphi\to 2\pi}\hat{\theta}_k(\rho e^{i\varphi}),\ \lim_{\varphi\to 0}\hat{v}_k(\rho e^{i\varphi})=\lim_{\varphi\to 2\pi}\hat{v}_k(\rho e^{i\varphi}),\ \rho>0. \tag{8}$$

Then, using (7) and (8), we can state:

$$\frac{\hat{v}_k(\rho)+\xi_{2k}}{\rho^2+a_k^2\left(\alpha^2+\rho^{\sigma-1}\right)}=\frac{\hat{v}_k(\rho)+\xi_{2k}}{\rho^2+a_k^2\left(\alpha^2+\rho^{\sigma-1}e^{2\pi i(\sigma-1)}\right)}. \tag{9}$$

The equality (9) is obtained using the following expression: $\hat{K}_2(\lambda)=\lambda^{\sigma-1}$. Let us suppose there is such a number ρ, for which $\hat{v}_k(\rho)+\xi_{2k}\neq 0$. Then the equality (9) is impossible, since the number $\sigma-1$ takes the values in the interval $(-1,0)$. If $\hat{v}_k(\rho)+\xi_{2k}\equiv 0$, for any $\rho>0$, then by the uniqueness theorem for holomorphic functions it follows that $\hat{v}_k(\lambda)+\xi_{2k}\equiv 0$, for any complex λ. The last identity is impossible, since point 2 of the space definition PW_+ is violated.

3. The problem of controlling vibrations of a beam with integral memory, in the case when the control action is distributed along the entire beam length

Let us consider the equation of vibration of a beam with integral memory:

$$\theta_{tt}(t,x)+G(0)\theta_{xxxx}(s,x)+\int_0^t G'(t-s)\theta_{xxxx}(s,x)ds=u(t,x),\ t>0,\ x\in(0,\pi), \tag{10}$$

where

$$G(t)=\sum_{j=1}^{N}\frac{c_j}{\gamma_j}e^{-\gamma_j t} \tag{11}$$

and c_j, γ_j are given positive constants such that $0<\gamma_1<\gamma_2<\ldots<\gamma_N$. It should be noted that in this equation the coefficient of the function θ_{xxxx}, which stands outside the integral term, is consistent with the kernel. This equation has the initial conditions:

$$\theta|_{t=0}=\xi_1(x),\ \theta_t|_{t=0}=\xi_2(x), \tag{12}$$

and also the conditions at the ends of the interval $(0,\pi)$:

$$\theta(t,0)=\theta_{xx}(t,0)=\theta(t,\pi)=\theta_{xx}(t,\pi)=0. \tag{13}$$

The task is to drive the solution of the system (10) - (13) to a state of rest. This means that it is required to construct such a control action $u(t,x)$, for which the corresponding solution $\theta(t,x)$ will be equal to zero identically for $t>T$, where $T-$ is some point in time. To drive the solution of the system to rest, it is necessary (but not enough) to satisfy the conditions:

$$\theta(T,x)=\theta_t(T,x)=0,\quad x\in(0,\pi). \tag{14}$$

The distributed control problems for similar systems in the case when there is no integral term in the equation (classical plate) were considered in [3].

We note that for the systems with integral memory the fulfillment of conditions (14) is not sufficient to stop the oscillations of the system, since after reaching the zero state the system can leave it (see, for example, [8]). In this section, we will prove that under certain conditions imposed on the initial problem data, the system (10) - (13) can be driven to a state of rest in finite time by a distributed control action arbitrarily small in absolute value.

The research methods in this work are close to those used in [10] and [11].

Let us consider the operator $A:=\dfrac{d^4}{dx^4}$ with the domain

$$D(A)=\left\{\varphi\in H^4(0,\pi):\varphi(0)=\varphi_{xx}(0)=\varphi(\pi)=\varphi_{xx}(\pi)=0\right\}.$$

Let $\{\psi_n(x)\}_{n=1}^{+\infty}$ be the corresponding orthonormal system of eigenfunctions and $\{n^4\}_{n=1}^{+\infty}$ defines the corresponding eigenvalues, i.e.

$$A\psi_n(x) = n^4\psi_n(x).$$

It should be noted that by virtue of the boundary conditions (13), the system of eigenfunctions $\{\psi_n(x)\}_{n=1}^{+\infty}$ coincides with the system of eigenfunctions of the Laplace operator with zero Dirichlet boundary condition.

Let us denote the linear function space $f : R_+ = (0,+\infty) \to D(A)$, by $W_{2,\gamma}^2(R_+, A)$ standardized with the norm:

$$\| \theta \|_{W_{2,\gamma}^2(R_+,A)} = \left(\int_0^{+\infty} e^{-2\gamma t} \left(\left\| \theta^{(2)}(t) \right\|_{L_2(0,\pi)}^2 + \left\| A\theta(t) \right\|_{L_2(0,\pi)}^2 \right) dt \right)^{\frac{1}{2}}, \quad \gamma \geqslant 0.$$

The function $\theta(t,x)$ is called a strong solution to the problem (10)–(13), if for some $\gamma \geqslant 0$ this function is an element $W_{2,\gamma}^2(R_+, A)$, satisfies the equation (10) almost everywhere in the variable t on the positive semiaxis R_+ and satisfies the initial conditions (12).

We define the function of a complex variable λ:

$$l_n(\lambda) := \lambda^2 + n^4\lambda\hat{G}(\lambda),$$

where

$$\hat{G}(\lambda) = \sum_{k=1}^{N} \frac{c_k}{\gamma_k(\lambda + \gamma_k)}.$$

Let us now write down the solution of the problem (10) - (13) in explicit form. To do this, we use the well-known results of [13], in which the corresponding formulas are given. Let us present the formulations of two theorems, the first one refers to the formula for the solution of equation (10) with zero right-hand side, the second theorem to the solution with zero initial conditions. Due to the linearity of the problem, the sum of these two solutions will be the desired solution.

Theorem 1. Let us suppose, that $u(t,x) \equiv 0$, $t \in R_+$, the function $\theta(t,x) \in W_{2,\gamma}^2(R_+, A)$, $\gamma > 0$, the strong decision (10)–(13), then for any $t \in R_+$ the formula is correct:

$$\theta(t,x) = \frac{1}{\sqrt{2\pi}} \sum_{n=1}^{\infty} \frac{(\xi_{2n} + \lambda_n^+ \xi_{1n}) e^{\lambda_n^+ t} \psi_n(x)}{l_n^{(1)}(\lambda_n^+)} + \frac{1}{\sqrt{2\pi}} \sum_{n=1}^{\infty} \frac{(\xi_{2n} + \lambda_n^- \xi_{1n}) e^{\lambda_n^- t} \psi_n(x)}{l_n^{(1)}(\lambda_n^-)} +$$
$$+ \frac{1}{\sqrt{2\pi}} \sum_{n=1}^{\infty} \left(\sum_{k=0}^{N-1} \frac{(\xi_{2n} - q_{k,n}\xi_{1n}) e^{-q_{k,n}t}}{l_n^{(1)}(-q_{k,n})} \right) \psi_n(x), \tag{15}$$

where $-q_{k,n}$ denotes the real function zeros $l_n(\lambda)$ ($q_{0,n} = 0$, $q_{k,n} > 0$, $k = 1,2,\ldots,N-1$), $\lambda_n^{\pm}$ – is a pair of complex conjugate zeros, ξ_{1n}, ξ_{2n} – are the Fourier coefficients of the initial data expansion in terms of eigenfunctions and series (15) converges in the norm of the space $L_2(0,\pi)$.

Theorem 2. Let us suppose, that $u(t,x) \in C([0,T], L_2(0,\pi))$ for any $T > 0$, $\theta(t,x) \in W_{2,\gamma}^2(R_+, A)$ – is the strong solution to the problem (10) - (13) for some $\gamma > 0$, $\xi_1 = \xi_2 = 0$. Then for any $t \in R_+$ the formula is correct:

$$\theta(t,x)=\frac{1}{\sqrt{2\pi}}\sum_{n=1}^{\infty}\omega_n(t,\lambda_n^+)\psi_n(x)+\frac{1}{\sqrt{2\pi}}\sum_{n=1}^{\infty}\omega_n(t,\lambda_n^-)\psi_n(x)+\frac{1}{\sqrt{2\pi}}\sum_{n=1}^{\infty}\left(\sum_{k=0}^{N-1}\omega_n(t,-q_{k,n})\right)\psi_n(x), \quad (16)$$

where $\omega_n(t,\lambda)=\frac{\int_0^t u_n(s)e^{\lambda(t-s)}ds}{l_n^{(1)}(\lambda)}$,

$u_n(t)$ – is the Fourier coefficient of the control action expansion in terms of eigenfunctions and series (16) converges in the space norm $L_2(0,\pi)$.

It should be noted that by increasing the smoothness of the right-hand side of the equation and the initial data, it is possible to achieve the required smoothness of the solution (see details in [13]). All zeros of the $l_n(\lambda)$ function (for each natural n) are known to be pairwise different; moreover, their asymptotics were studied in detail in [14].

For further reasoning, we need the following lemma, the proof of which for a similar problem is given in the paper [11].

Statement 1. For any natural number n , the following equality is true:

$$\frac{1}{l_n^{(1)}(\lambda_n^+)}+\frac{1}{l_n^{(1)}(\lambda_n^-)}+\sum_{k=0}^{N-1}\frac{1}{l_n^{(1)}(-q_{k,n})}=0.$$

Let the control action $u(t,x)$ satisfy the conditions of the theorem 2. As mentioned above, the solution to the problem (10) - (13) can be represented in the form:

$$\theta(t,x)=\frac{1}{\sqrt{2\pi}}\sum_{n=1}^{\infty}\frac{(\xi_{2n}+\lambda_n^+\xi_{1n})e^{\lambda_n^+t}\psi_n(x)}{l_n^{(1)}(\lambda_n^+)}+\frac{1}{\sqrt{2\pi}}\sum_{n=1}^{\infty}\frac{(\xi_{2n}+\lambda_n^-\xi_{1n})e^{\lambda_n^-t}\psi_n(x)}{l_n^{(1)}(\lambda_n^-)}+$$

$$+\frac{1}{\sqrt{2\pi}}\sum_{n=1}^{\infty}\sum_{k=0}^{N-1}\left(\frac{(\xi_{2n}-q_{k,n}\xi_{1n})e^{-q_{k,n}t}}{l_n^{(1)}(-q_{k,n})}\right)\psi_n(x)+\frac{1}{\sqrt{2\pi}}\sum_{n=1}^{\infty}\frac{\int_0^t u_n(s)e^{\lambda_n^+(t-s)}ds}{l_n^{(1)}(\lambda_n^+)}\psi_n(x)+$$

$$+\frac{1}{\sqrt{2\pi}}\sum_{n=1}^{\infty}\frac{\int_0^t u_n(s)e^{\lambda_n^-(t-s)}ds}{l_n^{(1)}(\lambda_n^-)}\psi_n(x)+\frac{1}{\sqrt{2\pi}}\sum_{n=1}^{\infty}\sum_{k=0}^{N-1}\left(\frac{\int_0^t u_n(s)e^{-q_{k,n}(t-s)}ds}{l_n^{(1)}(-q_{k,n})}\right)\psi_n(x). \quad (17)$$

As $q_{0,n}=0$, then for the derivative of the Fourier coefficient of the function θ , taking into account statement 1, we can write (for all the natural numbers n):

$$\theta_n'(t)=\frac{1}{\sqrt{2\pi}}\frac{\lambda_n^+(\xi_{2n}+\lambda_n^+\xi_{1n})e^{\lambda_n^+t}\psi_n(x)}{l_n^{(1)}(\lambda_n^+)}+\frac{1}{\sqrt{2\pi}}\frac{\lambda_n^-(\xi_{2n}+\lambda_n^-\xi_{1n})e^{\lambda_n^-t}\psi_n(x)}{l_n^{(1)}(\lambda_n^-)}+$$

$$+\frac{1}{\sqrt{2\pi}}\sum_{k=1}^{N-1}\left(\frac{(-q_{k,n})(\xi_{2n}-q_{k,n}\xi_{1n})e^{-q_{k,n}t}}{l_n^{(1)}(-q_{k,n})}\right)\psi_n(x)+\frac{1}{\sqrt{2\pi}}\frac{\lambda_n^+\int_0^t u_n(s)e^{\lambda_n^+(t-s)}ds}{l_n^{(1)}(\lambda_n^+)}\psi_n(x)+$$

$$+\frac{1}{\sqrt{2\pi}}\frac{\lambda_n^-\int_0^t u_n(s)e^{\lambda_n^-(t-s)}ds}{l_n^{(1)}(\lambda_n^-)}\psi_n(x)+\frac{1}{\sqrt{2\pi}}\sum_{k=1}^{N-1}\left(\frac{(-q_{k,n})\int_0^t u_n(s)e^{-q_{k,n}(t-s)}ds}{l_n^{(1)}(-q_{k,n})}\right)\psi_n(x). \quad (18)$$

Let us formulate the control $u(t,x)$, for which the conditions (14) are satisfied. We have

$$-\left(\frac{(\xi_{2n}+\lambda_n^+\xi_{1n})e^{\lambda_n^+T}}{l_n^{(1)}(\lambda_n^+)}+\frac{(\xi_{2n}+\lambda_n^-\xi_{1n})e^{\lambda_n^-T}}{l_n^{(1)}(\lambda_n^-)}+\sum_{k=0}^{N-1}\frac{(\xi_{2n}-q_{k,n}\xi_{1n})e^{-q_{k,n}T}}{l_n^{(1)}(-q_{k,n})}\right)=$$

$$=\frac{\int_0^T u_n(s)e^{\lambda_n^+(T-s)}ds}{l_n^{(1)}(\lambda_n^+)}+\frac{\int_0^T u_n(s)e^{\lambda_n^-(T-s)}ds}{l_n^{(1)}(\lambda_n^-)}+\sum_{k=0}^{N-1}\frac{\int_0^T u_n(s)e^{-q_{k,n}(T-s)}ds}{l_n^{(1)}(-q_{k,n})},\quad n=1,2,..., \quad (19)$$

$$-\frac{\lambda_n^+(\xi_{2n}+\lambda_n^+\xi_{1n})e^{\lambda_n^+T}}{l_n^{(1)}(\lambda_n^+)}-\frac{\lambda_n^-(\xi_{2n}+\lambda_n^-\xi_{1n})e^{\lambda_n^-T}}{l_n^{(1)}(\lambda_n^-)}-\sum_{k=1}^{N-1}\frac{(-q_{k,n})(\xi_{2n}-q_{k,n}\xi_{1n})e^{-q_{k,n}T}}{l_n^{(1)}(-q_{k,n})}=$$

$$=\frac{\lambda_n^+\int_0^T u_n(s)e^{\lambda_n^+(T-s)}ds}{l_n^{(1)}(\lambda_n^+)}+\frac{\lambda_n^-\int_0^T u_n(s)e^{\lambda_n^-(T-s)}ds}{l_n^{(1)}(\lambda_n^-)}+\sum_{k=1}^{N-1}\frac{(-q_{k,n})\int_0^T u_n(s)e^{-q_{k,n}(T-s)}ds}{l_n^{(1)}(-q_{k,n})},\quad n=1,2,.... \quad (20)$$

Let us introduce the short notation:

$$a_n=-(\xi_{2n}+\lambda_n^+\xi_{1n}),\ \bar{a}_n=-(\xi_{2n}+\lambda_n^-\xi_{1n}),\ b_{k,n}=-(\xi_{2n}+(-q_{k,n})\xi_{1n}),\ k=0,1,2,...,N-1.$$

Further, following the method of [10] and [11], we will equate the coefficients in front of the values

$$\frac{1}{l_n^{(1)}(\lambda_n^+)},\ \frac{1}{l_n^{(1)}(\lambda_n^-)},\ \frac{1}{l_n^{(1)}(-q_{k,n})},\ k=0,1,2,...,N-1.$$

on the left and right sides of the equations (19), (20). Therefore, for all natural n and $k=0,1,2,...,N-1$, we get the moment problem:

$$\int_0^T u_n(s)e^{\lambda_n^+(T-s)}ds=a_ne^{\lambda_n^+T},\ \int_0^T u_n(s)e^{\lambda_n^-(T-s)}ds=\bar{a}_ne^{\lambda_n^-T},\ \int_0^T u_n(s)e^{-q_{k,n}(T-s)}ds=b_{k,n}e^{-q_{k,n}T}. \quad (21)$$

Let us divide by a monomial factor in both sides of the equations (21). Then we get:

$$\int_0^T u_n(s)e^{-\lambda_n^+s}ds=a_n,\ \int_0^T u_n(s)e^{-\lambda_n^-s}ds=\bar{a}_n,\ \int_0^T u_n(s)e^{q_{k,n}s}ds=b_{k,n},\ k=0,1,2,...,N-1. \quad (22)$$

The system of integral equations (22) is a classical finite-dimensional moment problem for each natural n. Let us define that $\lambda_n=-\lambda_n^+$ and $\bar{\lambda}_n=-\lambda_n^-$.

We will seek for the solution (22) in the following form:

$$u_n(s)=D_{1,n}e^{\lambda_ns}+D_{2,n}e^{\bar{\lambda}_ns}+\sum_{j=0}^{N-1}C_{j,n}e^{q_{j,n}s},\ n=1,2,..., \quad (23)$$

where $D_{1,n}$ $D_{2,n}$ and $C_{j,n}$ are the unknown constants and $s \in [0,T]$. Let us note, that for $s > T$ control action u_n is equal to zero. Substituting (23) in (22), we obtain a system of algebraic equations for each natural n:

$$D_{1,n}\int_0^T e^{2\lambda_n s}ds + D_{2,n}\int_0^T e^{(\lambda_n+\bar{\lambda}_n)s}ds + \sum_{k=0}^{N-1} C_{k,n}\int_0^T e^{(\lambda_n+q_{k,n})s}ds = a_n,$$
$$D_{1,n}\int_0^T e^{(\bar{\lambda}_n+\lambda_n)s}ds + D_{2,n}\int_0^T e^{2\bar{\lambda}_n s}ds + \sum_{k=0}^{N-1} C_{k,n}\int_0^T e^{(\bar{\lambda}_n+q_{k,n})s}ds = a_n, \quad (24)$$
$$D_{1,n}\int_0^T e^{(q_{k,n}+\lambda_n)s}ds + D_{2,n}\int_0^T e^{(q_{k,n}+\bar{\lambda}_n)s}ds + \sum_{j=0}^{N-1} C_{j,n}\int_0^T e^{(q_{k,n}+q_{j,n})s}ds = b_{k,n}. \quad k = 0,1,2,\dots,N-1$$

Let Δ_n be the system determinant (24). This is the Gram determinant and therefore in this case it is not zero for any moment in time $T > 0$. Then we obtain the unique solution of the linear system (24).

If we choose the initial data of the problem (10) - (13) in such a way that they contain only a finite number of nonzero Fourier coefficients, then we obtain a finite number of finite-dimensional moment problems. Each of which has a solution in the form (23). It should be noted that this choice of initial data is "natural", since they can approximate the functions from more general classes. In this case, only a finite number of Fourier coefficients of the control $u(t,x)$ will be nonzero. In this case, the nonzero coefficients given by formula (23) tend to zero in modulus at T tending to infinity (for the proof of this fact, see [10] and [11]). This means that $u(t,x)$ can be made arbitrarily small in absolute value.

Let us now show that this choice of control brings the system to a state of rest, that is, the solution is identically zero starting from the moment $t = T$. For this, we will use the formula (17). Let us suppose, that $t = T$, then, as stated above, $\theta_n(T) = 0$. Let us prove that $\theta_n(t) \equiv 0$, at $t > T$. Let us suppose, that λ_0 – is a zero of the function $l_n(\lambda)$. Since the control $u(t,x)$ is equal to zero identically zero for $t > T$, then for such t it is possible to write:

$$\int_0^t u_n(s)e^{\lambda_0(t-s)}ds = e^{\lambda_0 t}\int_0^T u_n(s)e^{-\lambda_0 s}ds.$$

Next, we use the obtained equality, the moment system (22) and the formula (17). It is not hard to see that the solution $\theta_n(t)$, corresponding to the formulated control will be equal to zero identically for any $t > T$.

Summary

The present paper is devoted to the problems of distributed controllability for some models of plates and beams with integral memory. The goal of the control is to drive these systems to rest. The integral term of the plate equation contains the Abel kernel and a kernel (of the integral term) in the beam equation consists of the sum of a finite number of decreasing exponential functions. It is proved that by controlling the whole domain, the first system cannot be driven to rest but the second system is controllable to rest in finite time.

Financial support

The study was supported by a grant from Russian Science Foundation (project No. 16-11-10343).

References

IOP Conf. Series: Materials Science and Engineering 1083 (2021) 012041 doi:10.1088/1757-899X/1083/1/012041

[1] Gurtin M E and Pipkin A C 1968 Theory of heat conduction with finite wave speed *Arch.Ration. Mech. Anal.* **31** 113-126.

[2] Butkovkisky A 1965 The theory of optimal control of systems with distributed parameters (Nauka, Moscow.

[3] Chernousko F L 1992 Bounded control in distributed-parameter systems *Journal of Applied Mathematics and Mechanics* **56 (5)** 707-723.

[4] Lions J L 1988 Exact controllability. Stabilization and perturbations for distributed systems *SIAM Review* **30 (1)** 1-68.

[5] Romanov I and Shamaev A 2018 Exact Bounded Boundary Zero-Controllability for the Two-Dimensional Wave Equation *arXiv.* doi 1603.01212.

[6] Russell D L 1978 Controllability and stabilizability theory for linear partial differential equations: recent progress and open questions *SIAM Review* **20 (4)** 639-739.

[7] Pandofi L 2005 The controllability of the Gurtin-Pipkin equations: a cosine operator approach *Appl. Math. Optim.* **52** 143-165.

[8] Ivanov S and Pandofi L 2009 Heat equations with memory: lack of controllability to rest *Journal of Mathematical Analysis and Applications* **355 (1)** 1-11.

[9] Romanov I and Shamaev A 2016 Non-controllability to rest of the two-dimensional distributed system governed by the integrodifferential equation *Journal of Optimization Theory and Applications* **170(3)** 772–782.

[10] Romanov I and Shamaev A 2013 Exact controllability of the distributed system governed by string equation with memory *Journal of Dynamical and Control Systems* **19 (4)** 611-623.

[11] Romanov I and Shamaev A 2019 Exact controllability of the distributed system governed by the wave equation with memory *arXiv.* doi 1503.04461.

[12] Romanov I and Romanova A 2018 Some problems of controllability of distributed systems governed by integrodifferential equations *IFAC-PapersOnLine* **51 (2)** 132-137.

[13] Rautian N A 2011 On the structure and properties of solutions of integro-differential equations arising in thermal physics and acoustics *Mathematical Notes* **90 (3)** 455-459.

[14] Vlasov V V, Rautian N A and Shamaev A S 2011 Spectral Analysis and Correct Solvability of Abstract Integro-Differential Equations Arising in Thermophysics and Acoustics *Contemporary Mathematics. Fundamental Directions* **39** 36-65.

IOP Conf. Series: Materials Science and Engineering 1083 (2021) 012042 doi:10.1088/1757-899X/1083/1/012042

Calculation of a beam with external reinforcement. Contact layer model

N Yu Tsybin*, V I Andreev, R A Turusov

Moscow State University of Civil Engineering (National Research University), 26 Yaroslavskoye Highway, Moscow, 129337, Russia

E-mail: science@nikitatsybin.ru

Abstract. The article is devoted to the stress-strain state study of the beam, with external reinforcement in the tensile zone. The classical Euler-Bernoulli hypotheses are used for the beam. The interaction of the beam with the reinforcement layer is carried out using the contact layer. In order to simplify the previously obtained solution, the flexural stiffness of the reinforcement layer is not taken into account. As a result, the solution of the resolving equations system for this model has been obtained. It is shown that the obtained solution allows satisfying all boundary conditions, including the equality to zero of the shear stresses at the corner points. The influence of various mechanical and geometric parameters on the stress-strain state of the beam, the reinforcement layer and the contact layer is analyzed. The influence of the introduced simplifications on the solution results is analyzed.

Introduction

Increasing and restoring the bearing capacity of the existing beams is one of the most common construction tasks. One of the solutions to this problem is an external amplification device. In this case, as a rule, the stretched zone of the beams is strengthened. The figure below shows a general view of the model of a beam with external reinforcement in the tension zone.

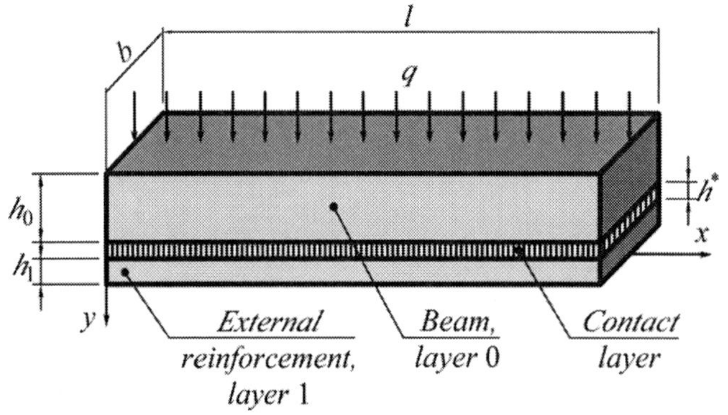

Figure 1. General view of the beam model with external reinforcement

The classical Euler-Bernoulli hypotheses are used for the beam taking into account axial displacements $u_0(x,0)=u_{0,0}(x)$.

The contact layer is considered as an orthotropic medium, with such properties that it can be represented as a set of elastic short rods not connected to each other. In such a medium, only shear stresses can occur in the shear plane and normal stresses perpendicular to the shear plane. The main characteristics of the contact layer to be determined from the macro experiment are its rigidity and thickness [5, 6, 13].

Reinforcement layer height h_1 is generally much less than the height of the beam section h_0, i.e., $h_0 \gg h_1$. This allows to neglect bending stiffness D_1 of the reinforcement layer, as well as the bending moments and shear forces arising in it. As a result, a uniaxial stress-strain state appears in the reinforcement layer.

The influence of this simplification on the stress-strain state of the model as a whole (in comparison with the model taking into account the bending stiffness of the reinforcement layer) will be analyzed below.

1. Resolving Equations

The system of resolving equations (initial) for a two-layer beam was obtained earlier in [1]. Let us rewrite it without changes.

$$\left.\begin{aligned} D_0\frac{d^4v_0}{dx^4}+\frac{\eta G^*}{h^*}(v_0-v_1)+B_0\left(\frac{h_0}{2}+\frac{h^*}{2}\right)\frac{d^3u_{0,0}}{dx^3}=q;\\ D_1\frac{d^4v_1}{dx^4}-\frac{\eta G^*}{h^*}(v_0-v_1)+B_0\left(\frac{h_1}{2}+\frac{h^*}{2}\right)\frac{d^3u_{0,0}}{dx^3}=0;\\ \frac{d^4u_{0,0}}{dx^4}\frac{B_0(h^*)^2}{12\eta G^*}-\frac{d^2u_{0,0}}{dx^2}\frac{B_0}{G^*}+u_{0,0}\frac{B_0+B_1}{B_1h^*}-\frac{dv_1}{dx}\frac{h^*+h_1}{2h^*}-\frac{dv_0}{dx}\frac{h^*+h_0}{2h^*}=0, \end{aligned}\right\} \quad (1)$$

wherein $D_0=E_0h_0^3/12;\ D_1=E_1h_1^3/12;\ B_0=E_0h_0;\ B_1=E_1h_1;\ E^*=\eta G^*$. The beam width b is taken equal to one. Shear stresses in the contact layer are related to the longitudinal force N_0 by the expression:

$$\tau^*=-\frac{dN_0}{dx}. \quad (2)$$

Longitudinal forces are related to axial displacements by the equality:

$$N_0=B_0\frac{du_{0,0}}{dx}. \quad (3)$$

Based on the static equilibrium conditions of the system as a whole $N_1=-N_0$.

To solve the problem formulated in this work, the following assumptions should be made in the original system of equations (1).

1) To neglect the bending stiffness of the layer 1 (reinforcement layer), compared to the beam bending stiffness, i.e. to accept $D_1=0$.
2) In view of assumption 1, the vertical displacements of layer 1 are equal to the layer 0 displacements $v_1=v_0=v$.

Taking these assumptions and equality (3) into consideration, the system of equations (1) will take the form:

$$\left.\begin{aligned} D_0\frac{d^4v}{dx^4}+\left(\frac{h_0+h_1}{2}+h^*\right)\frac{d^2N_0}{dx^2}=q;\\ \frac{(h^*)^2}{24G^*}\frac{d^4N_0}{dx^4}-\frac{1}{G^*}\frac{d^2N_0}{dx^2}+\frac{1}{h^*}\left(\frac{1}{B_0}+\frac{1}{B_1}\right)N_0-\frac{1}{h^*}\left(\frac{h_0+h_1}{2}+h^*\right)\frac{d^2v}{dx^2}=0. \end{aligned}\right\} \quad (4)$$

2. The problem solution

The system of equations (4) can be reduced to one resolving equation for the function v.

$$\frac{d^8v}{dx^8}-2\lambda_1\frac{d^6v}{dx^6}+2\lambda_1\lambda_2\lambda_3\lambda_5\frac{d^4v}{dx^4}=2\lambda_1\lambda_2\lambda_3\psi. \qquad (5)$$

This equation uses the notation below.

$$\lambda_1=\frac{12}{(h^*)^2};\ \lambda_2=\frac{G^*D_0}{h^*};\ \lambda_3=\frac{B_0+B_1}{B_0B_1D_0};\ \lambda_4=\frac{h_0+h_1+2h^*}{D_0};\ \lambda_5=\frac{\lambda_4^2}{4\lambda_3}+1;\ \psi=\frac{q}{D_0}.$$

The solution to the equation (5) is the following expression:

$$v=\frac{1}{s_1^4}\left(C_1e^{s_1x}+C_3e^{-s_1x}\right)+\frac{1}{s_2^4}\left(C_2e^{s_2x}+C_4e^{-s_2x}\right)+\frac{C_5x^3}{6}+\frac{C_6x^2}{2}+C_7x+C_8+\frac{\psi x^4}{24\lambda_5}, \qquad (6)$$

in which

$$s_1=\sqrt{\lambda_1-\sqrt{\lambda_1(\lambda_1-2\lambda_2\lambda_3\lambda_5)}};\ s_2=\sqrt{\lambda_1+\sqrt{\lambda_1(\lambda_1-2\lambda_2\lambda_3\lambda_5)}}.$$

Longitudinal force N_0 expressed by the displacement function v from the system of equations (4).

$$N_0=\eta_1\frac{d^6v}{dx^6}-\eta_2\frac{d^4v}{dx^4}+\eta_3\frac{d^2v}{dx^2}+\eta_2\psi, \qquad (7)$$

in which

$$\eta_1=\frac{1}{\lambda_1\lambda_2\lambda_3\lambda_4};\ \eta_1=\frac{2}{\lambda_2\lambda_3\lambda_4};\ \eta_3=\frac{\lambda_4}{2\lambda_3}.$$

After substitution (6) in (7), we get the final expression for the longitudinal force N_0.

$$N_0=\frac{\chi_1}{s_1^2}\left(C_1e^{s_1x}+C_3e^{-s_1x}\right)+\frac{\chi_2}{s_2^2}\left(C_2e^{s_2x}+C_4e^{-s_2x}\right)+\eta_3C_5x+\eta_3C_6+\frac{\eta_3x^2-2\eta_2(1-\lambda_5)}{2\lambda_5}\psi, \qquad (8)$$

in which

$$\chi_1=\eta_1s_1^4-\eta_2s_1^2+\eta_3;\ \chi_2=\eta_1s_2^4-\eta_2s_2^2+\eta_3.$$

Shear stresses in the contact layer are determined from the expression (2) taking into account (8).

$$\tau^*=-\frac{\chi_1}{s_1}\left(C_1e^{s_1x}-C_3e^{-s_1x}\right)-\frac{\chi_2}{s_2}\left(C_2e^{s_2x}-C_4e^{-s_2x}\right)-\eta_3C_5-\frac{\eta_3x}{\lambda_5}\psi. \qquad (9)$$

Stresses and displacements in the beam and reinforcement layer are determined from the expressions

$$\left.\begin{aligned}
&u_0=u_{0,0}-y_0\frac{dv}{dx};\ u_{0,1}=-\frac{B_0}{B_1}u_{0,0};\ \sigma_{x,0}=E_0\left(\frac{du_{0,0}}{dx}-y_0\frac{d^2v}{dx^2}\right);\ \sigma_{x,1}=E_1\frac{du_{0,1}}{dx};\\
&\tau_{xy,0}=\frac{E_0}{2}\left(y_0^2-\frac{h_0^2}{4}\right)\frac{d^3v}{dx^3}-E_0y_0\frac{d^2u_{0,0}}{dx^2}+\frac{\tau^*}{2};\\
&\sigma_{y,0}=\frac{E_0}{2}\left(\frac{h_0^2y_0}{4}-\frac{y_0^3}{3}\right)\frac{d^4v}{dx^4}+\frac{E_0}{2}\left(y_0^2-\frac{h_0^2}{4}\right)\frac{d^3u_{0,0}}{dx^3}-\frac{y_0}{2}\frac{d\tau^*}{dx}-\frac{q}{2},\ y_0=-\frac{h_0}{2}\ldots\frac{h_0}{2}.
\end{aligned}\right\} \qquad (10)$$

The statically equivalent internal forces in the beam and the reinforcement layer are determined from the stresses (10), by integrating over the area of the layer. Let us write them down.

$$N_k=\int_{-h_k/2}^{h_k/2}\sigma_{x,k}dy;\ M_0=\int_{-h_0/2}^{h_0/2}y_0\sigma_{x,0}dy;\ Q_0=\int_{-h_0/2}^{h_0/2}\tau_{xy,0}dy. \qquad (11)$$

Considering (10), we find:

$$N_0=B_0\frac{du_{0,0}}{dx};\ N_1=-N_0;\ M_0=-D_0\frac{d^2v}{dx^2};\ Q_0=-D_0\frac{d^3v_0}{dx^3}+\frac{\tau^*h_0}{2}.$$

The total forces in the beam are recorded below:

$$N_{beam}=0;\ M_{beam}=-D_0\frac{d^2v}{dx^2}-N_1\frac{h_0+2h_k+h_1}{2};\ Q_{beam}=-D_0\frac{d^3v}{dx^3}-\frac{dN_1}{dx}\frac{h_0+2h_k+h_1}{2}. \quad (12)$$

The total forces in the beam calculated using the expressions (12), should correspond to the forces occurring in a single-layer beam (due to static equivalence). The fulfillment of this condition is one of the most important checks of the obtained solution.

3. Border conditions

Unknown constants of integration included in (6), (8) and (9) are determined from 8 boundary conditions. As an example, let us consider a hinged-supported beam with edges free from the longitudinal forces.

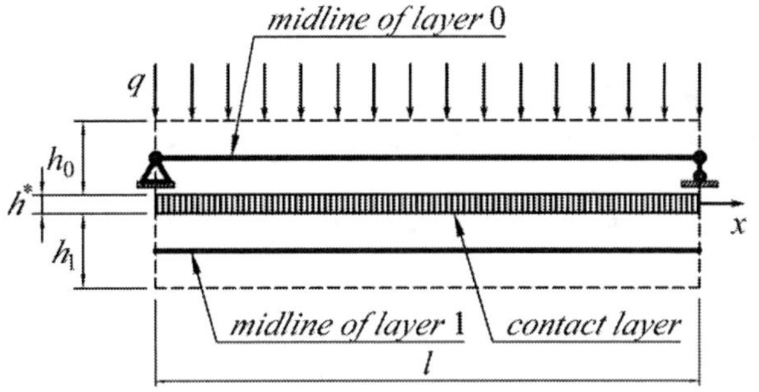

Figure 2. Model boundary conditions

Below are the boundary conditions of the model.

$$\left.\begin{aligned} v_0\left(\pm\frac{l}{2}\right)=0;\ M_0\left(\pm\frac{l}{2}\right)=0;\\ N_0\left(\pm\frac{l}{2}\right)=0;\ \tau^*\left(\pm\frac{l}{2}\right)=0.\end{aligned}\right\} \quad (13)$$

These boundary conditions correspond to the symmetric stress-strain state of the beam.

Analyzing the boundary conditions (13) and the expressions (6), (8) and (9) we come to the conclusion that to satisfy the boundary conditions it is necessary to take

$$C_1=C_3;\ C_2=C_4;\ C_5=0;\ C_7=0. \quad (14)$$

Taking into account (14) the expressions (6), (8) and (9) take the form:

$$\left.\begin{aligned} v=\frac{2}{s_1^4}C_1\cosh(s_1x)+\frac{2}{s_2^4}C_2\cosh(s_2x)+\frac{C_6x^2}{2}+C_8+\frac{\psi x^4}{24\lambda_5};\\ \frac{-M_0}{D_0}=\frac{d^2v}{dx^2}=\frac{2}{s_1^2}C_1\cosh(s_1x)+\frac{2}{s_2^2}C_2\cosh(s_2x)+C_6+\frac{\psi x^2}{2\lambda_5};\\ N_0=\frac{2\chi_1}{s_1^2}C_1\cosh(s_1x)+\frac{2\chi_2}{s_2^2}C_2\cosh(s_2x)+\eta_3C_6+\psi\frac{\eta_3x^2-2\eta_2(1-\lambda_5)}{2\lambda_5};\\ \tau^*=-\frac{2\chi_1}{s_1}C_1\sinh(s_1x)-\frac{2\chi_2}{s_2}C_2\sinh(s_2x)-\frac{\psi\eta_3x}{\lambda_5}.\end{aligned}\right\} \quad (15)$$

Constants C_1, C_2, C_6, C_8, included in the formulas (15), are determined from the boundary conditions (13) on the right (or left) edge of the beam. In matrix form, the system of linear algebraic equations to determine the integration constants is as follows:

$$\mathbf{A}\cdot\mathbf{c}=\mathbf{b}\Rightarrow\mathbf{c}=\mathbf{A}^{-1}\mathbf{b}, \quad (16)$$

in which

$$\mathbf{A}=\begin{pmatrix} \frac{\cosh(\omega_1)}{s_1^4} & \frac{\cosh(\omega_2)}{s_2^4} & \frac{l^2}{16} & \frac{1}{2} \\ \frac{\cosh(\omega_1)}{s_1^2} & \frac{\cosh(\omega_2)}{s_2^2} & \frac{1}{2} & 0 \\ \frac{\chi_1\cosh(\omega_1)}{s_1^2} & \frac{\chi_2\cosh(\omega_2)}{s_2^2} & \frac{\eta_3}{2} & 0 \\ \frac{\chi_1\sinh(\omega_1)}{s_1} & \frac{\chi_2\sinh(\omega_2)}{s_2} & 0 & 0 \end{pmatrix};\ \mathbf{c}=\begin{pmatrix} C_1 \\ C_2 \\ C_6 \\ C_8 \end{pmatrix};\ \mathbf{b}=\begin{pmatrix} -\frac{\psi l^4}{768\lambda_5} \\ -\frac{\psi l^2}{16\lambda_5} \\ -\frac{\psi\left[\eta_3 l^2-8\eta_2(1-\lambda_5)\right]}{16\lambda_5} \\ -\frac{\psi\eta_3 l}{4\lambda_5} \end{pmatrix},$$

$$\omega_1=\frac{s_1 l}{2};\ \omega_2=\frac{s_2 l}{2}.$$

Solving a system of linear algebraic equations (16) for given values of geometrical and physical-mechanical parameters of the model is not difficult. However, the general form of this solution is quite cumbersome, so it is not presented here. The numerical solution examples of a similar problem can be found in [4].

4. Calculation example

Let us consider a model of a two-layer beam with the following geometrical and physical and mechanical parameters:

$$E_0=1\cdot10^4\text{MPa};\ E_1=2\cdot10^5\text{MPa};\ G^*=200\text{MPa};\ \eta=2;$$

$$h_0=20\text{mm};\ h_1=2\text{mm};\ h^*=1\text{mm};\ l=200\text{mm}.$$

These parameters will be called basic. We take the value of the applied load equal to one.

Figure 3 shows the graphs of the displacements and internal forces distribution in the beam, the reinforcement layer and the contact layer, calculated by the formulas (10).

To assess the fulfillment of the conditions for static equilibrium of the beam as a whole, Figure 4 shows the graphs of the internal forces distribution, calculated by the formulas (12).

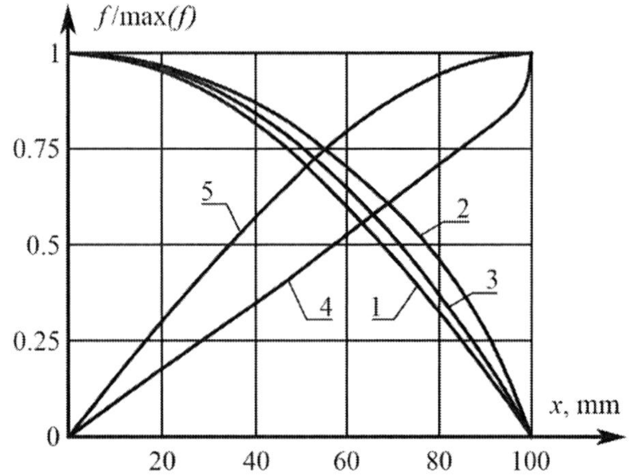

Figure 3. Relative displacements and internal forces in the beam.

$1-\frac{v_0(x)}{v_0(0)};\ 2-\frac{M_0(x)}{M_0(0)};\ 3-\frac{N_0(x)}{N_0(0)};$
$4-\frac{Q_0(x)}{Q_0(0.5l)};\ 5-\frac{u_{0,0}(x)}{u_{0,0}(0.5l)}.$

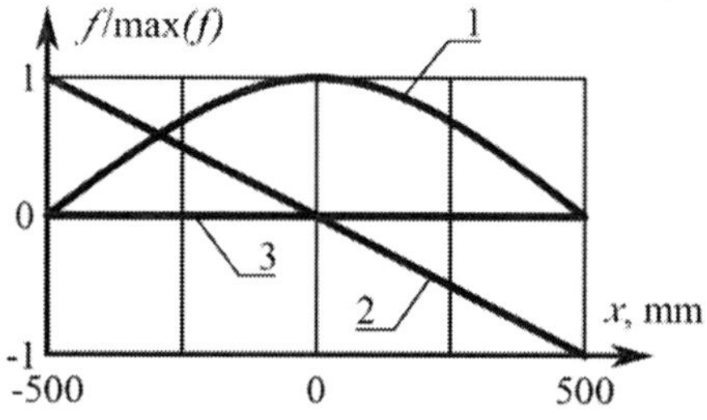

Figure 4. Total forces in the beam.

$1-\frac{8M_{beam}(x)}{ql^2};\ 2-\frac{2Q_{beam}(x)}{ql};$
$3-N_{beam}(x).$

The calculation results for three-layer plates and shells are given in the works [17, 18, 19].

The greatest difference in the results obtained from solving the system of equations (1) and (4), observed for tangential stresses and transverse forces in a narrow zone of the edge effect, presented in the figures below.

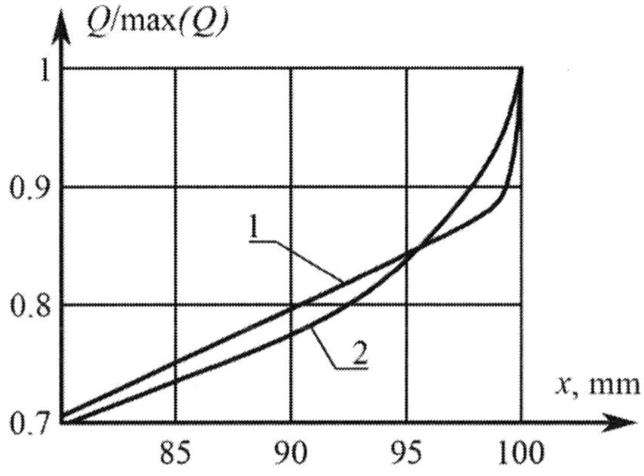

Figure 5. Transverse forces in the beam.
1 – from system solution (1).
2 – from system solution (4)

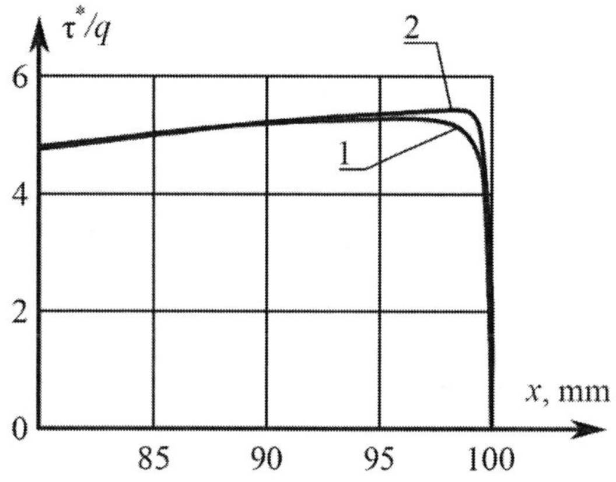

Figure 6. Shear stresses.
1 – from system solution (1).
2 – from system solution (4)

It is obvious that the decisive influence on the differences in the stress-strain state determined from the solution of the system of equations (1) and (4), the thickness (and, accordingly, bending stiffness) of the reinforcement layer. Analysis of the influence of the reinforcement layer thickness on the error in calculating the vertical displacements of the middle of the beam and the maximum value of shear stresses in the contact layer are shown in Figures 7 and 8.

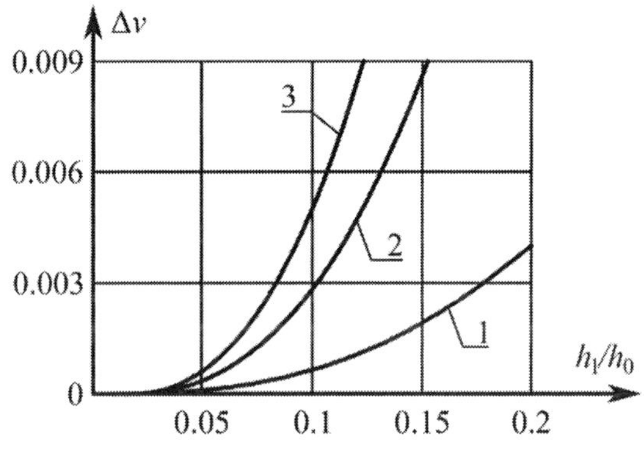

Figure 7. Calculation error of vertical displacements v.

$$1-\frac{E_1}{E_0}=1;\ 2-\frac{E_1}{E_0}=10;\ 3-\frac{E_1}{E_0}=20.$$

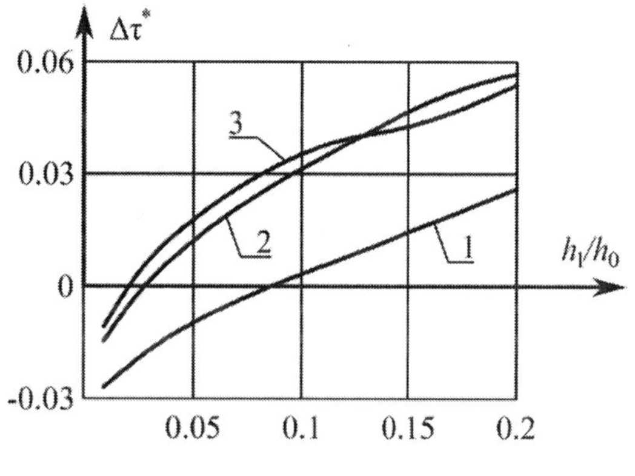

Figure 8. The error in calculating the maximum shear stresses.

$$1-\frac{E_1}{E_0}=1;\ 2-\frac{E_1}{E_0}=10;\ 3-\frac{E_1}{E_0}=20.$$

The graphs of the calculation error dependence for bending moments, longitudinal forces and displacements have a character that qualitatively and quantitatively coincides with the graphs shown in Figure 7.

Summary

A solution that makes it possible to predict the stress-strain state of beams with external reinforcement in the stretched zone based on the contact layer model has been obtained in this work.

Other models of interaction of layers (including elastic-plastic) can be found in [7, 14, 15, 16].

It is shown that the simplifications introduced into the original system of equations (1) (taking into account the bending stiffness of the reinforcement layer) do not significantly affect the results of determining the forces and stresses in the structure with the reinforcement layer thickness not exceeding 10% of the beam height. In this case, all boundary conditions are satisfied and the conditions for static equilibrium of the beam as a whole are satisfied.

References

[1] Turusov R, Andreev V and Tsybin N 2020 *IOP Conference Series Materials Science and Engineering* **913** 032053.
[2] Andreev V, Tsybin N and Turusov R 2019 *E3S Web of Conferences* **97** 04071.
[3] Tsybin N, Turusov R and Andreev V 2018 *IOP Conference Series Materials Science and Engineering* **456** 012063.
[4] Andreev V, Turusov R and Tsybin N 2016 *Procedia Engineering* **153** 59.
[5] Turusov R, Bogachev E and Andreev V 2019 *J. Phys.: Conf. Ser.* **1425** 012188.
[6] Turusov R, Bogachev E and Elakov A 2016 *Mechanics of composite materials and structures* **22** **3** 430-51.
[7] Glagolev V and Markin A 2018 *J. Phys.: Conf. Ser.* **973** 012003.
[8] Korol E 2018 *IOP Conference Series Materials Science and Engineering*, **456** 012075.
[9] Korol E, Tho V 2020 *IOP Conference Series Materials Science and Engineering* **775** 012115.
[10] Chepurnenko A, Andreev V, Beskopylny A and Jazyev B 2016 *MATEC Web of Conferences* **67** 06059.
[11] Krivoshapko S, Shambina S and Hyeng C 2017 *Materials Science Forum* **895** 45.
[12] Andreev V, Turusov R and Tsybin N 2018 *MATEC Web of Conferences* **251** 04066.
[13] Turusov R 2019 *Materials Science Forum* **974** 638.
[14] Sebestianiuk P, Perkowski D and Kulchytsky-Zhyhailo R 2020 *International Journal of Mechanical Sciences* **182** 105734.
[15] Anasiewicz K, Kuczmaszewski J 2019. Materials. **12(23)** 3911.
[16] Glagolev V, Markin A and Fursaev A 2016 *PNRPU Mechanics Bulletin* **2** 34.
[17] Chepurnenko A, Mailyan L and Jazyev B 2016 *Procedia Engineering* **165** 990.
[18] Yazyev B, Chepurnenko A and Savchenko A 2018 *Materials Science Forum* **935** 144.
[19] Chepurnenko A 2017 *Magazine of Civil Engineering* **76** 156.

Free oscillations of a beam with installation (1st part)

Kh P Kulterbaev*, V A Vodakhova, L A Baragunova, M M Shogenova, A Kh Tsipinova

Kabardino-Balkarian State University named after H.M. Berbekov, 173, Chernyshevsky St., Nalchik, 360004, Russia

E-mail: kulthp@mail.ru

Abstract. The transverse oscillations of a beam carrying a discrete mass in a span are studied. The mathematical model of oscillations is presented as a boundary value problem from the main partial differential equation of the fourth order hyperbolic type in the spatial coordinate, of the second order in time, the boundary conditions and the conditions for the sections' conjugation. The technical theory of the rods bending oscillations, based on Bernoulli's hypothesis, is used. The spectral problem of determining the eigenvalues and eigenmodes of oscillations (the Sturm-Liouville problem), necessary for analyzing the problems of forced oscillations, is considered. It is argued that the solution by analytical methods is impractical due to the large volume of transformations and cumbersome calculations. Methods of separation of variables and finite differences are used. An algorithm for solving the problem, implemented in the Matlab software environment in the form of high-precision graphic-analytical calculations, is developed. The conclusions for practical applications of the results are made.

1. Introduction

Beams oscillations have currently an extensive bibliography, since they are the widespread elements of buildings and structures and are used in installation in almost all the economy sectors. In recent years, the widespread use of robots, automated factories and workshops has led to the fact that beams experience non-traditional, non-classical technogenic external influences of a dynamic and kinematic nature. To date, transverse and longitudinal oscillations of beams and, in general, rods have an extensive bibliography [1-6]. But the scientific and technological revolution of recent decades as well as the requirements of the digital economy in recent years require more and more accurate calculations of structures, including beams.

Seismic and technogenic loads are special and very dangerous for beams, due to their random nature, which is very much tied to uncertainties at the earthquakes' occurrence place and the multiparametric intensity of their flow. Studies show that the dynamic behavior of structures depends significantly on the beams and actions' parameters, which form a large area in the multidimensional parameters space. Due to the limited volume of the given publication, in this case there is no way to move on to such problems. But the authors intend, having laid the foundation for this development, to continue the research, devoting it to the identified problems.

This work is devoted to the free oscillations of the beams carrying installation that is a source of dynamic loads. Its value lies in the fact that in such a spectral problem such beam parameters, which are necessary for the study of forced oscillations, such as frequencies and modes, will be obtained.

2. Mathematical model of oscillations

The meaningful statement of the problem is based on the design diagram of the beam, carrying a discrete mass and shown in Figure 1. The beam is acted upon in the transverse direction by static forces from its own distributed m and discrete - M mass.

In the transition to a mathematical model of free oscillations, we will use the technical theory of bending rods oscillations. Then the main equation has the form of a homogeneous differential equation of hyperbolic type in partial derivatives [2].

$$bu''''(x,t) + m\ddot{u}(x,t) + \varepsilon m\dot{u}(x,t) = 0, \qquad b = EJ, \qquad x \in (0, l), \qquad t > -\infty. \tag{1}$$

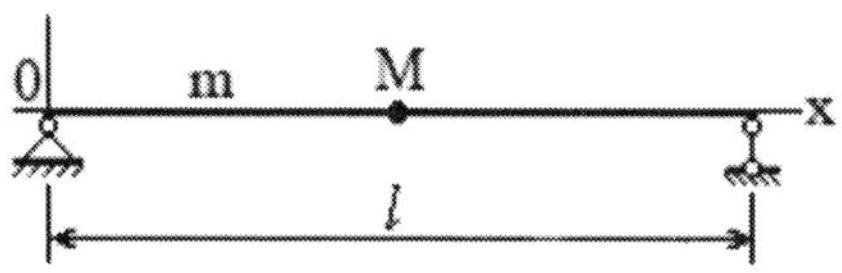

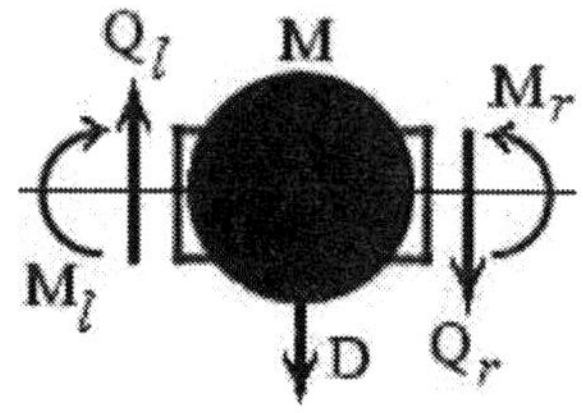

Figure 1. Design scheme of the beam **Figure 2.** Sections conjugation

Here $u(x,t)$ - is the desired function of beam deflections during motion, $b = EJ$ – is the bending stiffness of the beam, E – is an elastic modulus of the beam material, J – is an axial moment of the cross section inertia, m – is the linear beam mass, ε - is a specific coefficient of linear viscous internal friction. The dashes in the superscript denote the deflection function differentiation along the coordinate x, dots over the symbols denote time differentiation t. Static beam deflections will not be considered; they are calculated using the trivial methods of structural mechanics. In this case, it will be assumed that the oscillations occur around the equilibrium position.

Note that the presence of a discrete mass M will lead to the breakup of the second and third derivatives of the function $u(x,t)$ at the point of its location, as a result of which it is necessary to include the conditions for conjugation of sections to the left and right of it in the mathematical model of oscillations. Let us consider the situation near the mass M (Figure 2). It should be borne in mind that M is a material point with zero dimensions, so the left and right edges of the element will be considered drawn through one point.

The main equation (1) is joined by the boundary conditions corresponding to Figure 1 and the conditions for conjugating the sections to the left and right of the discrete mass. The boundary conditions for the hinged ends have the form:

$$u(0,t) = 0, \qquad u''(0,t) = 0, \qquad u(l,t) = 0, \quad u''(l,t) = 0. \qquad t > -\infty. \tag{2}$$

Some conditions for sections conjugation are kinematic and are that the function $u(x, t)$ and its first and second derivatives should be smooth at the points where the discrete masses are located x_k

$$u(l/2 - 0, t) = u(l/2 + 0, t), \qquad u'(l/2 - 0, t) = u'(l/2 + 0, t). \tag{3}$$

The need to fulfill the conditions (3) is obvious from the continuity and structure considerations. Other conditions are dynamic and arise due to the presence of force factors at the adjacent sections' conjugation (Figure 2): bending moments on the left and right M_l, M_r, lateral forces Q_l, Q_r, d'Alembert force of inertia D. The cross-sections on the left and right at this point rotate the same way. Figure 2 also implies that $M_l = M_r$. Equality of bending moments in the sections on the left and on the right corresponds to the function smoothness with respect to the second derivatives

$$bu''(l/2\text{-}0,\ t) = bu''(l/2+0,\ t). \tag{4}$$

Transverse forces Q_l, Q_r and d'Alembert force of inertia D act in a vertical direction and create a translational movement of the installation. The d'Alembert force of inertia is

$$D = Q_l - Q_r \tag{5}$$

and is expressed by the time derivative

$$D = M\ddot{u}(l/2,\ t).$$

From the above-said it follows that the equation (5) should be included in the conjugation conditions, but at the same time there is no such need for (3) and (4), since the smoothness and continuity of the function are ensured by them. Let us turn to expressions for the shear forces in terms of the deflection function and get

$$M\ddot{u}(l/2,\ t) = b[u'''(l/2-0,\ t) - u'''(l/2+0,t)]. \tag{6}$$

As it is possible to see, the smoothness of the function $u(x,t)$ on the third spatial derivative is not provided, and this should be taken into account in the forthcoming calculations. From the point of view of structural mechanics, this means that the transverse forces have jumps by the value of the d'Alembert force.

Equation (1), boundary conditions (2) and conjugation condition (6) form a mathematical model that makes it possible to study free, forced deterministic and stochastic oscillations of a beam carrying installation with a source of impact.

3. Free oscillations

The problem of free oscillations is mandatory before considering forced oscillations, since without such its results as natural frequencies and forms, a consistent consideration of forced oscillations is impossible. Therefore, we move on to setting and solving the problem.

Note that free oscillations are described by homogeneous basic equations and additional conditions and do not require initial conditions in time. We use the variables separation method [7] and represent the deflection function of the longitudinal axis of the beam in the form of a product

$$u(x,t) = X(x)e^{\lambda t}, \tag{7}$$

where $X(x)$ – is the desired eigenform function, $\lambda = -\varepsilon + j\omega$ – is a characteristic indicator, ε, ω – are damping coefficient and angular frequency of free oscillations, j – is an imaginary unit. This representation is based on the fact that it should describe the process of damped oscillations. Coefficient ε will be determined by bibliographic sources, frequencies ω arbitrary numbers in the class of real numbers R. Substituting (7) in (1), (2), (6), we can reduce the result by a common factor and obtain a system of homogeneous linear algebraic equations

$$B(\lambda)X = 0, \tag{8}$$

where $B(\lambda)$ - is a square matrix of coefficients that contain the desired frequency ω. Their determination by analytical methods is associated with cumbersome calculations and the need to solve a system of transcendental equations. A more versatile and simpler way to identify a spectral pair (ω, X) consists in using the numerical methods [8] and computer calculating systems. Let us apply the finite difference method (FDM) and software package Matlab. FDM requires a transition from the domain of a continuous argument $0 \le x \le l$ to the domain of a discrete argument (mesh)

$$L_h = \{x_i = (i-1)h, \quad i = 1,2,\dots,n\}$$

with the step $h = l/(n-1)$. After applying the method of variables separation, we get

$$bX''''(x) + \Lambda X(x) = 0, \qquad \Lambda = \mathrm{m}\lambda(\lambda + \varepsilon). \tag{9}$$

$$X''(0) = 0, \quad bX'''(0) - cX(0) = 0, \quad X''(l) = 0, \quad bX'''(l) + cX(l) = 0. \tag{10}$$

$$b\left[X'''(x_K - 0) - X'''(x_K + 0)\right] - M_i \lambda^2 X = 0, \quad i = 1,\ 2. \tag{11}$$

Hereinafter $x_k = l/2$ – is the coordinate of the beam sections' conjugation point. It is necessary to note that the conjugation conditions (3), (4) are not indicated because of the smoothness of the function itself *X(x)* and its first and second derivatives on the boundary of the sections.

We will perform the necessary procedures FDM by replacing the derivatives with their finite-difference analogs with accuracy $O(h^2)$ and instead of the main equation we obtain (1)

$$X_{i-2} - 4X_{i-1} + \mu X_i - 4X_{i+1} + X_{i+2} = 0, \qquad \mu = 6 + \Lambda h^4/b. \tag{12}$$

We carry out the similar actions for the boundary and conjugation conditions.

Left end:

$$X_1 = 0, \qquad 2X_1 - 5X_2 + 4X_3 - X_4 = 0, \tag{13}$$

Right end:

$$-X_{n-3} + 4X_{n-2} - 5X_{n-1} + 2X_n = 0, \qquad X_n = 0. \tag{14}$$

Conjugation condition at the point x_k:

$$3X_{k-4} - 14X_{k-3} + 24X_{k-2} - 18X_{k-1} + \eta X_k - 18X_{k+1} + 24X_{k+2} - 14X_{k+3} - 14X_{k+4} = 0, \eta = 10 - m\lambda^2/b. \tag{15}$$

The mathematical model of free oscillations is now represented by a system of algebraic equations (12) - (16), which we write in a matrix-vector form

$$A(\lambda)Y = 0. \tag{16}$$

where $A(\lambda)$ is a square coefficient matrix of the n – th order, Y – is a vector of discrete arguments replacing an eigenfunction X(x). It should be borne in mind that instead of exact values X approximate values $Y = \{y_1, y_2, \dots, y_n\}$, $y_i \approx X(x_i)$ will be determined, i – is the finite difference mesh node number. The coefficient matrix has the form

$$A(\lambda) = \left(\begin{array}{ccccccccccccccc}
1 & & & & & & & \vdots & & & & & & & \\
2 & -5 & 4 & -1 & & & & \vdots & & & & & & & \\
1 & -4 & \mu & -4 & 1 & & & \vdots & & & & & & & \\
 & 1 & -4 & \mu & -4 & 1 & & \vdots & & & & & & & \\
\cdots & \cdots & \cdots & \cdots & \cdots & \cdots & \cdots & \cdots & \cdots & \cdots & \cdots & \cdots & \cdots & \cdots & \cdots \\
 & & & & 1 & -4 & \mu & -4 & 1 & & & & & & \\
 & & & 3 & -14 & 24 & -18 & \eta & -18 & 24 & -14 & 3 & & & \\
 & & & & & & 1 & -4 & \mu & -4 & 1 & & & & \\
\cdots & \cdots & \cdots & \cdots & \cdots & \cdots & \cdots & \cdots & \cdots & \cdots & \cdots & \cdots & \cdots & \cdots & \cdots \\
 & & & & & & & \vdots & & & 1 & -4 & \mu & -4 & 1 \\
 & & & & & & & \vdots & & & & -1 & 4 & -5 & 2 \\
 & & & & & & & \vdots & & & & & & & 1
\end{array}\right).$$

Zero elements are not written here. Elements of the main diagonal η contain eigenvalues, which can be determined from the condition for the nonzero solutions existence to the system of equations (16), which is in the fact that

$$\det A(\lambda) = 0\,. \tag{17}$$

This equation is a high-order algebraic equation, and for the large n values its compilation and solution [9] presents significant difficulties. For the cases like this spectral problem, the problem of finding the eigenvalues and eigenvectors can be significantly simplified. There are two reasons for this:

1. There is no need to solve the complete eigenvalue problem. The practical need in the overwhelming majority of cases is to use only a few of the smallest eigenvalues and corresponding vectors for the beams, i.e., to solve the partial eigenvalue problem.
2. The used numerical finite difference method in conjunction with the Matlab computer complex and the ability to build a high-precision graph of the characteristic equation with subsequent implementation on a computer screen makes it possible to set the eigenvalues in very simple ways.

This method of solving the problem of eigenvalues and vectors was used by the authors in a number of problems and shows a high degree of accuracy in solving the problems of rods stability and oscillations [3].

The applied algorithm consists of the following sequential actions:

1. The square matrix is compiled according to the numerical method A(ω) *n* order, the elements of the main diagonal of which contain natural frequency.
2. The function graph detA(ω) is built and displayed on the monitor screen.
3. By the points detA(ω) = 0 the eigenvalues are read from the screen ω_k, where k = 1, 2 .... in the required quantities.
4. At the zero points of the determinant, the rank of the matrix A is decreased to the value R = n – 1. Therefore, one of the nonzero components of the vector is necessary Y to take an arbitrary number like $Y_k = 1$.
5. The column containing Y_k, is transferred with a change of signs to opposite to the right side of equation (16) and it becomes inhomogeneous.
6. To make a matrix A square again, any of its lines is deleted and thus the equation (16) becomes square of the *n-1* order.

$$AY = d. \tag{17}$$

7. The vector Y of the *n-1* dimensions is determined from (17). It is supplemented by the newly appointed $Y_k = 1$ and the desired order *n* eigenvector is obtained.

Example 1. Initial data: The beam consists of two steel I-beams No.30, l = 8 m, E = 210 GPa, J = 14160 cm^4, m = 73 kg/m, M = 4000 kg, $\varepsilon = 0,25$ s^{-1}, n = 1001, spatial step $h = 1/(n-1)$, frequency step $h_\omega = 0,25\ s^{-1}$

The result of the solution is displayed on the monitor screen (Fig. 3) as a graph *ω - det,* where the dot marks the first and second natural frequencies. The values of two elements of the natural frequencies' spectrum are determined by reading from the monitor screen

$$\omega_k = \{25,51 \quad 393,69\}\ s^{-1}.$$

These numbers are highly accurate, since the monitor screen makes it possible to increase the picture size and view the fragments and neighborhoods of points in a very enlarged form.

The use of regulatory documents [10, 11] shows that the first frequency of free oscillations corresponds to the highest dynamic factor under seismic impacts on the beam $\beta = 2,5$ (Fig. 4) for the structures on soils of all categories. From the point of view of this beam reliability, it is advisable to reduce it by changing the oscillatory system parameters. A qualified solution to such a problem should be based on methods of dynamic oscillation damping, methods of dynamic mechanical systems' optimal design. The arsenal of techniques and special means for such cases is very extensive and the choice depends significantly on the specific circumstances.

Oscillations in and around the second natural frequency are high-frequency with 63 revolutions per second and a dynamic coefficient equal to one. They are not dangerous for the beam, since they do not fall into the dangerous zone for seismic oscillations. The third natural frequency was not determined due to its very significant value, which practically does not occur for installation.

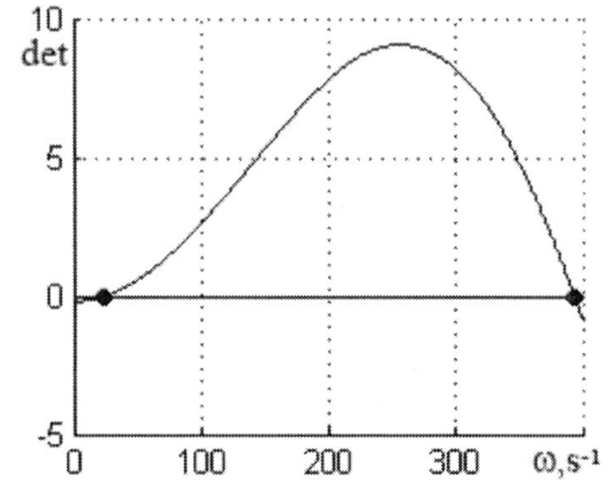

Figure 3. Free oscillations frequencies

1. I, II category soils,

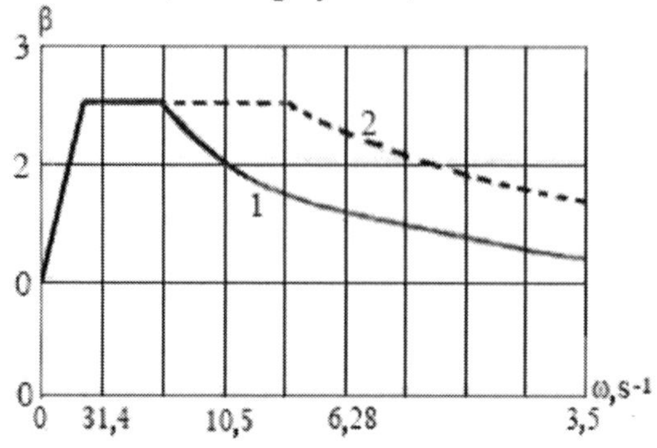

2. I, II category soils.

Figure 4. The dynamic coefficient dependence on oscillation frequencies

The next task is to find the eigenvectors Y_k $(k = 1, 2, 3... n)$ matrices A. Those that correspond to the frequencies of free oscillations found in the example are of interest. They can be determined from the system of equations (16) with known eigenvalues ω_k. In this case, the determinant of the matrix A becomes zero, which means that its rank decreases and is equal to $R = n\text{-}1$. Hence it follows that the components of the vector Y can be calculated only up to a factor. In this case, one of the nonzero components of the vector Y can be taken equal to an arbitrary number and excluded from the required unknowns. Then the system of equations (16) turns from homogeneous into inhomogeneous, and the remaining components of the vector Y are determined from it.

The resulting eigenforms, normalized to unity and corresponding to eigenvalues, are shown in the graphs in Figure 5.

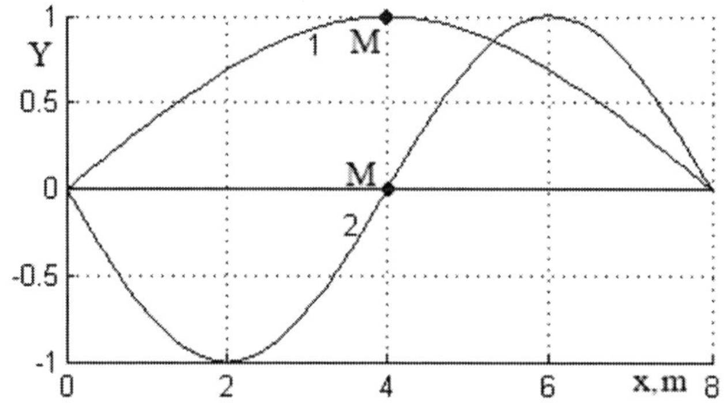

Figure 5. Free oscillation shapes

The first eigenform almost coincides with the half-wave of the sinusoid. Similar half-waves are contained in the second form. Moreover, they correspond to the boundary conditions and conjugation conditions: the ends are hinged supported, therefore, have zero deviations; the heavy discrete mass M remains stationary during oscillations in the second form. Based on these features, it can be assumed that the results obtained for determining the frequencies and eigenforms are reliable.

4. Summary

1. Mathematical models to study free oscillations of a continuous-discrete beam structure have been created.
2. A high-precision graphical-analytical method for solving the spectral problem for beams has been proposed.

3. An effective numerical method for determining natural frequencies and forms of free oscillations has been developed.
4. It has been proposed to remove the first natural frequency from the large dynamic coefficients zone according to Building Codes.

References

[1] Svetlitskiy V A 1987 *Rod mechanics. In two parts. Part II. Dynamics.* (Moscow, Higher school) 304.
[2] Bolotin V V 1978 *Oscillations in technology. Reference book in 6 volumes. Volume 1. Oscillations of linear systems* (Moscow, Mechanical engineering) 352.
[3] Kulterbaev Kh P, Baragunova L A, Shogenova M M, Shardanova M A 2019 The Solution of a Spectral Task on Variable Section Compressed Beams Oscillations by Numerical Methods *Materials Science Forum* **974** 704-710. ISSN: 1662-9752.
[4] Kulterbaev Kh P, Alokova M H, Baragunova L A 2015 Mathematical modeling of flexural oscillations of a vertical rod of variable section *The News of higher education institutions. North - Caucasion. region. Techn. sciences* **4** 100-106.
[5] Kulterbayev Kh P 2011 Kinematically excited oscillations of a continuous-discrete multi-span beam *Bulletin of the Nizhny Novgorod University. N.I. Lobachevsky. Proceedings of the X All-Russian Congress on Fundamental Problems of Theoretical and Applied Mechanics. Publishing house of NNSU named after N.I. Lobachevsky* **4 (2)** 198-200.
[6] Ilyin M M, Kolesnikov K S, Saratov Yu S 2003 *Oscillation theory* (Moscow, MSTU) 272.
[7] Tikhonov A N, Samarskiy A A 1977 *Equations of mathematical physics* (Mosocw, Nauka) 736.
[8] Samarskiy A A 1977 *The theory of difference schemes* (Moscow, Nauka, Ch. ed. physical-mat. lit.) 656.
[9] Verjbitskiy V M 2005 *Numerical methods (linear algebra and nonlinear equations)* (Moscow, LLC "Onyx 21st Century" Publishing House) 432.
[10] SNiP II-7-81. *Construction in seismic areas. Design standards* (Moscow, Gosstroy of the USSR, Stroyizdat) 48.
[11] BC 14.13330.2014 *Construction in seismic areas SNiP II-7-81* (updated SNiP II-7-81* "Construction in seismic areas" (with Change N 1).*

IOP Conf. Series: Materials Science and Engineering 1083 (2021) 012044 doi:10.1088/1757-899X/1083/1/012044

Forced oscillations of beams with installation (Part II)

Kh P Kulterbaev*, V A Vodakhova, L A Baragunova, M M Shogenova, A Kh Tsypinova

Kabardino-Balkarian State University named after H.M. Berbekov, 173, Chernyshevsky st., Nalchik, 360004, Russian Federation

E-mail: kulthp@mail.ru

Abstract. This article (Part II) is a continuation of Part I, published in the same edition, where the spectral problem is considered. Part II examines two types of transverse beam oscillations: forced harmonic and forced random. The beam carries its own distributed mass and discrete mass. The source of oscillations is operating equipment with an element moving during technological operations. The mathematical model of oscillations is presented as a boundary value problem from the main partial differential equation of the hyperbolic type of the fourth order in the spatial coordinate, the second order in time, the boundary conditions and the conditions of the beam sections' conjugation. The technical theory of the rods' bending oscillations is used, based on Bernoulli's hypothesis about the beam plane cross-sections' invariability. The methods of variables separation and finite differences are applied. The algorithms for solving the problems have been developed, implemented in the Matlab software environment. Verification of the proposed mathematical models is demonstrated using the specific examples. The particular examples have been carried out and the practical conclusions have been outlined.

Introduction

Beams are the widespread elements of buildings and structures of heavy and light industry, petrochemical industries, road transport highways, processing enterprises. Beams are the parts of machines, technological equipment, machine tools, robots and manipulators. The requirements of digital economy, automation and computerization of production required a more accurate and adequate design and construction of beams that experience non-classical man-made and seismic external influences of a dynamic and kinematic nature [1-6]. Beams are subjected to technogenic oscillations, which can be both deterministic and stochastic, are very dangerous and difficult to design. These two types of oscillations are very closely related, since random influences (processes) consist almost always of an infinite set of harmonics [7].

Forced oscillations' mathematical model

The design diagram of the beam, shown in Figure 1, carries the equipment with discrete mass M in the middle of the span. The equipment contains mass m_0, moving plane-parallel in the vertical direction or rotating uniformly around the axis. It is the source of the concentrated force F (t). In addition to them, the beam itself has its own linear mass *m*.

There is a wide variety of beams by material, different shapes of cross-sections, constant or variable cross-sections in length, etc. The most suitable in the case, when heavy equipment is installed on the beam and it should be in the same plane with the intermediate floor, is an *I*-beam paired section.

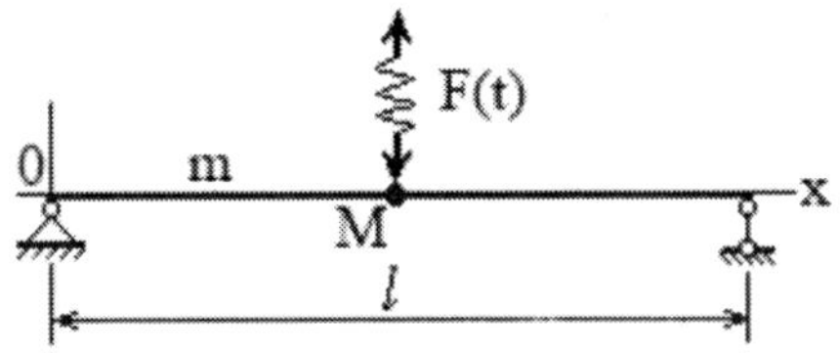

Figure 1. Calculation scheme.

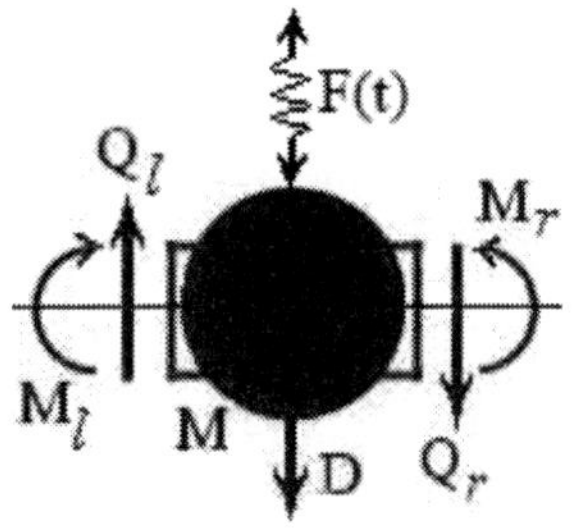

Figure 2. Sections' pairing.

We will use the technical theory of bending oscillations of rods and write out the basic equation of the inhomogeneous differential equation of hyperbolic type in partial derivatives [2].

$$b\,u''''(x,t) + m\ddot{u}(x,t) + \varepsilon m\dot{u}(x,t) = 0, \qquad b = EJ, \qquad x \in (0,\ l), \qquad t > -\infty. \tag{1}$$

Traditional notations are used here: $u(x,\ t)$ – bending function of the curved longitudinal axis of the beam; $x,\ t$ – coordinates in space and time; $b = EJ$ – constant bending stiffness of a beam along the axis; E – Young's modulus of the beam material; J – axial moment of a beam cross-section inertia; m - linear density of the beam material; ε – specific coefficient of internal friction linear viscous forces.

Dashes in the superscripts correspond to the derivatives with respect to the spatial coordinate, the dots above the letters correspond to the derivatives with respect to time. The number of lines and dots corresponds to the differentiation order.

Boundary conditions and a condition for mating beam sections are added to the main equation. Boundary conditions at the left and right ends of the beam are:

$$u(0,t) = 0, \qquad u''(0,t) = 0, \qquad u(l,t) = 0, \qquad u''(l,t) = 0. \qquad t > -\infty. \tag{2}$$

The condition for mating left and right sections:

$$b[u'''(l/2-0,\ t) - u'''(l/2+0,t)] - M\ddot{u}(l/2,\ t) = F(t). \qquad t > -\infty. \tag{3}$$

The function $F(t)$ represents harmonic or random effects of equipment on a beam.

Together (1) - (3) form a mathematical model of forced harmonic and random steady-state oscillations. It should be noted that in such problems the initial conditions are not required.

Harmonic oscillations

Harmonic oscillations of the beam are considered in view of the fact that technogenic oscillations are often harmonic or poly-harmonic. In addition, the results obtained for the harmonic oscillations give an approximate idea of random oscillations and greatly facilitate the understanding and solution of the problem about them.

In equation (3), in this case, we assume that the force is described by the function

$$F(t) = fe^{\lambda t},$$

where f – is the harmonic force amplitude, $\lambda = j\omega$, j – is an imaginary unit.

In the method of variables separation [8], which will be further applied, the output function has the form of a harmonic oscillation

$$u(x,t) = X(x)\ e^{\lambda t}.$$

Here $X(x)$ – is the oscillation amplitude. Similarly to the case of free oscillations, using the finite difference method [9, 10], it is possible to obtain the inhomogeneous system of linear algebraic equations

$$A(\lambda)\cdot Y = d,$$

IOP Conf. Series: Materials Science and Engineering 1083 (2021) 012044 doi:10.1088/1757-899X/1083/1/012044

where $A(\lambda)$ – is a square matrix of order coefficients n, $Y = \{y_1, y_2, \dots, y_n\}$ – is a vector of discrete arguments replacing a continuous function $X(x)$. The coefficient matrix and the vector on the right side have the form

$$A(\lambda) = \begin{pmatrix} 1 & & & & & & & \vdots & & & & & & & \\ 2 & -5 & 4 & -1 & & & & \vdots & & & & & & & \\ 1 & -4 & \mu & -4 & 1 & & & \vdots & & & & & & & \\ & 1 & -4 & \mu & -4 & 1 & & \vdots & & & & & & & \\ \dots & \dots & \dots & \dots & \dots & \dots & \dots & \dots & \dots & \dots & \dots & \dots & \dots & \dots & \dots \\ & & & & 1 & -4 & \mu & -4 & 1 & & & & & & \\ & & & 3 & -14 & 24 & -18 & \eta & -18 & 24 & -14 & 3 & & & \\ & & & & & & 1 & -4 & \mu & -4 & 1 & & & & \\ \dots & \dots & \dots & \dots & \dots & \dots & \dots & \dots & \dots & \dots & \dots & \dots & \dots & \dots & \dots \\ & & & & & & & \vdots & & & 1 & -4 & \mu & -4 & 1 \\ & & & & & & & \vdots & & & & -1 & 4 & -5 & 2 \\ & & & & & & & \vdots & & & & & & & 1 \end{pmatrix}.$$

$$d^T = \{0, \dots, d_k, \dots, 0\}.$$

Here zero elements of the matrix and vector are not inscribed, the vector of the right-hand side contains a single element with the number $k = (n+1)/2$, corresponding to the conjugation point of the left and right sections of the beam, the T sign in the superscript corresponds to the operation of vector transposition. The designations for the matrix elements are introduced:

$$\mu = b + \Lambda h^4/b, \qquad \eta = 10 - m\lambda^2/b, \qquad \Lambda = m\lambda(\lambda + \varepsilon).$$

Let us carry out the calculations for a specific beam with the parameters.

Example 1. The beam is composed of two I-beams No.30, $l = 8$ m, $E = 210\ GPa$, $J = 14160\ cm^4$, $m = 73\ kg/m$, $M = 4000\ kg$, $\varepsilon = 0{,}5\ c^{-1}$, $n = 4001$, $f = 100\ kN$, $\omega = \{1;\ 15;\ 50;\ 100\}$. c^{-1}, the spatial coordinate step $h = l/(n-1)$. For use in analyzing the results, we indicate the eigenvalues $\Omega = \{25{,}51;\ 393{,}69\}\ c^{-1}$. It is required to make the calculations and build a graph of the oscillation amplitudes function along the length of the beam $Y(x)$.

The counting results are shown in Figure 3 as graphs $Y(x)$ at the specified frequencies of exposure to equipment ω. Frequency numbers are signed in order.

At low oscillation frequencies, the force F, lower than the first natural frequency, the oscillations of the beam have the form of the first form (curves 1, 2). With a further increase in the frequency of disturbances (curves 3, 4), the first resonance oscillations with large amplitudes occur, then they decrease and change the phase of oscillations (curves 3, 4) to the opposite force phase. At high ω values, the oscillations almost disappear and become imperceptibly small and take place in the second and higher modes of oscillations. These results are of no practical importance and therefore are not shown. At the same time, they may be of theoretical interest.

The results obtained for the harmonic oscillations make it possible to predict the dynamic behavior of a beam under random influences.

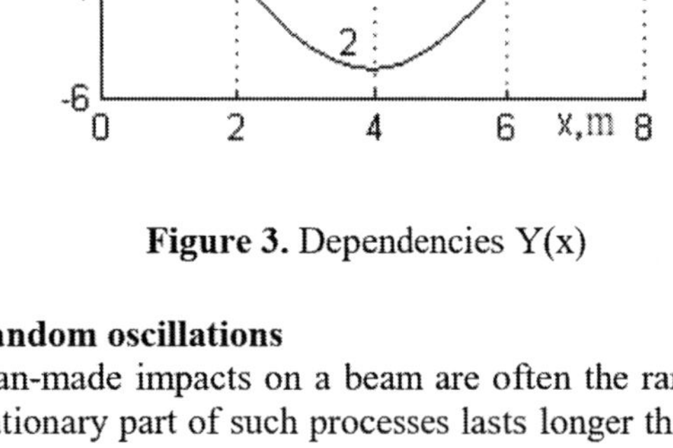

Figure 3. Dependencies Y(x)

Random oscillations

Man-made impacts on a beam are often the random processes in time. Moreover, in most cases, the stationary part of such processes lasts longer than the others and poses the greatest danger to building structures. Therefore, the corresponding oscillations will be further considered as stochastic within the

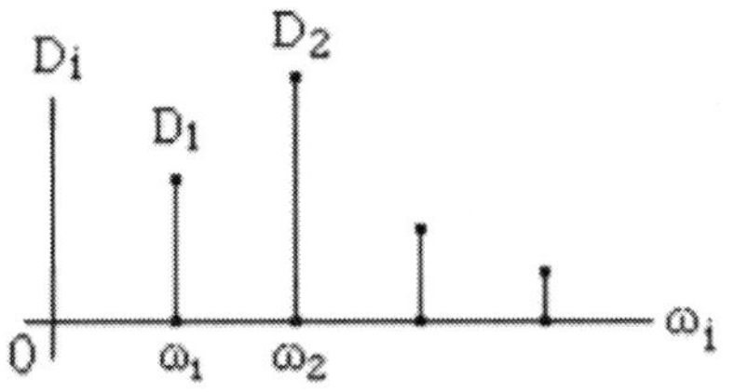

Figure 4. Discrete line spectrum

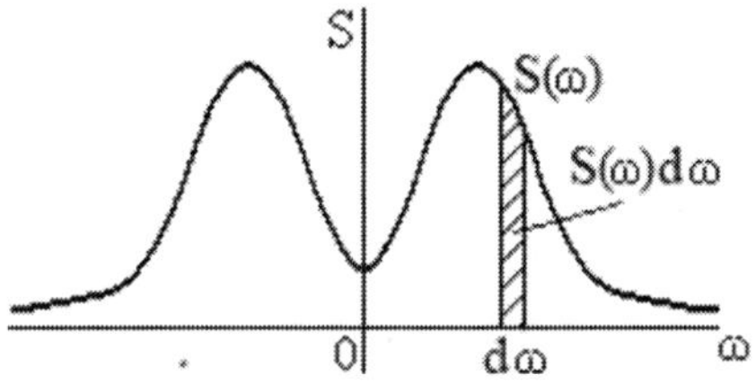

Figure 5. Spectral density

framework of the stationary random processes' theory. The results on harmonic oscillations obtained above will help to solve such problem in relatively simple ways, without resorting to complex and cumbersome existing methods. For this purpose, in the mathematical model for harmonic oscillations given above, we will make the necessary changes for the random oscillations. They will only concern the input force $F(t)$. According to the above-mentioned, the output process $u(x, t)$ will be a centered spatio-temporal random field, stationary in time and inhomogeneous in the spatial coordinate. The task is to find the standard deviations of the output process from the given spectral density of the input random process. Spectral density functions of the input stationary technogenic processes for more frequent cases have the form:

$$S(\omega) = \frac{4\theta^2 \alpha\sigma^2}{\pi\left[\left(\omega^2 - \theta^2\right)^2 + 4\alpha^2\omega^2\right]}, \quad \omega \in [0,\infty], \quad \theta^2 = \alpha^2 + \beta^2. \tag{4}$$

Here α, β - are the broadband and dominant frequency parameters, σ - is the process standard deviation $F(t)$. Processes with spectral density (4) are called the narrow-band random processes with a characteristic frequency. As it is known, a stationary random process can be represented as a Fourier series [7]

$$F(t) = \sum_{i=1}^{\infty}[U_i cos\omega_i t + V_i sin\omega_i t]. \tag{5}$$

The simple transformations taking into account the relationship between the variance and the correlation function of a stationary random process and the series coefficients correlation (5) give the force variance

$$D[F(t)] = \sum_{i=1}^{\infty} D_i.$$

This result is shown in Figure 4 in the form of the so-called discrete line spectrum variances. It turns out that there is a binary correspondence between dispersions and frequencies, that is, each elementary dispersion can be associated with a certain frequency of the spectral density. In this case, the random process variance of the input process $F(t)$ seems to be an improper integral [10]

$$D_F = \int_{-\infty}^{\infty} S(\omega)d\omega$$

Figure 5 shows that the elementary variance $S(\omega)d\omega$, shaded in Figure 5 is associated with each frequency ω. If the elementary variance is taken instead of the driving force amplitude, the output will be the elementary variance of the discrete mass deviations $dD_x(\omega_k)$. Subsequent summation over $\omega_k (k = 1, 2 \dots, n)$ gives the variance of deviations D_x. A very effective method for obtaining variances for stationary random oscillations at the output, which ultimately will consist in calculating the sum can be obtained from it:

$$D_x \approx h_\omega \sum_k S_k .$$

Here h_ω – is a frequency axis split step. Further calculations are carried out as in Example 1, namely, instead of the amplitude of this harmonic $F(t)$ the elementary variance of disturbances is inserted into the algorithm and computation program $S_k h_\omega$. The results of their summation give the output displacement process variance.

Example 2. The beam consists of two steel I-beams No. 30, $l = 8$ m, $E = 210$ *GPa*, $J = 14160$ *cm*^{4}, $m = 73$ *kg/m*, $M = 4000$ *kg*, $\varepsilon = 0{,}5$ *c*$^{-1}$, $n = 2001$, $\beta = \{4;\ 6;\ 90;\ 98\}$ c^{-1}, $\sigma_f = 141kN$, the spatial coordinate step $h = l/(n-1)$. It is required to calculate and plot the function of the standard deflections deviation of the beam oscillations $\sigma_y(x)$.

The calculations shown in the graphs in Figures 3 and 6, specially carried out in the most common frequency range of the equipment operation ω ϵ [1, 100]. When considered together, taking into account that standard deviations can only take positive values, they have many quantitative and qualitative similarities.

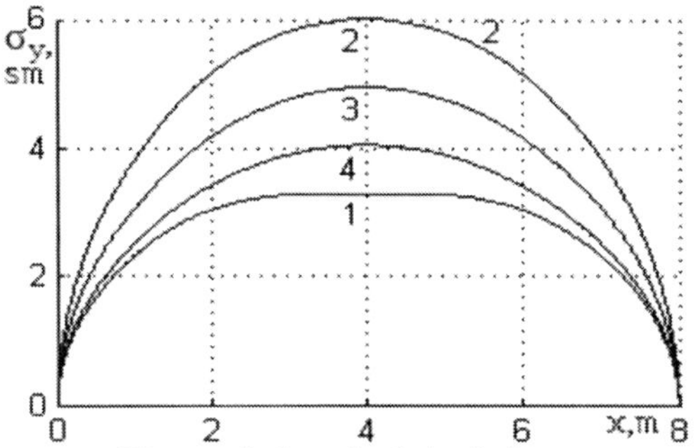

Figure 6. Standard deviations.

Summary

1. The standard deviations of random oscillations have the values and shapes close to the harmonic oscillations' amplitudes.
2. Small values of the disturbance frequency (ω =1, …, 3) and large (ω > 150) c^{-1} are not dangerous for this beam, since the values of the beam dynamic deflections are insignificant (0, ..., 2) mm.

References

[1] Bolotina V V 1978 *Oscillations in technology. Reference book in 6 volumes. Volume 1. Oscillations of Linear Systems* (Moscow, Mechanical engineering) 352.
[2] Panovko Ya G 1964 *Introduction to the theory of mechanical oscillations* (Moscow, Nauka) 276.
[3] Kulterbayev Kh P, Pshenichkina V A 2006 Random processes and fluctuations of building constructions and structures *Volgograd state university of architecture and civil engineering*, (Volgograd, VolgSACEU) 335.
[4] Kulterbayev Kh P 2012 Oscillations of a discrete-continuous beam, kinematically excited by harmonic and random vector processes *Bulletin of the universities. North Caucasus region. Tech. science* **3** 54-59.
[5] Ilyin M M, Kolesnikov K S, Saratov Yu S 2003 *Oscillation theory* (Moscow, MSTU) 272.
[6] Maslennikov A M 2016 *Dynamics and stability of structures. Textbook and workshop for universities* (Moscow, Yurayt Publishing House) 366.
[7] Khruscheva I V 2009 *Fundamentals of Mathematical Statistics and Theory of Stochastic Processes: Textbook* (SPb., Publishing house "Lan"') 336.
[8] Tikhonov A N, Samarsky A A 1977 *Equations of mathematical physics* (Moscow, Nauka) 736.
[9] Akimov P A, Belostotsky T B 2017 et.al. *Informatics in construction (with the basics of mathematical and computer modeling): Tutorial* (Moscow, KNORUS) 420.

[10] Kulterbayev Kh P 2019 On the influence of the correlation of seismic effects on the oscillations of a vertical rod *XII All-Russian Congress on Fundamental Problems of Theoretical and Applied Mechanics. Collection of reports of the Congress in 4 volumes. Vol. 1: General and applied mechanics* (Ufa, RIC BashSU) 344-346.

[11] Kulterbayev Kh P, Shogenova M M, Baragunova L A 2020 On the Influence of the Characteristic Frequency and Broadband of Seismic Effects on the Vertical Rod Oscillations *International science and technology conference "FarEastCon-2019"/ IOP Conf. Series: Materials Science and Engineering* **753 042040.** doi:10.1088/1757-899X/753/4/042040

IOP Conf. Series: Materials Science and Engineering 1083 (2021) 012045 doi:10.1088/1757-899X/1083/1/012045

Application of the software package in the improvement of floors made of hollow core slabs

S V Ovchinnikova[1*], E M Schneider [2], G V Kukinova[2]

[1] Kuban State Agrarian University named after I.T. Trubilin, 13, Kalinina st., Krasnodar, 350044, Russia

[2] Nevinnomyssk State Humanitarian and Technical University, 19 Bulvar Mira str., Nevinnomyssk, 357108, Russia

E-mail: svetlana.swetlana-ov@yandex.ru

Abstract. An overview of the rationale for the unweighted hollow-core floor slabs use is given. The technological and design features of hollow-core floor slabs are considered. The possibilities of rational use of unweighted hollow-core floor slabs are presented. The calculation of monolithic floor slabs was carried out in the Lira software package on the basis of existing technologies, current standards and the methods for calculating the unweighted hollow-core slabs. The performance of floor slabs with spans equal in two directions is considered. It was found that regardless of the span of the floor slab, the effect of reducing its own weight exceeds the effect of reducing the section rigidity. The total deflection due to the own weight of the floor slab and the payload on it for the hollow core slabs is lower than for solid slabs of the same span. The use of unweighted monolithic slabs provides a reduction in the dead weight of the floor slab up to 35%. Reducing the load on the slab also provides a reduction in the reinforcing steel materials application rate. The use of unweighted monolithic floors by reducing the weight of structures and reducing their deflections will significantly optimize the structural schemes of buildings. The performed numerical experiment makes it possible to establish the boundaries of the effectiveness of the unweighted hollow-core monolithic floor slabs' use, upon reaching which the use turns out to be significantly more profitable than the use of traditional beamless floors.

Introduction

The last decade of the twentieth century is characterized by the significant achievements in construction industry [1-2]. Reducing the dead weight of monolithic floor slabs while maintaining the required strength, reliability and durability using reinforced concrete structures is an urgent task of modern construction. The use of unweighted floors can significantly reduce the cost of vertical supporting structures and foundations [3-5]. Reduction in the slabs' weight can significantly reduce the cost of the entire structural frame of a building. In addition, in some cases, the use of modern methods for the unweighted monolithic floor slabs' construction can significantly increase the construction production industrialization, increase the speed of construction of buildings, and reduce the total construction duration [6].

The use of hollow-core floor slabs is one of the possible ways to reduce the material application rate and the mass of buildings erected from monolithic concrete, practically without reducing their reliability

[7-11]. Reduction in the structures' weight is an important advantage when constructing the facilities in seismically active regions.

Main part

A model of a hollow-core monolithic floor slab, for which there are experimental data on the bearing capacity and deformations of the structure under load was adopted as an object of research. The computational model of a hollow-core slab modeled by an anisotropic slab was verified with the results of the experimental study. The performed comparison confirms the adequacy of the proposed variant of modeling hollow-core floors with anisotropic slabs with reduced stiffness parameters. The numerical model of the slab provides relatively accurate data on the hollow-core slab deformations. The discrepancy between the results at low loads is explained by the absence of cracks in concrete at this stage, and, accordingly, by the high section rigidity. A significant difference in the data at loads close to the ultimate is explained by the nonlinear work of reinforced concrete.

The general theory of calculating slabs can be used for flat unweighted slabs, provided that the loads of their dead weight and rigidity are reduced, due to which the capabilities of the LIRA-SAPR program "Design" tab can be used to select reinforcement in the section. The flat slabs' thickness of solid sections is recommended to be taken at least 16 cm; at least 1/30 of the largest span's length and not more than 25 cm. Thus, the geometric characteristics of the flat slabs with a thickness equal to the thickness of hollow-core slabs differ significantly from slabs.

A numerical study of the hollow-core slabs and solid slabs' stress-strain state of the same span, the thickness of which was adopted according to the recommendations, was carried out, a comparison of their main parameters characterizing the cost of construction was made, as well as the strength and reliability of the floors (deflections, dead weight, reinforcement application rate). The geometric characteristics of floor slabs are given in Table 1.

Table 1. Geometric characteristics of slabs.

Span L, m	Solid cross- section slab thickness h_{sl}, m	Empty cross-sectional slab thickness h_{emp}, m
6	0.2	0.24
7	0.23	0.28
8	0.25/0.27	0.32
9	0.25/0.3	0.35
10	0.25/0.33	0.42
11	0.25/0.37	73.33
12	0.25/0.40	0.46
13	0.25/0.43	0.50
14	0.25/0.46	0.54
15	0.25/0.50	0.58

For the solid cross-section slabs with a span of more than 8 meters, 2 options are considered:

1) the section height is equal to 1/30 of the span;

2) the height of the span section is 25 cm.

When designing the slabs, the issues of ensuring the punching shear strength of the slab were not taken into account; in the future, the application rate of concrete for capitals and the steel application rate for transverse reinforcement were not taken into account. The use of capitals in a non-girder ceiling deprives it of its most important advantages over the slab and beam floors. Capitals increase the overall floor slab and floor heights in a similar way to beams. The lower surface of the slab has height differences, which requires the installation of suspended ceilings, and also creates the restrictions for architectural and planning solutions and the installation of engineering systems. For girderless capital less slabs, the shear strength of the slab (against the action of lateral forces near the support) is possible due to the construction of hidden steel capitals (distribution systems). The issue of their organization within the framework of the conducted technical and economic analysis has not been considered, but their arrangement only increases the total cost of the floor slab.

The general view of the design scheme used in the numerical experiment is similar to the scheme shown in Figure 1.

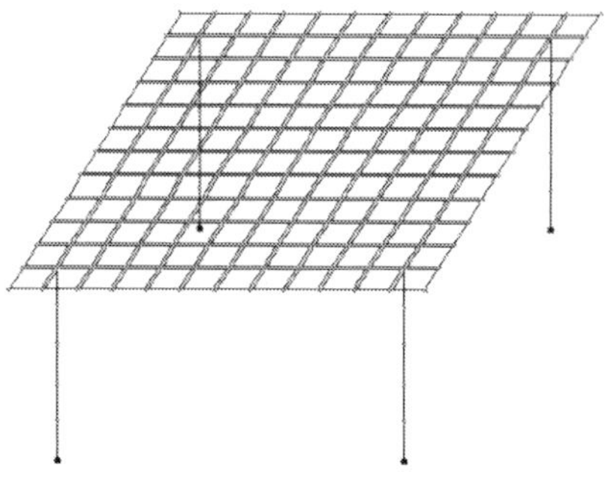

Figure 1. General view of the design scheme

Reinforcement application rate was determined based on the accepted background reinforcement of the floor slab. There are no absolutely rigid bodies in the design model that simulate the real bearing area of the floor on the columns. The spacing and diameter of the background reinforcement was determined based on the required reinforcement area over the entire area of the floor slab, the reinforcement zones of the upper reinforcement above the supports and minor bursts of the lower span reinforcement were neglected. When determining the reinforcement flow rate, the implementation of local reinforcement zones was not taken into account. In the case of designing a girderless solid floor slab with capitals, their deflection turns out to be slightly lower than that obtained from the experiment results, due to a decrease in the thin bending part, however, the own weight of the floor and the reinforcement application rate for it is only increased. Since the criterion for optimizing the structure was the dead weight of reinforced concrete structures, the assumptions made during the technical and economic analysis and modeling of the structure's operation only improve the situation for solid non-unweighted slabs.

The results of numerical experiments are shown in the form of graphs in Figure 2, 3 and 4.

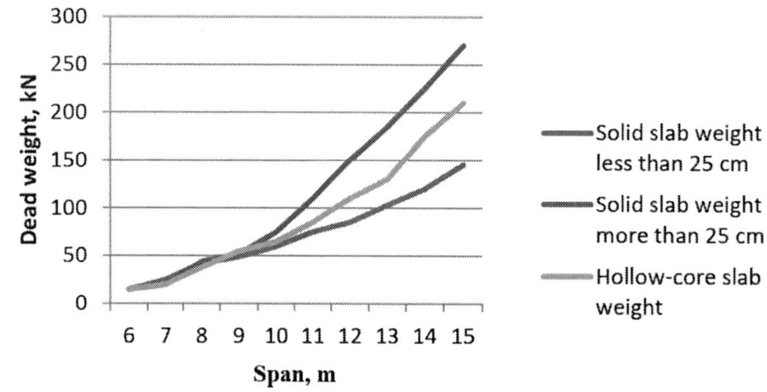

Figure 2. Graph of the floor slabs' dead weight depending on the span size

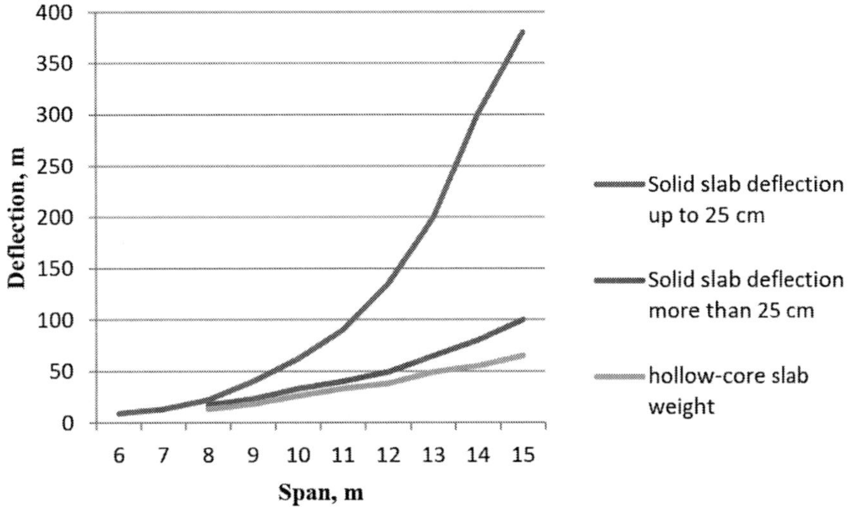

Figure 3. Graphs of the floor slabs' deflection depending on the span size

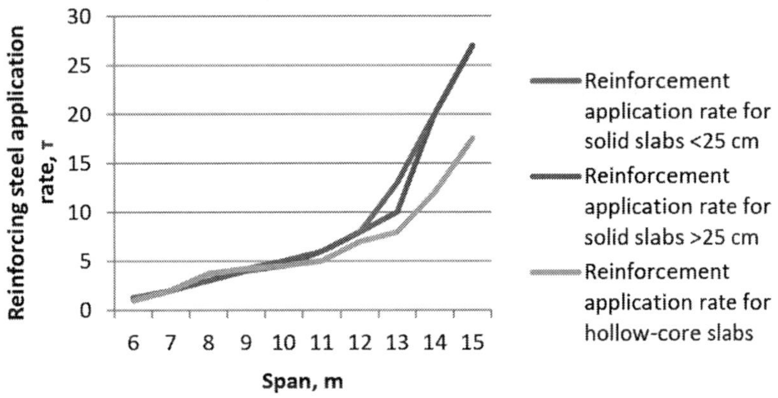

Figure 4. Graphs of the reinforcing steel application rate dependence on the span size

Comparison of calculated slab deflections with permissible values has been performed. The values of the ultimate deflections are taken according to the criterion of aesthetic and psychological requirements, as well as with the structural constraints associated with the presence of elements prone to the floors cracking. Excessive deflections in floor slabs can damage ceilings, finished floors and other elements. The requirements related to the possible location of partitions under the ceilings are not taken into account, since other design solutions are possible to prevent a decrease in the gap between the load-bearing structural elements and the partitions located under the elements. Information on the maximum permissible deflections is given in Table 2.

Table 2. Maximum deflections.

Span *L*, m	Maximum deflection (aesthetic and psychological limitations), mm	Maximum deflection (design constraints), mm
6	30	40
7	33	46.67
8	36	53.33
9	39	60

10	42	66.67
11	45	73.33
12	48	80
13	50.7	86.67
14	53.3	93.3
15	56	100

From the presented graphs it can be seen that with the spans of 10 meters or more, the dead weight of unweighted hollow-core slabs becomes greater than the weight of thin solid slabs up to 25 cm high, excluding capitals. However, when the span length of 10 meters is reached, the deflections of thin floor slabs up to 25 cm in height become greater than the limiting ones, therefore their use is unacceptable. With a span of more than 11 meters, the requirements for limiting the maximum deflections are no longer met by the solid slabs with a height h=L/30.

According to graph 4, the application rate of reinforcement for the hollow ceiling is less than for solid slabs with the same span. Thus, with a span of 10 meters or more, the use of unweighted floor slabs not only ensures the required rigidity, but also increases the economic efficiency of the structures used.

It can be noted that the hollow-core slabs with the cross-sectional characteristics adopted, with a span of 14 and 15 meters, also have deflections greater than the limiting ones. However, exceeding the permissible values is not as significant as for the solid slabs, a decrease in displacements can be achieved by increasing the section of columns and / or solid sections at the supports, or by slightly increasing the floor slab thickness.

With the spans up to 10 meters, the application rate of reinforcing steel for solid and hollow floor slabs is approximately the same. With an increase in the span, the application rate of reinforcing steel decreases. With a span of 13 meters or more, the difference in the accepted reinforcement diameters is more than one line of the standard assortment of reinforcing bars.

Summary

The maximum economic effect when using hollow-core floor slabs is achieved when several factors are combined:

1. Concrete application rate for a hollow floor slab is lower than the application rate of concrete for a continuous floor with the same spans. This efficiency criterion is achieved for the spans over 10 meters.
2. The application rate of reinforcing steel for a hollow ceiling is lower than the application rate for a solid floor with the same spans. This efficiency criterion is achieved for the spans over 13 meters.

Thus, it is rational to use the hollow-core monolithic slabs in the design of floors with spans of more than 10 meters, with an increase in the span, their efficiency increases.

The decrease in the cost of construction and operation of the facility from the use of unweighted floors due to their influence on other design solutions (reducing material application rate for vertical supporting structures and foundations, reducing the volume of the building, etc.) was not taken into account in the study.

The reduction in steel application rate due to the use of hollow core slabs is more than 35%.

In the general case, regardless of the applied design solutions, the required strength and stiffness of the floor slabs can be achieved by increasing the sections or reinforcing the elements. The key challenge is to design sustainable structures with the lowest cost.

The performed numerical experiment makes it possible to establish the effectiveness boundaries for the unweighted hollow-core monolithic floor slabs' use, when their use turns out to be significantly more profitable than the use of traditional girderless floors.

References

[1] Orlov D V 2010 Monolithic structures made of unweighted modified nano-concrete on the example of a five-story multi-level parking lot *Engineering and Construction Journal* **5 (15)** 12-15.
[2] Ovchinnikova S, Kalinichenko M, Markina N, Schneider E 2020 Energy modernization of housing stock *E3S Web of Conferences* **157, 06028**. doi.org/10.1051/e3sconf/202015706028
[3] Tsorayeva E N 2020 To the question of the economic efficiency of land use in the municipality *Omsk Scientific Bulletin. Society Series. History. Modernity* **5 (3)** 141-146.
[4] Tsorayeva E N, Baeva N A 2019 Problems of land legislation violation in the Krasnodar Territory *Bulletin of the Kyrgyz-Russian Slavic University* **19 (3)** 114-116.
[5] Tsorayeva E N, Ivanov A S, Gagarinova N V 2018 Land management as a mechanism for ensuring effective agricultural land use in the Krasnodar Territory *Bulletin of the Kursk State Agricultural Academy* **8** 256-261.
[6] Ovchinnikova S V, Plakhotnyuk E N 2019 Justification of the unweighted hollow-core floor slabs use *Actual problems of engineering sciences Materials of the VII-th (64) annual scientific-practical conference of teachers, students and young scientists of the North Caucasus Federal University "University science for the region"*, Stavropol, 222-228.
[7] Ovchinnikova S V, Bobkov D V 2019 Application of modified nano-concrete in construction *Actual problems of engineering sciences Materials of the VII (64) annual scientific-practical conference of teachers, students and young scientists of the North Caucasus Federal University "University science for the region"* Stavropol, 220-222.
[8] Dokucheva L M, Dolina E V, Yurkova R E, Stratinskaya E N, Shalashova O Yu 2011 Techniques excluding negative processes in the soils of irrigated agricultural landscapes of the chernozem zone of southern Russia *Scientific journal of the Russian Research Institute of Melioration Problems* **1 (1)** 4.
[9] Gagarinova N V, Tsorayeva E N, Bakumenko N S 2018 Problems of effective land management in Russia *Bulletin of the Adyghe State University. Episode 5: Economics* **3 (225)** 114-120.
[10] Sekisov A N, Varvarkina A V, Gura D A, Savenko A A, Shikhovtsov A A 2019 Pit design taking into account water level reduction using the software package midas gts nx *IIOAB Journal* **10 (52)** 93-100.
[11] Mikheev G V, Sekisov A N, Gura D A, Abazyan A G, Kuznetsova O A 2019 Economic and marketing adaptation of business processes in the modern russian real estate market *Revista Inclusiones* **6 (52-5)** 119-124.

IOP Conf. Series: Materials Science and Engineering 1083 (2021) 012046 doi:10.1088/1757-899X/1083/1/012046

Methodological foundations for the system analysis principles application for modeling the phenomena of heat transfer in the technological cleaning process of tanks for oil products.

S V Fedosov[1, 2], A G Azovtsev[3*], S A Syrbu[3]

[1] National Research Moscow State University of Civil Engineering, 26 Yaroslavskoe highway, Moscow, 129337, Russia

[2] Volga State Technological University, block 3 Lenin sq., Yoshkar-Ola, Republic of Mari El, 424000, Russia

[3] Ivanovo Fire and Rescue Academy of State Fire Service EMERCOM of Russia, 33, Stroiteley ave., Ivanovo, 153040, Russia

E-mail: asovtsev121@yandex.ru

Abstract. The article is devoted to the determination of the methodological foundations for the application of the system analysis principles for modeling the phenomena of heat transfer in the technological cleaning process of tanks for oil products. The article deals with the spontaneous combustion problem of pyrophoric deposits in oil and gas industry and a method for solving it using the principles of system analysis. The process of spontaneous combustion of pyrophoric deposits has several stages, therefore, decomposition of this process and further analysis of its structure are relevant. On the basis of a system analysis, the process stages of the pyrophoric deposits' spontaneous combustion have been determined and general and specific tasks have been formulated.

1. Introduction

Vertical steel tanks for storage of oil and oil products are potentially dangerous objects due to the circulation of flammable substances in them. In addition to the main causes of a fire, such as the presence of an open flame, the manifestation of atmospheric and static electricity, etc., this system has another potential hazard - spontaneous combustion of pyrophoric deposits - substances capable of spontaneous combustion in contact with atmospheric oxygen. Pyrophoric deposits consist of sulfides, polysulfides, oxides, iron hydroxides, elemental sulfur, its oxides, and heat formation occurs from the sulfides oxidation, the danger lies in the complex action of the components. In the article [1], the oxidation process of iron sulfides is divided into three stages: at the first stage, electrochemical reactions predominate, due to the presence of water. Presumptive reactions at this stage are [2, 3]:

$$FeS_2 + O_2 \rightarrow FeS + SO_2 + 222.5 \text{ kJ},$$

$$2FeS_2 + 7.5O_2 + H_2O \rightarrow Fe_2(SO_4)_3 + H_2SO_4 + 2777 \text{ kJ}.$$

Further, in the second stage, upon reaching 70-80°C an increase in temperature is reduced presumably due to an increase in the water evaporation intensity and a decrease in electrochemical reactions. The influence of water on this stage still needs to be checked. The third stage starts with temperature 100-110°C, where chemical reactions start prevailing. The first and third stages are characterized by a rather rapid rise in temperature, at the second stage even a slight decrease in temperature is observed.

Iron sulfides by themselves, when interacting with air, can heat up to a temperature 600-700 °C, however, it was shown in [4] that ignition occurs already at a temperature 200-210 °C, which corresponds to the self-ignition temperature of sulfur vapor. According to the article [5], spontaneous combustion of pyrophoric deposits was an ignition source in 12.8% of cases of fires on tanks. Most often, fires from spontaneous combustion of pyrophoric deposits occur during daylight hours, in sunny weather, in the presence of through holes in the tank, during the tank's long-term operation without cleaning, as well as when pumping oil [6]. Therefore, the consideration of heat transfer processes in the reservoir is an urgent task.

2. System analysis of the pyrophoric deposits' heating process

2.1. Building a hierarchical structure

When considering the problem of modeling the phenomena of heat transfer in the process of technological cleaning of reservoirs for oil products when heating pyrophoric deposits, it is necessary to use the system analysis methods. The system analysis strategy implies building a hierarchical structure of the process [7, 8] - breaking the problem into the levels, where the chemical reaction kinetics will be the lower level. The second level will be a chemical reaction in a layer of pyrophoric deposits, the third level will be heat transfer throughout the entire thickness from the air outside to the air inside the tank.

The whole process of heating a layer of pyrophoric deposits can be described by a generalized schematic-graphic model in the form of a structural diagram (Figure 1), where the input data will be:

- thickness of outer paint layer, steel tank wall, pyrophoric deposits (δ_P, δ_{ST}, δ_D respectively);
- heat capacity, density, thermal conductivity of the above layers (c, ρ, λ respectively);

perturbation:

- heat flux from solar radiation q_S;
- heat generated by oxidation of pyrophoric deposits q_D;

output data:

- temperature distribution $T(x, \tau)$ in time by thickness x.

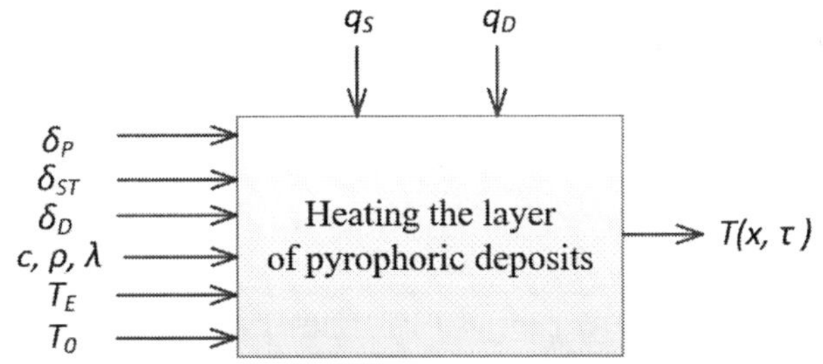

Figure 1. Block diagram of the object

At the heat transfer level, the problem of modeling heat transfer in the layer of pyrophoric deposits is formulated.

2.2. Physical model description

To formulate a mathematical problem, it is necessary to describe a physical model. As mentioned above, most often spontaneous combustion of pyrophoric deposits occurs in sunny weather, therefore, it is necessary to take into account the sun radiation effect in the model. The very wall of the reservoir with pyrophoric deposits can be considered as a one-dimensional three-layer endless plate (Figure 2), since the outer part of the tank, as a rule, is covered with a light paint for thermal insulation, and pyrophoric deposits are formed on the inside due to interaction with hydrogen sulfide. For simplicity, it is assumed that pyrophoric deposits have a homogeneous structure throughout the entire thickness.

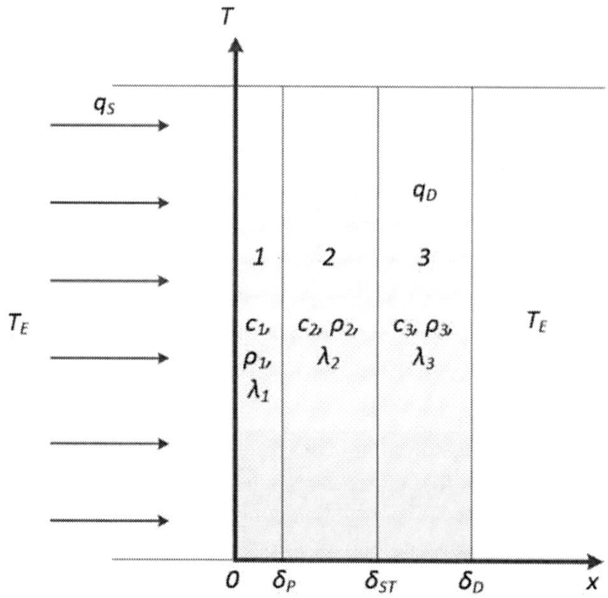

Figure 2. Illustration of a model for heating a tank wall by solar radiation and oxidation of pyrophoric deposits
c, ρ, λ – heat capacity, density, thermal conductivity, respectively,
1, 2, 3 – indices of paint layers, steel wall and pyrophoric deposits, respectively;
q_S – heat flux from solar radiation; q_D – heat generated by oxidation of pyrophoric deposits; T_E – environment temperature and the temperature inside the tank;
x – coordinate; T – temperature

2.3. A mathematical problem statement

Let us formulate a mathematical problem and define the boundary conditions. Heat propagation in the plate will be carried out according to the differential heat conduction equation:

$$\frac{\partial T_1(x,\tau)}{\partial \tau} = a_1 \frac{\partial^2 T_1(x,\tau)}{\partial x^2} \quad (\tau > 0; \quad 0 \le x \le \delta_P); \tag{1}$$

$$\frac{\partial T_2(x)}{\partial \tau} = a_2 \frac{\partial^2 T_2(x)}{\partial x^2} \quad (\tau > 0; \quad \delta_P \le x \le \delta_{ST}); \tag{2}$$

$$\frac{\partial T_3(x,\tau)}{\partial \tau} = a_3 \frac{\partial^2 T_3(x,\tau)}{\partial x^2} + \frac{q_D(T,x)}{c_3 \rho_3} + \frac{\varepsilon \rho_q}{C_q} \frac{\partial u(x,\tau)}{\partial \tau} \quad (\tau > 0; \quad \delta_{ST} \le x \le \delta_D); \tag{3}$$

where: a – is a thermal diffusivity;
ε – is a phase transition criterion;
ρ_q – denotes the specific heat of vaporization;
u – is a moisture content.
Initial conditions:

$$T_1(x,\tau)|_{\tau=0} = T_{10}(x), \tag{4}$$

$$T_2(x,\tau)|_{\tau=0} = T_{20}(x), \tag{5}$$

$$T_3(x,\tau)|_{\tau=0} = T_{30}(x). \tag{6}$$

Border conditions:
on the left border

$$q_S = \lambda_1 \frac{\partial T_1(x,\tau)}{\partial x} \quad (x=0), \tag{7}$$

on the right border

$$\alpha_3\left[T_E - T_3(x,\tau)\right] - (1-\varepsilon)\rho_q \rho_0 \left[u(x,\tau) - u_E\right] = \lambda_3 \frac{\partial T_3(x,\tau)}{\partial x} \quad (x=\delta_D); \tag{8}$$

where: ρ_0 – is medium density;
u – is transfer potential;
u_E – is a substance potential at the interface of pyrophoric deposits with the environment.
at the contact point between the layers 1 and 2

$$\lambda_1 \frac{\partial T_1(\delta_P,\tau)}{\partial x} = \lambda_2 \frac{\partial T_2(\delta_P,\tau)}{\partial x}, \tag{9}$$

$$T_1(\delta_P,\tau) = T_2(\delta_P,\tau); \tag{10}$$

at the contact point between the layers 2 and 3

$$\lambda_2 \frac{\partial T_2(\delta_{ST},\tau)}{\partial x} = \lambda_3 \frac{\partial T_3(\delta_{ST},\tau)}{\partial x}, \tag{11}$$

$$T_2(\delta_{ST},\tau) = T_3(\delta_{ST},\tau). \tag{12}$$

The formulated mathematical problem (1) - (12) is a complex system and it is necessary to use such a method of system analysis as decomposition, to break a complex system into the subsystems for its solution.

2.4. Decomposition of a mathematical problem as a system

The solution of the mathematical problem (1) - (12) will be carried out numerically and analytically after it is divided into the subsystems as follows: we will simultaneously consider the effect of the heat flow from the sun q_S and the heat generation from an internal source during the pyrophoric deposits' oxidation q_D.

The decomposition of the problem (1) - (12) for the subsequent solution on the heat flux effect from the sun on the left boundary is shown in Figure 3 (the conditions (13) - (16)), according to the generation of heat from an internal source during the pyrophoric deposits' oxidation in the layer in Figure 4 (the conditions (17) – (20)).

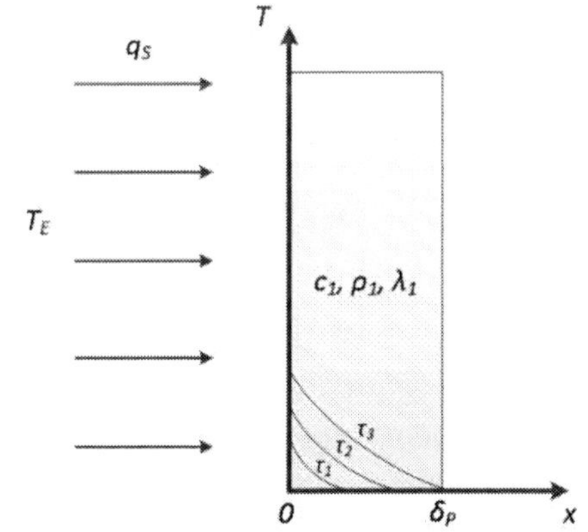

Figure 3. Illustration of a model of heating the outer paint layer of a tank by solar radiation and temperature profiles at different time values (at $\tau_i < \tau_{i+1}$)

$$\frac{\partial T_1(x)}{\partial \tau} = a_1 \frac{\partial^2 T_1(x)}{\partial x^2} \quad (\tau > 0; \quad 0 \le x \le \delta_P), \tag{13}$$

Initial conditions:

$$T_1(x,\tau)|_{\tau=0}=T_{10}(x). \tag{14}$$

Border conditions:
on the left border

$$q_S=\lambda_1\frac{\partial T_1(x,\tau)}{\partial x} \quad (x=0), \tag{15}$$

on the right border

$$\lambda_1\frac{\partial T_1(x,\tau)}{\partial x}=0 \quad (x=\delta_P); \tag{16}$$

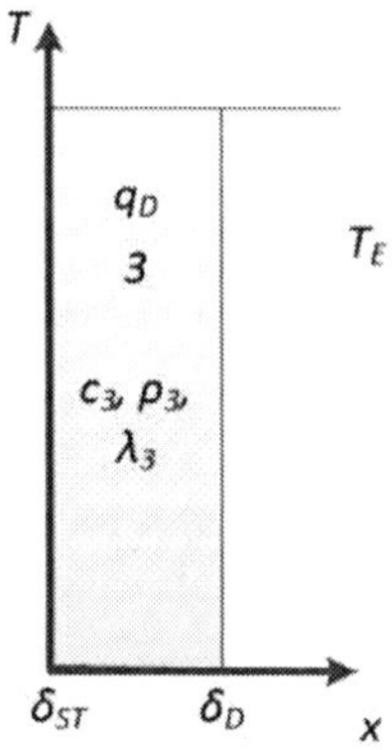

Figure 4. Illustration of a heat generation model in the pyrophoric deposits layer

$$\frac{\partial T_3(x)}{\partial\tau}=a_3\frac{\partial^2 T_3(x)}{\partial x^2} \quad (\tau>0;\quad \delta_{ST}\le x\le\delta_D), \tag{17}$$

Initial conditions:

$$T_1(x,\tau)|_{\tau=0}=T_{10}(x). \tag{18}$$

Border conditions:
on the left border

$$\lambda_3\frac{\partial T_3(x,\tau)}{\partial x}=0 \quad (x=\delta_{ST}); \tag{19}$$

on the right border

$$\alpha_3\left[T_E-T_3(x,\tau)\right]=\lambda_3\frac{\partial T_3(x,\tau)}{\partial x} \quad (x=\delta_D); \tag{20}$$

2.5. Methodology for solving the problem.

The subsystems obtained in Section 2.4 will be solved sequentially at small values of time until the temperature profile of the other layers' boundaries is reached. When the temperature profile reaches the boundary of another layer, a temperature gradient arises and a boundary condition of the second kind appears in the problem (which corresponds to the conditions (9), (10) of the general problem).

Let us consider the case when the temperature inside the tank is kept constant, and outside it rises due to heating by solar radiation. Then the temperature fields (Figure 5) will be characterized by the curves for τ_1, τ_2, τ_3.

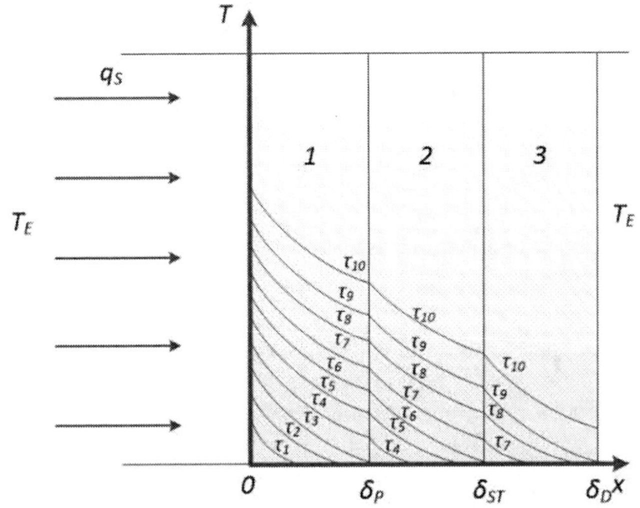

Figure 5. Illustration of temperature profiles at different time values τ

When $\tau_3 = \tau_1$* the temperature profile reaches the border δ_P. Prior to this, the temperature field varied along the curves for τ_1, τ_2. In this case, the problem was determined by the equation (13) with the initial condition (14) and the boundary conditions (15)-(16).

From the moment in time τ_3 the calculation system includes zone 2 - the steel wall of the tank. At a point with a coordinate $x = \delta_P$ the heat conduction equation will be replaced in the form of expression (2). The initial condition for the first step of calculating the temperature field is the equation (5). The boundary condition on the left boundary is the condition of heat transfer at the contact point of the layers between the layers 1 and 2 (9), taking into account the achieved temperature profile (21):

$$-\lambda_1 \frac{\partial T_1(\delta_P, \tau_3)}{\partial x} = q_1(\tau_3) = \lambda_2 \frac{\partial T_2(\delta_P, \tau_3)}{\partial x} \tag{21}$$

Boundary conditions on the right border (x = δ_{ST}):

$$-\lambda_2 \frac{\partial T_2(\delta_{ST}, \tau_3)}{\partial x} = 0 \tag{22}$$

Changing the temperature fields at a time step τ_4 will have the form of curves at τ_4, τ_5, τ_6.

From the moment in time τ_6 the layer 3 is additionally switched on - pyrophoric deposits and the calculation is carried out in the same way as for the layer 2.

When considering the case of the internal heat source q_D appearance in the layer 3, the calculation will be performed from the left and right boundaries. The temperature profile from the left border will look the same as when considering the case of heating the left border by solar radiation. As there will be an internal heat source in the layer 3, then heat will spread towards the left boundary.

The analytical solution of the system of equations (13) - (16) for a similar case is presented in the work [7] (23):

$$\Theta(\bar{x}, Fo) = \frac{T_1(x,\tau) - T_E}{T_E} = Ki\left\{(1-\bar{x}) + 2\sum_{n=1}^{\infty}\left[\frac{(-1)^n}{\mu_n^2}\sin\left[\mu_n\left(1-\bar{x}\right)\right]\exp\left(-\mu_n^2 Fo\right)\right]\right\} + \\ +2\sum_{n=1}^{\infty}\left[\cos\left[\mu_n \bar{x}\right]\exp\left(-\mu_n^2 Fo\right)\frac{(-1)^n}{\mu_n^2}\int_0^1 T_0(\xi)\cos(\mu_n \xi)\,d\xi\right], \tag{23}$$

where the following notation is introduced in a dimensionless form (24):

$$\Theta(\bar{x},Fo)=\frac{T_1(x,\tau)-T_E}{T_E};\quad Fo=\frac{a_1\tau}{\delta_P^2};\quad \bar{x}=\frac{x}{\delta_P};\quad Ki=\frac{q_S\delta_P}{\lambda_1 T_E} \tag{24}$$

μ_n – denotes the characteristic equation roots and can be defined as: $\mu_n = n\cdot\pi$.

Ki – is a Kirpichev criterion;

$\bar{x}$ – denotes the relative coordinate;

Fo – is a Fourier test;

$\Theta(\bar{x},Fo)$ – is dimensionless temperature as a function of relative coordinate and Fourier number.

Summary

The analysis results show that to determine the temperature distribution in the vertical tank wall thickness and pyrophoric deposits, it is necessary to split the general problem into the smaller ones.

As a result of the work performed, it is possible to assume:

- homogeneous layers, which will be used for the calculation have been defined;
- a method for calculating temperatures by a numerical-analytical method (by sequentially determining the temperature profiles along the layer thickness) has been proposed;
- the analytical solution for the sequential calculation of the temperature profile over the layers thickness has been determined. The resulting equation gives a possibility to obtain the temperature profiles (as shown in Figure 5) at the stage of heating three layers by the sun's rays. When an internal heat source occurs in the layer 3, the temperature profile will be different. The subject of further research is to obtain the temperature profiles taking into account the occurrence of an internal heat source in the pyrophoric deposits' layer.

References

[1] Dou Z, Jiang J C, Zhao S P, Mao G B, Zhang M G, Wang L, Wang Z R 2015 Experimental investigation on oxidation of sulfurized rust in oil tank *Journal of Loss Prevention in the Process Industries* **38** 156-162. DOI: https://doi.org/10.1016/j.jlp.2015.09.009.

[2] Beilin Yu A, Niselson L A, Begishev I R, Filimonov L I, Shishkanov B A, Ashcheulova I I, Podobaev A N, Reformatorskaya I I 2007 Corrosive pyrophoric deposits as promoters of spontaneous combustion of reservoirs with sulfurous oil *Journal of Metals Protection* **43 (3)** 290-295.

[3] Boyarov A N, Sumarchenkova I A 2009 Ensuring safety in the operation of oil and oil products storage tank farms *Youth and Science: Reality and the Future. Materials of the II Int. scientific-practical. conf. March 3, 2009 – Nevinnomyssk, VIII: Natural and applied sciences* 565-566.

[4] Boyarov A N, Karamyshev V G 2008 Investigation of the pyrophoric deposits behavior when heated under various conditions *STJ Interval* **9** 37-41.

[5] Petrova N V, Cheshko I D, Galishev M A 2016 Analysis of the practice of expert research of fires at oil and oil products storage facilities *Journal "Bulletin of the St. Petersburg University of the State Fire Service of the Ministry of Emergency Situations of Russia"* **3**. Information on http://vestnik.igps.ru/wp-content/uploads/V83/ 7.pdf

[6] Volkov O M 1984 *Fire safety of tanks with oil products* (Moscow, Nedra) 151.

[7] Kafarov V V, Dorokhov I N 1976 *System analysis of chemical technology processes. Basics of strategy* (Moscow, Nauka) 500.

[8] Natareev S V 2007 System Analysis and Mathematical Modeling of Chemical Technological Processes *Ivanovo state chemical-technological university: study guide Ivanovo* 80.

[9] Fedosov S V, Aloyan R M, Ibragimov A M, Gnedina L Yu, Aksakovskaya L N 2005 *Freezing of wet soils, basements and foundations* (Moscow, ASV Publishing House) 277.

IOP Conf. Series: Materials Science and Engineering 1083 (2021) 012047 doi:10.1088/1757-899X/1083/1/012047

"Scheme of the planning organization of the land plot" section development based on information modeling

M M Zheleznov, A O Rybakova*

Moscow State University of Civil Engineering, 26, Yaroslavskoe highway, Moscow, 129337, Russia

E-mail: ryibakovaao@mgsu.ru

Abstract. Improving the design efficiency of capital construction projects is one of the promising areas for the science and production development. Today, innovative methods exist and are being developed for rationalization at all stages of construction production. During the design phase, there are certain parts which work needs modernization. Section No. 2 of the project documentation "Scheme of the planning organization of the land plot" was not subjected to automated influence and its design has been carried out in an exclusively conservative way for a long period of time. With the development of design automation tools, including information modeling technologies, new prospects for work and overall rationalization have become opened up for developers. The aim of the work is to analyze the existing methods of computer-aided design for the section "Scheme of the planning organization of the land plot" and their prospects, and the task is to determine the methodology for information modeling, taking into account the existing tools, software systems and the requirements of the legislation in the Russian Federation. This article outlines the features of the automated development in the section, the relationship between the software functionality and the specifics of projects, as well as the prospects for the development and integration of information modeling technologies in the field of land plots organization and planning.

1. Introduction

Planning organization of a land plot (POLP) – is one of the sections of project documentation (PD) for capital construction objects, section No. 2 of project documentation in accordance with the Government Decree of the Russian Federation of February 16, 2008 No. 87 "On the composition of sections of project documentation (GD RF No. 87) and requirements for their content". POLP is required to obtain a positive conclusion of the state PD examination, a permit for the construction of an object, during reconstruction and additional development of a site, during land surveying and other changes in the project.

POLP development is carried out in accordance with the urban planning of the land plot (UPLP), the requirements of BC 42.13330.2016 "Urban planning. Planning and development of urban and rural settlements", BC 4.13130.2013 "Fire protection systems. Limiting the spread of fire at protected objects", BC 18.13330.2019 "Production facilities. Planning organization of the land plot", BC 59.13330.2016 "Accessibility of buildings and structures for people with limited mobility" and other regulatory documents, taking into account the specifics of the object.

The POLP section includes text and graphics. Each part consists of sections and subsections, the composition of which is determined by the RF Resolution No. 87 and the relevant regulatory documents. Over the past few years, the composition of the POLP section has expanded significantly, and the number of requirements has increased accordingly. This phenomenon is due to the regular amendments to the regulatory legal acts on the land relations regulation and changes in the organization of the Unified State Register of Expertise Conclusions of Design Documentation of Capital Construction Objects (USREC). (Annual report of USREC in Russia for 2019)

Therefore, the number of comments to the POLP section becomes more every year, which generally improves the quality of the section preparation, but at the expense of additional design organization. In most cases, adjustments based on POLP expert comments entail additional changes in other PD sections. As a result, it slows down the passage of the examination and obtaining a positive conclusion, which complicates the obtaining of a building permit and the process of the facility construction itself.

The issue of improving the quality of the POLP section at the design stage is one of the most relevant among designers. In addition to maximum compliance with the current requirements of regulatory documents, it is necessary to use modern computer-aided design tools. These tools can be implemented through information modeling technologies (BIM technologies). The main work of BIM technologies is initially aimed at the tasks of architecture, constructive and engineering support, however, with the right approach and the most effective use of tools, it is possible to develop a POLP section.

2. Methods

2.1 POLP section design according to Autodesk standard

Today, in most cases, specialists in development use the Autodesk AutoCad software package. However, AutoCad is only a tool for 2D drawings of the graphic part, without additional information content and a high level of visualization. If necessary, the specialized graphic editors for 2D and 3D modeling can be used. This conservative design method is the longest, most ineffective and without the possibility of prompt adjustments. The phased implementation of design tasks in this case corresponds to the composition of the POLP project.

The most modern and efficient design method is POLP based on the tools of the 2D and 3D Autodesk programs: AutoCad-Infraworks-ReCap-Civil3D-Navisworks. According to the BIM standard for programmatic interaction, Autodesk forms development POLP tasks with a big difference from the composition of the project, but in the end the result is obtained in full compliance with the RF PD № 87 (Table 1).

Table 1. List of POLP basic design works

No.	POLP Tasks	Result
1	Creation of space-planning solutions (if necessary, several options)	Space-planning solution
2	Consideration of options for the placement of sites, taking into account the earthworks volume	Selected site placement option
3	Development of several options for communication between the sites	Selected communication option between the sites
4	Uploading data for cost estimation	Assessment of capital costs for planning a land plot (PLP).
5*	The laser scan results processing: registration and cleaning of point clouds, cutting of point clouds into smaller parts for better control and use	Point clouds ready to use
6	Get engineering survey data and upload it to InfraWorks	Engineering survey data uploaded to InfraWorks

7	Designing the main part of the master plan, external water supply and sewerage networks, heating networks - BIM models' creating	Developed models of the master plan, external water supply and sewerage networks, heating networks
8	Creation of a BIM model of the site and on-site networks	BIM model of the site and intra-site networks

As a result of analyzing the Table, it is possible to see that this design method provides much more possibilities for design, in contrast to the standard 2D approach:

1. Ability to develop several design solutions and choose the best;
2. Ability to export and import data, including those from devices and gadgets;
3. The most accurate application of the engineering surveys' results;
4. Accelerated drafting of a master plan of networks with the ability to search for collisions;
5. Simultaneous development of the basic BIM model and the POLP model.

Along with the advantages of this approach, there are a number of disadvantages that complicate the work: a large amount of software and differences in data exchange formats. The set of software products involved may not always be at the disposal of the design team. Not every project company can purchase several diverse products, train employees or find specialists who own a large number of programs.

Despite the information exchange formats versatility, data conflicts arise at the time of exporting models. Due to the high level of Autodesk software development, format collisions occur less frequently and with the possibility of correction, however, the time spent on solving the problem may not be replenished in the future. Scheme of software interaction when designing a POLP section according to the design stage is shown in Figure 1.

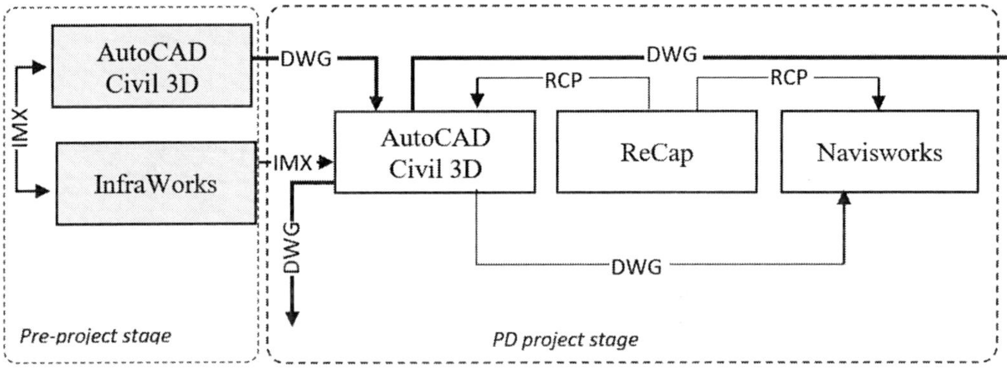

Figure 1. Interaction of programs by section POLP

The presented development method POLP the most efficient and commensurate with labor costs for a large-scale project, where the customer is interested in visual presentation and development of options in a short period of time. Taking into account the specifics and dimensions of the object, it is not always necessary and possible to design according to the above-mentioned scenario. In the case of overestimating the goals of the project, the advantages of this method may turn out to be disadvantageous.

2.2 POLP section design based on the software package GeoniCS

GeoniCS is one of the developing software systems in the field of horizontal design. GeoniCS is designed to automate the engineering surveys, design topographic plans, general plans, external engineering networks and roads, as well as to create a digital elevation model. GeoniCS is a Russian development, the complex is initially focused on the regulatory requirements of the Russian Federation in the field of POLP and documentation.

GeoniCS works on the AutoCAD platform, was originally developed for the surveyors, but now its functionality is regularly expanded for the work of related specialists in the form of separate modules.

The software complex consists of six modules, each of which gives an opportunity to solve a range of tasks in terms of engineering surveys, POLP section, construction organization project and master plan. In rare cases, the working model in the program can be used for rendering. The functionality of each module is presented in Table 2.

Table 2. List of basic POLP design works

№	Module GeoniCS	Function	POLP
1	Topoplane and Terplan	Creation of topographic plans, maps and relief based on archival materials, geodetic marks, data from GIS systems.	In terms of the relief organization diagram and the situation plan
2	Master plan	The module consists of groups of tools responsible for performing the tasks of designing the master plans: horizontal and vertical planning, landscaping.	In terms of the area improvement scheme
3	Networks	Tracing engineering communications, placing and combining networks with the ability to automatically build profiles and monitor compliance according to the standards. Automated documentation.	In the part of the engineering networks' master plan
4	Circuits	Design of linear objects using tools for editing geometric elements, plan, horizontal axes. Automated building of profiles and drawings, their editing and export.	In terms of the area improvement scheme
5	Sections	Automated construction of integrated profiles from the Topoplan and Trails modules, roads and volumes of earth masses movement.	In the part of the earth masses plan
6	Geomodel	Preparation of a 3D model complex, automated construction of all types of drawings, adaptation of the model for export, visualization.	In accordance with the reference terms

Along with the specialized functions of the GeoniCS modules, there are fundamental and general tools that help solving the following tasks:

1. Building a 3D model of a full-fledged scenario within the site;
2. Tracing and construction of on-site and access roads in three-dimensional format (3D model of the road);
3. Using and updating the symbol library;
4. Generation of drawings and preparation of documentation in accordance with Russian legislation.
5. Minimal visualization

Based on the maximum adaptation to the design features in our country, there is no need to create and configure templates, create additional libraries and administration. Model Detail Level POLP in each module may be different depending on the purpose and specifics of the object, the requirements of the technical task and the time for the section development. 3D model can be imported from other software systems for correction or revision, and also exported as initial data or assignments for other specialists. (Figure 2)

Figure 2. GeoniCS Situational Plan Model

3. Results

As a result of considering the design automation POLP section, the following general advantages can be highlighted: development time reduction due to the use of software tools, a high level of visualization due to graphic capabilities, processes multitasking and delegation, design solutions' variability, integration with other software systems, effective use of geodetic and topographic information. Despite the presence of local flaws in development methods, in most cases the advantages of software systems dominate. Depending on the purpose of the project, each tool can be used to varying degrees.

Designing based on the Autodesk standard using a suite of programs is most suitable for the large-scale projects for which visual component, cross-platform and flexibility are important. At the same time, there are all the necessary resources for the highest possible level of implementation: temporary, software and hardware, intellectual and financial.

Working in the GeoniCS software package does not provide a high level of functionality compared to the Autodesk bundle, however, it has a large number of high-quality tools that cover a wide range of works and significantly increase the POLP design efficiency. GeoniCS interface is made on the basis of simple algorithms, which makes learning and mastering the program easier. Working on the Autodesk AutoCAD platform provides fast integration and accelerated work.

4. Summary

Thus, all the design automation tools of the POLP section rationalize development, thereby providing a positive effect not only within the section, but also in the project as a whole. The ability to work out the several options at the same time and promptly make adjustments without routine actions ensure a high level of project development at the initial stages, as well as an accelerated examination of project documentation and obtaining a positive conclusion. Total efficiency from POLP development automation has a positive impact on the entire life cycle of the project. Research in information modeling and land management is shaping new directions for the construction industry.

Acknowledgements
The reported study was funded by RFBR, project number 20-37-90064.

References

[1] Ginzburg A 2017 *Bulletin of construction technique* **12 (1000)** 33-35.
[2] Information on https: www.gge.ru/analytics/analytics_and_statistics/
[3] Lua Y, Wub Z, Changa R and Lib Y 2017 *Automation in Construction* **83** 134-148.
[4] Ilhan B and Yaman H 2016 *Automation in Construction* **70** 26–37.

[5] Makisha E 2019 *Scientific journal «Modeling, Optimization and Information Technology»* 7(2) 339-350.
[6] Information on https://www.docs.cntd.ru/
[7] Garyaev N and Rybakova A 2018 *MATEC Web of Conferences* **170** 1076.
[8] Rybakova A and Kagan P 2019 *E3S Web of Conferences* **97** 01008.
[9] Mohamed M, Ahmed O 2020 *Sustainable Cities and Society* **57** 102120.
[10] Behnam A, Mohsen K, Abbas R, Serene H 2017 *Computers, Environment and Urban Systems* **61** Part A 24-38.
[11] Serene H, Abbas R 2016 *Land Use Policy* **57** 1-10.

The aggregation method of inhibiting factors in construction

A O Alekseev[1], V S Gladkikh[1], U V Aftahova [1], A G Aftahov[1], O V Takhumova[2*]

[1]Perm National Research Polytechnic University, 29, Komsomolsky Ave., Perm, 614990, Russia
[2] Kuban State Agrarian University named after I.T. Trubilin, 13, Kalinina Street, Krasnodar, 350044, Russia

E-mail: takhumova@rambler.ru

Abstract. The article presents a new methodology for the complex assessment of factors that restrict working conditions, which is based on matrix mechanisms of complex assessment, allowing to take into account heterogeneous complexity factors. As criteria for determining the cramped working conditions, the following are used: the mass and dimensions of the inventory, their number and the complexity of work to prepare the workplace for construction and installation work. The presented approach is able to take into account any factors' combination, which makes it possible to abandon the existing fixed number of coefficients. Illustrated examples are shown and that clearly demonstrates the methodology for assessing the complexity of construction and installation work. The proposed method can be used in the management of construction projects, to adjust the cost of work, etc. The screen forms of the software module created specifically to provide a comprehensive assessment of complexity factors are demonstrated.

1. Introduction

Construction and installation work can be performed under various conditions, including the non-standard ones, in which the work is carried out at a slower pace, or the complexity of the work increases, this may be due to the need to prepare the workplace, for example, due to poor lighting, clutter of the room with equipment or furniture, etc. Such conditions are usually called constrained, and the factors leading to delays in work are the factors that constrain working conditions.

For example, according to J. Leea, in the construction of multi-storey and low-rise buildings in Taiwan, the builders are faced with the problem of the wind influence of the upper floors, that is much higher than on the lower ones. This leads to an increase in construction time and the accidents' likelihood. In addition, in view of the growing number of atypical high-rise buildings, it is difficult to quickly assemble the traditional existing form of shuttering, additional work appears due to the changes that occur as the number of floors increases, thereby, again, delaying the construction period and increasing its cost [1]. According to O. Oyeyipo, the construction time and, as a result, its cost is influenced by the number of stakeholders. The more complex a project is, the more it needs to be coordinated with a large number of stakeholders. This means that stakeholders expose construction projects to varying degrees of risk and uncertainty during the construction process [2]. M. Adamu believes that if the principles of "total cost, efficiency and schedule" are followed during the construction of the project, then the implementation of the project can be carried out with great risks. That cannot be said about the inclusion of the lean production principles in the construction process, thereby it is possible to achieve a ratio of price and quality, i.e., to take into account all the factors that may affect the construction course before the project is finally approved for work [3]. The workers of a culvert in central Feldafing faced the constrained working conditions when they were forced to repair it. On the one hand, public transport should have an unrestricted passage over this structure, on the other hand, from the point of view of water management, the cross section of the canal can only be slightly reduced. In addition, private land passes through the canal, so the owners should experience as few restrictions as possible. All these factors undoubtedly affected the cost and time required for the repair [4]. S. Helmstädter believes that local conditions, environmental requirements, technical and economic aspects, changes in logistics and regional experience should be taken into account when choosing the construction methods and their options. This will undoubtedly leave its mark on the cost of construction and its timing [5]. The authors agree with the opinions of experts that when planning the construction work, it is necessary to take into account local conditions, including regional programs for the sustainable development, which is confirmed by the studies in the field of justification of directions for increasing the environmental sustainability of construction production [6, 7].

It is important to note that at present, both in Russia and in other countries, as a rule, in the regulatory and technical documents and standards, a fixed list of conditions restricting working conditions is defined, but in practice, the work can be performed with a simultaneous combination of several complexity factors. Therefore, it is required to develop a universal methodology for the comprehensive assessment of factors constraining working conditions. This determined the goal and objectives of this study.

2. Methods

To take into account modern complexity factors, it is proposed to use matrix mechanisms of complex assessment (hereinafter – MMCA). MMCA advantage among other mechanisms is that it is possible to evaluate and compare with each other heterogeneous criteria, by transferring them from the phase space to qualimetric (criterial), and it is also possible to take into account the nonlinear influence of factors. With regard to the problem under consideration, it is proposed to take into account the following factors: the mass and dimensions of the equipment required to prepare the workplace; its quantity, as well as the complexity of the preparatory work.

For a comprehensive assessment of heterogeneous criteria, it is necessary to draw up a graph in the form of a binary tree, which will show the sequence of particular criteria aggregation. So, it is proposed to first roll up such criteria as the mass and dimensions of the inventory, as the most common. Then the generalized result is rolled up with a criterion corresponding to the amount of inventory used, as a result of which a generalized assessment is formed - the complexity of the equipment operation, which in turn is reduced with the laboriousness of preparatory work and as a result, a single comprehensive assessment is formed, reflecting the complexity level of preparing a workplace for construction installation work (Figure 1).

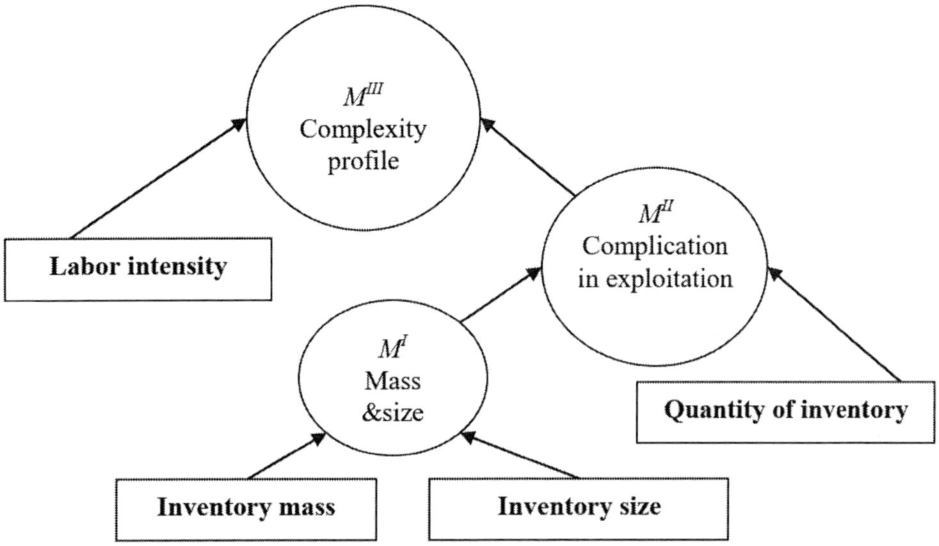

Figure 1. Criteria tree for complex assessment of factors constraining working conditions

The nodes of the criteria tree contain the convolution matrices (Fig. 2), reflecting the rules for aggregating the particular criteria. It should be noted that when developing MMCA, as a rule, a single descriptive rank scale from 1 to 4 is used, where 1 is an assessment reflecting the worst qualitative state of the criterion, and 4 is the best. In this study, an inverse scale is adopted, but with a similar number of gradations (Table 1). Therefore, the convolution matrices will be 4 × 4. Each convolution matrix is a set of compound inference rules "If, then", and takes into account the influence of each criterion, depending on the priorities set in the matrix. In the works devoted to the MMCA study [8,9] such convention is accepted that the convolution matrices' origin is located in the lower right corner, the left criterion corresponds to the rows, and the right criterion corresponds to the columns. In this study, we will take the same matrices arrangement.

Table 1. Descriptive scales of factors constraining working conditions and their criterion presentation in a rank scale from 1 to 4

№	**Criterion value (point)**	**Qualitative interpretation grade**	**Weight, [kg]**	**Dimensions, [m^3]**	**Quantity, [set.]**	**Labor intensity, [man-hours]**
1	1	The factor that complicates the work takes place	Low	Small	Small	Low
2	2	Moderately complicating factor	Average	Average	Average	Average
3	3	Factor that significantly (significantly) complicates the work	High	Large	Large	High
4	4	Critically Complicating Factor	Very high	Very big	Very big	Very high

Note: specific values for the criteria are determined for the specific complexity conditions. A detailed example is presented in Table. 2

Composite inference rules are formed in the form of causal statements, for example, if the mass is "high" (k_m=3), and the dimensions are "average" (k_s=2), then the generalized factor "Weight and dimensions" is one of the factors that averagely complicates the work ($k_{m\&s}$ =2). For the specific conditions of complexity, the set of rules can also be adjusted.

Figure 2 shows the criteria tree with convolution matrices describes 256 possible combinations of factors constraining working conditions.

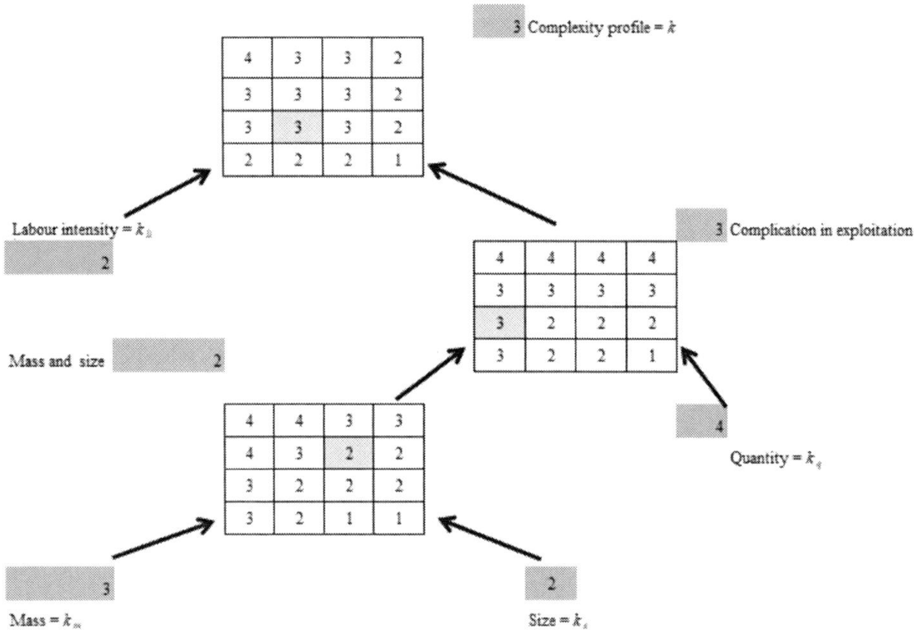

Figure 2. An example of complex assessment of complexity factors in Microsoft Excel®

Complex assessment is carried out sequentially, folding a pair of criteria in each convolution matrix, the resulting value of the convolution matrix is used for the matrix convolution of the next level until we reach the only complex indicator interpreted as the level of construction and installation work complexity. So, in Figure 2, the elements of the matrix are highlighted in yellow, corresponding to the values of the criteria rolled up in the matrix. For example, inventory weight is high (k_m=3), and the dimensions are average (k_g=2), then the third row and the second column are selected in the matrix, respectively, relative to the lower right corner. In the given example, the value 2 corresponds to the matrix element in the third row and column 2, which means that row 2 will be selected on the matrix of the next level. Let the amount of inventory be "very large" (k_q=4), then in the convolution matrix of weight and size with quantity, 2 rows and 4 columns are selected, which in the considered example corresponds to grade 3. Since this generalized grade is on the right according to the criteria tree, then on the top-level matrix, column 3 will be selected. In the demonstrated example (see Fig. 2), the complexity in the criterion space is 2 (k_{li}=2), then the row 2 should be selected. In the final matrix, at the intersection of 2 rows and 3 columns, the matrix element has an estimate of 3 (k=3). Thus, the complexity level, taking into account the considered factors in the complex, is defined as significantly complicating the work.

The works [8, 9] show the approaches to continuous and fuzzy complex assessment, using which it is possible to carry out a comprehensive assessment of any combination of factors that constrain the working conditions. Thus, the use of continuous mechanisms of complex assessment will allow taking into account the combination of any constraining factors, which ultimately makes it possible to adequately assess the real complexity conditions.

For continuous complex estimation, it is proposed to use the additive-multiplicative approach [8, 9], which is written using the expressions (1) – (7):

$$k^i = j_3^{M^i} + \gamma_{k_l}^{M^i} \cdot \left(j_4^{M^i} - j_3^{M^i}\right) + \gamma_{k_r}^{M^i} \cdot \left(j_5^{M^i} - j_3^{M^i}\right) + \gamma_{k_l}^{M^i} \cdot \gamma_{k_r}^{M^i} \cdot \left(j_6^{M^i} + j_3^{M^i} - j_4^{M^i} - j_5^{M^i}\right), \quad (1)$$

$$\gamma_{k_l}^{M^i} = \left\lceil k_{M_l^i} \right\rceil = mod(k_{M_l^i}, 1), \quad (2)$$

$$\gamma_{k_r}^{M^i} = \left\lceil k_{M_r^i} \right\rceil = mod(k_{M_r^i}, 1), \quad (3)$$

$$j_3^{M^i} = m_{rc} \in M^i \,|\, r = \left\lfloor k_{M_l^i} \right\rfloor ; c = \left\lfloor k_{M_r^i} \right\rfloor, \quad (4)$$

$$j_4^{M^i} = m_{rc} \in M^i \,|\, r = \min(\lfloor k_l + 1 \rfloor; 4); c = \left\lfloor k_{M_r^i} \right\rfloor, \quad (5)$$

$$j_5^{M^i} = m_{rc} \in M^i \,|\, r = \left\lfloor k_{M_l^i} \right\rfloor ;\ c = \min(\left\lfloor k_{M_r^i} + 1 \right\rfloor ; 4), \quad (6)$$

$$j_6^{M^i} = m_{rc} \in M^i \,|\, r = \min\left(\left\lfloor k_{M_l^i} + 1 \right\rfloor ; 4\right); c = \min(\left\lfloor k_{M_r^i} + 1 \right\rfloor ; 4), \quad (7)$$

where M^i – is a convolution matrix i-level of the criteria tree (see Fig.1), k_l – is the left test grade, k_r – is an assessment of the right criterion.

To find the value of a comprehensive assessment of the constrained working conditions at the 1st level, the following formula should be used:

$$k^I(k_m; k_s) = j_3^{M^I} + \gamma(k_m) \cdot \left(j_4^{M^I} - j_3^{M^I}\right) +$$
$$+\gamma(k_s) \cdot \left(j_5^{M^I} - j_3^{M^I}\right) + \gamma(k_m) \cdot \gamma(k_s) \cdot \left(j_6^{M^I} + j_3^{M^I} - j_4^{M^I} - j_5^{M^I}\right). \quad (8)$$

The value of the comprehensive assessment at the 2nd level will be determined by the following formula:

$$k^{II}\left(k^I(k_m; k_s); k_q\right) = j_3^{M^{II}} + \gamma\left(k^I(k_m; k_s)\right) \cdot \left(j_4^{M^{II}} - j_3^{M^{II}}\right) + \gamma(k_q) \cdot \left(j_5^{M^{II}} - j_3^{M^{II}}\right) +$$
$$\gamma\left(k^I(k_m; k_s)\right) \cdot \gamma(k_q) \cdot \left(j_6^{M^{II}} + j_3^{M^{II}} - j_4^{M^{II}} - j_5^{M^{II}}\right). \quad (9)$$

The calculation of the final comprehensive assessment at the 3rd level of the tree is determined by the formula:

$$k = k^{III}\left(k_{li}; k^{II}\left(k^I(k_m; k_s); k_q\right)\right) =$$
$$= j_3^{M^{III}} + \gamma(k_{li}) \cdot \left(j_4^{M^{III}} - j_3^{M^{III}}\right) + \gamma\left(k^{II}\left(k^I(k_m; k_s); k_q\right)\right) \cdot \left(j_5^{M^{III}} - j_3^{M^{III}}\right) +$$
$$+\gamma(k_{li}) \cdot \gamma\left(k^{II}\left(k^I(k_m; k_s); k_q\right)\right) \cdot \left(j_6^{M^{III}} + j_3^{M^{III}} - j_4^{M^{III}} - j_5^{M^{III}}\right), \quad (10)$$

Substituting the expression (8) in (9), and it, in turn, in (10), we obtain the desired equation, which determines the complexity level of construction and installation work:

$$k = k^{III}\left(k_{li}; k^{II}(k^I(k_m; k_s);\ k_q\right) = j_3^{M^I} + \gamma(k_{li}) \cdot \left(j_4^{M^{III}} - j_3^{M^{III}}\right) +$$
$$+\gamma(j_3^{M^{II}} + \gamma(j_3^{M^I} + \gamma(k_m) \cdot \left(j_4^{M^I} - j_3^{M^I}\right) + \gamma(k_s) \cdot \left(j_5^{M^I} - j_3^{M^I}\right) +$$
$$+\gamma(k_m) \cdot \gamma(k_s) \cdot \left(j_6^{M^I} + j_3^{M^I} - j_4^{M^I} - j_5^{M^I}\right)) \cdot \left(j_4^{M^{II}} - j_3^{M^{II}}\right) + \gamma(k_q) \cdot \left(j_5^{M^{II}} - j_3^{M^{II}}\right) +$$
$$+\gamma(j_3^{M^I} + \gamma(k_m) \cdot \left(j_4^{M^I} - j_3^{M^I}\right) + \gamma(k_s) \cdot \left(j_5^{M^I} - j_3^{M^I}\right) +$$
$$+\gamma(k_m) \cdot \gamma(k_s) \cdot \left(j_6^{M^I} + j_3^{M^I} - j_4^{M^I} - j_5^{M^I}\right)) \cdot \gamma(k_q) \cdot \left(j_6^{M^{II}} + j_3^{M^{II}} - j_4^{M^{II}} - j_5^{M^{II}}\right)) \cdot \left(j_5^{M^{III}} - j_3^{M^{III}}\right)$$
$$+ \gamma(k_{li}) \cdot \gamma(j_3^{M^{II}} + \gamma(j_3^{M^I} + \gamma(k_m) \cdot \left(j_4^{M^I} - j_3^{M^I}\right) +$$
$$+\gamma(k_g) \cdot \left(j_5^{M^I} - j_3^{M^I}\right) + \gamma(k_m) \cdot \gamma(k_g) \cdot \left(j_6^{M^I} + j_3^{M^I} - j_4^{M^I} - j_5^{M^I}\right)) \cdot \left(j_4^{M^{II}} - j_3^{M^{II}}\right) +$$

$$+\gamma(k_q)\cdot\left(j_5^{M^{II}}-j_3^{M^{II}}\right)+\gamma(j_3^{M^{I}}+\gamma(k_m)\cdot\left(j_4^{M^{I}}-j_3^{M^{I}}\right)+\gamma(k_s)\cdot\left(j_5^{M^{I}}-j_3^{M^{I}}\right)+$$
$$+\gamma(k_m)\cdot\gamma(k_s)\cdot\left(j_6^{M^{I}}+j_3^{M^{I}}-j_4^{M^{I}}-j_5^{M^{I}}\right))\cdot\gamma(k_q)\cdot\left(j_6^{M^{II}}+j_3^{M^{II}}-j_4^{M^{II}}-j_5^{M^{II}}\right))\cdot$$
$$\left(j_6^{M^{III}}+j_3^{M^{III}}-j_4^{M^{III}}-j_5^{M^{III}}\right) \quad (11)$$

For quantitatively measurable indicators, it is necessary to determine the reduction functions, which in the simplest case are linear. Casting functions are usually divided into monotonically increasing and monotonically decreasing. In this study, all the considered criteria (see Table 1) with an increase in their values lead to an increase in the complexity of working conditions, so the reduction functions will be monotonically increasing:

$$k_i= 3(x_i–x_{i\min})/(x_{i\max}–x_{i\min})+1 \quad (12)$$

where x_i –is the actual value of the difficulty factor in natural units, $x_{i\min}$–is the value of the complexity factor, below which the conditions are considered uncomplicated, $x_{i\max}$–is the maximum permissive value of the complexity factor, k_i–is the reduced value of the complexity factor to a scale of 1-4 (complexity criterion).

3. Results

Let us consider an example of continuous complex assessment of the working conditions complexity, using the example of low illumination of the workplace. In the case of low light, in preparation for work, the equipment is required, shown in the table below (Table 2):

Table 2. Criteria and their parameters for a comprehensive assessment of the complexity factor "Weak illumination of the workplace"

№	Criterion name	Incoming items	calculation formula	Total
1	**Inventory weight**	Cable	0.315 kg * 3m	0.945 [kg]
		Staples	0.005 kg * 24 pc.	0.12 [kg]
		Cord-grip lampholder	0.05 kg * 1 pc.	0.05 [kg]
		Lamp	0.04 kg * 1 pc.	0.04 [kg]
		Lamp	3 kg * 1 pc.	3 [kg]
	Total			4.155 [kg]
	MIN	4.155 [kg]		
	MAX	33.24 [kg]		
2	**Inventory dimensions**	1 box	0.42*0.33*0.27	0.038 [m³]
	Total			0.038 [m³]
	MIN	0.038 м3		
	MAX	0.15 м3		
3	**The amount of equipment required to prepare the workplace**	Cable	1 [set]	1 [set]
		Staples	1 [set]	1 [set]
		Cord-grip lampholder	1 [pc.]	1 [pc.]
		Lamp	1 [pc.]	1 [pc.]
		Lamp	1 [pc.]	1 [pc.]
	Total			5 [pc.]

	MIN	from 1 to 2 sets		
	MAX	over 7 sets		
4	**Labor intensity of preparation of the workplace**	Uniform rates and prices 23.1-3-1-02-G	0.16 person-hour for 1 m*5 pc.	0.8
		Uniform rates and prices 23.1-3-2-01-B	0.125 person-hour for 1 m*5 pc.	0.625
		Uniform rates and prices 23.1-17-1-01	0.14 person-hour for 1 pc*5 pc.	0.7
		Uniform rates and prices 23.1-17-2-01	0.03 person-hour for 1 lamp *5 pc.	0.15
		Uniform rates and prices 23.1-17-2-03	0.03 person-hour for 1 lighting fixture*5 pc.	0.15
	Total			2.4 [person-hour]
	MIN	2.4 [person-hour]		
	MAX	8 [person-hour]		

As it can be seen from the Table. 2, each factor is determined by the minimum and maximum permissible values in natural units. Substituting the actual values into the reduction function (12) for each criterion, respectively, we obtain the values on a single scale from 1 to 4 (in the criterion space):

1.1) Inventory weight - actual inventory weight 6.39 (used 3m cable weighing 0.945 kg, LSP 42-2x40-001 lamp weighing 5 kg, 2 36W lamps 82500 TL-D STANDARD 36W / 33-640 T8 G13 2850Lm 4000K weighing 0.152 kg each (in view of the peculiarities of the structure of the lamp, an additional cartridge is not required), staples in the amount of 28 pieces weighing 0.141 kg) with a minimum value of 4.16 kg and a maximum value of 33.24 kg will have an estimate in the criterion space of 1.23:

$$k_m = 3(6.39–4.16)/(33.24–4.16)+1 = 1.230$$

1.2) Inventory dimensions - the actual dimensions of the inventory 0.1365 m3 (for the previously listed type of inventory, a box of 1.3 * 0.7 * 0.15 was required) with a minimum value of 0.038 m^3 and a maximum value of 0.15 m^3 will be assessed in the criterion space 3.638

$$k_g = 3(0.1365–0.038)/(0.15–0.038)+1 = 3.638$$

1.3) The number of required inventory for the workplace preparation –the actual number of required inventory is 2 pieces, with a minimum value - 1 piece and the maximum - 8 pieces will have an estimate in the criterion space 1.429

$$k_q = 3(2–1)/(8–1)+1 = 1.429$$

1.4) Labor intensity of the workplace preparation - the actual time for laying out the equipment is 3.2 man-hours, with a minimum value of 2.4 people-hours, and a maximum value of 8 man-hours will have an estimate in the criterion space 1.429

$$k_{li} = 3(3.2–2.4)/(8–2.4)+1 = 1.429$$

Substituting the found values into formula (11), we obtain a complex estimate describing the level of complexity (k=1.429). The figure below shows the intermediate and final results of continuous complex estimation of complexity conditions:

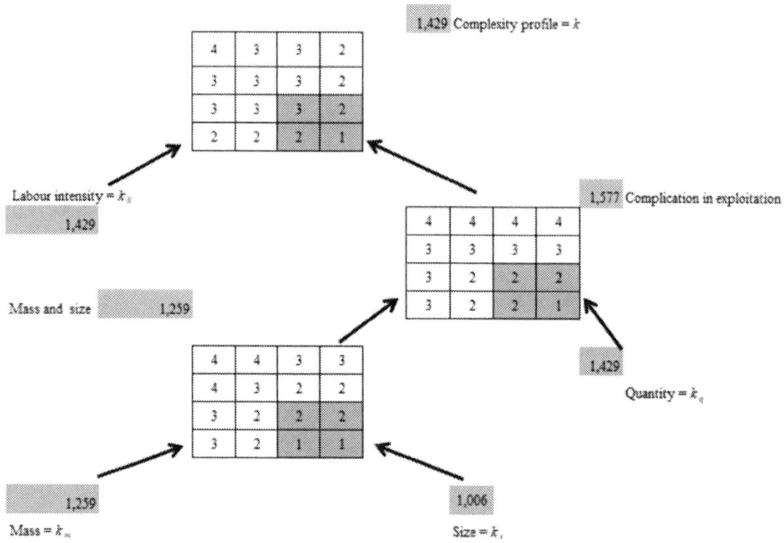

Figure 3. An example of the continuous complex assessment of complexity factors on the example of "Low illumination of the workplace" in Microsoft Excel®

Note: the elements corresponding to the parameters are grayed out on the convolution matrices j_3, j_4, j_5 and j_6

4. Discussion

The methodology for assessing the construction and installation works complexity proposed in this work can be used in the management of a construction project, as well as for adjusting the cost of construction and installation works if there are such facts in the construction organization project.

The existing groups of cramped working conditions and the corresponding correction factors are shown in Table 3:

Table 3. existing groups of cramped working conditions

№	Group of constrained working conditions	Range of coefficient values
1	Installation work in mountainous areas	from 1.25 till 1.5
2	Harmful working conditions (t air, MPC content in the air)	from 1.25 till 2.3
3	Installation work in underground conditions	from 1.68 till 3.0
4	Average working conditions	from 1.1 till 1.35

In this study, the average statistical working conditions, for which the range of values of the correction coefficient for cramped working conditions is 1.10–1.35 are considered. At the same time, the proposed methodology can be adapted to specific conditions, such as performing work in mountainous areas, in underground conditions and under harmful working conditions (see Table 3).

To convert the resulting value of the difficulty level from a scale of 1-4 to the range of values of the correction coefficient, the following expression [10] should be used:

$$K_j = \left(\frac{k_j - k_1}{k_2 - k_1}\right) \cdot (K_2 - K_1) + K_1 \quad (13)$$

where: K_j – is the required complexity factor for j- type of work;
k_j – is the comprehensive assessment reflecting the level of complexity for j- type of work;
K_1 and K_2 – are the minimum and maximum values of the difficulty coefficient for j- type of work, planned for use (in this study 1.10 and 1.35, respectively);
k_1 and k_2 – are the minimum and maximum values of the complex complexity estimate for j- type of work (in this study 1 and 4, respectively);

So, for the considered example, the following coefficient is obtained:

$$K_j = (1.35–1.10)\,(1.814–1)/(4–1)+1.10 = 1.168$$

Despite the fact that such calculations can be performed in already existing software products of the Decon family (Business-Decon [11], web-Business-Decon [12]), the authors have created an analogue of the software module based on Microsoft Excel® spreadsheets (Fig. 4):

Low object illumination

№	Criteria that determine the complexity of the preparation of the workplace	measure	value	Quantity regulator	Reduced value
1	Mass	kg	6,39	< >	1,231
2	Size	m^3	0,038	< >	1,006
3	Quantity	pcs	2	< >	1,429
4	Labour intensity	labor hour	3,2	< >	1,429

Minimum value	Maximum value
4,16	33,24
0,038	0,15
1	8
2,4	8

Complexity profile 1,429 k
Complexity factor 1,136 $K=(K_2-K_1)(k-1)/3+K_1$

K_1= 1,10
K_2= 1,35

Figure 4. Calculation of the complexity factor value, with the actual values of the integrated assessment in Microsoft Excel®

As it can be seen in Figure 4, for the given values of the criteria with an accuracy of two decimal places, the complex assessment is equal to 1.964, when transferred into the values from 1.10 to 1.35, the complexity factor coefficient is 1.18.

5. Summary

To test the proposed methodology, a series of computational experiments, in which the simplest constraining conditions were assessed were carried out. For these experiments the complex assessment was 1.0, and the complexity factor value was 1.1, respectively, and vice versa, a comprehensive assessment of really very difficult working conditions was carried out, according to the results of the complex assessment of which the maximum complex grade of 4.0 was obtained, and, as a consequence, the maximum value of the complexity factor was 1.35.

At the moment, using the methodology proposed by the authors, the complexity factors have been determined for such conditions as "Low illumination of an object", which is shown in this article, as well as "Dustiness, corrosion deposits, deposits of natural origin and other formations leading to preliminary cleaning of the object before performing work","The need for a second and subsequent visits of specialists due to the impossibility of providing complete and reliable initial data to determine the work scope","The need to collect and provide documents (technical passports) for the equipment (machines, mechanisms) used at the facility, in connection with the specifics of the object". Prior to this study, the complexity coefficients for the above-mentioned conditions have not been determined.

References

[1] Leea J, Leea D, ChoaH, KangaK 2017 Inhibiting Factors and Improvement Plan of Table formwork Method in High-Rise Building Construction *34th International Symposium on Automation and Robotics in Construction. 2017 Jun 28 - 2017 Jul 1* 571-576. DOI: 10.22260/ISARC2017/0079.

[2] Oyeyipo O, Odeyinka H, Owolabi J and Afolabi A 2019 Factors Inhibiting Stakeholder Management of Mega Construction Project *International Journal of Civil Engineering and Technology (IJCIET)* **10 (03)** 1261-1269.

[3] Adamu M, Olufemi S T 2019 Identification of the Inhibiting Factors Affecting Implementation of PPP Models using Lean Technology in Construction *International Journal of Engineering and Technical Research* **80 (7)** 140-143. DOI: 10.17577/IJERTV8IS070046.

[4] Al-Jamous A, Werner M, Kaiser M, Thyroff R, Gehrke B 2020 Instandsetzung der Verrohrung des StarzenBachs in FeldafingmitCarbonbeton *Zeitschrift fur den gesamtenIngenieurbau: Bautechnik, April 2020* 279-285. DOI: 10.1002/bate.201900073 (in German).

[5] Helmstädter S, Lorenzl H 2018 Die neuen Richtlinienfür den LändlichenWegebau (RLW) *Fachbeitrag, 143 Jg., Juni 2018* 349-353. DOI 10.12902/zfv-0217-2018 (in German).

[6] Degtyarev G V, Takhumova O V 2019 Designing an additional freshwater source infrastructure to ensure the environmental sustainability of coastal areas. IOP Conference Series: Earth and Environmental Science, 395, [EESE3951001.pdf]

[7] Bondarenko S V, Takhumova O V 2019 Economic and organizational mechanisms of forming business networks in the construction industry *IOP Conf. Ser.: Mater. Sci. Eng. 698 077040 Alekseev A., Salamatina A., Kataeva T. Rating and Control Mechanisms Design in the Program Research of Dynamic Systems. 21st IEEE Conference on Business Informatics* 96-105. DOI 10.1109/CBI.2019.10103.

[8] Alekseev A O 2019 Rating and control of complex objects in uncertainty cases *Applied Mathematics and Control Sciences* **2** 103–131. DOI: 10.15593/2499-9873/2019.2.06 (in Russian).

[9] Gureev K A, Gureeva E G 2016 Methodology and method development of employee labor intensity assessment regard complexity factors *Economics and Entrepreneurship* **11-2 (76-2)** 560-564 (inRussian).

[10] Alekseev A O, Melekhin M I, Shaidulin R F, Kharitonov V A 2014 Automated system for complex assessment of objects with the ability to choose a fuzzy convolution procedure in accordance with the degree of uncertainty of expert information about the parameters of their state. Patent license EVM No.2014660537.

[11] Alekseev A O, Melekhin M I, Korgin N A, Korepanov V O, Spirina V S, Shaidulin R F 2016 Software module for experimental study of the stability of the matrix anonymous generalized median mechanism to the strategic behavior of agents. Patent license EVM No.2016660758.

Modern architectural - planning, organizational, technological, constructive solutions at green roofing

L A Seferyan*, V E Chubarov, K V Chubarova

Don State Technical University, 1, Gagarin, Sq., Rostov-on-Don, 344000, Russia

E-mail: Luiza.seferyan@mail.ru

Abstract. The article discusses one of the most urgent problems of Russian cities - the lack of green areas in urban environment. In urbanization process with the growth of large cities and an increase in the population, the ecological state of the environment is transformed, and the concepts of "dwelling - environment" are also changed. One of the effective and rational steps towards solving the problem of the lack of green areas in Russian cities is the green roofs use. At the moment, there is a certain typology, as well as many different design solutions using modern building materials for the organization of green roofs. The article analyzes the advantages and disadvantages of roof gardening in detail, also considers the existing regulatory and legal framework governing their structure. Describes the functions of normative documentation on green roofs at all building life cycle stages.

Introduction

Population growth and long-term urbanization will inevitably lead to the compaction of buildings in modern cities. Less and less land remains within the city limits, free from buildings and highways, not bound by asphalt, paving slabs or other hard surfaces. The air is polluted by the emissions from cars in traffic jams. The lack of elementary culture and the thirst for momentary profit turned undeveloped territories and green plantings into unauthorized dumps. In large cities, organizational and technological solutions established by norms and rules are violated, since due to an increase in the density of urban development, the density of population is also increased, and at the same time, the area of green spaces per person is decreased. The problem of reducing green areas is becoming one of the most urgent in the cities of the Russian Federation (RF), since there is practically no free space for them [1,2].

There is a growing shortage of urban land and one of the possible and rational solutions in this situation is the use of roofs and facades of buildings, as well as all kinds of urban infrastructure surfaces, such as overpasses, underground garages, and so on, to create architectural and landscape solutions.

Materials and Methods

In modern multi-storey buildings, especially the high-rise buildings, 25 floors (approximately 75 m), a person feels himself in the "stone jungle". He absolutely does not feel the connection with nature, which, naturally, has a depressing effect. In order to somehow compensate for this negative feeling, resting places with elements of landscaping are needed, the so-called recreational spaces.

At its core, a green recreational space is an architectural solution with the use of small architectural forms that imitate a serene nature, the only purpose of which is to create an additional recreation area that succinctly fits into the structure of an architectural element, for example, a roof, a balcony or a transitional gallery. In the form of greenhouses, winter and stone gardens. The use of these architectural and landscape techniques increases the aesthetic appeal of the structure. And a variety of techniques for landscaping buildings with living plants on artificial foundations in particular on operated roofs, halls, or floor areas, has a beneficial effect on the quality of living [2].

Currently a program with the same name is being implemented in the Rostov region in order to provide the population with affordable and comfortable housing. Its operation started in 2014 and its main directions are not only the construction of new housing stock, but also the reconstruction and renovation of the existing one. Since with the growth of cities, not only their appearance, not only the ecological situation, but also the very state of modern housing as an environment is changed. But everything that surrounds a person, the area and height of the premises, the layout of the apartment, what he sees from his window, the structure of the living environment determined by the number of storeys and the shapes of the surrounding buildings, has a physiological and psychological effect on him. A feeling of connection with nature or rather the lack of it.

Living in modern apartments and comfort zone ensure the creation of an individual microclimate, possibility of physical, emotional and psychological rehabilitation of a modern person. It is impossible to imagine a modern apartment without indoor plants. The ability of plants to convert carbon dioxide into oxygen through photosynthesis changes the quality of indoor air in a direction favorable to humans. Plant leaves have the ability to absorb heavy metal particles from the atmosphere. In addition, plants are able to absorb harmful emissions. Thus, the arrangement of green roofs, vertical gardening and the arrangement of adjacent territories during construction and reconstruction can significantly improve the living conditions and life of people in modern cities [3,4].

Results

The technology, as well as the organization of solutions for landscaping, is a good option for the reconstruction of multi-storey buildings due to the fact that today modern cities abound in high-rise buildings, the level of landscaping is so little or none at all. The clear advantage of high-rise buildings landscaping is an aesthetic appeal. The variety of shapes, colors and shades of plants can be successfully used to create a perfect architectural look.

In modern megalopolises, where the main buildings occupy about 80% of the entire territory, the issue of landscaping is relevant for urban planners. The need to provide the human environment with green spaces is dictated by the very fact of human existence, in his own created environment.

Vertical gardening is an alternative type of gardening, in which vegetation is formed on the facade of buildings and structures using special structures. The French phrase "vegetable façade" or the English "green wall" translated as "green wall", can be understood as vertical landscaping of the building facade surface. Climbing plants are used for this. Their ability to grow is especially valuable, braiding vertical structures adjacent to wall fences. There are also such plants as grapes, that can grow on their own without additional supports. The main components in "green walls" construction are: the substrate (soil), the irrigation system (from pipes and pumps) that delivers water and fertilizers, the plants themselves, as well as the supporting elements along which green spaces grow and climb. A big plus of facade landscaping is a decrease in the energy consumption level of the building by reducing heat loss [5,6].

A green roof is a green space created through the use of soil layers and plants growing in it, located on top of a regular roof. This type of green space is, to a certain extent, an alternative to the objects of classical landscape gardening architecture, which can be conveniently located on the roofs of both ground and underground structures.

Green roofing is not an invention of modern architecture. Similar methods were used in the countries of Scandinavia and Central Asia. It was only in this way that they tried to solve another issue, namely, a decrease in temperature exposure. If in the North they tried to insulate their homes by using turf and moss, in the South, on the contrary, they escaped from the heat and nighttime temperature drops. There are also references to the gardensэ arrangement on rooftops and terraces, both in Babylon and the cities of Assyria. In Russia, the fashion for arranging roof gardens came from the European countries, somewhere in the late 19th, early 20th centuries, and even now it is widespread in many countries of the world.

Modern green roofs are the technically sophisticated and carefully balanced systems. They differ mainly in root systems into extensive and intensive.

The extensive green roof has a big plus as it does not require any special maintenance. On the waterproofing layer top of the roof, there is a drainage, in turn covered with a shallow layer of soil, undersized plants growing in it and grass cover corresponding to local climatic conditions. This type of roof gardening consists in the use of light porous soil, with a layer thickness of 5 to 15 cm. Planted unpretentious plants that do not require regular watering. Typically, hardy evergreen species are used for planting, forming a continuous carpet. The mass of the soil layer, together with the planted plants, is on average 20 kg per 1 m^2. Therefore, there is no need for additional base strengthening. In addition, extensive roofs are a very simple way to create a self-contained ecological system to protect roofs [7,8].

Intense green roofs are like a small square or park, more difficult to set up and require more maintenance than the extensive ones. For their device, a deeper layer of soil is required, usually over 30 cm, necessary for the full development of the root system of trees and shrubs.

Both types of modern green roofs are the well thought out and carefully calculated roofing systems, the cost of which, taking into account their life cycle, as a rule, does not greatly exceed the cost of a conventional roof.

a) b)

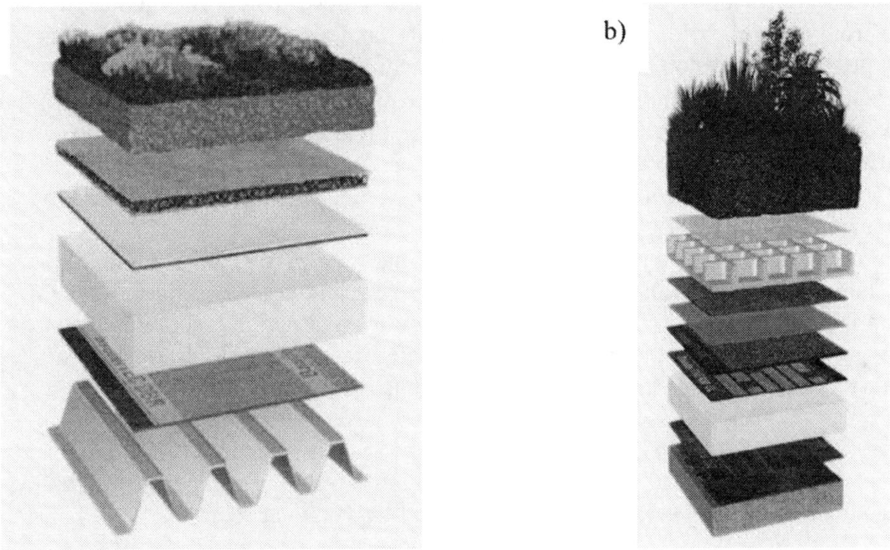

Figure 1. Extensive (*a*) and intensive (*b*) green roofs.

There are two types of green roofs - extensive and intensive. Extensive greening is the closest to natural conditions and gives insignificant loads on the roof structure (70 - 150 kg/m^2). The composition of such a system includes a layer of substrates that accumulate moisture and nutrients for plant roots and drainage, which removes excess water. They are separated from each other by a filtering layer that prevents small particles of substrates from entering the drainage plates, thus increasing the service life of the system. The extensive roof greening option is more economical and requires almost no maintenance, unlike the intensive greening systems.

Intensive greening offers great opportunities for roofing in terms of landscape design, for example, gardening, planting shrubs, trees and organizing recreation spaces. However, intensive green roofs are considered environmentally less efficient and more expensive compared to the extensive green roofs, since they create a significantly greater load on the supporting structures (from 300 kg/m^2) and require more mechanical support. But it is necessary to consider the low operating costs when using such facilities. Since the green roof is much more durable than the usual one. This economically justifies the costs at the design and construction stage of any facility. Since vegetation protects the membrane layers and supporting structures from the harmful effects of ultraviolet radiation and weather conditions. That is, the subsequent operation covers the initial costs of landscaping [9,10].

Regardless of what type of green roof will be organized, during its construction it is necessary to ensure the presence of the following layers:

- Thermal insulation - insulation made of mineral wool boards, foam glass, extruded polystyrene foam and other materials. An important condition for choosing a heat-insulating material is its non-hygroscopicity, as well as its ability to withstand the pressure of plant roots.
- Waterproofing membrane on bitumen or synthetic base. It is necessary both to protect the supporting structures of the building from atmospheric waters and to ensure an optimal microclimate for the green spaces' growth. Also, in addition to the membrane, a root film can be installed to protect the plant roots as well as aluminum or copper foil for additional protection of the structures from moisture.
- Drainage layer is required when installing green roofs to drain excess moisture and maintain a comfortable environment for plants. Drainage is made of rolled perforated polyethylene or high-pressure polystyrene.
- A filtration layer made of geotextiles or special heat-bonded material, which protects the drainage system from clogging with pieces of soil substrate, and also prevents the nutrients necessary for normal growth and development of plants from leaching out of the soil.
- Soil substrate is the basis of the vegetation layer life, which ensures the functioning of the green roof for many years. The composition and proportions of ingredients in the soil depend on the type of vegetation that will be planted.
- Green spaces - lawn, flowers, shrubs, trees, any plant that can withstand exposure to large amounts of direct sunlight.

The introduction of green roofing technologies has many advantages, both the most obvious environmental and image-building, as well as engineering and economic ones. Greening ensures the prolongation of the life cycle of the roof by protecting it from ultraviolet radiation, mechanical damage, as well as protecting the waterproofing layer from high temperatures. In addition, a green roof has pronounced thermal and sound insulation properties, thus providing passive energy savings and acoustic comfort.

However, the installation of green roofs requires additional investment from the customer and the involvement of designers and contractors with a high level of proficiency. When designing the supporting structures, it is imperative to take additional loads into account.

Discussion

As the analysis showed, the direct preparation of the roof surface for landscaping costs about 45 cu/m^2 if the roof is flat, and 65 cu/m^2 – in case it is oblique. On the inclined roofs to fix plants, it is necessary to install the additional honeycomb cells, which will cost the customer an additional 20 cu/m^2.

At the moment, the main regulatory document governing the design and organization of green and operated roofs in Russia is GOST R 58875-2020 "Green" standards. Greened and maintained roofs of buildings and structures. Technical and environmental requirements". The specified document entered into force on June 1, 2020 and contains the requirements for the production of work, operation and safety of green roof structures. It should be noted that for the first time, the clear definitions are given in GOST to three types of roof greening: intensive, semi-intensive and extensive, as well as the detailed design solutions for the design of these types of roofs. Establishing regulatory requirements for different types of green roofs plays an important role at all stages of a building's life cycle (Figure 2).

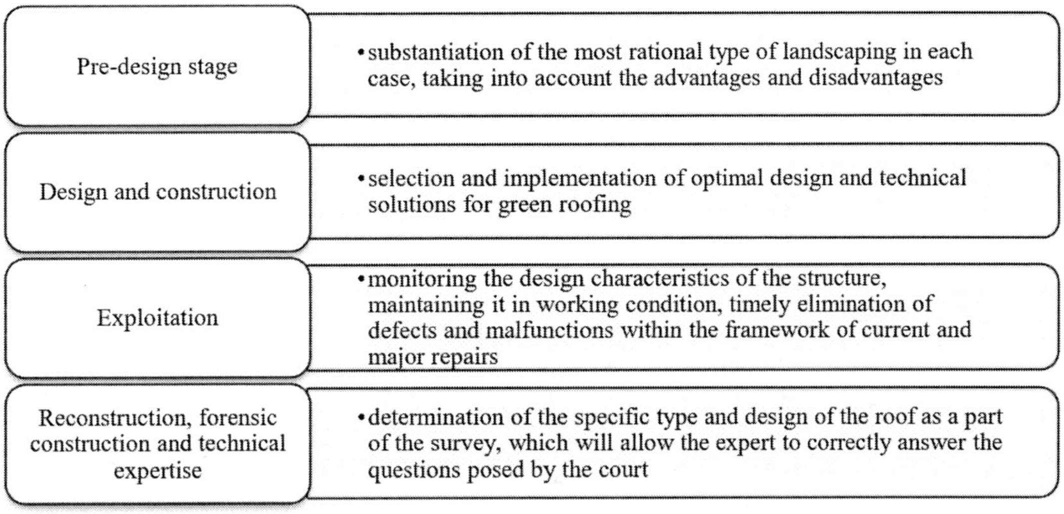

Figure 2. The main functions of regulatory documents in the green roofs design throughout a building's life cycle

From the point of view of calculating the load-bearing elements of the structure used, it is necessary to take into account the weight of the soil layer and the plants planted in it, as well as other operational loads, such as soil unevenly distributed over the roof and wind loads, which is especially important in the case of intensive landscaping. All drainage and other materials of the roofing structure layers should be calculated in a moisture-saturated state. The approved GOST standard does not contain any special recommendations for determining permanent and temporary loads on green roof operation, all loads are calculated in accordance with BC 20.13330.2016 "Loads and Impacts". Maximum weight loads are not standardized and are calculated in accordance with the project requirements.

It is worth noting that green roofs are considered a construction offset measure and are accounted for in the American and European green standards LEED, BREEAM and DGNB as supporting biodiversity, reducing air pollution, dust and greenhouse gases. In the Russian standard NAS NOSTROY 2.35.4-2011 "Green construction" greening of a building, including roof and vertical gardening, as well as "winter gardens", is taken into account in the rating of habitat sustainability calculated for residential and public buildings.

Measures for the area improvement, organization of an ecologically comfortable zone around the house (the creation of parks, playgrounds in the yard, etc.) and provision of landscaping standards in the absence of conditions for conventional long-term landscaping can be carried out using mobile landscaping.

Summary

Consequently, the use of green recreational space in urban landscaping makes it possible to solve the problem of the deficit of green areas and will have a beneficial effect on the city appearance [11,12].

Economic and engineering benefits of a green roof:

1. With the help of rainwater absorption, there is a saving of water resources and a decrease in the urban storm sewage cost;
2. Thermal insulating qualities - in hot season, the roof does not overheat and in the winter period the energy is saved, which guarantees passive energy saving;
3. The soil layer and plants reduce the level of sound waves, which significantly increases the level of sound insulation;

4. The effect of cost savings (compared to the life cycle of a conventional roof in the future for 25 years). The vegetation layer on the roof is a natural protection against mechanical damage, ultraviolet radiation and the effects of temperature fluctuations, which significantly increases the service life;
5. No need to create special operating conditions as well as additional investments;
6. Due to the constant moisture content of the soil layer, in case of fire, the rapid spread of fire over the roofing surface;
7. Possibility of placing various types of communications, both during installation and after it;
8. Installation work simplicity;
9. The root system of plants prevents soil slipping;
10. Local waterproofing availability;
11. Lack of serious investments and special operating conditions;
12. Versatility, location anywhere, if possible, vegetative soil cover arrangement.

The main advantages of a green roof from the environmental point of view:

1. Oxygen source;
2. By absorption, neutralizes harmful gases and dust in the environment;
3. Regulates air humidity;
4. Aggregation of rainwater;
5. Obtaining the new natural conditions for the life of fauna:

Image and social benefits:

1. Ensures a high rating during certification for compliance with green standards;
2. Increases the image attractiveness of the object for potential buyers and tenants;
3. Increases public appreciation of the building
4. Serves as an additional recreation area with vegetation and the possibility of plant care, which is also an open public space for human interaction.

Various rooftop greening endeavors take place in over 40 countries around the world. Active research and work on roof greening is carried out in Germany and France. In some countries, roof greening is supported by the legislation. Work in this direction is in progress in Moscow and St. Petersburg.

Greening of roofs of buildings during reconstruction currently becomes one of the most dynamically developing areas of landscape architecture and becomes increasingly popular in the large cities of the Russian Federation [13,14].

"The scent of mowed lawns creates the illusion of a village, awakens old pictures, taking you somewhere far away" - D. Hovhannes.

To solve the problem of organizing the technological solutions for greening multi-storey buildings in conditions of reconstruction, ecologically comfortable zones within the city without the alternative landscaping introduction seems extremely problematic or not possible at all.

References

[1] Seferyan L A 2010 State support for the reform in the housing and communal services sector *Resources Information Supply Competition*" **3** 254 – 256.

[2] Seferyan L A, Zilberova I Yu 2013 Reforming the housing and communal services of the region *"Construction-2013": Materials of the International Scientific and Practical Conference. Rostov-on-Don: RSCEU* 125 – 127.

[3] Seferyan L A, Vorontsyova O V, Shvets Yu S 2018 Methods for improving the energy efficiency of residential buildings *Engineering Bulletin of the Don* **2**. Information on ivdon.ru/ru/magazine/archive/n2y2018/4847

[4] Vasilyev S E 2010 An integrated approach to the effective use of financial resources for the reproduction of the housing stock *Property relations in the Russian Federation* **1** 50-57.

[5] Babayan K D, Bobkina K O, Seferyan L A 2019 Statistics of improvement programs in the city of Rostov-on-Don *Engineering Bulletin of the Don* **3**. Information on ivdon.ru/ru/magazine/archive/n3y2019/5845

[6] Sheina S G 2008 *Methodology for strategic management of the technical condition of the housing stock through modeling and optimization of organizational and technological solutions* (Rostov-on-Don) 13.
[7] Seferyan L L, Bulanova A S, Morozov V E 2017 Analysis of waste treatment plants in Europe *Don's Engineering Gazette* **2**. Information on ivdon.ru/ru/magazine/archive/n2y2017/4249
[8] The main tasks and problems of intermunicipal ecological waste processing complexes *XXVII International Scientific Conference of the Eurasian Scientific Union "Strategies for Sustainable Development of World Science"* Collection of scientific works of the XXVII International Scientific Conference of the Eurasian Scientific Union (Moscow, May 2017) 164.
[9] Yazyeva S B, Seferyan L A, Oparina L A, Golubeva A Y 2018 Greening. technology organizing of multistorey buildings, when reconstructing architectural and planning decisions with the use of modern building materials *Materials Science Forum* **931** 883-888.
[10] Seferyan L A, Kondrateva T N, Morozov V E, Leusenko I V 2019 Qualitative state improvement of public and courtyard areas in terms of Rostov-on-Don *IOP Conference Series: Materials Science and Engineering* **698 (5) 055004.**
[11] XXVII International Scientific Conference of the Eurasian Scientific Union "Strategies for the Sustainable Development of World Science" *Collection of scientific works of the XXVII International Scientific Conference of the Eurasian Scientific Union.*
[12] Sheina S, Fedorovskaya A, Yudina K 2018 "Smart City": Comfortable Living Environment *IOP Conf. Series: Materials Science and Engineering* **463 (2)** 032095.
[13] Sheina S G, Yudina K V 2018 The method of spatial analysis of the density of the distribution of landscaping during the reconstruction of urban development *Engineering Bulletin of the Don* **3 (50)** 107-119. Information on ivdon.ru/ru/magazine/archive/n3y2018/5083
[14] Aloyan R M, Seferyan L A, Mailyan A L 2017 Factors of sustainable development of the housing stock of a large city *Engineering Bulletin of the Don* **1**. Information on ivdon.ru/ru/magazine/archive/n1y2017/3992

IOP Conf. Series: Materials Science and Engineering 1083 (2021) 012050 doi:10.1088/1757-899X/1083/1/012050

Development of technological measures to ensure the safety of production facilities in petrochemical industry

V L Adamyan[1], G A Sergeeva[1], L A Seferyan[1*], N Y Gorlova[2]

[1]Don State Technical University, 1, Gagarin sq., Rostov-on-Don, 344002, Russia
[2]Rostov branch of the state treasury educational institution of higher education "Russian Customs Academy", 20, Budennovsky ave., Rostov-on-Don, 344002, Russia

E-mail: Luiza.seferyan@mail.ru

Abstract. This article, using the example of the production of methane oxidative ammonolysis, discusses the problems of safe process control. This is achieved by a set of precautionary measures. The industrial safety plan prevents one of the causes of technological equipment depressurization. The production of hydrocyanic acid by oxidative ammonolysis of methane is considered, in which, to prevent the vapor-gas mixture formation, it is necessary to ensure complete sealing of the apparatus in which the combustible substance is stored, and also to maintain the pressure in the apparatus equal to the design pressure.

A number of technical solutions, with the help of which it is possible to ensure the operation of devices with ammonia without the formation of explosive concentrations in them, are presented. These include: the presence of a mixture of combustible gas with an oxidizing agent with a working concentration in the apparatus above the upper or lower ignition limits, taking into account the safety margin; a system for supplying a non-combustible gas having a phlegmatizing property to the reaction zone; automatic regulators of the oxidizing agent and reducing agent ratio, as well as gas pressure regulators; protection against atmospheric and static electricity; automatic shutdown of the pipeline as a result of automatic shutoff of the valve; stationary gas analyzers that automatically signal a deviation from the norm.

Introduction

In the chemical and petrochemical industries, flammable liquids are the most of the starting materials and reaction products. This leads to the complexity of the design of technological processes that provide not only the raw materials conversion and the reaction products selectivity, but also the facilities' operation safety.

The purpose of this work is to analyze a set of measures to ensure the safety of technological processes, using the example of an ammonolysis plant for methane.

A source of danger in the production of hydrocyanic acid by the oxidative ammonolysis of methane are the areas that have the following characteristics:

- presence of flammable, toxic, explosive substances;
- availability of equipment operated under pressure more 0,07 MPa;
- storage and processing of the raw materials and products prone to spontaneous combustion, explosive properties' formation.

The whole complex of precautionary measures taken is a plan of measures to ensure industrial safety [1,2].

Materials and methods of research

Such anthropological factors as violations of technological parameters and, accordingly, the operating mode of technological equipment and related damages and accidents pose a great danger to the production of hydrocyanic acid. In this case, within a short period of time, a flammable concentration can be formed both inside and outside due to the release of FL in significant amount. So, for example, let us consider the possibility of preventing the formation of a vapor - air - gas mixture in the production of hydrogen cyanide (hydrocyanic acid), indicating the need:

– to ensure complete sealing of devices in which a flammable substance is stored;
– to maintain the pressure in the apparatus equal to the design [3,4].

The reaction of oxidative ammonolysis is carried out in a contact apparatus, which is a tubular apparatus with pipes 22x3.5 mm in diameter made of fused corundum (Figure 1). The catalyst - a metal plate - is applied to the inner surface of the tubes. The reagents, passing through the contact tubes, enter the cooling element, which consists of a large number of tubes washed with water, which makes it possible to very quickly cool the reaction products to 400 ^{0}C. Then the vapor-gas mixture enters the common collector and is directed to the extraction of hydrocyanic acid by the conventional methods.

Ammonia-methane
mixed air

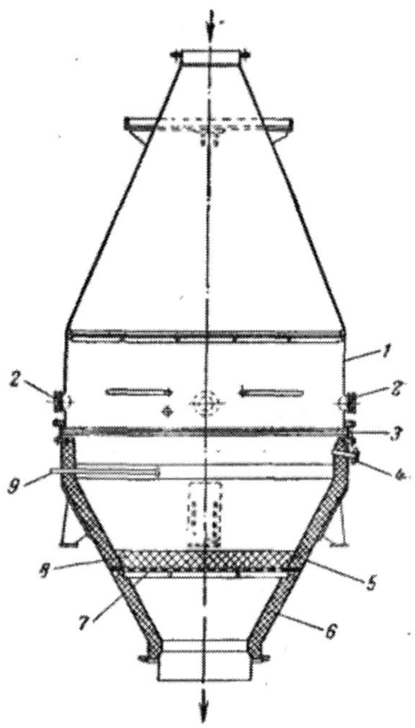

Hydrocyanic acid
HCN

Figure 1. Contact apparatus for the hydrocyanic acid synthesis by oxidative ammonolysis: 1 - apparatus body; 2 - observation windows; 3 - catalyst grid; 4 - thermowell for thermocouple; 5 - asbestos gasket between the body and the lining; 6 - lining; 7 - mesh; 8 - nozzle; 9 - sampling connection.

Research results and their results

The catalyst grid is made of a platinum-rhodium alloy (10% rhodium). In the apparatus, several meshes of wire of different thickness are placed one above the other, and the lower support mesh is made of the thickest wire. The service life of the catalyst is more than 400 hours; no loss of platinum due to evaporation [5,6].

Ensuring complete sealing of devices in which a combustible substance is stored is the exclusion of a vapor-air space in the device. However, in the production of hydrocyanic acid by the oxidative ammonolysis method, oxygen is one of the main components of the initial reaction medium supplied to the contact apparatus. In the synthesis by the Andrussov method, the reaction proceeds according to the following scheme (the process is similar to the catalytic oxidation of ammonia):

$$CH_4 + NH_3 + 1{,}5O_2 \rightarrow HCN + 3H_2O \qquad \Delta H = -114{,}9kkal$$

In addition to HCN formation, numerous side reactions occur, the most notable of which are:

1) interaction of methane with water formed in an intermediate reaction:

$$CH_4 + H_2O \rightarrow CO + 3H_2 \qquad Q_{ch.r.} = 48{,}87kkal$$

2) incomplete combustion of methane:

$$2CH_4 + O_2 \rightarrow 2CO + 4H_2 \qquad Q_{ch.r.} = -17{,}9kkal$$

3) complete combustion of methane:

$$CH_4 + 2O_2 \rightarrow CO_2 + 2H_2O \qquad Q_{ch.r.} = -192{,}5kkal$$

4) decomposition of ammonia with the release of a very large amount of heat:

$$4NH_3 + 3O_2 \rightarrow 2N_2 + 6H_2O \qquad Q_{ch.r.} = -302kkal$$

Due to the occurrence of side reactions, the consumption of ammonia and methane increases: in contact gases, except for hydrogen cyanide and unreacted feed gases (CH_4, NH_3, O_2, N_2), contained CO, CO_2 and H_2.

The overall process runs on a platinum catalyst autothermer. The required reaction temperature (≈1000 ^{0}C) is maintained by adjusting the composition of the reaction mixture, which most often has approximately the following composition: CH_4 12-13 %, NH_3 11-12 %, air 72-75 %.

Thus, in order to prevent an explosive concentration even in the devices, in which oxygen is the oxidizing agent for the vapor-gas-air mixture formation, it is necessary to take into account the limits of its explosiveness in comparison with the project one [7].

One of the common causes of damage to process equipment is corrosion, which can be the reason for its depressurization. Carbonyl corrosion is a major hazard in the production of oxidative ammonolysis of methane. As it can be seen from the above-shown reaction equations, one of the reaction products is carbon monoxide (II) – CO, which at low pressure and ambient temperature is inert towards metals. However, under the conditions of the ammonolysis technological process, stable iron pentacarbonyl is formed.

$$Fe + 5CO \rightarrow Fe(CO)_5$$

The reaction of iron carbonyl formation takes place at the interface between the solid and gaseous phases. The presence of water vapor in the reaction phase has a beneficial effect on the carbonyl formation, and the presence of ammonia increases the rate of carbonyl formation.

Iron pentacarbonyl is a flammable liquid that burns in an oxygen atmosphere to form iron oxide and carbon dioxide, or magnetite and carbon monoxide. Oxygen, passed through the liquid iron carbonyl, causes its ignition, accompanied by detonation.

The structural formula of iron pentacarbonyl is expressed as follows:

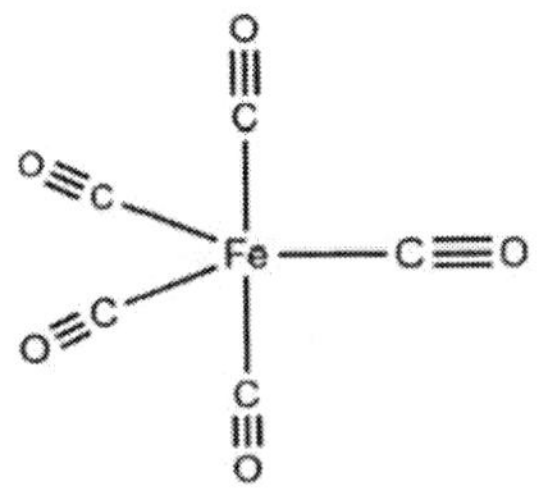

The loosening and destructive effect of iron pentacarbonyl on the metal surface layer is usually prevented by making a contact apparatus from chromium-nickel steel containing 20% nickel, 23% chromium, up to 30% chromium compounds and manganese bronzes.

When ammonia is decomposed, hydrogen is released, which, combining with air, forms an explosive gas, which spontaneously explodes with great force.

It is possible to ensure the operation of devices with ammonia without the formation of explosive concentrations in them using the following technical solutions:

- in the presence of a combustible gas mixture with an oxidizer, the working concentration in the apparatus should be set above the upper or below the lower ignition limits, taking into account the safety margin;
- in case of violation of the ratio automatic regulation of the oxidizing agent and the reducing agent components, the system for supplying a non-combustible gas with a phlegmatizing property to the reaction zone is connected [8];
- on the technological lines included in the apparatus, automatic regulators of the oxidizing agent and the reducing agent ratio, as well as gas pressure regulators are installed;
- the use of explosion-proof electrical equipment, protection against atmospheric and static electricity;
- automatic pipeline shutdown is established as a result of automatic valve shut-off;
- for continuous monitoring of the gases' mixture working concentration with an oxidizer, the devices are equipped with stationary gas analyzers that automatically signal a deviation from the norm.

The characteristics of ammonia and its effect on the human body are given in Table 1.

Table 1. Characteristics of ammonia and its effect on the human body

Characteristics of ammonia	unit of measurement	Indicators
Boiling temperature	^0C	-33.4
Density	g/cm^3	0.68
Permissible concentration of ammonia in air	mg/l	0.02
Danger to human life	%%	0.5-0.6
Explosion hazard with flame	%%	15.6-28.6
Density of concentrated solution	g/cm^3	0.91

Technological equipment operating mode violation can lead to unpredictable emergency situations, the most dangerous of which is fire [9].

Summary

On the basis of the studied and presented material, in order to protect equipment from fires, thermal shields, which are used to localize radiant heat, reduce irradiation at workplaces, are used (Figure 2). Alfol (aluminum foil), aluminum sheet, tinplate, aluminum paint are the materials used for heat-reflecting shields. The shield consists of a supporting frame, a reflective surface and attachments to the shielded equipment.

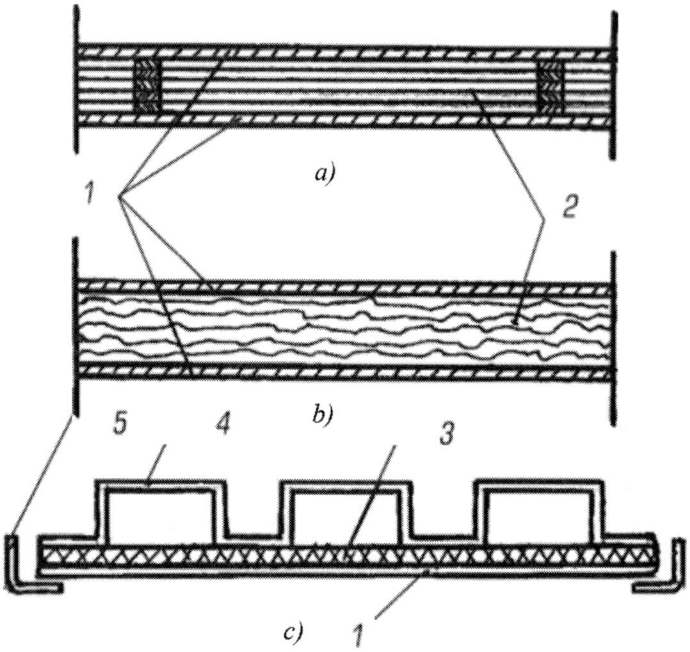

Figure 2. Heat shields: *a* - shield made of alfol, laid in rows in air spaces; *b* - shield made of crumpled alfol in the air spaces; *c* - combined shield;
1 - a metal sheet; *2* - alfol layer; *3* - thermal insulation layer; *4* - profiled aluminum sheet; *5* - frame

The weakening of the thermal flux behind the screen is due to its absorption and reflectivity.

References

[1] Adamyan V L 2015 Reduction of fire and industrial hazard of technological processes *Science perspective* **10** 72-74.

[2] Information about accidents at hazardous production facilities *QRZ: Official site Federal Environmental, Engineering and nuclear Supervision Agency.* Information on http://www.gosnadzor.ru (application date: 26.01.2020).

[3] Adamyan V L, Shikhamadov Sh M 2018 Improvement of fire extinguishing at the spill of volatile liquids *International student scientific bulletin* **5** 169. Information on http://eduherald.ru/ru/article/view?id=19218 (application date: 26.01.2020).

[4] Adamyan V L, Sergeeva G A, Drokin A V, Galushkin O S, Elmanov A A 2017 Problems of phlegmatization of a gas-air mixture in storage tanks for petroleum products *Science perspective* **4 (91)**7- 9.

[5] Sergeeva G A, Adamyan V L 2017 Muddle Areas In The Pshekha River Basin *Ecology and life safety. Collection of articles of the XVII International Scientific and Practical Conference* 135-138.

[6] Sergeeva G A, Jijin K S 2018 Geological cataclysms of the North Caucasus *Scientific journal "Globus", materials of XXXIII Intern. scientific-practical conf. "Achievements and problems of modern science" St. Petersburg* 17-25.

[7] Yazyeva S B, Seferyan L A, Oparina L A, Golubeva A Y 2018 Greening. technology organizing of multistorey buildings, when reconstructing architectural and planning decisions with the use of modern building materials *Materials Science Forum* **931** 883-888.

[8] Seferyan L A, Kondrateva T N, Morozov V E, Leusenko I V 2019 Qualitative state improvement of public and courtyard areas in terms of Rostov-on-Don *IOP Conference Series: Materials Science and Engineering* **698 (5) 055004.**
[9] Sheina S G, Yudina K V 2018 Development technique of urban-environmental passport of the reconstruction territory *Materials Science Forum* **931** 822-826. Information on https://www.scientific.net/MSF.931.822.

Optimal ways to develop ecological and energy-efficient construction

S V Ovchinnikova[1*], A Yu Kalinichenko[2], A A Lyamina[2] and E M Schneider[3]

[1] Kuban State Agrarian University named after I.T. Trubilin, 13 Kalinina st., Krasnodar, 350004, Russia
[2] North Caucasus Federal University, 1, Pushkin st. Stavropol, 355017, Russia
[3] Nevinnomyssk State Humanitarian and Technical University, 19 Bulvar Mira str., Nevinnomyssk, 357108, Russia

E-mail: svetlana.swetlana-ov@yandex.ru

Abstract. The construction industry has a significant impact on our environment. When erecting buildings and structures, building materials, energy, water and other resources, the production of which violates the ecological balance, are used. The construction process itself has also negative effect on environment and the nearby objects with noise, vibration exposure, waste incineration, groundwater and soil pollution in general. However, the time period required for the construction of the building is insignificant compared to the period of operation. Experts estimate the service life of modern monolithic buildings at 125 - 150 years, and in some cases this index reaches 300 years. During this time, a new problem arises - the ecological balance violation: insufficient insolation, changes in natural wind flows, hydrogeology, vegetation reduction, soil and vegetation layer pollution, air masses stagnation, etc. All this creates the prerequisites for the creation of special environmental protection measures aimed at maintaining ecological balance. Thus, at each stage of the life cycle of a building, it is necessary to carry out the measures to protect the environment and minimize harmful effects on the environment. Today the prerequisites for the green building development are being laid in Russia. The main idea is to reduce energy intensity and increase energy efficiency in construction.

Introduction

Currently, the environmental issue is acute. The industrial development of countries has led to a large-scale depletion of limited natural resources. Society was faced with the choice of finding more alternative energy sources [1-2].

Cities are the centers of economic development in Russia, more than 70% of the country's population is concentrated in them, therefore, the question of the need to preserve a favorable living environment arises most acutely in cities [3-5]. The high density of population, transport and industrial enterprises in limited areas is the main reason for the environmental problems of the cities, the main ones of which are:

- air pollution;
- surface water sources pollution;
- groundwater pollution;
- disturbance and destruction of the fertile soil layer, salinization, waterlogging and lands desertification;

- increase in the area occupied by landfills with solid domestic waste;
- lack of green space [6-8].

As a result, about 40 percent of the territory in the Russian Federation is experiencing environmental problems, ranging from water pollution to radioactive emissions into the atmosphere [9-10]. This, in turn, affects the life quality of the population and is health dangerous [11].

The most popular standards are: BREEAM (Building Research Establishment Environmental Assessment Method, Great Britain), LEED (Leadership in Energy and Environmental Design, USA), DGNB (Deutsche Gesellschaft fur Nachhaltiges Bauen Germany). Russia has recently introduced its own certification system "Green Standards".

Main part

"Green Standard" is effective in solving the following tasks:
- reducing the consumption level of energy and material resources by the building;
- reducing the adverse impact on natural ecosystems;
- ensuring a guaranteed level of comfort for the human environment;
- creation of new energy-efficient and energy-saving products, new jobs in the manufacturing and operating sectors.

In the countries where Green Building is developing, national standards that take into account the socio-economic and natural conditions of the country are being created: for example, it is inappropriate to introduce such recommendations in the northern regions of Russia as autonomous power generation by wind generators and solar panels, but in the south, this will give great savings energy resources.

Construction in accordance with international environmental standards makes it possible to minimize the consumption of natural resources, improve the energy efficiency of the system and improve comfort for the consumers. Figure 1 shows the percentage by the type of real estate properties in the world.

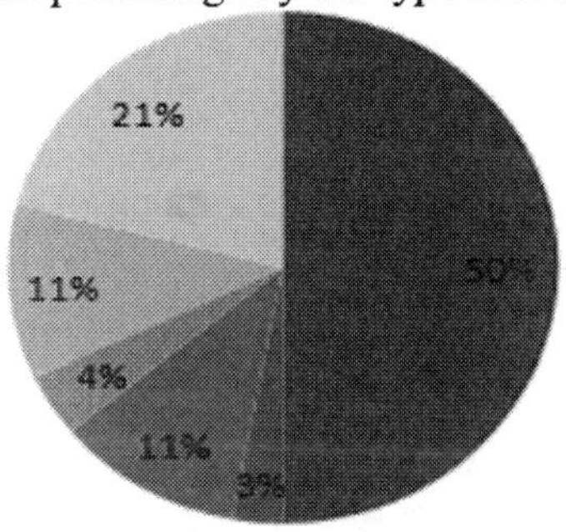

Figure 1. Percentage by the type of real estate commissioned "green" objects in the world: offices – 50%, hotels – 3%, production warehouse space -11%, production warehouse space –4%, retail real estate –11%

In the Russian market, most of the buildings that have any of the environmental certificates belong to the office segment - 59% of the total number of "green" buildings. Warehouse and industrial facilities are in second place in Russia - 23%. Hotels are in third place - 5% respectively (Figure 2).

In Russia, the construction of Olympic facilities in Sochi has become unprecedented in terms of energy efficiency. In addition to efficient energy saving thanks to new thermal protection materials, environmentally friendly production of electricity using solar panels was launched. In addition, rainwater was used in technical processes to preserve the hydraulic system of the Imeretinskaya Lowland. And the points for processing construction and wood waste made it possible to use about 90% of construction waste in the construction of highways, sidewalks, land reclamation on the Sochi Park territory.

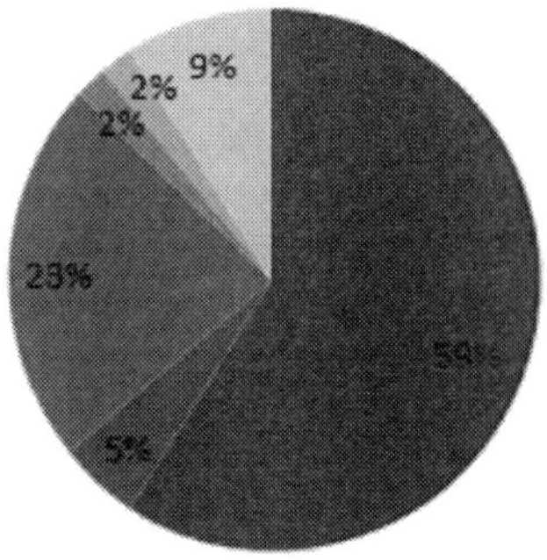

Figure 2. Percentage by the type of real estate put into operation "green" objects in Russia: offices - 59%, hotels - 5%, industrial and warehouse space - 23%, residential real estate - 2%, retail real estate - 2%

Today the prerequisites for the development of green building are being laid in Russia. The main idea is to reduce the energy intensity of the gross domestic product of the Russian Federation by at least 40%. The lack of energy resources slows down economy development, therefore this law should push the real estate market towards energy conservation, which in turn will entail the economy development of the Russian Federation. According to the law, the reduction in energy intensity will be carried out mainly by replacing the incandescent lamps, installing electricity consumption meters and reducing the cost of heating premises. However, the law is still not supported by the normative acts and real mechanisms of its implementation and incentives to use.

Green certification is voluntary. The certified property becomes more attractive and competitive in the real estate market, because guaranteed to have a high quality of construction and excellent performance of the facility, its energy efficiency and environmental friendliness. The development and implementation of environmental construction standards will intensify business expansion, the development of innovative technologies and the economy. We get the environmentally friendly and energy efficient buildings that comply with the following principles:

- quality improvement of real estate objects;
- reducing the cost of building the facility and its subsequent operation;
- caring for residents and creating comfortable conditions for them to be in buildings both within the framework of work and in ordinary life;
- creation of conditions for the minimum impact of facilities under construction on the environment.

The economic benefits of green building:

1. The energy consumption of the building is reduced by 25%, therefore the energy costs are reduced.
2. Water consumption is reduced by 30%, thus reducing the cost of water supply.
3. "Smart" controls and controls save on building maintenance.
4. Increased tenant satisfaction as well as reduced rental refusals, which in turn minimizes costs.
5. Green construction is an excellent marketing ploy to attract potential tenants and owners, increases the rate of return on leased space.
6. The characteristics of "green" buildings already meet the requirements of the law on energy efficiency and the expected tightening thereof, which will save on refurbishment and reconstruction of these facilities.
7. Environmental technologies will soon become an effective method of reducing the cost of a building. At the moment, the rise in the cost of construction of "green" buildings is 4-8%, in cases of particularly complex design solutions this index reaches 18%, however, the costs of construction of many "green" buildings do not exceed the cost of their "non-green" counterparts. In any case, a quick payback over several years by reducing the operating costs should not interfere with the introduction of eco-technologies into the real estate market.

Therefore, green building is undeniably more cost effective when considering the full life cycle of a building.

Today, Russia has all the necessary prerequisites for the green building market development:

- the global economic crisis has significantly slowed down the construction pace. It is necessary to introduce innovative technologies, reduce costs and improve the quality of facilities under construction to gain competitive advantages in the construction market;
- Russia has announced its commitments to improve energy efficiency in construction in accordance with the Kyoto Protocol;
- Russia plans to improve energy efficiency by 40% by 2022;
- in the long-term plans, Russia will introduce free pricing on the domestic energy market, which will soon lead to a sharp rise in prices for connecting and using electricity;
- the growth of social, environmental and economic problems in Russia is largely due to the ineffective use of outdated technologies and low-quality materials. Unreasonably large waste of natural resources is also detrimental

The growth in demand for green buildings is confirmed by the dynamics of the number of certified facilities in recent years.

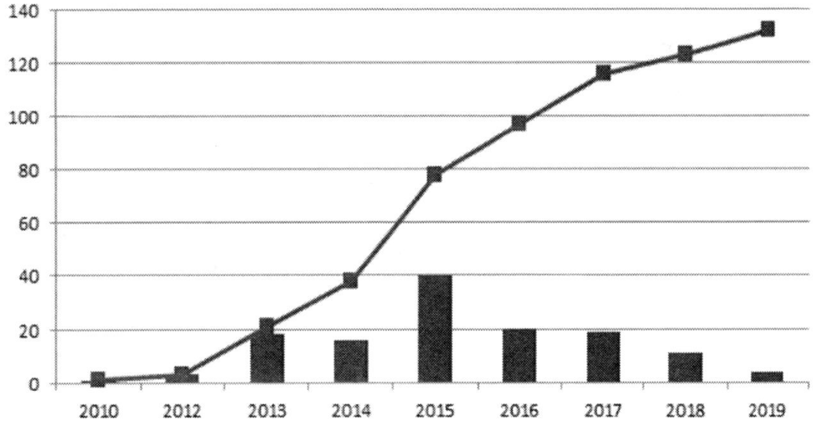

Figure 3. Dynamics of the number of certified objects according to BREEAM, LEED and DGNB standards: ▬ number of new certified facilities, -■- total number of certified facilities

Figure 3 shows the growth dynamics of "green" objects in the Russian real estate market for 2010-2019.

On the part of the state at the legislative level, a number of changes are already being made to spread the energy-efficient construction methods in the territory of the Russian Federation.

Summary

Green technologies are the most promising solution to the environmental and economic problems Russia is facing now. A large number of conferences are being held on the topic of "green" construction, scientific papers are published, projects of energy-efficient buildings are being developed. A legal framework is being prepared and introduced to help introduce green technologies to the Russian real estate market. The construction pace of energy efficient facilities is increasing every year, which shows great interest in this area of construction. Today Russia is at the beginning of a long way leading to sustainable development, both in economy and in social and environmental aspects of life.

References

[1] Ovchinnikova S, Kalinichenko M, Markina N, Schneider E 2020 Energy modernization of housing stock *E3S Web of Conferences* **157 06028**. doi.org/10.1051/e3sconf/202015706028

[2] Tsoraeva E N 2020 To the question of the economic efficiency of land use in the municipality *Omsk Scientific Bulletin. Society Series. History. Modernity* **5 (3)** 141-146.

[3] Kalinichenko M Y, Stojanov N I, Abornev D V, Ovchinnikova S V 2020 Optimization of the regeneration of the sodium-cation ion-exchange filter *IOP Conference Series: Materials Science and Engineering* **905 (1) 012034**. doi.org/ 10.1088/1757-899X/905/1/012034
[4] Tsoraeva E N, Baeva N A 2019 Problems of land legislation violation in the Krasnodar Territory *Bulletin of the Kyrgyz-Russian Slavic University* **19 (3)** 114-116.
[5] Tsoraeva E N, Ivanov A S, Gagarinova N V 2018 Land management as a mechanism for ensuring effective agricultural land use in the Krasnodar Territory *Bulletin of the Kursk State Agricultural Academy* **8** 256-261.
[6] Dokuchaeva L M, Dolina E V, Yurkova R E, Stratinskaya E N, Shalashova O Yu 2011 Techniques excluding negative processes in the soils of irrigated agricultural landscapes of the chernozem zone in southern Russia *Scientific journal of the Russian Research Institute of Melioration Problems* **1 (1)** 4.
[7] Gagarinova N V, Tsorayeva E N, Bakumenko N S 2018 Problems of effective land management in Russia *Bulletin of the Adyghe State University. Episode 5: Economics* **3 (225)** 114-120.
[8] Ovchinnikova S, Abornev D, Kalinichenko M, Kalinichenko A, Sekisov A 2021 Optimizing the temperature stress for the furnace volume of a fire-tube boiler *Advances in Intelligent Systems and Computing* **1259 AISC** 601-610. doi.org/10.1007/978-3-030-57453-6_57
[9] Sekisov A N, Ovchinnikova S V, Norenko M S, Matevosyan V S 2020 Prospects for the solar energy development in housing construction: An assessment of economic efficiency *IOP Conference Series: Materials Science and Engineering* **913 (4) 042051**. doi.org/ 10.1088/1757-899X/913/4/042051
[10] Degtyarev G V, Belokur K A, Sokolova I V 2018 Modeling of the building by numerical methods at assessment of the technical condition of structures *Materials Science Forum* **931** 141-147.
[11] Kondrachuk O E, Petrenko Ya I 2015 Prospects for the development of ecological and energy efficient construction in Russia (Basic research) **11-3** 579-583.

Application of photogrammetric methods in architecture, construction and land management

E N Tsoraeva[1*], R K Gadzhiev [2], S E Kuchiev [2], A A Pekh[2], S A Mezhyan[1]

[1]Kuban State Agrarian University named after I.T. Trubilin, 13 Kalinina str., Krasnodar, 350044, Russia

[2] Gorsky State Agrarian University, 37, Kirova str., Vladikavkaz, 362040, Russia

E-mail: elionora@list.ru

Abstract. One of the important tasks is the photogrammetry application study in construction with the aim, namely, to determine the engineering structures' deformations, to study the engineering structures' models and to solve other problems. With the use of modern technologies, it is possible to make photogrammetric surveys of buildings and structures using certain methods of performing these works. Photogrammetry plays a significant role not only in architecture, construction and land management. Photogrammetric methods make it possible to solve some applied problems directly from the images economically and fairly accurately, for example, to measure the terrain areas, determine their slopes, obtain quantitative characteristics of erosion processes, and perform vertical planning with the determination of the earthwork amount. In the article, the main goal is to consider the implementation of architectural and construction measurements for the purpose of reconstruction and restoration of buildings, as well to complete visualization of the area of interest, performed using photogrammetric methods on the basis of aerial photographs. The analysis made allows us to conclude that both classical and modern digital photogrammetry can be used in architecture, construction and land management.

Introduction

The relevance of the work is due to the fact that modern digital photogrammetry makes it possible to reveal its shape, size and spatial position in a given coordinate system, as well as its area, volume, various sections at the time of survey and changes in their values after a certain time interval from the images of the object under study.

Photogrammetry is used in a wide variety of science and technology branches [1, 2]:

– in geodesy and cartography (for creating plans and maps);

– in construction (for control measurements and structures' deformation research);

– in architecture (for shooting historical monuments);

– in astronomy and cosmonautics (for determining the position of space objects and mapping planets);

– in military engineering (to determine the target coordinates, trajectory).

All building structures are constantly exposed to the destructive forces of nature and civilization. If to consider the architectural monuments, the necessary work should be prepared with special care and carried out taking into account a number of points [3]. For example, such work requires mandatory inventory surveying, accurate and in accordance with the applicable standards. The same applies to the ruins, which must be preserved for humanity as an architectural heritage. At the same time, there is often a need for special scientific research, which in turn puts forward special requirements.

Discussions and results

Digital photogrammetry is based on the use of computer technologies for processing photographs. It should be noted that due to the high performance of the laser scanner, it is possible to quickly obtain complete information about the objects' geometry. Such scanners use coordinate points to provide data on object reflection. This reflected object can be perceived as these three-dimensional discrete models. Thus, today scanning can be performed using digital photogrammetric survey. That is, as a result of combining laser scanning data and digital photogrammetric survey, a realistic color three-dimensional model is obtained (Figure 1).

Figure 1. An example of a colored three-dimensional model of the monument to Nicholas I.

The main methods of photogrammetric processing technology are the digital methods of image processing. By scanning, the image is divided into a number of tiny equal areas called pixels. Each such area contains sufficient information regarding color and color density. In digital photogrammetry, the accuracy of obtaining results increases with increasing scan resolution. The smaller the pixel size, the more accurate the result.

In construction, photogrammetry is used to determine the deformations of engineering structures, control the large-panel buildings' installation accuracy, study the engineering structures models, calculate the volume of earthworks and select the optimal road route as well as to solve other problems. Engineering structures' deformation occurs under the influence of an external force - load. If the load does not exceed the limit established for a given material, then after the load termination, the structure returns to its original shape due to the material particles' interaction. Under heavy loads, permanent deformations appear [4, 5].

To study the engineering structures' deformation, photogrammetric and stereophotogrammetric methods are used. The photogrammetric method gives a possibility to determine the deformations that occur in the plane, and is used to study flat objects. The essence of the method is that several images of the object under study are obtained from the same fixed point, for example, the first one before the load, the second during the load, and the third after the load. In this case, the camera is installed so that the plane of the applied frame is parallel to the plane of the object and the orientation elements of the images are preserved. The photogrammetric method makes it possible to quickly measure architectural structures and get their projections on the horizontal and vertical planes. The application of this method is especially effective for studying hard-to-reach and complex architectural structures. The stereophotogrammetric method gives an opportunity to determine the deformation that occurs in space, and is used to study spatial objects. Several stereopairs of the studied object are obtained from the same basis, for example, the first one before loading, the second during loading and the third after unloading. To improve the accuracy of determining the deformation, special marks attached to the object under study are used.

With the market economy development in our country, the requirements for cadastral registration have changed, which is inextricably linked with the territory mapping [6, 8]. Orthophotomaps created on the basis of remote sensing data were selected as a cartographic basis. Land management includes aerial survey, topographic and geodesy, soil, geobotanical and other surveys and measurements, boundary surveying, development of proposals for the rational land use.

Digital 3D terrain modeling based on aerial photographs gives a complete territory visualization. The interactivity of 3D aerial photography is convenient for spatial analysis. For this, drones are used for aerial photography and this is justified, especially in the small areas and as a result, a full range of ready-made materials is created, including a 3D relief model. For the greatest detail, aerial point objects photography must be carried out both in nadir and with a difference from nadir, and the UAV in this case is the most optimal, in addition, cross photography is required on large areal objects, and from a financial point of view, the drone in aerial photography is the most profitable.

As the results of 3d aerial photography of the relief, the following data are possible are shown on figure 3.

Digital terrain model (DTM) applications:

- forecasting taking into account the landscape characteristics; calculation of geometric characteristics for engineering surveys, routing - navigation, architectural planning; possible risks assessment, taking into account the characteristic features of the relief - flooding areas at the time of floods, landslides and avalanches, etc. environmental monitoring taking into account the steepness of the slope, wind balance, illumination; - 3D visualization for laying power lines, designing highways, pipelines (gas, oil). 3D relief model makes it possible to eliminate possible errors in design and construction at an early stage.

Architectural design includes the design of new areas, buildings, structures, the development of new architectural forms, interior architecture. This also includes the spatial arrangement of objects, the choice of design and building materials, the organization of workplaces and other tasks. And here photos, photomaps, orthophotomaps can be used. Optical editing of photographic and stereo perspectives is used very often. Photos of the future structure are obtained through drawings or models. Architectural measurement is necessary in the absence of drawings during the restoration, reconstruction and restoration of structures, as well as when studying architectural monuments, making models and dummies, during archaeological excavations. Both photogrammetric and stereophotogrammetric images are used.

Table. 1. Measurement accuracy characteristics.

Measurement type	**Limit errors, cm**		**Scale**	**Type of work**
	major	**subsidiary**		
High precision	0,3-0,5	1-1,5	1:20	Drawings
Exact, II	1-2	3-5	1:50	Drawings
Exact, III	3-5	10-15	1:100	Drawings
Technical, IV	10-15	20-30	1:200	Sections, general drawings
Technical, V	20-30	30-50	1:500	Sections, schemes

Monuments of architecture and history, preserved in photographs, are restored using stereo models of single photographs received from different stations. Not only parts of the structure visible in photographs are restored, but also the invisible ones, using such indirect signs as shadows, profiles, color nuances and architectural style features. In this area, all activities are considered in accordance with the current laws on the historical and cultural monuments' preservation.

a

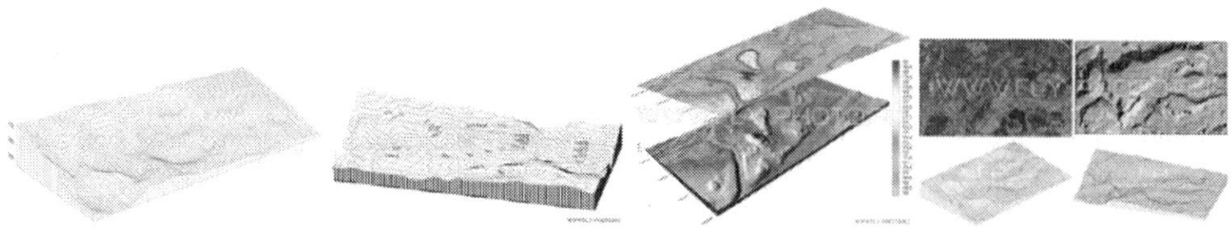

b

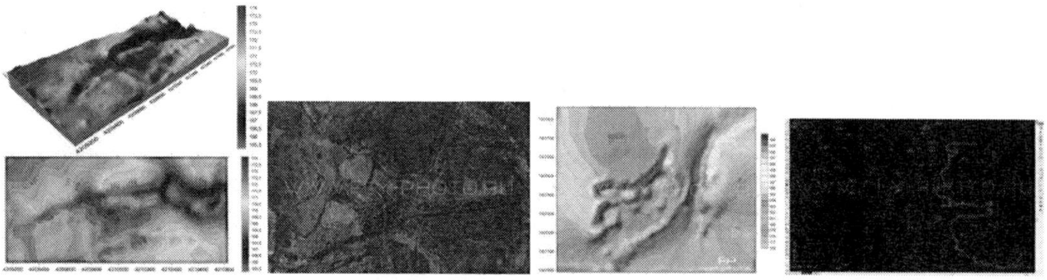

c

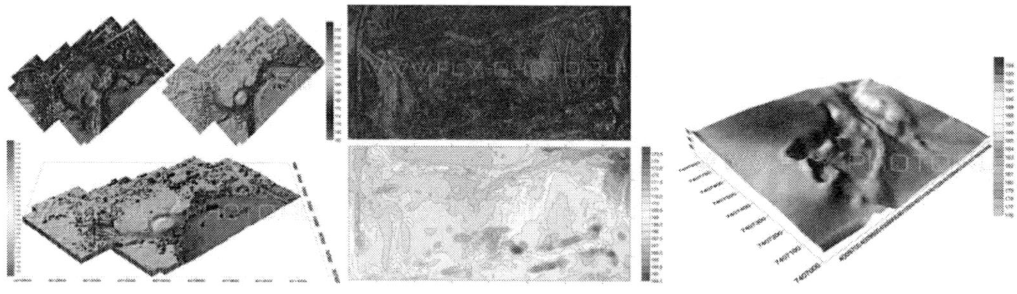

d

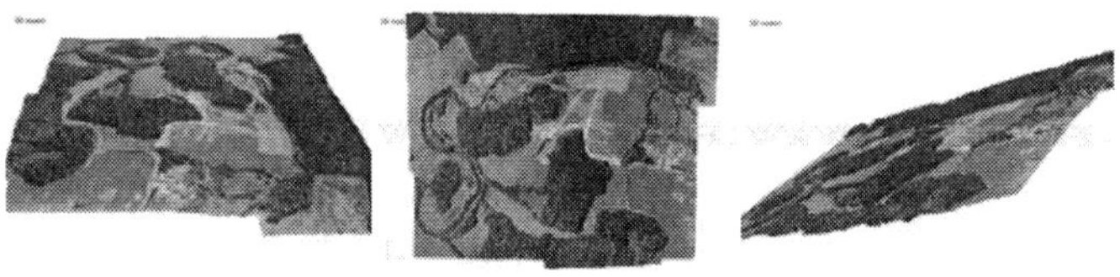

Figure 2. As the results of 3d aerial photography of the relief, the following data are possible; a - "relief shading" and wireframe; b - generalization of heights isolines with a possible step of up to 0.2 meters (data can be converted to available formats for engineering programs, for example, Autocad); c - DTM transformation based on aerial photographs in color layout and gradation of heights by colors; d - aerial photography and digital terrain model (DTM) with "pulling" aerial photography on digital wireframe

Figure 3. An example of work as a result of architectural measurement.

And these activities are associated with inventory, research of the state of monuments, including the deformations' determination, restoration of drawings during restoration, creation of an archive, promotion of national monuments, organization of exhibitions and expositions.

Of particular importance is the systematic nature of the documentation of existing monuments of architecture, not related to the needs of construction, but rather necessary for inventory. The resulting materials form the basis for repair or restoration work in case of damage. Finally, art research studies are based on such materials, if certain conclusions can be drawn from the shape and size of the monuments.

Part of the materials required to solve the above-mentioned problems, especially the images of monuments, can be quickly obtained using the ground-based photogrammetric methods with little manual labor and can be financially profitable. Compared to surveying by classical methods, photogrammetry, despite the high cost of the instruments, has numerous advantages: no need to climb the building structures; there is no need for the construction of forests; the duration of field work is reduced.

The development of modern photogrammetry and methods of photogrammetric processing of images is based on the use of new hardware and software.

New technical means include:

- GPS – to determine the elements of external orientation in flight;
- INS – to determine the corner elements of external orientation in flight;
- Latest precision aerial cameras;
- High and medium resolution space imaging equipment for multispectral survey;
- Scanner imaging systems (including laser survey);
- Digital filming cameras.

The software includes:

- GIS- systems;
- Computer graphics systems e.g. Autocad;
- Multi-area imagery analysis tools e.g. Erdas Imagine;
- Network software.

Summary

The use of photogrammetric methods has significant economic advantages, which, however, largely depend on the nature of the object and, above all, on the technical characteristics of the used survey equipment. Thus, digital systems allow performing all the processes of creating maps after carrying out the corresponding preliminary work (aerial photography and determining the control points' coordinates). In addition, the images obtained with digital systems can be corrected based on the satellite measurements taken both in the air and on the ground.

Based on the technical capabilities of modern technology of photogrammetric survey methods, a classification of the accuracy of performing measurement work is established, which is indicated in the Table. When creating photomaps of building facades, compiled for the survey purposes, the perspective displacements of minor parts (cornices, balconies) exceeding the tolerances are allowed.

Thus, one of the main tasks of photogrammetry in architecture is the performance of architectural and construction measurements for the purpose of reconstruction and restoration of buildings, as well as for the research purposes. The scale of the dimensional drawings, plans and individual fragments, the requirements for the completeness and accuracy of their compilation are determined in the terms of reference, depending on the purpose of the architectural and construction measurements. Photogrammetric work is the main component of modern technologies for creating and updating topographic maps, making photo maps, creating and updating topographic and specialized plans, i.e. play an important role in solving the problems of land management, land cadaster and monitoring, construction and architecture.

References

[1] Bulavitskiy V F, Jukova N V 2016 *Photogrammetry and remote sensing of the territory* (Pacific Publishing House, Khabarovsk).
[2] Limonov A N, Gavrilova L A, Obiralov A I 2006 *Improvement of technological processes for creating digital terrain models based on aerial photography materials* (Monograph, Moscow).
[3] Kalinichenko M Y, Stojanov N I, Abornev D V, Ovchinnikova S V 2020 Optimization of the regeneration of the sodium-cation ion-exchange filter *IOP Conference Series: Materials Science and Engineering* **905 (1)** 012034. https://doi.org/ 10.1088/1757-899X/905/1/012034
[4] Ovchinnikova S, Abornev D, Kalinichenko M, Kalinichenko A, Sekisov A 2020 Optimizing the temperature stress for the furnace volume of a fire-tube boiler *Advances in Intelligent Systems and Computing Springer* **1259 (2021)** 601-610. https://doi.org/10.1007/978-3-030-57453-6_57
[5] Ovchinnikova S, Kalinichenko M, Markina N, Schneider E 2020 Energy modernization of housing stock *E3S Web of Conferences* **157** 06028 (2020). doi.org/10.1051/e3sconf/202015706028
[6] Mezhyan S A, Tsoraeva E N 2020 On the state cadastral registration of land. *In collection: Land management, real estate cadaster and monitoring of land resources. Materials of the international scientific and practical conference. Ulan-Ude* 35-39.
[7] Mezhyan S A, Tsoraeva E N 2020 Mapping the lands of the cadastral quarter in the Mapinfo Pro software *Science Diary* **7 (43)** 17.
[8] Lyubitskaya E V, Ufimtseva Yu E, Tsoraeva E N 2019 State cadastral registration of real estate objects *Epomen* **25** 160-167.
[9] Pekh A A, Hugayeva L M, Katayeva M V 2019 The effectiveness of the application of the data of the state cadastre of real estate in the territory of the municipality of the city of Beslan, RNO-Alania *In the collection: Geodesy, land management and cadastres: problems and development prospects, dedicated to the 100th anniversary of Soviet geodesy and cartography. Collection of materials of the I International scientific and practical conference* 320-324.
[10] Gadzhiev R K, Pekh A A, Kuchiyev S E 2020 Prospects for the development of perennial plantations of common hazel (Corylus avellana l.) On the lands of the Irafsky district of the North Ossetia-Alania *Land management, cadastre and land monitoring* **6** 24-28.
[11] Kuchiev S E, Rogova T A, Basiyeva L J 2019 Protection of lands from erosion processes and the formation of environmentally sustainable agricultural landscapes for the mountainous zone of North Ossetia *Land management, cadastre and land monitoring* **8 (175)** 54-59.
[12] Pekh A.A., Hugayeva L M, Katayeva M V 2019 Problems of cadastral work in the Republic of North Ossetia - Alania. In the collection: Geodesy, land management and cadastres: problems and development prospects, dedicated to the 100th anniversary of Soviet geodesy and cartography *Collection of materials of the I International scientific and practical conference* 325-328.
[13] Sekisov A N, Ovchinnikova S V, Norenko M S, Matevosyan V S 2020 Prospects for the solar energy development in housing construction: An assessment of economic efficiency *IOP Conference Series: Materials Science and Engineering* **913(4)** 042051
[14] Sekisov A, Ovchinnikova S, Grebneva V, Chernyshova M 2020 Modern directions of low-rise housing construction development in the world: Economic and technological aspect *IOP Conference Series: Materials Science and Engineering* **913(4)** 042035.

General methodological foundations of measurement accuracy control

A R Gubeladze, E N Yagovkina* and I O Gubeladze

Don State Technical University, 1, Gagarin sq., Rostov-on-Don, 344010, Russia

E-mail: yagovkinaalena@yandex.ru

Abstract. The article discusses the issues related to quality control of the construction of engineering structures, the approach to the accuracy of performing geodetic measurements at the current pace of construction and installation work, as well as during operation.

Introduction

The role of geodetic works in the construction of engineering structures and installation of technological equipment has recently significantly increased. Particular attention is paid to quality control of construction and installation works, increasing the requirements for the structural geometric parameters' accuracy, its structural elements. At the same time, a sequence of technical measurements is required at all stages of both manufacturing and installation of equipment. Deviation from the design dimensions should not exceed the established tolerances. The accuracy of geodetic measurements is regulated by SNiPs and other regulatory documents for the work production.

However, the compliance with the requirements established by the regulatory documents on the required accuracy of performing geodetic measurements at the current pace of construction and installation work is not justified. Sometimes it is necessary to use such a measurement technique that allows to ensure the required accuracy, which complicates the process of construction and installation work, i.e. they either slow down or pause the survey duration. Therefore, when using geodetic methods for monitoring the technical condition of a structure or equipment both during construction or installation, and during operation, a different approach to the measurements' accuracy is possible: necessary and sufficient. Moreover, two different approaches can be used to the object itself:

1) when its geometry is formed;
2) when the technical condition of the structure is determined.

We will proceed from the assumption that, starting work, we know nothing about the object from the point of view of its state.

In accordance with the stages of the engineering objects' existence, their geometry can be represented as:

1) nominal;
2) design;
3) actual;
4) deformative.

Thus, the technical condition of any object at various stages can be represented by the following graph (Figure 1).

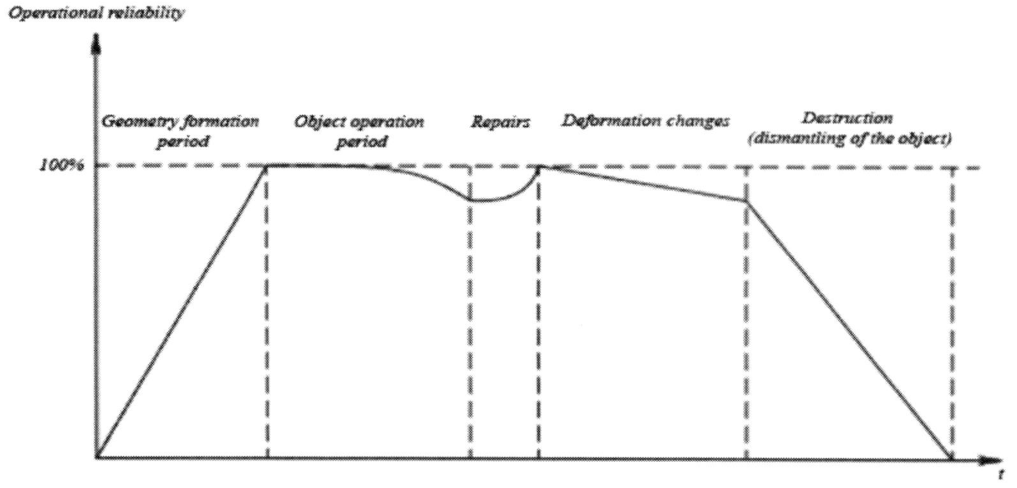

Figure 1. Technical condition stages of an object.

At the initial stage, during the period of construction and installation work, the geometry of the object is formed [1]. Its geometry changes from zero to 100% of the operational reliability. It should be borne in mind that the general state of the object consists of its actual state, taking into account the deviations arising from the technology of manufacturing individual structures and the production of construction and installation works, as well as deformation changes occurring due to an increase in the base loads, deformations of the Earth's surface on the site and other factors (Figure 2).

The next stage can be described as the operational period. At this stage, during the operation of an engineering structure, its geometry changes as a result of deformations arising from its elements' aging, equipment wear and a number of other reasons [2]. Consequently, its operational reliability is reduced. In this case, it becomes necessary to improve its reliability, i.e. a recovery period arises, as a result of which the object again reaches 100% operational reliability [3]. The last two periods can alternate repeatedly during the entire operation time of the structure. Ultimately, the object can undergo significant deformation changes, and, therefore, its geometry changes. In this case, elastic deformations transform into plastic ones, which leads to the destruction of the object.

According to Figure 2 an increase in deformation characteristics is observed at the initial stage. After the object has reached operational reliability, its deformation characteristics somewhat stabilize, although over time they continue increasing. Therefore, the technical condition of the object due to the resulting deformations will be a function of time t, i.e.

$$\xi_d = f(t). \quad (1)$$

While the actual geometry is formed up to 100% of operational reliability, after its formation it remains constant, i.e.

$$\xi_f = const. \quad (2)$$

The overall geometry is composed of the actual geometry and the geometry formed under the influence of the arising deformations, i.e.

$$\xi_0 = \xi_f + \xi_f. \quad (3)$$

IOP Conf. Series: Materials Science and Engineering 1083 (2021) 012053 doi:10.1088/1757-899X/1083/1/012053

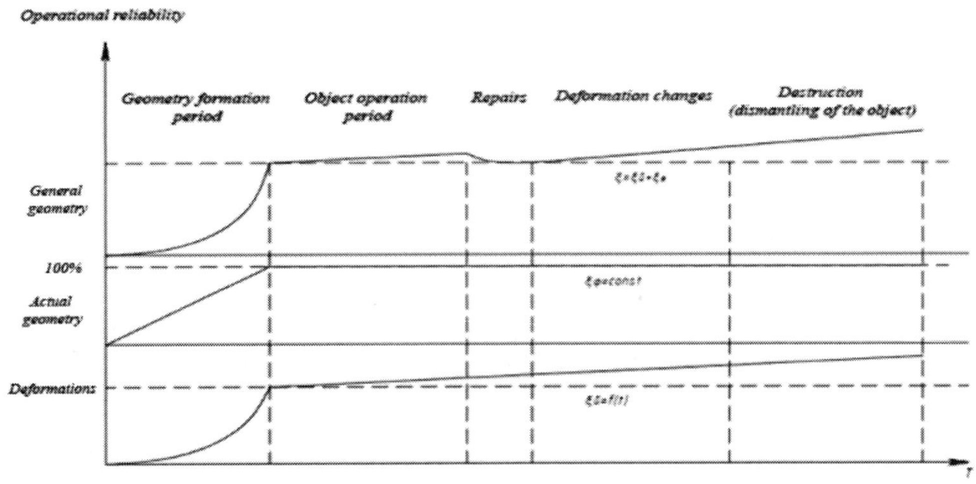

Figure 2. Forming the geometry of objects.

By establishing tolerances, and, consequently, determining the accuracy of geodetic measurements, it becomes necessary to assess the readiness of a structure for operation. Let us take the following approach to the assessment. Depending on the ratio of the structure parameters to the technological tolerances δ, let us consider the readiness of the structure, i.e. if the object is not ready for the operation, then the degree of its readiness ξ can be taken as zero, and in the case when the structure is completely ready, $\xi = 1$. Moreover, all its parameters correspond to the established tolerances. In all other cases, it is possible to talk about a partial readiness degree, estimated in fractions of a unit in the interval from 0 to 1. Therefore, depending on the readiness degree of the object, it is possible to establish sufficient and necessary accuracy of geodetic work (Figure 3).

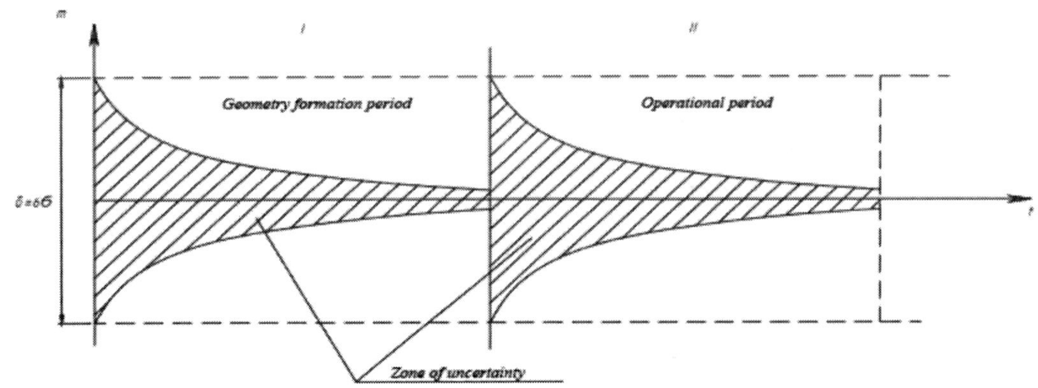

Figure 3. Graph of measurement accuracy m dependence on the operational reliability degree: I - when forming the geometry; II - when determining the technical condition of an object during operation.

Considering that at each stage of the state of the structure, a different approach to the accuracy of the production of geodetic works is carried out, we will establish the following mean square errors for measurements.

Let us consider the attitude to the accuracy of geodetic control at the initial stage, when the geometry of any technological equipment is still being formed [4]. At this stage of geodetic measurements, it is possible to use sufficient accuracy, i.e. below the norm. The root mean square error of the measurement result in this case will be a function of the maximum deviation of the parameters of the structure from the design:

$$m = C \cdot \Delta_{des}, \tag{4}$$

where m – is a mean square error of the measurement result;
C – is a precision factor;
Δ_{des} – is the maximum deviation of the parameters of the structure from the nominal.

As it can be seen from the graph (Fig. 3) when forming the object's geometry, the closer the technical state of the structure to the readiness level, the more reliable the information about the geometric parameters of our object should be, the less uncertainty, therefore, the measurement accuracy should increase.

After the facility commissioning, the accuracy requirements are sharply reduced, since in this case the necessary and sufficient information that the technical condition meets the operational standards, all the parameters of the structure do not go beyond the maximum permissible values, is required. Over time, the object is subjected to deformation measurements for a number of reasons, since the deformations occurring during this period are superimposed on its general geometry (Fig. 2). When the state of an object approaches a critical stress-strain ratio, the level of uncertainty should be significantly reduced, therefore, the requirements for the accuracy of geodetic measurements increase.

This approach will significantly reduce the time spent on construction and installation work. An example of this is the performance of geodetic control of the controlled object radial overhead travelling crane installation.

At the initial stage of the geodetic works' production, the accuracy coefficient was taken equal to 0.1. With the maximum deviation of the track axis points from the design 30 mm, the mean square error was 3.0 mm. As a result of the reassembly work, the largest deviations (15-30 mm) were first eliminated.

With further refinement of the track position, the accuracy of the work increased. At the second stage, the calculation of the root mean square error was made from the technological tolerance $\hat{o}_{max}$, i.e.

$$m = C \cdot \hat{o}_{max}. \tag{5}$$

The accuracy coefficient at this stage of work was taken from 0.3 to 0.1 of the tolerance, therefore, the root-mean-square error was 0.7 mm.

This made it possible to obtain the actual state of the rail track axis, which differs from the design one within the established tolerances. At the same time, the time for stopping installation work for the production of geodetic observations has significantly decreased.

Thus, the object after installation has reached a state where its readiness degree $\xi = 1$. The object is ready for operation. In the future, during operation, the structure or equipment is periodically subjected to geodetic control, taking into account the regulatory documents' requirements. According to the graph (Fig. 3), in this case, the technical condition of the object should be within the permissible values and the production of geodetic measurements can be performed with a lower accuracy in order to assess the technical condition of the object. In this case, to calculate a sufficient root mean square error of measurement, it is possible to use the dependence (5), taking the accuracy coefficient equal to 0.5. The root mean square error of measurement with a tolerance of 2.0 mm is 0.1 mm. To judge the object state, and in this case, the position of the overhead travelling crane path axis, such accuracy is quite enough.

Further operation of the structure leads to the fact that its elements, structures, individual parts are subject to wear, there is a displacement of the axes relative to the original position. The parameters deviate from the designed ones; therefore, the structure changes its technical condition. The degree of its readiness from 1 takes on the value $\ll 1$. If earlier it was stated that it was possible to use a sufficient degree of accuracy, now the more the structure deviates from its design position, the higher the accuracy of geodetic control over its condition. This is due to the fact that there is a limit of both the stress-strain state and the deviations of its structures from the designed ones for any structure. Further, even a slight movement of structural or equipment elements can lead to its destruction. Therefore, in such a period, the most reliable, most detailed information about the slightest changes occurring with the state of the structure is needed. To establish the accuracy of geodetic works, it is also possible to use the dependence (5), while setting the accuracy coefficient as $C = 0,1$.

A similar approach was taken when determining the technical condition of residential buildings. At the preliminary stage, when it was required to determine their technical condition, the definition of the rolls was carried out by the theodolite by projecting the upper corners of the buildings onto a horizontal ruler. The accuracy of determining the rolls was from 5 to 10 mm. This accuracy is quite enough to assess the state of the structure when the rolls reached 10, 20 or more centimeters.

When leveling the buildings, when it was required to set the values of the single point displacements on the plane, the accuracy of geodetic control over the rise of houses should be much higher. In this case, an electronic total station was used [5], the accuracy of work on determining the points coordinates on the building was about 1 mm. Especially such accuracy provided a high-quality rise of houses at the final stage.

Summary

As it can be seen, using the example of two objects, the proposed approach to the accuracy of geodetic control of the structure geometry during its formation and determination of its technical state is fully justified. Moreover, on the one hand, it facilitates the surveyors' work, on the other hand, it reduces the construction and installation work downtime during the period of geodetic measurements.

References

[1] Pimshin Yu I, Gubeladze A R, Arsenyev D M 2014 Technical control during the technological equipment installation *Engineering Bulletin of Don* **4 (2).** Information on http://ivdon.ru/ru/magazine/arcliive/n4p2y2014/2741
[2] Zabaznov Yu S, Gayrabekov I G, Pimshin Yu I 2010 Geodetic control of the geometry of the leveled building *Engineering Bulletin of the Don* **4**. Information on ivdon.ru/magazine/archive/n4y2010/250.
[3] Pimshin Yu I, Naugol'nov V A, Pimshin I Yu, Zabaznov Yu S, Yakovlev V V 2011 Method for determining the deformation characteristics of structures, Patent 2426089 RF., IPC G01 M 99/00. publ. 10.08.2011, bul. №22. http://www.fips.ru, http:// bankpatentov.ru/node/117239.
[4] Gayrabekov I G 2011 Method for determining the deformed state of buildings and structures *Engineering Bulletin of the Don* **1**. Information on ivdon.ru/magazine/archive/n1y2011/368.
[5] Pimshin Yu I, Zotov M V, Gubeladze A R 1998 Control of the dynamics of the spatial geometry of engineering structures by electronic tacheometry. *Sat. "Modern problems of mechanics and protection of the geological environment". Proceedings of the VII seminar. Rostov-on-Don: RSU.*

IOP Conf. Series: Materials Science and Engineering 1083 (2021) 012054 doi:10.1088/1757-899X/1083/1/012054

Investigation of the process organization of identifying an invalid test during construction expertise

V Mischenko, Yu Sergeev, A Sergeeva*, Yu Myasischev, R Myasischev

Voronezh State Technical University, 84, 20-letiya Oktyabrya ave., Voronezh, 394006, Russia

E-mail: 933947@mail.ru

Abstract. This article discusses the mechanism for studying the functional position of building structural units of a structure and identifying symptoms if there is a suspicion that the identifier influencing the building structure destruction was determined incorrectly. Approaches to determining the degree of damage to a building structure can use the methods based on mathematics and statistics. Calculated determination of a building structure failure degree cannot replace an expert-civil engineer, but provides significant assistance to him. The paper discusses the mathematical methods for determining the building structures' failure degree from both mathematical and technical points of view. Defect dynamics trend in a building structure in time is taken into account - a cardinal indicator when establishing the problem of finding the amount of working capacity loss. The stochastic methods of probability theory used with the fund recuperation of information about the work carried out to restore the building structures and their return to a working state provide a high level of determining defects in building structures. It has been proved that the policy of choosing a further examination reduces the total number of the studies performed, as well as reduces the examination duration, the moment of establishing the destruction degree determination of the building structure, which is important in some cases. The developed algorithm reduces the subjective moment both when establishing further examination and when determining the destruction degree of a building structure.

1. Introduction

Over the course of a certain period of building operation, under the climatic factors' dangerous moments influence and the technological processes distinctiveness, a regression of the structures' work occurs and there is a threat of deviation, in relation to the quality level set by the project [7]. Atmospheric phenomena, technological environment, man-made influences, factors influencing the building from the underground part are the reasons influencing the loss of the primary technical and operational properties of a construction. These processes increase the mechanical wear of the structure, thereby reducing the operational reliability reserve of the building, which can lead to the limiting state of the structure. The listed moments can cause an accident and destruction of the entire building [2, 3]. The performance of the survey, as a part of the technical and construction expertise, provides an opportunity to obtain technical information about the reliability level of building structures [11].

2. Approaches to determining a damage degree to a building structure

The intention of the technical survey is the synectics of the technical condition on a construction site, as well as the general operational activity, the synectics of the extensional absolutely necessary recreational developments, the pragmatization and feasibility of restoration, the express analysis of the root causes of defectiveness and malfunction or accidental damage to the construction site [4]. Obligatory nature of the work on the construction objects' functioning inspection, their extension, technical composition and purposefulness are determined by certain established intentions. Each stage of the survey includes a certain scope of work and a sequence of actions.

In approaches to determining the degree of damage to a building structure, discriminant analysis, Bayesian, neural networks, calculation trees based on the definitions of statistics and mathematics can be used [1]. These methods do not take into account the competence of an expert-civil engineer in matters of a building structure behavior during its destruction, the correlation of technological processes and identifiers that have an impact on destructuring. In the format of connecting stocks of technical and construction information into entirely mathematical methods, it is recommended to use the above-mentioned technical knowledge to ensure the detail and diversification of the statistical data, concretized by the accumulated knowledge base.

The information available for the civil engineer performing the examination makes it possible to supplement and concretize the distribution data. By such concretization and replenishment, in the mathematical approach, the statistics about the work of a building structure and the laws of the occurrence, development and outcome of technological processes that affect its reliability; timely assess the technical condition of the building structures of the technical facility; carrying out an express forecast about the existing possibility of the formation of malfunctions in construction facilities; developing procedures to eliminate or prevent these defects are included in the stock of information of an expert civil engineer [5, 6].

The maximum likelihood method is injected into the concept of probabilities as a generalized method for calculating the dominant complexity values by creating maximizing accepting of perception. The conclusion on the building structure reliability state is based on the time of the most probable destruction, and the probability of destruction is calculated according to the signs affecting this destruction.

In order to detect the certain parameters' probabilities of the present values that have an effect on the destruction of a building structure in any appearance of a flaw, the primary basis of information took place, and on this basis, rankings of a separate identifier that have an effect on destruction have been developed. The rankings of discrete properties influencing the destruction and dislocation consistency of these properties during fracture have been established. That is, the ranking and features affecting the structure destruction are well known $a_0 j = 1, 2, 3, \dots, n$

$$Q_{j\gamma}(a_j) = \{Q_{j\gamma}^{t1}(a_j), Q_{j\gamma}^{t2}(a_j), Q_{j\gamma}^{t3}(a_j), \dots, Q_{j\gamma}^{\tau}(a_j)\}$$

where $Q_{j\gamma}(a_j)$ – is the probability density value of the j- th feature affecting the destruction of a building structure with its value a_j and γ - m defect;

$t = 1, 2, 3, \dots, \tau(j)$ – define the moment (day) counted from the beginning of the change in the structure's properties.

Thus, according to the above-mentioned initial data for a separate identifier that influenced the failure of a building structure, it is possible to calculate the probability of an identifier that affects the building structure destruction - $Q_{j\gamma}^{t}(a_{0j}^{t})$ [10].

We will use the Bayes formula to determine one once-defined identifier affecting the destruction of a building structure [9]:

$$d_\gamma = \frac{d_\gamma Q_{j\gamma}^{t}(a_{0j}^{t})}{\sum_\gamma d_\gamma Q_{j\gamma}^{t}(a_{0j}^{t})} \quad (1)$$

where d_γ – is the probability distribution of detected defects in a building structure,

$\sum_\gamma$- defines the summary note of generalization $\sum_{\gamma=1}^{n}$ for each defect identified by an expert.

In the highlighted Bayes formula d_γ the objective possibility of a building structure failure in the right and left sides of the equation is interpreted in the same way.

With regard to d_γ the formula is solved (1).

To make the overview of this formula clearer, we will write it down for a pair of signs that destroy the structures: $\gamma = 1$ and $\gamma = 2$. Then the probability of the first sign of the building structure destruction has the form:

$$d_1 = \frac{d_1 Q_{j1}^t(a_{0j}^t)}{d_1 Q_{j1}^t(a_{0j}^t)+d_2 Q_{j2}^t(a_{0j}^t)} \tag{2}$$

Let us provide that in an expanded union of a pair of features $d_1 + d_2=1$. It is noticeable that the equality (2) has only two ways to resolve:

$d_1 = 0,\ d_2 = 1$ и $d_1 = 1,\ d_2 = 0$

The intention is to find a way to solve the problem, adequate from the point of view of finding a clear identifier that influences the destruction of the building structure. An iterative combination is recommended for the solution method. This iterative combination was used for the empirical formulation of the identifier determined by the formula (1) written in the form

$$d_\gamma^{m+1} = \frac{d_\gamma^m Q_{j\gamma}^t(a_{0j}^t)}{\sum d_\gamma^m Q_{j\gamma}^t(a_{0j}^t)} \tag{3}$$

The correlation index *m* used in the format of the quantitative parameter of the join of two variables. This indicator has the parameters from - 1 to + 1. If the indicator approaches zero, this indicates that there is no linear relationship. The linear relationship of the variables will be much closer than the indicator will approach $m + 1$ or $m - 1$. The result of the recommended iterative algorithm in practical application invariably converged; accordingly, the conditionality of determining the identifier that affects the destruction of the building structure was high [8].

The prerogative of the method lies in the fact that the iterative algorithm is unambiguously converged and this makes it possible to establish the sought probabilities with a range of calculus to a lesser extent.

3. Revealing the wrongly defined identifier

However, it happens that after the construction expertise and after the repair and restoration work, the defects similar to those that were found during the survey, continue to appear. This may mean that the cause of the defects was identified incorrectly [12]. In order to identify an invalid test, it is necessary to solve the following problem. One by one, tests or a group of tests are gradually removed from the previous surveys. After that, the most required survey is searched.

Let us introduce the S- operating time notation of a building structure without destruction. S_{max} can be achieved by applying the preventive measures for the existing technologies for the elements' recovery. The optimization problem is S_{max}, and the variable indicator is the number of the criterion (identifier), which influences the building structure destruction. The analogue of the functional purpose of the benchmark is ΔS, and a similar goal of improving - *max* ΔS, where ΔS – defines increase S as a result of the next survey; d – is a probability distribution of a diagnosed reduction in bearing capacity in a structure; γ =1, 2, 3,..., n – denote the number of the process destroying the structure; α – is a survey value, is interpreted as a parameter *k* (variable) and is an identical expression of any of the probable parameters of a feature that destroys the structure. The initial segregation of building structure defects was accepted by the consequential appraisal based on b – the previously diagnosed studies. The numbers of properties that destroy the structure are denoted by the letter $j:\ 1 \leq j \leq b$. α_{oj} – is a specific meaning of the j-th identifier influencing the destruction of the building structure, obtained as a result of the survey. Let us suppose the conversion range is also known $\forall a_m$, i.e a_m^{min} and a_m^{max}. The idea is to bring the non-destructive time of the building structure (S) to the maximum value. To do this, we (as a result of surveys) have to identify the signs that affect a building structure destruction.

If there is a suspicion that the identifier influencing the destruction of the building structure was determined incorrectly, then the data from the already carried out studies are removed. Its assumed parameter determines the total probability formula.

$$ac_{oj}^{j=dl} = \int_{a_{j\,min}}^{a_{j\,max}} a_j q(a_j) f a_j, \tag{4}$$

where

$$q(a_j) = \Sigma_\gamma p_\gamma q_\gamma(a_j). \tag{5}$$

With the given a_{oj}^{dl} there is d_{γ}^{new}.

$$a_{oj}^{j=dl} + a_{oj} \;\; j = 1,2,3, \dots, b, b-1, j \neq dl \rightarrow d_{\gamma}^{new}. \tag{6}$$

Following this, we will rename $d_{\gamma} = d_{\gamma}^{new}$ and circularly reproduce the calculations by the formulas (4) - (6) until determining a_{oj}^{dl} and d_{γ}. Found a_{oj}^{dl} and $d_{\gamma j}$ (more precisely, a comparison of these values with the verified test values and with the resulting failure probabilities) convincingly explain the reason and need for re-verification of the tests $j = dj$.

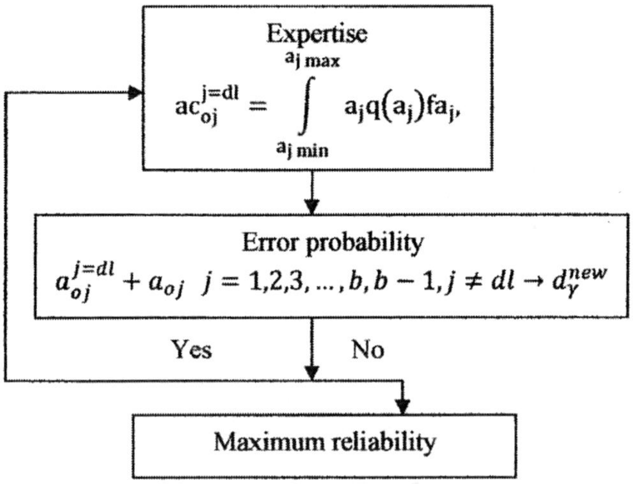

Figure 1. Survey control algorithm.

Rechecking is performed when specifying if the criterion S changes dramatically. For example, provided that after the suspicious estimate was excluded, and the S value (calculated by this method) has changed significantly, in this case it is necessary to double-check the survey. Whether the tests need to be repeated or not, can be decided in the same way as it had been found out, which test is the most important to perform. The survey control algorithm (Figure 1) has been designed to remove stochastic errors and to increase the surveys' reliability.

4. Calculating the likelihood of re-examination

The probability of re-examination j is calculated taking into account the previous surveys j. In some examinations, the tests that need to be repeated are identical to the primary ones, in other examinations the tests that need to be carried out are completely different. When answering the question of what further research should be done, those that have not been conducted yet and those that have already been completed, are taken into account. For those that have already been fulfilled, the need to repeat them again is determined. This is done for control, to remove stochastic errors, as well as to increase the surveys' reliability. A problem, which consists in answering the question at what point it is necessary to stop the measurement process of the next survey, or it should be continued, appears. Provided that the number of checks is not fully used, it is possible to develop the conditions for interrupting the procedure for finding the k calculation. In the event that the number of surveys is completed, and the specified criteria are incomplete, all this raises doubts whether the full composition of the applied surveys was really selected.

Sometimes the developed criteria depend on the price of the survey, as well as on what profit we will have after the building structure is restored. The recommended criteria do not depend on prices. In the form of a mandatory requirement to terminate the search for further examination, let us consider the correlation:

$$\sum_{\gamma} \frac{|d_{\gamma}^{new} - d_{\gamma}|}{d_{\gamma}} \leq \varepsilon_{d}, \tag{7}$$

$$\Delta S \leq \varepsilon_{\Delta S}. \tag{7a}$$

Inequality (7) reflects the stabilization process of refining the destruction degree determination (7a) and demonstrates that the additional examinations' profitability turned out to be small.

Since S has the moment dimension, therefore, we can solve the problem of the validity of the quantities ΔS. Another limitation of the smallest value ΔS and other criteria represents the correctness of the S, d_γ calculation and other parameters. The limiting conditions for the completion of calculations can be obtained by comparing with a building structure that does not have defects:

$$\frac{S^{new} - S_{11}}{S_{11}} \le \varepsilon_S. \tag{7b}$$

Where S_{11} – is the S value for an ideal structure, since the initial index of the first matrix refers to the fact that there are no defects, and the second index is one to the building structure restoration.

Another requirement is the greater reliability of the diagnosis:

$$(1\text{-max } d_\gamma) < \varepsilon_d. \tag{7c}$$

A series of features can be obtained by comparison with the ideal restoration of a building structure. A stricter rule is applicable to the fulfillment of the inequality conditions (7) - (7c) in a series of, for example, 3 calculations k performed one after the other. It is important that the k value found in a fragment of these calculations, for example, in 2 calculations, should be used.

Summary

The developed policy for selecting the next survey reduces the number of surveys performed, reduces the survey period, the duration of determining the destruction degree of a building structure, makes it possible to determine the identifier affecting the destruction of a building structure, which is fundamentally important for preventing a building accident. The developed algorithm allows minimizing the subjective factor when appointing the next examination, as well as when determining the destruction level of the examined structure.

References

[1] Mischenko V Ya, Emelyanov D I, Tikhonenko A A, Startsev R V 2012 Stochastic algorithms in solving multicriteria problems of optimization of resource allocation when planning construction and installation works *Scientific bulletin of the Voronezh State University of Architecture and Civil Engineering. Construction and architecture* **1** 92-97.
[2] Mischenko V Ya, Drapalyuk D A, Solntsev E A 2009 Monitoring of defects and accounting of the building structures' aging of of the housing stock *Scientific Bulletin of the Voronezh State University of Architecture and Civil Engineering. Construction and architecture* **4** 118-123.
[3] Mischenko V Ya, Golovinskiy P A, Drapalyuk D A 2009 Forecasting the rate of depreciation of the housing stock based on the monitoring of defects in building structures *Scientific Bulletin of the Voronezh State University of Architecture and Civil Engineering. Construction and architecture* **4** 111-117.
[4] Mischenko V Ya, Sergeev Yu D, Myasischev R Yu, Sotnikov O A 2015 A systematic approach to improving the operational quality of industrial and civil buildings *Scientific Bulletin of the Voronezh State University of Architecture and Civil Engineering. Construction and architecture* **4 (40)** 49-57.
[5] Myasischev Yu V, Sergeeva A Yu, Sergeev Yu D, Myasischev R Yu 2018 Development of a model for monitoring industrial and environmental safety for an objective assessment of the state of loads and bearing capacity of structures *Construction and real estate* **1-1 (2)** 63-67.
[6] Sergeeva A Yu, Sergeev Yu D, Krupenko S E 2014 Assessment of the system proximity to a crisis state *Economics and Management of Control Systems* **2.1 (12)** 215-218.
[7] Sergeeva A Yu, Myasischev Yu V, Myasischev R Yu, Sergeev Yu D 2017 Investigation of emergency state signs of load-bearing constructions of buildings and structures *Modern trends in the construction and operation of real estate, a collection of scientific articles based on the materials of the scientific and practical conference* 218-223.
[8] Sergeev Yu D, Mischenko V Ya, Myasischev Yu V, Myasischev R Yu 2019 Optimization of the

inspection process of load-bearing structures of pre-emergency buildings *FES: Finance. Economy* **16** 52-56.
[9] Hanson K M, Cunningham G S 1996 A computational approach to Bayesian inference *Computing Science and Statistics. VA 22039-7460* 202– 211.
[10] Cadez I V, Smyth P, McLachlan G J, McLaren C E 2002 Maximum Likelihood Estimation of Mixture Densities for Binned and Truncated Multivariate Data *Kluwer Academic Publishers* 7-34.
[11] Reza E, Richard E 2009 Knowledge management in construction companies in the UK *Al & Society* **24 (2)** 197-203.
[12] Liao J G, Lipsitz S R 2002 A type of restricted maximum likelihood estimator of variance components in generalized linear mixed models *Biometrika* **2** 401-409.

IOP Conf. Series: Materials Science and Engineering 1083 (2021) 012055 doi:10.1088/1757-899X/1083/1/012055

Engineering in modern construction

E Yu Romanenko[1*], N V Vodolazskaya[2]

[1] Don State Technical University, 162, Socialisticheskaya str., Rostov-on-Don, 344000, Russia
[2] Belgorod state agricultural university named after V. Gorin, 1, Vavilova Street, Mayskiy, Belgorod region, 308503, Russia

E-mail: Romanenko-RGSU@mail.ru

Abstract. The article highlights the issues related to the quality assurance of products and structures manufactured by the building complex subdivisions. The problems of this type complex objects' engineering are revealed, and an assessment of the technological production features as a sequential, interdependent process is given. The article considers the introduction effectiveness of the innovative technologies with digitalization elements in the process of manufacturing the reinforced concrete products and structural elements and the construction of buildings and structures on their basis, as well as the importance of engineering in modern construction in general. Based on the theoretical analysis, the recommendations for the use of engineering methods for reorganizing a business process with the controlled parameters technological operations and with the established quality indicators for the finished products are proposed.

Introduction

One of the most important components of a quality assurance system is improving the industrial products' reliability. At the same time, the methods for achieving the appropriate level of reliability are specific for each stage of the life cycle for the products and structures [1, 2]. This direction is especially important when considering the operational reliability issues, which is determined by the maintenance organization and includes such categories as maintainability, reliability, preservation, safety and durability. A fairly large number of works related to various management areas, including construction, are devoted to the problems of researching operational reliability, which in modern conditions is the most developing industry. The launch of new capacities for the mechanical engineering enterprises, instrumentation, the device of new transport interchanges and an increase in the volume of housing construction is impossible without activating the work of enterprises in the construction industry.

Construction industry plays a significant role in the Russian economy. Construction ensures the reproduction of fixed assets in all economy sectors, has numerous intersectoral connections and a high multiplier effect. More than 50% of total investment in fixed assets falls on the construction of buildings (residential and non-residential) and structures. The construction industry accounts for 6% of the total gross value added (GVA) for the economy as a whole. And, at the same time, the measures aimed at ensuring safety, both during the production of work and during the operation of buildings and structures, are gaining more and more speed. These actions are aimed at implementing the requirements of the Federal Law "On Technical Regulation" [3], in terms of the safety issues reflected in the Federal Law of the Russian Federation of December 30, 2009 No. 384-FL "Technical Regulations on the Safety of

Buildings and Structures" [4].

The main provisions on the construction industry and the strategy for its development are set out in the Draft Strategy for the Development of the Construction Industry in the Russian Federation until 2030 [5]. This industry is defined as a complex of participants in urban planning activities. It includes state and municipal authorities and administrations, organizations, enterprises, individuals engaged in the creation, preservation, improvement and utilization of buildings and structures (in the form of territorial planning, urban zoning, architectural and construction design, construction, overhaul, reconstruction, technical re-equipment, renovation, demolition), and the system of interaction between them at all the life cycle stages of capital construction objects and territories to form a comfortable and safe environment for people's life and activities [5]. The composition of the industry participants includes: citizens, developers and construction companies, contractors, design and expert organizations, engineering research organizations, financial and development institutions, government bodies that exercise regulation and supervision, self-regulatory organizations, educational and scientific organizations. The first and main principle of the Strategy is to focus on satisfying the consumers' rights, regardless of who they are represented by.

The implementation of this principle is possible only if there is a competent strategy aimed primarily at the innovative activity and labor productivity growth in the industry, which will be ensured by the improvement of legal and technical regulation and take into account the territorial specifics and the constituent entities' socio-economic development diversification high degree in the Russian Federation. This is impossible without expanding the information modeling technologies' use at all stages of the capital construction facility life cycle. The transition to such regulation of all urban planning processes, the criterion of which is coherence and algorithmicization, will make it possible to transfer all administrative procedures and technological processes into electronic form without exception. At the same time, in order to implement the Federal Law [3], the Federal Agency for Construction and Housing and Communal Services of Russia developed and agreed with the interested bodies a draft system of normative documents [6], which will ensure the transition to new methodological principles based on the technical regulations (TR) that find more and more widespread in the practice of building regulation and standardization of the developed countries [7].

Similar methodological principles are implemented in the process of scientific research. The level of such research work aimed at solving the scientific, technical, technological solutions and improving the quality of the source material, the level of the organizational and production process, product quality control at all stages begins with research and development work and ends with the work on implementation and transition for serial manufacturing of the products with the required quality parameters and the device based on them, durable and reliable buildings and structures [8]. For the widespread implementation of the scientific developments' results, there is no information base that makes it possible to evaluate, implement and ensure the positive results and possible advantages of innovative methods for the materials and products' manufacturing in the conditions of factory production and the construction of buildings and structures for various purposes on their basis. The withdrawal of the existing capacities for the production of reinforced concrete products and structures and the entire construction complex to a fundamentally new technical level is economically expedient both in time and in cost, which justifies the prospects of this direction in the construction industry development.

The problem analysis, goals and objectives of the work

The construction complex work is a diverse complex process aimed at meeting the consumer demand for finished products - buildings, structures, structures. For a fairly long period of time, the construction industry worked on the basis of the process ultimate goal fastest implementation principle - the construction of buildings and structures. Little attention was paid to the actual conditions, demand in the consumer market, meeting the urgent needs of the market and issues and the problem associated with the final stage - the disposal of products from the construction industry. Modern conditions of the consumer demand market have changed the very approach to this problem. More and more attention is paid not to production itself, but to the issues related to the justification of the need for this step, the prospects for finding the consumer demand, efficiency of use, time of the main life cycle and disposal complexity. And insofar as modern living conditions dictate new requirements, it is necessary to take a new approach to the organization of construction production as a whole.

For a more complete understanding of this problem, it is proposed to consider the life cycle of the construction products manufacturing.

The life cycle is a set of interrelated processes of sequential changes in the state of a product from the formation of initial requirements for it to the end of its operation or application. The life cycle of construction products includes a number of stages:

1) marketing, search and market research;
2) design and development of technical requirements, product development;
3) material and technical supply;
4) preparation and development of production processes;
5) production;
6) control, testing and inspection;
7) installation and operation;
8) technical assistance and mantainance;
9) disposal after operational reliability loss, etc.

Research, design and development of any product is based on promising marketing research, fundamental and predictive SRW and scientific research, applied marketing research and rationing of requirements for product quality, taking into account regional characteristics [9 - 11]. This should also include the conduct of the applied research, development of technical specifications and assessment of the design and technical level of products; drawing up a business plan for the development and manufacture of the products with the required quality parameters, attracting investments; development of technical documentation and expert work, marketing and commercial testing of construction products, study, analysis and experience use in the creation and operation of similar products at all stages of its life cycle, the selection of target markets, which is carried out in parallel with the design and development of products; development of a set of marketing activities.

The stage of the product life cycle is its conditionally allocated part, which is characterized by the specifics of the work performed in this area and the final results. A conditional representation of the life cycle of construction products can be obtained from the diagram, shown in Figure 1.

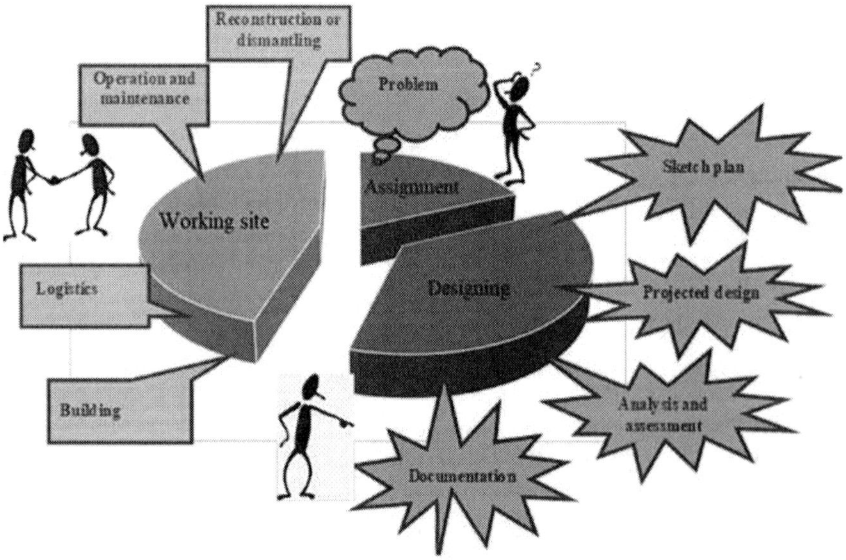

Figure 1. Stages of the construction products' life cycle

It is proposed to analyze some aspects of this situation from the point of view of the phased formation of problems, design and technical solutions. First of all, when designing, it is necessary to keep up with the times and provide the building complex with the products that meet the requirements of consumer demand in modern market. From the modern construction projects' features analysis, it is obvious that this market for meeting consumer demand is dominated today by high-rise buildings, which in general is inherent in densely populated megalopolises. Evaluating modern buildings and structures, we can unequivocally say that when designing them, all modern requirements are taken into account, designed to ensure the technical safety and operational reliability of newly erected buildings and structures. However, this situation forces us to take a completely new look at the problem of forming consumer demand, the problem of meeting it and the economic feasibility of this step as a whole.

As already noted, the security requirements and the operational reliability of buildings and structures have led to the need to increase the requirements for the quality and technical properties of both materials and products, as well as the process of erecting buildings and structures using them. These requirements form other approaches to ensure compliance with the original regulatory requirements for the materials' quality, for the preservation of quality parameters throughout the entire technological cycle, for the parameters of the technological process, as a guarantee of the operational reliability of buildings and structures erected and put into operation.

Research part, recommendations for manufacturing enterprises

Modern requirements for the organization of technological processes are, first of all, the introduction of individual operations' automation, as well as the devices for operational control and non-destructive methods of the technological parameters' quality control of the process and finished products. But first of all, we should talk about the possibility of providing the industry with high-quality raw materials. This problem is dealt with by the engineering service, which ensures the internal and external circulation of material resources.

Any production has the ability to make a decision to use finished products from other industries in its work or organize its own production, which will ensure the predicted quality of both materials and products made on their basis.

In this situation, the process of selecting suppliers is becoming increasingly important.

The main function, however, lies with the marketing services. Their main function becomes: search for the information about suppliers; request for information on prices; processing of commercial offers; holding a tender and selecting the suppliers based on tender results.

The block diagram (process diagram) is shown in Fig. 2.

In the presence of the raw materials' reliable source, a building complex enterprise can quite effectively use the opening opportunities. This is, first of all, the provision of materials of the required quality and volume, the absence of disruptions in the supply of concrete mixtures, the ability to control the process of concrete strength gain, since all technological parameters of the concrete mixture are known and the development of hardening acceleration modes will not cause great difficulties. At the same time, it is possible to maximize the time of formwork removal by providing continuous control over the strength of concrete in the formwork and thereby not only increase the efficiency of its turnover, but also ensure that the concrete reaches the required strength, which will avoid the appearance of defects in the supporting structures. Quality control of finished products and structures, in such a situation, involves checking the geometric dimensions and shapes of products, the quality of the outer surface, the size of the protective layer, the position of reinforcement and embedded parts, which will ensure that buildings and structures comply with the requirements of the project as a whole [8, 12, 13, 14]. At the same time, a number of the related tasks of the enterprise (organization) are being solved, namely:

- increasing production efficiency by selling excess concrete mix to third parties,
- the ability to realize surplus equipment, formwork, etc. using the three-level system: construction equipment - sale of building material and sale of unused construction equipment on the one hand and a branded landing page for the organization's construction services: rental of construction

equipment - general construction work - external and internal finishing - engineering and technical support on the other side.

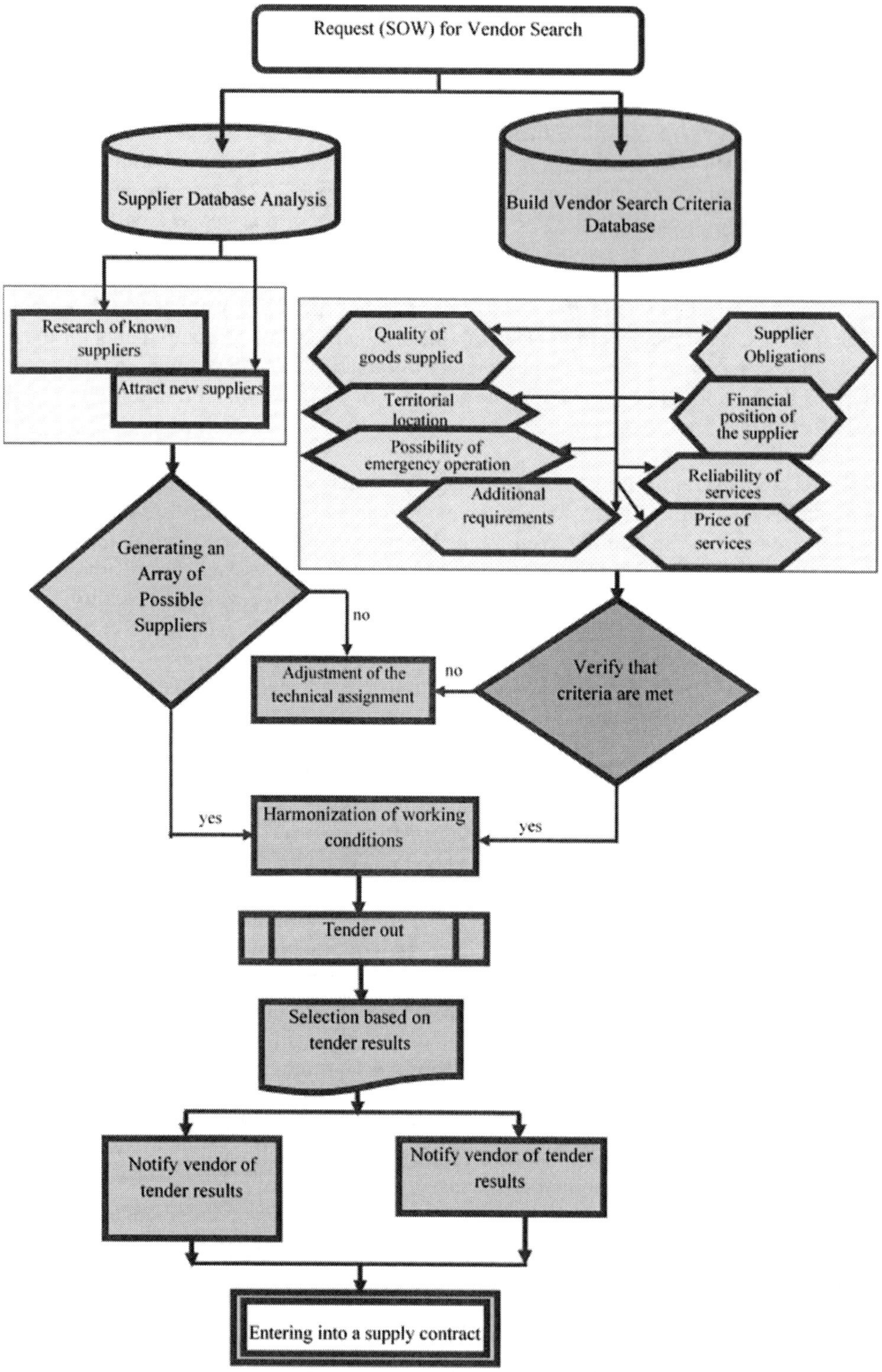

Figure 2. Process approach to supplier selection

Summary

Analyzing the above-said, we can conclude that the business process reorganizing effective engineering methods' introduction into production according to the operational form is possible by transferring the technological process from the category of assessing the quality of the finished product to the category of organizing production with the controlled parameters of the technological process and with the established quality parameters of the finished products. At the same time, digitalization and automation of the technological processes and product quality operational control, transfer of their management to an electronic platform, will significantly improve the production culture at enterprises, establish the causes of defects and improve the quality of products.

References

[1] Patuk I, Hasegawa H, Borodin I and Pavlyuk R 2020 Measurement of Strain Induced on a Deep Placement Fertilizer Applicator by Static load and Finite Element Simulation // *IOP Conference Series: Materials Science and Engineering, 1st International Conference on Innovative Informational and Engineering Technologies (IIET-2020) 28-29 May 2020, Stavropol, Russian Federation* **873** 012033. doi: 10.1088/1757-899X/873/1/012033.

[2] Kravchenko I N, Tretyakov A M, Erofeev M N 2003 Mathematical description of the approximating function of the process of wear of the blades of concrete mixing machines *Construction and road machines* **1** 36-38.

[3] Federal Law of the Russian Federation of December 27, 2002 No. 184-FL "On Technical Regulation" (November 28, 2018) [Electronic resource] // Electronic fund of legal and normative-technical documentation. Information on http://docs.cntd.ru/document/901836556 (date of access: 23.10.2020).

[4] Federal Law of the Russian Federation of 30.12.2009 No. 384-FL "Technical Regulations on the Safety of Buildings and Structures" (July 2, 2013) [Electronic resource] // Electronic Fund of Legal and Normative and Technical Documentation. Information on http://docs.cntd.ru/document/902192610 (date of access: 23.10.2020).

[5] Project from 06.12.2019. Development strategy for the construction industry of the Russian Federation until 2030. [Electronic resource]. Information on https://www.garant.ru/files/3/7/1334573/strategiya-razvitiya-stroitelnoy-otrasli-rossiyskoy-federacii-do-2030-goda.pdf (date of access: 25.10.2020).

[6] Resolution of the Government of the Russian Federation of 04.07.2020 N 985 "On approval of the list of national standards and sets of rules (parts of such standards and sets of rules), as a result of which, on a mandatory basis, compliance with the requirements of the Federal Law" Technical Regulations on the Safety of Buildings and Structures " and on recognition as invalid of some acts of the Government of the Russian Federation "[Electronic resource] // Consultant Plus: reference legal system. Information on http://www.consultant.ru/document/cons_doc_LAW_356807/ (date of access: 25.10.2020).

[7] Vodolazskaya N 2019 Types and ways of modernization in a context of the international experience *Virtual Economics, London, England* **2** **2(1)** 81–93. https://doi.org/10.34021/ve.2019.02.01(5).

[8] Romanenko E Yu, Nalimova A V 2020 The engineering of the reinforced concrete products manufacturing business process as a guarantor of the buildings and constructions' operational reliability *IOP Conf. Series: Materials Science and Engineering* **913** 032044. IOP Publishing doi:10.1088/1757-899X/913/3/032044.

[9] Vodolazskaya N V 2013 Problems and Prospects for Improving Regional Marketing Strategies *Eastern European Journal of Advanced Technologies* **1/10(61)** 95 – 98.

[10] Purwanto M R, Mukharrom T, Zhilyakov D I, ect. 2019 Study the importance of Business Ethics and Ethical Marketing in Digital Era JCR **6(5)** 150-154. doi: 10.22159/jcr.06.05.26

[11] Vodolazskaya N 2009 Models of network planning and management of power-consuming industries *Application of new technologies in management. ANTiM 2009. Proceedings.Vrnjačka*

Banja. Serbia **2** 811 – 818.

[12] Romanenko E, Ryabichenko S 2019 Construction sector enterprise product quality as a result of strategic innovation management *IOP Conf. Ser.: Mater. Sci. Eng.* **698** 022074. doi: 10.1088/1757-899X/698/2/022074.

[13] Drozdov Anatoliy 2018 Improvement of assembly technique efficiency for installation of buildings and related structures in light building constructions *MATEC Web Conf.* **251** 03042. doi: https://doi.org/10.1051/matecconf/201825103042.

[14] Kim, Hyun Tae, et al. 2017 The Measurement of SVOC Emission Rates from Building Materials *Materials Science Forum, Trans Tech Publications, Ltd., Mar. 2017* **893** 369–374. Crossref, doi:10.4028/www.scientific.net/msf.893.369.

A non-standard mark's center recognition correctness by a 3D laser scanner

D A Gura[1*], G G Shevchenko[1], N A Dyakova[1], S K Pshidatok[2]

[1] Kuban State Technological University, 2, Moskovskaya Street, Krasnodar, 350072, Russia

[2] Kuban State Agrarian University, 13, Kalinin Street, Krasnodar, 350044, Russia

E-mail: gda-kuban@mail.ru

Abstract. Terrestrial laser scanners have sophisticated specialized software that makes it possible to measure three-dimensional coordinates of a huge number of points in a short period by registering vertical and horizontal angles. To determine the spatial position of the scanner and the coverage of the area of interest on one scan, special marks are placed around the scanner. These marks are used for further high-precision scan matching. The problem of the process of recognizing the center of black and white marks during field shooting, as separate from the scanning process itself, is a separate task, which has not been sufficiently studied. The purpose of the experiment is to establish the correctness dependence of the mark center recognition on its various characteristics. The technique is the experimental component and theoretical analysis of the obtained material. The experiment was carried out under the same conditions, so that the influence of the distance from the device to the mark, a change in the angle between the laser beam and the mark plane could not affect the recognition correctness of the mark center. In the course of the experiment, 8 marks were prepared with a non-standard view and size, as well as one standard black and white mark, to check the reliability of the data obtained (diameter 74.1 mm, two opposite sectors are black, and the other two are white). This experiment showed that the process of recognizing the mark center depends on the combination of the following characteristics: mark geometry, mark color, the opposite sectors of the mark.

1. Introduction

One of the most widespread geodetic instruments in the XXI century is a ground laser scanner, which is being replaced by an air laser scanner [1]. Laser scanners are increasingly used in various branches of human activity, including laser scanning data that can be used to create geographic information systems [2] and monitoring [3-5]. This is due to the growth in the number of scanners among the contractors and understanding of using three-dimensional laser scanning method possibilities among the potential customers. Laser scanners can directly measure three-dimensional coordinates of a huge number of points in a short period [6]. Such features of the method give an opportunity to talk about the possibilities of its application in any areas where information on the objects' size and their spatial position is needed [7].

To determine the spatial position and coverage of the area of interest on one scan, special marks are placed around the scanner [8]. These marks are the points of the working space, and for the accuracy of the subsequent comparison of point clouds, multiple measurements of the centers of special marks are performed [9].

IOP Conf. Series: Materials Science and Engineering 1083 (2021) 012056 doi:10.1088/1757-899X/1083/1/012056

Marks are required for stitching scans based on the laser scanning results. Such marks can solve two problems:

- difficulties during scanning (possible scanner orientation loss during scanning);
- problems associated with achieving high accuracy of matching point clouds.

By merging point clouds across marks, the ground-based laser scanning software organizes all scans using the unique geometry of the object as a reference. To do this, the complex calculation algorithms are used.

In [10], the author notes that when performing work with a ground laser scanner and the subsequent merging of the obtained point clouds, the optimal number of special marks is equal to 8, located in a circle around the scanner. The distance of marks from the scanner should be no more than 100 m. Due to violation of these requirements, stitching by marks does not always give an accurate result. Then, on the final merged point cloud, some defects may appear (ghosting, reversal of scans) [11].

The location of the marks and their distance from the scanner is of great importance, but first it is necessary to recognize the data center of the marks. The center mark recognition process can be done in two ways:

- in software [12,13];
- directly in the field when performing laser scanning.

For example, in [10], the author describes the stages of laser scanning. One of them is the identification and determination of the approximate coordinates of the special marks' centers. The author also writes that, depending on the scanner model, these operations are performed either on the received scan, or on a digital image, if the scanner is equipped with a digital camera, but the process of recognizing the mark center itself is not considered.

In [14], the author divides special marks into two groups, flat and three-dimensional. At the same time, he claims that the determination of the center of special flat marks (Figure 1) is due to its contrast with respect to the surrounding background, and the volume mark (Figure 2) is based on their shape and size data.

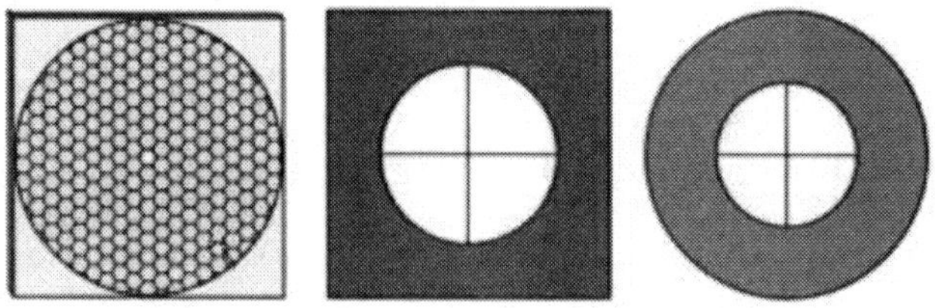

Figure 1. Types of flat special marks.

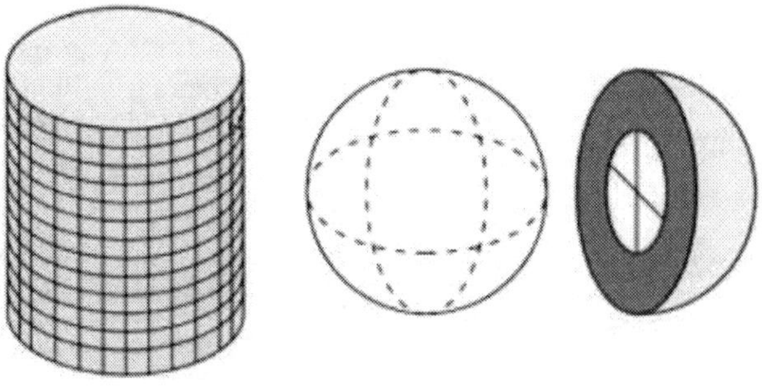

Figure 2. Types of three-dimensional special marks.

However, production experience has shown that often surveyors have to use a more budgetary version of marks - flat black and white marks (Figure 3), however, it is not available to everyone. The process of recognizing the data center of marks during the field survey, as an individual task, separate from the scanning process itself, is not sufficiently studied.

Figure 3. Standard Mark Result

To study the process of recognizing the center of marks during field work, it is necessary to set the following experiment goal - to establish the dependence of the correct recognition of the mark center on its various characteristics, for the possibility of further use of marks of own production.

From the goal setting, the following tasks follow, which need to be solved:

- to check the dependence of the recognition correctness of the mark center on its size;
- to determine the recognition correctness dependence of the mark center on its geometry;
- to establish the correctness dependence of the mark center recognition on the color of the mark sectors.

2. Materials and methods

To achieve the set tasks, the process of recognizing the mark center was investigated using three-dimensional laser scanning technologies and a ground-based laser scanner Leica ScanStation C10. The experiment was based on the classic black and white mark used in production. This is a standard 74.1 mm black and white mark, two opposite sectors of which are black and the other two are white (Figure 3).

Before performing the experiment, the scanner was installed at a distance of 5 meters from the wall, the marks were glued to the wall so that the sighting beam was perpendicular to the mark plane. The experiment was carried out under the same conditions, without changing the scanner position and with the same instrument settings. Thus, the influence of the distance from the device to the mark, a change in the angle between the laser beam and the plane of the mark could not affect the correct recognition of the mark center.

3. Results

During the experiment with mark 1, no influence on the size of this mark was found (Fig. 4).

The influence of the mark geometry on the recognition of its center was tested on mark 2, mark 3 and mark 4 (Figure 4). Scanning these marks showed that center recognition is not based on orientation to the center of the circle (mark 2). Changing the geometry of the opposite sectors also does not affect the recognition of the center (mark 3). When initially aiming at the areas remote from the center, recognition passes without errors (mark 4).

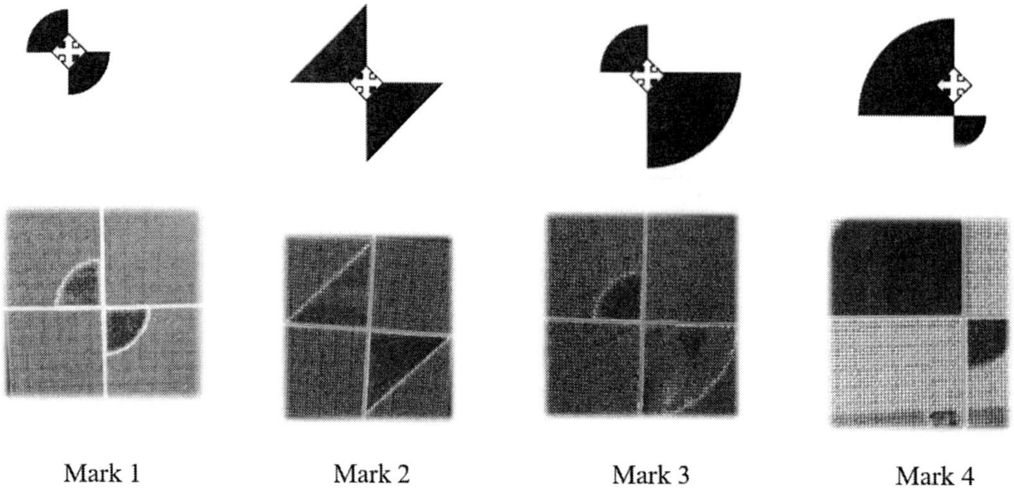

Figure 4. The result of experiments with marks

From the work [15], it can be concluded that the process of recognizing the mark center depends on the wavelengths, by finding the maxima of the reflected signals in the mark center, or simply on the color. In this regard, a study of mark 5 was carried out, the center of which was absent, but the device could determine its center (Figure 5).

Since white surfaces have nonselective reflectivity (reflectance coefficient more than 60%), and black surfaces have nonselective absorptive power (reflectance coefficient less than 10%), it was assumed that it is the contrast of white and black that is necessary for the program to recognize the mark center.

To check this relationship, the color of the standard mark was changed to blue and yellow (mark 6), but the scanner recognized the center of this mark (Figure 5). Making an assumption about the possibility of the opposite white sectors' influence, an experiment was carried out with mark 7 (Figure 5). In the experiment with mark 7, all sectors were replaced with primary colors with an optical spectrum: red 625 - 780 nm, yellow 565 - 590 nm, green 500-565 nm, blue 440 - 485 nm, the scanner could not determine the center of mark 7 (Figure 5).

Replacing the sectors of the standard mark with yellow, blue and two opposite ones with red, it was concluded that 2 identical opposite sectors are of great importance. This dependence was noticed earlier on marks 1-6, where it was necessary to find dependence on the size, geometry and color of the sectors (Figure 5).

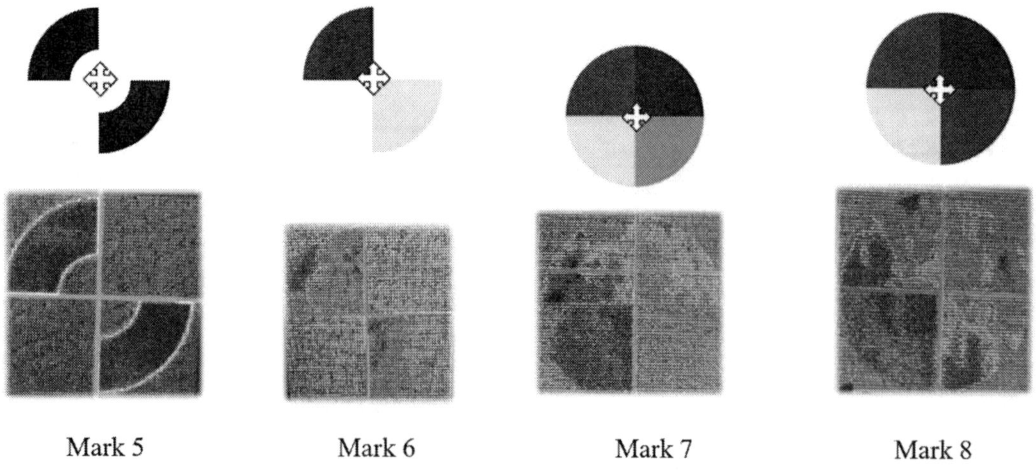

Figure 5. Result with mark experiments

4. Discussion

Thus, for correct recognition of the mark center by the scanner, all marks given during the experiment (with the exception of the mark 7) can be applied.

It should be noted that marks 1- 4, 6 and 8 had the same opposite colors of the sectors, and based on these marks, a false assumption had initially been made. This assumption stated that the process of recognizing the mark center depends only on the wavelengths, by finding the maxima of the reflected signals in the mark center, or simply on the color of the opposite sectors, which could later be confirmed by an experiment with mark 7. However, the study showed that in the case of mark 5 having excluded the mark center with the assumed maxima, the scanner correctly recognized its center, refuting the false assumption. Thus, the use of special or standard black and white marks is not required. Since the changes in some characteristics do not affect the correct recognition of the mark center, and therefore the correctness of the point clouds' comparison. However, this approach can only be relevant for determining the mark center in the field using the Leica ScanStation C10 scanner.

Summary

On the basis of the experiment carried out with a ground laser scanner Leica ScanStation C10, we checked the dependence of the correct recognition of the mark center on its size, determined the dependence of the correct recognition of the mark center on its geometry, established the dependence of the correct recognition of the mark center on the color of the opposite sectors of the mark. This experiment showed that in the process of recognizing the center of the mark, not individual characteristics of the mark are taken into account, but their totality.

For completeness of the study on this issue, it is useful to perform the additional studies on other terrestrial laser scanners. In the course of the research, the following tasks are required: determining the capabilities of scanners and further determining the dependences of the mark center identification during field shooting from the changes in the mark characteristics, as well as the possibility of using non-standard marks in work with these scanners.

References

[1] Benjamin Brede, Kim Calders, Alvaro Lau, Pasi Raumonen, Harm M Bartholomeus, Martin Herold, Lammert Kooistra 2019 Non-destructive tree volume estimation through quantitative structure modelling: Comparing UAV laser scanning with terrestrial LIDAR *Remote Sensing of Environment, November 2019* **233** 111355.

[2] Gribkova I S, Dyakova N A, Tsarkova Yu A 2020 The use of GIS for managing a single real estate complex on the example of the oil and gas industry *In the collection: Youth Science. Collection of the best scientific works of young scientists. Krasnodar* 33-35.

[3] Gura D A, Dubenko Yu V, Markovskiy I G 2020 Monitoring of transport infrastructure facilities using scanning technologies *Technosphere safety technologies* **2 (88)** 74-86.

[4] Gura D A, Dubenko Yu V, Dyshkant E, Pavlyukova A, Akopya G 2019 3D Laser scanning for monitoring the quality of surface in agricultural sector *IOP Conference Series: Earth and Environmental Science* 012184.

[5] Han W, Zhao S, Feng X, Chen L 2014 Extraction of multilayer vegetation coverage using airborne LiDAR discrete points with intensity information in urban areas: A case study in NanjingCity, China *International Journal of Applied Earth Observation and Geoinformation* **30 (1)** 56-64. DOI: 10.1016/j.jag.2014.01.016.

[6] Schmitz B, Kuhlmann H, Holst C 2020 Investigating the resolution capability of terrestrial laser scanners and its impact on the effective number of measurements *ISPRS Journal of Photogrammetry and Remote Sensing, January 2020* **159** 41-52.

[7] Seredovich V A, Seredovich A V 2011 Features, problems and perspectives of GLS application *Geo-Siberia* **1(2)** 223-224.

[8] Liu L, Lim S 2018 A voxel-based multiscale morphological airborne lidar filtering algorithm for digital elevation models for forest regions *Measurement* **123** 135-144.
[9] Komissarov A V 2015 Theory and technology of laser scanning for spatial modeling of territories (Dissertation for the degree of Doctor of Technical Sciences).
[10] Seredovich V A, Komissarov A V, Komissarov D V, Shirokova T A 2009 Terrestrial laser scanning (monograph, SSFU, Novosibirsk).
[11] Voronova K V, Akopyan G T, Schevchenko G G 2019 Experience in using the Cloud Constraint function in Cyclone to increase the links between scans when they are oriented relative to control marks *Electronic network polythematic journal "Scientific works of KubSTU"* **7** 1-8.
[12] Gura D A, Kuziakina M V, Dubenko Yu V, Pshidatok S K, Shevchenko G G, Granik N V, Markovskii I G 2019 Classification and automatisation of laser reflection points processing in the detection of vegetation *Proceed. of the International Symposium "Engineering and Earth Sciences: Applied and Fundamental Research" dedicated to the 85th anniversary of H.I. Ibragimov. Groznyi. Atlantis Press Publ.* 593-596. DOI: 10.2991/isees-19.2019.117.
[13] Ma H, Zhou W I, Zhang L 2018 DEM refinement by low vegetation removal based on the combination of full waveform data and progressive TIN densification *ISPRS Journal of Photogrammetry and Remote Sensing* **146** 260-271.
[14] Vistrchil M G 2014 Substantiation of methods for external orientation of digital models of mine workings obtained from the results of surveys by laser scanning systems (Dissertation for the degree of candidate of technical sciences).
[15] Gusyev V N, Aleksenko A G, Volokhov E M, Golovanov V A, Zverevich V V, Kiselev V A, Pravdina E A 2016 (St. Petersburg Mining University, SPb).

IOP Conf. Series: Materials Science and Engineering 1083 (2021) 012057 doi:10.1088/1757-899X/1083/1/012057

Geodesic provision for cadastral works: development prospects

D A Gura[1,2*], A P Nedyakina[2]

[1] Kuban State Technological University, Department of Cadaster and Geoengineering, Krasnodar, 350072, Russian Federation

[2] Kuban State Agrarian University, Faculty of Land Management, Krasnodar, 350044, Russian Federation

E-mail: gda-kuban@mail.ru

Abstract. The article analyzes the geodesic provision development when performing a cadastral survey. It has been established that improvement will be manifested in the use of modern geodetic technologies - three-dimensional laser scanning. The analysis of literary sources showed that there is a significant number of scientific works on the use of laser scanning for obtaining cadastral data and creating a three-dimensional cadaster. Methods for performing cadastral surveys, criteria for choosing a measuring instrument, methods for assessing the accuracy of coordinates, algorithms for processing the scan data, building models have been developed. However, the issues of further use of three-dimensional information for the cadastral documentation preparation are practically not covered, while the cadastral documentation is the basis for introducing the information into the real estate cadaster. The article presents the results of performing the ground-based laser scanning of a linear object. Additional information that can be introduced to the real estate cadaster in connection with the use of three-dimensional technology - height, technical condition of a linear object has been established. The sections of the technical plan and the changes made to them have been given. In addition, it was revealed that when using three-dimensional data in the preparation of technical plans, it would be inevitable to develop the software that can create boundary, technical plans based on three-dimensional data.

1. Introduction

In the Russian Federation, the information about real estate objects, about rights to them, about restrictions, encumbrances, boundaries of administrative-territorial entities, is contained in the Uniform State Register of Real Estate (USTRE), which is a global database.

The real estate cadaster is a part of this database and consists of information entered in the process of cadastral registration regarding land plots, buildings, premises, structures, single real estate complexes, parking spaces and objects of unfinished construction.

USTRE and its component part of the real estate cadaster are of great importance in the following areas:

- state and municipal management of land resources and real estate, manifested in territorial planning and forecasting land use, land protection, monitoring of land and real estate, land management, provision, withdrawal, conservation of land and other disposal of real estate;

- the formation of an objective payment for the real estate use, which is expressed in the cadastral value determination, which affects the payment of land tax and the amount of lease payments;
- ensuring market turnover of land and objects related to them;
- recognition by the state of the existence or the existence termination of a real estate object, without which the implementation of the above-mentioned actions is impossible.

Based on the cadastral data influence significance, absolutely objective requirements are imposed on them: accuracy, relevance, reliability, availability.

Information entry is carried out during the state cadastral registration, on the basis of the following documents:

land plot plan - for land plots;

technical plan - for buildings, structures, premises, parking spaces, objects of construction in progress.

Work in relation to the above-mentioned objects, carried out to draw up either a boundary or technical plan, are called the cadastral works. The work technology can be divided into two stages: field and office.

At the field stage, geodetic work is carried out, during which the location, coordinates of characteristic points of the real estate objects boundaries in the local coordinate system are determined.

Currently, the field stage of cadastral work is mainly carried out using electronic total stations or satellite receivers.

However, there is a noticeable tendency to integrate another measurement tool - a laser scanner.

Laser scanning as a method of photographing the Earth's surface and the objects on it was known back in the 80-90s of the XX century, but did not receive wide distribution in the Russian Federation, due to the high technology cost and the existing socio-political tension in that historical period [1]. Recently, there has been a tendency for the increasing introduction of laser scanning in many sectors of production and economic activity, such as environmental monitoring, monitoring of the state of infrastructure facilities, modeling, construction, architecture, mapping, geodetic and geological surveys, archeology, agriculture, certification of roads, including the real estate cadaster.

An example of the technology integration into the Russian real estate cadaster is a pilot project on the use of laser scanning for complex cadastral work in three municipal districts of the Republic of Dagestan, carried out until December this year. Moreover, it is planned to analyze the results of this work for the concept and proposals' development for scanning real estate in other constituent entities of the Russian Federation.

The use of laser scanning technology allows obtaining high quality geoinformation data about an object. In addition, the use of laser scanning makes it possible to speed up the process of collecting geodata due to a wide range of shooting and automation of aiming a laser beam at the terrain objects, increases the information saturation of the material [2]. However, a large amount of data obtained as a set of data points requires optimization of their processing [1].

The study objective is to analyze the possibility and mechanism of using laser scanning data for the cadastral documentation preparation, to identify the changes in its structure in terms of adding the new objects' characteristics.

To solve it, it is necessary to conduct an analytical review of the main sources of scientific literature on the use of laser scanning technology for cadastral work; search for the practical application of laser scanning technology in cadastral works; consider the structure of cadastral documentation; substantiate the improvement of geodetic works for the real estate cadaster formation.

2. The research question state analysis

A significant part of research in the field of laser scanning belongs to Professor V.A. Seredovich. In his works, the author describes the methods of performing work using scanning for a particular process, assesses the accuracy of the data obtained, analyzes the possibility and results of using ground, air and mobile laser scanning.

In the study [3,4], the authors reveal the concept and highlight the significance of the “height” parameter in the legal aspect of using, owning, disposing of real estate. In the work [3], the authors propose to use ground or air laser scanning for determining the powers of the owners of buildings structures, premises in the vertical parameter ("in height"); supervision over the use of land plots; clarification of information in the cadaster; bug fixes; cadastral assessment; in the procedures for monitoring the accuracy when registering objects. At the same time, it will be possible to expand the information content of the inventory due to accurate forecast and monitoring data in relation to the AUC technical condition and land conditions.

A significant part of the research [3, 5-7] highlights the principles of ground, air, mobile scanning, their advantages and disadvantages, scanning technology, the obtained accuracy of determining coordinates in plan and height, depending on the scanning type. In addition, the tasks of the cadaster that are supposed to be solved using the technology are indicated: correcting cadastral errors, performing supervision, monitoring land, adding a parameter to the volume of buildings, structures, premises in the cadastral data and accounting for it in cadastral valuation. The main task in these studies is the formation of a three-dimensional cadaster, where laser scanning acts as a means of geodesic provision.

In works [5,6] the characteristics of scanning systems are given. In the study [5], the most famous firms producing terrestrial laser scanners are indicated - Leica Geosystems (Switzerland), Topcon (Japan), Trimble (USA), Zoller + Fröhlich (Z + F) (Germany), Riegl (Austria), Callidus (Germany), Optech (Canada). The article [6] provides the technical characteristics of the scanners produced by the above companies, according to the parameters "maximum measured distance", "accuracy of distance determination" depending on the measurement principle: impulse, phase, distance determination method, optical triangulation method. The study concludes that distance measurement determines the accuracy of distance determination. The features of scanners Topcon GLS-2000, IMAGER 5010, Leica HDS8800, Trimble TX6, RIEGL VMX-1HA are considered.

The research [8] is characterized by a greater degree of the scanning performance issue study. As a result, the criteria for choosing a scanning method when performing measurements to form a three-dimensional inventory were developed. It was revealed that when creating separate 3D models of real estate objects, the method of ground laser scanning should be used. In addition, a technique for obtaining a three-dimensional model of a real estate object with the required accuracy has been developed. The relationship between the removal of scanning stations and the degree of three-dimensional model detail is shown. A detailed study of the accuracy issue of constructing such models is described in [9]. Accuracy assessment can also be performed by comparing the coordinates of the control points of the corners of the roofs of buildings, determined by the total station with the coordinates of the same points of the model [10].

The study [11] indicates that according to surveyors, after 10 years, the scanner is considered as one of the best tools for obtaining the vertical cadastral information of apartments with shared ownership units. In this regard, in this work, we compared the use of lidars (Callidus and Faro) and a rangefinder when preparing floor plans in both two-dimensional and three-dimensional formats. As a result, it was revealed that lidar survey is more efficient for the production of 3D models than for drawing up 2D plans. The transition to a two-dimensional image to create a 2D plan from a three-dimensional one occurs with errors associated with an ambiguous definition of the object geometry.

The results of the practical application of three-dimensional data for the inventory are given in articles [10, 4].

In [10], much attention is paid to foreign experience. The Netherlands and China, Singapore, Sweden, Australia are rightfully considered to be the most successful pioneers in the three-dimensional cadaster geodesic provision issue. The authors identify the trend of transition from a two-dimensional inventory to a three-dimensional one, which consists in the formation of three-dimensional models from the existing documents (for example, a floor plan). In addition, the dynamics of the BIM technologies introduction at the stage of designing the objects is traced in order to obtain a model of an object without scanning, and correcting the model at the stage of accounting for an object. It is assumed that the application of scanning will be carried out for some objects [10].

The study [4] describes the process of the first cadastral registration of multi-level rights in relation to a railway zone project, which is a combination of infrastructure, mainly a railway, and urban reconstruction based on a 3D act (3D PDF). Such an act contains a three-dimensional representation of rights; rights legend, object location coordinates, 2D cadastral map. The 3D act was obtained by exporting a title deed from specialized software CAD (Rhino 3D).

In the Russian Federation, back in 2010, a joint Russian-Dutch project "Creation of a model of a three-dimensional real estate cadaster in Russia" was launched. When creating three-dimensional objects for the prototype, the GoogleSketchUp software package was used. It should be noted that in the course of the project, floor plans were used to prepare three-dimensional models of pilot objects (buildings), and not newly performed geodetic measurements [10]. However, in 2020, the implementation of complex cadastral works, in which the location and characteristics of the object are obtained from the data of three-dimensional laser scanning is being tested.

The order of data processing is a "connecting bridge" between receiving the data and receiving the result. [12]. In this regard, a large number of scientific works are devoted to processing technology, accuracy assessment, equalization of point clouds, classification and recognition of objects, algorithms for obtaining a three-dimensional model, analysis of software required for data processing, depending on the required tasks [13-17]. An overview of the most common programs used in office processing is considered in [15]. The authors distinguish three groups of software: "software from laser scanner manufacturers" (for example, LeicaCyclone 9, RiScanPro, RiProcess TrimbleRealWorx) "Independent solutions" (TerrasolidTerraScan, GeoPlusVisionLidar 2018, Technodigit 3DReshaper 2017), "Solutions inside large CAD» (AutodeskAutoCADCivil 3D 2018, BentleyPointools, Descartes, Credo 3D Скан 1.4, «IndorSoft» IndorCAD 2018) [15].

The authors of [18] point out that lidar technology provides a fast, continuous and cost-effective opportunity to obtain 3D geospatial information for cadastral, urban planning, infrastructure mapping and security analysis in urban environments. The study focuses on the issue of data processing for the classification and retrieval of cadastral information. The authors have developed a technique for such actions using digital images and lidar data. The first source is needed to obtain information about the brightness of targets, and the second - for the parameter of the height and intensity of the reflected signal, on the basis of which the recognition of infrastructure objects is carried out.

The analysis of literature sources showed that there is a significant number of scientific works on the application of laser scanning technology to obtain cadastral data and create a three-dimensional cadaster. Methods for performing cadastral surveying, criteria for choosing a measuring instrument, methods for assessing the accuracy of coordinates, algorithms for processing scan data, building models have been developed.

The tendency of large-scale use of laser scanning in cadastral works is revealed. However, the issues of preparing cadastral documentation based on scanned data are little studied. This topic from the studied works is devoted to the articles [19, 4]. The possibility of using three-dimensional data for the cadastral documentation (technical plan) development is discussed in the article [19]. However, the work is characterized by an insufficient degree of the methodology elaboration for carrying out such a process, the required software or the necessary parameters for its creation is not specified, the analysis of the existing form of the technical plan and the information specified in this document is not carried out, it is not indicated what information will be added to the technical plan by using a fundamentally different shooting method.

In this regard, we can conclude that the further improvement of geodesic provision of cadastral works will be associated with the three-dimensional data processing improvement, the modernization of cadastral documentation for the real estate objects' cadastral registration and the development of software that carries out the most automated preparation of such cadastral documentation.

3. Improvement of geodesic provision for cadastral works

The operation principle of a laser scanner is to measure the angles and distances to terrain points in the vertical plane up to 270 degrees and in the horizontal plane up to 360 degrees with fixing XYZ coordinates for each point [20]. The collection of such points forms a three-dimensional digital geospatial image of the scanned array, presented initially as a point cloud [21].

Depending on the measurement technologies used, a distinction between ground, air and mobile laser scanning is made. The choice of the shooting method is determined based on the goals, objectives and types of scanned objects. If it is necessary to scan the small objects, then ground scanning is used, for the large areal objects - aerial, and for the extended linear ones - mobile (Figure 1).

Based on the conclusions drawn from the analytical review results, the scope of laser scanning data processing will be considered in more detail, in terms of drawing up the cadastral documentation for a linear object.

Figure 1. Point cloud obtained from the ground laser scanning results

Figure 2. Three-dimensional model of the object after data processing

The concept of improving geodesic provision can be schematically represented in the form of the following directions (Figure 3).

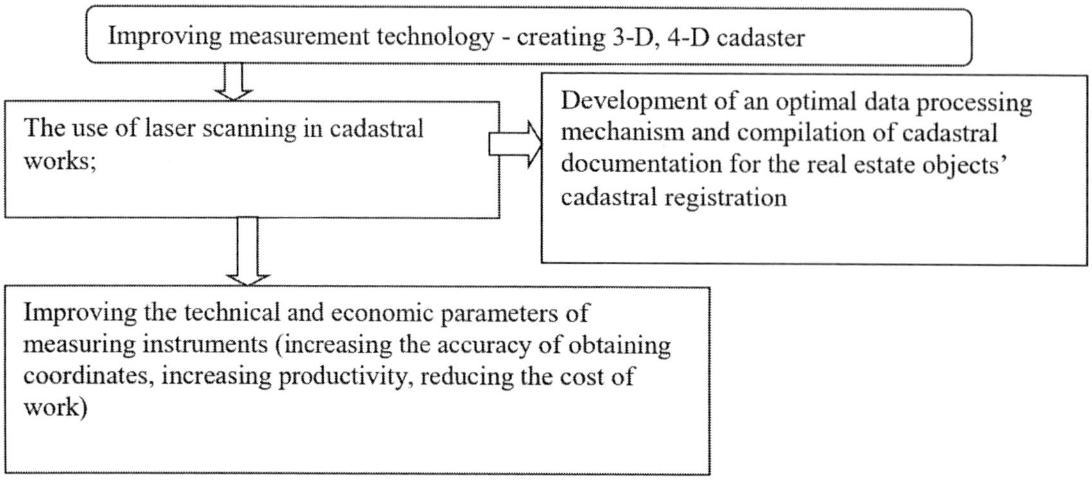

Figure 3. Concept for the geodesic provision development for cadastral works

The basis for cadastral registration of a linear object is a technical plan, consisting of the sections: general information, initial data, information on the measurements and calculations performed, characteristics of the real estate object, the conclusion of a cadastral engineer, a diagram of geodetic constructions, a layout on a land plot, a drawing. When using laser scanning in some sections of the technical plan, new information will appear (Table 1).

Table 1. New information of the technical plan when performing work with a laser scanner.

Section	Section	New information
Initial data	Information about the geodetic base	Z
	Information about measuring instruments	Laser scanner ________
Measurement and calculation information	Coordinate determination method	Ground-based laser scanning method; Aerial laser scanning method; Mobile laser scanning method
Object location description	Description of the location of the building, structure, object of construction in progress on the land plot	Z
Real estate object's characteristics	Characteristics' description	Height, technical condition [21]

Scheme of geodetic constructions	Scanner points (for terrestrial scanning); Flight plan, points of standing of satellite GNSS receivers (for aerial scanning); scheme of the route of movement, points of standing of satellite GNSS receivers (for mobile scanning)
Layout on the land Outline drawing	Three-dimensional model of the object

Currently, the programs most used by the cadastral engineers to create a technical, boundary plan are: «Polygon», «PCZO», «TechnoCAD», «ARGO», however, these programs do not have functionality for working with laser scanning data.

4. Summary

Laser scanning is a high-tech, fast and accurate method of shooting. The integration of this method for obtaining spatial information has also affected the real estate cadaster, an integral part of which is the cadastral documentation - landmarks and technical plans. When drawing up a technical plan for a linear object according to three-dimensional laser scanning data, in our opinion, the information will be added to the document structure - the value of the Z coordinate (in the information on the geodetic base and the requisite " the location description of a building, structure, object of construction in progress on a land plot"), the name of the new measurement method is the method of ground-based laser scanning; aerial laser scanning method; the method of mobile laser scanning, the value of the height of the technical state of the object (in the attribute "name of the characteristic"). In the graphic part (in the geodetic constructions scheme), an image of the scanner positioning points (for ground scanning), the flight plan, the positioning points of GNSS satellite receivers (for air scanning) will be added; scheme of the movement route, the location points of GNSS satellite receivers (for mobile scanning); a three-dimensional model of the object will be added to the object layout diagram on the land plot and the outline drawing.

When introducing laser scanning as a geodesic provision for cadastral work, it is necessary to pay attention to the development of software that allows the formation of cadastral documentation based on three-dimensional data. The following basic requirements for the program can be distinguished:

- modularity;
- data format support LAS/LAZ;
- laser scanning point classification and real estate recognition;
- automated reconstruction of a three-dimensional model of an object using point clouds;
- the presence of a graphic editor;
- formation of cadastral documentation;
- generating XML files.

Acknowledgements

The reported study was funded by Russian Foundation for Basic Research and Administration of Krasnodar Region of the Russian Federation according to the research project № 19-48-233020 Study of the possibility of using the complex of three-dimensional laser scanning for monitoring and ensuring the safety of infrastructure facilities in the city of Krasnodar and the Krasnodar Territory

References

[1] Gura D A, Pavlyukova A P and Solodunov A A 2020 Updating the algorithm for processing laser scanning data using linear objects as an example *IOP Conf. Ser.: Mater. Sci. Eng.* **913** 042041. doi:10.1088/1757-899X/913/4/042041
[2] Matikainen L, Lehtomäki M, Ahokas E, Hyyppä J, Karjalainen M, Jaakkola A, Kukko A and Heinonen T 2016 Remote sensing methods for power line corridor surveys 10-31. doi.org/10.1016/j.isprsjprs.2016.04.011
[3] Parhomenko D V and Parhomenko I V 2016 Laser scanning in the state real estate cadastre: technological and legal aspects *Vestnik Sibirskogo Gosudarstvennogo Universiteta Geosistem i Tekhnologij* **1 (33)**114-123.
[4] Stoter J, Ploeger H, Roes R, Riet E, Biljecki F and Ledoux H 2016 First 3D Cadastral Registration of Multi-level Ownerships Rights in the Netherlands, 5 th International *FIG 3D Cadastre Workshop, Athens, Greece.* Information on http://www.gdmc.nl/3DCadastres/literature/3Dcad_2016_37.pdf
[5] Markovskij I G and Gura D A 2018 Application of ground-based laser scanning for the purpose of forming a 3-D real estate cadastre *Development i Innovacii v Stroitel'stve* 143-145.
[6] Hahulina N B and CHerkasov A A 2018 Laser scanning as a method of collecting spatial information for the real estate cadastre *Kadastrovoe i Ekologo-landshaftnoe Obespechenie Zemleustrojstva v Sovremennyh Usloviyah* 260-264.
[7] Khoo V, Low E and NG Z H 2014 3D laser scanning to detect property encroachment, FIG Congress, Engaging the Challenges *Enhancing the Relevance, Kuala Lumpur, Malaysia* 6993. Information on https://www.fig.net/resources/proceedings/2014/2014_3dcadastre/3Dcad_2014_03.pdf.pdf
[8] Altyncev M A and CHernov A V 2018 Application of laser scanning technology for modeling real estate objects in 3D cadastre *Geodeziya i Kartografiya* **79** (9) 52-63. DOI: 10.22389/0016-7126-2018-939-9-52-63
[9] Snezhko I I 2014 Method for calculating the accuracy of building models of real estate objects in the 3D cadastre (dissertation for the degree of candidate of Sciences, Moskva: MIIGAiK).
[10] Avrunev E I and Giniyatov A I 2019 Current state of geodesic provision for creating and maintaining a 3d cadastre *Neftegazovyj Kompleks: Problemy i Resheniya* 51-56.
[11] Pouliot J and Vasseur M 2014 Terrestrial lidar capabilities for 3D data acquisition (indoor and outdoor) in the context of cadastral modelling: a comparative analysis for apartment units *4th International Workshop on 3D Cadastres, Dubai, United Arab Emirates*. Information on http://www.gdmc.nl/3DCadastres/literature/3Dcad_2014_17.pdf
[12] Sarychev D S 2014 Processing laser scanning data *SAPR i GIS Avtomobil'nyh Dorog* **1 (2)** 16-19.
[13] Seredovich A V and Ivanov V A 2013 Studies of the accuracy of measurements made by a ground-based laser scanner *Intereskpo Geo-Sibir'* **1 (3)** 134-143.
[14] Altyntsev M A and Antsifirov E S 2013 Research on the accuracy of mobile laser scanning data equalization *Interekspo Geo-Sibir'* **1 (3)** 90-95.
[15] Medvedev V I, Sarychev D S and Scvorzov A V 2014 Pre-processing of mobile laser scanning data in the Indorcloud system *SAPR i GIS Avtomobil'nyh Dorog* **2 (3)** 67- 74
[16] Nevolin A G and Medvedskaya T M 2014 Processing of ground-based laser scanning results taking into account the signal reflection coefficient *Vestnik Sgugit (Sibirskogo Gosudarstvennogo Universiteta Geosistem i Tekhnologij)* **1 (25)** 47-53.
[17] Medvedev V I and Raikova L S 2017 Software for processing data from laser scanning of terrain *SAPR i GIS Avtomobil'nyh Dorog* **2(9)**10-31. doi: 10.17273/CADGIS.2017.2.2.
[18] Kumar P, Rahman A A and Buyuksalih G 2017 Automated extraction of buildings from aerial lidar point cloud and digital imaging datasets for 3D cadastre *Cadastre: Geo-Information Innovations in Land Administration* 159-165. DOI: 10.1007/978-3-319-51216-7_14
[19] Styagov A S and Gura D A 2018 Method of drawing up a technical plan for linear structures using laser scanning *II International Research Competition* 44-47.
[20] Pavlyukova A P and Gura D A 2018 Determination of the location of the border of water bodies

using laser scanning technologies *Nauchnoe Obespechenie Agropromyshlennogo Kompleksa* 434 – 436.

[21] Gura D, Dubenko Y, Dyshkant E, Pavlyukova A and Akopyan G 2019 3D laser scanning for monitoring the quality of surface in agricultural sector *IOP Conference Series: Earth and Environmental Science* **403** 012184. DOI: 10.1088/1755-1315/403/1/012184.

Organizational and technological approaches to the construction and technical expertise of cultural heritage sites

I Yu Zilberova, I V Novoselova* and AL-Fatla Tareq Neamah Mohsen

Don State Technical University, 1, Gagarin Square Rostov-on-Don, 344000, Russia

E-mail irina1000000@gmail.com

Abstract. Construction and technical expertise plays an important role in the preservation of cultural heritage sites. Based on the data obtained in the course of construction and technical studies the technical condition of cultural heritage objects or their parts is determined as well as the possibility and methods of carrying out repair and restoration work. Construction and technical expertise of cultural heritage sites allows for control ensuring the prevention of damage, destruction, reshape or complete destruction of historical and cultural monuments. The special feature existence of such objects determines a special order and approaches to conducting construction and technical research attached to them.

1. Introduction

Construction and technical expertise as one of engineering and technical expertise is carried out on the basis of knowledge in the field of capital construction, technologies and features of building design,building materials production etc. [1]. It plays an important role in investigating events that have caused accidents, deformations or the external signs of hidden defects appearance in structures and units of capital construction objects.

The list of objects to research by a construction expert is very wide and differs in functional purpose, natural and technical characteristics as well as in procedural position [2, 3].

Work with historical and cultural monuments is one of the most difficult expertise types. The fate of such objects directly depends on the expert's conclusion and his assessment. Construction and technical expertise makes it possible to prove the facts that give rise to include capital construction objects in the cultural heritage register as well as withdrawal from the state register, to identify the projects compliance of protected zones of historical and cultural monuments with provisions on the protection of cultural heritage and town planning regulations, to determine on the basis of historical and cultural significance and other signs, to which monuments category should one or another object be attributed [4]. According to the examination results, repair and restoration work can be assigned, and the composition of measures for the object preservation can be determined.

2. Materials and methods

Work with monuments of history and culture in Russia implies the availability of both traditional certificates of admission for construction and design, issued by self-regulatory organizations, and additional approvals and permits from the Ministry of Culture and other organizations responsible for the state of the historical and cultural heritage of the Russian Federation. Relations arising in the process of preservation and use of cultural heritage objects in Russia are regulated by Federal Law No.

73 of June 25, 2002 "On cultural heritage objects (historical and cultural monuments) of the peoples of the Russian Federation".

Cultural heritage objects are divided into the following categories of historical and cultural significance:

- cultural heritage site федерального значения;
- objects of cultural heritage of regional significance;
- objects of cultural heritage of local (municipal) significance.

The status of an object does not determine property rights, and monuments of federal significance can be owned by regions, as well as vice versa.

Cultural heritage objects stand out among other real estate objects with their historical value as well as architectural, planning and constructive solutions, which determine the presence of some peculiarities in the construction and technical examination of such objects [5].

When carrying out construction and technical expertise of historical and cultural monuments, the objects of research can be:

- capital construction objects, building structures and building materials;
- unfinished repair and restoration work;
- construction production technologies;
- results of previous construction examinations;
- accident reports
- design and estimate documentation
- contracts.

Work with these sites requires deep knowledge in many related areas so specialists must have a high level of qualifications and professional training, this is a distinctive feature of the construction and technical expertise of cultural heritage sites.[6]. When carrying out construction and technical expertise directly on historical and architectural monuments as well as various sculptural compositions not only technical but also creative methods are used.

3. Results and Discussion

The result of the expert's work can determine the further object fate - reconstruction, restoration or other work. Expert work with objects of cultural heritage is designed to solve a wide variety including highly specialized tasks, which include [7, 8]:

- determination of the object age;
- setting the terms of restoration work;
- fixation of changes in the hydrogeological regime;
- identification of deformations and defects of supporting structures;
- determination of the upporting structures reliability;
- monitoring of the technical condition of the building;
- determination of the redevelopment or reconstruction possibility;
- recommendations formation for increasing the elements bearing capacity;
- determination of restoration methods for the most complete and effective object restoration;
- implementation of adaptation projects of historical and cultural monuments for operation in accordance with modern requirements;
- impact assessment of various technogenic factors on the cultural heritage site;
- verification of the performed repair and restoration works compliance with the requirements of regulatory and contractual documentation;
- verification of the design estimatesvalidity;
- quality control of finishing, repair, restoration work, etc.

The content of the engineering and technical expertise is determined by the customer's assignment. Construction expertise is appointed in controversial issues, it can be pre-trial or carried out in court. Basically, the following types of construction expertise are distinguished at cultural heritage sites:

- examination of the building technical condition, its destruction degree, normal operation time without restoration or reconstruction work;
- cause examination of accidents, destruction, defects;
- examination of design and estimate documentation
- examination of the quality, volume and cost of repair and restoration work;
- estimation of the work cost in progress during the conservation of capital construction objects;
- damage determination of building or structure .

Cultural monuments, as a rule, do not have standard design documentation. In this case during the construction and technical expertise, archives are studied and materials are examined for a specific historical period or for similar objects, therefore, the study of design documentation for cultural heritage sites is a rather complicated and time-consuming process. When performing construction and technical expertise of cultural heritage objects, it is necessary to apply special measures:

- study of materials from various archives related to the object under study (search for archival information, analysis of materials on similar buildings and structures built at the same time as the object under study);
- visual and instrumental examination, as well as diagnostics using modern technology;
- geological surveys to determine the foundation durability and possible geological and hydrogeological risks;
- determination of the material strength used in construction by means of load simulation;
- object measurements to determine the exact geometry of buildings, including three-dimensional scanning and volume modeling;
- examination of all structures, defect identification (the strength of the supporting structures is determined as well as possible ways to strengthen them);
- performing calculations required for strength checks taking into account detected defects, damages and changes in material characteristics;
- drawing up a report based on the results of the examination containing if necessary comprehensive recommendations for strengthening and restoring the object.

The status of historical and cultural monuments stipulates a more detailed study. Significant part of buildings and structures related to cultural heritage sites are under state protection. Their technical condition is examined in order to preserve such objects.

Construction and technical expertise production in relation to cultural monuments often involves the use of methods and devices for non-destructive testing.

Non-destructive methods imply the determination of the structure strength characteristics of the building under study without their direct destruction while the results obtained are quite reliable [9]. This control method makes it possible to inspect critical structures of buildings, for which the selection of individual samples is not possible.

Despite the presence of quite significant drawbacks, the non-destructive testing method is actively used in the activities of construction experts. Any weakening of the surveyed objects, which is inherent in destructive methods, leads to a decrease in the bearing capacity not only of individual structures but also of the building as a whole. Also subsequent repair of damaged areas requires additional costs. Method use of non-destructive testing makes it possible to avoid such negative consequences thereby reducing not only the time, but also the cost of the technical inspection of buildings and structures [10].

Modern devices for non-destructive testing allow to check various construction objects, analyze the internal state of building structures, elements and assembly units without violating their integrity. The current selection of devices with different operating principles is applicable to a wide range of materials, including metal, concrete and brick. Also during the examination ordinary measuring instruments are used to check the geometry of structures and to measure damage: rulers, tape measures and levels. Non-destructive testing devices allow to determine the characteristics and data of building structures, which are necessary in the examination process. Such possibility not only reduces the examination time but also increases the accuracy and precision of the research carried out. The use of any devices during construction and technical inspection implies the need to check them because if the examination is carried out using measuring instruments and instruments that have not been verified but are subject to verification its results will be inadequate evidence.

Inspections of the technical condition of cultural heritage sites in Russia are regulated by GOST R 55567-2013 "Procedure for organizing and conducting engineering research at cultural heritage sites. Monuments of history and culture ". The document contains provisions on the scope of work in terms of the engineering and technical conditions of cultural heritage objects necessary to establish the state of cultural heritage objects, to carry out conservation, restoration, adaptation of such objects for operation in modern ones, as well as to recommend their preservation.

Engineering and technical research of a complex nature is provided by research, design and production work, including additional examinations in the process of repair and restoration and other works, the study of archival, historical, expert and other information.

When carrying out work on the survey of historical and cultural monuments the status of the objects should be taken into account. In some cases the protected part of a building or structure (facade, walls, any fragments of columns, etc.), however the safety of a separate part depends on the integrity of the entire object. During a full-scale instrumental examination, especially valuable fragments of monuments of cultural history are examined and evaluated most carefully. Increased requirements are imposed on the determination of the strength and elements durability [11].

Inspection of the technical condition of the cultural heritage object is carried out before the start of design work as well as the period of work on the preservation of cultural heritage after the structures opening that are inaccessible during the main survey.

The completeness and accuracy of the full-scale examination of the object is of great importance for further research and drawing up a conclusion [12]. When performing the survey the features of the object and the specifics of its environment should be taken into account, while the expert must use personal protective equipment that meets the conditions of the inspection, comply with the requirements of labor protection and safety.

All data obtained as a result of inspection of a cultural heritage site must be recorded in detail, including through the use of modern photo and video equipment. The use of special technical means and devices in the process of examination allows obtaining more accurate information about the object of research [13].

Further to the object inspection all available materials should be studied and if necessary, supplemented with comments. After completing the full-scale inspection, the expert needs to check the equipment and aids for completeness.

When performing construction and technical studies experts should have access to an appropriate laboratory and material and technical base or the ability to involve specialized laboratories in carrying out certain work types.

The use of information technologies in the process of construction and technical expertise allows more efficient work in this area. Optimization of information support for construction and technical expertise through the use of the latest generation software allows to solve a large number of repetitive processes at high speed, which in turn makes it possible to increase the efficiency of construction experts[14].

Based on the data obtained as a result of the construction and technical research the technical condition of the building or its individual parts is determined, as well as the possibility and methods of restoring the cultural heritage object.

4. Summary

As a result of the examination the customer receives an expert opinion, which indicates all the main characteristics of the cultural heritage object, determined as a result of the examination. Based on the data obtained, the expert can determine the technical condition, the destruction degree and the time of possible operation of the facility without repair and restoration work. Also the conclusion may contain the professional opinion of an expert on what methods can be used during the planned repair and restoration work to increase their efficiency, recommendations for restoring the original building appearance, the methods description and materials that are closest to those used in its construction.

References

[1] Petrenko L, Manzhilevskaya S, Shilov A and Abdullah Saleh A H A 2018 The far-seeing planning systems and models for the construction management *MATEC Web of Conferences* **170.** doi: 10.1051/matecconf/201817001007.

[2] Korol E, Ostyakova A and Perfilowa E 2018 The operational characteristics of the housing stock as a result of renovation *IOP Conference Series: Materials Science and Engineering* **365.** doi: 10.1088/1757-899X/365/6/062027.

[3] Barkalov S, Kurochka P and Nasonova T 2018 Optimal placement of maintenance facilities *MATEC Web of Conferences* **170.** doi: 10.1051/matecconf/201817001124.

[4] Ivanova D G, Ivanova O E and Sukhinin S A 2018 Features of Economic Activity on Lands of Historical and Cultural Heritage *Materials Science Forum* **931** 851-855.

[5] Zilberova I Y, Nikulina O V and Novoselova I V 2019 Main principles of choice of multi-purpose design solutions for construction operations when conducting judicial construction and technical expertise (JCTE) *IOP Conference Series: Materials Science and Engineering* **698.** doi: 10.1088/1757-899X/698/5/055012.

[6] Samosudova N V 2017 Modern leadership and management methods for development organizations *MATEC Web of Conferences* **106.** doi: 10.1051/matecconf/201710608062.

[7] Mishchenko V Ya, Gorbaneva E P and Sevryukova K S 2020 Foreign and Russian experience conducting major repairs of housing fund taking into account energy-efficient measures Russian *Journal of Building Construction and Architecture* **1** 28-41.

[8] Sheina S G, Minenko E N and Sakovskaya K A 2018 Complex Assessment of Resource-Saving Solutions Efficiency for Residential Buildings Based on Sustainability Theory *Materials Science Forum* **931** 870-876.

[9] Beskopylnyi A N, Vernezi N L, Veremeenko A A and Arakelyan R M 2019 Experience non-destructive testing of the metal of bridge structures *IOP Conference Series: Materials Science and Engineering* **698** doi: 10.1088/1757-899X/698/6/066001.

[10] Bovteev S, Kanyukova S, Okrepilov V and Rezvaia A 2016 Construction work tasks duration: new method of estimation and quality control *Journal of Applied Engineering Science* **1** doi: 10.5937/jaes14-10443.

[11] Zelentsov L B, Mayilyan L D and Shogenov M S 2019 Organizational and technological simulation of the construction organization activity in the complex infrastructure projects implementation *IOP Conference Series: Materials Science and Engineering* **698.** doi: 10.1088/1757-899X/698/7/077048.

[12] Zilberova I Yu, Petrov K S and Al Fatla Abdullah Neama Mohsen 2020 Features of the planning system for repair and construction production as a part of the urban programs' implementation *IOP Conference Series: Materials Science and Engineering* **913.** doi: 10.1088/1757-899X/913/4/042020.

[13] Kluchnikova O and Pobegaylov O 2017 Application of Strategic Management Principles as Tool to Improve Quality of Construction Technological Process *IOP Conference Series: Materials Science and Engineering* **262.** doi: 10.1088/1757-899X/262/1/012075.

[14] Sheina S and Chubarova K 2019 Formation the construction cost for residential buildings at the design stages *Advances in Intelligent Systems and Computing* **983** 216-223.

Opportunities for the drinking water resources' export-oriented commercialization as a potential element of the economic framework in mountainous areas

R O Kalov[1*], D D Kiloev[2], R S Elmurzaev[3]

[1] Kabardino-Balkarian State Agrarian University, 1 Lenin str., Nalchik, 360030, Russia
[2] Grozny State Oil Technical University named after Academician M.D.Millionshchikov, 100 Kh. Isaeva, Grozny, 364061, Russia
[3] Chechen State University, 17 Boulevard Dudaeva ave., Grozny, 366007, Russia

E-mail: calov.r@yandex.ru

Abstract. The predicted reserves of fresh groundwater in the Kabardino-Balkarian Republic (KBR) reach 7200 thousand m^3/day, which is 31.2% of similar resources in the North Caucasus Federal District, ranking first in the region.
In terms of mineralization in general, groundwater in the KBR belongs to the class of "fresh groundwater"; in terms of chemical composition they are classified as "drinking"; in terms of total hardness - to "moderately hard water". Water does not contain significant concentrations of heavy metals and inorganic nitrogen compounds. However, the use of the indicated potential is limited only to the creation of local lines for bottling water with a small capacity, mainly focused on the republican and regional markets of Russia.
The purpose of this study is to substantiate the hypothesis about the sufficient resources' availability of hydrogeological materials suitable for industrial production and bottling as ecologically clean drinking water.
Within the framework of the designated goal, the following tasks were solved:
- search for mountain landscapes, in the first approximation localizing the balance reserves of fresh groundwater;
- environmental monitoring of the sites previously designated as favorable for the drinking water extraction;
- the productive aquifers' assessment nature and the conditions of their natural protection from pollution;
- identification of natural components that maintain the balance of groundwater and factors of its change over time;
- establishing the specifics of groundwater discharge and the dynamics of regime-forming processes in the vertical zoning conditions in the study area.

Introduction

The provision of the world's population with high-quality drinking water is recognized as an important condition for the high life expectancy of people, in connection with which the demand for bottled drinking water is steadily increasing. The latter has become the most important product of industry and world trade.

Assessing the chances of the republic's participation in the international division of labor in the sphere of drinking water circulation, it should be noted that the predicted potential of underground drinking water is 18 times higher than needed in KBR. 28 deposits of underground drinking sources with balance reserves of 1041 have been tested thousand m³/day in Kabardino-Balkaria. The content of normalized components in them in the overwhelming majority of deposits is within the Russian standards and regulations' requirements, as well as the European standards according to WHO. At the same time, only 15% of the reserves of the explored part of groundwater resources have passed the state examination. The share of underground sources in the structure of drinking water use does not exceed 7-8%, which indicates a weak degree of their involvement (Figure 1).

Meanwhile, having significant opportunities for export-oriented industrial bottling of ecologically clean drinking water of the highest quality, the republic is practically not involved in the process of the current dynamic distribution of the formed markets for its sales.

Results

Stimulated by the dynamic development of the soft drink industry giants, relatively low capital costs for organizing production, the natural balance of the reserves volume and the groundwater chemical composition, the potential budgetary price for products, Kabardino-Balkaria may also have a strong chance of entering this market.

A comparative characteristic (Table 1) of the chemical composition of some bottled drinking water bottled in a number of European countries and Kabardino-Balkaria is presented below.

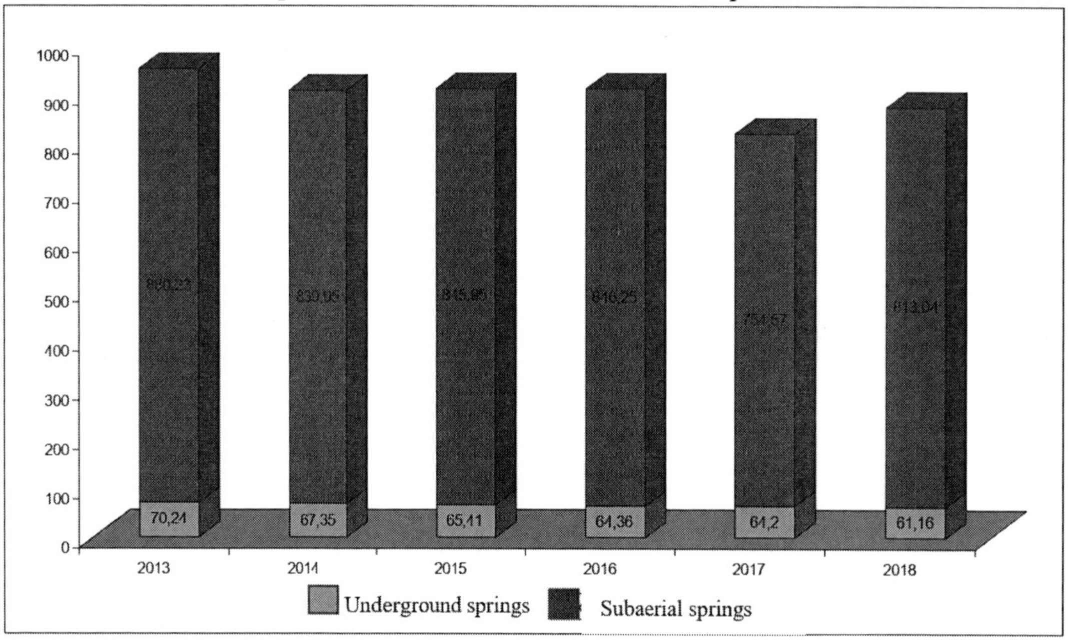

Figure 1. Structure of fresh water consumed in Kabardino-Balkaria by sources

Table 1 shows that the maximum permissible concentrations (MPC) of chemical elements in different countries differ significantly. At the same time, KBR bottled drinking water do not exceed the upper limits of chemical compounds' permissible concentrations than the quality indicators of the high-quality imported analogues. According to the solid residual, "Crystal Valley", for example, is several times less saturated with the dissolved inorganic and organic substances than a number of well-known brands. The fact of the high-quality republican water products is confirmed by the certificates of the relevant services of the World Health Organization (WHO), the experts from the Bodicoal laboratory (Great Britain). According to their data, drinking water "Bezengi", "Azau", "Elbrus", "Vertical", "Crystal Valley" are chemically and microbiologically flawless and comply with the Codex of Hygienic Practice for Bottled or Packaged Drinking Water Bottled / Packaged Drinking Water), as well as the Russian SanPIN 10-113 RB 99. The initial mineralization of the base underground basins of the drinking water listed brands does not exceed 01-0,3 g/dm³. This water has a composition and physical characteristics that are favorable for the human body and do not require chemical water treatment, with the exception of aeration, filtration, etc.

With a view to the export-oriented commercialization of drinking water use in KBR the authors have identified a number of presumably promising mountain areas within the Baksan, Irik, Adyl-Su, Malka, Zhapyr-Tala, Karasu, Kurkuzhin tracts using large-scale hydrogeological maps. Within their limits, a field sanitary and hydrogeological survey has been carried out:

- the actual sanitary and ecological state (for May 2020) of the proposed areas for the operational reserves' formation;
- the groundwater natural protection level from external pollution.

It was revealed that the natural elements feeding groundwater are natural precipitation, melted glacial and partly surface water, from which it follows that the deposits are of exclusively infiltration origin. Moreover, the ratio of the elements feeding the deposit is noticeably corrected in space and time.

The mineral composition of groundwater in the designated areas is determined by their hydrogeological structures: the rocks' composition, the deposits' depth and the unloading water' migration nature. The latter circulates in veins and fissures of Proterozoic crystalline rocks (shales, gneisses, quartzites, etc.), often broken by granite intrusions, which are widespread in the geology of high-altitude levels within 4 thousand meters (Figure 2). Taking into account the low solubility of crystalline rocks, it can be predicted with a high probability that the water in the indicated cavities are weakly mineralized. Consequently, the substances dissolved in water will have an infiltration genesis, from which it follows that the predominant glacial feeding of most deposits should limit the downward components' migration of anthropogenic origin [1].

Table 1. Characterization of the chemical composition of some bottled drinking water

Indicators quality, mg/l	Category of high-quality drinking water						Azau (KBR)	Crystal valley (KBR)	Standards quality water by WHO	Council directive EC № 80/778
	Misia (Italy)	Val-vert (Bel-gium)	Catha-rein (Nether-lands)	Cris-talroc (Fran-ce)	High-land Spring (Scot-land)	Thonon (Swit-zeland)				
1	2	3	4	5	6	7	8	9	10	11
pH	—	7.7	8.09	7.8	7.8	7.4	7.2-7.4	7.5	6.5-8.5	6.5-9.5
Solid residual	216.8	201	160	223	136	342	200-310	8.0	1000	1500
Minerali-zation	314	302	212	320	200	509	<500	285.1	—	200-500
HCO_3	202.1	204	147	195	136	350	80-120	152.5	—	30-400
SO_4	20.39	18	1.9	21	6	13	40-70	16.9	250	250
CI	6.71	4	7.4	11	7.5	9	5-10	5.2	250	250
NO_3	1.73	4	0.4	<1	<1	12	<5	<5	50	50
Ca	68.93	67.6	42.4	70	35	108	25-50	60	—	—
Mg	4.01	2	2.8	3	8.5	14	4-12	6.7	—	50
Na	3.51	1.9	9.66	4.8	6	3	1-30	<20	—	—
K	0.55	0.7	0.5	1.4	0.6	<1	1-45	3.2	—	2-20

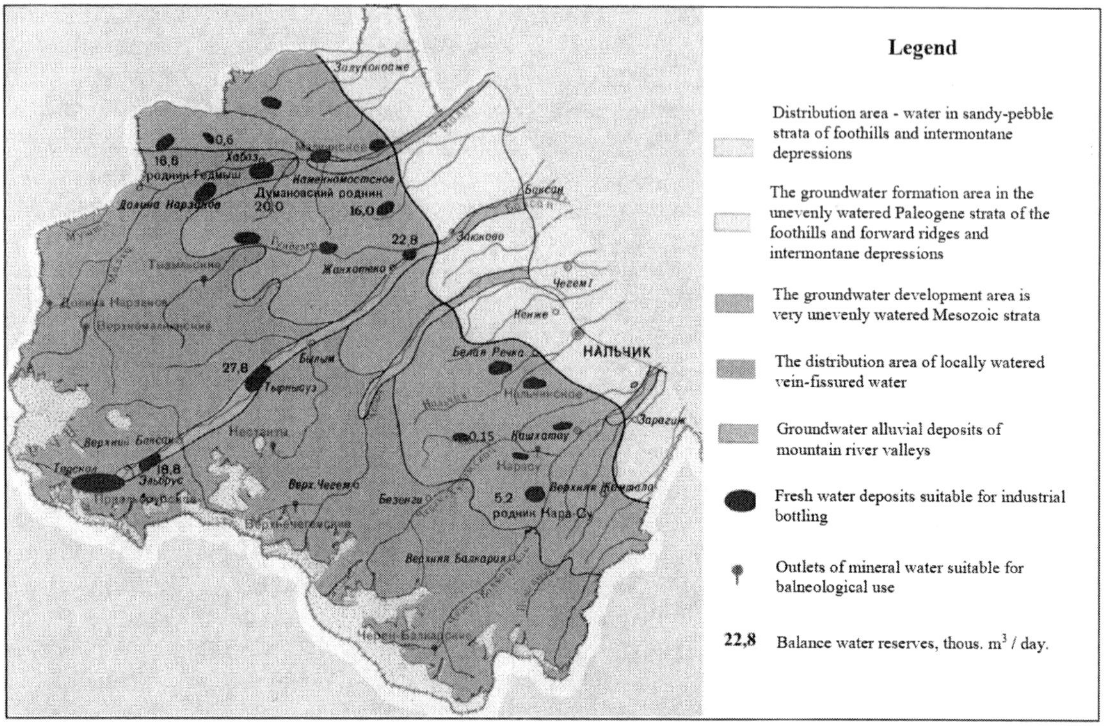

Figure 2. Schematic map of fresh water deposits presumably suitable for industrial bottling

A more significant problem of involving vein-fissured groundwater in drinking water use is mainly the local watering of ecologically clean high-altitude areas. The exception is the relatively large deposits within the catchment areas of the Malka, Karasu, Kurkuzhin rivers, the localization of which has a basin character.

The low forward ridges of the Greater Caucasus are predominantly composed of soluble Jurassic limestones. Judging by the nature of the numerous springs' discharge and their chemical composition, free-flow underground water horizons with a small thickness of the aeration zone are localized here [2]. Confirmation of this hypothesis is the constant of the drinking water quality, which will allow, if it is involved in nature management, to switch from the purification of produced water to their pollution prevention. Therefore, for example, the water of the Kara-Su spring in the basin of the Cherek River can be packaged in their natural state without preliminary water treatment, except for the need to comply with strict sanitary-preventive measures during their intake and bottling. This water is not only harmless, but biologically valuable and retains high consumer properties.

The underground water of the Ciscaucasia can be represented by an artesian basin with the reservoir-type water accumulations in significantly dislocated and lithified terrigenous-carbonate complexes. Even lower are the layers of sedimentary rocks of the North Jurassic depression, which, together with aquifers, are submerged under the cover of younger layers, creating a significant head of high-quality hydro-carbonate calcium water [3]. The increased content of hydro-carbonates in the component composition of the Ciscaucasia underground water predetermined their weak mineralization. This feature is associated with the fact that limestones represent an active sorption barrier that reduces and neutralizes the pollution migrating inward. Consequently, groundwater with such protection represents an unfavorable environment for the accumulation of standardized micro components and is epidemiologically safe.

In the course of this study, an assessment of the current character of groundwater discharge has been carried out. The water content of the hydrogeological folded region in Central Caucasus is due to the weathering crust of crystalline rocks. At the depths from 60 to 100 m, the fault zones have been formed, penetrating into the body of the hydrogeological massif. Within the carbonate karst strata, open discharge of the basins in the form of springs with a flow rate of up to several hundred liters per second prevails. Intrusive and metamorphic massifs are naturally less water-abundant, since fresh water circulates in a limited space of fractured and fractured-vein horizons [4].

Groundwater within river valleys is fed from the subaerial springs, including precipitation through the accelerated infiltration of alluvial deposits. The aquifers here are often not overlapped and not underlain by mature, low-permeable sediments, therefore, their natural protection is insufficient. For water springs which depth is less than 5 m, the risks of increased mineralization are higher. The indicated hydrodynamic horizon is a zone of free water exchange, due to which groundwater is directly affected by the modern climatic dynamics and is drained by the mountain rivers. Therefore, the salinity of the horizon water is likely to vary within 1 g/l, which meets the quality requirements for water for centralized water supply.

At the same time, this aquifer has a large safety margin in terms of uninterrupted drinking water supply due to the high infiltration parameters of pebbles, sands and sandy loam formations above the floodplain terraces. However, the high productivity of these basins dictates the need for systematic hydrogeochemical studies of bottom sediments, soils, snow throughout the water reserves formation area [3].

The groundwater operational resources' calculation can be made according to the formula [5]:

$$Q = \frac{2VKHS}{\ln R - \ln r}$$

Where Q – is water intake capacity, m³/day;

V – is volumetric fluid loss;

K – defines the rock filtration coefficient, m/day;

H – is the thickness of flooded rocks, m;

S – denotes the level lowering, m;

R – is the influence radius, m;

r – is the well radius, m.

After the groundwater basins' potential productivity calculation, it is possible to proceed to the specific locations' substantiation for the production wells. It is advisable to draw up a water intake scheme with grid steps that ensure a stable water resources' flow rate. In the presence of extensive complex groundwater recharge, as, for example, in the subniveal zone, the volume of water withdrawal in the autumn-summer period is practically not limited. In the areas with low infiltration capacity of horizons, the drainage size should be controlled. The dependence of the well flow rate on the groundwater level lowering is shown in Figure 1.

Figure 3 makes it possible to judge the well productivity dynamics at different lowering of the groundwater level. The presented graphs allow to determine the limit of the permissible level lowering for the stable obtaining of the given flow rate with free-flow and pressure water. $^1/_3$ thickness of the watered strata, at which the consumer can count on 95% stability of the field's fluid loss can be taken as the value of the limiting water level lowering from its initial indicator in the pool.

The final stage of the work was the recording of the actual ecological state of the promising areas, drawing on the topographic map of all the detected water manifestations (springs), identifying the potential channels for possible groundwater pollution. Thus, in the process of field studies of the upper reaches in the Malka River, early drilled but abandoned wells that were used for distant pasture farming and for milk processing in the mountains, have been discovered. Sluggish contamination of groundwater can occur through the unsealed wellbores [2].

The potential of mountain springs is practically not used. Most of them have not been studied, their cadastral registration has not been conducted, the qualitative characteristics of the sources have not been studied, the exit places are not recorded on the topographic map. The applied value of the springs is limited only by the occasional use of some unsettled sources in tourism.

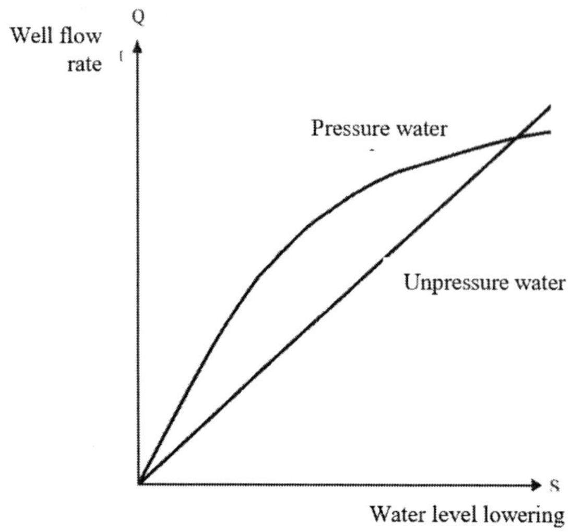

Figure 3. Graph of the well flow rate dependence from water level lowering

Meanwhile, a number of springs with the current production rate can become a self-sufficient source of water supply to the mountain settlements, which have not yet been provided with drinking water reliably protected from pollution. Mountain-folded areas are characterized by good washing, dense fracturing, in the middle mountains - karst, therefore the springs have a quite acceptable level of mineralization [2]. However, before a decision is made to fill the water supply network with them, it is necessary to establish the observation of candidate springs for involvement for several years to identify the size of their minimum discharge in a dry season. The springs with discharge less 10 l/s cannot supply water networks without interruption. In this case, the minimum flow rate ratio to its maximum level should be in the ratio 1:2. This type of discharge allows the spring to be classified as a permanent water source [5].

The current geography of fresh groundwater distribution and the nature of their mineralization can be corrected by post volcanic carbon dioxide migrating in the form of a solution from the Elbrus magma chamber. Therefore, in the short term, isotopic studies of surface and ground water should be included in the general complex of prospecting works to assess the nature and intensity of water exchange [1].

The appraisal work should be completed by the pilot-filtration testing of deposits, determination of the estimated production rate of wells, analysis of all significant groundwater basins' quality in accordance with the international standards. But for this complex of works, a potential subsoil user should obtain a state license (permission for geological subsoil study).

Summary

The underground water of the mountainous zone of Kabardino-Balkaria are represented by the locally watered vein-fissure, bed-block and cover-flow deposits. The most significant of them can be considered 13 deposits not involved in water use with total forecast reserves in 144 thous. m^3/day. Judging by the chemical composition of the springs associated with a number of these deposits, the level of mineralization in none of the flooded areas does not exceed 1g/dm^2.

Due to the overcrowding of unused underground storage facilities, tens of thousands of cubic meters of fresh water, in particular, especially valuable melted ice water, irretrievably flows into the sea every day. In the conditions of the republic's truncated tax base and extremely low incomes of the population, this is an impermissible luxury.

Meanwhile, this sphere of nature management can become an important element of the ecological and economic zone's design in the mountains of the KBR, proposed by the authors. The revealed regularities of the space-time dynamics of groundwater made it possible to outline the specific areas for the commercialization of drinking water use. The described potential is a serious reserve for Kabardino-

Balkaria to acquire the status of a reliable exporter of drinking water in world economic relations. To do this, it is necessary to create a network of bottling plants with a modern production infrastructure. The motives for the users activating potential subsoil should be relatively low capital costs for organizing production; rather high profitability of the enterprises producing, bottling and selling drinking water; the prospect of an ever-increasing demand for high-quality drinking water.

At the stage of entering the world drinking water market, the main task of the aquatic products' manufacturers should not be to maximize profits, but to persistently build the trust of domestic and foreign consumers, as well as loyalty of control and regulatory bodies in republican brands. In case of water use controlled commercialization civilized organization from the point of view of preventing the new property redistribution in the mountains and entrepreneurial selfishness, the brand of drinking water of Kabardino-Balkaria can take a stable place in the world market of soft drinks.

Acknowledgements
The study was carried out with the financial support of the Russian Foundation for Basic Research within the framework of a scientific project № 19-010-00882.

References
[1] Abaykhanov U I 2010 Features of the carbonic mineral water formation in Central Caucasus and Ciscaucasia *Bulletin of the Southern Scientific Center of the Russian Academy of Sciences* **6 (1)** 41-51.
[2] Zhinzhakova L Z, Cherednik E A 2019 The analysis results of the spring water composition in Kabardino-Balkarian Republic *International scientific research journal. Earth sciences* **2 (80)** 65-68.
[3] Krainov S R, Ryzhenko B N, Shvets V M 2004 *Geochemistry of groundwater. Theoretical, applied and environmental aspects* (Moscow) 667 p.
[4] Kalov R O, Toguzaev T Kh, Bekarov G A *Ecological and economic zone as a balanced natural and economic structure with desired properties* (Moscow, ASV) 138 p.
[5] Bekarov G A 2019 Natural preconditions for the marketing of water resources (on the example of the Kabardino-Balkarian Republic) *Russian Economic Internet Journal* 1-3.

To the problem of determination the aerodynamic resistance coefficient value to the swirled dust and gas flow in the air ducts of engineering and environmental systems

D P Borovkov[1,*], K O Chicherov [2]

[1]Volgograd State University, 100, Universitetsky avenue, Volgograd, 400062, Russia
[2]Penza State University of Civil Engineering, 28 Titova street, Penza, 440028, Russia

E-mail: friggate@yandex.ru

Abstract. The experimental studies' results of the aerodynamic drag of a swirling dust and gas flow passing through a network of air ducts of engineering and environmental systems are presented. The limits of the method applicability for determining the aerodynamic resistance coefficient of a gas-dust swirling flow passing in a round aspiration duct, without taking into account the solid phase presence in it, have been determined.

Introduction

Flow swirling in air ducts is an effective way to combat the formation of dust deposits that worsen the aspiration systems' efficiency [1-5]. The coefficient value of specific aerodynamic resistance to movement in the channel of a swirling gas flow containing solid dust-like particles is known to exceed that for a pure gas flow with the same swirl intensity at lower values of mass concentrations of solid particles than those observed in the absence of the flow rotational ($\varepsilon > 0{,}01\ mg/mg$) [6 - 8].

The reason for this discrepancy is the interaction of dust particles moving into the near-wall flow zone with the channel walls (air duct), which causes an increase in friction, and, as a consequence, additional expenditures of flow energy for moving along the channel.

The verification possibility of using the technique for determining the pressure loss of a swirling dust and gas flow when passing through the air ducts of engineering and environmental systems was carried out using the experimental methods. When conducting the experimental studies, the task to determine the solid phase mass concentration value effect on the specific aerodynamic resistance coefficient value to the movement of a flow with a rotary motion passing in a ventilation (aspiration) duct, circular cross-section, was set.

Giving rotational motion to the dust-gas flow was carried out by using a tangential-type swirling device, which provides the possibility of smooth control of the swirl intensity [9, 10]; the device is shown in Fig. 1. Regulation of the swirl parameter at the outlet of the turbulizer is carried out by adjusting the cross-sectional area of the tangential branch pipe. For this purpose, the filling device is equipped with a movable wall 3, which can be moved longitudinally by means of an adjusting screw 4 along the section of the tangential branch pipe 2. The range of the filling device Φ * shape parameter values varies within 0.5 ... 5.

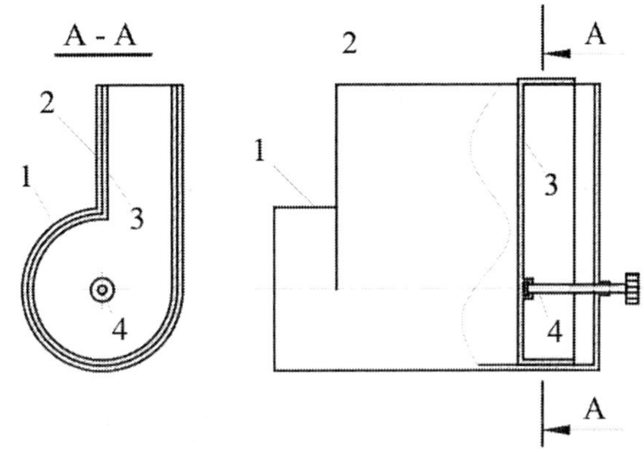

Figure 1. Dependence of the specific aerodynamic resistance coefficient values to the swirling gas-dust flow movement on the intensity of the flow swirling averaged along the length of the duct $\lambda_{sf}\left(\overline{\Phi^*}\right)$, at $Re = 1\times10^5$ and various values of the mass dust concentration: $1 - \varepsilon = 0{,}01$; $2 - \varepsilon = 0{,}009$; $3 - \varepsilon = 0{,}008$; $4 - \varepsilon = 0{,}007$; $5 - \varepsilon = 0{,}006$; $6 - \varepsilon = 0{,}003$.

The schematic diagram of the experimental setup created for the research is shown in Fig. 2. The unit is a horizontal air duct with a relative length of 50 ... 100 diameters, along which a dusty stream moves by means of a draft device. To ensure the possibility of generalizing the experimental data during the experimental studies, air ducts with diameters of 100 ... 400 mm were used. This range is most typical for most small and medium-sized suction and dust extraction systems. To impart a rotational motion to the flow, a tangential-type swirler is installed at the end of the air duct with the ability to change the initial swirl parameter. The flow dustiness was controlled by a metering device. The gas flow rate flowing through the installation was measured using a micromanometer and a pneumometric tube, the aerodynamic resistance was estimated as the difference in static pressures on the walls. The static pressure values were measured using a battery micromanometer connected to the fittings (8) located at equal distances of ten air duct diameters.

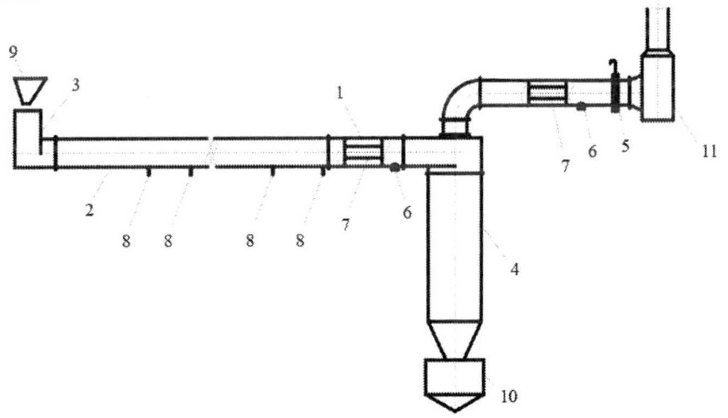

Figure 2. The experimental setup diagram: 1,2 - glass air duct sections; 3 - adjustable curler; 4 - inertial dust collector; 5 - control damper; 6 - metering air inlet; 7 - flow straightener; 8 – air inlets for removing static pressure; 9 - loading device; 10 - dust collection bin; 11 - fan.

The pressure loss of the swirling flow was determined as the difference between the static pressures in the initial section and the final section of the air duct. The position of the initial section is conventionally taken as a distance of four diameters of the duct, in order to achieve a damping flow behind the swirler. The final section is located at a distance of ten duct diameters in front of the flow damper (7), to avoid taking into account its influence on the kinematic flow pattern.

The dust particles' mass concentration value in the gas flow was varied with the help of nozzle metering devices within the limits $\varepsilon = 10^{-3}...10^{-2}$ *kg/kg*. The aerodynamic operating mode of the experimental setup was determined based on the Reynolds criterion values (calculated from the axial components of the gas flow velocities averaged over the channel cross section), varied in the range $Re = 10^5...5\times10^5$, which is typical for the environmental engineering systems' ducts.

For the experimental studies, pre-prepared quartz sand, which has a true density $\rho = 2600$ *kkg/m*^{3} and median diameter $\delta_{50} = 90$ *mkm*, was used. When using fibrous materials, as well as the materials with increased adhesion, as well as when significant electrostatic forces occur (for example, when using polymer materials for the air ducts' manufacturing), the resistance to flow movement, both swirling and axial, can significantly increase.

Based on the data obtained during the experimental studies, the specific aerodynamic resistance coefficients values to the swirling flow λ_{sf} were determined, corresponding to different values of the flow swirl intensity (Φ^*) and the mass concentration of dust particles in it (ε).

The results of determining the resistivity coefficient values to the movement of a two-phase swirling flow, obtained in the course of experimental studies, are shown in Figure 1.

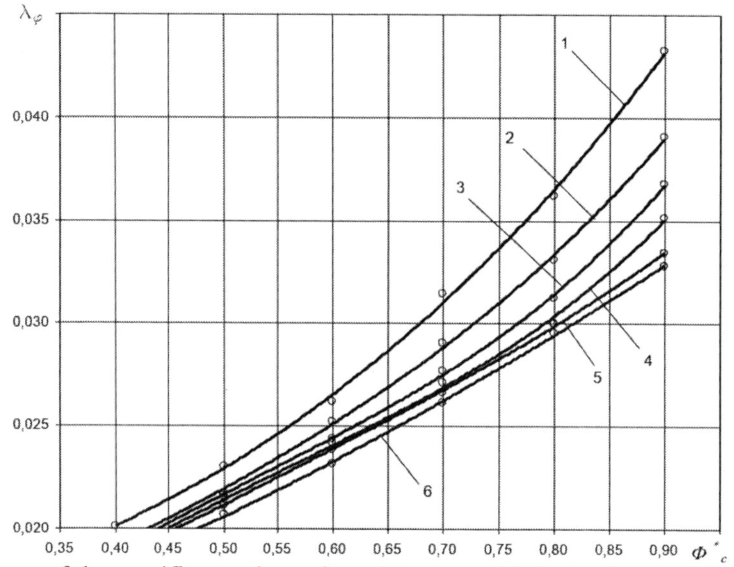

Figure 3. Dependence of the specific aerodynamic resistance coefficient values to the swirling gas-dust flow movement on the intensity of the flow swirling averaged along the length of the duct $\lambda_{sf}\left(\overline{\Phi^*}\right)$, at $Re = 1\times10^5$ and various values of the mass dust concentration: $1 - \varepsilon = 0{,}01$; $2 - \varepsilon = 0{,}009$; $3 - \varepsilon = 0{,}008$; $4 - \varepsilon = 0{,}007$; $5 - \varepsilon = 0{,}006$; $6 - \varepsilon = 0{,}003$.

To improve the clarity and convenience of the comparison, the experimental data are presented in the form of the specific aerodynamic resistance coefficient values ratios of dusty and clean swirling gas-dust flows corresponding to certain combinations of the experimental factors λ_{sf}/λ_0 (Figure 2)

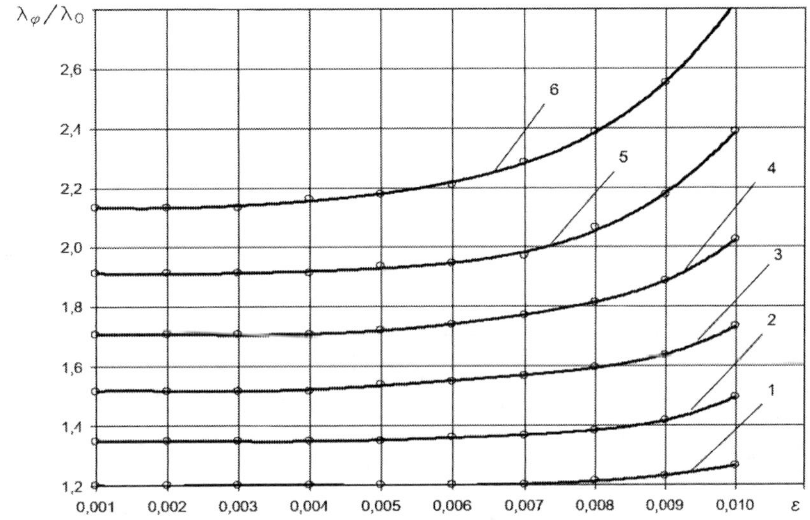

Figure 4. Dependence of the specific aerodynamic resistance coefficients ratio of the two-phase and single-phase rotating flows on the mass concentration of dust $\lambda_{sf}/\lambda_0(\varepsilon)$, at different values of the flow swirl intensity averaged over the duct length: 1 – $\Phi^* = 0{,}4$; 2 – $\Phi^* = 0{,}5$; 3 – $\Phi^* = 0{,}6$; 4 – $\Phi^* = 0{,}7$; 5 – $\Phi^* = 0{,}8$; 6 – $\Phi^* = 0{,}9$.

As follows from the experimental data, the specific aerodynamic resistance coefficient value to the axial flow movement is practically independent of the solid phase particles' mass concentration over the entire region of the experiment, which fully corresponds to the data [6 - 8].

However, in the presence of a rotational flow motion, this situation changes. An exponential growth of the specific aerodynamic drag coefficient is observed with an increase in the flow swirl intensity value. In this case, the values obtained at mass concentrations of the solid phase in the range of values $\varepsilon = 0{,}001...0{,}004$ with practically sufficient accuracy correspond to the results of calculations performed according to the methods [6 - 8].

When the indicated values of ε are exceeded, the values of the coefficient of specific aerodynamic resistance to the movement of the swirling flow begin to exceed those obtained by the calculation (see Figure 2). Figure 3 graphically shows the dependence of the ratio of the coefficients λ_{sf}/λ_0 on the intensity of the flow swirl Φ^*.

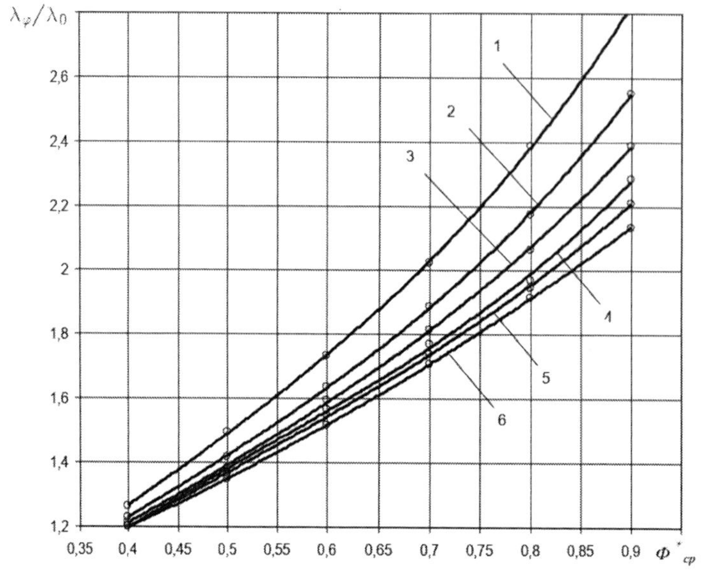

Figure 5. Ratio dependence of the specific aerodynamic drag coefficients of two-phase and single-phase rotating flows on the swirl intensity $\lambda_{sf}/\lambda_0\left(\Phi^*\right)$, at various values of the dust mass concentration: *1 – ε =0,01; 2 – ε =0,009; 3 – ε =0,008; 4 – ε =0,007; 5 – ε =0,006; 6 – ε =0,003.*

According to the shape of the curves shown in Figures 2 and 3, it can be concluded that the high degree of influence exerted by the mass concentration on the ratio λ_{sf}/λ_0value. In addition, it follows from the data presented that the influence exerted on the sought ratio by the mass concentration value is noticeably increased with an increase in the flow swirl Φ^* intensity.

From a practical point of view, the most important result of the described experimental studies is the dust particles mass concentration values' determination, at which it is possible with sufficient accuracy to use the calculated dependences intended to determine the specific aerodynamic resistance coefficient values of a single-phase swirling flow passing through a circular steel ventilation duct. It follows from the results obtained that this value does not exceed the value $\varepsilon = 0{,}005$, which is two times lower than the similar values typical for the axial dust flows.

However, it should also be noted that the obtained value $\varepsilon = 0{,}005$ corresponds to the volumetric concentration values of the reference dust (quartz sand) with = *600 mg/m³* under normal conditions, which is quite enough for the overwhelming majority of currently operated engineering and environmental systems.

Summary

1. Practically significant differences in the aerodynamic resistance values of dusty and pure swirling flows start becoming obvious at lower values of the solid particles' mass concentrations than those characterizing the axial flows.
2. The studies carried out confirm the possibility of using the calculated dependencies to determine the values of the specific aerodynamic resistance coefficient to the movement of a swirling gas flow to the movement of a dusty flow in steel circular air ducts of engineering and ecological systems at the dust particles' mass concentration values $\varepsilon < 0{,}005$.

References

[1] Azarov V N, Borovkov D P, Redhwan A M 2014 Application of swirling flows in aspiration systems *International Review of Mechanical Engineering* **8 (4)** 750-753.

[2] Azarov V N, Borovkov D P 2003 Application of swirling flows in aspiration systems in the construction industry *United scientific journal* **5** 102-104.
[3] Azarov V N, Borovkov D P 2012 Application of swirling flow in aspiration systems at enterprises of the construction industry *Construction Materials* **5** 65-67.
[4] Azarov V N, Borovkov D P 2012 On the use of swirling flow in aspiration systems at the construction industry enterprises *Bulletin of the Central Regional Branch of the Russian Academy of Architecture and Construction Sciences* **16** 12-17.
[5] Borovkov D P, Chicherov K O 2013 Aspiration systems with swirling flow in air ducts *Regional architecture and construction* **1** 115-121.
[6] Borovkov D P, Chicherov K O 2012 Aerodynamic calculation of aspiration systems when organizing flow swirling in air ducts *Regional architecture and construction* **3** 145-148.
[7] Schukin V K, Khalatov A A 1982 *Heat transfer, mass transfer and hydrodynamics of swirling flows in axisymmetric channels* (Moscow, Mechanical engineering) 200.
[8] Khalatov A A 1989 *The theory and practice of swirling flows* (Scientific thought) 192.
[9] Azarov V N, Zheltobryukhov V F, Borovkov D P A 2003 Device for cleaning air ducts of aspiration systems under excess pressure. Utility model patent RUS 35325 05.06.2003
[10] Azarov V N, Borovkov D P, Martyanov V N, Azarov D V 2003 Device for cleaning air ducts from dust. utility model patent RUS 33755 05.06.2003

Procedure for digitalization of measures to create a gas distribution network

D A Yedukov*, V A Yedukov

Samara state technical university, 244, Molodogvardeyskaya street, Samara, 443100, Russia

E-mail: inbox163@inbox.ru

Abstract. The article describes the possibility of measures digitalization to create a gas distribution network. The description of processes at definition of the contract category on connection, performance and the statement of the design documentation is resulted. The analysis of possible reduction of actions terms at the expense of electronic geodetic database use, a uniform platform on coordination of gas distribution networks projects is carried out. As a result of the introduction of measures digitalization to create a gas distribution network, the contractual obligations term on the part of the gas distribution organization will be reduced to 3 calendar months.

Introduction

At present, the Russian Government is working on an instructions list based on the audit results of the legislative execution aimed at developing regional gasification [1]. This instruction was received from the President of the Russian Federation as a result of unsatisfactory inspection by the Federation Council of the country gasification rate, which currently does not exceed 1% per year [2]. Instructions were also received on the need to ensure digitalization of gasification measures at all levels of executive power and other measures that will have to be phased out by 2024 and 2030 to complete gasification of the country [3, 5, 8].

It should be noted that the new rules for connecting to gas distribution networks are constantly being criticized [3, 6] and adjusted by the Government which number reached 15 in seven years.

According to the current regulatory acts of the constituent entities of the Russian Federation and the requirements for construction and installation work in public areas the time period for construction of a gas distribution network to the border of the applicant land plot can be up to nine months [4]. During this period it is required to carry out numerous approvals of the geodetic survey material and design documentation with operating organizations and local authorities. It is also necessary to obtain numerous construction and installation permits in order to legally comply with the possibility of building a gas distribution network in constricted conditions in the vicinity of existing engineering networks. Upon completion of the gas distribution network construction it is necessary to obtain a permit to commission the capital construction facility and make a technological connection with gas commissioning.

Materials and methods

Consider the gasification stages from the application to the approval of the project documentation in order to identify how to reduce the time for interaction responsible decision-making and activities digitalisation.

At the first stage the applicant submits an application for connection to the gas distribution network (Figure 1). At present, this process is ensured by the employees of the gas distribution organisation (GDO) who ensure further review and conclusion of contractual obligations after reviewing the package of documents attached to the application.

The GDO then analyses the possibility of connecting the applicant's capital construction facility in accordance with the requested gas consumption to the gas distribution network and determines the connection point. These data are decisive indicators when drawing up the contract for technological connection and its further conclusion between GDO and the applicant. After conclusion of contractual obligations on connection of the capital construction object to the gas distribution network, exploration and design works are carried out by GDO. In this process the Department of Urban Development, operating organisations and third parties that interests are affected in the course of the gas distribution network construction are involved.

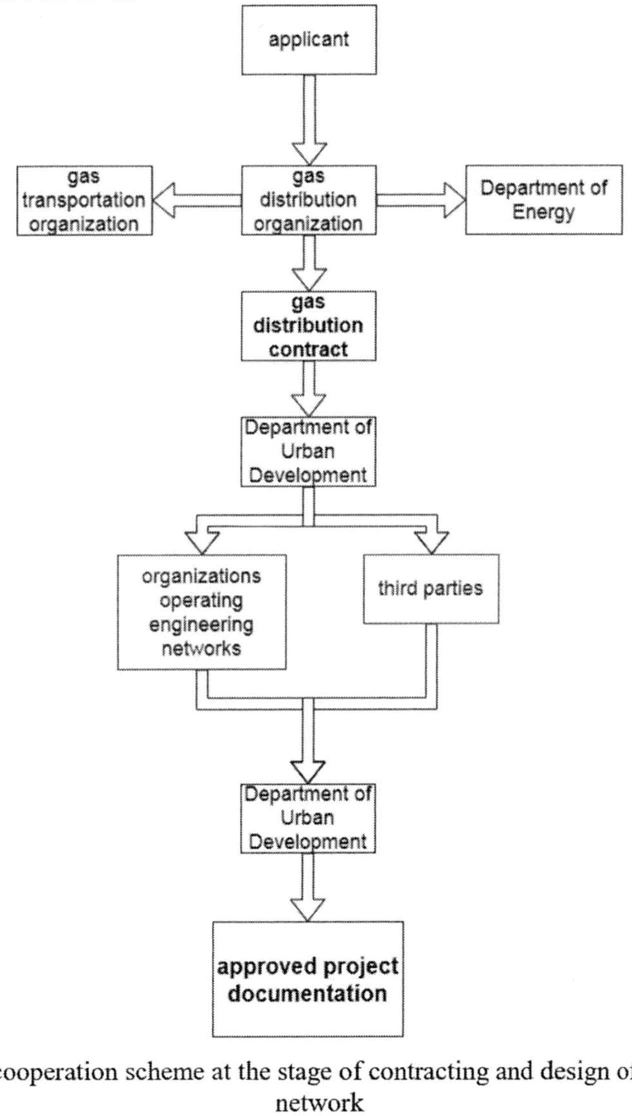

Figure 1. Existing cooperation scheme at the stage of contracting and design of the gas distribution network

It should be noted that if in the existing scheme (Fig.1) the declared volume of consumption does not exceed 500 m^3/h such measures as determination of the connection point and the consumption volume when connecting the future consumer before signing the contract for technological connection are not agreed with the Ministry of Energy and the gas distribution organisation (GDO). These data are transmitted to these authorities by the general register as quarterly reports.

Based on the existing model of interaction between the GDO and the main regulator of energy relations (Ministry of Energy) and the natural gas supplier (NGS), it can be seen that the latter two agencies receive information after concluding contractual obligations. This fact reduces the interaction degree at the stage of determining the connection point to the existing gas distribution network.

Results

In order to address the above mentioned shortcomings in the interaction of interested organisations and agencies it is necessary to strengthen the relationship and create a single digital space where all participants in the process will have to work: from identification of the connection place to the existing/reconstructed gas pipeline to construction and installation work in the city.

It is proposed to consider one of the possible options of the procedure for digitalization of the gas distribution network in terms of preliminary definition of the contract category (definition of the gas pipeline connection point) and performance of design works on construction of the gas distribution network (Figure 2).

On the instructions of the Head of State of 31 May 2020 it is necessary to ensure the possibility of applying for connection not only in GDO, but also in numerous multifunctional centres in the country [1]. The instruction implementation will allow increasing the number of centres accepting gasification applications, but will also make it possible to centrally send data on the total number of applications and the corresponding load on the gas distribution network in a particular area to the Ministry of Energy and the Gas Transport Organisation (GTO). This condition will make it possible to monitor in real time the dynamics of possible connections to the existing gas distribution network for further forecasting of measures to increase the capacity of existing gas distribution stations.

In such a way, at the first stage the analysis of possible connection options to the existing gas distribution network is performed. It is necessary to create a single digital space where interaction between the applicant - GDO - GTO - municipal authorities - Department of Urban Planning - operating organisations will be possible.

It should be noted that at present the process of interaction between the abovementioned organisations and the executive authorities is not sufficiently established. Absence of a short normative term for consideration of the laying possibility in the public territory and coordination of design documentation increases the approval time for the construction of gas distribution networks. In the case of constrained construction, many design solutions are not coordinated due to the proximity to existing engineering networks and possible subsequent difficulties in operation, maintenance and repair.

To meet this condition it will be necessary to create a single site to agree the design documentation for the construction of engineering networks including the gas distribution network. Each responsible person of each of the organizations involved will be given a certain period of time to agree or reasoned refusal with recommendations on adjustment of the design decision. In order to increase the degree of responsibility and accuracy in making decisions on coordination of the gas pipeline route it is necessary to create the site basis for coordination of project documentation - current digital tablets of the Department of Urban Planning. These tablets should display capital construction objects and existing engineering communications on the site. Providing information to site participants in electronic form (DWG format) will allow using the possibility of timely changes to the existing location of engineering communications and strengthening the organisation degree when considering design solutions.

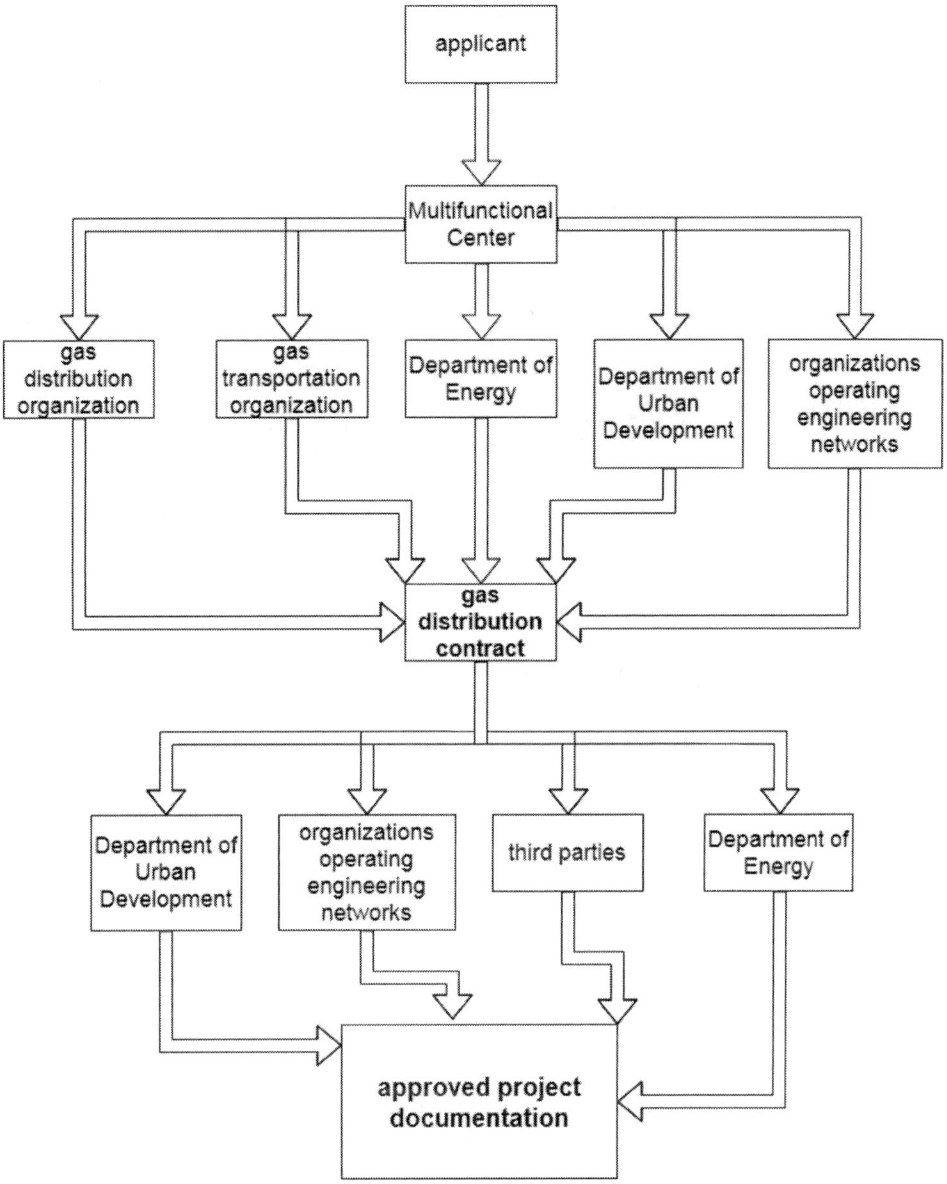

Figure 2. Proposed interaction scheme for digitization of measures to conclude the contract and design the gas distribution network

In this way, to implement the digitalisation of the first stage where the contract for connection and execution of design works for creation of a new gas distribution network is concluded it is necessary to create an electronic surveying database containing up-to-date data on location of all existing facilities, capital construction objects and engineering communications. This electronic topographic database must include all interested participants in the process of designing and building the gas distribution network and must be updated in a timely manner.

In addition to coordinating and approving the design documentation for the construction of the gas distribution network the availability of a single digital space will make it possible to discuss the following issues that need to be resolved jointly with the Ministry of Energy and executive authorities:

– construction of a new low or medium pressure gas distribution network to connect new applicants for the area under consideration;

- necessity and options for reconstruction of the existing gas distribution network or parallel laying of the new network;
- necessity to carry out measures to increase pressure at the end points (connection, installation of additional gas reduction points (GDP) on the medium and high pressure network, etc.);
- possibility of alternative routes for the gas pipeline in the presence of forest land, constricted conditions, third party territories.

The adoption of the above decisions when creating new or reconstructing existing sections of the gas distribution network will directly affect the dynamics of gasification of the region as a whole [9, 10, 11, 12, 13].

The need to implement the Government Decree on the establishment of security zones along the route of the gas distribution network does not have a good effect on the gasification development dynamics. Very often in practice it turns out that the owner of a land plot whose interests will be affected during the design of a gas pipeline or GDP near its land plot refuses to agree the design documentation. For this reason, the design organisation has to repeatedly change the route of the gas distribution network or force the installation of a GDP in the applicant's territory due to the presence of a 10-metre security zone. In most cases, the Department of Urban Planning and executive authorities are not involved in resolving such disputed points and the creation of a common electronic platform for the approval of gas pipeline routes will allow monitoring and timely resolution of such failures and optimisation of construction and installation costs.

Table 1. List and timing of measures from the submission time of an application for gasification to approval of project documentation for construction of the gas distribution network

№	Name of event	Duration before digitalization, days	days Duration after digitalisation, days
1	Determining the possibility of connection (preparing the technical conditions) to the existing networks taking into account the development perspective	5	-
2	Identification of connection point options together with the Department of Urban Planning and executive authorities for the possible development prospects of the region	-	5
3	Approval of increase in supply volume with GTO	-	1
4	Preliminary approval of the pipeline route with operating companies	14	7
5	Coordination of the pipeline route with third parties	14	-
6	Coordination of the gas pipeline route with third parties involving the Department of Urban Planning and executive authorities	-	7
7	Execution of surveys	14	14
8	Reconciliation of survey results with interested organisations and the Department of Urban Planning	44	14
9	Design of the gas distribution network	14	14
10	Approval of the gas pipeline route with interested organisations and the Department of Urban Planning	44	14

IOP Conf. Series: Materials Science and Engineering 1083 (2021) 012061 doi:10.1088/1757-899X/1083/1/012061

11	Examination of industrial safety	14	7
12	Transfer of the project work results to the applicant	7	1
13	Approval of project documentation for the gas consumption network	14	1
Amount		184	85

Upon completion of the development stage of project documentation for connection to the gas distribution network it is necessary to obtain in electronic form an approved project with all interested organisations. Then the route plan of the gas distribution network should be sent to the applicant in order to be able to link the connection point at the border of the land plot to the projected gas consumption network.

Discussion

The result of the introduction of the stage digitalization concluding an agreement for the connection and design of a gas distribution network (Table 1) there will be a 53% reduction in the time for implementation of measures allowing for more efficient decision-making on the prospective development of the gas distribution network and an increase in the consistency degree between disparate operating organisations. In order to ensure timely decision-making on approval of project documentation for construction of the gas distribution network it is necessary to ensure a short time for approval at a single electronic site. In this case, the design team of the gas distribution organisation will be able to perform its duties in a short time as the number of approvals with interested organisations operating external engineering networks may reach 12 within the city limits. In addition, in case of simultaneous involvement of the executive authorities in this time-consuming process it will be possible to solve problems by agreement with third parties that significantly regulate the cost and timing of measures for the construction of the gas distribution network.

Summary

The construction dynamics of the gas distribution network largely depend on the consistency degree of the interested organisations involved in making decisions on the location of the connection point and coordinating design decisions. Creating a single digital space and changing the existing approach to project approval measures will reduce the time required to connect capital construction projects to the gas distribution network.

References

[1] Information on http://www.kremlin.ru/acts/assignments/orders/63454.
[2] Information on https://iz.ru/982258/2020-03-02/matvienko-raskritikovala-tempy-gazifikatcii-v rossii.
[3] Information on https://www.rbc.ru/economics/02/03/2020/5e5d036f9a794719b25b8e47
[4] Information on http://council.gov.ru/events/news/114180/.
[5] Information on https://www.rbc.ru/society/12/03/2019/5c87981c9a79472d96955924
[6] Information on https://tass.ru/ekonomika/7880275.
[7] Information on https://rg.ru/2020/03/02/matvienko-predlozhila-propisat-v-zakone-usloviia-podkliucheniia-gaza.html.
[8] Information on http://kremlin.ru/events/president/news/60052.
[9] Urabe, Yasuhiko&Kawamura, Toshio&Sakanoue, Takashi&Uno, Osamu&Matsuzaki, Yoshio 2016 *Gas Supply Infrastructure.*
[10] Perrow, Melissa 2019 Securing competitive gas supply - a manufacturing *The APPEA Journal*
[11] Novopashina N A, Yedukov V A, Yedukov D A 2013 Prospects for the development of gas distribution networks after the adoption of Resolution No. 1314 of the Government of the Russian Federation dated 30.12.2013

[12] Balzannikov M I, Galitskova K S, Strelkova A K 2016 Traditions and innovations in construction and architecture *Construction Technologies. Collection of articles, Samara State University of Architecture and Civil Engineering* (Samara) 364-369.
[13] Brünje H &Corsten, H-D &Weßing W 2003 From natural gas heating to the "all-gas supply" house 674-678.
[14] V N Melkumov [et al] 2015 Design of urban and village gas supply distribution systems [Electronic resource] (a training manual/essay, Voronezh State University of Architecture and Civil Engineering, EBS ASV).

Road incidents detecting

A R Feizullaev

Don State Technical University, 1, Gagarin sq., Rostov-on-Don, 344010, Russia

E-mail: f.artur1987@gmail.com

Abstract. Timely detection of traffic incidents largely determines the traffic management algorithms' effectiveness. To date, an accurate method for detecting road incidents has not been developed. As a result, different approaches to determine the incidents' duration are used. This paper observers currently available approaches to identifying road incidents and suggests dividing incidents into four stages: detection time, reaction time, clearance time and recovery time. Determining the impact degree of various factors on the incident duration is important for designing an efficient traffic management algorithm under incident conditions. Four-stage split can be used for developing an algorithm minimizing the incident duration as well as the time needed for road traffic recovering. The article presents the results of traffic flows' modelling, which makes it possible to estimate the road incidents' detection time as well.

1. Introduction

One of the topical areas of research in the field of traffic management is development of the systems that automatically monitor the traffic state on the roads. Such systems should not only monitor the current state at the road sections and intersections, but also predict the traffic flow state, warn the drivers and reduce the number of accidents. Collecting, analyzing and managing the data obtained from the detectors during incidents is an important task that should be addressed in order to manage traffic effectively.

2. Material and methods

Incident duration determining is one of the main tasks because the timely response to an occurred incident allows minimizing both the duration of its existence and the time required to return the situation to normal. In early studies, the incident duration was directly included the deviations cause lifetime from the normal situation [1]. Then it was proposed to determine the incident duration as the time difference between the incident detection and the first reaction to it (the arrival of response services, the decision to redistribute traffic flows, change the traffic light object operation mode) [2]. However, this approach eliminates the time required to detect an incident. Subsequently, the lifetime of the incident started being measured by the duration between the time when the incident was detected and the time when the incident was eliminated. Thus, the incident duration can be divided into the following stages:

- detection time: the period from the incident occurrence to its detection;
- reaction time: the period from the detection moment to the first exposure time;
- time to eliminate the incident's causes: the period from the first exposure time to the incident's causes elimination;
- recovery time: the period from the incident's causes elimination to the normal traffic conditions' restoration.

At the moment, incidents are identified based only on the traffic video surveillance results. However, this method is limited, since it is impossible to control the road network in the operator mode. In addition, two types of congestion are possible: periodic congestion and non-repeating congestion, the occurrence place of the latter is not known in advance. The main solution is to determine the possibility of promptly detecting congestion and dangerous situations on the road network according to various types of technical means for monitoring the traffic flows' characteristics in conjunction with the traffic mathematical modeling methods.

There are various techniques for detecting incidents. The following methods can be distinguished as the main ones:

- methods based on comparing the traffic flow characteristics on a given section of the road in various situations [5];
- statistical methods for determining the incidents' probability [6];
- incident detection methods based on time series models [7];
- methods of filtering data on the traffic flows' characteristics [8];
- methods based on traffic modelling [9];
- methods based on artificial intelligence [10];
- image-processing methods [13].

Taking a limited set of information on the characteristics of traffic flows in real-time into account, it is necessary to develop the algorithms and models that are capable to identify deviations from normal traffic conditions in the event of dynamic bottlenecks, traffic accidents and other incidents on highways.

The incident detecting system's effectiveness is determined at two levels: the level of data collecting and the level of data processing. Data collecting involves detecting and surveillance technologies. Data processing includes algorithms to detect and classify the incidents by analyzing the movement parameters registered by the detectors or sensors in order to notify the participants of an incident. These two levels, when combined, provide a technical platform for developing and implementing various algorithms.

Most of the algorithms were developed to detect incidents on highways, where relatively low intensity is observed at a sufficiently high speed of vehicles.

Developing accurate algorithms for controlling each interval will make it possible to act effectively during the incident, reduce its duration and minimize a negative impact. Figure 1 shows the incident evolution over time.

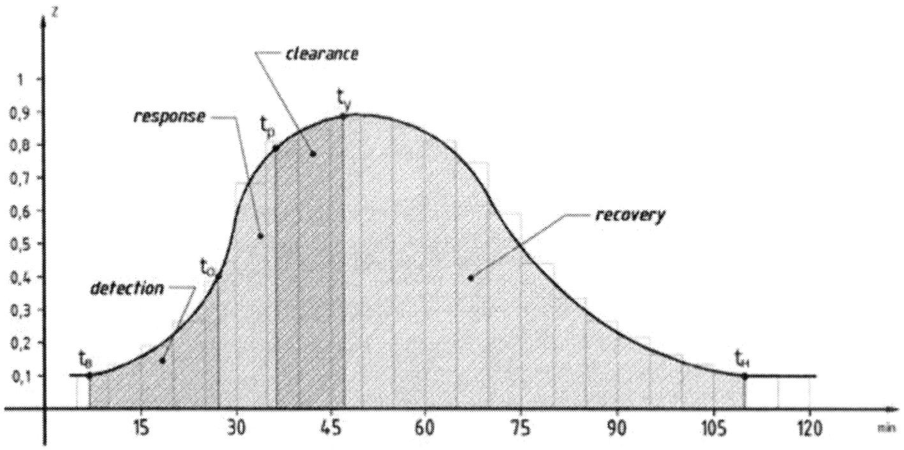

Figure 1. Incident development in time

Load factor is given by the formula:

$$Z = \frac{q}{q_{max}} \tag{1}$$

q – is the traffic flow, veh/h;

q_{max} – denotes road capacity, veh/h.

Developing an effective traffic management algorithm in an incident, it is necessary required to determine the influence degree of various factors on the incident duration.

3. Result and discussion

As part of the study, transport models of incidents in various conditions have been developed. The modeling of traffic flows was carried out on a linear section of the road with the allowed speed of 60 km / h and a traffic intensity close to the carrying capacity (about 80%). Figure 2 shows a two-lane section of the road with an incident at a distance of 50, 100 and 150 meters from the detector. The difference in traffic density in the normal state and during an incident was used as a criterion for detecting an incident.

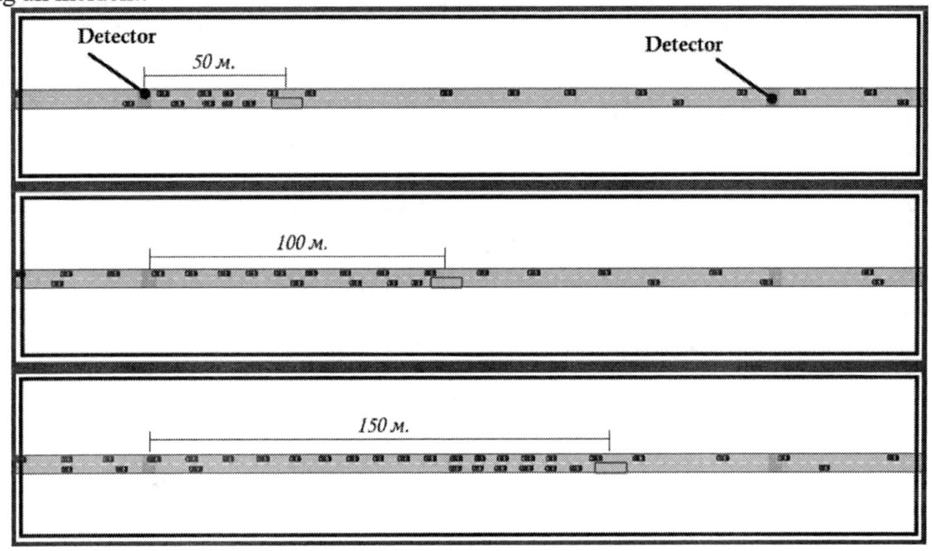

Figure 2. Two-lane section, single lane incident

Figure 3 shows a three-lane section of the road also with an incident at a distance of 50, 100 and 150 meters from the detector.

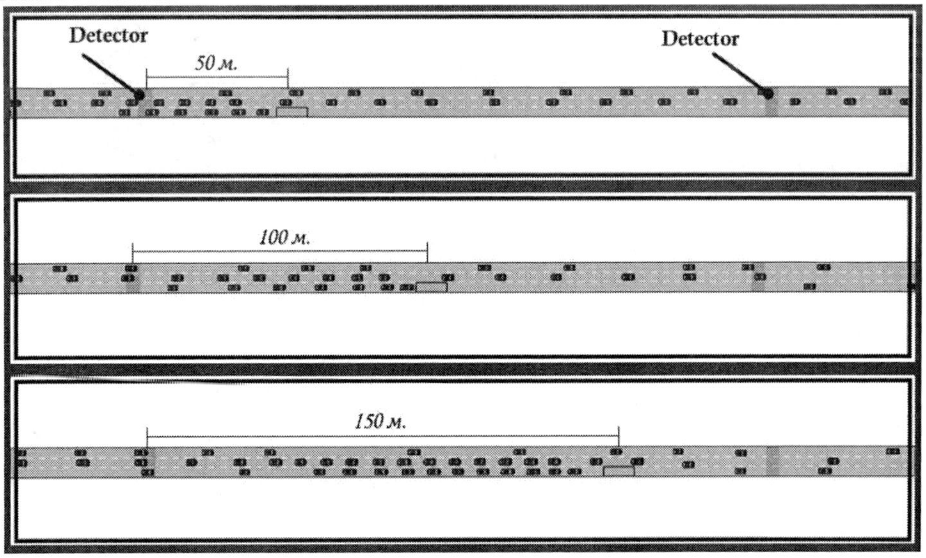

Figure 3. Three-lane section, single lane incident

Finally, Figure 4 shows a section with three lanes and the limited movement in two lanes

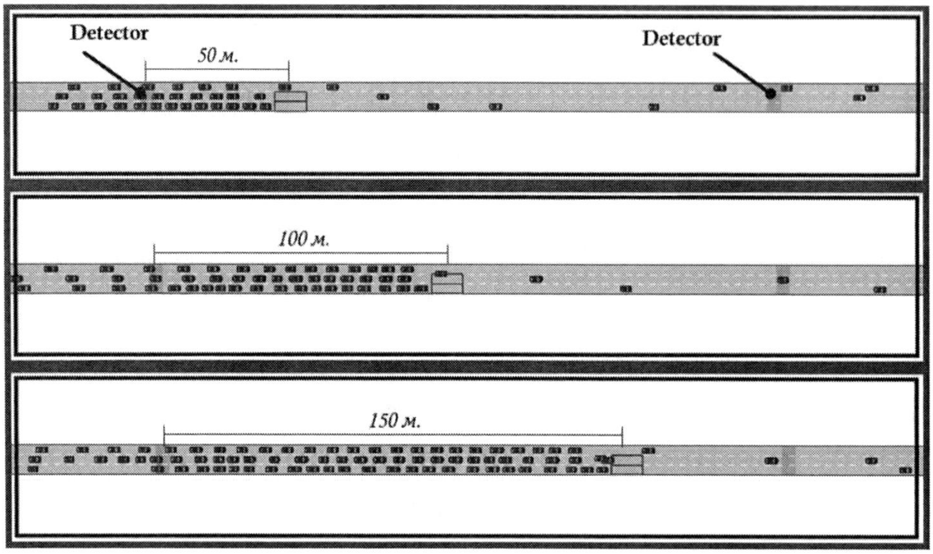

Figure 4. Three-lane section, two-lane incident

The period time from the moment the incident occurred until the moment it was detected was established for each experiment. Figure 5 shows the simulation results.

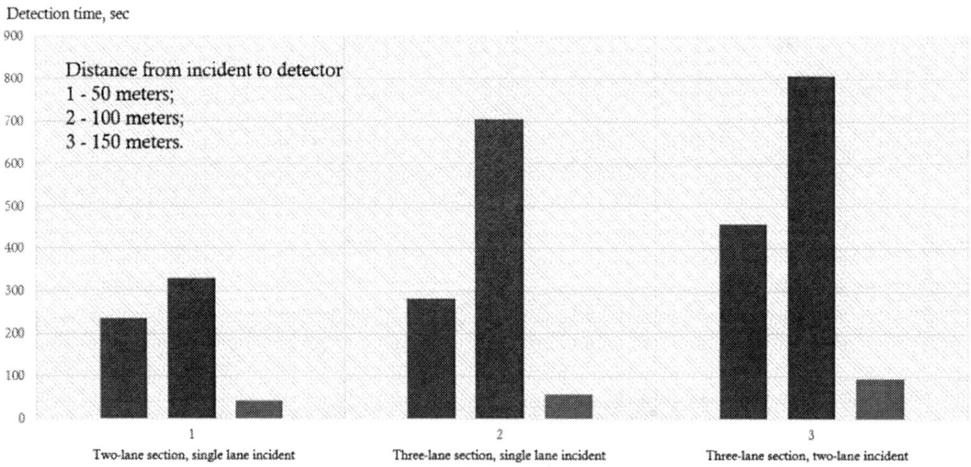

Figure 5. Simulation results

Based on the simulation results, we can state that the fastest detecting of the incident was recorded on a three-lane stretch of road. Thus, we can conclude that when the incident is detected shortly, the situation on the road is the most difficult. The rest of the data show that small disturbances in the traffic flow cause a low speed of the emerged shock wave and the incidents are detected for much longer.

4. Summary

Developing effective methods and models for detecting road incidents is of crucial scientific and applied value and can be used to solve the following problems:

- reducing the number of accidents on roads;

- reducing travel time for drivers;
- increasing the transport stream speed;
- minimizing expenses for organizing transport process;
- reducing in fuel consumption and environmental pollution.

Instead of the current practice of analysing global flow, automatic monitoring systems should be based on the detailed analysis of the behaviour of each vehicle at intersections. The systems should be able to identify each vehicle, track its behaviour and recognize dangerous situations or the events leading to them. Tracking intersection behaviour is often difficult due to a large number of vehicles and walking inclusions. This problem's solution is possible by developing an algorithm that is resistant to such problems.

References

[1] Derose J 1964 Analysis of reandom freeway traffic accidents and vehicle disabilities *Highway Research Record* **59** 53-65.

[2] Jones B, Janssen L and Mannering F 1991 Analysis of the Frequency and Duration of Freeway Accidents in Seattle *Accident Analysis and Prevention* **23(4)** 239-255.

[3] Peterfreund N 1999 Robust tracking of position and velocity with kalman snakes *IEEE Trans. Pattern Anal. Machine Intell.* **21** 564–569.

[4] Deniz O, Celikoglub H B 2011 Overview to some existing incident detection algorithms: a comparative evaluation *Procedia – Social and Behavioral Sciences.*

[5] Payne H J and Knobel H C 1976 Development and testing of incident detection algorithms, Volume 2: users guidelines *Report No. FHWA-RD-76-21, FHWA, Washington D.C., April 1976.*

[6] Dudek C L, Messer C J and Nuckles N B 1974 Incident detection on urban freeway *Transportation Research Record, TRB, National Research Council* **495** 12-24.

[7] Ahmed M S and Cook A R 1980 Time series models for freeway incident detection Journal of Transportation Engineering, ASCE **106 (6)** 731-745.

[8] Cook A R and Cleveland D E 1974 Detection of freeway capacity-reducing incidents by trafficstream measurements *Transportation Research Record, TRB, National Research Council* **495** 1-11.

[9] Black J and Sreedevi I 2001 Automatic incident detection algorithms *ITS Decision Database in PATH.* Information on http://www.path.berkeley.edu/~leap/TTM/Incident_Manage/Detection/aida.html>, February 2001.

[10] Adeli H and Samant A 2000 An adaptive conjugate gradient neural network-wavelet model for traffic incident detection *Computer-Aided Civil and Infrastructure Engineering* **15 (4)** 251.260.

[11] Lin C K and Chang G L 1998 Development of a fuzzy-expert system for incident detection and classification *Mathematical and Computer Modeling* **27 (9-11)** 9-25.

[12] Hsiao C-H, Lin C-T and Cassidy M 1994 Application of fuzzy logic and neural networks to automatically detect freeway traffic incidents *Journal of Transportation Engineering,* ASCE **120(5)** 753-772.

[13] Michalopoulos P G, Jacobson R D and Anderson C A 1993 Field implementation and testing of a machine vision based incident detection system *Pacific Rim TransTech Conference Proceedings, ASCE, Seattle, WA* **1** 69-76.

[14] Kim H J, Choi H-K 2001 A comparative analysis in incident service time on urban freeways *IATSS RESEARCH* **25 (1)** 62-72.

[15] Fleyeh H 2004 Color detection and Segmentation for road and Traffice signs *Proceedings of the 2004 IEEE Cybernetics and Intelligent Systems* **2** 809-814.

[16] Miller R, Sun Z, Bebis G 2004 On Road vehicle Detection *IEEE Transactions on Pattern Analysis and Machine Intelligence* **28** 694-711.

[17] Kass M, Witkin A, and Terzopoulos D 1988 Snake: Active Contour Models *Int. J. Computer Vision* 321-331.

IOP Conf. Series: Materials Science and Engineering 1083 (2021) 012063 doi:10.1088/1757-899X/1083/1/012063

Automated transport and storage systems for road transport companies

E V Marchenko*, Ju V Marchenko, S I Popov, S G Kuren, I V Marchenko

Don State Technical University, 1, Gagarin sq., Rostov-on-Don, 344000, Russia

E-mail: daedwardrambler.ru@mail.ru

Abstract. The technologies of automated transport and storage systems' work for motor transport enterprises, placement, storage and parts release from the warehouse, as well as labor organization in the warehouse are considered. The calculation and design of the warehouse is given.

Introduction

The work performed in different warehouses is approximately the same. This is due to the fact that for different logistics processes, the functions performed by warehouses are similar: temporary placement and storage of inventories; providing logistics services in service system; material flows' transformation. However, specific warehouses require their own work organization with their own characteristics, which depend on the purpose of the warehouse and its size. The work organization for large warehouse complexes is not suitable for a small, highly specialized warehouse. Transportation, storage and warehousing of foodstuffs requires different conditions and approach in contrast to industrial products [1-11].

The structure of the warehouse for automotive spare parts has a number of features that should be taken into account for the efficient operation of the warehouse. The spare parts warehouse is characterized by a wide range and heterogeneity of the sizes of the parts stored in the warehouse.

Theoretical section

Parts acceptance

Acceptance is an important part of the process. Acceptance is carried out by the responsible persons in accordance with the internal rules of the company.

Parts for the warehouse are accepted in terms of quality and quantity. Acceptance technology is determined depending on how the goods are delivered and packed. Acceptance by quantity is carried out by comparing data from the accompanying transport documentation (waybills, specifications, invoices, etc.) with the actual parts' availability.

Acceptance by quality is performed according to the requirements of standards, contracts, technical conditions and accompanying documents. Products that have not passed mandatory certification are not subject to acceptance as well as parts with defects. If a non-compliance of the supplied parts with the company's requirements is found, acceptance is suspended.

Parts are accepted by warehouse workers (storekeepers) unilaterally.

If a shortage or non-compliance of the goods with the company's requirements is revealed, an act is drawn up with the signatures of the persons who carried out the acceptance. This act is sent to the supplier, it indicates the parts' names, the invoice number, the type of shortage and the amount of missing goods.

Parts placement and storage

Parts that have passed acceptance are moved to storage areas to create the necessary parts stock in the warehouse. This technological operation is the basis for performing further functions of the warehouse technological process related to the rational supply organization of the company with spare parts.

The warehouse uses a rack method for storing spare parts. The parts themselves are in a container that has its own unique number (address), with the help of which, using the database of the system for accounting for parts in the warehouse, you can quickly determine the required cell in which the part required for release is located. Such targeting of storage provides operational management of warehouse stocks.

To maintain the proper quality and condition of the parts, the storage regime is observed. The warehouse is provided with control over maintaining the optimum air humidity and temperature. These microclimate parameters are controlled by ventilation and heating.

Parts release from the warehouse

To carry out repair operations or maintenance, it is necessary to provide the workers with the necessary spare parts and materials that are stored in the warehouse. Parts release is the distribution of parts from the warehouse directly for work and services.

In accordance with the established norms of the enterprise, the parts are distributed from the warehouse directly to the sites, but only to strictly defined employees who have the right to perform this kind of actions.

For release registration, a limit card can be used.

A quick search for the necessary parts is carried out by virtue of the warehouse accounting system database, which is based on bar coding. The part is written off from the warehouse automatically when scanning the barcode on the documentation and additional confirmation of the operation by the storekeeper carrying out the release.

After performing the above-mentioned operations, the storekeeper takes the required number of the specified parts from the specified cell and takes them to distribution, having previously completed all the accompanying documentation for the release.

Labor organization in the warehouse

First of all, with the introduction of a robotic loader, the warehouse organizational structure changes. If there is no need to work manually, the need for storekeepers disappears, the loader will do the work for them. But there is a need for periodic maintenance and adjustment of the loader; for these purposes, a person who will take the position of an operator-setter is needed. Otherwise, labor organization in the warehouse remains the same.

The duties of the commissioning operator include the following:

monitoring the technical condition of the loader;
carrying out scheduled maintenance and running repairs, as well as unscheduled repairs;
software adjustments;
manual control of the loader if it is necessary to transport goods outside the warehouse;
loader monitoring;
acceptance and placement of parts in the warehouse using a loader;
filling out the accompanying documentation related to the loader.

Parts acceptance

The technological process of accepting the supplied parts remains the same, except that there are no storekeepers in the warehouse who previously accepted the goods. Now these responsibilities are assigned to the operator.

Calculation and design of a warehouse

In connection with the inclusion of a robotic loader into operation and, as a result, a change in the warehouse operation technology, it is necessary to calculate the overall parameters of the racks on which the containers will be stored.

When calculating the parameters, the dimensions of the container, which has dimensions suitable for the loading mechanism of the electric loader used in the warehouse are taken as basic. Accordingly, the cell should have the dimensions slightly larger than the overall dimensions of the container for free loading and unloading on storage racks [7-15].

Taking into account the robotic equipment operation in the warehouse, and, accordingly, the absence of a human factor, we can assume minimal errors when loading and unloading containers. In this case, it is possible to allow minimal displacement of the loading mechanism to the sides; and in order to increase the density of the warehouse layout, allow an increase in the storage rack cell relative to the overall dimensions of the container by only 20 mm on the sides (40 mm in total).

Then, the cell width will be:

$$B_c=B_{cont}+40,$$

where B_{cont} – is the container width, mm.

$$B_c=1080+40=1120 \text{ mm.}$$

Free height for lifting the container should be slightly larger than the side gaps, first, for the ability to free load and unload the container when it is not fully loaded, and second, when filling the container with large-sized parts, there should be a space between the shelf of the storage rack and the parts exceeding the overall size of the container.

Then, the cell height will be:

$$H_c=H_{cont}+85$$

where H_{cont} – is the container height, mm.

$$H_c=915+85=1000 \text{ mm.}$$

The depth of the shelf corresponds to the depth of the container and is 920 mm.

Thus, we get a rack cell with the following parameters: 900x1120x1000 mm.

Taking into account the dimensions of one cell of the main rack, it is possible to calculate the total number of the main storage racks' cells that will be in the warehouse. With a height from floor to ceiling of the warehouse equal to 4900 mm, it is possible to fit the following number of cells in height (excluding the storage racks' shelves height):

$$n = H_{wh} + /H_c,$$

where H_{wh}= 4900 – is the warehouse height, mm.

$$n = 4900/1000 = 4{,}9 \text{ pc.}$$

The resulting number n=4,9 rounded down due to the impossibility of increasing the warehouse height. Then we get the number of cells in height, n=4 pc.

Taking into account the storage racks' shelves height and the height of the resulting number of cells, we obtain the following value for the main storage racks' height:

$$H_{sr} = n^* (H_c + h_{sh}),$$

where h_{sh} – rack shelf height, h_{sh}=50 mm.

$$H_{sr} = 4^*(1000 + 50) = 4200 \text{ mm.}$$

Taking into account the storage racks' length in the warehouse and the width of one cell, the calculation of the number of cells of the main racks is as follows:

$$N_{mr} = (a + b + c + d)*n/B_c,$$

where a, b, c, d – denote the main racks' length, mm (Figure 1).

$$N_{mr} = (4480 + 13440 + 5600 + 10080)*4/1120 = 120 \text{ pc.}$$

Thus, the width and height of the cells of the additional racks are the same with the width and height of the main racks' cells, the difference is only in depth. It follows from this that for additional racks it is necessary to select the containers with the following parameters: 560x1080x915 mm. In turn, the cell of the additional rack will have the following parameters: 560x1120x1000 mm.

In this case, the number of additional racks is calculated using the following formula:

$$N_{add} = e*n*2/B_c,$$

where e – is the additional rack's length, mm.

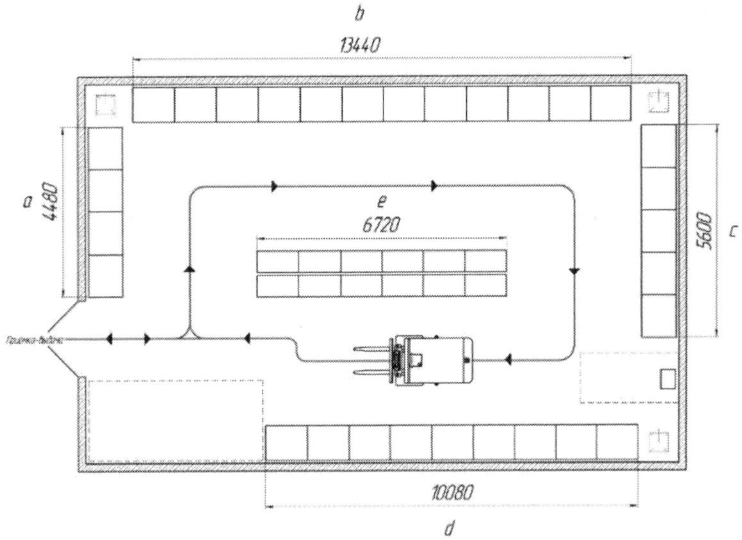

Figure 1. Plan of storage racks' arrangement

$$N_{add} = 6720*4*2/1120=48 \text{ pc.}$$

As a result, a total of 168 cells with the containers are available for storing spare parts in the warehouse, of which 120 are large and 48 additional containers are smaller.

For fast and energy efficient operation of a robotic electric loader, the passage width should be sufficient for free maneuvering around the warehouse, as well as for loading and unloading goods. In this case, the steering control of the loader allows it to turn on the spot 90°. Based on this, for free maneuvering and performing operations, the loader needs a passage width a little more than its overall length. If such a maneuver is impossible, it is possible to use the U-turn algorithm in several steps; to execute this algorithm, the collision and route orientation parameters are included in the loader software.

Based on the above-noted parameters of the loader and its overall length, which is 3210 mm, we obtain the following dimensions of the passage width (Figure 2):

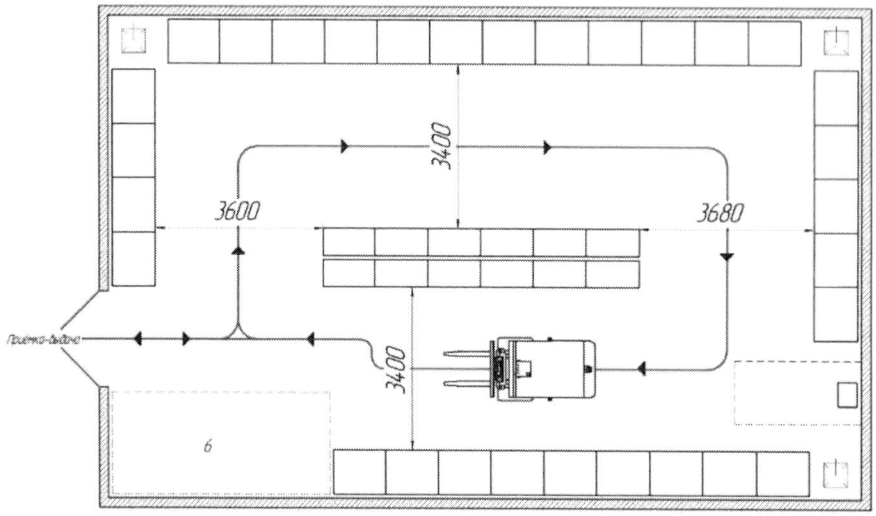

Figure 2. Loader passage width

Parts' placement and storage

Upon completion of the parts acceptance, they should be sorted into places. This process is carried out using a loader. The electric loader receives the address of the container into which the sorting will be carried out and brings it to the point of acceptance and release. For storing parts, large plastic containers of various sizes are used on runners (Figure 3).

Figure 3. Plastic container for storing spare parts

These containers have a volume of up to 700 liters and can withstand dynamic and static loads on the bottom of up to 1000 kilograms. They also have a long service life with proper handling and are suitable for transportation by loader and do not overload the racks and loader with their own weight, unlike similar metal containers.

All containers available in the warehouse have a certain placement order. First of all, the containers are determined by two criteria:

1. mass;

IOP Conf. Series: Materials Science and Engineering 1083 (2021) 012063 doi:10.1088/1757-899X/1083/1/012063

2. demand.

These criteria were not chosen as the main ones by chance, the heaviest and most demanded containers are located as close as possible to the release point, this is done to reduce the load on the loader, save energy and increase the distribution speed of the necessary goods. In accordance with these criteria of containers, the racks in the warehouse are divided into three zones for placing the containers (Figure 4):

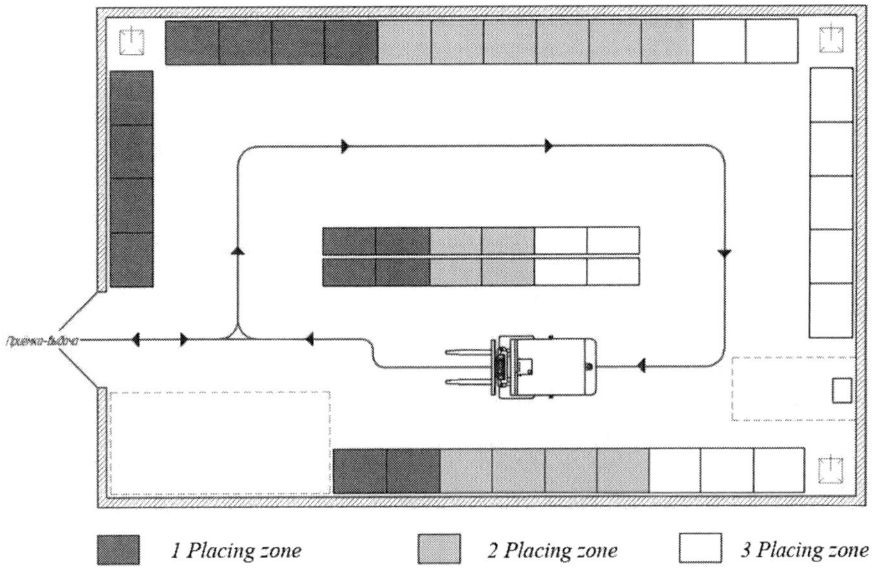

Figure 4. Layout of disposition areas on storage racks

1. disposition area for heavy, medium and light high-demand containers;
2. disposition area for heavy, medium and light medium-demand containers;
3. disposition area for heavy, medium and light low-demand containers.

Within these zones, the distribution of containers by weight is also carried out, according to the principle, the further from the acceptance and release, the easier. Weight designation is numerical and looks like this:

1 – heavy;
2 – average weight;
3 – light.

The demand for containers is also divided into 3 groups, but indicated by letters:

X – high-demand;
Y – medium-demand;
Z – low-demand.

Further distribution is made based solely on the contents of the container. All containers are distributed according to specific parts and their belonging to specific groups. Membership in a group is indicated by the letters (A, B, C, etc.), and belonging to a particular part is indicated numerically (1-valves, 2-pistons, etc.). This distribution includes the following parts' groups:

A – engine;
B – body (all attachments such as lights, handles, wiper blades, etc. are stored in a warehouse);
C – transmission;
D – running gear;
E – steering control;
F – brake system;
G – electrical equipment;

H – cooling system.

The result of such parts sorting by containers is the following example of a unique code for each container with spare parts (Figure 5):

1ZA136

Figure 5. Unique cipher of the container

After deciphering this designation, it turns out that the container is heavy (1), contains little-demanded (Z) engine parts (A), namely camshafts (136).

Parts release from the warehouse

Parts are also released to persons who have the right to receive them from the warehouse, but the receipt process changes radically. The warehouse is fully automated and does not require human participation to carry out the release process. To obtain the necessary spare parts, the person to whom the release will be made comes to the pick-up and drop point with accompanying documentation, which contains a list of spare parts to be dispensed and there is a barcode for each part, which has the detailed information in which the container it is located for the loader. This information is read by the scanner and sent wirelessly to the loader via WI-FI. To locate the necessary container in the warehouse at the received address, RFID technology is used [16-20].

RFID – it is a technology that uses radio waves to transmit information. This system consists of tags (data keepers), a reader, an antenna (to amplify the signal) and a system that processes data.

The reading distance of an RFID tag will vary depending on the equipment and environment used, and is customizable based on needs. Data in the form of a cipher (Figure 5) is recorded on the RFID tag. The data received from the scanner and the data from the tag are compared by the loader and the route to the container with the necessary spare parts is determined. Following the algorithms, the loader loads the container and takes it to the acceptance point, where the customer picks up the necessary part.

Accounting for material assets in the warehouse

With the general automation of the warehouse, the processes of accounting for material assets are partially automated. So, when ordering goods from a warehouse according to the process described above, the barcode carries, in addition to the container address, information about the order, and specifically the number of released parts, their code numbers, recipient data. All data is automatically recorded by the computer when releasing and the details are entered into the database as shipped (released).

Summary

As a result of this work, the calculation and the warehouse design has been carried out, the plan for the storage racks arrangement has been calculated, the width of the passage for the loader has been determined, the containers' disposition areas have been highlighted and a unique marking code for each container with spare parts has been proposed.

References

[1] Teplyakova S V, Kotesova A A, Popov S I, Kotesov A A 2020 The transition from the sample data to the total aggregate of the final volume and the analysis of this transition laws *International Scientific Conference «Construction and Architecture: Theory and Practice for the innovation Development» (CATPID-2020): IOP Conference Series: Materials Science and Engineering* **913** 042054. doi:10.1088/1757-899X/913/4/042054

[2] Ivanov V V, Popov S I, Dontsov N S, Kotesova A A 2020 The oxide film formation under vibration processing in the high-resource parts manufacture in transport engineering *International Scientific Conference «Construction and Architecture: Theory and Practice for the innovation Development» (CATPID-2020): IOP Conference Series: Materials Science and Engineering* **913** 042056,. doi:10.1088/1757-899X/913/4/042056

[3] Korotky A A, Marchenko E V, Popov S I, Marchenko Ju V, Dontsov N S 2020 Theoretical foundations of modeling the process of transport vehicles steel ropes structural defects formation *XIII International Scientific and Practical Conference «State and Prospects for the Development of Agribusiness – INTERAGROMASH 2020»: E3S Web of Conferences* **175** 05018. doi.org/10.1051/e3sconf/202017505018

[4] Ivanov V V, Popov S I, Dontsov N S, Ekinil G E, Oleynikova Ju A, Denisenko Ju N 2020 Mechanical coating formed under conditions of vibration exposure *XIII International Scientific and Practical Conference «State and Prospects for the Development of Agribusiness – INTERAGROMASH 2020»: E3S Web of Conferences* **175** 05023. doi.org/10.1051/e3sconf/202017505023

[5] Korotky A A, Popov S I, Galchenko G A, Marchenko Ju V, Drozdov D S 2020 The use of SmartBox container for agrobusiness logistic processes optimization *XIII International Scientific and Practical Conference «State and Prospects for the Development of Agribusiness – INTERAGROMASH 2020»: E3S Web of Conferences* **175** 13019. doi.org/10.1051/e3sconf/202017513019

[6] Ivanov V V, Popov S I, Selemeneva E M, Babazhanov N T 2019 Study of technological characteristics of the process of formation of vibration mechanochemical oxide coating *XV International scientific-technical conference «Dynamics of technical systems» (DTS-2019): AIP Conference Proceedings* **2188** 020015. doi.org/10.1063/1.5138389

[7] Ivanov V V, Popov S I, Marchenko Ju V, Marchenko E V, Dontsov N S, Timofeev S A 2019 Thickness of vibrational mechanochemical solid-lubricant coating in friction pairs of transport engineering products *XII International Scientific Conference on Agricultural Machinery Industry (INTERAGROMASH 2019): IOP Conference Series: Earth and Environmental Science* **403** 012115. doi:10.1088/1755-1315/403/1/012115

[8] Korotky A A, Marchenko E V, Ivanov V V, Popov S I, Marchenko Ju V, Dontsov N S 2019 Model of forming vibration mechanochemical solid lubrication coating on surface of steel rope *XII International Scientific Conference on Agricultural Machinery Industry (INTERAGROMASH 2019): IOP Conference Series: Earth and Environmental Science* **403** 012116. doi:10.1088/1755-1315/403/1/012116

[9] Kotesova A A, Teplyakova S V, Popov S I, Kopylov F C 2019 Ensuring assigned fatigue gamma percentage of the components *International Scientific Conference «Construction and Architecture: Theory and Practice of Innovative Development» (CATPID-2019): IOP Conference Series: Materials Science and Engineering* **698** 066029. doi:10.1088/1757-899X/698/6/066029

[10] Ivanov V V, Popov S I, Kirichek A V 2018 Investigation of optimal chemical composition of cast aluminum alloys for vibrational mechanical-chemical polishing and deposition of protective and decorative coatings *XI International Conference on Mechanical Engineering, Automation and Control Systems (MEACS 2017): IOP Conference Series: Materials Science and Engineering* **327** 032026. doi:10.1088/1757-899X/327/3/032026

[11] Ivanov V V, Popov S I, Kirichek A V 2017 Qualitative Characteristics of MoS_2 Solid-Lubricant Coating Formed by Vibro-Wave Impact of Free-Moving *Indenters Key Engineering Materials* **736** 18-22. DOI:10.4028/ www.scientific.net/ KEM.736.18

[12] Nikolaev N N, Marchenko Yu V, Filatov S K 2019 Research and modeling of the taxi service in small towns *International Scientific Conference «Construction and Architecture: Theory and Practice of Innovative Development» (CATPID-2019): IOP Conference Series: Materials Science and Engineering* **698**. doi:10.1088/1757-899X/698/6/066027

[13] Kalmykov B Yu, Stradanchenko S G, Sirotkin A Y, Garmider A S, Kalmykova Y B 2016 *ARPN Journal of Engineering and Applied Sciences* **11(17)** 10205-10208.

[14] Kalmykov B Yu, Ovchnnikov N A, Kalmikova O M, Guguyev I K, Kushnariva I V 2015 *ARPN Journal of Engineering and Applied Sciences* **10 (10)** 4366-4371.

[15] Ovchinnikov N A, Kalmykov B Yu, Stradanchenko S G, Kozyreva E A, Chefranova O V 2015 ARPN Journal of Engineering and Applied Sciences **10 (22)** 10511-10522.

[16] Kalmykov B Yu, Ovchinnikov N A, Kalmikova O M, Jigulskii V I, Yurshin Y G 2015 *ARPN Journal of Engineering and Applied Sciences* **10 (8)** 3793-3797.

[17] Shcherbakov I N, Ivanov V V, Korotkiy A A 2018 *Solid State Phenomena* **284** 1140–1143. doi:10.4028/www.scientific.net/ssp.284.1140

[18] Shcherbakov I N, Korotkiy A A, Egelskaya E V 2020 *Lecture Notes in Mechanical Engineering* 1279-1285. doi.org/10.1007/978-3-030-22041-9_133

[19] Shcherbakov I N, Korotkiy A A and Shulga G I 2019 *AIP Conference Proceedings* **2188** 020011. doi.org/10.1063/1.5138385

[20] Demyanov Alexander, Demyanov Alexey, Pavlitskiy Boris, Sherbakov Igor, Shapsha Alexander and Shapshal Sergey 2019 *IOP Conference Series: Earth and Environmental Science* **403 (1)** 012015. DOI: 10.1088/1755-1315/403/1/012015

Integral assessment of anti-corrosion materials' quality

V V Deryushev [1], M M Zaitseva [1]*, I V Kundryutskova [2], O S Ovchinnikov [3]

[1]Don State Technical University, 1, Gagarin sq., Rostov-on-Don, 344002, Russia
[2] Kamensk Technological Institute (branch), Platov South-Russian State Polytechnic University (NPI), St. Saprygina, 6, Kamensk-Shakhtinskiy, 347800, Russia
[3]Platov South-Russian State Polytechnic University (NPI), St. Prosvescheniya, 132, Novocherkassk, 346400, Russia

E-mail: marincha1@rambler.ru

Abstract. In modern conditions, protective paint and varnish coatings are used everywhere to ensure the safety of metal structures and include the materials of various composition and application. The paints and varnishes market offers conventional hard and thixotropic anticorrosive coatings applied by the devices using airless spray technology. The article describes the experiment results, which consists in conducting a comparative analysis of protective materials in the framework of laboratory studies. The main physical and mechanical properties are determined. Modeling of anticorrosive coatings' indicators, which determine the degree of change in material properties in terms of adhesion, thickness, uniformity, flexibility and weather resistance of the treated surface of a metal structure, has been carried out. The values of the indicators were formed within the framework of the existing database, created and replenished using the real data obtained during the maintenance and repair of the existing facilities. The considered method is applicable for any number of analyzed surfaces and parameters of anticorrosive materials. As a result, it was found that the traditional coatings within the considered sufficiency indicators meet the specified requirements by 85-90%. However, the primer "Tectyl BT Coat" is the best option for using anticorrosive materials to protect the steel structures exposed to aggressive environmental influences. The sufficiency level for this primer is 92%, which determines its first place in the rating. The proposed method of experimental research and subsequent integral assessment of satisfying the requirements adequacy is versatile and is a tool that can be used when solving the problems of multi-criteria selection or ranking of a finite number of different anti-corrosion coatings or other objects.

Introduction

Natural-climatic and man-made factors negatively affect metal structures and materials during operation, which leads to corrosion destruction. The impact of temperature changes, its sharp fluctuations lead to a violation of the protective coatings' integrity, their cracking and peeling due to the occurrence of internal stresses due to the difference in the thermal expansion coefficients of the film and the substrate. Moisture leads to swelling, coating softening, washing out the water-soluble impurities from it, and accelerating the binders' destruction process. Salt water has a particularly destructive effect on protective coatings and metal structures in general.

Therefore, one of the directions for solving the urgent problem of ensuring the efficiency and safety of the metal structures' operation in modern conditions is the search for the most rational ways to protect them from corrosion. Of course, the operating conditions should be taken into account, focusing on the maximum environment aggressiveness, when choosing a protective coating.

1. Methods and materials

In modern conditions, protective paint and varnish coatings are used everywhere to ensure the safety of metal structures and include the materials of various composition and application.

The surface to be protected is processed in a technological sequence. The first coat provides good adhesion between the coatings and is called "primer". At this stage, primer that combines not only the paint and varnish material applied directly to the metal structure is used, but also the coating film formed from this material can be used. Primer prevents potential differences at the metal-coating interface, reducing the possibility of static electricity on the metal structure surface. In addition, the use of priming materials practically excludes diffusion development in metal [1-4]. The properties of this protective coating are influenced by the chemical composition of the primer (Figure 1), the type of metal structure surface, as well as the thoroughness and adherence to the technology of preparing the processed metal before painting.

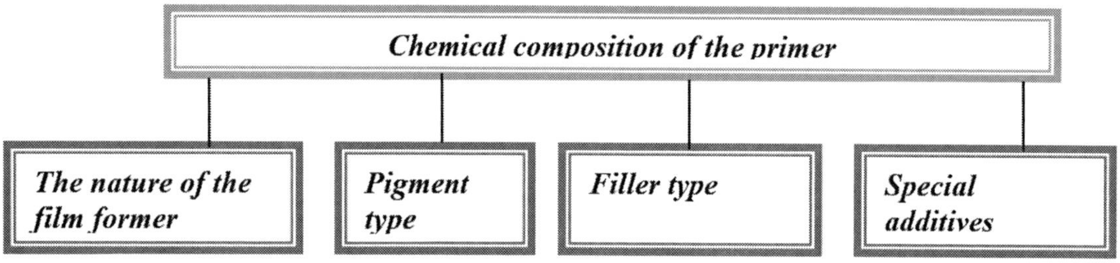

Figure 1. The chemical composition of the primer affecting its properties

The next step in protecting the metal structure surface is the application of anticorrosive and coating enamels designed to create a physical barrier in the form of a film that prevents the penetration of corrosion agents, in particular, the streams of salt water. These coatings provide high values of the weather resistance of the metal structure surface.

Anti-corrosion primers FL-03k, EF-065, VL-02 and enamels MS-17, HS-5226, EP-46u were selected for the study. The materials in question are popular among the consumers and are used not only to protect the metal structures operating in difficult conditions, including salt water, but also to provide a decorative effect.

In addition, there are thixotropic anticorrosive coatings on the paints and varnishes market, applied by the devices using airless spray technology. Due to the application of this processing method, such paints and primers change the viscosity parameters already in a single-layer application, which makes it possible to create the coatings with a thickness of 80-200 microns. The primer "Tectyl BT Coat", having a thixotropic composition based on mineral oil, was chosen for the analysis. This material is designed to protect the metal structures operating in the aggressive environment of fresh and sea water. It differs by the peculiarity that during surface treatment it creates a thick, oily, translucent film with a recommended thickness of 80 microns and a coating 12,5 m^2/l.

The experiment, which consists in carrying out a comparative analysis of protective materials, was carried out in the framework of laboratory research. The main physical and mechanical properties were determined. In article [5] adhesion, thickness and weather resistance were considered. This paper evaluates the uniformity and flexibility of the coating.

Coating uniformity is the ability of the applied material to fill the entire required volume, excluding the formation of voids or other composition defects.

The coating flexibility is assessed by checking the ability of the treated surface in bending, exposure to negative temperatures not to form cracks and breaks.

The study was carried out on the pre-made samples in the form of plates 150x100mm in size and 5mm thick, made of materials St45 and aluminum alloy D16T. Airless spray "Fakel-4" was used to

IOP Conf. Series: Materials Science and Engineering 1083 (2021) 012064 doi:10.1088/1757-899X/1083/1/012064

apply anti-corrosion coating. The physical and mechanical properties were evaluated under normal conditions (temperature 22^{0}C and relative humidity 60%) after the time required for the protective dry film formation on the treated surface. Further, the samples were exposed to various types of effects (Table 1), carried out independently of one other.

Table 1. Types of anti-corrosion coating impact and the necessary equipment.

№	Impact type	Equipment
1	Low and high temperatures	Climate heat chamber "Nema"
2	Impact impact	Setting the pendulum tumblebug type
3	Vibration loads	Vibration stand
4	Influence of sea water salinity	Container with sea water

The change in the physical and mechanical parameters of the coating was assessed by comparison with the indicators obtained under normal conditions after each type of exposure.

2. Results and discussion

The first step was to study the temperature effect on the protective coating quality. When the temperature changes from -10^{0}C till 50^{0}C with a step equal 5^{0}C the parameters of physical and mechanical properties were determined. The experiment was carried out by placing the coated samples in a Nema climate heat chamber for 20 minutes. The time was chosen taking into account the sufficiency for complete cooling or heating of the paint and varnish material layer. The results of the experiment were used to plot the uniformity dependences S and flexibility h anti-corrosion coatings from the temperature (Figure 2). Numerical values for common primers and enamels were given to average values.

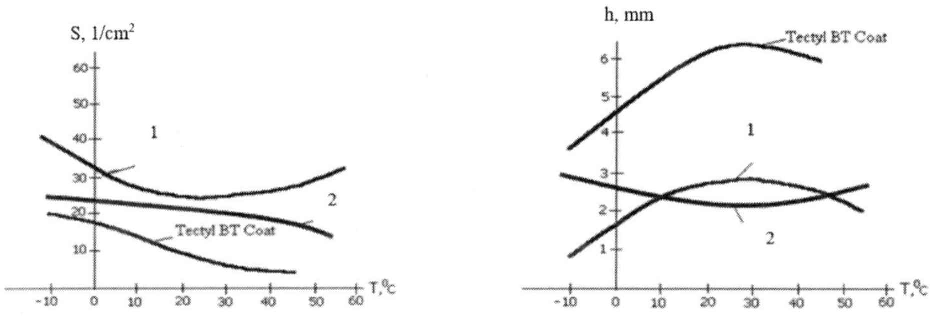

Figure 2. Coating characteristics' dependence on temperature:

1- Primers EF-065, FLK-03k, VL-023; 2- Enamles EP-46u, MS-17

At normal temperature, the number of microcracks per unit surface of the "Tectyl BT Coat" coating was observed as twice less than that of conventional coatings, due to the viscosity of the oily film, which very densely covers the sample surface. With increasing temperature "Tectyl BT Coat" becomes even more viscous, and all microcracks seem to "collapse", while the number of microcracks in hard coatings increases. At low temperatures, cracking of both conventional coatings and "Tectyl BT Coat" occurs.

The flexibility of hard coatings is half that of "Tectyl BT Coat", which is due to the presence of viscous binders in the latter. When investigating the effect of low temperatures, it was found that cooling significantly affects the limiting value of the sample bending. Moreover, the flexibility of "Tectyl BT Coat" and conventional hard coatings are reduced by 50 and 60% respectively. It is worth noting that the flexibility of a thixotropic coating, even at low temperatures, is higher than that of hard coatings at normal temperatures.

The second step was to assess the effect of the number of freeze-heating cycles on the properties of conventional and thixotropic primers. The coated samples were cooled from 22⁰C (normal temperature) to -10⁰C. Then they heated up to 50⁰C and cooled back to normal temperatures. Thus, the real operating conditions of the metal structure were simulated by passing the temperature through the zero value. The experimental dependences of the uniformity S and flexibility h on the number of cycles n are shown in Figure 3. Each coating was first subjected to 10 cycles, then the number of cycles was increased in increments of 5. Maximum n was 70 cycles, which is equal to the average value of the number of days per year with temperature changes during the metal structures operation in a moderately cold climatic region.

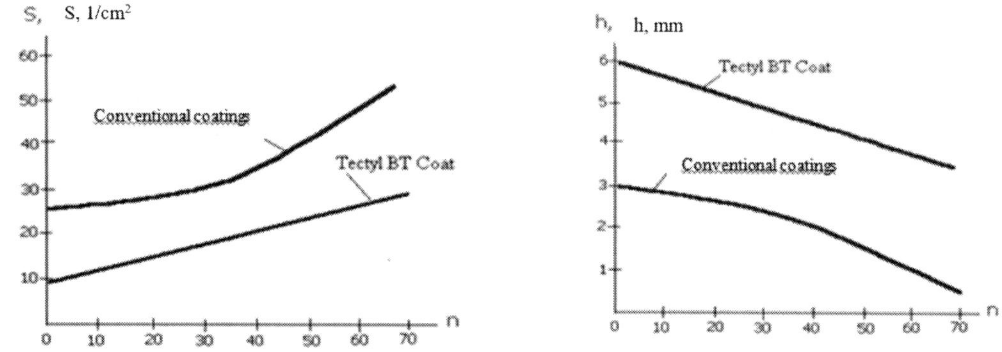

Figure 3. The dependence of the primers' characteristics on the number of "Freeze-heating" cycles.

The number of microcracks in hard coatings starts increasing sharply after 30 cycles and doubles at the maximum number of cycles. The number of microcracks on "Tectyl BT Coat" covering increases linearly without sharp jumps and reaches a value equal to the initial number of microcracks in a conventional hard coating.

Flexibility with an increase in the number of cycles is reduced by 2 times for all coating types.

The third stage was to study the vibration loads effect on the anti-corrosion coating properties. To carry out the experiment, the painted samples were fixed in the installation in such a way that the blow of the pendulum tumblebug head hammer fell on the sample's back. The magnitude of the kinetic energy of impact E varied in the range from 1 to 6 J with a step of 1 J. The results are shown in Figure 4 in the form of uniformity dependences S and flexibility h from impact energy E.

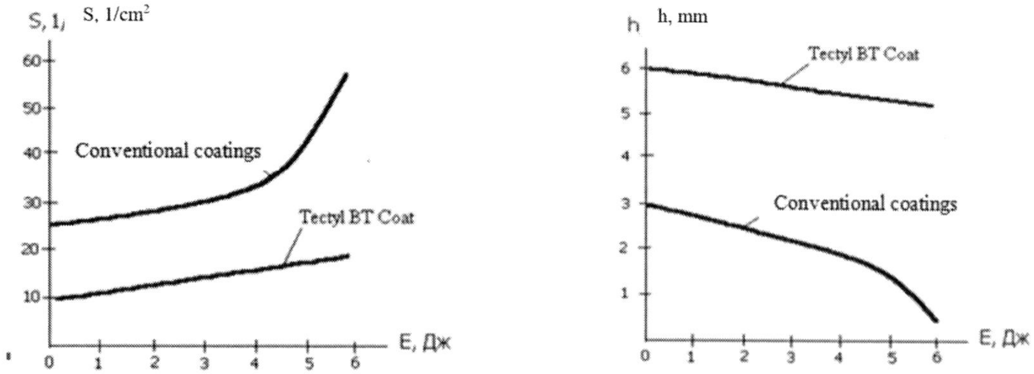

Figure 4. The primer characteristics' dependence on impact energy

The number of microcracks in conventional coatings increases exponentially more than twofold with an increase in the impact energy, which is explained by cracking and spalling of the material. The quantity S increases linearly on "Tectyl BT Coat" covering.

At the fourth stage of the experiment, the coated samples were placed for 10 days in containers with sea water of varying degrees of salinity. It is known that sea water salinity can reach 40%, which certainly has a significant effect on the physical and mechanical properties of anti-corrosion materials. During the experiment, the samples were exposed to aggressive action of water with varying salinity degrees. The water salinity content varied from 0 to 50% with an increase of 10% in each container. The influence of the seawater salinity is especially typical for metal structures, some of which are in salt water or in contact with it, as well as for the tanks of various values filled with water. The obtained dependences of the physical and mechanical properties of the paint coatings under study on the percentage of salts in water are shown in Figure 5.

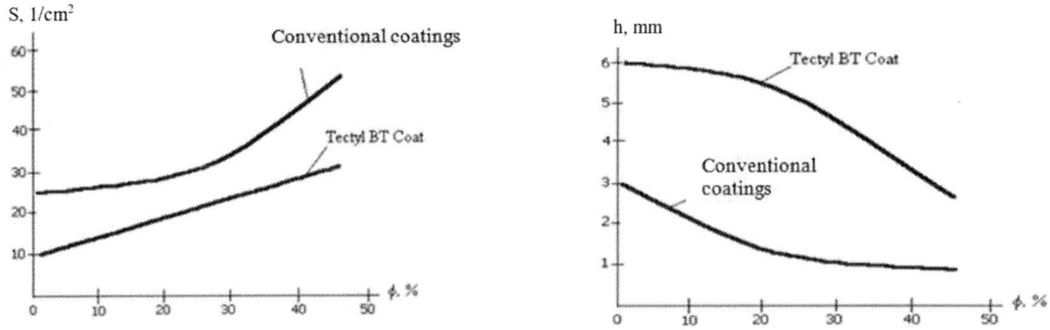

Figure 5. The primer characteristics' dependence on water salinity

Prolonged exposure to salt solution leads to cracking of all coatings, that is, to a decrease in uniformity. This increases the permeability of the anti-corrosion material. The flexibility of the coatings is also reduced. Deflection value *h* at "Tectyl BT Coat" is reduced by half, for conventional coatings – by almost three times.

As the study on the assessment of the main physical and mechanical properties of conventional and thixotropic anticorrosive materials has shown, when considering various operating conditions, the characteristics of coatings have both advantages over others and disadvantages. The problem of choosing a rational option for the protective treatment of metal structures is solved using the methods described in [6-8], by applying a multi-criteria selection (ranking) of technical objects, the concept of sufficiency and the criterion that implements it. The option of corrosion protection, which has the maximum value of the sufficiency indicator will be considered the "best" option. In this case, the scatter degree of the integral assessment of particular indicators from the criteria (levels of sufficiency) will be minimal, providing maximum protection of the processed surface of the metal structure from the aggressive effects of various environmental factors [9-12].

The basis for obtaining the sufficiency criterion for the anti-corrosion coatings analysis (ACC) is to find the weighting factors that establish the advantage of some criteria over others. A functional method is used to determine the weighting factors. The main physical and mechanical properties of anticorrosive materials are selected as estimated indicators y_j: adhesion, thickness, uniformity and flexibility, weather resistance. Their minimum y_{0j}, maximum y_{Mj} and sufficient y_{tj} values in accordance with the technical characteristics (Table 2).

Table 2. Modeling indicators.

Indicator characterizing the parameters ACC	**Property ACC**	**Minimum indicator y_{0j} value**	**Maximum indicator y_{Mj} value**	**Sufficient indicator value y_{tj}**	**Units**
y_1, σ	adhesion	10^{-2}	2	0,3	MPa
y_2, H	thickness	1	100	30	μm
y_3, S	uniformity	0,2	50	20	$1/cm^2$
y_4, h	flexibility	1	5	3	mm
y_5	weather resistance	0.03	3	1.0	year

Modeling of indicators of anticorrosive coatings, which determine the degree of change in material properties in terms of adhesion, thickness, uniformity, flexibility and weather resistance of the treated surface of a metal structure, has been carried out. The indicators values were formed within the framework of the existing database, created and replenished with the help of real data obtained during the maintenance and repair of the existing facilities [14, 15]. This method is applicable for any number of surfaces under consideration and parameters of anti-corrosion materials.

The values of the sufficiency indicator obtained as a result of modeling z_i are ranked, determining the place in the rating for each considered protective material (Table 3).

Table 3. Anti-corrosion coatings.

No.	**Name ACC**	**The value of the sufficiency indicator z_i**	**Ranking place**
1.	Primer FL-03k	0.88	4-5
2.	Primer EF-065	0.85	7
3.	Primers VL-02 and VL-023	0.91	2
4.	Primer MS-17	0.88	4-5
5.	Enamel EP-46u	0.87	6
6.	Enamels HS-5226	0.9	3
7.	Primer "Tectyl BT Coat"	0.92	1

3. Summary

Thus, we can conclude that traditional coatings within the considered sufficiency indicators satisfy the specified requirements by 85-90%. However, the primer "Tectyl BT Coat" is the best option for using anticorrosive materials to protect the steel structures exposed to aggressive environmental influences. The sufficiency level for this primer is 92%, which determines its first place in the rating.

The proposed method of experimental research and subsequent integral assessment of the adequacy of satisfying the requirements is versatile and is a tool that can be used when solving the problems of multi-criteria selection or ranking of a finite number of different anti-corrosion coatings or other objects.

References

[1] Ivanov V V, Popov S I, Selemeneva E M, Babazhanov N T 2019 Study of technological characteristics of the process of formation of vibration mechanochemical oxide coating *XV International scientific-technical conference «Dynamics of technical systems» (DTS-2019): AIP Conference Proceedings* **2188** 020015. doi.org/10.1063/1.5138389.

[2] Ivanov V V, Popov S I, Marchenko Ju V, Marchenko E V, Dontsov N S, Timofeev S A 2019 Thickness of vibrational mechanochemical solid-lubricant coating in friction pairs of transport engineering products *XII International Scientific Conference on Agricultural Machinery Industry (INTERAGROMASH 2019): IOP Conference Series: Earth and Environmental Science* **403** 012115. doi:10.1088/1755-1315/403/1/012115.

[3] Korotky A A, Marchenko E V, Ivanov V V, Popov S I, Marchenko Ju V, Dontsov N S 2019 Model of forming vibration mechanochemical solid lubrication coating on surface of steel rope *XII International Scientific Conference on Agricultural Machinery Industry (INTERAGROMASH 2019): IOP Conference Series: Earth and Environmental Science* **403** 012116. doi:10.1088/1755-1315/403/1/012116.

[4] Ivanov V V, Popov S I, Kirichek A V 2018 Investigation of optimal chemical composition of cast aluminum alloys for vibrational mechanical-chemical polishing and deposition of protective and decorative coatings *XI International Conference on Mechanical Engineering, Automation and Control Systems (MEACS 2017): IOP Conference Series: Materials Science and Engineering* **327** 032026. doi:10.1088/1757-899X/327/3/032026

[5] Deryushev V V, Zaitseva M M, Kosenko E E, Kosenko V V 2020 The quality analysis of the anti-corrosion coatings metal structures operating in difficult conditions *IOP Conference Series: Materials Science and Engineering* **913(4)** 042059.

[6] Deryushev V V, Sidelnikova E G 2013 Analysis of metal structures protection against corrosion *Engineering Bulletin of the Don* **27(4)** 268.

[7] Deryushev V V, Kosenko V V, Cherpakov A V, Zaitseva M M, Kondratova L E 2015 Modeling of stress state of welded joint in pipeline taking into account uneven distribution of mechanical propertie Physics and Mechanics of New Materials and Their Applications *Proceedings of the 2015 International Conference on Physics and Mechanics of New Materials and Their Applications, devoted to 100-th Anniversary of the Southern Federal University* 501-507.

[8] Kasyanov V E, Deryushev V V, Kosenko E E, Kosenko V V, Pshenichnaya A Yu 2018 Synthesis of methods and principles of ensuring the reliability of one-off and serial production machines *International Conference on Modern Trends in Manufacturing Technologies and Equipment (ICMTMTE 2018). electronic edition. "MATEC Web of Conferences"* 02106.

[9] Panfilov A V, Deryushev V V, Korotky A A 2020 Recommended safety systems for a risk-based approach *Labor safety in industry* **5** 48-55.

[10] Deryushev V V, Sidelnikova E G 2013 A generalized indicator of sufficiency for assessing the technical condition of construction and material handling equipment *Scientific Review* **9** 164-167.

[11] Deryushev V V, Sidelnikova E G 2013 The structure and model of building an integral indicator for assessing the quality of construction and material handling equipment *Scientific Review* **9** 311-313.

[12] Deryushev V V, Kosenko E E, Kosenko V V, Zaitseva M M 2019 Making technical decisions in conditions of uncertainty in the presence of risk *Safety of technogenic and natural systems* **2** 56-61.

[13] Deryushev V V, Sidelnikova E G 2014 Determination of the range of quality indicators for construction and material handling equipment *Scientific Review* **11(3)** 775-777.

[14] Deryushev V V, Zaitseva M M, Evseev D Z, Kosenko E E 2020 Concentration of thermal stresses in metal materials and constructions under local heating *Materials Science Forum*.

[15] Deryushev V, Zaitseva M, Megera G, Fedyanov A 2019 A method for the identification of dynamic characteristics car suspension materials *IOP Conference Series: Materials Science and Engineering* **698** 066031.

IOP Conf. Series: Materials Science and Engineering 1083 (2021) 012065 doi:10.1088/1757-899X/1083/1/012065

Improving the reliability of the truck frame, taking into account maintenance frequency

V V Deryushev [1], M M Zaitseva [1*], D N Kundryutskov[2], O S Ovchinnikov [3]

[1]Don State Technical University, 1, Gagarin sq., Rostov-on-Don, 346780, Russia
[2] Kamensk Technological Institute (branch), Platov South-Russian State Polytechnic University (NPI), St. Saprygina, 6, Kamensk-Shakhtinskiy, 347800, Russia
[3]Platov South-Russian State Polytechnic University (NPI), St. Prosvescheniya, 132, Novocherkassk, 346400, Russia

E-mail: marincha1@rambler.ru

Abstract. In this work, the reliability of the KamAZ-4308 truck frame is increased, taking into account its maintenance frequency. This brand was chosen due to the fact that it is one of the best-selling cars in Russia among the cars with a gross weight of 14 to 40 tons in 2019. A method for optimizing the value of the intervals between the planned car maintenance (MAINT) is proposed. When calculating the reliability, the problem of minimizing the amount of initial data was solved in order to reduce the experiment time, its labor intensity and the financial testing costs. As one of the options for reducing unit costs, a change in the material for manufacturing a truck frame is considered. The empirical and approximating functions of the general totality of the finite volume (GTFV) of the Tp frame of the KamAZ-4308 car service life are obtained. At a given failure rate λ = 10–7 h-1, period values τ = 10 h and cost U/C = 100 the optimal frequency of maintenance is Topt = 993 h ≈ 1,4 month taking into account the minimum technogenic risk Smin = 0,0513.

Introduction

Reliability issues are closely related to finding the ways to reduce labor, time and economic losses as a result of equipment failure. The development and application of methods for reducing the duration of tests, financial costs for conducting an experiment and obtaining the basic data is an urgent task in modern conditions. Minimization of costs for production, operation and prediction of the service life of a part can be achieved by reducing the volume of samples for strength, load and service life when calculating the vehicle reliability indicators.

Improving the reliability of truck components is inextricably linked to ensuring timely and high-quality maintenance. Optimization of the frequency and duration of MAINT time remain relevant.

1. Justification of the technical object choice

In this work, the reliability of the KamAZ-4308 truck frame is increased, taking into account its maintenance frequency. This brand was chosen due to the fact that it is one of the best-selling cars in Russia among the cars with a gross weight from 14 to 40 tons in 2019. Information on the position of the considered and other brands on the Russian market is presented in Figure 1.

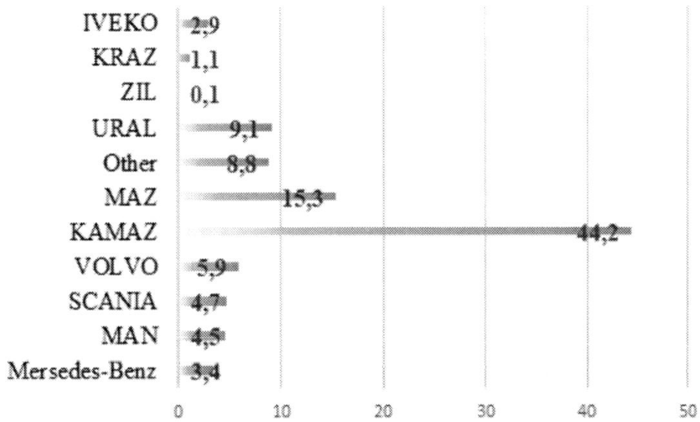

Figure 1. Percentage share of OJSC "KamAZ" and other manufacturers in the Russian market at the end of 2019 according to Autostat

OJSC "KamAZ" is the undisputed leader in the trucks supply to the Russian market. According to the company's data, 35.5 thousand trucks were sold in 2019, more than 5 thousand vehicles of which were exported. Among the company's buyers are such large enterprises as Gazprom, Rossiiskie Seti, Iteko-Russia, PC Rosneft, AM GROUP and others. In order to ensure the competitiveness of cars, the company needs to solve the problem of reducing the risks of equipment failure during operation, as well as increasing the service life of parts. Increasing the service life of the vehicle can be achieved by optimizing the time between the scheduled maintenance.

The KamAZ-4308 truck frame was chosen as the technical object of the study, since it is its basic part, a highly loaded supporting structure, which ensures the reliable operation of the machine as a whole in many respects by attaching the engine, chassis, transmission elements, body, cab and additional special equipment. This part is considered as a construction of increased responsibility, the operating features of which should be studied and improved to increase its reliability and efficiency of using the car as a whole.

It is known that the car frame is subjected to alternating bending and torsional loads, mechanical and human-caused influences, as well as the aggressive environmental influences (moisture, chemical agents to prevent road icing, vibration, and others). All these factors together lead to the appearance of defects in the part and, as a consequence, to a decrease in the durability of the part, a decrease in its service life.

2. Optimizing reliability

To ensure the optimal reliability of equipment and its components (Figure 2), it is recommended to strive to reduce the time of the experiment, its labor intensity, the number of tests and the financial costs of their implementation [1, 2]. This task can be solved by developing the methods for minimizing the amount of initial data for calculating the reliability indicators [3, 4].

In this paper, it is proposed to optimize the probability of failure-free operation of the truck frame *P(t)* by the criterion of unit costs *Ci* (Figure 3) taking into account the time length between maintenance. Reliability calculation is carried out using the method described in [5] and based on the use of small samples of strength, loading and service life. To model the general totality of a finite volume, the distribution of absolute ranges is used $W_{r_s} = x_r - x_s$. $(r,s=1,\dots,n)$

$$F(W) = \int_0^W f(W)dW,$$

where $f(W_{rs})$ – is the shear density of the ranges (in general):

$$f(W_{rs}) = C_{rs} \int_{-\infty}^{+\infty} P^{r-1}(x)p(x)[P(x+W_{rs}) - P(x)]^{s-r-1} p(x+W_{rs}) \times$$
$$\times [1 - P(x+W_{rs})]^{n-s} dx,$$

$C_{rs} = \dfrac{n!}{(r-1)!(s-r-1)!(n-s)!}$; $P(x)$ – is the characteristic shear function; $p(x)$ – is the characteristic probability density.

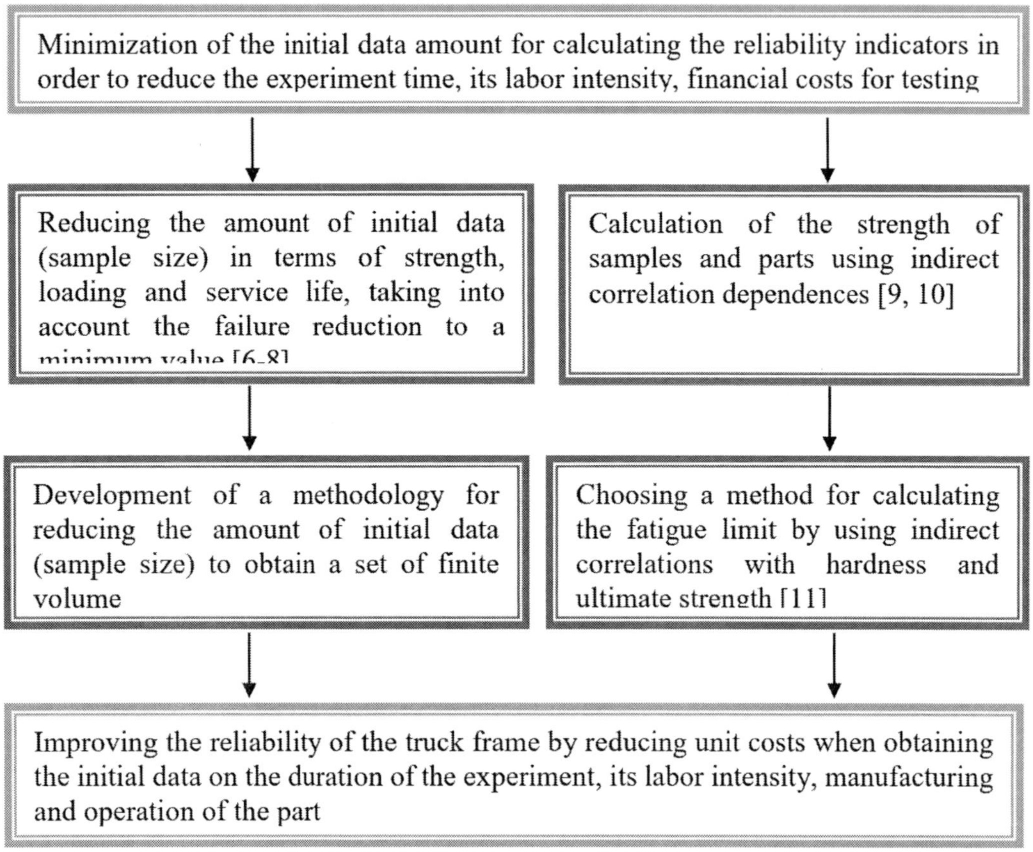

Figure 2. Ensuring optimal reliability of equipment and its components

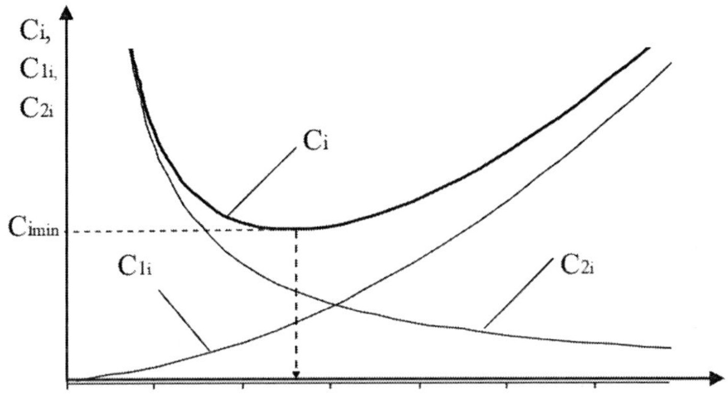

Figure 3. Optimization of the vehicle frame manufacturing option (and, accordingly, the likelihood of the vehicle frame failure-free performance)

In the course of modeling, empirical and approximating functions of the general totality of finite volume (GTFV) *Tp* frame of the KamAZ-4308 car service life were obtained (Figure 4).

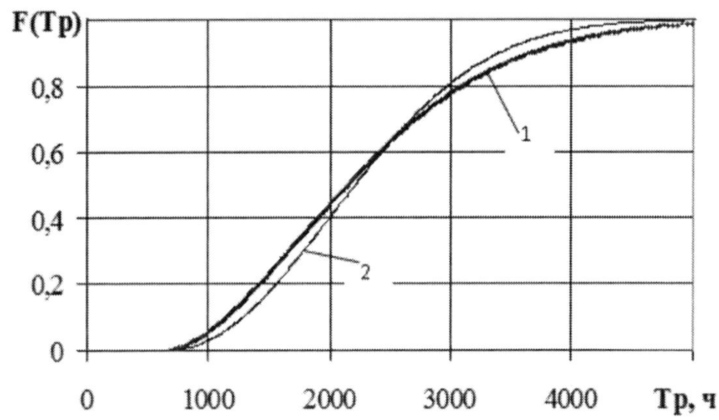

Figure 4. GTFV for the KAMAZ -4308 truck frame service life distribution functions: 1 – empirical; 2 – approximating

A change in the steel grade, wall thickness, an increase in the cross-sectional area of the frame and others were applied as the options for the frame manufacture.

An important aspect of this study is the accounting for maintenance, since maintaining the vehicle frame operability has a significant impact on the reliability of the entire vehicle as a whole. Timely maintenance in specialized automotive services makes it possible to reduce the risk of sudden failures, which leads to an increase in the vehicle operation safety and a decrease in the cost of extending its service life [12, 13]. It should be borne in mind that each element of the car may experience the implicit failures not determined visually or during conventional diagnostics, but can only be detected using the special monitoring and diagnostics tools. Having determined the functions of the probability dependences of no-failure operation and the failure rate on time $P(t)$ and $\lambda(t)$ accordingly, it is possible to solve the problem of optimizing the duration of the service life period of an object (car frame) between the scheduled maintenance. The financial costs of eliminating the hidden failures are denoted by u, then the unit costs for time are denoted by τ.

$$S(\tau)=Up(t<\tau)+Cp(t\geq\tau)+u\int_0^\tau tf(\tau-t)dt=U[1-P(\tau)]+CP(\tau)+u\int_0^\tau[1-P(t)]dt=$$
$$=U+(C-U)P(t)+u\left[\tau-\int_0^\tau P(t)dt\right].$$

The average unit costs over time τ

$$s(\tau)=\frac{1}{\tau}\left\{U+(C-U)P(t)+u\left[\tau-\int_0^\tau P(t)dt\right]\right\}.$$

Differentiating the expression in time and equating the derivative to zero, we obtain

$$P(\tau)-\tau\frac{dP(t)}{dt}-\frac{u}{U-C}\left[\tau-\int_0^\tau P(t)dt\right]=\frac{U}{U-C},$$

Having solved this equation, we find the desired value of the optimal value of the maintenance frequency *Topt*.

Determining the size of the period between the scheduled MAINT, it is necessary to strive to minimize the quantitative risk value [14, 15], taking into account financial losses for the car restoration and repair, simple equipment and the entire mechanized complex of machines during work.

We represent the risk function as

$$S=CP_2(t)+UP_3(t)$$

To obtain the practical results, let us consider the frame of a KamAZ-4308 truck. The choice of this technical object is justified at the beginning of this work.

3. Results of solving the optimization problem

To present the results obtained (Table), we denote the minimum amount of risk by S_{min}, *U*/*C* – *is* the cost ratio of a sudden failure to the maintenance cost.

Table 1. The results of solving the problem of optimizing the intervals between the planned MAINT for the truck KamAZ-4308.

τ, *h*	***U*/*C***	**Optimal maintenance intervals T_{opt}, [h]**		
		$\lambda=10^{-8}\,h^{-1}$	**$\lambda=10^{-7}\,h^{-1}$**	**$\lambda=10^{-6}\,h^{-1}$**
10	10	9972	3164	982
	100	3252	993	317
	1000	995	304	93
	10000	326	92	24
20	10	14136	4457	1396
	100	4463	1394	426
	1000	1384	427	121
	10000	429	133	27
50	10	22331	7032	2195
	100	7032	2187	665
	1000	2188	658	173
	10000	658	178	36

100	10	31544	9930	3081
	100	9922	3065	905
	1000	3044	906	235
	10000	907	233	40

Analyzing the table, we find that, for example, for a given failure rate $\lambda = 10^{-7}\ h^{-1}$, the period values $\tau = 10$ h and cost U/C = 100 the optimal maintenance frequency is $T_{opt} = 993$ h ≈ 1,4 month taking into account the minimum technogenic risk $S_{min} = 0{,}0513$.

4. Summary

In this article, the problem of improving the reliability of the frame of the KamAZ-4308 vehicle has been solved, taking into account the optimization of the intervals between scheduled maintenance. The values of the MAINT optimal frequency, depending on the failure rate, the costs of restoring the facility in case of a sudden failure, the size of the period and human-caused risk have been obtained.

References

[1] Deryushev V V, Zaitseva M M, Kosenko E E, Kosenko V V 2020 The quality analysis of the anti-corrosion coatings metal structures operating in difficult conditions *IOP Conference Series: Materials Science and Engineering* **913(4)** 042059.

[2] Deryushev V V, Kosenko E E, Kosenko V V, Zaitseva M M 2019 Making technical decisions in conditions of uncertainty in the presence of risk Safety of technogenic and natural systems **2** 56-61.

[3] Kotesova A A, Teplyakova S V, Popov S I, Kopylov F C 2019 Ensuring assigned fatigue gamma percentage of the components *IOP Conference Series: Materials Science and Engineering* **698(6)** 066029

[4] Deryushev V, Zaitseva M, Megera G, Fedyanov A 2019 A method for the identification of dynamic characteristics car suspension materials *IOP Conference Series: Materials Science and Engineering* **698** 066031.

[5] Kasyanov V E, Rogovenko T N, Zaitseva M M 2013 Providing a given fatigue life of machine parts using small samples of initial data *Mechanical Engineering Bulletin* **5** 10-15.

[6] Evseev D Z, Kotesova A A, Kosenko V V, Golubeva A Y 2019 Ways to Improve the Durability of Pneumatic Tires *IOP Conference Series: Materials Science and Engineering* **698(6)** 066032.

[7] Deryushev V V, Zaitseva M M, Evseev D Z, Kosenko E E 2020 Concentration of thermal stresses in metal materials and constructions under local heating *Materials Science Forum* **974** 729-734.

[8] Kasyanov V, Deryushev V, Shulkin L, Kosenko E, Kotesova A 2018 Endurance tests of single machines production *MATEC Web of Conferences* **224** 02107.

[9] Veremeenko E G, Rogovenko T N 2020 Handling equipment reliability effect on grain terminal risks *IOP Conference Series: Materials Science and Engineering* **913(4)** 042062.

[10] Teplyakova S V, Kotesova A A, Popov S I, Kotesov A A 2020 The transition from the sample data to the total aggregate of the final volume and the analysis of this transition laws *IOP Conference Series: Materials Science and Engineering* **913(4)** 042054.

[11] Kasyanov V, Kosenko E, Kosenko V, Krymsky V 2020 Method for processing experimental strength data using statistical methods *E3S Web of Conferences* **157** 02019.

[12] Kasyanov V, Deryushev V, Kosenko E, Kosenko V, Golubeva A Y 2018 Synthesis of methods and principles of ensuring the reliability of one-off and serial production machines *MATEC Web of Conferences* **224** 02106.

[13] Ivanov V V, Popov S I, Marchenko Ju V, Marchenko E V, Dontsov N S, Timofeev S A 2019. Thickness of vibrational mechanochemical solid-lubricant coating in friction pairs of transport engineering products *XII International Scientific Conference on Agricultural Machinery Industry (INTERAGROMASH 2019): IOP Conference Series: Earth and Environmental Science* **403** 012115. doi:10.1088/1755-1315/403/1/012115.

[14] Kasyanov V E, Korotky A A, Egelskaya E V, Kosenko E E 2020 The algorithm for ensuring the parts' (assembly nodes, machines) reliability based on the theory of risk *IOP Conference Series: Materials Science and Engineering* **913(4)** 042052.
[15] Panfilov A V, Deryushev V V, Korotky A A 2020 Recommended safety systems for a risk-based approach *Labor safety in industry* **5** 48-55.

Checking the adequacy of the technical and economic model for ensuring a target life of an individual production machine

V E Kasyanov , E E Kosenko*, V V Kosenko, D B Demchenko, R V Khvan

Don State Technical University, 1, Gagarin sq., Rostov-on-Don, 344010, Russia

E-mail: a123lok@mail.ru

Abstract. A number of principles to ensure the reliability of individual production machines have been formulated. A block diagram of the model for ensuring the gamma-percentage target life of the individual production machine parts has been built on their basis. To assess the model adequacy, the accelerated operational tests without a test stand base have been proposed. A machine (units, parts) is considered as a system with mutually loaded elements, i.e. self-stand.

Introduction

The machines of individual production (multi-bucket stripping wheel overburden excavator for mining in a specific quarry, a rolling mill for a certain type of rolled metal in metallurgical production, a car of the President of the country, a helicopter and a plane of the President, a lunar rover, a Mars rover), as well as the cars of foreign production (Maybach (2005), Rolls-Royce (2008), Cadillac DTC (2009), Ferrari (2010), Bugatti (2011), Lamborghini (2012), Cadillac Elmiraj (2013)) unlike serial machines require a different approach to ensure their reliability [1,2].

If for serial machines of domestic production, great experience in solving the problems to ensure reliability [3-5] has been accumulated since the end of the 60s of the last century, and for the past 50 years they have been engaged in ensuring, improving the reliability of serial machines [6-8], then for the machines of individual production these tasks started being solved from about 2010 to 2016 [1,2].

These or those models (mathematical, technical and economic) are known to be used in scientific research, from the practical point of view, and they make it possible to perform calculations and obtain the necessary results rather quickly [9,10].

In this case, it is required to determine the calculation error for making a decision.

When ensuring the reliability of machines of individual production, a model, the block diagram of which is shown in Figure 1, is used.

A number of principles (2, 3, 4) for ensuring the reliability of serial machines, which are used in the development of a new method for ensuring the reliability of machines of individual production, have been formed. The principles 1,5,6 are formulated for a new method (individual production) [1].

The basic principles, used in the method, are as follows.

1. A homogeneous sample is formed from the parts not for their intended purpose, but for a gamma-percentage resource with an acceptable error [11].
2. A set of finite volume is used.

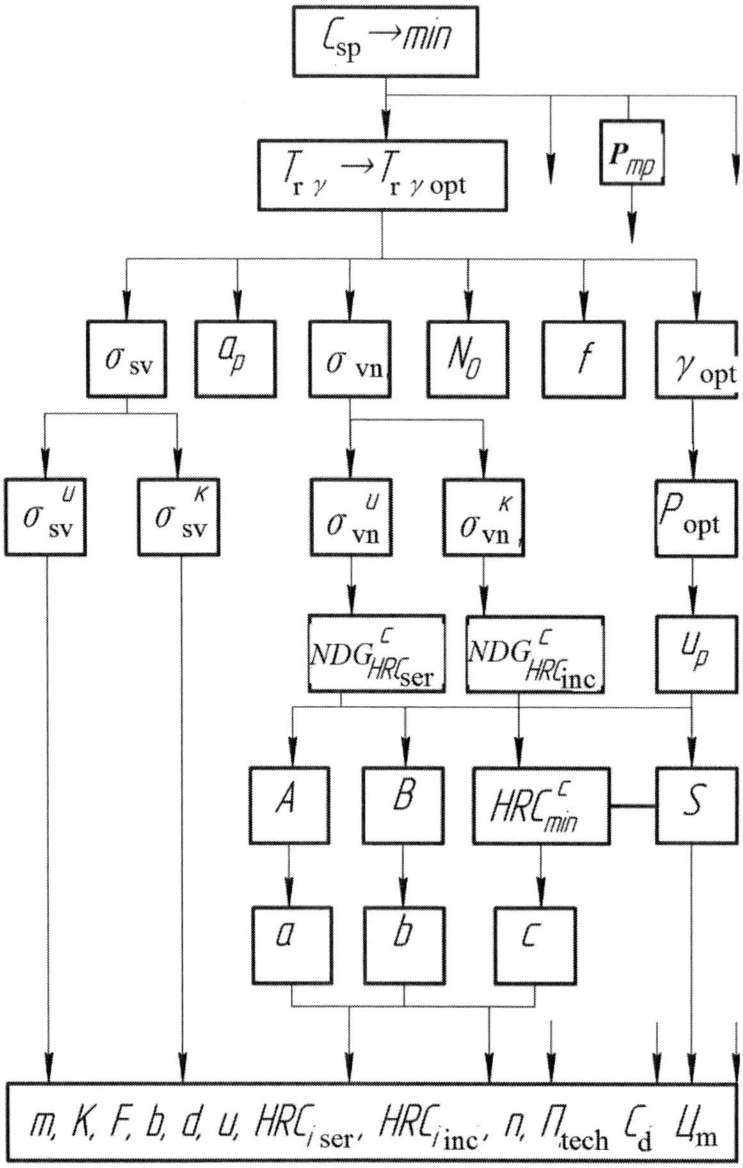

Figure 1. Block diagram of the model for ensuring the gamma-percentage resource of the machine parts of individual production

3. For the set, Weibull's law is applied without the shift parameter; it is replaced by the minimum first value of the variation series. This replacement is due to the fact that the shift is less than the minimum value of the variational series of the sub-total, which is not correct.
4. The parts with sudden fatigue failures are designed for a target life with a certain margin.
5. Adjacent parts are the parts of the test stand; parts of the whole machine form a stand.
6. For the period of scheduled and application repairs and maintenance, it is necessary to have a backup machine.

A feature of the last 50-60 years of the mechanical engineering development and the applied probabilistic and statistical methods was the processing of significant amounts of information, for example, the samples of 50, 100 and more units (machines, assemblies, parts).

Such methods were widely used to determine the reliability indicators of serial machines [10, 12].

IOP Conf. Series: Materials Science and Engineering 1083 (2021) 012066 doi:10.1088/1757-899X/1083/1/012066

Main part

Methodological materials of the state and industry levels were developed, allowing to process the statistical data on the various products' reliability properties.

At the same time, the issues of ensuring the reliability of the most important products of individual production (the car and helicopter of the President of the country, a unique rolling mill and overburden excavator, a lunar rover, a rover, etc.), as well as foreign-made passenger cars [2] and the statistical methods necessary in these cases require their own solutions.

The attempts to find the recommendations in this area of science and practice in the technical literature have not showed any results yet (perhaps this is a closed topic).

The key questions in the formation of samples are their representativeness (most often used) and homogeneity.

A sample is called representative if it sufficiently well represents the characteristics of the general sub-total.

The sample is homogeneous if each sample object has a main parameter that is the same (or almost) with other sample objects [11].

The sub-total or a sample from it will be homogeneous if the main parameter, for example, the target life is the same (with some error) with other details.

Figure 2 shows the examples for the machine parts.

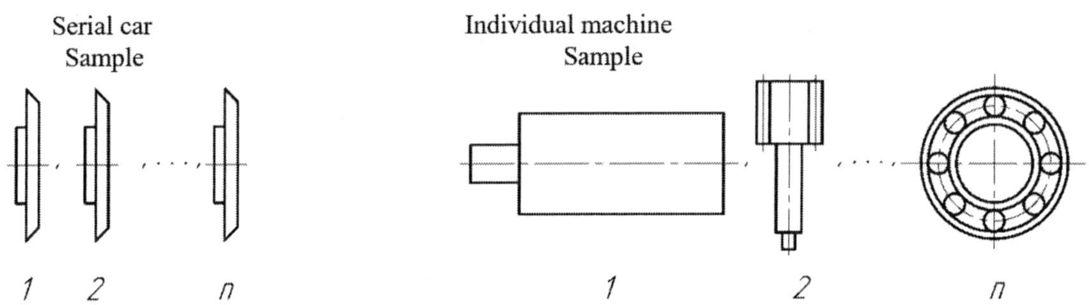

Identical parts (gears) by appointment and the main parameter

Identical details on the main parameter, but different by appointment:
1 – shaft;
2 – gear shaft;
n – frictionless bearing.

Figure 2. Machine parts of serial and individual production

In case of using one gear (Figure 3), the sample size will be determined by the number of its teeth. Then these teeth form a sample both by the main parameter and by the purpose.

If there are two or more gears in the unit (gearbox) (Fig. 3, b, c), the sample can be formed by the teeth of two and three gears.

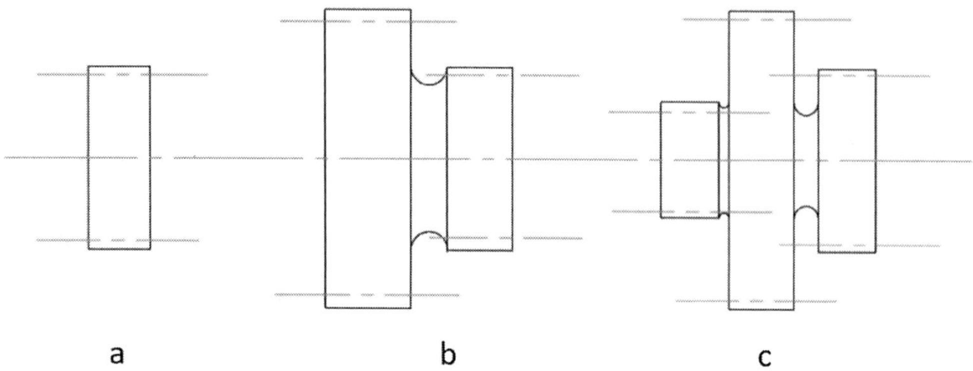

Figure 3. Gears of individual production: a - one gear, b - double block of gears, c - triple block of gears.

At the same time, the gamma percentage resource of the part should be used as the main parameter.

Then the homogeneity of the sample (sub-total) will be determined by the same gamma percentage resources of the details (within the acceptable error)

$Tl\gamma 1 \approx Tl\gamma 2 \approx \ldots \approx Tl\gamma\, n \approx Tl\, t$,

where Tl t – is the target life to the limit state of the part.

The main condition for ensuring the same main parameters of parts with sudden failures - resources (single-individual production machines) - is the calculation by the Wohler-Sorensen-Kogaev formula based on the experimental initial statistical data of fatigue strength and loading.

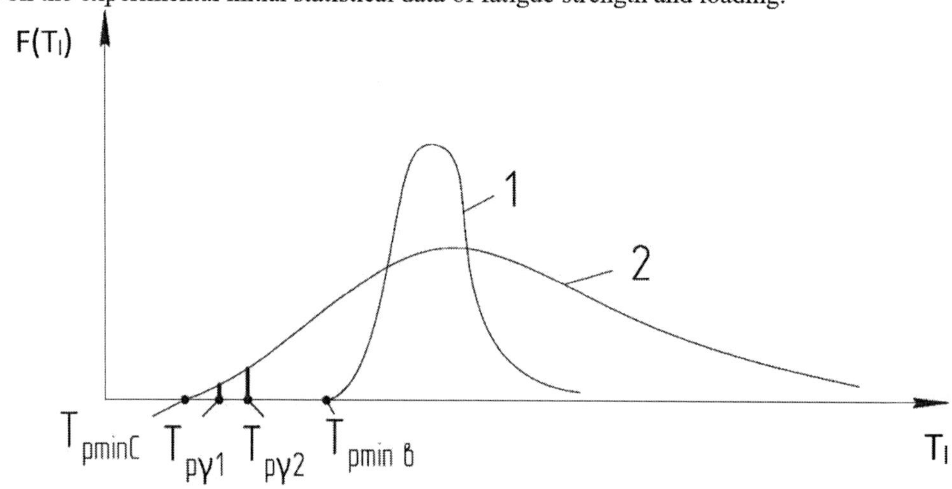

Figure 4. Probabilistic resource distributions of the samples and sets of parts with fatigue sudden failures: 1-sample; 2-set

Figure 4 shows that for the distribution of the sub-total, the resource span is greater than the sampling span.

The algorithm for determining the parameters of the sample and the sub-total is shown in Figure 5.

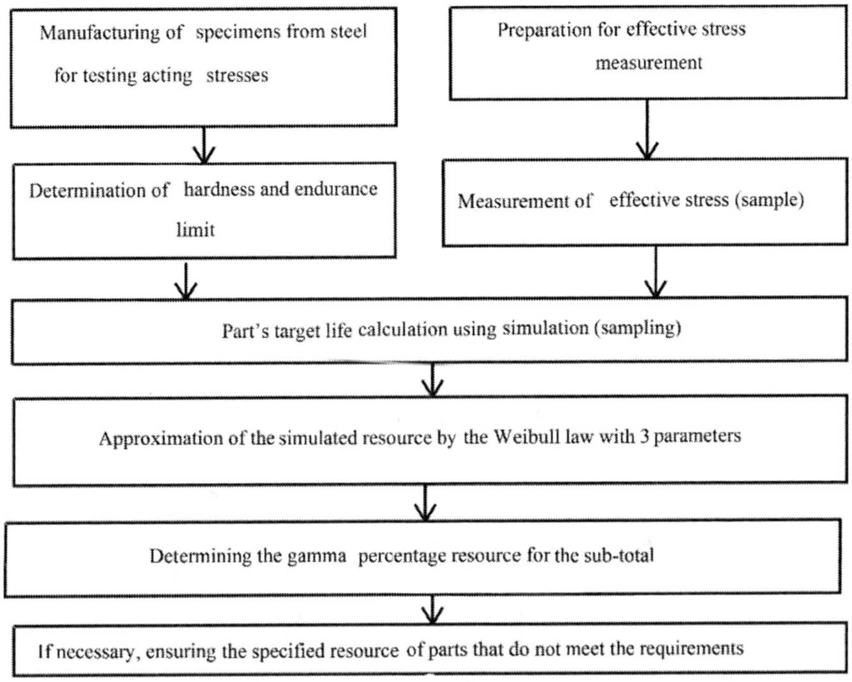

Figure 5. Algorithm for determining the parameters of the sample's probability distributions and the totality of the fatigue life of a part.

Information for calculating the resource of the part is shown in Figure 6.

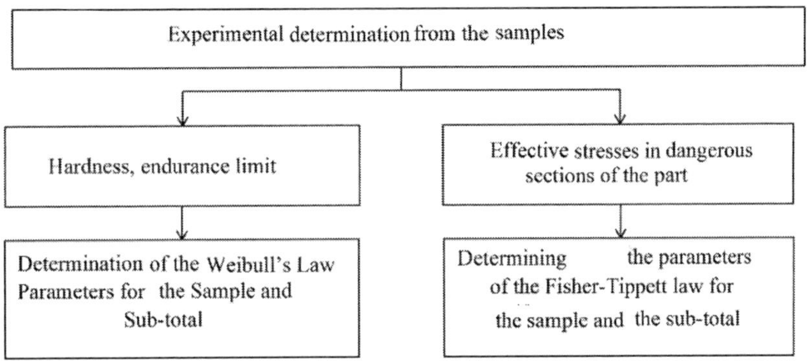

Figure 6. Information for calculating the target life of a part

The entire composition of machine parts should be divided taking into account the effects of part machine failures (Table 1).

Table 1. Influence of types of damage to parts on the reliability of the machine (unit).

Parts with sudden failure	Parts with drift failure	Unloaded details
Fatigue	Wear, fatigue	No failure

Not diagnosed	Are diagnosed, the failures are prevented by the planned replacements	
Cause machine failure	Do not cause machine failure	

With regard to a serial machine, the indicators for its reliability properties are used, both for the machine as a whole and for the units and parts. In addition, the complex indicators are used: availability and technical utilization factors. The same indicators are determined for an individual production machine.

An important place in the system of ensuring the reliability of individual production machines is occupied by the tests, which should confirm the calculated resources of the machine parts (assembly).

Analysis of various types of tests showed that the most effective type is the operation of a machine (unit), almost every part is tested and at the same time serves as a loading element for another part (self-stand machine).

In this case, it is necessary to assess the adequacy of the model to the experimental data.

A certain difficulty is the duration of the experiment (10-20 years) to confirm the adequacy of the calculated results to the experimental data.

To solve this problem, it is advisable to use the accelerated tests without a stand base and increased loads, but only under operating conditions and by reducing the calendar operating time of the tested machine [7].

The positive side of such accelerated tests in time is the fact that the strength and loading correspond to the design modes of units and parts. Such tests are operational, and the results obtained (in contrast to stand tests with increased loads) do not require recalculations and are more reliable [7].

In this case, the machine (units, parts) is considered as a system with mutually loaded adjacent elements, i.e. as a stand [1].

This is the first time we have come to this conclusion, having been involved in machine reliability for over 50 years [3].

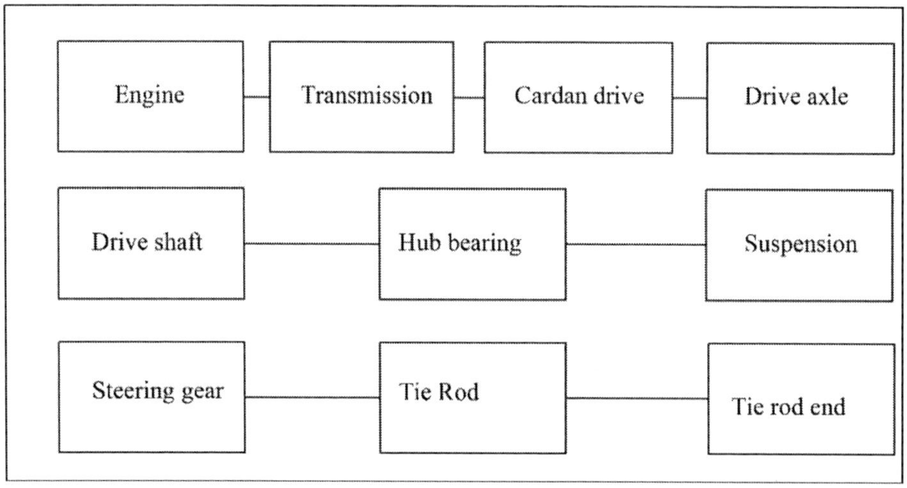

Figure 7. Variants of neighboring nodes for the self-stand

Figure 7 shows that, for example, for a part of a car (a), the transmission is a loading element of the engine; for the drive axle, the loading unit is a cardan drive.

Calculating the machine's target life:

Typical case,	Typical case,

for 10 years of operation:

T_{tl} = 10 years.

D_y = 253 days a year.

T_h = 8 h work in 1 day.

$T_{tl} = 10 \cdot 253 \cdot 8 = 20240$ h;

2530 days in 10 years.

Accelerated in time

testing machine:

D_y = 360 days a year.

T_h = 23 h work in 1 day.

$T_{sy} = 23 \cdot 360 = 8280$ h a year.

T_{st} = 20240: 8280 = 2,45 year.

Acceleration: 10: 2,45 = 4,1 times.

for 20 years of operation:

T_{tl} = 20 years.

D_y = 253 days a year.

T_h = 8 h work in 1 day.

$T_{tl} = 20 \cdot 253 \cdot 8 = 40480$ h;

5060 days in 20 years.

Accelerated in time

testing machine:

D_y = 360 days a year.

T_h = 23 h work in 1 day.

$T_{sy} = 23 \cdot 360 = 8280$ h a year.

T_{st} = 40480: 8280 = 4,9 year.

Acceleration: 20: 4,9 = 4,1 times.

Consequently, one of the two manufactured machines according to the same drawings and technologies is subjected to the operational tests for 2.5 or 5 years with the recording of operational information about failures, repairs, etc. [12].

Perhaps during these five years of operational tests of one (second) machine, it will be necessary to ensure an increase in the target life of parts, machine components by changing the design and manufacturing technology [9,13].

Figure 8 shows a diagram of ensuring the reliability of the test machine No. 1 and the main machine No. 2.

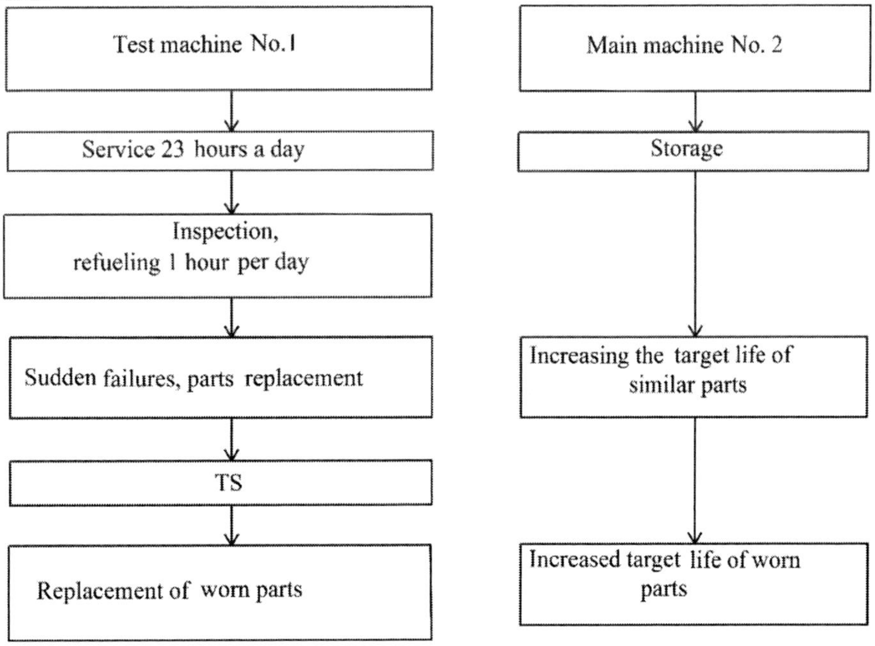

Figure 8. Scheme of ensuring the reliability of the main and backup machines

Summary
Thus, to ensure the reliability of aa individual production machine, when at first glance there is no possibility of using the proven probabilistic-statistical and other methods of ensuring the reliability of serial machines, there is a real opportunity to apply the above-mentioned method.

References
[1] Kasyanov V E 2015 Method of ensuring the reliability of machines of individual production *Construction and architecture – 2015: materials int. scientific-practical conf., RSCEU* 111-112.
[2] Lurye P V, Nazarov R A 2016 100 cars that changed the world (Publishing house «E», M.).
[3] Kasyanov V E, Belenkiy D M 1969 Reliability control system for mining machines *RISS: Reports for the 26th Scientific and Technical Conference*.
[4] Kasyanov V E, Schulkin L P, Trembach I A 1971 Reliability control of hydraulic supports *Collection of RECI*.
[5] Kasyanov V E 1976 Tests of excavators in operation and calculation of their reliability indicators on a digital computer *Reliability and quality control* **6** 15-19.
[6] Kasyanov V E 1980 Choosing the optimal strategy for increasing the reliability of machines *Reliability and quality control* **5** 52-58.
[7] Kasyanov V E 1986 On the issue of standardization of methods for testing machines for reliability *Reliability and quality control* **8** 57-59.
[8] Kasyanov V E 1988 Principles of creating practically trouble-free machines // Standards and quality **7** 39-42.
[9] Kasyanov V E 1990 Integral assessment, increase and optimization of machine reliability // Bulletin of mechanical engineering **4** 7-8.
[10] Kasyanov V E, Rogovenko T N 1999 Probabilistic-statistical assessment of the gamma-percentage resource of the machine frame (on the example of the Don-800 power unit) **6** 10-12.
[11] Handbook of probabilistic calculations 1970 (Military Publishing, M).
[12] Kasyanov V E 2014 A method for assessing the reliability for a sample and a set of finite volume *Scientific Review* **11** 5.
[13] Kasyanov V E, Rogovenko T N, Schulkin L P 2003 Fundamentals of the theory and practice of creating reliable machines [Text] *Mechanical Engineering Bulletin* **10** 3-6.

Theoretical substantiation of ensuring the machine parts' reliability for the assigned target life

S V Teplyakova*, A A Kotesova, A A Kotesov

Don State Technical University, 1, Gagarin sq., Rostov-on-Don, 344010, Russia

E-mail: svet-tpl@yandex.ru

Abstract. Increasing the machines' reliability remains an urgent task of the world mechanical engineering, the solution of which is aimed at significantly reducing the number of the machines' critical, basic and other elements (parts) failures. To solve this problem, a need to move from the sample data to a general sub-total of a finite volume is indicated, the theoretical foundations for ensuring the machine parts' reliability for the assigned target life are formulated and the economic feasibility of increasing the machine parts' reliability is analyzed.

Introduction

The first machines, produced at the beginning of the twentieth century in rather small volumes, had a very low target life, which entailed frequent failures and, as a result, significant costs for maintenance and repair. Therefore, the idea of creating reliable machines have worried many scientists for a long time.

Every year the number of equipment produced by the world manufacturers is growing inexorably and reaches hundreds of thousands a year. Such an increase in production also leads to an increase in the cost of maintaining a working condition against the background of a decrease in machine operation efficiency. This provision has led to the need for scientific research devoted to the creation of machines without failures, despite the prevailing stereotype that "if a machine works, then it necessarily breaks down and requires repair".

The analysis showed that this issue in the scientific literature is not sufficiently sanctified, since there is a minimum number of publications on this topic, and there is also a trade secret that provides machine manufacturers with the necessary competitiveness.

Main part

A distinctive feature of reliability as a science is the operation of the probability theory laws and mathematical statistics in predicting and ensuring a certain level of machine parts and assemblies' reliability. This is due to the need to use the significant amounts of statistical information: from 100 to several million units and various sample sizes from them.

The sample size has a great influence on the results' accuracy when processing the collected information, which can be in the range of 5 - 20 or 100 - 200 units.

Not many sources justify the need to apply the laws with restrictions in the dispersion of parameters and fixing the sample sizes relative to the general sub-total of a finite size [1, 2, 3]. To approximate the values of sample data with a volume of 50 - 100 units, the Mises criterion was applied in the calculations and the probability laws were used with the restrictions on the left (three-parameter Weibull's law) or right (Fisher-Tippett law) part of the distribution curve.

There are situations when a high probability of failure-free operation of parts (P=0,95) does not allow solving the problem of the required level of machine reliability. In this case, a different scientific approach is needed to exclude failures, ensure trouble-free operation, inadmissibility of accidents and significant costs for repairing machines. It is known that the failure-free operation probability values' optimization can reduce the number of failures. However, as the experience of operating machines shows, failures occur before the calculated parts' target life.

The studies and calculations made it possible to establish the following:

- the calculation of the gamma-percentage target life of parts and assemblies during design is carried out according to the sample strength and loading data in volume N=10-100 data;

- at the same time, thousands of machines are in operation, forming sub-totals of these details 10^4 parts (if about a thousand parts (machines) are produced per year, and the production period of this series without changing the design and manufacturing technology is 10 years).

On probabilistic paper (Figure 1) for the three-parameter Weibull's law in general form, the line segment AB represents a sample of volume n=100 and the segment CD represents the sub-total volume Nc=10^3-10^6 [4, 5].

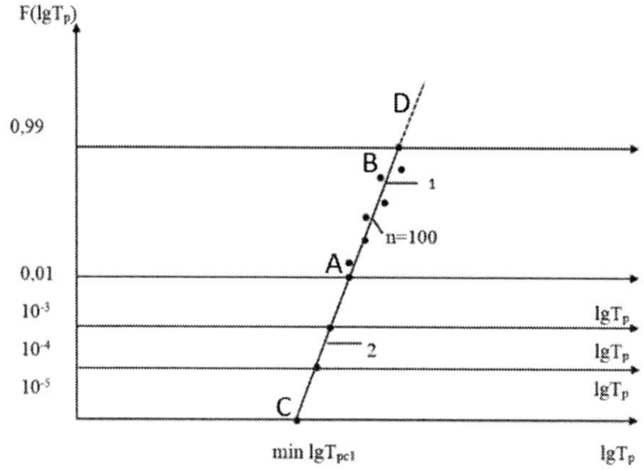

Figure 1. Distribution functions of the sample and the sub-total target life of parts according to Weibull's law.

Figure 1 confirms the fact that the value of the minimum target life according to the sample data is significantly greater than the same value according to the sub-total data to which the sample belongs. The calculations have shown that for the range R=100 the discrepancies are from 4 to 294 times [6, 7].

The study of this issue showed that the use of the selective initial data in the calculations significantly reduces the number of actual failures, which entails an early underestimation of the predicted costs of their elimination. Moreover, this situation actually causes early failures of a set of parts in comparison with failures for a sample (Figure 2).

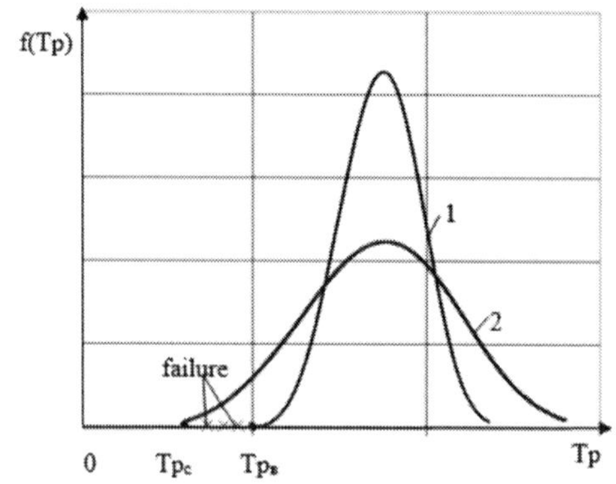

Figure 2. The target life values' distribution: 1 - by the sample; 2 - by the sub-total.

Figure 2 shows the statistical data approximation by the Weibull's probability law with a shift (curves 1 and 2), while it can be seen that the failures according to the data of a finite volume set will begin earlier than according to the sample data.

Extrapolating the sub-total curve to obtain the sub-total shift value Cc according to Weibull's law is incorrect. This is confirmed by the fact that the shift parameter Cc for the sub-total is less ($Cc < Tp_{min}$) the minimum value of the sub-total variation series (Figure 3), and this contradicts the theory [8, 9].

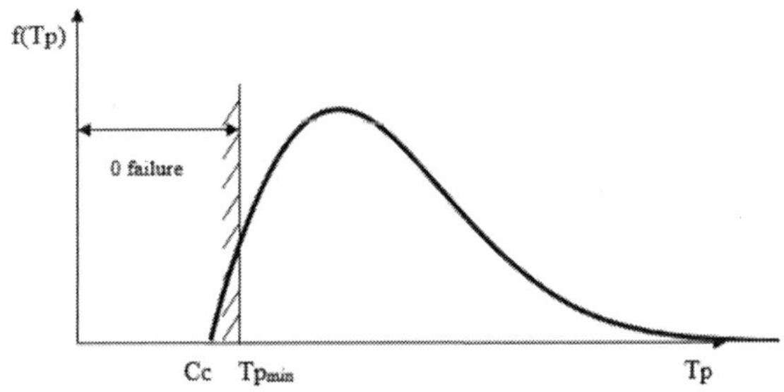

Figure 3. Extrapolation of the Weibull distribution curve to the region of "zero" failures.

So, instead of shifting Cc it is recommended to use the first value of the sub-total variation series Tp_{min}. Then, in the interval $0 \ldots Tp_{min}$ there are no failures, i.e. reliability is ensured [10], and the probability of failure-free operation tends to unity, and the probability of failure Q=0 respectively.

Naturally, such a statement is usually not taken seriously by the opponents and is disputed by the fact that in this case the cost of parts will increase and the machines will be too expensive. This reaction is spontaneous, and the term reliability is a fairly common case.

Having built the sample distribution in logarithmic coordinates in order to determine the value Tp_{min} (Figure 1) [8] we get a graph in the form of a straight line of distribution, common for the sample and for the sub-total. This assumption was verified by modeling the samples from the sub-totals with the size №=10^3; 10^4; 10^5. Discrepancy between the obtained values Tp_i and the direct distribution of sub-totals did not exceed 1-3% [11, 12].

Extension (Figure 1) of the line is a direct sample extrapolation that allows an approximate result with some probability. Therefore, for guaranteed reliability in the interval 0 - Tp_{min} the assigned target life of the part should be increased by 15-30% (Figure 4).

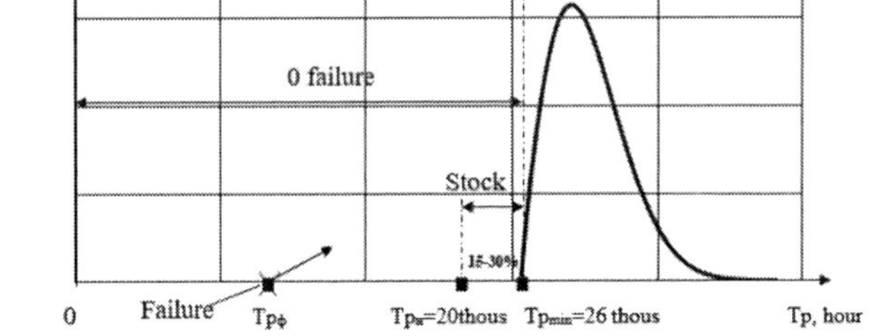

Figure 4. Zero-failure principle.

As previously indicated, the sample line on the logarithmic graph of the Weibull distribution coincides with the direct distribution of the sub-total, then the parameters of the scale and shape (except for the shift) of the Weibull's law for the sub-totals are taken the same as for the samples from them (Figure 5).

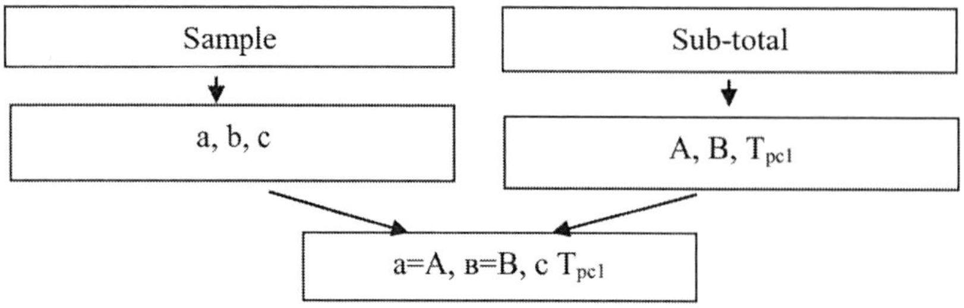

Figure 5. Correspondence of the distribution parameters' values of the sub-total and the sample.

Having analyzed the approximate composition of the parts and their direct impact on the reliability indicators, for clarity, a classification table, indicating failures, was formed (Table 1).

Table 1. Relationship between failure modes and machine uptime.

The part is not diagnosed	With sudden failure	With sudden failure
	Sudden car stop	Sudden car stop
	Safety loss	Significant economic damage
	Absolute reliability required	Absolute reliability required
The part is diagnosed	With gradual failure	Gradual failure (not diagnosed)
	Application after diagnosis or planned stop for TS and car repair	No repair base
	Some economic damage	Some economic damage
	Ensuring Optimum Availability	Absolute reliability required

After analyzing the above-said, the theoretical foundations of ensuring the machine parts' reliability for the assigned target life are formulated:

1. In calculating the reliability reserve, as a quality indicator, it is necessary to operate with the data of the finite volume general sub-total;
2. For the transition from sample, research data to the general sub-total data of a finite volume, it is necessary to use the probabilistic laws with a shift: Weibull's law with three parameters; the three-parameter Fisher-Tippett law; log-normal law with shift.
3. To determine the values of the parameters of the distribution of the sub-total according to sample data obtained experimentally.
4. To consider in the calculations that the target life of the part must be greater than the assigned target life.
5. To set a guarantee of reliability (zero failures for the assigned target life) using a margin of 15-30%, increasing the target life.
6. For the key parameters (especially those having a power-law relationship with the target life), it is necessary to apply tolerances that provide an acceptable dispersion of the target life.

For perception clarity of the above-said, a block diagram of ensuring reliability has been built (Figure 6).

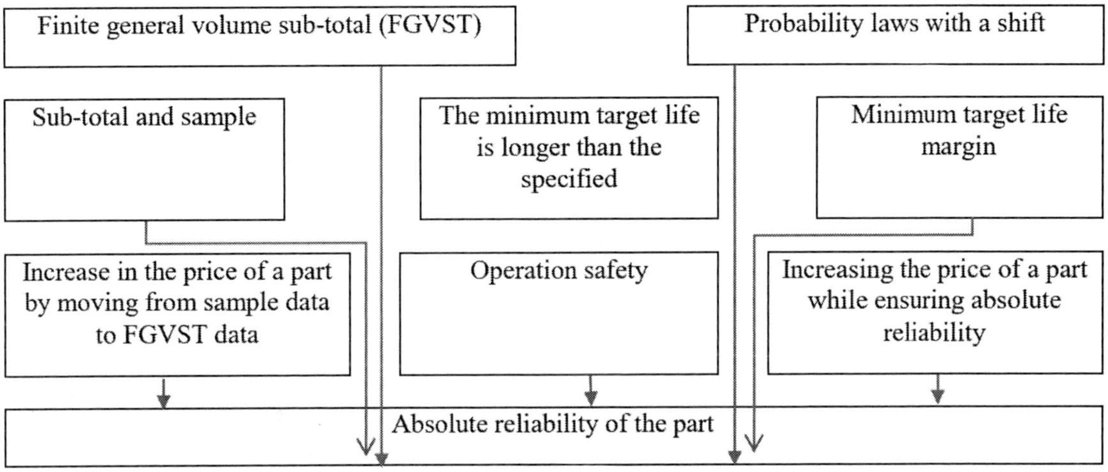

Figure 6. Structural diagram of reliability assurance.

With an increase in reliability (an increase in strength characteristics and (or) a decrease in stresses acting in a dangerous section) and ensuring the required reliability, there is a certain rise in the cost of manufacturing a part (assembly, machine). However, at the same time, there is a significant reduction in repair and maintenance costs, a decrease in the need for spare parts, their storage and transportation. As a result, the unit costs per unit of output will be significantly lower (Figure 7).

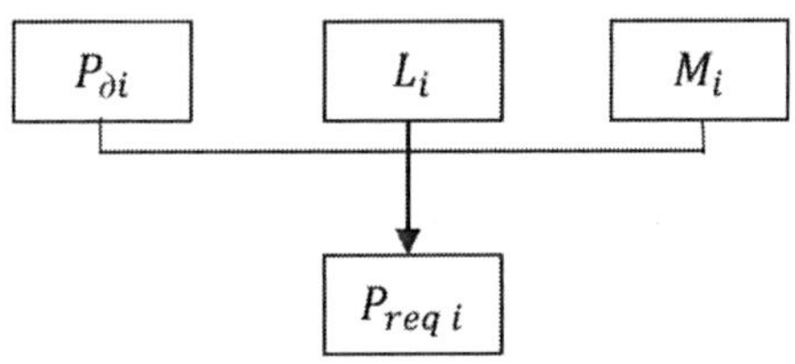

Figure 7. The relationship of proposals for increasing the target life with the cost of the part: $P_{pi}-$ is the initial part price; L_i- denotes the suggestions for improving the strength characteristics; M_i- denotes the proposals for reducing the effective voltage in the dangerous section; $P_{reqi}-$ is the part's price, taking into account the cost of ensuring reliability.

The calculations have been carried out using the approximate data: the proportion of fatigue (without diagnostics) failures ~50%; fatigue and wear failures with diagnostics ~40%; unloaded parts make up about 10%, the assigned target life of parts and machines declared by the manufacturers is 20 thousand hours; fatigue slope indicators m_2=15; for wear m_2=4; increase in fatigue life - 17 times; target life including wear - 9 times.

Table 2. Data used and results obtained.

m_2=15; 1 cheap part 20 thous.rub.	Fatigue and wear m_2=15; 20% parts, wear m_2=4, 20%
To enlarge min target life 17 times, taking into account an increase of 15-30% to guarantee absolute reliability	To enlarge min wear life 9 times
To increase strength $\sqrt[15]{17} = 1{,}206$ times	To increase wear strength $\sqrt[4]{9} = 1{,}73$ times
Part price increase (P_P=20 thous.rub) on 20,6%	Part price increase with wear 300*0,73=219 thous.rub.

Increase in car price $\frac{1{,}504-1{,}5}{1{,}5}100\%=0{,}27\%$	Increase in machine prices with 20% wear $\frac{1{,}719-1{,}5}{1{,}5}100\%=14{,}6\%$
1 expensive part 120 thousand rubles price increase 120*0,2=24 thousand rubles car price increase $\frac{1{,}524-1{,}5}{1{,}5}100\%=1{,}6\%$	
5 parts cost 300 thousand rubles An increase in their price is 300 thousand rubles. *0,2=60 thousand rubles Increase in car price $\frac{1{,}56-1{,}5}{1{,}5}100\%=4\%$	All parts: 50% fatigue and 40% wear. Increase in car price 150+60+219=429 thousand rubles. The new price of the car is 1.929 million rubles. $\frac{1{,}929-1{,}5}{1{,}5}100\%=28{,}6\%$ Check 10%+4+14,6=28.6%

Table 4. Increase in the machine cost with an increase in the parts' reliability.

Option no.	Ensuring absolute reliability	Increase in car price in %
1	1 piece is cheap	0,27
2	1 expensive detail	1,6
3	5 average cost parts	4
4	50% parts with fatigue failures are not diagnosed	10
5	40% parts with fatigue and wear failures are diagnosed	18,6
6	100% machine parts	28,6

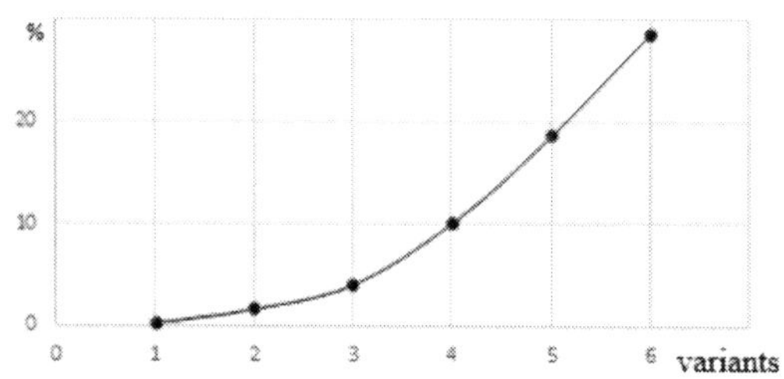

Figure 8. Increase in the machine cost with an increase in the parts' reliability .

Figure 8 shows the increase in the machine cost with an increase in reliability from 1 part to 100% of parts. The situation with complete reliability of the entire machine takes place in cases where there is no possibility of maintenance or repair.

Summary

Despite the fact that the above calculations are approximate, it can be considered that the fact of obtaining too expensive parts or machines with an increased cost several times can be refuted.

Thus, the main opportunity to exclude the onset of failures during a certain (assigned or given) target life is to increase the minimum target life of each part to a value greater than the specified target life, taking into account the guaranteed stock.

References

[1] Teplyakova S V, 2016 Ensuring the reliability of machine parts during the assigned fatigue life (dissertation of the candidate of technical Sciences).

[2] Kasyanov V E, Rogovenko T N, Topilin I V 2003 Determination of the minimum target life of the basic elements of machines *Bulletin of mechanical engineering* **3** 9.

[3] Kasyanov V E, Kotesova A A, Teplyakova S V 2013 Simplified determination of discrepancies between the minimum target lifes of samples and sub-totals for critical machine parts Don Engineering Bulletin **2 (25)** 119.

[4] Kotesova A A, Teplyakova S V, Popov S I, Kopylov F C. 019 Ensuring assigned fatigue gamma percentage of the components International Scientific Conference «Construction and Architecture: Theory and Practice of Innovative Development»: IOP Conference Series: Materials Science and Engineering **698**. doi:10.1088/1757-899X/698/6/066029

[5] Ivanov V V, Popov S I, Marchenko Ju V, Marchenko E V, Dontsov N S, Timofeev S A 2019 Thickness of vibrational mechanochemical solid-lubricant coating in friction pairs of transport engineering products *XII International Scientific Conference on Agricultural Machinery Industry (INTERAGROMASH 2019: IOP Conference Series: Earth and Environmental Science* **403**. doi:10.1088/1755-1315/403/1/012115

[6] Teplyakova S V, Kotesova A A, Popov S I, Kotesov A A 2020 The transition from the sample data to the total sub-total of the final volume and the analysis of this transition laws *International Scientific Conference «Construction and Architecture: Theory and Practice of Innovative Development»: IOP Conference Series: Materials Science and Engineering* **913**. doi:10.1088/1757-899X/913/4/042054

[7] Korotky A A, Marchenko E V, Ivanov V V, Popov S I, Marchenko Ju V, Dontsov N S 2019 Model of forming vibration mechanochemical solid lubrication coating on surface of steel rope *XII International Scientific Conference on Agricultural Machinery Industry (INTERAGROMASH 2019): IOP Conference Series: Earth and Environmental Science* **403**. doi:10.1088/1755-1315/403/1/012116

[8] Teplyakova S V 2014 Graphic-analytical method of definition parameters of Weibull distribution is offered *Scientific journal «Scientific review»* **11(2)**.

[9] Teplyakova S V, Kotesova A A 2014 To ensure the practical availability of items of construction machinery *Scientific Internet – journal «Science World»* **2.**

[10] Kasyanov V E, Kosenko E E, Demchenko D B, Teplyakova S V, Hwan R V 2020 Managing the reliability of technical systems *Truck: transport complex, special equipment* **3.**

[11] Kasyanov V E, Kosenko E E, Demchenko D B, Teplyakova S V 2020 Managing the reliability of technical systems *Safety of technogenic and natural systems* **2.** Information on http://www.bps-journal.ru/publications/2020/1-2020

[12] Teplyakova S V, Kotesova A A, Nikolaev N N 2020 Modeling the vehicle quality integral indicator *Advanced Engineering Research* **20 (2)**. Information on https://doi.org/10.23947/1992-5980-2020-20-2-150-154

[13] Deryushev V V, Zaitseva M M, Kosenko E E, Kosenko V V 2020 The quality analysis of the anti-corrosion coatings metal structures operating in difficult conditions *IOP Conference Series: Materials Science and Engineering* **913(4)** 042059.

[14] Deryushev V V, Zaitseva M M, Evseev D Z, Kosenko E E 2020 Concentration of thermal stresses in metal materials and constructions under local heating *Materials Science Forum MSF* **974** 729-734.

[15] Teplyakova S V, Kopylov F S, Krimsky V S 2019 Constructive features of robots-manipulators production with adjustable bend *Young researcher of the Don region* **5.** Information on http://mid-journal.ru/upload/iblock/3e8/17_871-Teplyakova_94_97.pdf

IOP Conf. Series: Materials Science and Engineering 1083 (2021) 012068 doi:10.1088/1757-899X/1083/1/012068

The Internet of Things in Transport Technology Improvement and Project Learning

N N Nikolayev

Don State Technical University, Gagarin sq., 1, Rostov-on-Don, 344000, Russian Federation

E-mail: nnneks@yandex.ru

Abstract. A fairly new field of science and technology - the Internet of Things (IoT) and Industrial Internet of Things (IIoT) are considered in relation to the development and digital transformation of the transport industry and higher education. At the same time, the influence of this technology on both transport processes and transport education is presented. An example of the industrial Internet of things use in the development of a package product 'smart container-transformer' and an exhibition complex for presenting this product is given. The means and results of the project-based learning implementation based on the Internet of Things in the course of teaching the discipline 'Transport systems' among graduate students are presented.

1. Introduction

Today, we can regretfully state that a fairly large number of technical specialists and teachers of technical universities are not familiar with such concepts as the Internet of Things (IoT) and the Industrial Internet of Things (IIoT). However, the Internet of Things is one of the main trends in technology development. The world's leading industrial and IT companies work in this direction.

It is of no secret that the Russian economy has been heavily dependent on energy prices for several decades. At the same time, there is a decline in production in all industries. Not to mention the extremely small number of new technologies being developed and introduced into production. In order to change the situation, it is necessary to track and apply the global development trends. However, this requires technicians who are aware of these trends, and who also know how to find the new development opportunities and apply them. The emergence of a sufficiently large number of such technical specialists is impossible without the educational system transformation in universities, especially at the second stage of obtaining higher education - in the programs for Master's degree.

Today, many actions that previously could only be done by a person are performed by robots, and what previously could only be done on our own, it has become possible to perform online. For example, ordering food online, paying bills, and so on. If these capabilities of automated subjects are combined with the capabilities of Internet networks, then this is a simplified concept of the Internet of Things.

The idea of the Internet of Things was born in 1999 [1]. Kevin Ashton from the Massachusetts Institute of Technology put forward the idea of automating the management of the supply of goods to the warehouse at his presentation. If to combine all warehouse terminals into one network, then it will be possible to regulate many issues without the participation of people. For example, if the warehouse runs out of some product, then the machines could figure out what is missing and also order replenishment on their own. The concept is dubbed the 'Internet of Things'. The idea is based on the use of ordinary things in human life, but not as static objects, but as things that can analyze the environment and make the appropriate decisions. Almost any thing can be done smart no matter the application area is, and the new IPv6 protocol makes it possible to assign a network address to absolutely every 'smart' thing.

Modern technology is inevitably changing the education concept. Now the information accessibility level is such that a person with a certain educational base, but without special education, is able to master quite complex technologies at a practically significant level in the shortest possible time, for example, programming microcontrollers and creating 'smart' things on this basis. However, to create a base that will enable the bulk of specialists to use the information products, it is necessary to rebuild the education system. This requires radical transformations and is fraught with certain problems.

2. The problem analysis

The problem of introducing the Internet of Things into the economy and higher education needs to be analyzed from different points of view. On the one hand, there are technical and economic complexities. On the other hand, there are the problems of modern pedagogy and higher education. Let us analyze these problems.

2.1. Analysis of publications on this issue

At the moment, there is a fairly large number of studies and publications on the IoT topic. A group of scientists from Croatia, Spain and Italy [2] have made a detailed survey study reflecting the development of the Internet of Things at the moment. The study divides the direction of IoT development into developments in the field of energy, the creation of smart cities, healthcare and transportation. That is, the relevance of the transport topic from the point of view of the Internet of Things is confirmed by other studies.

The publication [2] also noted that an obstacle to the development of the Internet of Things in Africa, Latin America and some countries in the Middle East is the low speed of Internet networks.

The authors of [2] also considered an issue that is usually not thought about: the expenditure of the production resources for IoT products, as well as the subsequent disposal of products that have expired. It is suggested that resources and minerals may not be enough at the current pace of IoT devices' development. However, it is noted that the IoT technologies' growth is associated with the growth of the world's population, urbanization and resource scarcity. Growth is driven by the need to improve urban infrastructure.

IoT development is also taking place in the field of improving the environmental friendliness and autonomy of road transport. Smart cars learn to do without a person in matters of traffic control, registration for maintenance, diagnoses when diagnosing faults, finding parking, active and post-crash safety as well as many other options.

A group of Polish and American authors presented the support system development for public transport passengers based on IoT technologies [3]. The publication highlights the development of a new model BIBO (Be-In / Be-Out), which allows, based on IoT devices (smart beacons and smartphones), to automatically pay for travel during the trip without the participation of a passenger. Self-learning models of passenger identification in transport were applied based on the statistical models and neural networks. According to the authors, in this case, the statistical models work better.

A group of Indian authors published [4] the information on the developed system for ensuring the passengers' safety on buses in case of emergency situations. It is proposed to equip buses with an intelligent system equipped with a set of sensors for the occurrence of various emergency situations (fire, increased vibration on wheels, over-speeding). The sensors send a signal to the microcontroller, which sends information to the data cloud and issues the commands to security systems. These can be the signals to alert the driver, passengers, a command to stop the engine by cutting off the fuel supply, opening emergency exits after stopping and starting the fire extinguishing system. In the event of an emergency, a signal will be sent to the rescue services automatically via GSM networks, accompanied by detailed information about the incident.

Other authors from India presented [5] the concept of an intelligent system for fast parking space search. This problem is especially acute in large cities and metropolitan areas around the world. The biggest difficulties arise when trying to find a parking space during business hours. As a result, a huge amount of time is spent searching for empty seats, while huge amounts of fuel are burned and large amounts of exhaust gases are emitted. The researchers proposed an IoT system that includes ultrasonic sensors for monitoring parking spaces, analog-to-digital converters based on Arduino microcontrollers, a cloud database and a Raspberry Pi-based processor. To track free parking spaces where it is not possible to install ultrasonic sensors, it was proposed to use quadcopters with video cameras. In the future, it is proposed to expand the project based on the tactile Internet and 5G networks new technology.

The authors from Malaysia, the United Kingdom and the United States [6] propose an integrated system for transport management, data collection and decision-making based on all existing IoT applications: radio frequency identification, Bluetooth and WiFi-based networks, vehicle communication networks, 4G standards and 4G, wireless sensors and cloud technologies. Combining the capabilities of these technologies into an IoT system, according to the authors, will improve the safety, economy and production efficiency of vehicles.

The analysis of publications on the topic of industrial technological development, including road transport, developing in the direction of IoT technology, makes it possible to conclude that this new technology is being actively implemented, but it still has a very large potential for further development. Accordingly, there is a need for training specialists of a wide variety of profiles who have the skills to use IoT in their professional activities.

The authors from Australia published a paper [7] devoted to the IoT introduction in the educational process in training specialists in the field of agriculture. In the course of the study, an analysis of the IoT electronics available on the market, services for the databases and data exchange formation for IoT purposes was carried out. As a result, the optimal microcontrollers and sensors were selected, and free, but highly efficient resources were chosen as the means of accumulating and processing information. As a result, a system for visualizing environmental parameters within an agricultural farm was created. This system is introduced into the educational process, and the concept of problem learning is implemented on it. The authors noted the lack of educational literature and methods for introducing IoT into the educational process of universities.

During the literature search, it was not possible to obtain the information about the implementation of IoT systems in the educational process of training transport areas. Therefore, the pedagogical developments in the field of IoT systems presented in this article are relevant and are designed to fill the existing gap.

2.2. Analysis of the problems in higher education

The author of this article has over 15 years of experience in teaching in higher education. During this time, educational standards have changed, the Bologna system of higher education (bachelor, master) was adopted, and a student-oriented competence-based approach to learning was proclaimed. However, in essence, these processes were carried out on the basis of the remnants of the Soviet education system, and inherited many of the shortcomings of this educational model. People who continue to lead the educational process often do not understand the essence of the transformations being carried out, and they follow them exclusively when preparing documentation. The educational process itself remains unchanged. Controlling structures are also interested, first of all, in properly formatted documentation.

Let us consider the shortcomings of the existing education system:

- Weak interdisciplinary communication. Most disciplines are studied by sporadically, especially in general education and general technical disciplines.
- Excessive overload of theoretical calculations. Often a teacher, especially the one who has been leading the discipline for many years and has learned the material well, gives students a huge number of long formulas, diagrams and graphs. And he sees the learning process as a thorough memorization of the material given out, and then - the reproduction of what was learned for the exam.
- Lack of many disciplines' connection to the professional field of graduates.

It should be noted that 15-20 years ago, these teaching methods were perceived by the students back as inevitable hardships of the process of obtaining higher education. Many students memorized the material without thinking about the meaning of this process, passed it successfully, and soon forget it. Today's generation of students is more pragmatic and such methods stop working. It is difficult to force a student to learn the material if he does not understand why it is needed.

We live in the information age, when access to a wide variety of knowledge can be obtained with a mouse click. But the information availability does not replace education, as many students believe. But memorizing formulas and definitions without thinking is not the way to get an education in the modern world. If earlier it was believed that a teacher in a discipline should know everything, then in the information age he should not and cannot do it. The task of the teacher is to teach how to search, screen out, analyze and apply the available information, to help students acquire skills and abilities in the professional sphere. Modern education is based on project activities that are part of the project-based learning system [8].

Despite the fact that project-based learning has been known for more than 100 years, in the information age it has only increased its relevance [9, 10]. For the employer, it does not matter what a university graduate knows, it is much more important what he knows how to do and if he knows how to master new skills on the basis of his existing educational base. such Modern technologies as IoT have given the learning opportunities of a higher order.

3. Development of IoT products and project learning methods based on them

3.1. Smart transformer container and educational and exhibition complex

Let us consider the 'SmartBoxCity' [11] project, in the course of which the 'Internet of Things' technology was applied. At the same time, both an industrial IoT product and an educational and exhibition complex, consisting of operating car models with a manipulator and reduced-size transforming containers, were obtained. The author of this article is a member of the project team and the direct developer of the remote-control system for the educational and exhibition complex.

'SmartBoxCity' is an innovative IoT package product for container shipping that consists of:

- Hardware - a smart transformer container (Figure 1), equipped with sensors for cargo weight, temperature, humidity, light, inclination, impact, video cameras for monitoring cargo online, a mechanism for folding and unfolding, raising and lowering doors and side walls (Figure 2, Figure 3), batteries and solar panels.
- Software part - server software, data cloud, website, mobile user and driver applications, API integration of the Yandex.Routing system for managing the transport process logistics.

Figure 1. Unfolded transforming container.

Figure 2. Convertible container in the process of folding.

Figure 3. Folded convertible container.

The 'SmartBoxCity' package product gives an opportunity to transport packaged cargo, as well as bulk cargo using 'flexy tank' flexible containers. The system is currently designed for the transport of goods in an urban environment. However, the design and technology can be transferred to the containers of any size and purpose. Folding, unfolding, formation of an order for transportation, acceptance of the order by the car driver with a manipulator and tracking the progress of the transport process occurs from a smartphone via a mobile application.

Figure 4. Manipulator operation when unloading a folded container layout.

Figure 5. Manipulator operation when unloading the expanded container layout.

Figure 6. Unfolding the container layout.

Figure 7. Unfolded container layout.

To obtain the opportunities to demonstrate the development functionality in rooms and on a small area, as well as for educational process use in the university, an educational and exhibition complex was developed and produced. It consists of a manipulator with a telescopic trailer (Figure 4, Figure 5) mounted on an electric vehicle and two fully functional miniature convertible containers (Figure 6, Figure 7). The manipulator and container drives are based on linear actuators powered by 12 V lithium-polymer batteries.

Let us consider the electronic circuit of power supply and control of the educational and exhibition complex using the example of the container-transformer exhibition layout, which is shown in Figure 8. The linear actuator (not visible in Figure 8, as it is located in the bottom of the container) is powered by battery 1. It is connected through the connectors 2 to the electromagnetic relay 3, which performs the function of switching the polarity of the supply on the linear actuator. The function of a load weight strain gauge, a GPS sensor, as well as the container software is performed by a simulator program built into a smartphone 4.

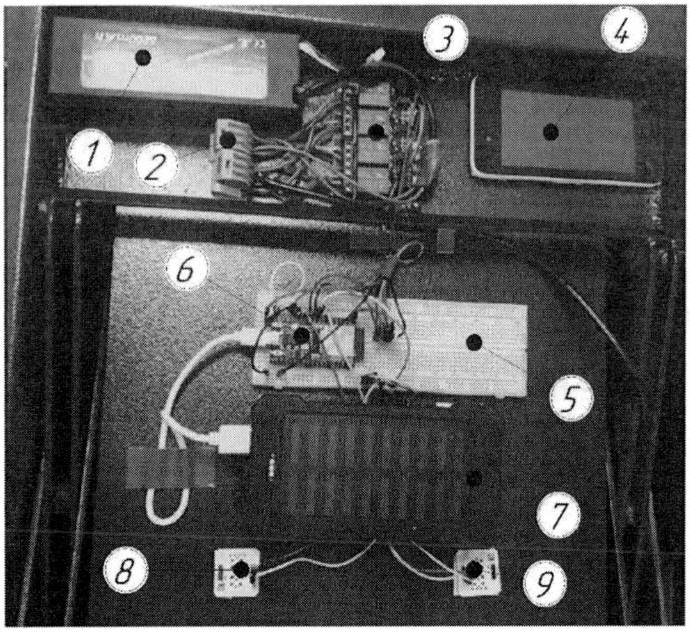

Figure 8. Electronic circuit of power supply and control of the transformer container exhibition layout

To ensure the possibility of correcting the connection diagram without soldering, a prototype board 5 was used. It has a microcontroller 5 of the ESP32 model with a built-in Wi-Fi antenna and a Bluetooth adapter that controls relay 3 depending on the commands sent from the smartphone. The power supply for the control system is separated from the power section and is supplied through the power bank 7, which simulates a solar battery. The microcontroller 6 is connected to humidity and temperature sensors 8 and light inside the container 9 to check the tightness of the door closing. Sensor data is sent to the server and then transferred from the server to the simulated smartphone, as well as to the main container management tracking application.

Management of the educational and exhibition complex is implemented in two ways: via Wi-Fi using the MQTT Internet of Things protocol via a free account of the MQTT broker Cloudmqtt and via Bluetooth via a free terminal application Dabble.

The educational and exhibition complex will be used in the process of training bachelors and masters of transport training areas of the Don State Technical University.

3.2. IoT-based project learning

The simplicity of the IoT technology electronic systems used to equip the educational and exhibition complex of the 'SmartBoxCity' project and the ease of mastering the programming of microcontrollers in the Arduino IDE based on numerous templates from the Internet allows creating a variety of projects, including in the field of transport. This was the basis for the introduction of project training in the Master's program 'Technology of transport processes' among the students.

In practical classes, the students were introduced to the IoT technology essence and demonstrated the possibilities and simplicity of creating smart devices based on the ESP32 microcontroller. This microcontroller has a built-in timer, Hall sensor, analog-to-digital converter for connecting analog sensors, built-in WiFi and Bluetooth, 36 pins, 18 of which are available for creating projects. The microcontroller is equipped with a dual-core processor.

The students were asked to familiarize themselves with the variety of devices and sensors that can be used with the ESP32. An overview of the main models of microcontrollers on the market was also given, with an analysis of the advantages and disadvantages.

Each graduate student has a Master's thesis topic that they must defend in order to obtain a Master's degree. The students are encouraged to develop a small IoT project connected with the dissertation topic. An IoT project is carried out in the form of a business plan, which should contain:

- Description of the IoT project essence with the disclosure of scientific novelty and practical value.
- Comparison with analogues.
- Description of the design, structure, software and hardware components of the project.
- Assessment of the potential sales market and areas of implementation.
- Assessment of economic efficiency and payback periods.

Project-based learning based on IoT technologies solves the problems of modern higher education described above:

- Knowledge from a variety of disciplines is actively applied, interdisciplinary connections are used.
- Students study the material of various disciplines from the point of view of practical applicability, and the mathematical apparatus is used as a tool for finding the IoT project parameters.
- The acquired skills and design results are used in the course of writing a Master's thesis and can be a part of professional activity. Students are encouraged to submit the projects for participation in competitions and grants.

In project-based learning, the student acquires many practical skills. Some projects contain elements that go far beyond the direction of student training (applied programming, electronics, etc.). In this case, it is recommended to involve students in specialized areas of training and organize teamwork. Organizing and managing a team is also a critical professional skill that project learning can provide.

4. Summary

The purpose of introducing IoT technologies into transport technologies and the educational process is to raise the technology level of any technical specialists to a new world level. 'Smart' technology makes life easier for users and gives new economic opportunities. However, it requires highly qualified personnel in a variety of specialties. Thus, such new technologies as IoT, are a factor in the need to transform the education system. On the other hand, they provide opportunities for such transformations, such as IoT-based project learning. It is planned to extend the experience of the Master's program to the Bachelor degree students as project training development, and then to establish a mechanism for organizing teamwork with the involvement of students in specialized training areas and the subsequent nomination of educational projects for competitions and grants.

References

[1] Ashton K 2009 That 'Internet of Things' Thing: In the Real World Things Matter More than Ideas *RFID Journal* **06** 4986.

[2] Nizetic S, Solic P, Gonzalez-de-Artaza D L, Patrono L 2020 Internet of Things (IoT): Opportunities, issues and challenges towards a smart and sustainable future *Journal of Cleaner Production* **274** 122877.

[3] Mastalerz M, Malinowski A, Kwiatkowski S, Śniegula A, Wieczoreke B 2020 Passenger BIBO detection with IoT support and machine learning techniques for intelligent transport systems *Procedia Computer Science* **176** 3780–3793.

[4] Vanitha N S, Karthikeyan J, Kavitha G, Radhika K 2020 Modelling of Intelligent Transportation System for Human Safety using IoT *Materials Today: Proceedings MTP* **06** 421.

[5] Kanakaraja P, Nadipalli S, Kumar S V A, Kumar K S, Kavya K C S 2020 An implementation of advanced IoT in the car parking system *Materials Today: Proceedings MTP* **06** 042.

[6] Rahim Md A, Rahman Md A, Rahman M M, Asyhari A T, Bhuiyan Md Z A, Ramasamy D 2020 Evolution of IoT-enabled connectivity and applications in automotive industry: A review *Vehicular Communications* **VC** 100285.

[7] Gunasekera K, Borrero A N, Vasuian F, Bryceson K P 2018 Experiences in building an IoT infrastructure for agriculture education *Procedia Computer Science* **135** 155–162.

[8] Sharma A, Dutt H, Sai Ch N V, Naik S M 2020 Impact of Project Based Learning Methodology in Engineering *Procedia Computer Science* **172** 922–926.

[9] Ivanov V V, Popov S I, Kirichek A V 2018 Investigation of optimal chemical composition of cast aluminum alloys for vibrational mechanical-chemical polishing and deposition of protective and decorative coatings *IOP Conference Series: Materials Science and Engineering, IOP Conf. Series: Materials Science and Engineering* **327** **(3)**.

[10] Kazun A P, Pastukhova L S 2018 The practices of project-based learning technique application: experience of different countries *The Education and Science Journal* **20** **(2)**.

[11] Korotkiy A A, Nikolaev N N Container transportation in the urbanized environment with using of technology 'internet of things' *Safety of technogenic and natural systems* **3** 47–53.

IOP Conf. Series: Materials Science and Engineering 1083 (2021) 012069 doi:10.1088/1757-899X/1083/1/012069

Some aspects of the bridges' functional qualities restoration

A V Makarov, S A Kalinovsky*, T V Ereshchenko, M A Pavlova

Volgograd State Technical University, 28, Lenin Ave., Volgograd, 400005, Russia

E-mail: sk0522@yandex.com

Abstract. The paper considers the reasons for reducing the load-bearing capacity of bridge structures. Factors affecting the feasibility of major bridge repairs are identified and some ways of their implementation are analyzed. An improvement of one of the modern methods of repairing bridge beams – pasting them with carbon lamellas is proposed. The improvement essence is the use of a device for tensioning reinforcing belts and the proposal to use carbon fiber lamellas as puffs, similar to Sprengel puffs. The method consists in stretching the lamella along the lower edge of the beam edge and fixing it using the developed anchor, not by gluing.

1. Introduction

In the modern world with the concentration of population and production in large cities, transport security issues are becoming more and more relevant. Not only security against terrorist threats and man-made disasters, but also security in the sense of ensuring sustainable, reliable, and uninterrupted operation of facilities and systems. Stable operation should be understood as trouble-free operation of the systems, including transport systems with possible external negative impacts on the system. Bridges are the most vulnerable link in the transport system. Restoring the consumer qualities of the bridge, its load capacity and safety, is disproportionately more expensive than restoring a highway section. And this process takes much longer. Therefore, the problem of restoring bridges that are in poor condition or have passed into poor condition is very urgent. The repair and reconstruction of a bridge in a state of disrepair, which makes it impossible to continue operating, usually requires serious investment comparable to new construction. However, due to irregular inspection, bridge structures are often in unsatisfactory condition and all measures for their repair are carried out even when their damage is already significant [1].

2. Factors influencing the choice decisions about restoring or dismantling bridge structures

The existing method of a bridge structure inspection involves a comprehensive study of its technical condition for all groups: the bridge deck, spans, supports and the bridge space [2,3]. This is done so that a decision about the future use of the bridge structure – major repairs, reconstruction or dismantling and new construction can be made. The determining factor here is the state of superstructures and their resource. Major repairs with the replacement of all load-bearing structures are not effective, as the price is comparable to the new construction, and sometimes exceeds it due to the old structures' dismantling. It is also important to note that it is not possible to use modern methods of design and construction in case of a major repair bridge because the old bridge is made in accordance with a certain scheme, which on the one hand not be significant changes, and the other does not fully meet modern requirements for this bridge structure. For example, the breakdown into spans, location and the obstacles' intersection angle may be less well chosen for modern conditions than for the conditions in which this bridge was built. The possibility of making certain structures from certain materials may also have changed, as well as the possibility of using other, more modern materials.

Promising in terms of major repairs with relatively low costs are the bridge structures with span beams, the resource of which is not exhausted, but only reduced due to changes in load capacity standards and removable defects. Load capacity is a characteristic of the technical condition of a bridge structure that corresponds to the maximum impact of a temporary vertical load, at which there is no transition to the limit state in any of the main load — the construction bearing structures. The load capacity of the structure as a whole is determined by the load capacity of the weakest of the main load-bearing structures [4].

3. Some methods of bridge structures' recovering

Figure 1 shows the bridges with reduced load capacity, mainly due to changed standards. The facade surfaces of the superstructure's extreme beams are in poor condition and require either repair or replacement. In the first case, it will not be possible to increase the load capacity of the structure, but only to prevent further development of the reinforcement and concrete corrosion defect. Moreover, the beam reinforcement was previously performed. If the end beams are replaced with new modern ones, the load capacity of the bridge in this part of the superstructure will be increased to the required standard. The second pair of adjacent beams does not require replacement, since the forces in it are lower than in the extreme ones. The beams in the superstructure are loaded in accordance with the transverse installation coefficient [5]. The ordinates of the pressure influence line for the extreme beams are determined by the expression:

$$y_1 = \frac{a_1^2}{2\sum a_i^2 + \frac{n \cdot l_p^2 \cdot G_b \cdot I_b}{3 \cdot E_b \cdot I_k}},$$

and for the adjacent beams (second beams), the expression:

$$y_2 = \frac{a_1 \cdot a_2}{2\sum a_i^2 + \frac{n \cdot l_p^2 \cdot G_b \cdot I_b}{3 \cdot E_b \cdot I_k}}.$$

The ratio of expressions will be $\frac{a_1}{a_2}$, if $a_1 = n \cdot a$ and $a_2 = (n-2) \cdot a$ then we get $\frac{n}{n-2}$. Usually in the span structures of bridges beams from 6 to 10 pieces. Then the coefficient of transverse installation for the beams' second pair is 40 percent less at n=6, and 20 percent less at n=10. The standard change in the time load of the mobile load from A11 to A14 corresponds to an increase in internal forces in the beam by 22 percent. Thus, this old part of the span structure is also out of danger.

Figure 1. Figure with short caption (caption centred).

This solution was proposed to restore the capacity of the overpass in the city of Volgograd on Mendeleev street. The technical condition of the overpass on Mendeleev street is far from the standard. The survey showed the presence of defects in load - bearing structures, destruction of the protective layer of concrete, corrosion damage to working fittings, unsatisfactory condition of sidewalks-there are no slopes, separation of the railing (Figure 1). Corrosion of bridge reinforcement occurs due to the penetration of chlorides and carbon dioxide into concrete [6,7]. The destruction of the beams was facilitated by the passage of vehicles exceeding their standard load [8]. The structures have cracks, supporting parts and expansion joints in unsatisfactory condition [9]. The destruction of the waterproofing layer contributes to the concrete reinforcement and carbonation corrosion. The structure has a load capacity limit of 10 tons, which does not meet modern standards.
Removing the end beams will allow you to use a new beam that meets all the requirements of the project. Replacing the pavement with reusable layers of asphalt concrete will reduce the constant load, restore the waterproofing layer and make traffic comfortable.

There is another solution to the problem of restoring the bridge's load capacity. They are less costly. If the span beams are in good condition and the physical reduction in load capacity is 10 percent, then it is possible to use a high –strength composite-carbon slats glued from carbon fibers. Due to the high strength characteristics of carbon fiber, it is possible to increase the load – bearing capacity of the structure almost without losing the useful volume and increasing the own weight of the structures-the thickness of the reinforcing elements is usually from 1 to 5 mm. Beams' reinforcement is performed by applying carbon fiber in the most stressed areas: for the perception of increased bending moment – usually in the center of the span along the lower face of the structure [10].

Composite reinforcement consists in the fact that the carbon lamella glued to the surface of the structure takes a part of the effort from the temporary load, thereby increasing the load - bearing capacity of the reinforced element (Figure 2). The disadvantage of the new type of reinforcement is its cost- it is more expensive than conventional reinforcement, but this is only at first glance. First, it is necessary to take into consideration that the durability and strength of the material increases the service life, and, therefore, potentially increases the inter-repair period and the cost of subsequent repairs. Second, the composite material itself is more expensive, and the technology of the amplification device is simpler. A small difference in the cost of metal and composite reinforcement can be beneficial later on.

a) *b)*

Figure 2. Reinforcement of the superstructure beams with carbon lamels.
a) execution of work; b) finished structure.

The disadvantages of reinforcement with carbon lamellas include: first, the possibility of detachment of the reinforcement tapes and difficulties with diagnostics; second, the inability to perceive a constant load acting on the structure; third, when installing the reinforcement without stopping traffic on the bridge, the non-hardened adhesive will allow the tape to shift slightly under the influence of a temporary load and not work for part of this load in the future.
The calculation of the required composite cross section is made from the condition $M \leq M_{ult}$, , where M – is the bending moment from the external loads; M_{ult} – is the maximum bending moment that can be perceived by the reinforced section of the element. M_{ult} value for bendable rectangular, t-shaped, and i-beam elements that have a shelf in the compressed zone with $\xi = \frac{x}{h} \leq \xi_{R,f}$ determine depending on the position of the neutral axis of the section.

If the neutral axis passes in the shelf (Figure 3), then the condition is met,

$$R_s A_s + R_f A_f \leq R_b b'_f h'_f + R_{sc} A'_s,$$

The bending moment should be determined by the formula:

$$M_{ult} = R_b bx(h_0 - 0{,}5x) + R_{sc} A'_s (h_0 - a') + R_f R_f a \quad (1)$$

The height of the compressed zone should be determined by the formula:

$$x = \frac{R_s A_s - R_{sc} A'_s + R_f A_f}{R_b b}$$

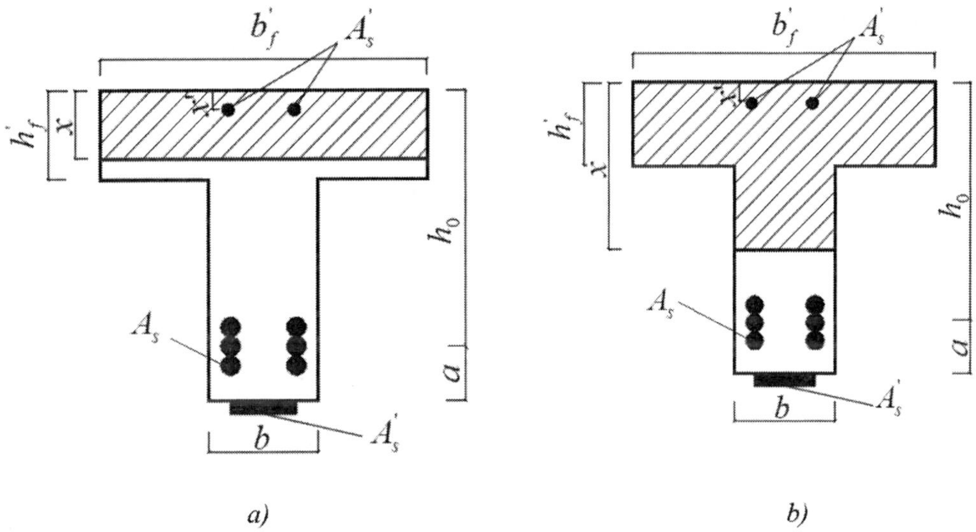

Figure 3. Cross-section of the beam: *a*) under normal loading (the neutral axis passes in the shelf); *b*) overloaded (the neutral axis passes through the edge).

If the neutral axis passes through the edge (Figure 4), then the condition is met,

$$R_s A_s + R_f A_f > R_b b'_f h'_f + R_{sc} A'_s,$$

The bending moment should be determined by the formula:

$$M_{ult} = R_b bx(h_0 - 0.5x) + R_b (b'_f - b) h'_f (h_0 - 0.5h'_f) + R_{sc} A'_s (h_0 - a') + R_f A_f a, \quad (2)$$

The height of the compressed zone should be determined by the formula:

$$x = \frac{R_s A_s + R_f A_f - R_{sc} A'_s - R_b (b'_f - b) b'_f}{R_b b}$$

When calculating the strength of the superstructures' bending beams in formulas (1) and (2) of the elements, it is recommended to observe the condition $\xi = \frac{x}{h} \leq \xi_{R,f}$. Then the maximum bending moment M_{ult} can be determined by replacing the calculated resistance of the external reinforcement made of composite materials R_f with the stress in it σ_f, determined by the formula:

$$\sigma_f = \left[\varepsilon_{b2}\left(\frac{\omega h}{x} - 1\right) - k\varepsilon_{bt}^{0}\right] E_f \quad (3)$$

The problem of eliminating the disadvantages of simply gluing carbon tapes with an adhesive composition is solved. For this purpose, a patented device for tensioning reinforcing belts is used [11,12]. It is proposed to use a carbon fiber lamella as a tightening by analogy with a Sprengel tightening. The method consists precisely in stretching the lamella along the lower edge of the beam edge, and fixing it with the help of a developed anchor, and not by gluing. The fixed anchor is a steel plate with a glued lamella clamped between two plates by the tension force of the bolts. The tension anchor is a welded part made of steel plates and a corner. The carbon fiber composite reinforcement device provides for the following technological operations: the span beams are repaired – the cracks are expanded, old destroyed concrete is beaten off, the reinforcement is cleaned of rust, the torn cross reinforcement is welded and treated with inhibitors, and then everything is sealed with a repair solution.

The fixed part of the anchor is fixed in the place determined by calculations. To install, you need to break off the protective layer of reinforcement and clean the reinforcement itself from concrete and rust. Two high-strength bolts are inserted into the metal plate, then the plate is welded by electric welding to the lower row of the reinforcement package. The fixed part of the tension anchor is attached in the same way. Places where anchors are welded to the reinforcement are sealed with a repair compound, and the protective layer of concrete is restored. The required length of carbon fiber tape is cut off with a margin of 80 - 100 cm by 40 - 50 cm on each side [13]. The steel plate and the movable anchor are covered with adhesive, wrapped with tape, crimped and left under the press until the glue is completely cured. After that, the plate is placed in place in the fixed part of the anchor by putting the holes on the bolts and fixing them with nuts. High-strength bolts and nuts are inserted into the existing holes in the tension anchor. The force created by tightening the nuts is pulled on the lamella, which makes the structure similar to the Sprengel tightening, but the bridge dimension is not engaged and the load-bearing bridge structure is not weighted. After creating forces, the metal parts of the anchors are protected from atmospheric corrosion by painting. The lamella top is covered with an adhesive to protect it from weather and mechanical damage.

4. Summary

1. Major repairs of reinforced concrete bridges with the replacement of all load-bearing structures are not effective, it is relevant only when to achieve the standard load capacity of the bridge, it is enough to replace only the extreme beams of the superstructure.

2. Effectively increase the load capacity of bridge structures, composite materials such as carbon fiber lamellas can be used, but the lamellas should not be glued, but stretched along the lower edges of the beam ribs, and fixed with the help of developed anchors. The use of a tensioning device increases the durability and reliability of the structure.

References

[1] Radic J, Bleiziffer J and Tkalcic D 2005 Maintaining safety and serviceability of concrete bridges in Croatia *Bridge Structures* **1(3)** 327-344.

[2] Makarov A V, Guluev G G and Shalaev S V 2017 Reconstruction of the overpass as a safety requirement *Engineering Bulletin of the don: an electronic journal* **2**. Information on http://www.ivdon.ru.
[3] Code of practice 79.13330.2012 2012 *Bridges and pipes. Rules for surveys and tests* (Ministry of regional development of the Russian Federation).
[4] ODM 218.4.026-2016 2016 Methodological recommendations for determining the load capacity of bridge structures in use on public roads. Concrete and reinforced concrete structures. (Moscow: Rosavtodor).
[5] Salamakhin P M, Makovsky L V and Popov V I 2007 Engineering structures in transport construction. In two books Book 1. (Publishing center "Academy" Moscow).
[6] Kamaitis Z 2002. Damage to concrete bridges due to reinforcement corrosion: Part II-Design considerations. *Transport* **17(5)** 163-170
[7] Migunov V N, Ovchinnikov I G 2010 Modeling the effect of transverse cracks in an aggressive environment on the physical-technical characteristics of reinforced concrete structures *Roads and bridges* **24**. Information on rosdornii.ru/UserFiles/File/dim/24-2/08.pdf
[8] Nikolenko M A, Golovan V Y 2016 Analysis of the causes of defects affecting the load-bearing capacity of artificial structures on the example of the bridge km 1009+279 (right) of the M-4 "Don" highway *Engineering Bulletin of the Don: an electronic journal* 2016 **4**. Information on ivdon.ru/ru/magazine/archive/n4y2016/3800/
[9] Osipov V O, Kozmin Y G, Anciperovsky V S and Kirsta A A 1975 Maintenance and reconstruction of bridges (Transport, Moskow).
[10] Chernyavsky V L, Axelrod E Z 2003 Reinforcement of reinforced concrete structures with composite materials *Housing construction* **3** 15-16.
[11] Makarov A V, Rekunov S S 2019 Strengthening bridge spans by composite materials *IOP Conference Series: Materials Science and Engineering* **687** 7. Information on https://iopscience.iop.org/article/10.1088/1757-899X/687/3/033038/pdf. - Doi:10.1088/1757-899X/687/3/033038
[12] Makarov A V, Kulbin S V 2017 Fastening and tensioning device *Useful model 173025 Russian Federation, MPK E01D 22/00.* (Volgograd: Volgograd State Technical University)
[13] Makarov A V, Kulbin S V 2017 A new way to strengthen bridge spans made of composite materials *Bulletin of the Volgograd state university of architecture and civil engineering. Series: Construction and architecture* **48 (67)** 140-149.

Modeling the stability of horizontal generations worked in parallel to the front of a uniform slope

O A Bogomolova[1*], A V Zhidelev[2], P V Ivanova[3]

[1]Volgograd State Technical University, 28, Lenin avenue, Volgograd, 400005, Russia
[2] Vzlet Branch of the Moscow Aviation Institute (National Research University) in Akhtubinsk, 5, Dobrolyubova Street, Akhtubinsk, 416501, Russia
[3] Don state technical university, 1 Gagarin sq., Rostov-on-Don, 344002, Russia

E-mail: banzaritcyn@mail.ru

Abstract. Improvement of modern urban infrastructure, expansion of transport networks, creation of innovative engineering communications – as well as other circumstances stimulate the development of underground construction, both in our country and abroad. Like any underground structure, it should have the ability not to collapse when exposed to external loads. For the successful operation of such structures throughout their entire service life, it is necessary to constantly assess the influence degree of all factors' totality, one way or another, affect its strength. The task, the purpose of which is to conduct the experimental studies on models made of equivalent materials to assess the geometric parameters' influence of the "slope – underground working space" system on the stability of the latter, has been set. Geometric parameters mean the shape of the underground working space cross-section and its position in the soil massif: the shape of the underground working space cross-section is round and semi-elliptical; the underground working spaces' bottom is at the level of the slope bottom, the underground working space position is determined by the shortest distance from the transition point of the bottom to the slope to the vertical axis of the cross-section symmetry. The criterion for long-term stability is a criterion, the qualitative feature of which is the absence of inelastic (plastic) deformation zones on the development contours, and the quantitative indicator is the value of the reduced pressure of connectivity σ_{con}. As a result of model experiments, safe distances d were determined, on which underground working spaces should be located in the near-slope area so that the stability of the "slope - underground working space" system is ensured. It turned out that these distances for underground working spaces of both circular and semi-elliptical cross-section are practically the same, that is, the underground working spaces of such a cross-section, all other sections being equal, were equally stable. Practical coincidence of the model experiments' results and numerical finite element analysis, calculations, allows us to speak about the experimental data reliability.

Introduction

The underground space development, associated with the transport infrastructure expansion in large cities due to population growth and an increase in the road traffic volume, is proceeding at a rapid pace, which ensures an increase in the underground construction volume.

The issues related to the underground space development are quite relevant, as evidenced by the large number of scientific publications on this topic and the growth of the planned volume of construction work. For example, in the city of Volgograd, due to the peculiarities of the relief, the tram lines are partially underground. Now the length of the underground section is more than 7 kilometers, and in the near future it will be extended by another 4 km.

Figure 1. Underground space of high-speed tram stations in the city of Volgograd

The cost of underground construction is quite high, which is associated with the need to implement a number of measures to ensure the stability of the underground complex and aboveground structures. It is possible to reduce the corresponding costs using the special techniques and calculation methods that allow determining the safe parameters of underground working spaces and ensuring the safe operation of facilities in the future. The stability of underground structures (underground working spaces) depends on a number of factors: geometric dimensions, cross-sectional shape, features of the geological structure and relief, external loads, physical and mechanical properties of the enclosing soil, and much more.

One of the important tasks in underground construction, in particular, in the construction of transport tunnels in mountainous terrain, is to determine the position of the underground working space in the slope region, which would guarantee the stability of the "slope – underground working space" system without taking any additional measures to strengthen it.

In this regard, we carried out the experiments aimed at determining such geometrical dimensions and such a position of unsecured horizontal underground working spaces of a circular and semi-ellipsoidal cross-section, in which the plastic deformation areas or destruction zones do not appear on their contour of underground working spaces. We call such geometrical parameters safe and consider the production itself stable when these parameters are maintained.

During the experimental studies, the slope models were formed in an experimental tray with the following dimensions of 1.0 × 1.79 × 0.105 m, the side faces of which were made of 8 mm thick plexiglass (see Figure 1*a*).

In the slope models, the voids were created (using the laid-out templates), simulating underground working spaces with a cross-sectional shape in the form of a circle and half an ellipse. The position of the underground working spaces was determined by the distance *L*, measured by the shortest distance from the slope transition line to its base to the vertical axis of the underground working space symmetry.

When forming the models from an equivalent material, the value L=0,15H; 0,2H; 0,25H; 0,3H, and the slope model angle β=60°. The base width of the semi-elliptical underground working space is taken equal to the diameter of the circular underground working space d=0,1H. The geometric parameters, dimensions of the model are given in fractions of its height H=0,6 m. (see Figures 2*c*, 2*d*).

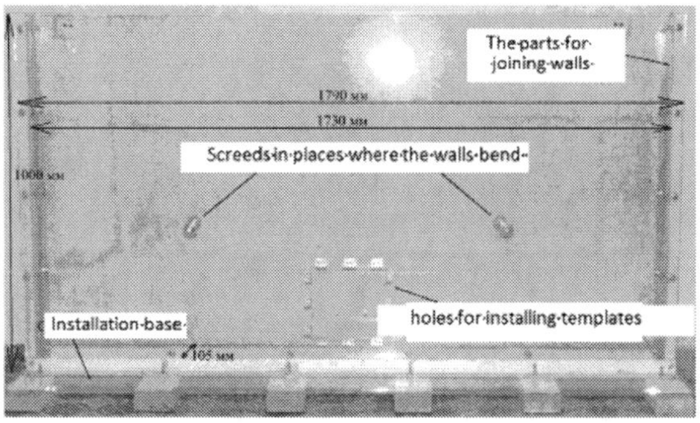

a)

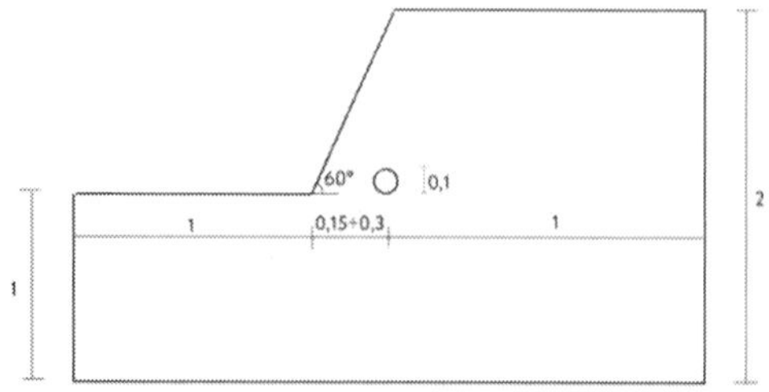

b)

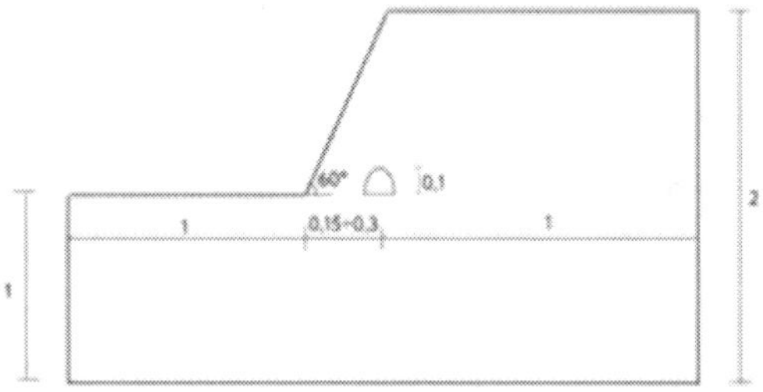

c)

Figure 2. Photo of the tray for carrying out model experiments (a); geometric dimensions and parameters of the model (in H shares) underground working spaces with a round cross-section (*b*) and semi-elliptical (*c*) shape

A sand-oil mixture was used as an equivalent material from which the slope models were formed, the shear characteristics of which in the unconsolidated and compacted state were established by testing at consolidated shear in the GGP-30 box shear apparatus (Figure 3a). The tests were carried out in the soil mechanics laboratory LLC "Radian".

Shear diagrams are shown in Figures 3 *b*, *c*, the numerical values of physical and mechanical properties - respectively in Tables 1 and 2.

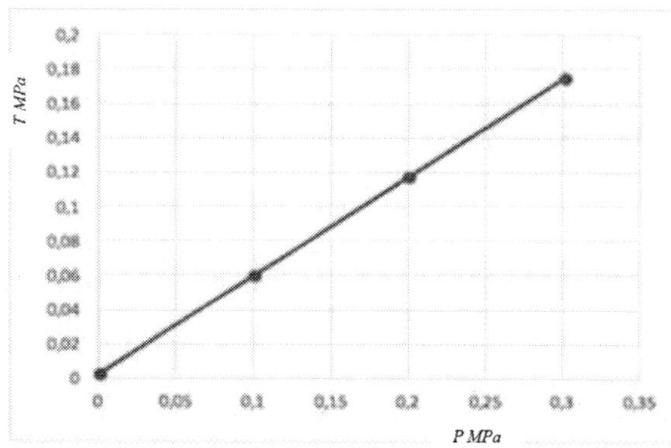

b)

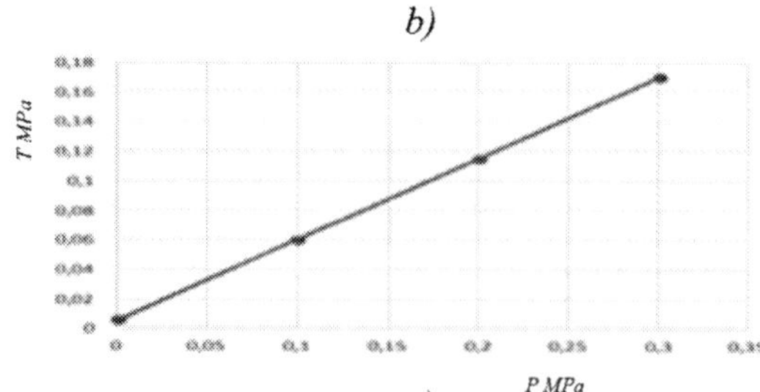

a) *c)*

Figure 3. Box shear apparatus GGP-30 *(a)* and shear diagrams of unconsolidated and compacted equivalent material (*b, c*)

Table 1. Physical and mechanical properties of an equivalent material in consolidated state

P, [MPa]	τ, [MPa] value		tg φ	φ	Sh, [kPa]
	experiment	calculation			
0.1	0.06	0.06	0.55	28°49'	5
0.2	0.115	0.115			
0.3	0.17	0.17			

Table 2. Physical and mechanical properties of an equivalent material in unconsolidated state

P, [MPa]	τ, [MPa] value		tg φ	φ	Sh, [kPa]
	experiment	calculation			
0.1	0.06	0.06	0.575	29°54'	2.5
0.2	0.118	0.118			
0.3	0.175	0.175			

The average value of the equivalent material bulk density of the models is determined to be γ=17,2kN/m^3; and the value of its lateral pressure coefficient, found on the basis of the method of K. Terzaghi [1-3] - ξ_o=0,4.

Slope models were generated as follows. On the front wall of the box, the contour of the future model was outlined with colored chalk, taking into account the geometric dimensions and parameters adopted during the modeling. At the first stage, a layer of equivalent material was laid to a height from the bottom of the chute to the lower boundaries of the underground working space templates fixed on the walls of the experimental chute. Subsequent layers, approximately 10 cm thick, were laid alternately with a time delay of about 30 minutes after the previous layer completion of the equivalent material for natural compaction. A tray filled in this manner was left for three hours to allow gravitational compaction of the equivalent material. After this time, a careful extraction of soil from a part of the chute was carried out (unloading, underground working space off the slope), and a slope model was formed. After the formation of the model, it was photographed, and it was again left for three hours to redistribute the stresses in connection with the model "unloading". After that, a careful extraction of the box-template was carried out, photographing of the model, violations' fixation of its integrity and the state of the underground working space imitation. Photos of the models taken before and after removing the box are shown in Figures 4 and 5.

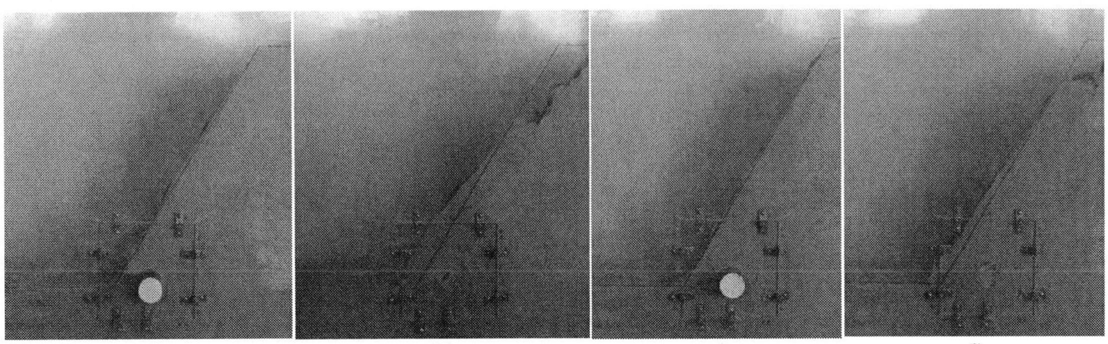

a) *b)* *c)* *d)*

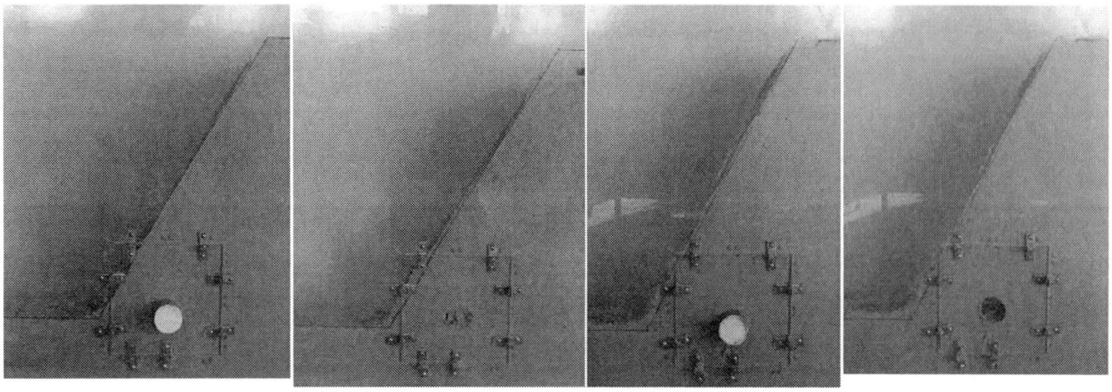

e) *f)* *g)* *h)*

Figure 4. Photographs of a slope model made of equivalent material, undermined by a round horizontal underground working space, before and after removing the template box at the distances L=0,15H (*a; b*), 0,2H (*c; d*), 0,25H (*e; f*), 0,3H (*g, h*).

The results show that for $L \geq 0.3H$ the stability of underground working spaces with a round and semi-elliptical cross-sectional shape is ensured with almost the same values *d* provided that the physical and mechanical properties of the equivalent material of the model and the transverse dimensions of the simulation underground working spaces are the same.

In addition to physical modeling of the stability loss process of the "slope - underground working space" system, computer modeling of this process was performed. For this, the computer program "Stability. Stress state" [4], developed with the participation of one of the authors of this work, was used. To calculate stresses, the program formalized the finite element method (FEM). The program gives a possibility to use the stability criterion [5; 6], the qualitative feature of which is the absence of inelastic deformations areas on the development contour, and the quantitative indicator is the value of the reduced pressure of connectivity σ_{con}=C$(\gamma H tg\varphi)^{-1}$, where: C, γ, φ – are respectively specific adhesion, angle of internal friction and volumetric weight of equivalent material; H – is the height of the slope model. To find the shear ranges (APD), the program uses the Coulomb strength condition [7] in the form presented by G. Cacko [8].

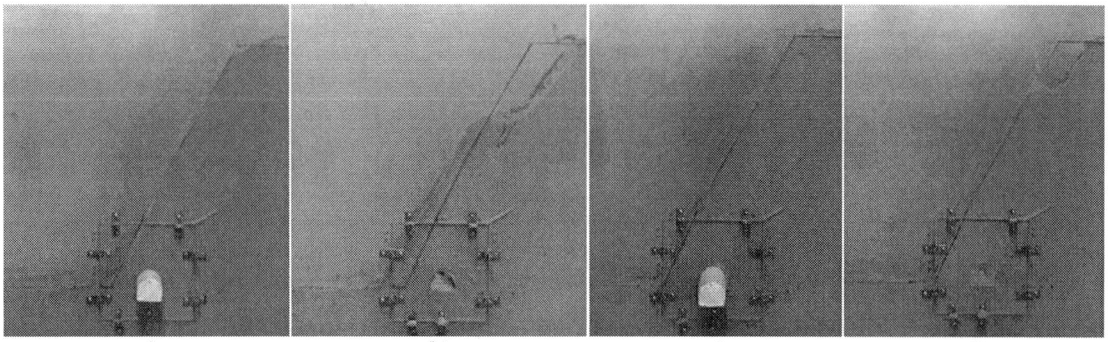

a) *b)* *c)* *d)*

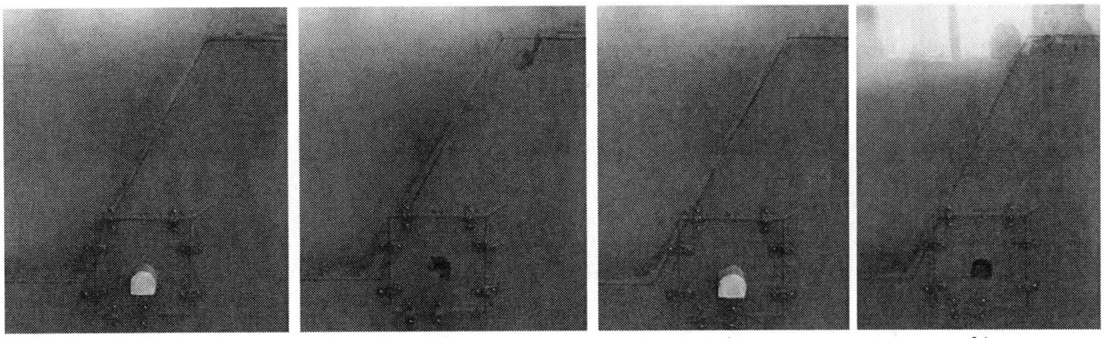

e) *f)* *g)* *h)*

Figure 5. Photographs of a slope model from an equivalent material, undermined by a semi-elliptical horizontal underground working space, before and after extraction of the template box at the distances L=0,15H (*a; b*), 0,2H (*c; d*), 0,25H (*e; f*), 0,3H (*g, h*)

Figures 5 and 6 graphically represent the computer simulation results. It turned out that the values *L*, calculated using a computer program [4] differ by no more than 7-15% from the similar distances obtained experimentally.

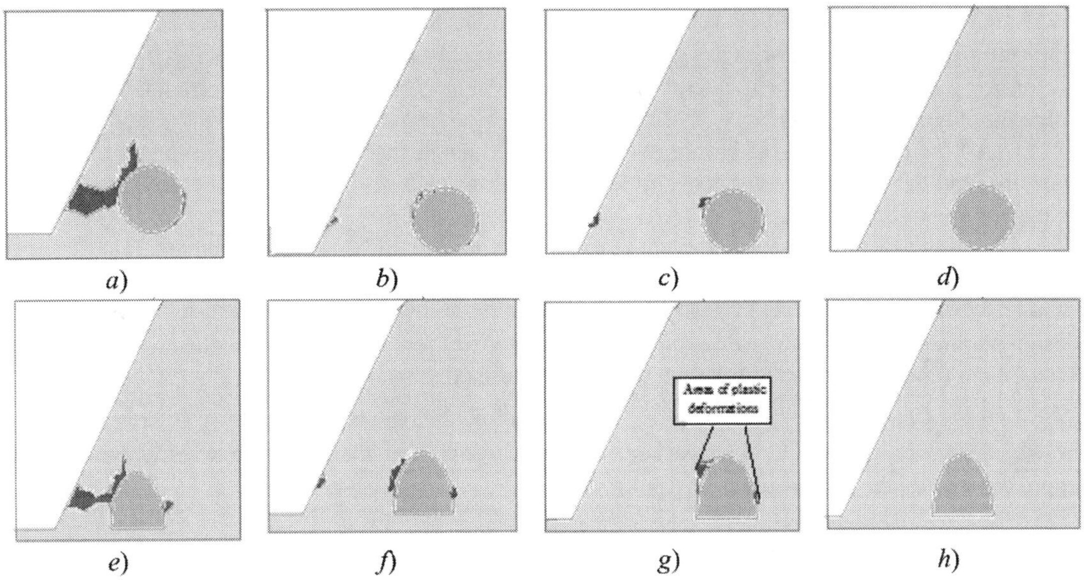

Figure 6. Pictures of plastic deformation areas of underground working spaces of round and semi-elliptical shape, for the distances L=0.15h (*a; e*), L=0.20h (*e; f*), L=0.25h (*c; g*), L=0.3h (*d; h*).

This circumstance gives an opportunity to conclude that it is possible to use the program [4] to solve similar problems.

Based on this premise, the buckling process computer simulation of single horizontal underground working spaces with a circular cross-section of variable diameter was performed *d*=(0,05; 0,15; 0,2; 0,25; 0,3)*h*, which are alternately located at the distances $L=h$; $L=h+((2h/tg\beta-h)/2)$; $L=2h/tg\beta$ from *A* points of the slope transition to the bottom (see the design diagram in Figure 7).

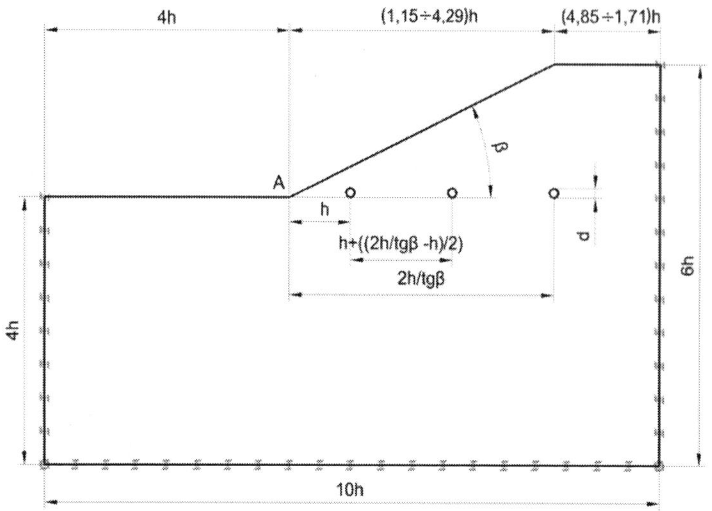

Figure 7. General view of the design FEM scheme, its geometric parameters and boundary conditions

The slope angle takes three values β=25°; 45°; 60°, and the value of the reduced pressure of connectivity in the calculation is assigned as six values σ_{con}=0,1;0,5;1; 2; 2,75; 3,5. All calculations are performed with the value of the lateral pressure coefficient ξ_o=0,75, which corresponds to the average value for clayey soils [9-11].

All the calculations, as before, were performed using the computer program "Stability. Stress state" [4] and the FEA computer program [12], which was also developed with the participation of one of the authors of this article.

The view of the FEM design scheme with the main geometric dimensions and boundary conditions is shown in Fig. 7.

As a result of the calculations performed, for all possible combinations of numerical values of variable design parameters, the "safe" values of the soil internal friction angles of the enclosing mass φ_s, which ensure the presence of a qualitative trait, have been defined.

Based on the calculation results analysis, the graphical dependencies of the form $\varphi_s = f(\sigma_{\text{con}})$, some of which are shown in Fig. 8, have been built. It was found that all the curves obtained can be approximated by the expression:

$$\varphi_s = a\sigma_{con}^{b}, \tag{1}$$

where: a and b – are the real coefficients, moreover, the coefficient a has the dimension [deg], and the coefficient b – is a dimensionless quantity.

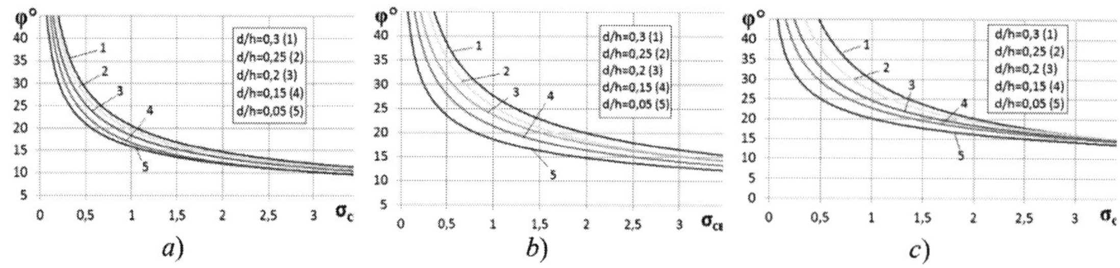

Figure 8. Graphical dependencies view $\varphi_s = f(\sigma_{\text{con}})$ at different values of the underground working space diameter β=25° (*a*); β=45° (*b*); β=60° (*c*) at a distance from the center of the underground working space to the point *A* *L*=*h* (*a*); *L*=*h*+((2*h*/*tgβ*–*h*)/2) (*b*) and *L*=2*h*/*tgβ* (*c*) respectively.

As a result of the numerical values' analysis of the coefficients included in the expression (4), the graphs were built to determine their numerical values, which are shown in Figures 9; 10. The curves' analysis shown in Fig. 9 shows that the change in the relative (in fractions *h*) underground working space diameter from 0.05 to 0.3 for all the considered distances from the center of the underground working space to the point *A*, entails a change in the value of the coefficient *b* for 7-18%.

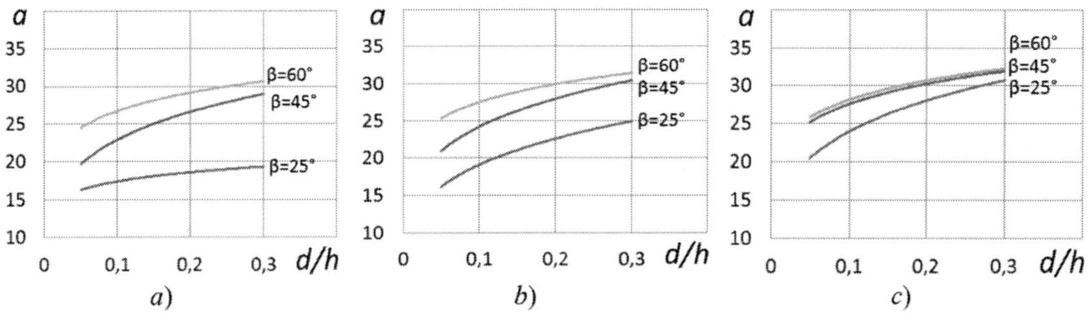

Figure 9. Graphs for determining the numerical value of the coefficient *a* depending on the relative underground working space diameter at *L*=*h* (*a*); *L*=*h*+((2*h*/*tgβ*–*h*)/2) (*b*) and *L*=2*h*/*tgβ* (*c*).

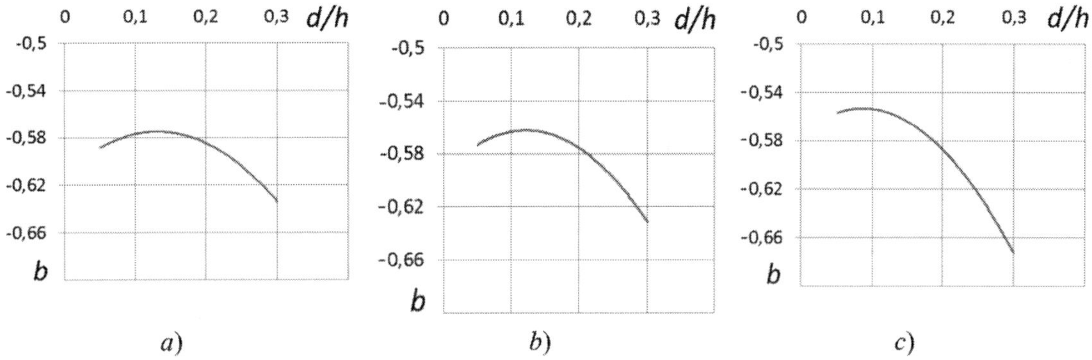

Figure 10. Graphs for determining the numerical value of the coefficient *b* depending on the relative underground working space diameter at *L*=*h* (*a*); *L*=*h*+((2*h*/*tgβ*–*h*)/2) (*b*) and *L*=2*h*/*tgβ* (*c*).

As a result of the additional calculations, it was established: if we take the value of the coefficient *b* constant and equal *b*=-0,62, then the numerical values of the quantity φ_s, obtained on the basis of formula (4) and the graphs shown in Fig. 9 and 10 will differ from the corresponding values φ_s, calculated by the formula

$$\varphi_s = a\sigma_{con}^{-0,62} \quad (2)$$

using the graphs shown in Fig. 8, no more than 10%, moreover, this difference goes into the stability margin.

Summary

1. The stability of a horizontal underground working space depends on the geometrical parameters of the slope and the underground working space, its location, physical and mechanical properties of the

IOP Conf. Series: Materials Science and Engineering 1083 (2021) 012070 doi:10.1088/1757-899X/1083/1/012070

constituent soil. The closer the underground working space is located to the base of the slope, the more intensively, all other things being equal, the process of destruction of its contour proceeds.

2. Graphical dependencies and analytical expressions, which together represent an engineering method for a preliminary assessment of the stability of a horizontal underground working space of a circular cross-section worked out in a homogeneous soil slope at the level of its bottom, have been obtained.

3. Verification calculations showed that the numerical values of the quantity φ_s friction obtained on the basis of the proposed engineering method differ by no more than 10.2% from the results obtained by direct use of the computer programs [4; 12], and the error goes into the stability margin.

References

[1] Terzaghi K 1961 Soil mechanics theory (Gosstroyizdat, M.).

[2] Terzaghi K, Peck R 1958 Soil mechanics in engineering practice (Gosstroyizdat, M.).

[3] Terzaghi K 1933 Structural soil mechanics based on its physical properties (Gosstroyizdat, Moscow, Leningrad,).

[4] Bogomolov A N et.al. 2009 Stability (stress-strain state): certificate of state. registration of computer programs № 2009613499. № 2009612297; statement 19.05.2009; registered in the Register of computer programs 30.06.2009. – 1 p.

[5] Bogomolov A N, Nestratov M Yu, Solovyev A V, Bogomolova O A 2009 A new criterion for assessing the long-term stability of homogeneous slopes based on the analysis of the stress-strain state *Bulletin of the Volgograd State University of Architecture and Civil Engineering. Series: Building and architecture. - Volgograd: Publishing house of VolgSACEU* **14 (33)** 5-12.

[6] Bogomolov A N, Bogomolova O A, Shiyan S I 2009 On two criteria of stability of soil slopes and slopes *Modern technologies in construction Theory and Practice: Materials of All-Russian international scientific-practical conference, dedicated to the 50th anniversary of the Construction department of the Perm State Technical University, 22—23 sept. 2009 г.). Perm: PSTU* 54 – 58.

[7] Coulomb C 1773 Application des riles de maximus et minimis a quelques problemes de statique relatifs a L`architecture (Memories de savants strangers de L`Academlie des sciences de Paris).

[8] Cagout G 1934 Eguilibre des massifs a frottemenet interne (Paris).

[9] Vyalov S S, Zaretsky Yu K 1962 Strength and creep of frozen soils, and calculations of ice-ground fences (USSR Academy of Sciences, Moscow).

[10] Vyalov S S 1978 Rheological foundations of soil mechanics (Higher School, Moscow).

[11] Vyalov S S 1959 Rheological properties and bearing capacity of frozen soils (USSR Academy of Sciences, M.).

[12] Bogomolov A N [et.al.] 2015 FEA: state registration patent for the computer program No. 2015617889. Registered. July 23, 2015.

Evolution of loading paths in a heterogeneous soil mass during the foundation pits' construction

Z G Ter-Martirosyan, A Z Ter-Martirosyan, Yu V Vanina*

National Research Moscow State University of Civil Engineering (NRU MGSU), 26, Yaroslavskoye highway, Moscow, 129337, Russia

E-mail: yuli.jullies@gmail.com

Abstract. This article shows the construction of loading paths for the soil mass points using the example of a phased construction of the subway structure underground part. The solution to the problem was obtained by the finite element method using a PC PLAXIS 2D. The graphs of changes in the points' stress state in coordinates p'/q are shown.

1. Introduction

The Moscow metro construction history has almost a century-long history. The stations of the first stage were built mainly shallow in pits up to 12 m deep. Along with the subway construction underground space development, the theoretical and applied science of soil mechanics, which today is the main analysis tool in the new station complexes and running tunnels' design, as well as other underground construction projects, has been developed.

At present, the formation and development of theories for describing the soils mechanical behavior mathematical models based on solutions of the theory of elasticity continues. The appearance of the CamClay mathematical soil model (the Cambridge clay model) in the 1950s made it possible to better predict the stress paths in comparison with the linear-elastic and elastic-plastic models Linear Elastic and Mohr-Columb. On its basis, a separate direction in the models' development was identified with the introduction of an independent law of soils behavior during deformation, while the shear hardening region develops independently of volumetric deformations. Such models include the soil model of Zaretsky Yu.K. [1], as well as the elastic-plastic model with isotropic Hardening Soil and its modifications. Hardening Soil and Hardening Soil Small are often used options for designing pit barriers. The Hardening Soil model includes deviatoric loading, which induces shear hardening, and oedometrique loading, which induces compressive hardening.

There are two main types of hardening: shear hardening and compression hardening. Shear hardening is used to model irreversible deformations resulting from primary deviatoric loading, while compression hardening is used to model the irreversible plastic deformations caused by primary compaction under oedometrique and isotropic loading. In the Hardening soil model, the shear strength is determined using a triaxial E_{50} deformation modulus, compressive hardening - oedometrique E_{oed}. The soil mass behavior is processed during numerical modeling by constructing stress paths. For the behavior of the laboratory sample, the stress paths are shown in Figure 1 [8].

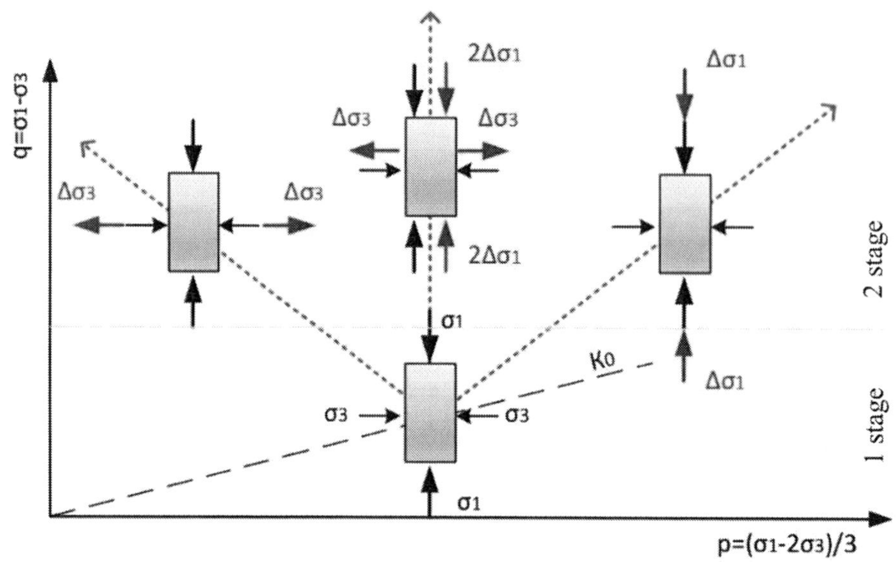

Figure 1. Basic paths of drained loading in a triaxial tool

According to Figure 1, after the stage of anisotropic loading with vertical pressure exceeding the lateral pressure, the further path of stresses can change in three ways [8]:

1) with a decrease in lateral pressure with a constant vertical - the path deviates to the left;

2) with an increase in vertical pressure with a simultaneous decrease in lateral pressure with a constant ratio of them (constant shear stresses) - the vertical position of the line (an increase in the vertical load is equal to a double decrease in the horizontal load);

3) with an increase in only vertical pressure with a constant lateral pressure - the path deviates to the right.

2. Formulation of the problem

To determine the stress state for some points of the soil massif, a geotechnical calculation was carried out for a stage-by-stage construction of a pit for the assembly-shield chamber (ASC) construction using the PLAXIS 2D software package [11].

The main parameters of the design scheme (Figure 2) are: the enclosing structures of the pit for the construction of the assembly-shield chamber (ASC) are provided in the form of a monolithic reinforced concrete "wall in the ground" of a trench type 800 mm thick. The length of the "wall in soil" is 39.5 m. The depth of the pit is 25.6 m. The stability and spatial invariability of the pit fence is ensured by a spacer system consisting of 5 tiers of spacers and distribution belts. Spacer installation step is 4 m.

The geotechnical model of the foundation was adopted on the basis of analysis and generalization of the materials from engineering and geological surveys carried out at the construction site. In mathematical modeling, the Hardening Soil and Mohr-Coulomb model were used to describe the mechanical behavior of the soil. The base is represented by sandy soils with water-saturated layers. At the excavation stage for the 4th tier, dewatering works are carried out for further development of the soil.

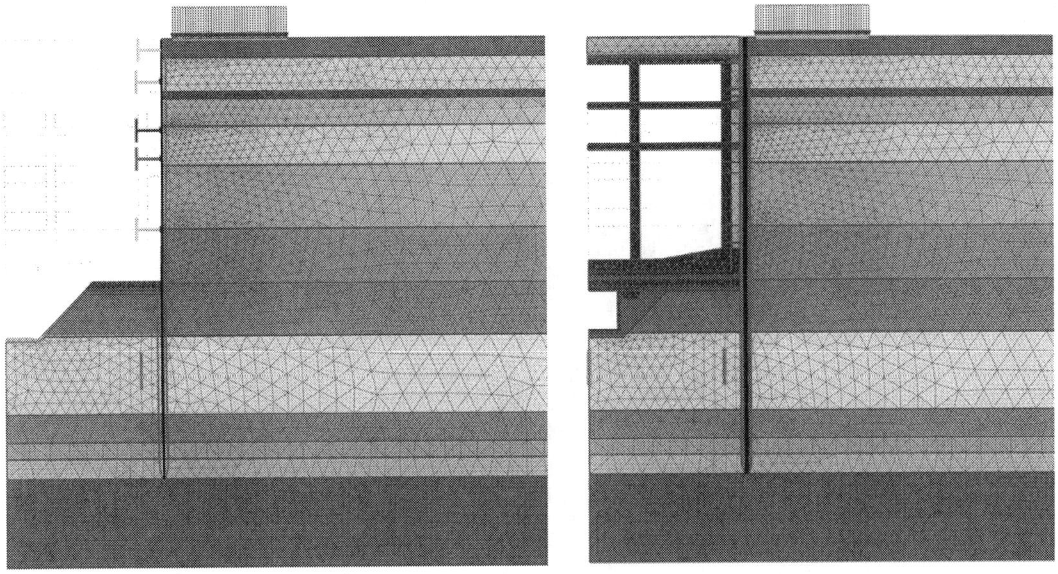

(a) (b)

Figure 2. Design scheme in PC PLAXIS 2D: a) at the stage of excavation development to the bottom with the installation of the 5th tier; b) at the stage of dismantling the 1st tier and backfilling

To study the stress state formation during step-by-step development with the installation of a spacer attachment and the erection of structures with the subsequent removal of the spacer system tiers, the most characteristic points that were within the collapse prism have been selected (Figure 3). The picture of the soil mass stress state is presented as follows (Figures 4-5.).

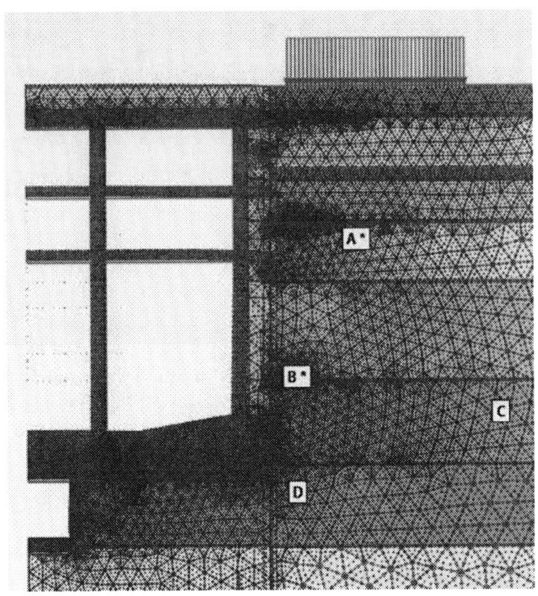

Figure 3. Characteristic points for studying stress paths.

In PC PLAXIS 2D the stress state is defined in the form of the following areas [11]:

1. **Failure points** (fracture point, red) - points of the Coulomb-Mohr strength condition violation, which is associated with the shear deformations development;

2. **Cap points** (hydrostatic reduction point, blue) – the points correspond to isotropic (hydrostatic) reduction;
3. **Hardening points** (hardening points, green) – the points correspond to shear hardening
4. **Cap + Hardening points** (brown) – the points of hardening on the shear hardening envelope and on the hip-roofed surface.

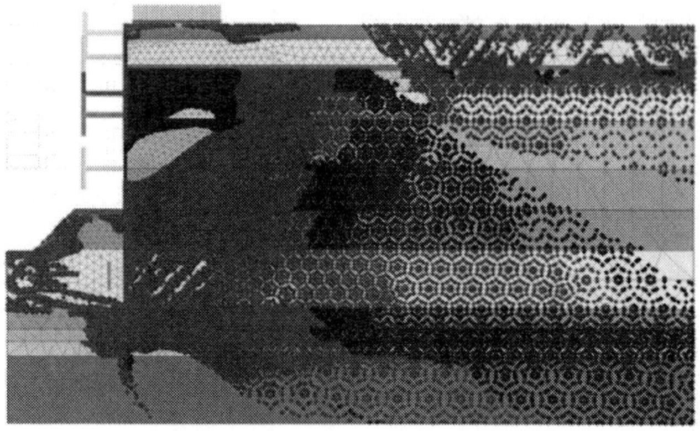

Figure 4. Areas of stress state at the complete excavation stage.

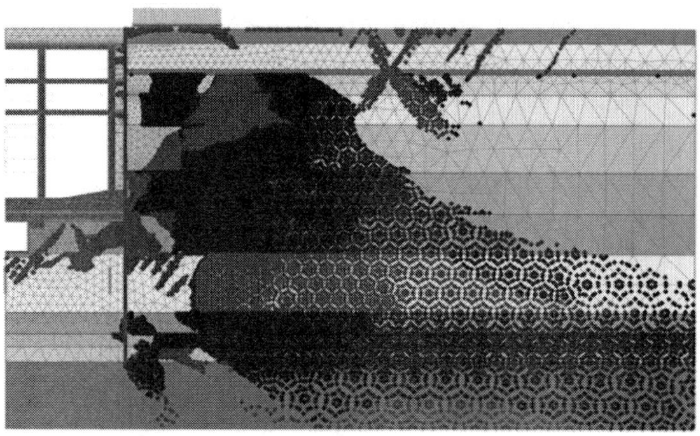

Figure 5. Areas of stress state at the stage of dismantling the 1st tier of the spacer attachment and backfilling.

Further, the paths of stresses for the selected points were constructed with a description of the model behavior in space p'-q (Figures 6-8). The beginning of each new phase is indicated by large markers. The construction stages are presented in the following sequence:

- Stage 1 - Recess for the 1st tier of the spacer attachment;
- Stage 2 - 5 - Consecutive cuts for the spacer fastening tier followed by the foundation pit installation;
- Step 6 - Berm Development and Temporary Tray Design;
- Stage 7 - Dismantling of the temporary tray, pit excavation to the pit level with the installation of the 5th tier.
- Stages 8-12 - Consecutive erection of structures with the subsequent dismantling of the spacer fastening tiers

3. Solution of the problem

Figures 6 to 9 show the loading paths for the points A-D.

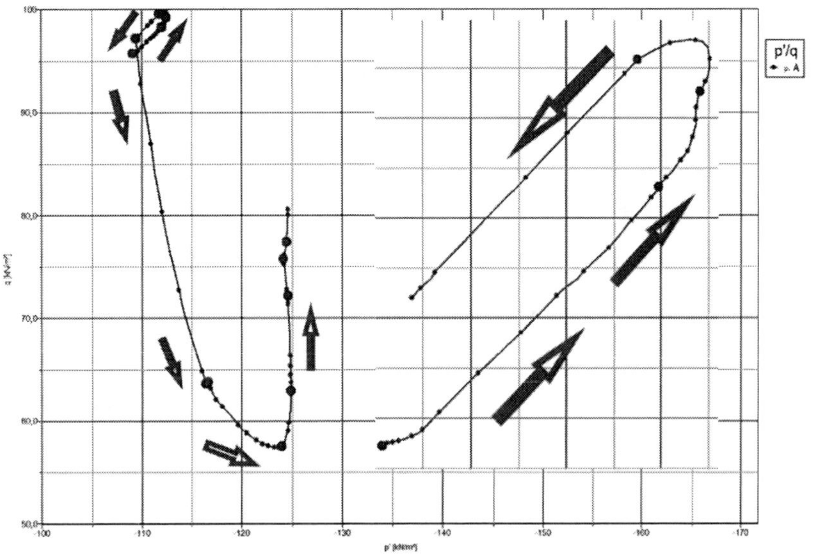

Figure 6. Loading paths of point A at the stages 1-12 in axes p'/q.

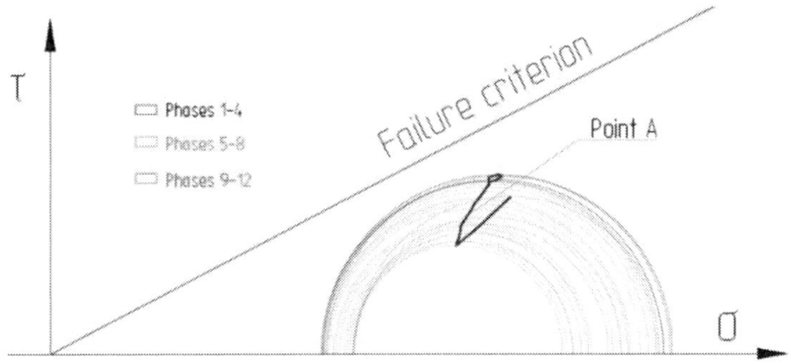

Figure 7. Theoretical construction of the loading path of point A at the stages 1-12.

At the stages 1-2, point A is characterized by an increase in vertical stresses and a slight increase in horizontal stresses, which leads to a deviation of the stress path to the right. At the stages 3-5, the lateral pressure decreases at a constant vertical pressure, which manifests itself in the form of a shift of the stress path to the left (since the point is at the level of the notch for the 3rd tier of the spacer). By stage 7, the stress path tends to the right with a decrease in the radii of Mohr's circles, as the foundation is unloaded to the level of the pit. At the stages 8-12, with repeated loading, an increase in lateral pressure occurs with a slight increase in vertical - the stress path tends vertically up.

Point A at all stages corresponds to isotropic (hydrostatic) reduction. As shown in Figure 6, the path of isotropic compression is not a strictly horizontal line, which is associated with the influence of the cuts' dimensions and shapes in the design model, the introduction of spacer fastening structures, as well as edge effects, boundary conditions. However, for point A, the displacement of the path to the right determines the type «Cap» in the stress state picture.

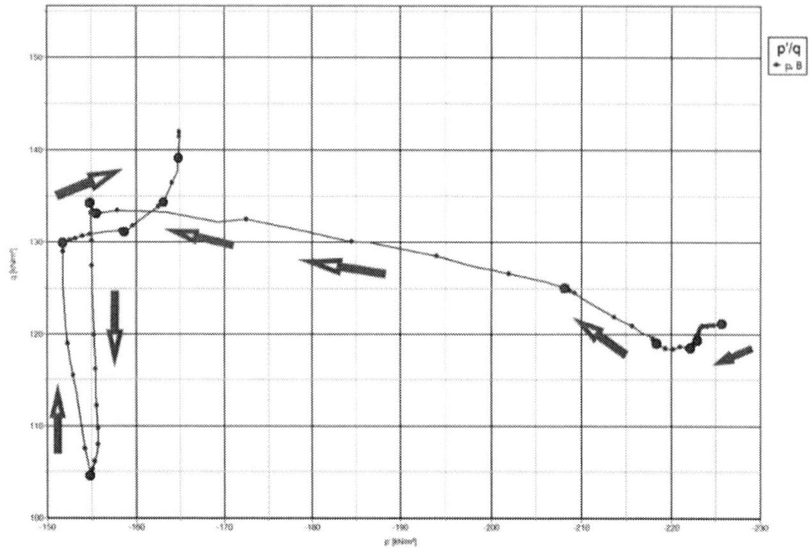

Figure 8. Load paths of point B at the stages 1-12 in axes p'/q.

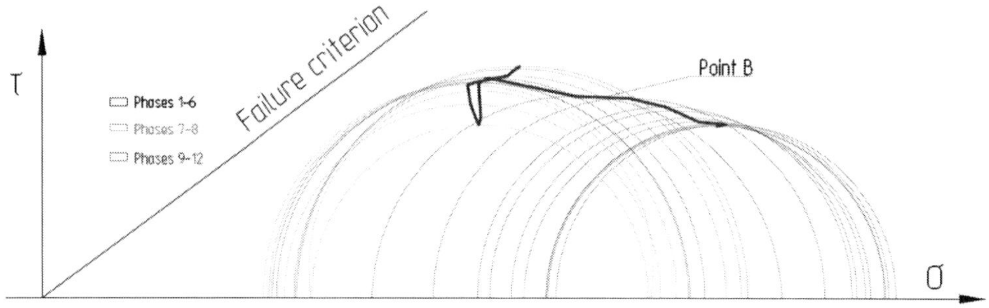

Figure 9. Theoretical construction of the point B loading path at the stages 1-12.

Point B is notable for the path displacement to the left at the stages 1-6, as evidenced by the decrease in vertical pressure with increasing lateral pressure (with a phased unloading of the foundation). At the stage 7, a change in the stress path to a vertically directed downward (unloading of the base to the level of the pit) is observed with a decrease in the radii of Mohr's circles. At the stage 8, the lateral pressure increases with a slight increase in the vertical pressure - the path deviates to the left. In general, for point B, the path displacement to the left at the stages of the forward stroke (stages 1-7) determines the type of "Hardening point" for it in the stress state picture, which corresponds to shear hardening. However, at the stages 9-12 with an increase in vertical and lateral pressures, the stress path deviates to the right, which determines the type of "Cap + Hardening point" for it in the stress state picture.

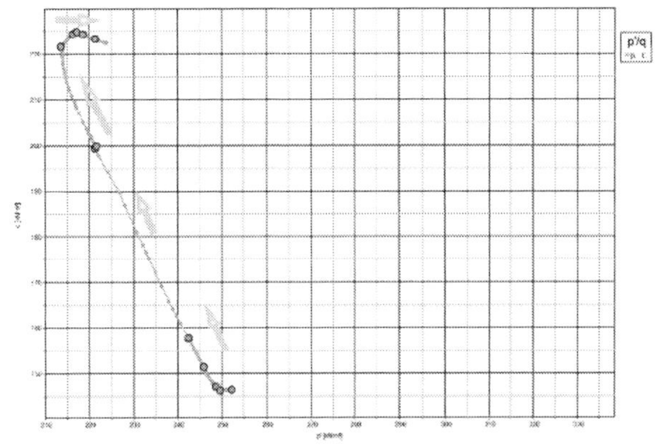

Figure 10. Trajectories of point C loading at the stages 1-12 in axes p '/ q.

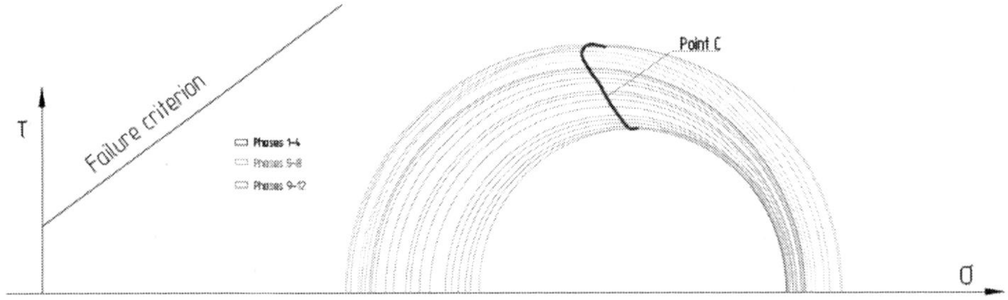

Figure 11. Theoretical construction of the point C loading path at the stages 1-12.

Point C, similar to point B, changes the stress state type during the forward and reverse stages. However, here there is a change from the type "Hardening point" (the path displacement to the left) to the type "Cap + Hardening point" (the path displacement to the left with the same ratio of increments of q and p ') at the stage 3, and at the stages 8-12 there is a change to the type Cap (the path displacement to the right) due to repeated subgrade loading.

Point D (Figures 12-13) is characterized by a shift of the loading path to the left. At the stages 9-12, with an insignificant increase in vertical and lateral pressures, the stress path deviates to the right, but in the stress state picture at all stages it is characterized by the "Hardening point" type.

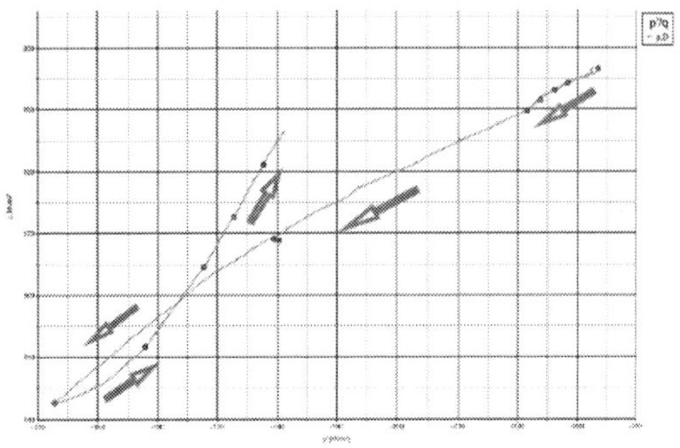

Figure 12. Loading paths of point D at the stages 1-12 in axes p'/ q.

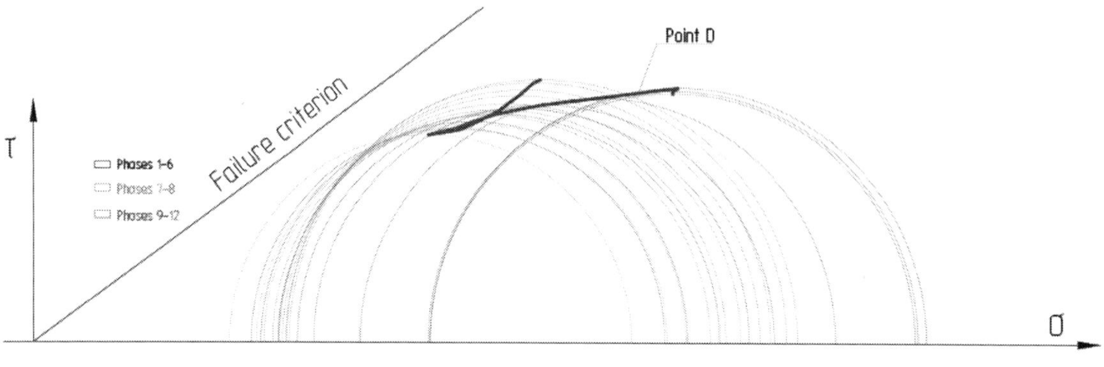

Figure 13. Theoretical construction of point B loading path at the stages 1-12.

4. Summary

As a result of numerical scheme modeling of the pit development stages with the installation of spacer tiers and the erection of structures for the underground part of the metro structure, the main loading paths for the points located in the collapse prism have been obtained. Analysis of the solution results shows that the stress state for the same point can be transformed at the stages 1-7 and 8-12, which illustrates the construction of loading paths.

Determination of the stress state fields at the soil mass points is necessary to assess the deformability of soil mass layers, to determine the effect on nearby buildings, underground structures and utilities [2].

References

[1] Zaretsky Yu K, Lombardo V N, Groshev M E 1979 Plastic flow of soil massifs *News of HEI. Construction and architecture. Novosibirsk* **2** 3-24.

[2] Zaretskiy Yu K, Kabantsev M I 2006 Influence of the sequence of erection of nearby high-rise buildings on precipitation and roll of foundation slabs *MGSU Herald* **1** 51-56.

[3] Karaulov A M, Korolev K V, Galter D S 2018 Ultimate pressure of a round stamp on a base in cramped conditions // News of HEI. Construction **12 (720)** 23-30.

[4] Paramonov V N 2012 The finite element method for solving nonlinear problems in geotechnics (SPb.).

[5] BC 22.13330.2011 Foundations of buildings and structures. Updated edition of SNiP 2.02.01-83 *. Moscow: Ministry of Regional Development of the Russian Federation, 2011.

[6] BC 120.13330.2012 Subways. Updated edition of SNiP 32-02-2003 (with Amendments N 1-4)

[7] Ter-Martirosyan Z G, Ter-Martirosyan A Z 2020 Soil mechanics in high-rise construction with a developed underground part (ASV, M.).

[8] Fedorenko E V 2016 Calculation of subgrade on a weak foundation in the PLAXIS program *Designing the development of a regional network of railways* **4** 393-398.

[9] Tsytovich N A, Ter-Martirosyan Z G 1981 Fundamentals of applied geomechanics in construction (Higher school, M.).

[10] Plaxis 2D. User guide. 2017 - 1021 p.

IOP Conf. Series: Materials Science and Engineering 1083 (2021) 012072 doi:10.1088/1757-899X/1083/1/012072

Improving the methodology of organizing the parking process in large cities of the world

Aleksandr Sekisov [1*], Leonid Ivanov [2], Alexander Begiashvili [3]

[1]Kuban State Agrarian University I.T. Trubilin, 13, Kalinina str., Krasnodar, 350044, Russia
[2]Russian Engineering Academy, building 4, 9, Gazetny per., Moscow, Russia
[3]Perm National Research Polytechnic University, 29, Komsomolsky ave., Perm, Russia

Email: alnikkss@gmail.com

Abstract. It has not been more than a century since cars became a global phenomenon, and the possibility of unobstructed parking has gone from being a worry-free situation to a big problem. At first, it was not considered an important matter, then it was a topic for discussion, and later most of the parking lots became an object of commercial and logistic interest, an object of profit by any means, even to the human environment detriment [7, 8, 13]. Currently, the ability to own and dispose a parking slot is perceived by the subjects of public and economic relations as a significant advantage, and often even more significant than the possession of a vehicle. In different states, the evolution stages of this issue were uneven, but in general the trends were the same - an increase in the number of parking zones and an increase in their capitalization.

Introduction

In the 21st century, during the active development period of modern information and automated technologies aimed at accelerating the productive effect, and, at the same time, closely related to the new humanistic principles of the post-industrial era, new approaches aimed at improving the life quality of people in every possible way while maintaining or even improving their habitat have emerged [1, 4, 5, 9]. One of such approaches was the tubing parking principles' development, which can significantly save the consumer's money, as well as optimize the use of city squares used in single-level parking areas. In this regard, this paper considers the issues of improving the methodology of organizing parking slot process in cities based on the tubing parking principles.

Main part

Every year there are more and more cars in the big cities of the world. Due to the outstripping growth rates of the population's motorization over the parking slots provision for cars, the problem of the parking slots' lack for parking the vehicles not only in the central part of the city, but also in its peripheral areas is becoming more and more urgent.

Our analysis of various studies in this area indicates the existence of a whole scientific archive for solving the problems associated with parking slot. In particular, in his research: "The high cost of free parking" by the professor at the University of California at Los Angeles Donald Shoup focuses on the fact that “if the problems with parking slot are not regulated, then after a certain period of time an area of an England size will be needed for the world car fleet location” [3].

The relevance of the parking slot problem is also proved by the results of the IBM Global Parking Survey [2], in which 9055 motorists in 25 cities of the world were interviewed. It was discovered that:

- at least 40% of traffic jams in large cities in the world arise from the fact that drivers in search of parking slot to park a vehicle create obstacles on the roadway;
- 8 out of 10 interviewed respondents are often forced to refrain from looking for a parking slot for a short-term parking of their vehicle and move to another place;
- more than 1/4 of the respondents came into conflict with other drivers over parking slot.

Due to the lack of space for short-term storage of vehicles, drivers use the carriageway of the city, the parking time ranges from 20-30 minutes to 6 or more hours. At the same time, the organization of “unregulated parking lots” has bad consequences for all road users: one parked vehicle on the stretch of the carriageway in high traffic conditions creates obstacles for 700-800 vehicles every hour. Traffic jams are the result of such chaotic parking. Also, due to the lack of parking slot in the courtyards, residents of the city park their vehicle on the roadways and thereby create obstacles for moving cars [11].

In Russia, the situation is aggravated by the fact that the Soviet designers of roads and residential areas could not predict that by the beginning of the 21st century there will have been many times more cars per 100 people. The number of cars per capita is growing, if not exponentially, then in arithmetic progression [7, 8, 11].

Analysis of the survey results carried out on city streets showed the following.

1. The carriageway is used by about 80% of all parking cars, partly on the carriageway and sidewalk - 15%, completely on the sidewalk - 5%.
2. The average density of passenger car parking reaches 390 vehicles / km on the trunk network, and 280 vehicles / km on the local network.
3. The average density of car parking on sidewalks is 90 vehicles / km.
4. The share of cars parked with traffic violations reaches 45% (the average number of offenders in the city is 130 vehicles / km).

The main types of parking violations are the following car placements:

- at an angle to the carriageway edge in the places where only longitudinal placement is allowed - 33% of all violations;
- on the sidewalk or with a partial drive onto the sidewalk in the absence of permissive signs and markings - 27% of all violations;
- closer than 5 m from the edge of the crossed carriageway or pedestrian crossing - 14%;
- within the coverage area "Stopping prohibited" – 7%.

The problem of parking for private vehicles has existed for a long time. Parking slots are sorely not enough, so the car owners are forced to park their vehicles where they can find free space. Because of parked cars on the roads, today it is punctual, but still there are serious obstacles in traffic. Public transport often cannot approach stops and passengers disembark from the second row. Pedestrians bypassing vehicles on the road run the risk of being hit by the wheels. Also, due to spontaneous parking in the courtyards of residential buildings, it is difficult to walk to entrances, playgrounds, schools and preschool institutions. Violation of traffic rules is often a compulsory measure for a car owner who decides to park his vehicle. As a result, pedestrians complain, while motorists shrug their shoulders, saying that they simply cannot park their car correctly [6, 10, 15, 16].

For example, in Moscow, 25% of the drivers surveyed noted that the longest time they spent looking for parking in the last year was from 11 to 20 minutes, and 15% named the figure from 21 to 30 minutes. Moscow is in second place in the number of fines received by drivers for parking in the wrong place (the average number of fines received per year is 8.5) and in fourth place among the world’s megacities. The main problem with parking lies in the huge number of cars on the roads, which were built decades ago and were not designed for such a number of cars. In addition, the construction of buildings in Moscow does not always allow for parking on the ground floor and below, as is done in most European cities. Among other things, Russia has a completely different driving culture: there are drivers who are not afraid of fines, as in Europe (possibly due to the low cost of Russian punishment). And one more factor is the mentality specifics [17, 18].

Also, for example, in Krasnodar at this time, as our studies have shown, the lack of parking slots, depending on the area, ranges from 35% to 65%. The average lack of parking slots in the city is 54%. The lack of parking zones in isolated cases of the city exceeds 65%. In addition, the impact that has on the efficiency of the organization and use of parking slot, the pricing policy of economic entities operating in this market should be noted. For example, the price for using paid municipal parking per hour in Krasnodar is 30 rubles. However, there are parking lots, the cost of which is 2-3 times higher. In particular, the price per hour of parking located on the territory of the Railway Station Square Krasnodar-1 is 100 rubles, at the Pashkovsky International Airport - 250 rubles, etc. In general, the study of parking slots in the city of Krasnodar revealed the fact that the situation that has formed in various zones of the city roadway is extremely difficult, which is constantly aggravated by an increase in the motorization level, the value of which significantly exceeds a similar indicator in other less traffic-loaded settlements of the Russian Federation.

Based on this, it can be concluded that there is a need for a significant increase in paid municipal parking zones, which will make it possible to free the road network from cars and establish a development vector for private parking slots. In addition, improving the state and conditions of public transport will provide the citizens with a worthy alternative to using individual vehicles. When implementing the decisions on the organization of traffic on roads, a special role belongs to the road and transport infrastructure development, as well as the use of various technical means and tools: road signs, road markings, road barriers, traffic lights, etc. [12, 14].

Based on the above-said, it should be noted that one of the most effective solutions to the existing problem is the use of automatic parking systems "Tubing parking".

The "Tubing parking" system (Figure 1) is a metal structure consisting of a supporting frame and six rows of mobile parking platforms, located in four tiers one above the other, on which cars are stored.

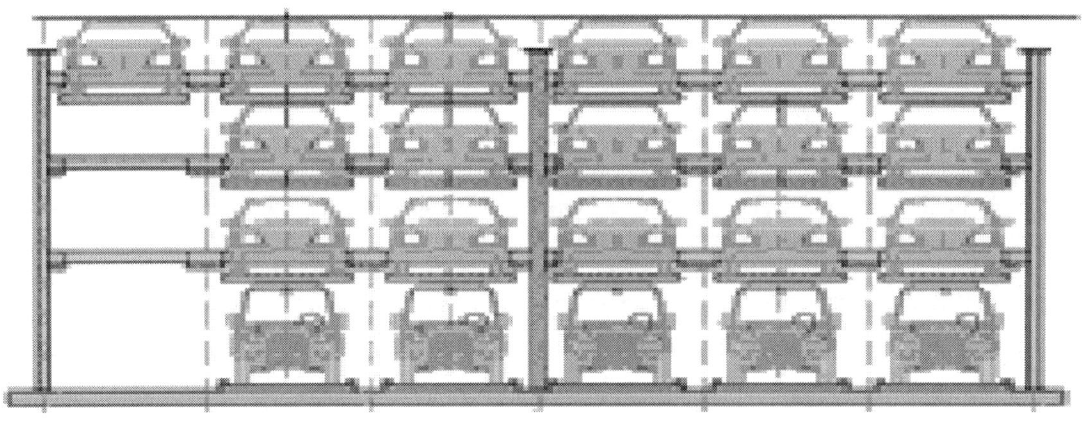

Figure 1. Diagram of the mechanical parking system "Tubing parking"

The capacity of this parking is up to 250 vehicles. Average daily parking load - 75%. The parking system makes it possible to save up to 35-50% of usable space, due to the multi-storey structure and the use of a hydraulic drive or an electric drive with great energy savings, equivalent to the power of an electric kettle (2.2 kV).

The hydraulic mechanism, which is equipped with each tubing parking system, performs the vertical movement of transport spaces on the parking levels in the absence of a ramp, in a state of limited space.

Benefits of an automated parking system:

– there is no need for additional construction work;

– the total number of parking slots increases to 50% due to the correct use of the space height;

– automatic system and operating mode;

– significant time savings for drivers;
– environmental friendliness - no harmful emissions into the atmosphere.
Segment of the parking lot under construction - people with an average income.
After commissioning, the plans include the following prices for parking services:
– rental price for one parking slot per year - 26,000 rubles;
– rental price for 1 month - 3500 rubles;
– rental price for 1 day - 150 rubles.

Also, the privileged places can be provided for the socially unprotected parts of the population in the amount of 25% of car places from the total. Consequently, this project can be paid off in 6-7 years, which is typical for this type of facility (Table 1).

Table 1. Calculation of net profit, thousand rubles.

Article title	0 year	1 year	2 year	3 year	4 year	5 year	6 year	7 year
Potential gross income	0	4695	5634.7	6198.1	6818	7499.8	8249.7	9074.7
Total direct costs	0	2810	2810	2810	2810	2810	2810	280
Gross profit	0	1885	2824.72	3388.1	4008.0	4689.8	5439.7	6264.7
Administrative costs	0	10	10	10	10	10	10	10
Total fixed costs	7845	1503	1548.3	1625.8	1707.1	1792.4	1882.0	1976.,1
Wearout, [%]	0	60	60	60	60	60	60	60
Total costs	7845	1747	1792.7	1870.2	1951.5	2036.8	2126.4	2220.5
Profit before tax payment	7845	137.9	1031. 9	1517.9	2056.5	2652.9	3313.3	4044.1
Income tax		8.274	61.9	91	123.3	159.1	198.7	242.6
Net profit		129.6	970	1426.8	1933.1	2493.7	3114.5	3801.5

Summary

Thus, the development of the tubing parking principles by the introduction of the "Tubing parking" system guarantees the provision of residential and public areas of the city with the necessary share of parking slots. At the same time, operational contracts can be implemented in the form of a public-private partnership both in previously built-up zones and in the territories included in the future development plans. The "Tubing parking" system will also significantly reduce consumer costs.

References

[1] Degtyarev G V, Takhumova 2019 Designing an additional freshwater source infrastructure to ensure the environmental sustainability of coastal areas *IOP Conference Series: Earth and Environmental Science* **395** [EESE3951001.pdf].
[2] Information on https://ru.newsroom.ibm.com – результаты международного исследования «IBM Parking Survey»
[3] Information on https://www.researchgate.net/publication/235359727_The_High_Cost_of_Free_Parking
[4] Ivanov L A, Bokova E S, Muminova S R, Katuhin L F 2020 Nanotechnologies: a review of inventions and utility models *Part I. Nanotechnologies in Construction* **12 (1)** 27–33. DOI: 10.15828/2075-8545-2020-12-1-27-33.
[5] Ivanov L A, Razumeev K E, Bokova E S, Muminova S R 2019 The inventions in nanotechnologies as practical solutions *Part V. Nanotechnologies in Construction* **11 (6)** 719–729. DOI: 10.15828/2075-8545-2019-11-6-719-729.

[6] Nehaj R, Molotkov G, Rudchenko I, Grinev A, Sekisov A 2017 Algorithm of composing the schedule of construction and installation works *IOP Conference Series: Earth and Environmental Science* 1-8. doi:10.1088/1755 1315/90/1/012019.
[7] Popov R A, Sekisov A N, Shipilova N A 2016 The Economics of Innovation in Modern Russia: Practice, Problems and Prospects Prospects *International Journal of Economics and Financial* **6 (8)** 184-188. Information on https://www.econjournals.com/index.php/ijefi/article/view/3727/pdf.
[8] Popov R A, Shipilova N A, Sekisov A N, Soloveva E V, Gura D A 2019 Innovative development of construction in russia: economics, technologies, management *Amazonia Investiga* **8(19)** 653-663. Information on https://amazoniainvestiga.info/index.php/amazonia/article/view/281/258.
[9] Bondarenko S V, Takhumova O V 2019 Economic and organizational mechanisms of forming business networks in the construction industry *IOP Conf. Ser.: Mater. Sci. Eng* **698** 077040/
[10] Alekseev A, Salamatina A, Kataeva T 2019 Rating and Control Mechanisms Design in the Program Research of Dynamic Systems *21st IEEE Conference on Business Informatics* 96-105. DOI 10.1109/CBI.2019.10103.
[11] Sekisov A, Ovchinnikova S, Grebneva V, Chernyshova M 2020 Modern directions of low-rise housing construction development in the world: economic and technological aspect *IOP Conference Series: Materials Science and Engineering* **913** 042035. doi:10.1088/1757-899X/913/4/042035.
[12] Sekisov A N 2019 Improving the efficiency of the organization of construction production based on the use of BIM-technologies *IOP Conference Series: Materials Science and Engineering* **698** 066005. doi:10.1088/1757-899X/698/6/066005.
[13] Sekisov A N 2019 Problems of organizing and conducting engineering surveys in construction *IOP Conference Series: Materials Science and Engineering* **698** 055016. doi:10.1088/1757-899X/698/5/055016.
[14] Sekisov A N, Degtyareva O G, Samsonova N V, Grigoryan M N 2018 Development of the Methods Improving the Production Costs Formation Process *International Conference on Construction and Architecture: theeory and practice of industry development 2018* 1210-1213. doi.org/10.4028/www.scientific.net/MSF.931.1210.
[15] Sekisov A N, Varvarkina A V, Gura D A, Savenko A A, Shikhovtsov A A 2019 Pit design taking into account water level reduction using the software package MIDAS GTS NX *IIOAB Journal* **10 (S2)** 93-100. Information on https://www.iioab.org/IIOABJ_10.S2_93-100.pdf.
[16] Vavrinchuk P A, Ryabkova E B 2014 Parking is the main solution to the deficit of parking slots *New ideas of the new century: materials of the international scientific conference IAD PNU* **2** 47-53.
[17] Gorilchenko M A 2013 Comparative analysis of modern systems of mechanized car parking *Mechanization in construction* **7 (829)** 35-38.
[18] Levashev A G 2013 Estimation of demand for parking in parking lots of various queuing facilities *Bulletin of Irkutsk State Technical University* **11 (82)** 211-216.
[19] Cherevko S N 2013 Construction of parking lots in cramped urban conditions *Engineering Bulletin of the Don* **26 (3) (26)** 156.

IOP Conf. Series: Materials Science and Engineering 1083 (2021) 012073 doi:10.1088/1757-899X/1083/1/012073

Interregional water transport cluster as a tool for economic convergence and integration

E N Rudskaya*, I A Eremenko, I V Popova

Don State Technical University, 1, Gagarin Square, Rostov-on-Don, 344000, Russia

E-mail: alenkaru09@gmail.com

Abstract. The aim of the study is to develop a conceptual project for creating an interregional water transport cluster along the Don River. The presented theoretical studies have been carried out using the methods of analysis and synthesis of the existing scientific developments in the field of territorial and inter-territorial transport infrastructure organization. Observation and static analysis methods were used in practice to determine the economic connectivity level of the regions along the North-South axis of the European part of Russia. According to the study results, the following results have been obtained:

- the role of inland waterway transport as a factor of convergence and integration of regional economies, territorial consolidation on the systemic cluster approach basis to the spatial organization of the economy has been proved;
- the effectiveness of the network-centric interregional cluster formation on the cooperation principles for the passenger and cargo shipping revival along the small rivers-confluents of the Don river in order to ensure maximum transport accessibility of remote areas and the corresponding multiplier effect for the development of all sectors of the economy has been substantiated;
- the idea of forming and maintaining a mega-regional ecosystem along the North-South axis along the Don River on the basis of dredging consequences' interregional monitoring, as well as the formation of water tourism embedded cluster along the rivers of the Don basin has been proposed.

1. Introduction

The problems of the spatial development of Russia are caused by the imbalance of interests at the national, regional and municipal levels. Power decentralization in the early nineties of the twentieth century laid the foundations for increasing independence in solving the territorial development problems. However, modern structural global transformations require the managerial actions' integration of regional and local authorities to solve common socio-economic problems. The regions' sectoral profiles tend to converge, which is confirmed by use of such a spatial organization form as a cluster.

The research subject is the concept of the interregional water transport cluster model.

The study showed that, a resource, network-centric, systematic approach has been formed in this area of economics. At the same time, the complexity and heterogeneity of the scientific research theoretical base in relation to the research subject still remains unchanged.

The above-mentioned reasons confirm the need to test the following hypothesis: the interregional cluster creation integrator can be such a geographical form factor as a river, which simultaneously acts as an economic resource. Small rivers form a fairly fast and cheap transit of goods and passengers, supporting the economy in the remote areas, providing employment for the population, preventing the extinction risk of villages and settlements.

The logical project toolkit with the use of the economic ties cooperation principles will allow to ensure the formation of a qualitatively new mega-ecosystem, improve the river resources regions-users infrastructure state, stimulate industrial development, realize a synergistic effect and increase the territorial connectedness and stability by means of the interregional sectoral interaction.

2. Methods and materials

The research was carried out using the methods of analysis, synthesis, based on integration theories and concepts, scientific developments on the modern economy spatial organization problems and its interregional transport infrastructure.

Most scientists agree that consolidating processes based on the network resources' use are decisive. Within the overall production chain, intra-industry interaction acquires new quality. The territories of Russia, occupying a huge area, are connected by a unique water network, which makes it possible to literally develop all the economy sectors on the cluster principles.

F. Perroux played an important role in the study of the regional economies' spatial organization problems, having developed the "Growth Poles" concept in the middle of the nineteenth century. He understood Poles as the industries dominating due to their significant market share, including in the field of innovation. Such "poles ... generate a chain reaction of the emergence and growth of industrial centers" [1].

A. Marshall also enriched the regional sciences with fundamental research in this area, introducing the concept of "industrial regions" and proving the synergistic effect of nearby enterprises through free access to suppliers, knowledge exchange and innovation [2].

The most significant contribution to the development of modern science about regional clusters was made by M. Porter, defining them as "... geographically concentrated interrelated types of economic activity ...". The scientist considered clusters as "a new progressive form of Inter-industry interaction of business", contributing to the country's competitiveness strengthening [3].

Let us present the most characteristic parameters of the cluster, which are noted by the majority of scientists:

- general geographic location;
- same or similar industry;
- vertical and horizontal functional communication of enterprises;
- coexistence of commercial and non-profit organizations to strengthen competitive positions;
- common technologies use;
- the cluster "core" presence.

These features of a regional cluster in total characterize the interaction effectiveness between economic entities - cluster members. Such market interaction between the members of cluster associations is based on competition and cooperation, making it possible to quickly adapt to changing external conditions.

For the purposes of our study, it is necessary to clarify the essence of interregional interaction. The most reasonable is the judgment of M.Yu. Makhotayeva and O. A. Bakumenko:

- it is "... a complex of resources flows' exchanges carried out within the framework of agreements between authorities, legal entities and individuals of different regions, adopted de jure or de facto, in order to represent common interests in the national and global economic space and increase the level of sustainable development in these regions ...";
- "business structures' interaction of the regions in solving common problems, creating production and technological chains, clusters, associations". [4]

To substantiate their position, the above-mentioned authors single out two most important properties of interregional interaction:

1. Economic integration, which is based on interregional production cooperation and interregional division of labor, ensuring the regional development strategic goals' achievement.

2. Economic connectivity, reflecting the interconnection of economic growth in the regions forming economic macro space.

In our opinion, the most promising form of interaction between territories is interregional clusters. Summarizing the various interpretations of the interregional cluster essence in scientific literature, we define this phenomenon as the integration of regions interconnected by a common resource base to achieve a certain common goal of socio-economic development [5].

Despite the vastness and depth of the research in the field of clustering the economy, insufficient attention is still paid to the methodological issues of organizing interregional sectoral clusters.

3. Research results

3.1 Transport-logical cluster as a basis for interregional interaction: system integration and network-centric approaches

Communicative interregional ties are based primarily on the transport territorial connectivity. Transport infrastructure ensures the redistribution of all factors of social reproduction and serves as an instrument of the territories' competitive struggle for a dominant position in the economy.

Transport and logistics clusters play a special role in these processes. A.B. Volynchuk notes the following basic features of these clusters: common spatial boundaries, the presence of economic entities with developed production potential, the presence of a territory center with a high density of interaction between its participants, transport lines with several transport modes [6].

Let us prove the validity of this statement using a systematic approach. So, the sign of the space unity means the possibility of the cluster's exclusive existence within the territory. With regard to the interregional quality of transport clusters, this means the length of water and land transport routes, useful storage areas, interaction intensity between the economic agents, taking into account regional specifics.

An important feature of the transport and logistics cluster is also a high level of concentration of the enterprises with developed production potential. This circumstance is associated with a decrease in the cost of moving raw materials and finished products. This feature correlates with the formation of the cluster core as a space-forming factor: effective economic contacts determine the real boundaries of the cluster.

Let us pay special attention to the types of transport in the transport and logistics cluster. As a system, the latter can have one basic type of transport, which should be dynamic, accessible and have a simple infrastructure. To ensure interregional connections and sufficient carrying capacity, water transport, road and rail are suitable. To form a cluster, at least two types of transport are needed: the first creates internal cohesion and unity of the system, the other - communication with other clusters and economic systems.

We believe that it is the inland water transport that is the strategic sector of interregional interaction with a multiplier effect. The river system of Russia acts as a natural network of transport and communications for most regions. We also emphasize the solution to the problem of greening transport by increasing the goods and passengers' transportation by water transport.

The essence of the network approach in the area under study is well explained by V.V. Tsyganov through the concept of supported development based on self-organization: "... this is harmonious ... development ... in which the use of natural resources, the investments direction, the orientation of scientific and technological development, personal development and institutional changes are coordinated with each other and strengthen the current and future potential for human satisfaction, needs and aspirations, ensuring the quality of people's life ...". The author develops the idea of a network approach to network centricity, the essence of which is to support the self-organization and development of capital centers, including those with the use of public-private partnership mechanisms [7].

Thus, the existing scientific developments confirm the validity of the proposed idea to consider inland waterways and water transport as the aforementioned strategic centers, taking into account the multiplicative effects of increasing investments in projects for the inland waterways' reconstruction and development. The ultimate goal is to increase the regions' integration level, develop the economic potential of the territories, increase the welfare of the population.

3.2 The river waterways and inland waterway transport state analysis

The territory of Russia is characterized by the world's longest exploited network of inland waterways - 101.7 thousand km, including the unified deep-water system of the European part of Russia is 6.5 thousand km and is part of the most important inland waterways of global importance.

In the Russian regions with operated inland waterways, 80% of the country's population is concentrated and about 90% of the gross domestic product is produced.

At the same time, the share of Russia's internal river transport in the total cargo turnover is less than one and a half percent. For comparison, in the European countries this index was: in Germany - 11%, the Netherlands - 34%, France - 10% of the freight turnover. Inland waterway transport in the European countries is complementary to rail and road transport, forms more than 35% of cargo deliveries to the largest seaports of Europe, is an environmentally friendly and safe mode of transport.

Passenger traffic by inland waterway transport has also steadily decreased since the early 90-s of the XX century, although in the regions of Siberia and the Far East, this type of transport remains highly important for the population. At the same time, in large cities the potential of water transport is practically not used to transport passengers in conditions of congested roads. Only 300-400 thousand people are transported per year on the tourist routes on small and large rivers.

The qualitative characteristics of inland waterways should be recalled here:

- with guaranteed dimensions of navigable passages and with conditions - 48.3% of the total tracks length, or 49 thousand km, of which with illuminated conditions - 25%, with reflective conditions - 11.3%, with unlit conditions - 11.9%.

Inland waterways with guaranteed dimensions of the passages make up slightly more than 26 thousand km or 53.5%;

- without guaranteed dimensions of navigable passages and with the situation, inland waterways make up about 3 thousand km or 3.1%, of which only 0.4% with illuminated conditions; with unlit conditions - 2.7%;

- without navigation barriers and guaranteed dimensions, the waterways length is almost 49.5 thousand km or 48.7% [8].

A significant problem is the infrastructural restrictions on navigation on the main waterways in the European part of Russia, where the sections length with a depth of less than 4 meters today is almost 1.8 thousand, or about 25% of the length of the deep-water system waterways.

Experts state that, the length of navigable routes with guaranteed depths has decreased by almost one and a half times over the past three decades, there is a dilapidation of navigable hydraulic structures, the turnover of inland waterway transport has decreased by 4 times, and the passenger turnover - by 6 times. Approximately 95% of self-propelled river vessels are over 20 years old and over 40 years old - 40% of ships.

We present the comparative characteristics of inland waterway transport with other modes of transport according to the following efficiency parameters:

- in the total volume of fuel consumption, inland waterway transport accounts for 2.9%, at the same time, railway transport has a share of 8.3%, road transport - 32.4%. In terms of energy efficiency, inland waterway transport is the second only to railroad transport: 113.6 against 158.7 thousand ton-km / t conv. fuel and much more efficient than road transport, for which this indicator is only 4.7 thousand ton-km / t conv. fuel

- total costs for the maintenance and development of public highways of federal and regional significance per 10 ton-km of road transport freight turnover are 14 times higher than inland water transport.

- the transport modes share in terms of their contribution to environmental pollution are distributed as follows: road transport - 94%, railway - 1.5, river transport - 0.6%.

- the cost of transporting dry cargo by inland waterway transport is reduced compared to the road transport at a distance of more than 200 - 300 km, and for oil products - at a distance of more than 600 km. Compared to the railway transport, the highest efficiency is achieved when transporting metal and metal products by water transport over a distance of more than 200 km, agricultural products - more than 250 km, oil products - more than 700 km [8].

The analysis made it possible to identify and generalize the following systemic problems of the Russian inland waterway transport:

- negative stable dynamics of a decrease in the share in the country's transport system;
- deterioration of the inland waterways' quality parameters, technical condition of hydraulic structures, high physical and moral deterioration of self-propelled vessels;
- decrease in the availability level of water transport services and their quality for remote rural areas, both in terms of cargo supply and passenger traffic;
- poor use of the small rivers' transit potential.

3.3 The water transport complex of the Rostov region state analysis

The Don River flows from the North to the South of the European part of Russia and has a length of 1870 km, the area of the river basin ranks 8th among the rivers in the country and is 421.7 thousand sq. km, of which 368.6 thousand sq. km are within Russia. The river, along with its confluents, adjoins 16 regions, including 390 thousand settlements. Only in the Rostov region there are 1960 settlements and directly in the river delta there are about 2.5 million water users.

The river basin is represented by a developed hydrographic network with 13 thousand streams and 8 thousand lakes, 10 thousand ponds and small reservoirs. There are 43 integrated reservoirs with a volume of 10 million cubic meters for seasonal flow regulation.

Let us highlight the most important problems of the water transport complex in the Rostov region, reflecting the general condition of the country's inland waterways:

- shallow waterways, caused by shallowing, drift and overgrowth. Low water leads to a decrease in the throughput and safety of vessels' navigation; therefore, the depths also decrease in the waters of the international seaports of Azov and Rostov-on-Don, the carrying capacity of incoming ships is limited to 3-3.5 thousand tons.
- a significant part of the "river-sea" fleet and marine serves under a foreign flag and have the right to work only in the waters of the Azov Sea, not going up the river. In this regard, the river ports operate as storage tanks and double transshipment of goods is required, which leads to an increase in logistics costs and a decrease in the competitiveness of products;
- the equipment of river ports and quay walls does not allow to properly service promising oil and chemical cargo;
- shallow depths do not allow passenger shipping development, and the infrastructure for it has fallen into disrepair. A significant number of vessels are 30-40 years old, the depreciation of the production assets of water transport exceeds 70%. All these factors hinder the development of water tourism in the region along the Don River and the Taganrog Bay of the Azov Sea, to Moscow, St. Petersburg, Volga cities;
- there is an increase in the processes of coastal destruction, silting, sewage pollution, the growth of blue-green algae.

The Azov-Don basin is an integral part of the European water transport ring and ensures the movement of goods to the southern European countries. A special role is assigned to the Volga-Don shipping canal, which passes through the regions that form half of the industrial potential and 65% of the country's agricultural potential. The importance of inland waterway transport in the region is increasing in connection with the international transport corridor Danube-Don-Volga development. The transport complex of the Rostov region integrates all types of mainline transport, warehouse and information logistics infrastructure.

3.4. Development of a project concept for an interregional water transport cluster along the Don River

The proposed project meets the requirements of the innovative scenario for the transport strategy of Russia for the period up to 2030. The project purpose is to prevent the internal water transport infrastructure degradation threat in the European part of the country, to increase the efficiency of transport enterprises and systems, and to bring population mobility closer to the developed countries' indicators.

The Don River flows directly through the territory of five regions: Tula, Lipetsk, Voronezh, Volgograd and Rostov regions.

3.4.1. Organizational and economic mechanism of interaction between the regions along the North-South axis along the Don River

- public authorities and local self-government of the territories participating in the water cluster ensure the solution of common problems for the infrastructure development along the entire river bank on the basis of agreements and cooperation programs;
- business structures' interaction in the adjacent regions to create production and technological chains for servicing transport facilities and structures of the water complex;
- public-private and municipal-private partnerships to attract investment resources in the interregional water cluster mechanism development.

The studies have shown that the interregional interaction implementation in practice is possible only if it is positioned as the goal of strategic management for the regional development. Very few regions set tasks for mutual partnership and integration.

3.4.2. The regions participating interaction tasks in the cluster being formed

- the territorial management process simplification - it is necessary to develop a regulatory legal act at the federal level to regulate the basic elements of interregional interaction; at the regional and local levels - the creation of interregional associations and foundations is required;
- reducing costs in the structure of regional gross products based on integrated transport and logistics infrastructure and information and telecommunications infrastructure;
- implementation of marketing projects of the territories - creation of an integrated water tourism interregional cluster on the water transport cluster basis.

3.4.3 Economic indicators of the main regions participating in the cluster being formed

Let us analyze the economic potential of the regions under study (see Table 1). The information sources were the data of the Ministry of Economic Development of the Russian Federation, the Rating Agency "RIA Rating", the Federal State Statistics Service.

Table 1. Indicators of the scale and efficiency of the regions participating in the projected interregional water transport cluster along the Don River

Indicators and regions	Tula region	Voronezh region	Lipetsk region	Volgograd region	Rostov region	Russian Federation
Industrial production index in January-May 2020, % to January-May 2019	124.1 (2nd place in the rating)	109.3 (10th place in the rating)	103.6 (29th place in the rating)	101.9 (36th place in the rating)	97.8 (59th place in the rating)	97.6
The volume of shipped goods of own production, work performed and services per resident, in January - May 2020,	300.9	200.8	344.5	201.2	156.5	281.2

thousand rub						
The volume of investments in fixed assets in the first half of 2020 mln ru per % to the similar period 2019	44368.3 /60.5	75548.8 /86.3	60452.2 /122.8	70222.1 /116.2	127853.9 /136.7	6916.6 /96.0
Surplus (deficit) of the consolidated budgets of the constituent entities of the Russian Federation, thousand rubles	- 3091134	8252365	-220072	248248	1709576	-
profitable enterprises' share, %	63.1	68.9	65.6	63.5	64.7	64.7
Dynamics of real money income of the population, in the first half of 2020 at % to the similar period 2019	100.1	95.1	96.1	90.9	92.0	96.9
Employment rate, at %	59.6	56.4	59.4	54.0	56.7	58.0
Unemployment rate, at %	4.8	4.5	4.4	8.7	5.1	6.2
Transport Infrastructure Development Index	3.00	2.96	3.28	3.08	3.06	3.24
The development level of the public-private partnership sphere, points	55.9	39.2	32.6	49.2	40.5	-
Tourism development level, points	77. 6 (24th place)	73.7 (28th place)	70.5 (31st place)	75. 5 (27th place)	89 (15th place)	-

The territories of the Don River basin are quite different in terms of the industrial production development level. So, the Tula and Voronezh regions are among the top ten leaders in the country. Although in terms of the absolute value of the shipped products volume per inhabitant, the Lipetsk region has the first position among the considered regions - 344.5 thousand rubles, the Rostov region - only 156.5 thousand rubles. At the same time, in terms of investment in fixed assets, the Rostov region is ahead of all the cluster regions. Also, a significant dissonance is observed in the budgetary sphere of the regions - two out of five regions have deficit budgets.

The share of profitable enterprises fluctuates between 63-65%, only the Voronezh region stands out from the total number with an index of 68.9%. But in terms of employment, the Tula and Lipetsk regions are in the lead, approaching 60%. The same two regions managed to practically keep the dynamics of real money income at the level of 2019.

Interestingly, only the Lipetsk region exceeds the value of the industry average index of transport infrastructure development, the lowest indicator is in the Voronezh region. At the same time, in terms of the public-private partnership mechanism use level, the Lipetsk region has the weakest positions in comparison with the rest of the regions participating in the sample.

This unevenness of territorial development confirms the need for integration and convergence on the cluster interaction basis in an interregional format based on the joint use of water resources, expansion of sales markets, ensuring economic connectivity and transport accessibility for the population.

3.4.4. Explanations of the concept individual elements

The current economic model of a federal state determines the competition of regions for federal funds and foreign investment. At the same time, the resource, systemic and network-centric approaches to the organization of interregional interaction are not used to level the living standards and develop the regions' economic potential [9].

The transport water cluster along the Don River will optimize the placement of the economic infrastructure of the studied regions on the cooperation basis, eliminate ineffective budgetary expenditures, unite the economic potential of the territories to solve large-scale investment projects based on the public-private partnership mechanism.

Particular attention should be paid to the creation of a monitoring and dredging system along the entire length of the Don River. Some regions carry out such work in case of emergency, when the depth goes beyond the maximum minimum and navigation becomes impossible. In connection with the high cost of equipment and work, we propose to create an interregional fund on terms of cooperation, purchase self-dredging dredgers, barges with opening hull, rope excavators, multi-bucket shells, anchor carriers for joint use on the basis of an interregional year-round schedule. The teams will be local, and in the absence of the necessary specialists in one region, it will be possible to use the principles of personnel leasing. At the same time, we propose to create an interregional service for monitoring the impact of dredging operations on the environment, since such works negatively affect fish resources.

High scores in the national tourism rating of the studied regions prove the feasibility of integrating into the interregional transport cluster and the water tourism cluster. Gravitation to the coastline will ensure the concentration of tourism industry objects evenly across all participating regions and, accordingly, increase employment in the social and technical infrastructure of the industry. In addition, all of these regions have significant cultural heritage. Thus, Tula is the third museum center after Moscow and St. Petersburg; Volgograd is the center of military-patriotic and historical and educational tourism, the cities of the Rostov region are included in the tourist route "The Golden Ring of the Bosporus Kingdom", the Voronezh region specializes in folk and event tourism, the Lipetsk region develops pilgrimage and archaeological tourism programs [10].

Thus, the results of the study confirm the need to move from a strategy of competition for resources to a strategy of interregional cluster interaction.

4. Discussion

The concept of integrating internal resources of territories based on the formation of cluster structures is of great scientific interest, since it determines the possibility of achieving competitive advantages and economic growth of the regions.

To study the role of resources in the activities of various clusters, it is necessary to discuss the provisions of the resource theory and the resource-oriented approach to managing the regional economy in conditions of limited resources. E. Penrose, J. Barney, B. Wernerfelt, C.K. Prahalad made a significant contribution to the development of this scientific field: G. Hamel, K. Ketels and others.

Synthesizing scientific developments on this issue, we can conclude that the territories' interaction is possible through the clusters for the purpose of their further development based on the joint use of available resources. For example, K. Ketels's idea of inter-cluster cooperation based on specialized platforms is interesting. In general, the scientists agree on the following positions:

- the cluster-network principle allows organizing effective integration interaction of regions, ensuring the alignment of their spatial development and economic cohesion;
- the use of natural resources, for example, inland water resources, forms a direct link between competitive advantages and effective management of strategic resources of territories.

According to the research results of the Harvard Business School, labor productivity at enterprises participating in the cluster is 20-40% higher than the industry average and, accordingly, the level of wages is one third higher than the national average. The clusters form up to 60% of GDP and more than 30% of employment in the US [11].

The European Union is pursuing a network transport policy aimed at stimulating the development of information, financial and social networks that contribute to the development of transport complex. Currently, such a policy is gradually transforming into a network-centric one, when network centers are selected and maintained as points of the transport system growth. In turn, a synergistic effect is created from directing the efforts of participants to a specific object or project.

So, there are more than 80 transport and logistics clusters, including 42 thousand km of inland waterways, 73 thousand km of roads, 215 thousand km of railways, 366 airports on the territory of the EU. The most prominent example of interregional transport clusters in Europe is the Frankfurt am Main cluster located in the five largest cities in the state of Hesse. This cluster, in addition to highly developed road and rail networks and the third airport in the world in terms of passenger traffic, includes a river port with access to the North and Black Seas [11].

Within the framework of the cluster policy, the state provides direct support in the format of financing programs: over the past decade, more than half a billion euros were allocated in Germany for the development of 32 clusters; in France, three billion euros were spent on the needs of 71 clusters [11]. Also, within the framework of the state cluster policy, specialized organizations with supporting functions (marketing, information, consulting, educational) are created.

5. Summary

The analysis showed that for the transit potential implementation in the region, the creation of an interregional water cluster is expedient. Inter-territorial interaction on the basis of cooperation will permit:

- to increase guaranteed navigable depths up to 4-4.5 meters through systematic dredging and cleaning works throughout the river by all the user regions on the basis of a joint coordinated schedule;
- to increase the share of the ports in the Rostov region in cargo processing, including container and chemical, among the ports of the Azov basin;
- to form a unified system of freight forwarding services on water transport with a logistics center in Rostov-on-Don;
- to create a favorable investment climate on the public-private partnership basis for the development and technical re-equipment of the river fleet and hydraulic structures on the Don River and its confluents;
- to improve interaction with other transport modes, to develop port rail and road approaches to river ports;

Restoration of inland navigation will reduce the transport costs share in the cost of intra-regional cargo flows of agricultural products, construction materials, fertilizers, and, accordingly, increase the transshipment of goods, according to experts, to 26-27 million tons per year. Full-fledged passenger river navigation and the construction of passenger port points, piers will ensure the development of water tourism, excursion and recreational services, cruise transportation to the Black Sea, and resorts including the Crimean.

References

[1] Perroux F 1954 *L'Europe sans rivages (Grenoble: Presses universitaires de Grenoble)* 859.
[2] Marshall A 1983 *Principles of Political Economy (Moscow, Progress)* 416.

[3] Porter M E 1998 *The Competitive Advantage of Nations (2nd. ed. — New York, Free Press)* 896.
[4] Mahotaeva M Y and Bakumenko O A 2017 Interregional sectoral cooperation of Pskov region:cluster approach *Services in Russia and Abroad* **11 (7)** 32-45.
[5] Klochko E N, Bogdanov I Y, Kobersy I S, Klimenkova M S and Litvinova V Y 2018 Research of organizational and economic resources, participants, cooperation of clusters in a certain territory *Espacios* **39 (31)** 10.
[6] Volynchuk A B and Krylova I A 2018 Theoretical approaches to the determination of the transport and logistic cluster *Azimuth of Scientific Research: Economics and Administration* **1 (22)** 77-80.
[7] Tsyganov V V, Borodin V A and Lemiashkova A V 2018 Consept of network-centric transport policy in Eurasia *Information technologies in science, education and management* **4** 45-48.
[8] Pantina T A and Borodulina S A 2018 Criteria and factors of inland water transport competitiveness growth *Bulletin of the Astrakhan State Technical University* **3** 68-77.
[9] Kushnarenko T V, Kaplina A V, Tabakov A N and Fomina I B 2018 Clustering as an integration pattern of business activity *European Research Studies* **21** 553-563.
[10] Kazaryanz I G, Eremenko I A, Tkachenko M Aand Suhovenko A V 2018 Institutional formation to develop tourism and recreation clusters *European Research Studies* **21** 167-177.
[11] Lebedeva A S and Rogavichene L I 2018 Innovative development of the transport services market: a cluster approach *Forecasting problems* **4 (169)** 83-91.

Programs implementation prospects for the large city territories development in Russia: institutional aspect

I A Bachurinskaya[1*], N V Vasileva[1], S N Maksimov[2], M N Yudenko[1]

[1] St. Petersburg State University of Economics, 21, Sadovaya, 191023, St. Petersburg, Russia

[2] St. Petersburg State University, 7-9, Universitetskaya emb., 199034, St. Petersburg, Russia

E-mail: basia@mail.ru

Abstract. The article shows that the development of large cities as centers and locomotives of economic growth requires efficient use of urban areas, ensuring their sustainable and balanced development. This necessitates improving the institutional framework and methodology for spatial planning. The authors note that the functioning of the institutions on which the rational land use in large cities depends does not meet modern requirements, primarily in the field of integrated sustainable territorial development (ISTD). The article analyzes the experience of St. Petersburg in the urban areas' development aspect, which made it possible to identify the main problems, including the lack of public consensus on the choice of territories for the programs' implementation. ISTD and insufficiently active participation of the city in the implementation of these programs, especially in terms of engineering infrastructure Preparation. Analysis of literature and practice in the study area showed the need to improve the mechanism for the formation and implementation of programs for the territorial development in large cities, the development of real estate market institutions, rules and regulations related to ISTD. The article presents the recommendations of the authors, to a certain extent filling the existing gaps. The author's definition of the "depressed territory" concept is proposed (to solve the problem of choosing the territories for ISTD), substantiated the need to form a unified city program for ISTD project development with their introduction into the General Plan of the city development, as well as the use of the economic-spatial model of the developed territory to determine the possible effect of its development and the costs necessary to obtain it. The article proposes a unified methodological basis for the territorial development programs' formation and implementation. The recommended methodological approach is aimed at improving the analysis and assessment quality oof development programs for the territories in a megapolis, at ensuring the effective use of urban land.

1. Introduction

The issues of spatial development, effective use of land resources, urban areas have recently been in attention focus of both state and municipal authorities and specialists in the field of land management, land use, territorial and urban planning. Land is recognized as the most important factor in economic growth when used efficiently. At the same time, an increasing concentration of economic, scientific, scientific, technical and innovative activities is presented in large cities and megapolises. The relevance of the article is determined by the fact that attention-attracting programs of territorial development, implemented in order to increase the urban land use efficiency and aimed at solving the socio-economic problems of society in providing housing and commercial real estate objects are closely related to the development of real estate market institutions and the construction of large cities. In this connection, the problems of institutional development in the field of territorial development and renovation have a significant impact on the prospects for the implementation of programs for the territorial development in a large city.

The authors of the article focus on the analysis of programs for the territorial development in large cities, which is due to the following reasons:

- such territories are most in demand for construction and development, they are actively involved in circulation, built up with housing and other objects, and are developed as public spaces;
- the use of land resources and the territorial development of large cities require special attention of the authorities, well-thought-out measures to regulate land use to ensure not only economic results and growth of city budget revenues, but also to create comfortable and safe living conditions in the city, to achieve the city development goals;
- the territories of large cities are the attractive objects for investors, which requires local authorities to actively work to create a favorable investment climate, prepare land plots for bidding, resolve issues of developing the necessary communal and other infrastructure and ensure an integrated approach to the urban areas' development. Russia has vast territories, the land fund of the country is 1,712,519.1 thousand hectares, or 17 125.2 thousand square km. The share of land in settlements in the structure of the country's land fund is insignificant - 1.2% (20.5 million hectares), of which the land of urban settlements occupies 41% (8.4 million hectares) [1].

With such a small share, these lands are densely populated and intensively used, especially in large cities. Currently, the share of the urban population is 74.6% [2].

These data show the relevance and importance of ensuring sustainable land use in large cities. At the same time, the development of institutions on which rational land use in large cities depends, although it is proceeding at a significant pace, but their current functioning is not enough for the modern development of society and the city's economy, primarily in the field of land use and construction for the purposes of integrated sustainable territorial development.

This means that the recommendations to improve the mechanism for the formation and implementation of programs for the development of large territories are needed, which would include the recommendations for the development of real estate market institutions, rules and regulations related to the integrated sustainable territorial development (ISTD), taking into account the characteristics and potential ISTD for large cities. But today these issues are considered sporadically within the scientific and professional community.

2. Literature review

In the Strategy for the Spatial Development of the Russian Federation until 2025, large urban agglomerations and megapolises are considered as the leading centers of economic growth, which requires the rational and efficient use of their potential, the most important component of which is the spatial resource - urban land. The solution to this problem involves improving the methodology of spatial planning, overcoming "unbalanced spatial development of large urban agglomerations and the largest urban agglomerations" [3]. However, increased attention to these issues has not yet led to the required qualitative changes in institutional aspects, both from the standpoint of analyzing the development of individual institutions, norms and rules, and from the standpoint of the impact of territorial development programs on the real estate market institutional structure transformation.

A large number of scientific works, both scientific researchers and practitioners in this field, are devoted to the problems of territorial development and the territorial development programs' implementation and are considered as general issues of sustainable regional development associated with the territorial development, including the use of green economy technologies , and certain issues of the territorial development programs' implementation from the standpoint of the individual institutional norms' development in the context of sustainable territorial development.

Natalia Klimenko (2011) analyzes the principles and instruments' development of the regional policy implemented by the European countries, shows the growing trend towards an increasing role of territorial conditions and geographical factors in the development of plans and strategies for balanced regional development [4].

Ferova, Lobkova et.al. (2019) the article presents the research issues aimed at finding and analyzing the existing approaches and tools for monitoring territorial development, as well as developing methodological and procedural foundations for assessing the stability of territories to solving socio-economic problems [5].

Vertakova and Plotnikov (2019) note that sustainable development is a priority in the modern world and one of its effective tools is the development of a "green" economy. The scientists propose the priorities and directions for the green economy development to regulate regional development in order to achieve the main goal - sustainable development in Russia [6].

In work "Green" technologies in housing: the experience of Russia authors I. A. Bachurinskaya, N. V. Vasileva, M. N. Yudenko and S. A. Nikolikhina (2020) an analysis of the institutional norms governing "green" construction in Russia at the present stage was carried out and the recommendations in the field of housing construction in the city were given [7].

Maksimov, S.N., Vasileva, N.V., Bachurinskaya, I.A (2015) consider the problematic aspects of the territorial development of St. Petersburg from the standpoint of analyzing the development trends of the territorial land and property complex [8].

The authors Maksimov S.N., Sirotnikova M.V. (2017) consider the projects for the urban areas' renovation on the example of St. Petersburg and Moscow. The foreign experience of residential areas' renovation is presented on the example of Great Britain and Japan. The main differences between Russian and foreign practice are noted. It is noted that the institutional aspect is decisive in the analysis of the prospects for the territorial development [9].

The listed scientists, researchers and practitioners agree that the territorial development is an important element in solving the socio-economic problems of the city, is closely related to the problems of sustainable development and the "green" economy and has a significant impact on the harmonious development of institutions and institutional norms for the formation and development of the real estate market in a large city, as well as the prospects for the implementation of territorial development programs as a mechanism for the functioning and development of the real estate market.

3. Analysis of the urban areas' developing practice in Russia on the example of St. Petersburg

The deployment of the practice of forming urban spaces in Russia is proceeding in the direction of the transition from the projects' implementation for the undeveloped territories' development, mainly in the form of projects for the integrated territorial development (ITD) to programs and projects for the built-up areas' development, which is reflected in the formal and informal institutional norms of the practice of using urban areas in large cities.

It is believed that the projects for the integrated development and development of the territory include those in which the construction of real estate with an area of more than 100 thousand square meters is declared. This includes housing, commercial real estate, infrastructure. According to the Consulting Center "Petersburg Real Estate", today about 130 such projects have been announced on the territory of St. Petersburg and suburban regional territories. This is about 65.4 million square meters of housing, and in St. Petersburg 98 projects (or 53.7 sq. m.), and in the Leningrad region - 34 projects (or 11.7 million sq. m.) were announced [10]. According to the Consulting Center, 29 of the announced projects involve the redevelopment of industrial territories, where the construction of 9.5 million square meters of real estate is possible; 23 objects are the projects for the residential blocks' renovation (such as "Khrushchev type real estate"). At the same time, according to the Office of the Federal State Statistics Service for St. Petersburg and the Leningrad Region (Petrostat), the investments in land improvement amounted to 66,226 million rubles in January-June 2020, which is 39.5% of investments in 2019, which

IOP Conf. Series: Materials Science and Engineering 1083 (2021) 012074 doi:10.1088/1757-899X/1083/1/012074

in our opinion is due to the objective conditions of the pandemic. The sources of financing the investments in fixed assets in St. Petersburg in January-June 2020 have a ratio of 15% - funds from the federal budget, 41% - funds from the budget of a constituent entity of the Federation and 44% - funds from organizations and population.

The built-up areas' development program implementation started in St. Petersburg in 2008:

– Law of St. Petersburg dated May 6, 2008 No. 238-39 "On the address program of St. Petersburg" Development of built-up areas in St. Petersburg ";

– Resolution of the Government of St. Petersburg dated July 24, 2008 No. 904 "On measures to implement the law of St. Petersburg" On the address program of St. Petersburg "Development of built-up areas in St. Petersburg" [11].

The experience of implementing the program over the past decade has revealed a number of problems in the development of institutional norms and rules, which have remained largely unresolved to date and hinder its implementation. It is possible to distinguish the following problems:

- lack of public consensus regarding the choice of territories for the territorial development projects' implementation;
- lack of an effective mechanism for resolving contradictions between the developers' interests and the population of built-up areas;
- relatively passive participation of the city in the program implementation (in particular, in solving the problem of residents' resettlement, in the engineering infrastructure preparation).

Currently, the number of projects for the built-up areas' development is expanding due to the appearance in the legislation of a new form - projects of integrated sustainable territorial development (ISTD). The legislative basis for its implementation was the aforementioned Federal Law of July 3, 2016 No. 373-FL "On Amendments to the Urban Planning Code of the Russian Federation, certain legislative acts of the Russian Federation in terms of improving the regulation of the preparation, coordination and approval of documentation for the planning of the territory and ensuring a comprehensive and sustainable territorial development and the recognition as invalid of certain provisions of legislative acts of the Russian Federation" [12]. ISTD project implementation practice makes only the first steps, despite the fact that more than four years have passed since the adoption of 373 - FL. The reasons for this situation in many respects overlap with those that we have already indicated above regarding the practice of implementing the program for the development of built-up areas in St. Petersburg. Compared with the above-mentioned program, as well as the Moscow programs for the housing stock renovation, the ISTD projects have two essential features.

First, these projects can and should affect the territories occupied by different types of development - industrial, public and business, residential.

At the same time, the result of the ISTD projects' implementation should become the multifunctional territories intended "for the placement of capital construction of residential, industrial, public-business and other purposes and necessary for the functioning of such facilities and to ensure the life of citizens of the objects of communal, transport, social infrastructure" [12]. This is their significant difference from the programs of emergency resettlement and dilapidated housing stock and programs for the residential buildings' renovation.

Second, the ideology of ISTD projects is based on the use of the public-private partnership principles with the presence of obligations both for investors and local governments, distributed between them, as well as the right-holders when they initiate the implementation of territorial development projects, i.e., formation and development of the institution of public-private partnership of ISTD projects.

In contrast, the programs for resettlement of emergency housing stock are funded from budgetary funds, while renovation programs for residential buildings are mainly funded from non-budgetary sources. Thus, the current Moscow program for the residential areas' renovation is built exclusively on budget funding, while the program for the residential areas' development in St. Petersburg, on the contrary, is almost exclusively based on private funding.

The option of implementing territorial development projects, especially if the initiator of such a project are the local governments based on the public-private partnership principles, is the most acceptable for a number of circumstances.

First, the use of budget funds, especially at the initial stages of the territory development project implementation, is evidence of the municipality interest in the successful implementation of the project and in itself is already a factor in increasing the investment attractiveness of the project for private investors.

Second, in the early stages of the project (during the pre-project research and development period), the active participation of the municipality makes it possible to more accurately determine the conditions for attracting private investment, taking into account the socio-economic feasibility and potential profitability of the territory development project.

In addition, the direct participation of the municipality in the development of regulatory documents for the implementation of development projects of the territory (defining the boundaries of the developed territory, developing and approving the project for planning the territory, determining the list of objects to be demolished or reconstructed) makes it possible to take into account the needs of the local community more fully and creates favorable conditions for holding public hearings on the project. The key to ISTD mechanism effective use in the interests of developing not only local territories, but also the entire territory of the city, has a choice of territories planned for development. Such a choice should be made at the level of the city as a whole, taking into account the state and significance of specific territories from the point of view of urban development as a whole. This task should be solved by forming a unified city program for the ISTD projects' development with their inclusion in the Master Plan of the city development.

An important institutional mechanism that regulates the territorial development is the Standard for the Integrated Territorial development, which is a methodological document that defines the main approaches to the formation and development of residential and multifunctional development territories in accordance with the actual needs and demands of urban residents [13]. This standard is advisory in nature. It was developed on behalf of the Government with the support of the Ministry of Construction and Housing and Utilities by the Unified Institute for Development in the Housing Sector, a subsidiary of Dom.rf JSC, and is aimed at a wide audience: residents, designers, developers, representatives of local government and government different levels. The standard gave certain directions for the provisions' implementation of the integrated territorial development into the regulatory framework. In June 2020, the amendments to the Code of Rules "Streets and roads of settlements" adopted in December 2019 came into effect. Thus, although the Standard for the Integrated Territorial development is of a recommendatory nature, it has begun to be actively implemented, which means that the problem of the relationship between formal and informal institutions is being actively addressed in order to determine the prospects for effective territorial development. However, not all problematic aspects have been resolved to date.

4. Main results

From a methodological point of view, it is fundamentally important for the territorial development effective institutions formation to identify the depressed territories, the state of which negatively affects urban development, and therefore in need of development using the mechanisms of integrated sustainable territorial development. Analyzing the concept of "depressive territory" the authors of the article came to the following conclusion. In general, a depressed territory, in our opinion, should be understood as an inefficiently used territory, that is, a territory characterized by ineffective and irrational land use (incompatible or outdated location and architectural and planning solutions of buildings), due to structural deterioration of the state and chaotic development of the objects located on it; real estate (high degree of physical and functional deterioration); unfavorable living conditions and economic activity, environmental degradation and underdeveloped infrastructure; which leads to the loss of income from land use (including budgetary income) due to inefficient use of land and real estate.

Several studies on the use of ISTD mechanisms, a list of characteristics of the territory depression is proposed. So, for example, in the study of the Institute for Urban Economics Foundation, the following option is proposed: "The following characteristics of residential and non-residential real estate objects (including land plots on which such objects) on such territory are: emergency condition; high level of wear, inadequate technical condition, impossibility or inexpediency of overhaul, low improvement level; the danger of use for the life and health of people, the environment, historical and cultural monuments, including due to non-compliance with the legislation requirements in the field of environmental protection, legislation in the field of ensuring the sanitary and epidemiological well-being of the population" [14].

The allocation of the above-mentioned characteristics seems reasonable, but, in addition to the first position, they are quite general and, in our opinion, need to be concretized and need more complete methodological and methodological justification.

Taking into account the importance in the field of real estate economics of such a theoretical structure as the concept and the basic principle of the best and most efficient use of real estate (in the Russian practice of the best use), it is proposed to use them as a methodological basis.

As it is known, the best use of real estate (land) is understood as such use that is physically possible, legally permissible, financially affordable and most cost effective. In this case, the economic efficiency criterion is the maximization of the land value.

This approach, in our opinion, is quite appropriate when choosing the options for the territorial development. At the same time, a development option is understood as a whole complex of sequentially considered options: from the choice of a territory for development to the functional filling of the developed territory.

At the initial stages of project implementation, when choosing a territory for development, the concept of the best use can be applied precisely when determining the depression signs. Taking this into account, in order to effectively operate the institution of territorial development, it is proposed to supplement and clarify the above-mentioned characteristics with the following norms:

- ineffective functional use of land plots and real estate objects on the territory, which differs from the best use option (in total - at least 50% of the territory);
- income from the use of state-owned commercial real estate located on the territory is lower than with the best use of the territory;
- the cost of maintaining state-owned non-commercial (social) real estate, located on the territory, is higher than in the case of the best use of the territory;
- the level of provision of the territory with engineering, transport and social infrastructure facilities is less than 50% of the standards established by legal acts, in accordance with existing and future needs;
- the aggregate physical depreciation of real estate objects on the territory is at least 50%, and functional depreciation is at least 75%;
- the delimitation of land plots does not meet the necessary requirements (improvement of the territory adjacent to the buildings) for at least 50%;
- the use of buildings and structures is carried out in violation of the existing norms and rules (for operation, fire safety, etc.) in at least 50% of cases.

A logical continuation in determining the project parameters is the development of an economic-spatial model of the territorial development: determination of the most appropriate boundaries of the developed territory, taking into account both the immediate and distant surroundings, taking into account the same principle of the best use. The land value included in the development area, as well as the surrounding area, may increase or decrease due to differences in the area and configuration of the land plot. The understanding of its functional content and structure, the definition of a set of functions, proportions between various types of development, commercial and non-commercial use of parts of the developed territory are organically connected with the definition of the boundaries of the territory. On the basis of the economic-spatial model of the territory under development, it is possible to determine the effect of its development and the level of costs required to obtain it. At the same time, it is necessary to consider several possible scenarios, including taking into account the changes in the boundaries of the developed territory.

Determination of the projected amount of costs and the planned effect is the basis for choosing the type and organizational and legal form of implementation of the development project (the project for the integrated territorial development, the program for the residential buildings' renovation, the program of emergency housing resettlement, etc.) to determine the role and contribution of the state, municipalities, property owners, investors in the initiated project.

On the basis of the developed economic-spatial and organizational-legal model, the formation of a system of contractual relations begins, ensuring the implementation of the project for the territorial development. The above-mentioned mechanism for the territorial development projects' formation includes, as we can see, spatial, economic, organizational and legal aspects and forms the new institutional norms and rules for the territorial development of urban areas.

The recommended methodological approach, in our opinion, improves the analysis quality and development programs' assessment for the territories in a large city and has a significant impact on the urban land use efficiency.

5. Summary

This paper presents the recommendations for improving the mechanisms for the formation and implementation of programs for the territorial development of a large city in order to increase the economic efficiency of using urban land. Taking into account foreign practice, it is necessary to note certain positive aspects of foreign practice, which, in our opinion, can be applied in Russia: development delegation and implementation of strategic territorial development specific projects to the level of municipal authorities; continuous monitoring of the implementation of projects, their goals and main parameters that affect the quality of the project and its results; consistency and continuity of programs and projects for the integrated territorial development. A unified methodological basis for the formation and implementation of programs for the territorial development based on the principle of the most efficient land use is proposed.

At the same time, it should be noted that socio-economic and social aspects are of paramount importance for the successful implementation for the built-up areas' development projects (and this is their fundamental difference from projects for the development of undeveloped areas).

The most important socio-economic problem is the problem of the property institution, namely, taking into account the interests of owners or holders of other rights to real estate in the territory subject to development. As the analysis of implemented or currently being implemented programs or projects for the territorial development shows, it is precisely because of the incorrect consideration of the interests of property rights that many programs have not been fully implemented, while others have significantly increased costs and implementation times.

Taking into consideration the importance of the issue of developing institutions that contribute to the development of the land market in a large city, the mechanisms to effectively resolve the problems that inevitably arise in the implementation of territorial development projects in order to form prospects for territorial development based on the effective use of land are needed, taking into account the principle of the most efficient land use. At the same time, it is advisable to amend the current legal norms that provide for the ISTD projects' implementation at the initiative of the municipality, the application of the procedure for the seizure of real estate for municipal needs in accordance with the Land Code of the Russian Federation in relation to all 100% of the objects within the boundaries of the ISTD territory. This procedure may cause reasonable objections from the property owners. More flexible mechanisms and norms are needed, based on agreements between the project initiators and real estate owners using the buyout procedure. At the same time, it is worth analyzing and using the experience of Moscow, where in the housing stock renovation program implementation, a mechanism to achieve the consent of the majority of owners and tenants under a social lease agreement in an apartment building (at least two-thirds) is used. Elaboration and creation of effective mechanisms for coordinating the interests of project initiators and owners will significantly reduce the risks associated with the owners' attempts to block the project implementation.

References

[1] Information on: https://rosreestr.ru/site/activity/gosudarstvennoe-upravlenie-v-sfere-ispolzovaniya-i-okhrany-zemel/gosudarstvennyy-natsionalnyy-doklad-o-sostoyanii-i-ispolzovanii-zemel-rossiyskoy-federatsii/
[2] Information on: https://showdata.gks.ru/report/278932/
[3] Information on: https://www.economy.gov.ru/material/file/a3d075aa813dc01f981d9e7fcb97265f/130219_207-p.pdf
[4] Klimenko N 2011 The territorial conesion as the basis for a balanced territorial policy in the Kaliningrad region. *Balt_region* **1(7)** 86-92.
[5] Ferova I S, Lobkova E V, Tanenkova E N and Kozlova S A 2019 Tools for Assessing Sustainable Territorial development Taking into Account Cluster Effects *Journal of Siberian Federal University. Humanities & Social Sciences* **4 (12)** 600–626.
[6] Vertakova Y V and Plotnikov V A 2019 Assessment of the economic activity greening level and the green economy development directions *IOP Conference Series: Earth and Environmental Science* **392.**
[7] Bachurinskaya I A, Vasileva N V, Yudenko M N and S A Nikolikhina 2020 "Green" technologies in housing: the experience of Russia *IOP Conference Series: Materials Science and Engineering* **913.**
[8] Maksimov S N, Vasileva N V, Bachurinskaya I A 2015 Land-property complex of Saint-Petersburg and its development in present situation *Asian Social Science* **11(14)** 1-9.
[9] Maksimov S N, Sirotnikova M V 2019 Comparative analysis of domestic and foreign programs for the renovation of urban areas Real estate: economics, management *Real estate: economics, management* **2** 67-72.
[10] Information on: https://www.novostroy-spb.ru/analitika
[11] Information on: http://docs.cntd.ru/document/8477823
[12] Information on: http:// www.consultant.ru/document/cons_doc_LAW_200754/3d0cac60971a511280cbba229d9b6329c07731f7
[13] Information on: https://дом.рф/urban/standards/printsipy-kompleksnogo-razvitiya-territoriy/
[14] Information on: https:// www.researchgate.net/publication/250277238_UK_Urban_regeneration_policies_in_the_early_twenty-first_century_Continuity_or_change

Development of an algorithm for increasing the construction organization efficiency in the context of using production outsourcing

E N Klochko*, **E A Britikova**

Kuban State Agrarian University named after I.T. Trubilin, 13 Kalinina, Krasnodar, 350044, Russia

E-mail: magadan.79@mail.ru

Abstract. Modern economic conditions faced by the construction organizations in Russia necessitate the use of outsourcing as a technology for transferring certain types of activities or functions of their business to third-party specialized organizations, together with responsibility for the result of the latter. The article proposes an algorithm for increasing the construction organizations efficiency in terms of the production outsourcing use. The main stages are considered as a part of the algorithm, being guided by which it is possible to make a decision about the outsourcing effectiveness in these organizations. The authors believe that the justification of outsourcing advisability should be based on a clear algorithmizing of the decision-making process on the issue under consideration. After determining the approach to the decision on the outsourcing use, the construction organization management should determine what type of outsourcing for the integrity of the business processes transferred to the side is necessary at this stage, conduct an economic and strategic assessment of the outsourcing use and develop the measures to adapt the construction organization organizational structure.

Introduction

The transformational processes taking place in the economy necessitate the search for the new tools to increase the modern enterprises' competitiveness. One of these tools is outsourcing. The issues of the formation and development of outsourcing in Russia are dealt with by A.A. Kizim, E.V. Kazakovtseva [1], A.A. Kirillova [2], E.E. Koba [3], I. D. Kotlyarov [4], T.F. Kutuzova, O. N. Rudenko [6], L.S. Churkina [7], N.V. Lutskaya [9], N.V. Mandrik [10], V.S. Poleshchuk [11]. The specifics of studying the features of the use of outsourcing in the construction industry, which significantly reduces the circle of researchers to several people are worth noting - I.V. Lineva [8], V.V. Sinyaeva V.E. [12-14], Solovieva, I.S. Lebedev [15].

When making a decision concerning the use of outsourcing, it is necessary to take into account possible risks, which requires the development of a special approach that facilitates the process of developing a managerial decision about an outsourcing partnership. It is necessary to assess the outsourcing implementation effectiveness by increasing the efficiency of the organization's activities, increasing labor productivity in advance, which will avoid possible problems when introducing production outsourcing into the construction organizations' activities. This presupposes the implementation of important analytical work to develop a solution to optimize the activities of a construction organization using production outsourcing, which precedes the stage of analyzing the existing outsourcing cooperation.

Materials and Methods

The theoretical and methodological basis of the study was the works of famous Russian scientists, economists and builders, devoted to disclosing the essence and content of outsourcing, its development trends, as well as the works devoted to analyzing the state and development of this management tool potential in the construction sector.

The article uses the methods of systemic, monographic, structural and logical research. Each of the methods will be used based on functionality.

Results and Discussion

Construction organizations face a large number of risks that need to be considered when deciding whether to use outsourcing as a tool to improve the construction business efficiency. Consequently, the construction organization management should substantiate in detail the approach to developing a decision on an outsourcing partnership. To make such an informed decision, it is necessary to rely on a set of criteria that will take into account possible risks.

Before making such decisions, it is necessary to act according to a certain algorithm, which is shown in Figure 1.

Initially, it is necessary to assess the own resources available to the construction organization, namely the property complex. Moreover, such an assessment should have a multifaceted nature, expressed in determining the quality level, use efficiency, and the property importance for a construction organization.

Let us consider the main stages of the algorithm presented in Figure 1.

At the first stage, it is necessary to assess the economic efficiency of the production business processes' implementation in a construction organization. Here it is necessary to take into account the economic analysis seriousness of the organization's activities. In economic terms, the emphasis should be on profit and profitability indicators. Analysis of these indicators will allow not only to understand how efficiently the construction organization works, but also whether it is possible to outsource a part of the organization's business processes or not.

At the second stage, the efficiency indicators' calculation using the fixed assets of the construction organization is carried out. Conducting this analysis will determine the factors that hinder the construction business effectiveness. Initially, it is best to focus on the capital productivity indicators, the deviation from the standard values of which will reveal the presence of a problem situation.

At the third stage, it is necessary to determine the adequacy of the construction organization potential to carry out the measures to improve the use efficiency of the fixed assets involved in the production business processes' implementation. In our opinion, it is best to use a combination of quantitative and qualitative indicators using the rating determination methodology based on a special test.

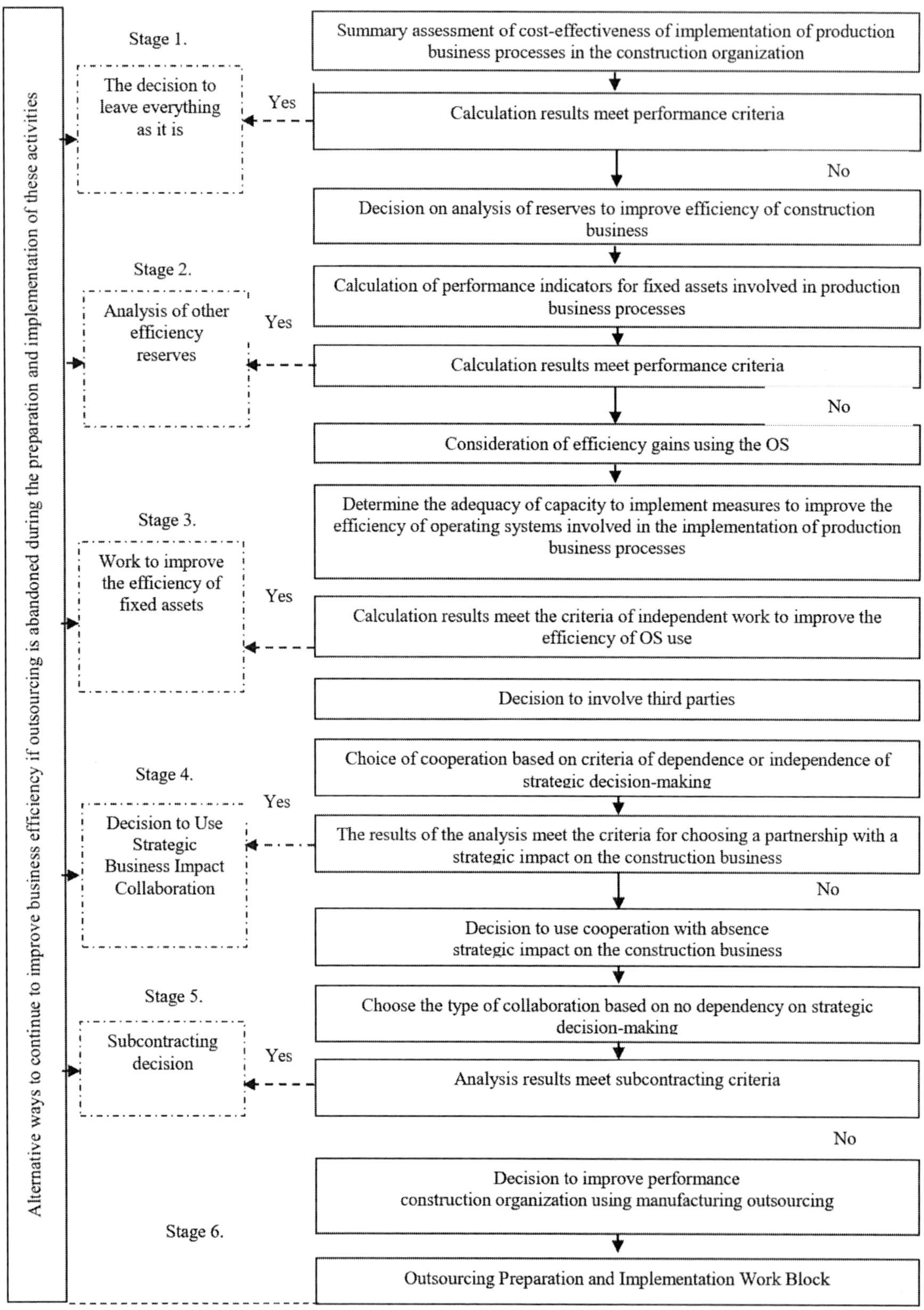
Alternative ways to continue to improve business efficiency if outsourcing is abandoned during the preparation and implementation of these activities
Stage 1.
The decision to leave everything as it is
Yes
Summary assessment of cost-effectiveness of implementation of production business processes in the construction organization
Calculation results meet performance criteria
No
Decision on analysis of reserves to improve efficiency of construction business
Stage 2.
Analysis of other efficiency reserves
Yes
Calculation of performance indicators for fixed assets involved in production business processes
Calculation results meet performance criteria
No
Consideration of efficiency gains using the OS
Stage 3.
Work to improve the efficiency of fixed assets
Yes
Determine the adequacy of capacity to implement measures to improve the efficiency of operating systems involved in the implementation of production business processes
Calculation results meet the criteria of independent work to improve the efficiency of OS use
Decision to involve third parties
Stage 4.
Decision to Use Strategic Business Impact Collaboration
Yes
Choice of cooperation based on criteria of dependence or independence of strategic decision-making
The results of the analysis meet the criteria for choosing a partnership with a strategic impact on the construction business
No
Decision to use cooperation with absence strategic impact on the construction business
Stage 5.
Subcontracting decision
Yes
Choose the type of collaboration based on no dependency on strategic decision-making
Analysis results meet subcontracting criteria
No
Decision to improve performance construction organization using manufacturing outsourcing
Stage 6.
Outsourcing Preparation and Implementation Work Block

Figure 1. Algorithm of a methodology for increasing the construction organization efficiency in terms of using production outsourcing

When using this technique, it is necessary to rely on the following indicators [4]:

1) provision of the organization with its own funds;
2) provision of the organization with borrowed funds;
3) sufficiency of the authorized capital of the organization;
4) investment attractiveness of the organization;
5) formation of the basic and working capital of the organization;
6) capital turnover of the organization;
7) liquidity of the organization's balance sheet;
8) the solvency of the organization;
9) return on assets, equity and sales.

Consequently, the construction organization should have property that, if necessary, should provide it with the possibility of obtaining a loan from the bank. Otherwise, it is possible to talk about the low importance for the construction organization of its property complex, which allows it to reasonably attract partners to increase the business efficiency.

At the fourth stage, it is necessary to choose the cooperation option for the construction organizations, relying on the dependence or independence criterion of strategic decision-making in the interaction process.

When choosing the need for cooperation with other construction organizations, it is necessary to take into account two possible options for the development of the situation in the future:

1) the ability to work, taking into account cooperation ties while maintaining the independence of decision-making of partners in the long term;
2) the ability to work taking into account cooperation ties without maintaining independence in the development of long-term solutions.

At the fifth stage, it is necessary to determine what type of cooperation will be used in making the long-term decisions.

The choice of cooperation based on the absence of dependence in making strategic decisions implies two options for subsequent cooperation within the framework of production outsourcing: it will be either a partnership based on the full responsibility of the outsourcing customer to the subcontracting construction customer, or based on outsourcing with distributed responsibility.

Let us designate the criteria for choosing between outsourcing interaction and subcontracting.

Subcontracting in this case can be considered as a means of temporary additional loading or addition of production capacities on a one-time basis, while the operational and production management of the business process remains under the subcontracting customer's jurisdiction. Distributed responsibility outsourcing implies a strategic decision on the further refusal to independently execute one or another production business process (with the possibility of liquidating the corresponding division and the freed-up property) and transferring it on a contract basis to an outsourcer, together with the authority for operational and production management of this business process. Outsourcing gives an opportunity to redistribute the risk for the final product between the construction organization and the cooperation partner, then the presence of a developed outsourcing market of the required qualifications contributes to the development of a decision on outsourcing cooperation.

So, first of all, the management of a construction organization needs to analyze the market for outsourcing services for the considered production business process' implementation. The purpose of analyzing the market for outsourcers capable of carrying out the execution of the desired production business process should be reduced to choosing the most appropriate partner. A guideline in choosing an outsourcer can be the results of evaluating the following criteria for a potential outsourcing partner [6]:

1) compliance of the outsourcer's qualifications with delegated powers;
2) application of new technologies;
3) the period of existence on the market;

4) a positive history of project implementation, the presence of positive customer reviews;
5) the rating of economic sustainability;
6) the cost of the services provided.

The listed criteria should be assessed based on the previously collected information using the method of expert assessments. In this case, the weight of each criterion can be determined by the experts using pairwise comparison. In turn, the most preferable will be the applicant who will eventually have the maximum total score for all criteria.

Thus, the fact of the presence of an outsourcer that meets the requirements of a construction organization, as well as a satisfactory assessment of the external partner's operational and production management efficiency can become the objective prerequisites for the implementation of outsourcing of a production business process in a construction organization.

At the sixth stage, it is necessary to carry out the work on the preparation and implementation of outsourcing in a construction organization.

After an informed approach to the decision to use outsourcing based on certain criteria, the management of the construction organization should:

1) to determine what type of outsourcing necessary at this stage for the integrity of the business processes transferred to the side (full or partial);
2) to give a reasonable economic and strategic assessment of the outsourcing use;
3) to develop measures to adapt the organizational structure.

In this case, the basis of the listed criteria can be a set of indicators calculated in the process of substantiating a decision on outsourcing. Thus, the assessment of economic efficiency at the stage of implementation of an outsourcing contract in the developed methodology will be an adequate component of the general approach to assessing the feasibility of using outsourcing in the construction organizations' activities.

Summary

We believe that it is necessary to carry out an examination of the need for both full and partial outsourcing forms implementation. Based on the results obtained, the management is able to draw a conclusion about certain advantages of the outsourcing process.

The conclusion and implementation of an outsourcing contract for a construction customer should be associated with an actual increase in the degree of its sustainability, or at least should guarantee the level that existed before outsourcing, since sustainability is a significant characteristic in the context of the interest of partner companies in the construction activities organizations.

References

[1] Kizim A A, Kazakovtseva E B 2016 Outsourcing of modern network technologies *Economy of sustainable development* **1** 192-198.
[2] Kirillova A A 2015 *Outsourcing as a tool to improve the efficiency of the organization RISK* **4** 329-332.
[3] Koba E E 2015 Outsourcing of the internal control system *Economy and entrepreneurship* **6** 1163-1165.
[4] Kotlyarov I D 2015 Outsourcing as a form of inter-firm cooperation: theoretical analysis *Bulletin of the Institute of Economics of the Russian Academy of Sciences* **5** 19-31.
[5] Kotlyarov I D 2013 Methodological problems of determining the price of outsourcer services *Problems of economics and management of the oil and gas complex* **11** 21-24.
[6] Kutuzova T F, Rudenko O N 2015 Outsourcing as a factor in increasing the efficiency of enterprises' activities *Actual problems of economics and management* **4** 32-36.
[7] Churkina L S 2016 Outsourcing of business processes of the *Economics enterprise* **12** 72-73.
[8] Linev I V 2016 Outsourcing in the construction industry *Herald of VSUET* **4** 453-459.
[9] Lutskaya N V 2016 Outsourcing: the level of services provided and models of interaction between the parties *Competence* **2** 28-34.

[10] Mandrik N V 2016 Outsourcing as a factor in the effective development of an enterprise in a competitive environment *Economy and entrepreneurship* **2-2** 978-981.
[11] Poleshchuk V S 2015 Outsourcing as a factor of the company's competitiveness *Professional project: ideas, technologies, results* **2** 34-37.
[12] Sinyaev V I 2008 Development of a commercial outsourcing system in construction *Bulletin of Belgorod University of Consumer Cooperatives* **1** 191-196.
[13] Sinyaev V V, Zhukov A A 2012 Outsourcing management in the construction sector of small and medium-sized businesses *Vector Science TSU. Series: Economics and Management* **4** 149-155.
[14] Sinyaev V I 2014 Outsourcing of marketing services in the field of construction entrepreneurship *Russian entrepreneurship* **22** 129-136.
[15] Solovyeva V E, Lebedev I S 2015 Outsourcing in construction *Scientific works of the Kuban State Technological University* **6** 240-249.

Increasing the economic efficiency of design and construction solutions due to the automated identification of construction works and structural elements of information models

I A Tyurin*, A V Ginzburg

Moscow State University of Civil Engineering, Yaroslavl highway, 26, Moscow, 129337, Russian Federation

E-mail: it3000@yandex.ru

Abstract. Currently in many countries around the world there is an active introduction of information modeling technologies in the construction industry. This process is largely aimed at optimizing approaches to the formation of design estimates for capital construction objects. However the established views on the development of project documentation do not always fit into the concept of new approaches. In an attempt to fit traditional processes of identification of construction works and structural elements into information modeling technologies modern engineers are faced with the problems of optimizing and automating established processes. Modern software for the identification of construction work and structural elements although is developed taking into account the interaction with information models, does not change the previously established design approaches. This article explores a methodology that can significantly increase the economic efficiency of design and construction solutions by automating the identification of construction work and structural elements using information modeling technologies.

1. Introduction

Undoubtedly the introduction of information modeling technologies (IMT) in the construction industry is a global trend. Everywhere in most countries of the world at the state level implementation plans, development strategies and other driving aspects are being developed. The use of the IMT for the design of construction projects receives all kinds of preferences from government customers. In Russia it is increasingly becoming mandatory to develop project documentation only by means of the IMT, which is clearly indicated in the technical specifications of State contracts. In 2019-2020 state expert organizations began accepting projects of construction projects only in electronic form with the possibility of presenting them in the format of information models. Several such projects have already successfully received positive opinions and despite the fact that this was the first experience the projects were examined on IMT e and in some cases even earlier than the deadline [1].

The advantages of the IMT using in the development of design documentation for construction projects over traditional design methods are obvious. The IMT have certain disadvantages as all new technologies especially those that are introduced so intensively. They are expressed in the use of foreign software IMT (software IMT), which is usually developed under the state standards of the development country. For example, at the output of the process of generating architectural and construction drawings documentation is generated that does not meet the design requirements established by the Russian

Federation. Design engineers have to modify the generated drawings in manual mode, which in some cases is comparable to the development of such drawings by the traditional method. Although solutions are provided for joint work in information spaces on the subject [2].

This trend can be traced throughout the design process using the IMT. This is the incompatibility of the data exchange formats between the IMT software developed by many vendors, the lack of feedback between the generated output information and the information model, the lack of the IMT solutions for the development of specialized sections of project documentation and so on. Despite this IMT allow not to be limited only to the development stage of design and estimate documentation but are successfully applied at all stages of the life cycle of the object under consideration [3].

The situation with the preparation of estimate documentation is a standing out from the whole mass of the IMT shortcomings. This is due to the fact that the formation principles of estimate documentation are of a purely territorial nature. The process of drawing up estimates differs not only between countries but sometimes also between regions of a single country. For example in the Russian Federation in different regions, different collections of estimated standards are used and if you recalculate the estimated cost of construction using different regional collections the total amounts will differ quite significantly. One of the options for solving the problem of estimated norms variability is the use of enlarged estimated standards in the process of creating an information model but this option for constructing a price model gives significant errors as a result of calculations [4].

In addition to differing territorial estimate standards, the goals - for which in fact the development is being carried out - have a strong influence on the principles of developing estimate documentation. In many countries including the Russian Federation, estimate documentation can be attached to the concluded contracts and have legal force along with the terms of the contract. In some countries the estimate documentation is only informative in nature and may not be drawn up at all [5]. In the Russian Federation the development of estimate documentation is mandatory for objects that are financed with the involvement of the state budget. For other capital construction projects, the preparation of estimate documentation is allowed but not mandatory [6].

According to state statistics in 2018 in the Russian Federation the share of state financing in investment and construction projects amounted to more than 70% of the total mass and if given the trend of commercial customers to establish requirements for the estimates inclusion in project documentation, moreover it was developed according to state budget standards and not according to firm prices, it can be assumed that in the Russian Federation in 80% of project documentation cases development estimates were also developed [7].

For many people the estimate documentation is presented in the form of ordinary calculations, sometimes not always objective sometimes the estimate starts to be studied only when the project has passed the state expert review.

This is due to well-established design principles, which imply the development of cost estimates as they say "post facto" that is engineering – technical solutions are put ahead of the economic component. It is connected with limited possibilities of automated design systems (CAE) and the estimated program complexes (EPC) used at traditional approaches of working out of the design-budget documentation. However with the IMT arrival in the construction industry the concept can and should change [8]. The IMT priority directions should be the technology of a construction object as the objects are built not for the construction or implementation of some engineering solutions but have real goals both in the development of the state and in commercial terms.

In most countries including Russia state collections of estimate standards are used for preparation of estimate documentation. The basis of the estimate rationing in Russia is the State elemental estimate standards (SEES) and methods of their preparation, updating and application. SEES are collections of tables with indication of resource requirements for building and construction works (BCW) and descriptive part (technical part of the collection) with indication of application peculiarities of this or that table positions depending on BCW performed. An example of a table from the collection of the SEES is shown in Figure 1.

IOP Conf. Series: Materials Science and Engineering 1083 (2021) 012076 doi:10.1088/1757-899X/1083/1/012076

Table SEES 06-01-012 Formwork (bottom) and supporting structures for high grillages

Scope of work:

01. Cutting and boards installation. 02. Installation of formwork panels. 03. Fastening of formwork elements with construction nails.

Meter: 100m^2 area of the grillages horizontal projection

06-01-012-01 Formwork (bottom) and supporting structures for high grillages

Resource code	Cost element name	Unit of measure	06-01- 012-01
1	Labor costs of construction workers	mhrs	95,92
1.1	Average work category		2,9
2	Labor costs of machine operators	mhrs	0,34
3	MACHINES AND MECHANISMS		
021141	Cranes on the road while working on other types of construction	mash.hour	0,07
030101	10 t	mash.hour	0,27
400001	Lifting loaders5t	mash.hour	0,1
	Flatbed vehicles, carrying capacity up to 5 t		
4	MATERIALS		
101-1805	Construction nails	t	0,0147
102-0053	Edged boards of coniferous species 4-6.5 m long, 75-150 mm wide, 25 mm thick, grade III	m^3	0,32
102-0061	Edged boards of coniferous species 4-6.5 m long, 75-150 mm wide, 44 mm thick and more, grade III	m^3	0,42
		m^2	5,44
203-0512	Board panels 40 mm thick	t	0,021
405-0253	Unhydrated lime for construction, grade I	m^3	0,061
411-0001	Water		

Figure 1. Example of a table from the SEES collection

As can be seen from Figure 1, the SEES table contains a brief description of the technological processes of this type of the BCW and departments with the need for manpower, mechanized equipment and construction resources. That is it contains information from which if desired it is possible to form resource sheets, to carry out calendar and network planning of works, to prepare schedules of materials delivery and so on [9-10], which is currently completely absent in the design process.

2. Formulation of the problem

In various IMT software the automation of identification of construction works and structural elements is implemented through the use of databases with “primary” and “secondary priorities”. The use of databases with "primary priority" is carried out by returning to the information model of elements containing estimated attributes. This method of automated identification of construction works and structural elements is widely used in large design and construction companies that are concerned about the economic and organizational components of their activities, and they have the ability to accumulate information on implemented projects and form their own corporate databases with "primary priority". This option of automated identification of construction work and structural elements is very effective but its application is possible only in the design and construction company of the database developer.

The approach to automation of construction works identification using databases of "secondary priority" is based on the application of collections of estimated standards post factum development of engineering solutions. This approach is the first attempt to automate the development of estimate documentation using the IMT [11].

As can be seen from the description of the current applied approaches to automate the identification of construction work and structural elements using the IMT although they fulfill the task but they have a number of significant disadvantages. This is due to the lack of methods for the selecting the most cost-effective design solutions and established technology for performing design work. Also an essential factor that limits the adoption of cost-effective design solutions is the lack of approaches to the abundance of information contained in the SEES.

3. Methods and techniques

The proposed methodology for automated identification of construction works and structural elements in information models of capital construction objects is based on the use of “secondary priority” databases as “primary priority” databases. This is based on the methods of forming SEES, which indicates that the tables are not tied to a specific construction organization and reflect the average demand for material, technical, construction and human resources for this type of the BCW.

The proposed methodology for the automated identification of construction works and structural elements is based on the automated assignment principle of attributes of the corresponding SEES tables to the information model elements. From there having such a powerful tool as the IMT it becomes possible, already in the process of developing project documentation to track changes in the estimated cost of construction when enumerating possible types of technical solutions [12] as well as to obtain related information that will help to track changes in construction time, compare the resource availability of the construction region with the need for construction resources of the projected capital construction facility and perform many other related calculations. In this way having the technical ability and working prototypes of the automated identification system for construction works and structural elements by a small number of attributes assigned at the stage of the conceptual project design it remains to figure out their use in the development of building information models.

Consider an example: when developing project documentation, the design engineer considers an element of the information model "wall" as a free constructive space that can be filled with various materials: brickwork, wooden beams, foam concrete or tongue-and-groove blocks (Figure 2).

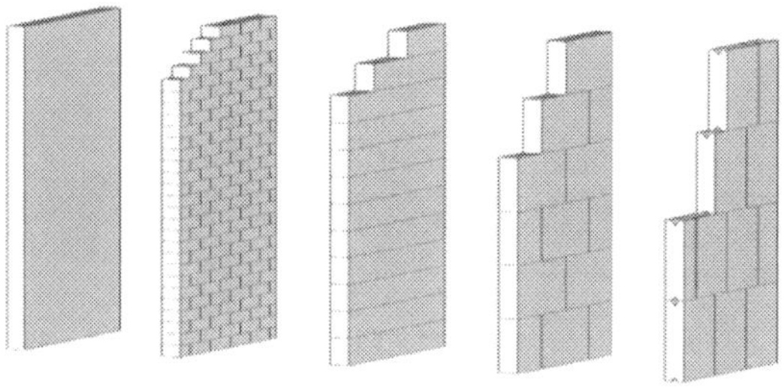

Figure 2. Visualization of the options selection for filling the element "wall" (from left to right: the element "wall", options for filling 1, 2, 3, 4).

Applying traditional approaches to the implementation of design work, the design engineer can choose any option since all four presented options are absolutely suitable for adoption into the project. The choice of a specific building material is usually based on the specification or the personal experience of the design engineer. In order to formalize the choice of one of the options the selection criteria are required, similar to the criteria of engineering and functional stability of life support objects [13]. From there having established the criteria for building materials it is possible to determine the indicator of their interchangeability.

To determine the interchangeability indicator, the following criteria are proposed, which can be obtained from the SEES collections:

- criterion C1: applicability in a specific filled space of the structure;
- criterion C2: the amount of human labor costs to perform the BCW
- criterion C3: the need for machines and mechanisms;
- criterion C4: the amount of associated construction resources.

For example during the parametric description of a "wall" element, the design engineer manually or using automated identification means assigns a "basic" building material - ordinary brick - to fill the considered structural space. Criterion C1 is determined by comparing the applicability of the considered BCW technology and if the value of the "partition" defined as a building element coincides criterion C1 is assigned a value equal to 100, otherwise -100. Criteria C 2, 3, 4 are calculated by comparing the difference value of the considered departments of the SEES tables to the "basic" building material and if the difference is negative, then the criterion is assigned a value of -10, otherwise the criterion is +10. The interchangeability index is calculated by the formula:

$$P(i) = C1 + C2 + C3 + C4;$$

The $P(i)$ values can range from -140 to 140. Significant proportion of C1 in the calculation is assigned to exclude building materials that are not suitable for filling the space of the building structure, thus the selected SEES tables with the value $P(i) < 60$ are discarded and not used in further identification. For the rest of the case the resulting rating table of interchangeable building resources of the SEES is compiled.

In the case of applying the proposed approach of automated identification of construction works and structural elements the design engineer receives more complete information about the decisions made (Figure 3).

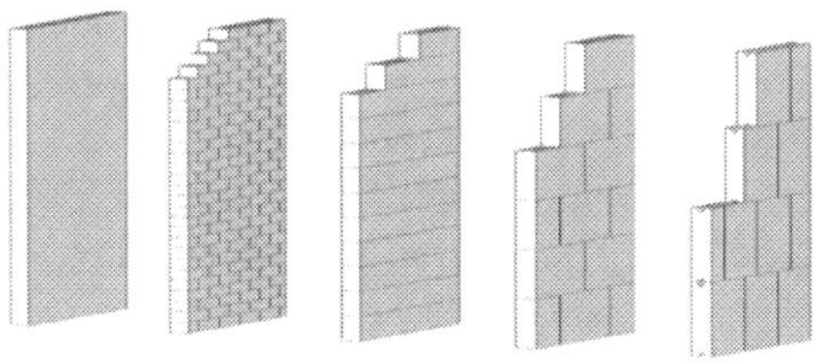

Construction space	Brick filling	Wood filling	Foamed concrete blocks filling	Tongue-and-groove plates filling
P(i)=	100	-120	120	90
Material cost, $/100m^2	256.00	345.00	225.00	750.00
The cost of work $/100 m^2	608.00	456.00	650.00	249.00
Total, $/100 m^2	864.00	801.00	875.00	999.00
Work time (team of 6 people), shift/100 m^2	3.5	2.6	3.8	1.5

Figure 3. Visualization of variants selection of the "wall" element filling when applying the proposed methodology (from left to right: the "wall" element, variants of filling 1, 2, 3, 4).

As can be seen from Figure 3, there is information both on cost and the duration of work, which allows to carry out the analysis of economic efficiency of accepted constructive decisions in a mode of real time instead of post facto drawing up of the estimate documentation. In addition to the interchangeability indicator the list of information returned from the SEES included labor costs based on the fact that the time factor may prevail over the economic component of a single construction object. The proposed methodology provides an opportunity to set appropriate priorities and perform invariant calculations.

Implementation of the proposed approach is also of variant nature. During the survey the approach was implemented as a plugin "Cost Assistant" for Autodesk Revit software (Figure 4).

IOP Conf. Series: Materials Science and Engineering 1083 (2021) 012076 doi:10.1088/1757-899X/1083/1/012076

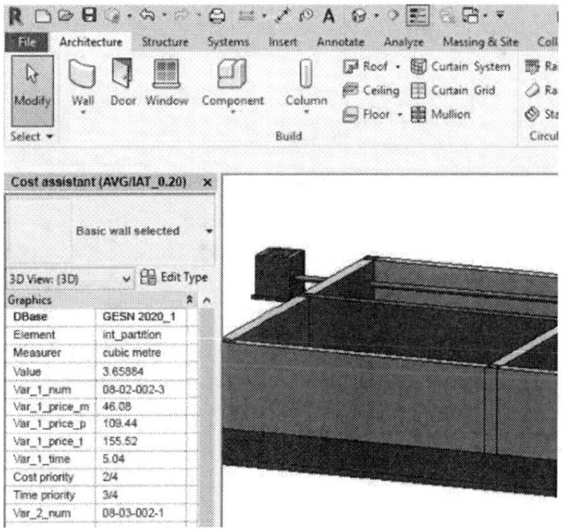

Figure 4. The approach implementation in the form of a plug-in "Cost Assistant" for the software IMT Autodesk Revit

When selecting an information model element and calling the "Cost Assistant" plugin, the following information is available to the user:

- DBase" - the current database of estimated standards (in the example, the SEES database as of Q1 2020 has been selected);
- Element" – is the selected element type ("int_partition" - internal partitions);
- "Measurer" – is the unit volume meter of the element;
- "Value" – is the volume value per meter;
- "Var_1_...", "Var_2_...", "Var_t_...". – is the result of identification of a construction element in the selected database;
- "Cost priority" – is the priority (rating) among the selected options set by the price parameter ("2/4" 2 of 4);
- Time priority" – is the priority (rating) among the selected options set by the time parameter.

As can be seen from the above the price and time ratings of filling the constructive space with the offered materials become available for the user. The plugin work can be provided both with manual assignment of one of the selected options and in automatic mode initially assigning priority to reduce the cost or duration of construction of the analyzed element of the information model.

Application of the proposed methodology already at the stage of sketch elaboration of volumetric planning solutions makes it possible to analyze the erection cost (filling the structural space) of various types of construction resources. As a result of the application of the proposed methodology for automated identification of construction works and structural elements it is assigning estimated attributes to the elements of the information model with the subsequent generation of estimated documentation in automatic mode.

The proposed approach significantly reduces the duration of the design documentation development by reducing the time for decision-making on the erected structural elements of the information model of the capital construction object.

4. Summary

The introduction of IMT in the construction industry is a breakthrough event that has brought huge opportunities for investment and construction projects. The potential of IMT tools in particular the selection of cost-effective technical solutions elements of information models of capital construction objects is very significant in comparison with the established technology of designing construction projects.

All this allowed to look at design processes in a new way and to develop methodology of automated identification of construction works and structural elements. The proposed methodology allows to make the most cost effective decisions already at the stage of sketch design development.

The proposed methodology is aimed at improving the economic efficiency of design and construction solutions through automated identification of construction works and structural elements of information models through the use of databases with "secondary priority" in the technology that provides for the use of databases with "primary priority". Assigning the elements of the information model attributes from the compendia of the SEES gives additional opportunities for planning the organization of construction processes, calendar and network construction schedules, as well as the implementation of related calculations of needs and the movement of manpower on the construction site, analysis of compliance with the resource availability of the region of construction and so on.

The proposed methodology has a significant potential, many times greater than currently available solutions to automate the estimated cost of construction as it provides optimization of the development of design documentation at all stages of design not only at the stage of making the estimated documentation.

References

[1] Lyapina A R, Borodin S I 2018 Information questions of technologies implementation of modeling in construction: state expertise of projects in Russia *Izvestia vuzov. Investments. Construction. Real Estate* **8 (2) (25)** 11-17.

[2] Anastasia Stepanova, Liubov Shilova 2020 Prospects of using Task management in design *23rd International Scientific Conference on Advance in Civil Engineering: Construction - The Formation of Living Environment, FORM 2020; National University of Civil Engineering55 Giai Phong RoadHanoi; Viet Nam; 23-26 September 2020* **869 (6)**.

[3] Ginzburg A V 2018 LE IM: Living Environment Information Modelling *International Scientific Conference Environmental Science for Construction Industry, ESCI 2018; Ho Chi Minh City; Viet Nam: MATEC Web of Conferences* **193** 05030.

[4] Pilyay A, Shilova L 2018 The use of normative basis for the construction cost for introduction of 5D BIM in Russia *21st International Scientific Conference on Advanced in Civil Engineering: Construction - The Formation of Living Environment, FORM 2018; Moscow State University of Civil Engineering (MGSU) Moscow; Russian Federation; 25-27 April 2018* **365 (6).**

[5] Rosasco P, Sdino L The regional price lists for estimating the costs of construction, Green Energy and Technology, ISSN:1865-3529, E-ISSN:1865-3537, Pages 213-229

[6] Town-Planning 2019 Code of the Russian Federation 29.12.2004 № 190-FL (edit от 02.08.2019) (add., entered into force 13.08.2019)

[7] Information on https://rosstat.gov.ru/storage/mediabank/Stroit-2018.pdf, accessed on: 15/11/2020.

[8] Ginzburg A V, Ryzhkova A I 2018 Possibilities of artificial intelligence to improve the organizational and technological reliability of construction production *Vestnik MGSU* **13 (1) (112)** 7-13.

[9] Ginzburg A V, Kagan P B 1999 CAD of construction organization *CAD and Graphics* **9** 32-34.

[10] Kagan P B, Ginzburg A V 1997 Automation of organizational and technological design in construction *Design Automation* **4** 36-45.

[11] Kuznetsov M A, Kokareva Ya A 2020 Automation of budgeting based on the BIM model *Collection of scientific articles of the XI International Scientific and Practical Conference "Improving the managerial, economic, social and innovative and technical potential of enterprises, industries and national economic complexes", Penza, May 21-22*

[12] Ilya Tyurin 2020 Automation identification of construction work and structural elements in BIM development *International Scientific Conference "Construction and Architecture: Theory and Practice of Innovative Development" (CATPID-2020) - Part 1 26-30 September 2020, Nalchik, Russian Federation* **913** 042010.

[13] Volkov A, Shilova L 2016 Some criteria of critical infrastructures stability *5th International Scientific Conference on Integration, Partnership and Innovation in Construction Science and Education, IPICSE 2016; Moscow, State University of Civil Engineering; Russian Federation; 16-17 October 2016* **86** 05009.

Comparative analysis of thermophysical characteristics for screen insulation and building heat-insulating materials

V I Lepilov, D G Usadskiy, A V Kovylin*

Volgograd State Technical University, 2 Degtyareva street, Volgograd, 400006, Russia

E-mail: Kovylin.andrei@mail.ru

Abstract. Methods for calculating massive screen insulation consisting of metal sheets with sufficient volumetric heat capacity and air layers are presented. The comparison of the thermophysical characteristics of a massive screen system with the steel screens with building thermal insulation materials is given.

Introduction

In stationary processes of heating or cooling bodies, the most important is the coefficient of thermal conductivity, which characterizes the ability of a substance to transfer heat [1]. When studying the non-stationary processes of heating or cooling bodies, along with the thermal conductivity coefficient, the thermal diffusivity of the substance is used, which characterizes the rate of temperature equalization over the body volume and the volumetric heat capacity of the substance, which is the product of the specific heat capacity by the substance volume.

Main Part

A setup diagram for studying the properties and characteristics of massive screen insulation is shown in Figure 1. It represents a structure of steel sheets and air spaces between them.

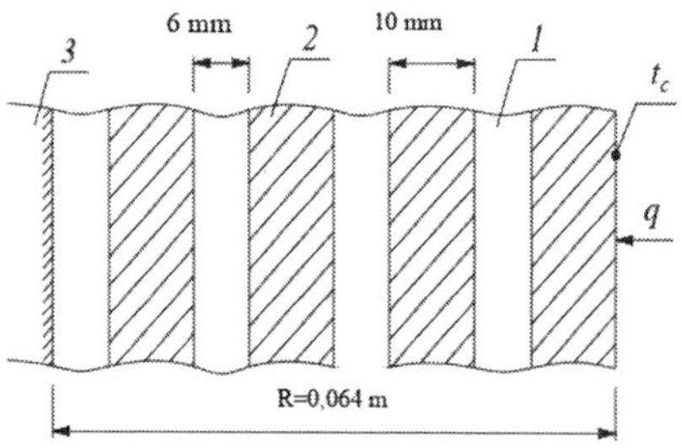

Figure 1. Screen scheme: q — specific heat flux, W / m^2; t_w — wall temperature, °C; *1* — air gap; *2* — steel sheet; *3* — sample (body)

To find the effective thermal conductivity coefficient (this coefficient means the thermal conductivity coefficient of the entire system of screens with air spaces) [2], let us consider the ratio:

$$\frac{R}{\lambda_{ef}} = n\frac{\delta_{\text{air}}}{\lambda_{\text{air}}} + n\frac{\delta_{\text{st}}}{\lambda_{\text{st}}}, \tag{1}$$

where R — is the total thickness of the screen system, m; λ_{ef} — is effective thermal conductivity, W/(m·K); $\lambda_{\text{air}} = 3{,}21\cdot10^{-2}$ W/(m·K) — is the air thermal conductivity coefficient; $\lambda_{\text{st}} = 45{,}24$ W/(m·K) — is the thermal conductivity coefficient of steel; δ_{air} and δ_{st} — denote the thickness of one air layer and one steel screen, respectively; m; n — is the number of layers, pcs.

From the relation (1), we express the effective thermal conductivity coefficient:

$$\lambda_{ef} = \frac{R}{n\frac{\delta_{\text{air}}}{\lambda_{\text{air}}} + n\frac{\delta_{st}}{\lambda_{\text{st}}}} \tag{2}$$

Substituting the corresponding values into the formula (2), we obtain, W/(m·K):

$$\lambda_{ef} = \frac{0{,}064}{4\frac{0{,}006}{0{,}032} + 4\frac{0{,}01}{45{,}24}} = 0{,}085.$$

To find the effective volumetric heat capacity (we mean the volumetric heat capacity of the entire system package of screens with air layers) [3], let us consider the equation of the sum of the volumetric heat capacities:

$$(c\rho)_{\text{ef}}\cdot V_0 = (c\rho)_{\text{air}}\cdot V_{\text{air}} + (c\rho)_{\text{st}}\cdot V_{\text{st}}, \tag{3}$$

where $(c\rho)_{\text{ef}}$ — is effective volumetric heat capacity of the screen system, J / (m^3·K); $(c\rho)_{\text{air}}$, $(c\rho)_{\text{st}}$ — volumetric heat capacity of air and material of screens included in the system, J/(m^3·K), respectively; V_0, V_{air}, V_{st} — volume of the entire screen system, m^3, total air space, m^3, screen material volume, m^3, respectively. It is clear that these volumes are determined by the formulas:

$$V_0 = RF;\ V_{\text{air}} = \sum_1^n \delta_{\text{air}} \cdot F;\ V_{\text{st}} = \sum_1^n \delta_{\text{st}} \cdot F, \tag{4}$$

where F — is the screen surface area, m^2.

Substituting the values of these volumes into the formula (3), we get:

$$(c\rho)_{\text{ef}}\cdot RF = (c\rho)_{\text{air}}\cdot \sum_1^n \delta_{\text{air}} \cdot F + (c\rho)_{\text{st}}\cdot \sum_1^n \delta_{\text{st}} \cdot F. \tag{5}$$

We divide both sides of the equation by the surface area of the screening system — F, m^2 and get:

$$(c\rho)_{\text{ef}}\cdot R = (c\rho)_{\text{air}}\cdot \sum_1^n \delta_{\text{air}} + (c\rho)_{\text{st}}\cdot \sum_1^n \delta_{\text{st}}. \tag{6}$$

This equation is the equation of the linear volumetric heat capacities sum for calculating the bulk screen systems. The equation characterizes the linear distribution of volumetric heat capacities included in the massive screening system [4, 5], referred to 1 m^2 screen area. In the future, for the calculations, we will use exactly this formula.

For our particular case, the formula will take the form:

$$(c\rho)_{\text{ef}}\cdot R = n(c\rho)_{\text{air}}\cdot\delta_{\text{air}} + n(c\rho)_{\text{st}}\cdot\delta_{\text{st}}. \tag{7}$$

Let us express the effective volumetric heat capacity from this equation

$$(c\rho)_{\text{ef}} = [n\,(c\rho)_{\text{air}}\cdot\delta_{\text{air}} + n\,(c\rho)_{\text{st}}\cdot\delta_{\text{st}}] / R. \tag{8}$$

We substitute the data into the formula (8) and obtain the effective volumetric heat capacity, J/(m^3·K):

$$(c\rho)_{ef} = [4(0{,}946\cdot1009)\cdot0{,}006 + 4(394\cdot7900)\cdot0{,}01] / 0{,}064 = 19{,}45\cdot10^5.$$

The thermal diffusivity is determined by the known expression:

$$a = \frac{\lambda}{c\rho}. \quad (9)$$

Substituting the data into the formula (9), we obtain, m²/s:

$$a = \frac{0{,}085}{19{,}45\cdot10^5} = 0{,}44\cdot10^{-7}$$

The results obtained are included in Table. 1.

Table 1. The comparison of thermophysical characteristics of a massive screens system (four flat steel screens) with air layers at t_c = 100 °C with monolithic: steel plate, concrete, cinder block, air with the same size *R*, m

material name	λ, W/(m·K)	(cρ), J/(m³·K)	*a*, m²/s
Massive screen system	0.085	$1.95\cdot10^6$	$0.044\cdot10^{-6}$
Steel plate	45.24	$3.1\cdot10^6$	$12.5\cdot10^{-6}$
Concrete	1.28	$2.6\cdot10^6$	$0.5\cdot10^{-6}$
Slag concrete	0.43	$1.9\cdot10^6$	$0.24\cdot10^{-6}$
Air	0.032	955	$33.6\cdot10^{-6}$

Table 1 shows a comparison of the thermophysical characteristics between massive screening system with steel screens with other materials.

From the comparison (see Table 1), it can be concluded that the effective thermal conductivity coefficient of the massive screening system (steel screens with air layers) is less than the thermal conductivity coefficients' values of all the materials under consideration, except for air and expanded polystyrene, and the system under consideration acquired the properties of a heat-insulating material [6, 7]. The value of the effective thermal conductivity coefficient of the screen system λ_{ef} is less than the thermal conductivity coefficient of steel by 532 times, concrete by 15 times, cinder concrete by 5 times, but more than the thermal conductivity of air by 2.66 times.

Thermal diffusivity a_{ef} screen systems are less than the thermal diffusivity of steel by 284 times, the thermal diffusivity of air by 764 times, concrete by 113 times, slag concrete by 5.5 times, expanded polystyrene by 5.7 times.

Also, at present time, an important role is played by the heat losses reduction from external barriers, therefore, it is important to use high-quality thermal insulation materials. Knowledge of the thermophysical properties (TPP) makes it possible to choose a heat-insulating material with the best heat-insulating characteristics, which will reduce heating costs and reduce heat losses of the building. To study the TPP, a sample was made from a mineral wool plate with the dimensions of 250 × 250 × 50 mm. The sample is a mineral wool heat-insulating plate, which contains basalt fiber, which has excellent fire-retardant and thermal insulation properties. A distinctive feature of the mineral plate is its high thermal insulation efficiency.

The experimental research procedure runs as follows. The test sample is placed in a device for implementing the method. The temperature of the test sample is measured using an additional thermocouple installed in the middle of the sample surface from the side of the heater to detect the temperature wave formed before the stationary mode onset. The coefficient of thermal conductivity, coefficient of thermal resistance, density of the stationary heat flux determination reveals the temperature wave on the surface of the material under study from the side of the heater. Using the obtained experimental data, the required thermophysical characteristics are determined by the formulas.

In the initial period, the sample temperature was 19 ° C. Before the stationary thermal regime onset the sample was heated from the initial temperature of 19 ° C to 40 ° C in 24 min. Starting from 24

minutes until the end of the measurement, the sample temperature from the side of the heater was T_h = 40°C.
The thermal conductivity coefficient is calculated by the formula:

$$\lambda = \frac{q \cdot \delta}{T_h - T_r} = \frac{16.6 \cdot 0.05}{40 - 17} = 0.036 \text{ W/m}^2$$

where: q – is the density of the stationary heat flux passing through the sample under study, W/m²; δ – is sample thickness, m; T_h – is the heater temperature, °C; T_r – is the refrigerator temperature, °C.

The thermal resistance coefficient is calculated by the formula:

$$R = \frac{T_h - T_r}{q} = \frac{40 - 17}{16.6} = 1.385 \text{ (m}^2K\text{)/W},$$

where T_h – is the heater temperature, °C; T_r – is the refrigerator temperature, °C; q – is the density of the stationary heat flux passing through the sample under study, W/m².

The oscillation amplitude of the temperature half-wave is calculated by the formula:

$$\vartheta_{h\text{-}w} = 0{,}5(T_{max} - T_{min}) = 0{,}5(41 - 17) = 12 \text{ °C}$$

where T_{max} – is the maximum temperature of the test sample from the side of the heater, °C; T_{min} – is the minimum surface temperature of the test sample from the side of the heater, °C.

Thermal activity is calculated by the formula: $B = \frac{q^{max}}{\vartheta_п} = \frac{85{,}6}{12} = 7{,}13$ W/(m²°C),

where q^{max} – is the maximum heat flux density, W/m²; $\vartheta_п$ – is the temperature half-wave amplitude, °C.

The volumetric heat capacity of the test sample is calculated by the formula:

$$c\rho = \frac{B^2 \cdot z}{\lambda \cdot 2\pi} = \frac{7{,}13^2 \cdot 1980}{0{,}036 \cdot 2 \cdot 3{,}14} = 445227{,}62 \text{ J/(m}^3\text{K)},$$

where B – denotes thermal activity, W/(m²°C); z – is the time of temperature measurement from the side of the heater before the onset of stationary mode, s.; λ – is the coefficient of thermal conductivity, W/m².

The thermal diffusivity is calculated by the formula $a = \frac{\lambda}{c\rho} = \frac{0{,}036}{445227{,}62} = 0{,}08 \cdot 10^{-6}$ m²/s,

where λ – is the coefficient of thermal conductivity, W/m²; $c\rho$ – is volumetric heat capacity of the test sample, J/(m³K).
An experimental study was also carried out at a different temperature regime of heating.

In the initial period, the sample temperature was 17 ° C. Before the stationary thermal regime onset, the sample was heated from the initial temperature of 17 ° C to 40 ° C in 25 min. Starting from 25 minutes until the end of the measurement, the sample temperature from the side of the heater was T_h = 40°C.

The growth of the thermophysical properties of the sample was carried out according to the method described above. The obtained thermophysical properties of the heat-insulating mineral plate using two experiments are shown in Table 2.

Table 2. Thermophysical properties of a heat-insulating mineral plate

Value	Experimental data	
	I experiment	II experiment
1. *Th*, °C	40	40
2. T_x, °C	17	19

3. λ, W/(m·K)	0.036	0.035
4. R, (m^2·K)/W	1.388	1.428
5. q, W/m^2	16.6	16.1

The obtained experimental values of the thermal conductivity coefficient λ = 0.036 W/(m·K) and λ = 0.035 W/(m·K) were compared with the passport data of the thermal conductivity coefficient of the test sample, which are equal to λ = 0.036...0.07 W/(m·K). The discrepancy is no more than 10%, that is, the proposed method for determining the TPP of a heat-insulating material makes it possible to accurately determine the thermal conductivity coefficient.

Summary

A significant decrease in the effective thermal diffusivity coefficient of a structure made of steel screens with air spaces is explained by the high thermal resistance of the air layer [8, 9] and the high volumetric heat capacity of the steel layer.

The performed physical and mathematical research confirms the possibility of using massive screen systems with steel screens as protection against thermal effects in stationary and non-stationary modes [10].

References

[1] Kovylin A V, Usadskiy D G, Lepilov V I 2019 The buildings innovative method fencing thermal regime study *International scientific conference on construction and architecture: theory and practice for the innovation development 2019, CATPID 2019 Kislovodsk,* 01-05 october 2019 **055026.**

[2] Lepilov V I 2007 Investigation of thermophysical properties and characteristics of systems of massive screens *author. dis. ... Cand. tech. sciences. Astrakhan.*

[3] Lepilov V I and Perfilov V A 2010 Energy-Saving Fencing Elements *VIII International Scientific Conference The quality of internal air and environment. Volgograd* 201-206.

[4] Mikheev M A, Mikheeva I M 1973 *Heat Transfer Basics* (Moscow, Energy) 319.

[5] Lepilov V I, Karapuzova N Yu, Kovilyn A V 2015 Theoretical foundations for determining the thermal properties of materials and heat and mass transfer processes in fences *a study guide* (Volgograd, VolgSACEU) 113.

[6] Fokin V M, Kovylin A V 2009 Theoretical foundations for determining thermal conductivity, volumetric heat capacity and thermal diffusivity of materials by thermal measurements on the surface by non-destructive testing *Bulletin of the Volgograd State University of Architecture and Civil Engineering. Series: Building and architecture* **14 (33)** 123—127.

[7] Fokin V M, Kovylin A V, Chernyshov V N 2011 *Energy efficient methods for determining the thermophysical properties of building materials* (Moscow, Spectrum) 155.

[8] Fokin V M, Kovylin A V Theoretical foundations for determining thermal conductivity, volumetric heat capacity and thermal diffusivity of materials by thermal measurements on the surface by the method of non-destructive testing *Bulletin of Volgograd State University of Architecture and Civil Engineering, VolgSACEU, Series "Construction and Architecture"* **14** 123-127.

[9] Kovylin A V and Fokin V M 2010 Method of Determining the Coefficients of Thermal Conductivity, Heat Absorption, Thermal Inertia, Thermal Diffusivity and Bulk Heat Capacity of Building and Heat-Insulating Materials by Non-Destructive Testing Method *VolgSACEU Bulletin. Series: "Building and Architecture"* Volgograd **19** 112-117 Pp.

On the question of increasing the purification efficiency of flue gases from the boiler unit of HEPS Volzhsky automobile plant

O A Balandina*, S M Puring, D N Vatuzov, E B Filatova

Samara State Technical University, Academy of Civil Engineering and Architecture, 194, Molodogvardeyskaya st., Samara, 443001, Russia

E-mail: balandinaolya88@rambler.ru

Abstract. The paper considers the measures to suppress emissions of nitrogen and carbon oxides emitted during natural gas combustion, used on the VAZ heat electro-power station territory in Togliatti. The analysis of the quantitative distribution of fuel consumption and pollutant emissions for 2015 - 2019 has been carried out. The description of the process flow diagram of the gas-air duct of the TGME-464 boiler unit (KA No.12) is presented. The possibility of using the method of selective catalytic reduction of nitrogen and carbon oxides on the honeycomb ceramic catalyst surface is considered. The graphical dependences of the volumetric-mass air flow rates, flue gas fuel, specific flow and speed energy on the change in the heat load of the boiler unit are determined. The possibility of using a mathematical model for calculating the change in concentrations and along the length of the catalytic layer depending on the flue gas flow temperature is considered.

1. Introduction

The process of sustainable energy development cannot be imagined without the development of solutions to environmental problems. For this purpose, the platform “Environmentally friendly heat energy of high efficiency” was approved in 2014 [1]. The strategic objectives of this platform include the development and implementation of measures for the flue gases purification from heat electro-power station plants, which ensure minimal harmful emissions of carbon oxides and nitrogen into the atmosphere [1]. The solution to this kind of problems is the development of plans for the existing boiler equipment modernization. For the boilers commissioned before 1995, the transition to the principle of "best available technologies" is being carried out [1, 2]. Therefore, it is necessary to make a quantitative assessment of the amount of harmful emissions into the atmosphere at the modernization stage for the existing equipment [1].

The HTEPS of the Volzhsky Automobile Plant of the Samara branch of PJSC T Plus is one of the largest thermal power plants in Europe, generates up to 40% of electric and 30% of thermal energy from all energy produced in the Samara region [3,4].

2. Materials and Methods

2.1. Analysis of the quantitative distribution of fuel consumption and pollutant emissions from the boiler unit TGME-464 (KA No. 12) VAZ HTEPS.

The VAZ HEAT ELECTROPOWER STATION is intended for power and heat supply of the Avtozavodsky residential area of Togliatti, as well as the Volzhsky Automobile Plant and other industrial enterprises. For the 2019 calendar year, the VAZ heat electro-power station generated electric and thermal energy, respectively - 2 838 561 thousand kW / h and 5 126 824 Gcal [3,4].

At all boilers installed at the VAZ heat electro-power station, the measures to suppress emissions were taken based on the use of flue gas recirculation and modernization of the injection cups. Information on fuel consumption and the amount of pollutant emissions for 2015–2019 is presented in Table 1 [5].

Table 1. Information on fuel consumption and the amount of pollutant emissions from VAZ heat electro-power station for 2015 - 2019.

year	Electricity generation, thousand kW / h	Fuel consumption			Emissions, tons					
		Total	including		Total	including				
			gas	fuel oil		NO_2	NO	CO	SO_2	other
2015	3 355 837	1 656 089	1 656 089	0	1755.9	1 503.2	244.6	8.13	0	0
2016	2 909 819	1 479 299	1 432 692	46 607	3178.0	1 305.9	212.2	7.90	1646.4	0
2017	2 892 578	1 444 708	1 444 708	0	1289.7	1 104.8	179.5	0.12	0	5.3
2018	3 017 595	1 550 730	1 550 730	0	1592.4	1357.8	220.6	8.72	0	5.3
2019	2 838 561	1 429 180	1 429 175	5	1397.6	1191.2	193.6	7.3	0.177	5.3

Environmental data on emissions from VAZ heat electro-power station (Table 1) demonstrate a decrease in emissions in 2019 compared to 2018. The trend towards a decrease in the amount of pollutant emissions into the atmosphere is associated with a reduction in electricity and heat generation, as well as in fuel consumption by the enterprise [5].

According to the recommendations of the Ministry of Energy of the Russian Federation "On Approval of Heat Supply Schemes for Settlements and Urban Districts" dated April 15, 2020, to local authorities when concluding the contracts for the development and updating of heat supply schemes for the settlements, urban districts with a population of 500 thousand people or more, as well as the federal cities values need to include the development of proposals to reduce the harmful substances emissions volume into the air [6].

To reduce CO and NO_x concentrations in the exhaust gases of the VAZ heat electro-power station, the authors of this study consider the possibility of using the method of selective catalytic reduction NO_x and CO to boiler type TGME-464 (KA No.12), installed on the territory of the TPP of the Volzhsky Automobile Plant of the Samarsky branch of PJSC T Plus.

2.2 Analysis of the technological scheme of the gas-air path of the boiler unit.

Boiler unit No. 12 is located in the boiler department with a length of 24 m. Regenerative air heater (RAH) 4 is installed outside the boiler house. On the same open area there is a vacuum deaerator (VD), a gas recirculation fan (GRF) and an exhaust fan (EF) [7].

The flue gas flue from the boiler convection shaft goes to the upper branch pipe of the RAH. The flue gas duct that removes flue gases from the RAH is connected to its lower branch pipe. To emit flue gases into the atmosphere, the boilers are connected by a brick gas duct to the chimney No. 4 with a height of 250 m [7].

The air supply path, the removal of boiler flue gases, as well as the flue gas recirculation path are made according to a single-line scheme, equipped, respectively, with one blower fan of the VDN-25x2-1 type, one RAH-88, one backup smoke exhauster of the DOD-28.5 GM-1 type and one exhaust fan for gas recirculation Gd-20-500U [7].

To reduce nitrogen and carbon oxides emissions, the boiler is equipped with flue gas recirculation. However, this method has the following disadvantages [8]:

- Instability of the combustion process caused by excessive cooling of the flame peripheral zone by recirculating gases in the combustion core;
- Increase in energy consumption for auxiliary needs caused by significant volumes of required recirculation gases;
- Increase in heat loss with flue gases [8].

To determine the possibility of applying the method of selective catalytic reduction of nitrogen and carbon oxides to a type boiler TGME-464 (KA No.12) it is necessary to determine the volumetric mass flow rates of the fuel and the procedure for calculating the catalyst length that provides the required degree of flue gas cleaning.

A ceramic honeycomb-type exhaust gas neutralization system is considered as a catalyst.

The determination of the volumetric-mass characteristics was carried out on the basis of the regime maps data, the volume data of air and flue gases for the TGME-464 boiler art. No. 12 (Table 2) operating on natural gas (Table 3) [7].

Table 2. Data of the regime card of the TGME-464 boiler art. No. 12 when burning natural gas (for 2019).

No.	Name		Index	Size	Boiler load (D_{cp}), [t/h]							
					220	260	300	340	380	420	460	500
1	Furnace pressure		P_F	[kgs/m^2]	75	83	93	96	117	130	180	185
2	Oxygen content behind the water economizer		$O_2{}^{"WE}$	%	5.1	4	3.6	2.7	2.5	2.2	2.1	2
3	GRF Load		I_{GRF}	A	20	25	26	28	32	35	38	42
4	Excess air ratio behind the water economizer		$\alpha^{"WE}$	-	1.29	1.21	1.19	1.13	1.12	1.11	1.1	1.09
5	Gas temperature	before RAH	t_r^{RAH}	°C	275	280	290	296	300	310	320	333
6		after RAH	t_{yx}	°C	106	107	110	113	116	117	121	123
7	Carbon monoxide concentration (α=1,4)		CO	[mg/nm^3]	no more than 80							
8	Nitrogen oxides concentration (α=1,4)		NO_x	[mg/nm^3]	80	88	89	95	104	115	166	205

Table 3. Data on air volume, flue gas and boiler concentration TGME-464 art. No.12 when burning natural gas (for 2019)

Name	Boiler TGME-464 art. No.12							
D_{pp}, [t/h]	220	260	300	340	380	420	460	500
Fact. volume exhaust gases, [m^3/h]	367322	406300	456495	500171	547539	573822	636230	687723
Fact. air volume, [m^3/h]	195614	215730	242302	263232	287963	302420	333947	361070
Concentration NOx, [mg/m^3] (at α =1,4)	80	88	89	95	104	115	166	205

The gas-air duct of the boiler unit No. 12 installed at the VAZ heat electro-power station is made according to a single-line scheme, equipped with one VDN-25x2-1 exhaust fan, one RAH-88, one DOD-28.5 GM-1 backup exhaust fan and one Gd-20-500U gas recirculation fan. Flue gases are taken after the water economizer and are fed by a recirculation fan (GRF) of the GD-20-500U type into the hot air ducts, from where they are mixed with hot air into the burners [7].

3. Results

3.1. Graphical representations of the results of changes in volumetric flow rates, specific flow energy and flue gas velocity in the load range KA No. 12.

The section of the gas-air duct from the RAH to the chimney is considered as the reaction zone. The verification of the mathematical model is carried out on a chimney section with a section of 2500x5500 mm, length 50700 mm (Figure 1) [7].

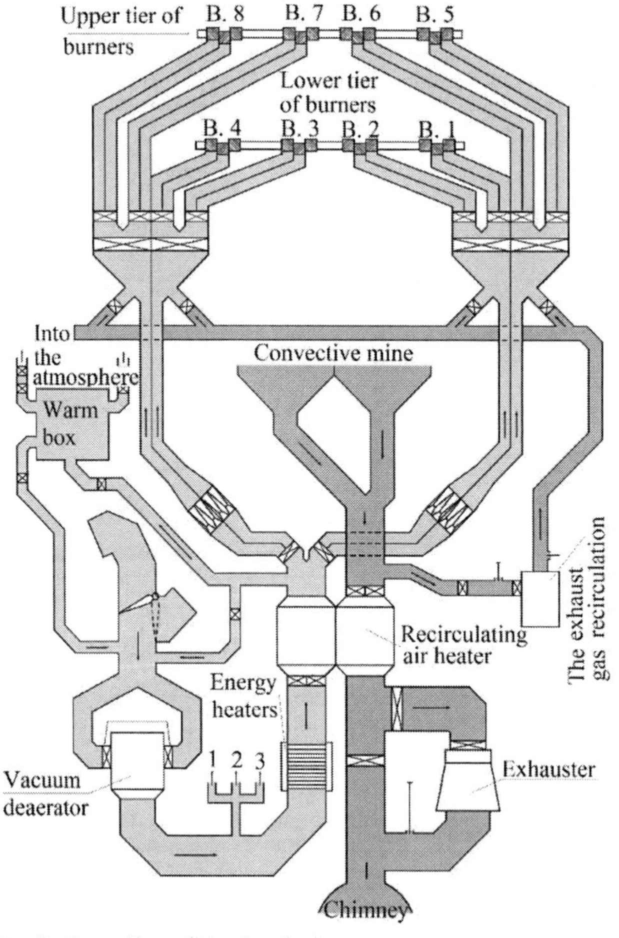

1 - On the sealing of blowing devices
2 -To seal bins
3 - To seal the shot trap

Figure 1. Gas-air duct KA No. 12 VAZ HEPS diagram

Diagrams of changes in volumetric flow rates, specific energy of the flow movement velocity of flue gases in the load range KA No. 12 (220-1500 t/h) are presented in Figures 2, 3.

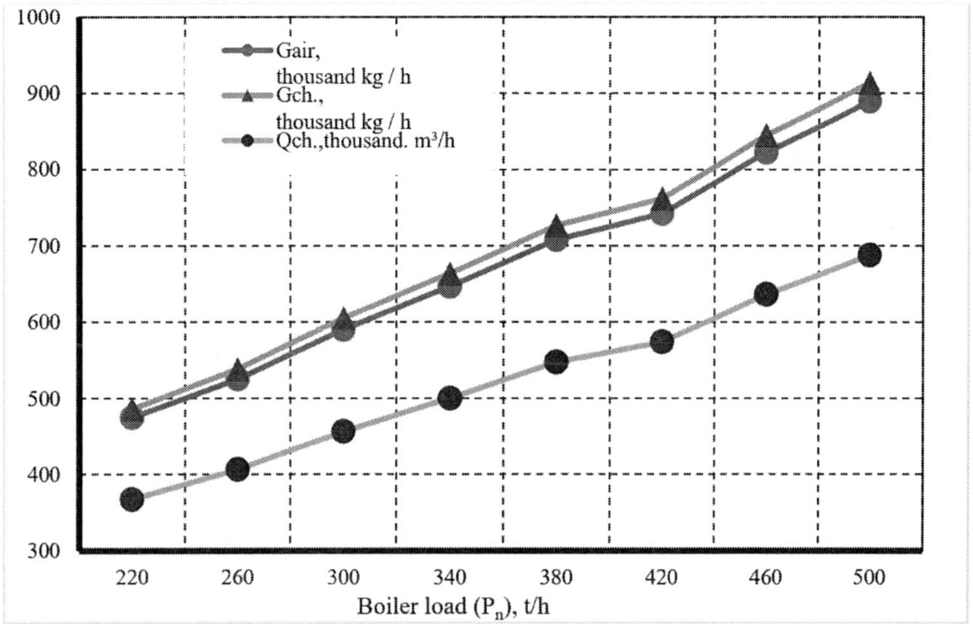

Figure 2. Graphical dependence of volumetric-mass flow rates of air and flue gases on changes in heat load KA No. 12 VAZ HEPS, where G_a– is the mass volume of air, thousand kg / h; $G_{f.g.}$– is the total mass flow rate of flue gases, thousand kg / h; $Q_{f.g.}$– is the total volumetric flow rate of flue gases, thousand m^3 / h.

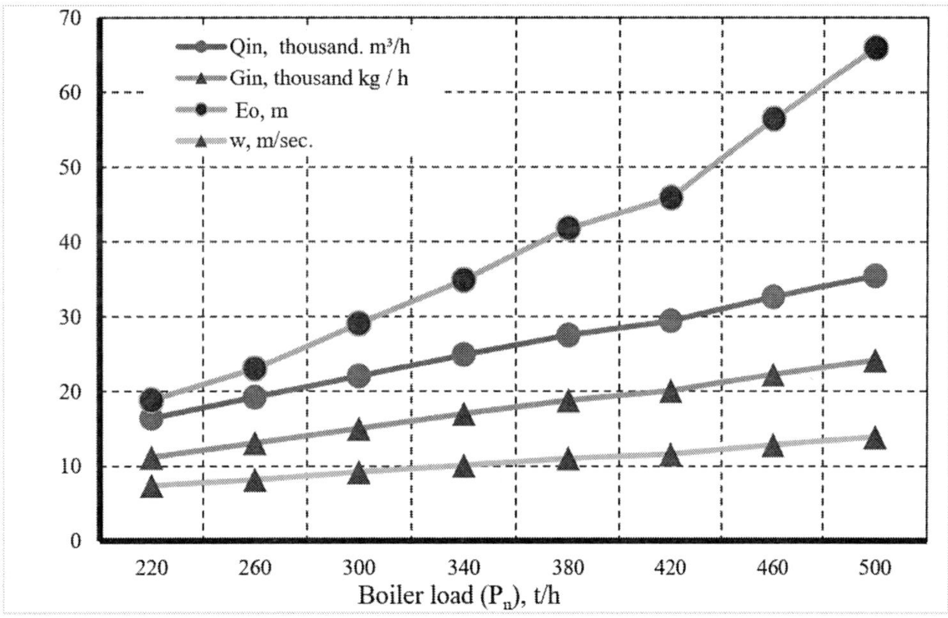

Figure 3. Graphical dependence of fuel consumption, specific flow energy and flue gas velocity on changes in heat load KA No.12 VAZ HEPS, where Q_F– is the fuel consumption by the device, thousand m^3 / h; G_F– is the mass fuel consumption, thousand kg / h; E_0 – is specific flow energy (J/kg); w – denotes flue gas velocity, m / s.

3.2. Systems of material balance equations for changes in CO and NO concentration along the gas path length and the gas flow temperature along the catalyst layer height.

To determine the required length of the catalytic layer, it is necessary to set the value of the required degree of flue gases purification from nitrogen and carbon oxides [9, p.135]. It is determined by the formula:

$$\eta = 1 - c_l^{NO,CO} \Big/ c_0^{NO,CO} \tag{1}$$

where $c_l^{NO,CO}$ – is the current concentration of nitrogen oxides, carbon along the catalytic layer length, $kmol/m^3$; $c_0^{NO,CO}$ – is the initial concentration of nitrogen oxides, carbon at the catalyst inlet, $kmol/m^3$ [9, p.135].
To determine the distribution of the concentrations of nitrogen and carbon oxides along the catalytic layer length, the authors proposed a mathematical model, describing the interaction CO and NO with atomic oxygen on the catalyst surface Pt [10-13]:

$$\text{For } NO \quad \begin{cases} \dfrac{\partial C^{NO}}{\partial h_к} = \dfrac{\sum_{m=1}^{z} k_m \cdot C_k^{NO}}{w}, \\ \dfrac{\partial T}{\partial h_к} = \dfrac{\sum_{m=1}^{z} k_m \cdot C_k^{NO} \cdot q_m}{w \cdot c_p}. \end{cases} \tag{2}$$

$$\begin{cases} \dfrac{\partial C^{NO}}{\partial \tau} = \beta \cdot \left(C_0^{NO} - C_к^{NO}\right) \cdot \alpha, \\ \sum_{m=1}^{z} k_m \cdot C_k^{NO} \cdot q_m = \alpha \cdot \left(T_г - T_k\right) \cdot \alpha. \end{cases}$$

$$\text{For } CO \quad \begin{cases} \dfrac{\partial C^{CO}}{\partial h_к} = \dfrac{\sum_{m=1}^{z} k_m \cdot C_k^{CO}}{w}, \\ \dfrac{\partial T}{\partial h_к} = \dfrac{\sum_{m=1}^{z} k_m \cdot C_k^{CO} \cdot q_m}{w \cdot c_p}. \end{cases} \tag{3}$$

$$\begin{cases} \dfrac{\partial C^{CO}}{\partial \tau} = \beta \cdot \left(C_0^{CO} - C_к^{CO}\right) \cdot \alpha, \\ \sum_{m=1}^{z} k_m \cdot C_k^{CO} \cdot q_m = \alpha \cdot \left(T_г - T_k\right) \cdot \alpha. \end{cases}$$

where τ – denotes reaction time, c; C_k^{NO}, C_k^{CO} – show the NO and CO concentration on the catalyst surface, respectively, $kmol/m^3$; C_0^{NO}, C_0^{CO} – is the NO and CO concentration in the flue gas stream, respectively, $kmol/m^3$ k_m – is the reaction rate constant m, $1/c$; z – denotes a number of reactions; α – denotes a specific catalyst surface area, m^2/m^3; T_k, T_G – is the catalyst surface temperature and flue gas temperature, respectively, K; – is a heat transfer coefficient, $kW/(m^2 \cdot K)$; q_t – – is the thermal effect of reaction t, kJ/mol; β – is the mass transfer coefficient,

$mol/(m^2 \cdot c \cdot mol/m^3)$; h_{cl} – is the catalytic layer height, m; w – is the gas flow rate, m/s; c_p – is the specific isobaric gas flow heat capacity, $kJ/(m^3 \cdot K)$.

4. Discussion

To solve the problem of determining the catalyst layer thickness at a given degree of flue gas purification, it is necessary to determine the sequence of determining the catalytic reduction process physical parameters.

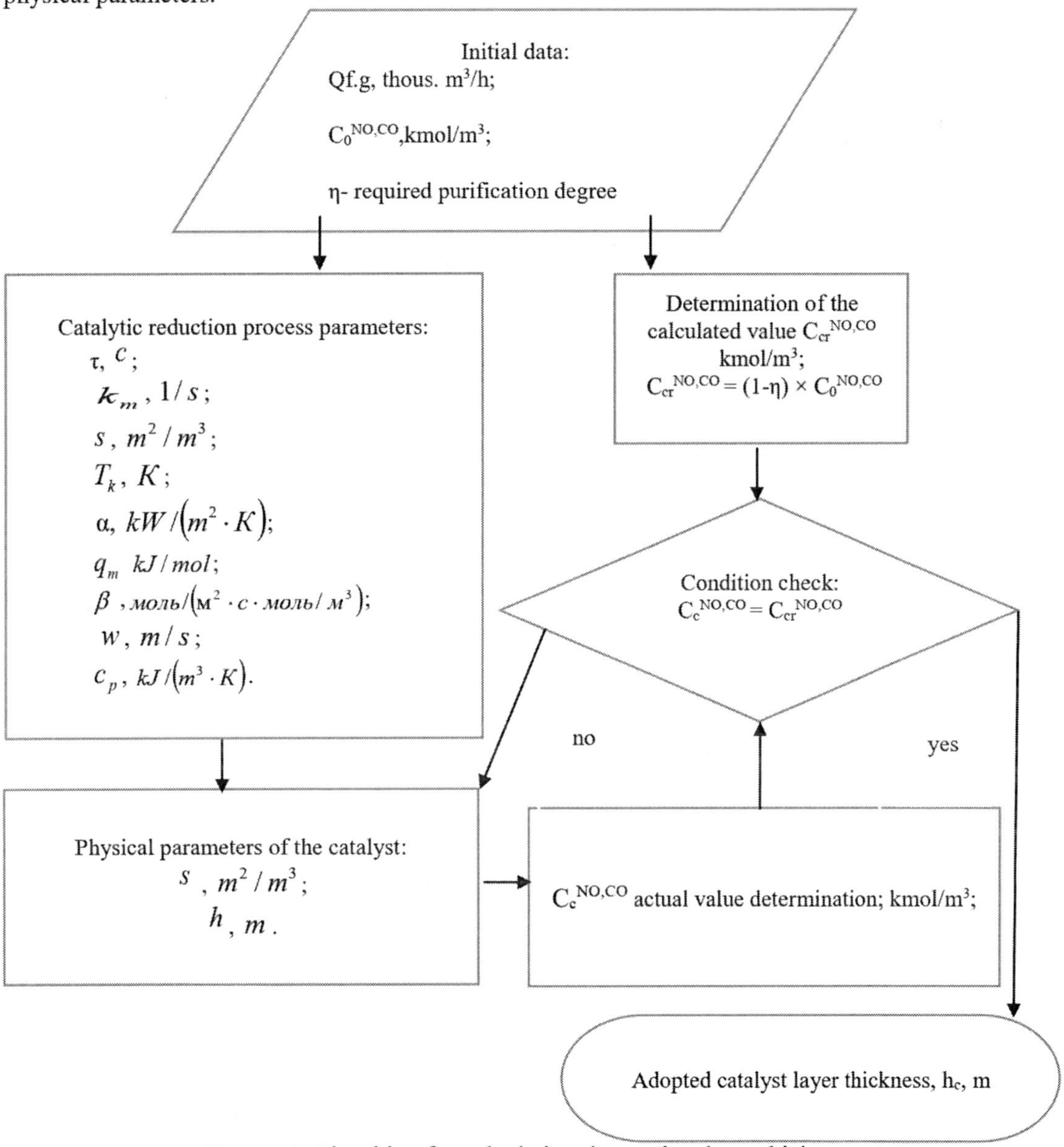

Figure 4. Algorithm for calculating the catalyst layer thickness.

5. Summary

The authors of the study considered the possibility of applying the selective catalytic reduction method to the TGME-464 boiler (KA No. 12) installed on the territory of the HEPS of the Volzhsky Automobile Plant of the Samarsky branch of PJSC T Plus. The volumetric-mass flow rates of the fuel and the algorithm for calculating the length of the catalyst providing a high purification degree of flue gases have been determined. The results of numerical calculations for the proposed algorithm can be used in environmental and economic calculations in order to optimize the systems for cleaning the exhaust gases of boiler plants from harmful emissions [14,15].

References

[1] Roslyakov P V, Chudopal Yu S Development of a program for calculating hazardous emissions from TPPs *Section 2 Heat power engineering: development of steam-gas and coal energy, renewable energy, energy saving, ecology* 29.

[2] Government, Medvedev D. 2015 Order of the Government of the Russian Federation of May 29, 2015 N 996-r Moscow "Strategy for the development of education in the Russian Federation for the period until 2025.

[3] Gromov V I 2010 Samara energetic histories.

[4] Alpatova N A 2011 Energy of the Volga region.

[5] Gobunova G V [et.al.] 2019 Report 2 TP-Air of the Production Enterprise of the VAZ heat electro-power station of the Samara branch of PJSC "T Plus" for 2019 [Text].

[6] Letter of the Ministry of Energy of Russia dated April 15, 2020 No. MJ-4343/09 "On approval of heat supply schemes for settlements, urban districts."

[7] Steam boiler E-500-13, 8-560GME (Model TGME-464) 1986 Technical description instruction manual 08.0302.377 (PA Krasny Kotelshchik).

[8] Ibragimov U Kh, Begimov O N 2015 Method of reducing nitrogen oxides with recirculation of flue gases *Young scientist* **4** 192-194.

[9] Belov S V 2013 2012 (25) Life safety and environmental protection (technosphere safety): a textbook for bachelors of universities (Yurayt, M.).

[10] Balandina O A, Filatova E B, Puring S M 2020 On the question of methods for determining the kinetic parameters of oxidative reactions on the surface of a Pt catalyst *International research journal* **7 (97) (1)** 39-44.

[11] Martsenyuk-Kukharuk M G [et al.] 1996 Development of the SCR process for cleaning waste gases from nitrogen oxides *Chemical industry* **4**.

[12] Koop J, Deutschmann O 2009 Detailed surface reaction mechanism for Pt-catalyzed abatement of automotive exhaust gases *Applied Catalysis B: Environmental* **91 (1-2)** 47-58.

[13] Chernova L E et al. 2017 On the question of the mechanism of carbon monoxide oxidation on a platinum catalyst *Advances in chemistry and chemical technology* **31 (8) (189).**

[14] Mantri D B, Aghalayam P 2007 Micro-kinetic study of reduction of NO on Pt group catalysts *International Journal of Chemical Reactor Engineering* 5 **(A1. 2)** Koop J., Deutschmann

[15] Mitrichev I I et al. 2016 Computer modeling and optimization of the process of removing nitrogen oxide and carbon monoxide on a highly porous catalyst *Advances in chemistry and chemical technology* **30 (4) (173).**

IOP Conf. Series: Materials Science and Engineering 1083 (2021) 012079 doi:10.1088/1757-899X/1083/1/012079

Automated monitoring systems for defects and damages of building structures and materials

S I Evtushenko[1], V A Lepikhova[2], M A Kuchumov[2], E G Scibin[2], T A Krachmalniy[2]

[1]National Research Moscow State University of Civil Engineering (NRU MGSU), 26, Yaroslavskoye highway, Moscow, 129337, Russia
[2]Platov South-Russian State Polytechnic University (NPI), 132, Prosveshcheniya, Novocherkassk, 346428, Russia

E-mail: evtushenkosi@mgsu.ru

Abstract. The automated systems for monitoring defects and damages of building structures and materials are considered. The presented systems make it possible to obtain information on the building structures cracks and joints' opening, the measurements of oscillation and shock acceleration, dynamic soil testing and defects control in composite materials. The "System for monitoring the state of cracks and joints of buildings and structures» is one of the presented systems. This system is necessary for collecting and transmitting information about the cracks and joints' opening width of the building structures as well as environmental conditions, can be used when inspecting buildings and structures, as well as in automated systems for monitoring the buildings and structures' technical condition. The modernization of the system sensor makes it possible to measure the humidity and temperature of the external environment and to determine the dependence of the crack opening width on weather conditions. The next element presented by the system is a sensor for measuring oscillation and shock acceleration, also based on an electrochemical principle of operation. The invention makes it possible to solve the problems of increasing the reliability of converting dynamic mechanical quantities into an electrical signal in a wide range of loads and frequencies. It can be used on the construction sites, for seismic measurements and experimental studies. Further, in the system under consideration, an element of which is a load cell sensor, designed to measure voltage in foundation soils, based on an electrochemical principle of action. Load cell allows performing dynamic soil tests, while the technical result is achieved in increasing the device reliability and the measurements accuracy. To determine the defects of composite materials, a non-destructive testing system based on the acoustic emission signals analysis is considered.

Introduction

A change in the performance characteristics and a decrease in the construction operational properties of buildings and structures occurs as a result of exposure to aggressive technological and natural influences. In addition, such negative phenomena are promoted by errors in design, damage due to violation of manufacturing and installation technologies, defects due to improper operation and untimely repairs, metal corrosion, biochemical effects, materials aging [1-5]. There are different types of composite materials, which differ in their composition and properties. The most widespread and used in construction are, for example, such types as sandwich panels, carbon fiber panels, laminated materials, textolites, fiberglass, which have high performance and decorative effect. Composite is used not only in the construction of residential buildings. It is difficult to imagine a bridge or dam that does not use CFRP panels. Various architectural elements, such as arches or domes, are also often created using composite materials. This is beneficial for the developers, since it provides them with significant savings in the construction of structures, installation, storage and transportation of material, and at the same time, the reliability, quality and other operational characteristics of the future building are not affected. Non-destructive products' testing is a necessary means of improving the quality of products made from composite materials. Of particular importance is the control of products operating in corrosive environments, experiencing high thermal stresses, oscillation loads, etc. Composite materials for a number of properties associated with the penetrating radiation propagation occupy an intermediate position between metals and concretes, for which non-destructive testing methods and industrial equipment for testing by these methods have been developed. Analysis and systematization of the current and periodic inspections' results, as well as the results of special instrumental examinations of industrial buildings, make it possible to analyze the change in the damage size to building structures in time, to identify the work dynamics of the structures, to calculate the residual life of an industrial building and to take timely measures to prevent its destruction. ... The development of devices and sensors for instrumental measurements of static deformations and stresses has been carried out in Platov South-Russian State Polytechnic University (NPI) for several decades [6 - 8]. The proposed work continues the study of static deformations and dynamic stresses on the same methodological principles.

Materials and methods

One of the first monitoring systems under consideration, consisting of one or more displacement sensors and a recording device, for example a computer or laptop, allows the crack opening width measurements of buildings and structures. The automated process of transmitting data via wireless communication is especially relevant if the controlled cracks are located in hard-to-reach places [9, 10].

Each displacement sensor (Fig. 1) consists of: a slide-wire and a current collector, an analog-to-digital converter, a microcontroller, a transmitting radio module with Zigbee protocol support, a chemical supply voltage source and a clock-alarm module.

The operation of the sensor is that when the controlled crack expands, the distance between the sensor body and the stop for the retractable spring-loaded dial indicator rod is increased, which shows the movement. For visual monitoring of the system, standard hour-type indicators are used, configured for the frequency of signal removal 1-2 times a day, followed by the sensor transition to sleep mode, which leads to significant energy savings. The slide-wire current collector is mechanically connected to the moving dial indicator rod, it receives a signal proportional to the sliding rod movement. Further, the signal from the current collector goes to the analog-to-digital converter, where the input analog signal is converted into a digital code and enters the microcontroller input and through the transmitting radio module transmits the displacement value to the receiving radio module and the recording device. In this regard, there is no need to use wired communication lines. If it becomes necessary to monitor several cracks located in different parts of the building at the same time, several sensors are used according to the number of cracks being controlled. Further, the receiving radio module alternately communicates with the transmitting radio modules that are part of each sensor. Each of them has its own number and transmits information over the radio channel to the recording device.

In the process of work, a second-generation sensor was developed, which includes additional means of measuring a temperature and humidity sensor, which makes it possible to expand the device functionality by increasing the number of determined parameters and to increase monitoring efficiency. The results obtained at the facility are automatically defined in the database for further analysis.

A sensor for measuring mechanical quantities is an element of a system that uses an electrochemical operation principle. The monitoring system is designed to convert dynamic mechanical quantities, including oscillation and shock acceleration, into an electrical signal and can be used in various industries, in particular in the construction field, for seismic measurements and experimental research. The mechanical value sensor consists of a support made of an electrically insulating material, an elastic conductive plate and an inertial element, two electrochemical cells, each of which includes a housing, an electrolyte, and a current-collecting counter electrode. When the sensor is exposed to oscillation, seismic oscillations and other shocks, they are perceived by an elastic metal plate with an inertial element, while the upper and lower surfaces of the metal plate are alternately stretched and compressed. Since the change in the area of the cells' deformable electrodes is close in magnitude, but opposite in sign, the output signal is doubled. Thus, the mechanical deformation direct transformation effect of the sensitive elastic element into an electrical signal is achieved. One of the advantages of sensors based on the " solid metal electrode interfacial layer elastic charging effect" is the direct transformation of an elastic element deformation (electrode of an electrochemical cell) into an electrical signal, which increases the reliability of the measurements under dynamic loads that change rapidly over time.

The next sensor is the load cell for measuring stress in soils. The invention relates to a measuring technique and is intended to measure the stress in soils under dynamic loads, in particular, during soil compaction, blasting, earthquake and can be used in construction, mining, experimental research [11]. The load cell works as follows (Figure 1). When dynamic stresses arise in soils, they are perceived by the elastic membrane of the working cover, while the hydraulic fluid deforms the elastic metal membrane, and the liquid electrolyte deforms the elastic metal membrane of the electrochemical cell [12].

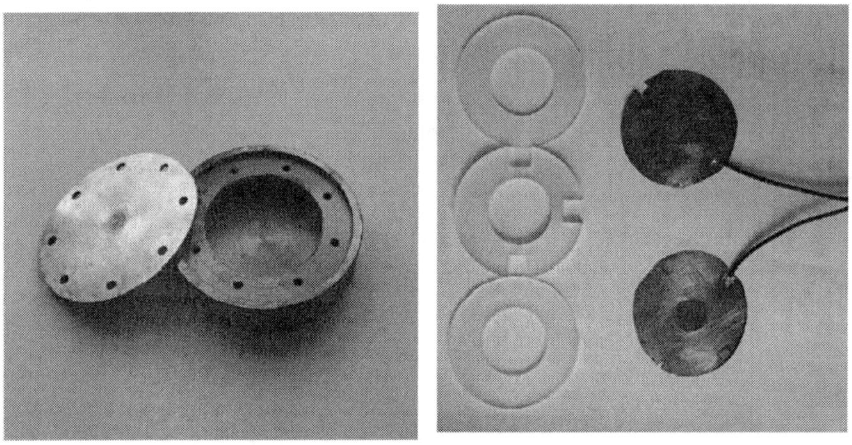

Figure 1. Constructive diagram of the electrochemical load cell: 1 - annular body and working cover; 2 - two round elastic membranes

Under elastic tension-compression of a flat plate, its initial area changes linearly. In the case of a circular membrane clamped along the contour, the design scheme is shown in Figure 2. It can be seen that in the area of the circular ring and the central part of the membrane, the same in magnitude, but opposite in sign deformations occur. When such a membrane contacts electrolyte, the electrode potential will not change during its deformation. In the proposed device, the use of an elastic electrically insulating protective coating in the form of a circular ring with an inner radius equal to half the electrode membrane radius makes it possible to select the central part of the membrane-electrodes, one of which will experience tension and the other compression and, therefore, the appearance of an electrical signal between them.

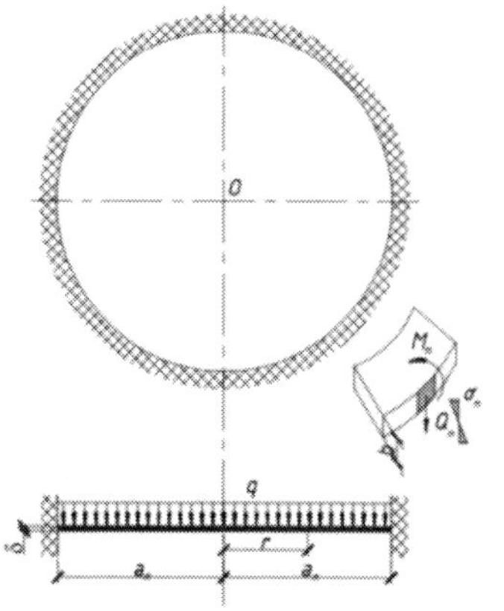

Figure 2. Design model, rigidly embedded along the contour of the membrane under load

The carried-out tests of the load cell showed its stability under shock loads.

The specific properties of composite materials require the development of non-standard methods and means of non-destructive testing. One of such methods is acoustic spectral-timbre non-destructive testing of products. On its basis, the authors have developed a stand for non-destructive testing of composite materials, the block diagram of which is shown in (Figure 3).

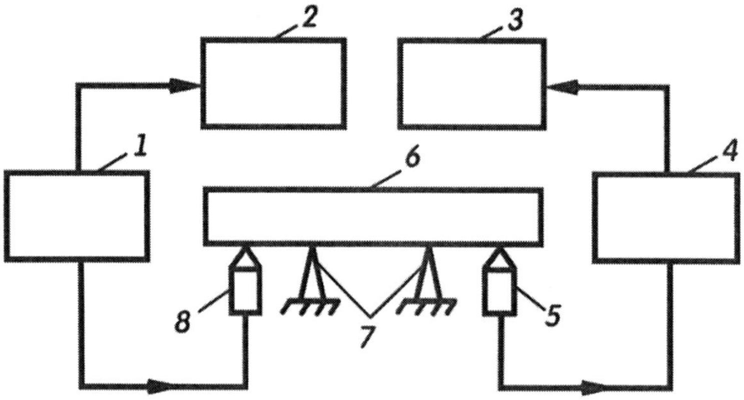

Figure 3. Block diagram of the stand for spectral-timbre non-destructive testing of composite materials.

1 - generator of electrical oscillations; 2 - frequency counter; 3-computer; 4 - amplifier; 5-piezoceramic sensor; 6 – tested item; 7 - supports; 8-sensor generator.

In the acoustic method, the emitting and receiving piezoelectric sensors are placed on different sides of the sample. A defect located in the ultrasonic signal propagation path reduces the signal energy transmitted from the transmitter to the receiver due to the waves' scattering in different directions. This happens when the acoustic wave impedances the material and the defective medium differ significantly. Therefore, it is most effective to detect the defects in the form of widespread air inclusions, single large cracks, systems of small cracks, cavities, delamination, etc. In the presence of such a defect, the signal level reaching the receiving sensor and recorded by a computer turns out to be significantly lower than the signal level in places without defects.

The discussion of the results

Considering the system, the element of which is a load cell sensor, we note the difference from analogues, which consists in the use of an electrochemical principle for the action.

When a metal electrode contacts an electrolyte solution at the metal-liquid interface, a potential shock φ is considered in electrochemistry to occur based on electrical double layer (edl). In the field of concentrated electrolyte solutions edl likened to a flat capacitor, one of the plates of which is formed by charges of a flat metal electrode q on the surface, and the other - by the charges of the opposite sign ions, adjacent from the solution to the electrode surface at a distance d, equal to the ion radius. The charge density will be proportional to the potential jump between the metal and the solution, and is determined by the ratio

$$q = \varphi \cdot c \tag{1}$$

where c – is edl capacity

The capacity of a flat condenser is found by the formula

$$c = \varepsilon \cdot S / 4\pi d \tag{2}$$

where in the case under consideration S – is the surface area size of the electrode in contact with the electrolyte; Ɛ – is a dielectric solvent constant.

The work of the load cell was based on the "effect of elastic charging of the interfacial layer of a solid metal electrode" [12].

The essence of this effect lies in the fact that the surface of a metal electrode in contact with an electrolyte increases or decreases by elastic deformation. In this case, according to (2), the edl capacity and if the electrode charge is maintained constant, for example, by introducing a damping resistance into the electrode circuit, then according to (1) the potential of the electrode itself is relative to the solution changes, which is more convenient for recording. The resistance is chosen so high that the interfacial layer does not have time to discharge during the deformation period (the discharge time constant is equal to the electrode capacity product by the indicated resistance). Analyzing the system for determining defects in composite materials using emitting and receiving piezoelectric sensors, the acoustic method is used. In this case, the fast Fourier transform method (FTM) can be used to analyze the acoustic emission signals generated by the formed defects. In this case, the transition from physical parameters to the frequency-amplitude energy spectrum parameters obtained from the time series of samples, the signal, by the discrete fast Fourier transform is carried out. Non-standard means of assessing the products quality, technology of its production using the analysis of acoustic emission (AE) signals arising from the diagnosed systems' generator. Decomposition of these signals into natural elementary components of the Fourier spectrum, based on higher timbre harmonics, makes it possible to identify the AE sources and evaluate their role in the product quality analysis. Let the acoustic emission signal be sampled in time in some way. Then the samples for N signal can be written (3) its following representation in the frequency domain:

$$X(k) = \frac{1}{N}\sum_{1}^{N-1} x(n)\left[\cos\frac{2\pi nk}{N} - j\sin\frac{2\pi nk}{N}\right], \quad 0 \le k \le N-1 \tag{3}$$

where n – denotes the sampling indices (discrete) of the signal in the time domain,

k – denotes the sampling indices (discrete) of the signal in the frequency domain,

$x(n)$ - denotes the sampling indices (discrete) of the signal in the frequency domain,

$X(k)$ - denotes the frequency domain samples.

Signal amplitudes X(k) can be calculated in accordance with the following relations (4):

$$\begin{cases} A(X(k)) = \sqrt{\mathrm{Re}(X(k))^2 + \mathrm{Im}(X(k))^2} - \textit{indication amplitude } X(k), \\ \mathrm{Re}(X(k)) = \dfrac{2}{N}\sum\limits_{n=0}^{N-1} x(n)\cos(\dfrac{2\pi nk}{N}) - \textit{FTM real part}, \\ \mathrm{Im}(X(k)) = \dfrac{-2}{N}\sum\limits_{n=0}^{N-1} x(n)\sin(\dfrac{2\pi nk}{N}) - \textit{FTM virtual part}. \end{cases} \tag{4}$$

To assess a defect in a particular sample, it is possible to use the method of studying the set of modal harmonics based on the FTM method [13]. In accordance with this method, a database containing information on subsets of modal harmonics corresponding to a particular defect in the sample under study is preliminarily formed. Subsequently, many harmonics of the Fourier spectrum of the acoustic emission signal of the sample are analyzed in real time, and modal harmonics corresponding to one or another defect are distinguished in it. The presence of these modal harmonics indicates the presence of a defect in the sample, data about which are stored in the computer database. To assess the shape and size of a defect contained in a test sample, it is necessary to use the information on timbre sub-spectra, i.e. on the sub-spectra of the total spectrum of the acoustic signal generated by one or another modal harmonic.

Summary

Automated monitoring systems have found application in the construction of buildings and structures. So, the installation of sensors on structural elements and the collection of information from them will make it possible to determine changes and deviations from the permissible values.

When analyzing the changes in the used building materials and the presence of zones of no connection determined by the system using piezoceramic sensors in them, the oscillations amplitude and spectrum will change. Closely located defects cause the appearance of lower frequencies in the spectrum of natural oscillations and an increase in the reflected signal amplitude in comparison with the defect-free regions of the sample. On the contrary, deep-seated defects cause the appearance of higher frequencies in the oscillation spectrum and a decrease in the amplitude of the signal reflected from the defect. Thus, complex monitoring using sensor systems will allow timely decisions on the building overhaul.

References

[1] Volosuhin V A [et.al.] 2013 Defects and damage to building structures of bridges on reclamation canals of the Rostov region: monograph (Platov South-Russian State Polytechnic University (NPI), Novocherkassk).

[2] Evtushenko S I, Krahmalny T A, Evtushenko A S, Krahmalnaya M P 2014 Monitoring of the technical condition of buildings and structures as the basis for integrated safety in construction *Building and architecture* **2 (4) (5)** 182-185. DOI 10.12737/17525.

[3] Evtushenko S I, Krahmalny T A 2019 Defects and Damages of Soil Foundations of Industrial Buildings *Construction and Architecture* **7 (3) (24)** 45-49. DOI 10.29039/2308-0191-2019-7-3-45-49.

[4] Evtushenko S I, Krakhmalny T A 2019 Defects and damage of columnar foundations of industrial buildings *Construction and architecture* **7 (4) (25)** 36-40. DOI 10.29039/2308-0191-2019-7-4-36-40

[5] Maximenko L A 2010 Formation of a database on typical damages and defects of building structures *SSGA, Novosibirsk* **6**.

[6] Krasnopolskaya N A 1969 Lighthouses for measuring the development of cracks in the walls of buildings. Experimental studies of engineering structures *Materials of the II Symposium, Leningrad, Sept. 1969 Editorial and publishing department NPI: Novocherkassk* 73-75.

[7] Murzenko Yu N 1969 Devices and sensors for remote registration of displacements and loads *In sb. Experimental studies of engineering structures. Materials of the II Symposium, Leningrad, Sept. 1969 Editorial and publishing department NPI: Novocherkassk* 43-51.
[8] Murzenko Yu N 1969 Load cells with a ring membrane for measuring stresses in soil *In collection of works. Experimental studies of engineering structures. Materials of the II Symposium, Leningrad, Sept. 1969 Editorial and publishing department NPI: Novocherkassk* 59-67.
[9] Krahmalny M P, Krahmalnaya T A, Evtushenko S I 2011 System for monitoring the state of cracks and joints of buildings and structures: patent for utility model 102035 Rus. Federation: E04G 23/00 (2006.01), application 2010140258, application no. 01.10.2010; publ. 02/10/2011, Bul. No. 4.
[10] Krahmalny M P, Krahmalnaya T A, Evtushenko S I 2012 System for monitoring the state of cracks and joints of buildings and structures: Pat. 2448225 Rus. Federation: IPC E04P 23/00 / / Application No. 2010140257/03 app. 01.10.2010; publ. 20.04.2012, Bul. No. 11.
[11] Evtushenko S I, Firsov V V, Skibin E G, Kuchumov M A 2017 Load cell for measuring stress in soils: Pat. 2657550 Rus. Federation: IPC E02D 1/00 (2006.01) / / No. 2017127438 App. 07/31/2017; publ. 06/14/2018, Bul. No. 17.
[12] Gokhshtein A Ya 1976 *Surface tension of solids and adsorption* (Moscow, Nauka).
[13] Puresev A I, Lepihova V A, Toropov O A, Malykh E A, Sorokin N P Pat. 2222807 RF, IPC-G01 N 29/02. A method of processing acoustic emission signals of generated disperse systems / - Applied. 02/12/01; Publ. 01.27.04, Bul. No. 3.

An approach for automatization of geotechnical monitoring in cryolithozone

A Alekseev[1,2], L Shilova[1*], E Mefedov[1]

[1]Moscow State University of Civil Engineering, 26, Yaroslavskoe highway, Moscow, 129337, Russia
[2]JSC "RCC "Construction", N. M. Gersevanov research Institute National Research, 6, 2nd Institutskaya str., Moscow, 109428, Russia

E-mail: ShilovaLA@mgsu.ru

Abstract. The article presents an analysis of publications in the field of research devoted to the problems associated with the construction of buildings and structures in areas of permafrost soils, which belong to the cryolithozone. At the same time, the analysis of the research works made it possible to conclude that currently the data obtained because of monitoring, as a rule, are processed manually. In the article the process of geotechnical monitoring is presented in the form of a generalized diagram, from which a sufficiently large array of data is collected during the process of geotechnical monitoring. According to this information graphs of changes in the movement of the building foundations in time and graphs of thermometric observations of the soil massif, etc. are made. That fact justifies the need of automatization of the collection and processing of geotechnical monitoring data. The article presents an approach for automatization of geotechnical monitoring. Systematization of the geotechnical monitoring data and ensuring the operation of equipment synchronously in an automated mode will provide the timely notification of stakeholders about the identified deviations of the controlled parameters and reduce risks and avoid accidents.

1. Introduction

Currently, more and more attention is paid to the problems associated with the construction of buildings and structures in the areas of permafrost soils, which belong to the cryolithozone - the upper layer of the earth's crust, which is characterized by a negative temperature of rocks and soils and the presence or the possibility of the existence of underground ice. This fact is confirmed by publications presented in the bibliographic and abstract database Scopus. At the same time, 255 publications are displayed in the system for the keyword "cryolithozone", most of which (78%) belong to authors from the Russian Federation, this is primarily due to the fact that about 60-65% of this country territory belongs to the regions of the eternal permafrost. If we consider on a global scale, then permafrost occupies about 25%. The figure 1 shows the distribution of permafrost depending on the ice content as a percentage of the area.

At the same time, the construction of buildings, structures, and elements of transport infrastructure in cryolithozone against the background of climate change, as well as their subsequent operation, necessitates solving a complex of engineering problems.

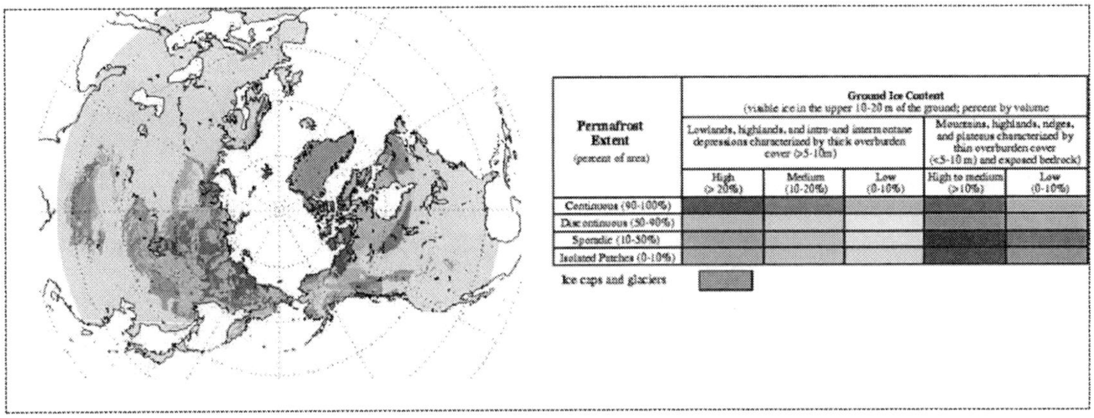

Figure 1. Length and types of permafrost [1]

It should be noted that the studies presented in the publications can be conditionally divided into several groups. Some of them are devoted to modeling and calculating the states of the cryolithozone under the operating conditions of construction objects [2-16], as well as others are devoted to the study of changes in the state of soils and ecosystems taking into account global climate change [17-21, etc.].

For example, research work [2] is devoted to the assessment of the state of the road network in the Magadan region, taking into account the changing climatic conditions of its operation. The assessment of the state of the roadway was carried out by modeling the temperature regime of the soil at the reference points, for which there were data sets on long-term values of climatic data. As a result of the study, the risk values for the considered reference points were formed.

The peculiarities of the interaction of the railway subgrade and permafrost in the subarctic part of the permafrost zone, as well as the reasons for the degradation of permafrost soils, are described in [3]. In the publication [4], variants of design solutions for self-cooling stand-alone columnar foundations of factory production in conditions of degradation of permafrost are proposed, considering the latest achievements in the area under consideration. A laboratory study of the bearing capacity of a metal pile depending on local geocryological conditions is presented in [5].

The work [6] presents the results of the analysis of data of long-term monitoring of accidents and safety of hydraulic structures in the permafrost zone from 1776 to 2015.

Interesting within the framework of this study are research works related to the use of modern information technologies to support the construction and operation of buildings and structures in the permafrost zone.

The article [7] presents an approach to solving the problem of maintaining the stability of the embankment slope due to its artificial freezing. At the same time, to calculate the heat flow in the embankment slope, a three-dimensional mathematical model was developed in the form of an application for Microsoft Windows. The application presented in the work allows you to form a long-term forecast of the temperature field in the embankment slope, depending on the design and technological features. As a result of the research, the efficiency of using seasonal cooling devices to increase the stability of embankment slopes due to their freezing was revealed.

The publication [8] describes an approach to detecting leaks from pipelines taking into account possible damage to the pipeline foundation due to thawing of permafrost; solving problems of automatic classification of defects that led to leaks; problems of prompt detection of corrosion spots in pipelines, as well as problems of determining the current state of the corrosion process in a pipeline using machine learning technologies.

The article [9] presents an updated algorithm for assessing the predicted risk of roadway operation for the climatic conditions of Yakutsk and Urengoy from the warming conditions characteristic of these zones.

The work [10] presents a set of programs for determining changes in the boundaries of permafrost zones in the conditions of the spread of thermal fields from various engineering objects operating in the Arctic regions. In the paper [11] algorithms and software are presented for determining the boundary conditions on the surface of frozen soil, taking into account the real process of freezing and thawing of the pore solution of the soil.

Intelligent management system of the automobile road's technical and operational condition in the cryolithozone are shown in the publication [13]. This system can be used for collecting information about the current state of the road, weather and climatic conditions, characteristics of traffic flows.

At the same time, the analysis of publications indicates that the most important issue in the construction and operation of structures in the areas where permafrost soils are spread is their preservation in a frozen state. The stability and durability of structures significantly depend on the temperature regime of the soils of the foundation, the provision of which is the main factor ensuring the safety of structures during construction and operation.

The control over the provision of the required temperature regime of the base soils is carried out by geotechnical monitoring. Geotechnical monitoring is understood as a set of works based on observations of the temperature and hydrogeological regimes of the permafrost soils, deformations of foundations and overground structures of buildings and structures.

Geotechnical monitoring also includes observations of the state of structural material, soil temperature at the base of structures; the air temperature in the underground; deformations of foundations; hydrogeological regime of the foundation, dynamic and seismic vibrations. Monitoring includes a wide range of multifaceted observations, performed both in automatic and manual modes.

Thus, the presented study is devoted to the development of an approach to the automation of geotechnical monitoring in the permafrost zone to reduce the timing of its implementation

2. Methods and methodology

Geotechnical monitoring is carried out to timely identify and prevent irreversible processes in soil foundations, deformation of buildings and structures, as well as to reduce risks and prevent emergency situations. In addition, the results of geotechnical monitoring can be used to study the operational suitability of buildings and structures, as well as assess the environmental safety of the territory, which is one of the priority tasks in the construction of buildings and structures in permafrost conditions.

At present time, the rules for conducting the geotechnical monitoring during the construction of construction sites are strictly regulated. In the Russian Federation, SP 305.1325800.2017 «Buildings and structures. Rules for conducting geotechnical monitoring during construction» was developed and put into effect. In addition, SP 25.13330.2012 «Bases and foundations on permafrost. The updated version of SNiP 2.02.04-88» is important and describes the design rules for foundations and foundations of buildings and structures erected in the territory of permafrost (permafrost) soils, which includes a chapter with requirements for geotechnical monitoring [22,23].

3. Results

According to the valid regulatory documentation in Russian Federation the generalized scheme for conducting geotechnical monitoring is made by authors and shown in Figure 2.

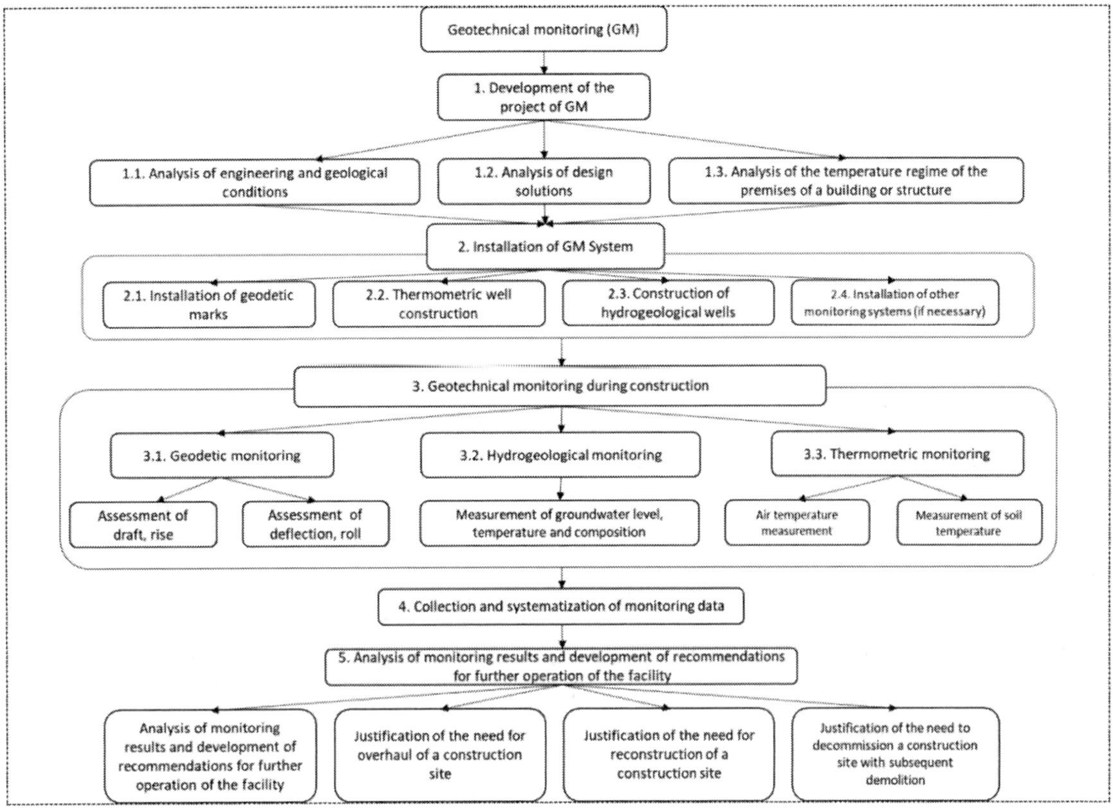

Figure 2. Generalized scheme for conducting geotechnical monitoring of a construction site in permafrost conditions

Figure 2 shows that in the process of carrying out geotechnical monitoring, a rather large array of data is collected, on the basis of which graphs of changes in the movement of the building foundations in time are plotted (Figure 3a), and graphs of thermometric observations of the soil massif (Figure 3b) are made and etc.

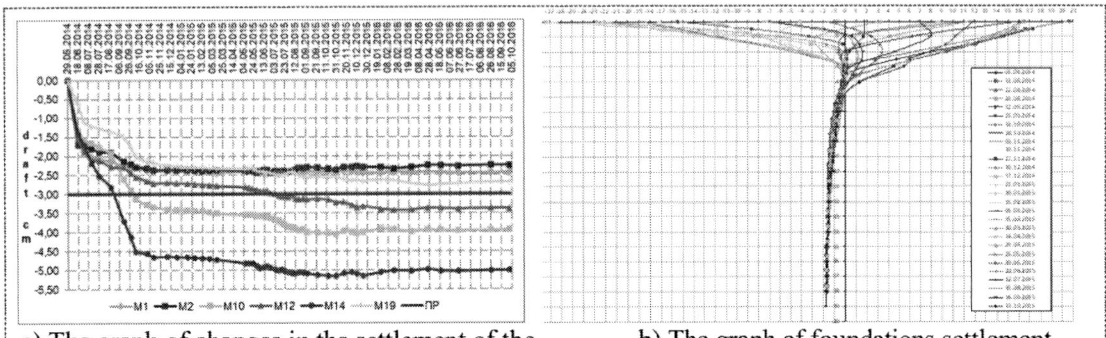

a) The graph of changes in the settlement of the foundations of the building in time

b) The graph of foundations settlement

Figure 3. The examples of graphs obtained as a result of geotechnical monitoring

All of the above leads to the need of developing an automated system for collecting and processing information obtained in the course of geotechnical monitoring, where the main task will be to automate the information processing and to present a results as a graph using a simple scheme presented in Figure 4.

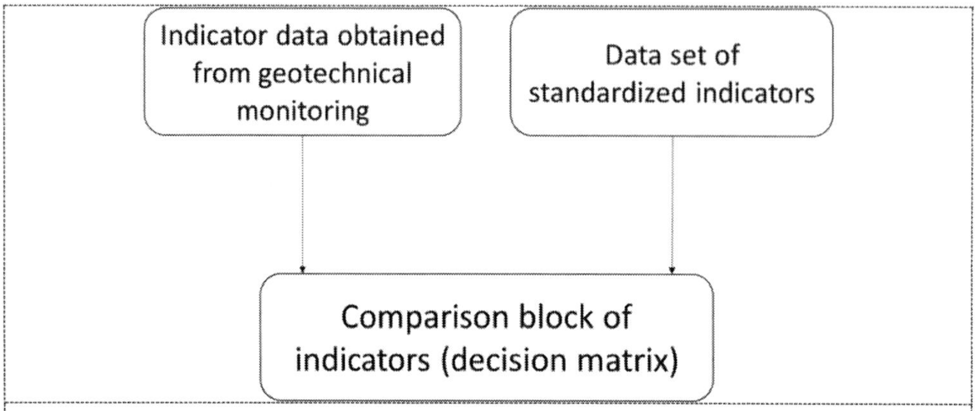

Figure 4. Generalized scheme for processing the results of geotechnical monitoring

At the same time, in the process of making decisions (figure 2) for the further operation of the construction object, it is necessary to consider various combinations of deviations of the controlled parameters (including the tendencies of their changes exceeding the expected ones) from the design values and the results of heat and geotechnical forecast. The algorithm of the program and its program code, as well as the software application itself will be presented in further works of the authors.

Summary

Systematization of the geotechnical monitoring data and ensuring the operation of equipment synchronously in an automated mode will provide the timely notification of stakeholders about the identified deviations of the controlled parameters and reduce risks and avoid accidents.

References

[1] Brown J, Ferrians O J, Heginbottom J A, Melnikov E S 2001. Circum-arctic map of permafrost and ground ice conditions. *Boulder, CO: National Snow and Ice Data Center/World Data Center for Glaciology*

[2] Yakubovich A, Yakubovich I 2021 Using the response surface to assess the reliability of the russian cryolithozone road network in a warming climate *Advances in Intelligent Systems and Computing* **1258 AISC** 486-495 DOI: 10.1007/978-3-030-57450-5_42

[3] Ashpiz E S 2020 The problems of the railway subgrade construction in the subarctic part of the Russian cryolithozone and the ways of their solution *Lecture Notes in Civil Engineering* **49** 295-302. DOI: 10.1007/978-981-15-0450-1_30

[4] Rozina V E, Komarov A K 2020 Main achievements and directions of scientific research in the field of foundation construction on sites composed of permafrost in the Russian Federation *IOP Conference Series: Materials Science and Engineering* **880 (1)** 012016. DOI: 10.1088/1757-899X/880/1/012016

[5] Kudriavtcev S, Shin E-C, Kovshun V 2020 The laboratory investigation of load-bearing capacity of the metal pile depending on the local geocryological conditions *16th Asian Regional Conference on Soil Mechanics and Geotechnical Engineering* ARC **159084.**

[6] Kronik Y A 2019 Analysis of the Data of Monitoring the Safety of Hydraulic Structures in the Cryolithozone Power Technology and Engineering **53 (2)** 191-199 DOI: 10.1007/s10749-019-01059-z.

[7] Romanova E, Khokhlova Y 2020 Maintaining thermal stability of a fill slope in cryolithozone *E3S Web of Conferences* **192** 01021. DOI: 10.1051/e3sconf/202019201021.

[8] Timofeev A V, Denisov V M 2020 Machine Learning Based Predictive Maintenance of

Infrastructure Facilities in the Cryolithozone *Studies in Systems, Decision and Control* **255** 49-74 DOI: 10.1007/978-3-030-31328-9_3

[9] Trofimenko Y V, Yakubovich A N, Yakubovich I A, Shashina E V 2020 Modeling of influence of climate change character on the territory of the cryolithozone on the value of risks for the road network *International journal of online and biomedical engineering* **16 (7)** 65-74

[10] Filimonov M Yu, Vaganova N A, Akimova E N, Misilov V E 2020 Supercomputer technologies for long-term modeling of permafrost changes *CEUR Workshop Proceedings* **2543** 362-369

[11] Vinokurova T A, Permyakov P P 2019 Identification of Heat Exchange Boundary Conditions at Various Natural and Technogenic Factors *Journal of Physics: Conference Series* **1392(1)** 012090. DOI: 10.1088/1742-6596/1392/1/012090

[12] Tumel N, Zotova L 2019 Diagnostics and mapping of geoecological situations in the permafrost zone of Russia *Geosciences (Switzerland)* **9(8)** 353. DOI: 10.3390/geosciences9080353

[13] Yakubovich A N, Yakubovich I A, Trofimenko Y V, Shashina EV 2019 Intelligent Management System of the Automobile Road's Technical and Operational Condition in the Cryolithozone *Systems of Signals Generating and Processing in the Field of on Board Communications* 8706742 DOI: 10.1109/SOSG.2019.8706742

[14] Yakubovich A N, Trofimenko Y V, Yakubovich I A, Shashina E V 2019 A forecast model for a road network's section traffic capacity assessment on a territory of the cryolithozone in conditions of the climate change *Periodicals of Engineering and Natural Sciences* **7 (1)** 275-280 DOI: 10.21533/pen.v7i1.380

[15] Debol'skaya E I, Gritsuk I I, Debol'skii V K, Ionov D N , Maslikova O Y 2018 Effect of Bank Deformations on Pollutant Transport in Rivers in Cryolithozone: Laboratory and Mathematical Modeling *Water Resources* **45 (4)** 542-552 DOI: 10.1134/S0097807818040073

[16] Legostaeva Y, Ksenofontova M, Gololobova A 2017 The stability of ecosystems cryolithozone in the extraction hydrocarbon feedstock *World Congress on Civil, Structural, and Environmental Engineering* 139598. DOI: 10.11159/awspt17.111.

[17] Desyatkin R V, Desyatkin A R 2019 The Effect of Increasing Active Layer Depth on Changes in the Water Budget in the Cryolithozone *Eurasian Soil Science* **52 (11)** 1447-1455 DOI: 10.1134/S1064229319110036.

[18] Butakov V I, Slagoda E A, Tikhonravova Y V, (...), Tomberg I V, Zhuchenko N A 2020 Hydrochemical composition and rare-earth elements in ice wedge of the Kara Region cryolitozone key areas *Bulletin of the Tomsk Polytechnic University, Geo Assets Engineering* **31(2)** 78-91 DOI: 10.18799/24131830/2020/2/2483.

[19] Goncharova O Y, Matyshak GV, Timofeeva M V, (...), Bobrik AA, Tarkhov MO 2019 Autotrophic and Heterotrophic Soil Respiration in Cryolithozone: Quantifying the Contributions and Methodological Approaches *The Case of Soils of the North of Western Siberia Contemporary Problems of Ecology* **12 (6)** 534-543 DOI: 10.1134/S1995425519060040.

[20] Pomortsev O A, Pomortseva A A, Trofimtsev Y I 2019 Cyclic Organization of Geological Environment: Permafrost Zone of Yakutia *IOP Conference Series: Earth and Environmental Science* **272 (2)** 022059 DOI: 10.1088/1755-1315/272/2/022059

[21] Kaverin D A, Lapteva E M, Shchanov V M, (...), Sharaya L S, Shary P A 2019 Climatic geoinformation analysis of the cryolithozone in the northeast of European Russia *Earth's Cryosphere* **23 (4)** 68-78 DOI: 10.21782/KZ1560-7496-2019-4 (68-78)

[22] Alekseev A G 2012 Geotechnical monitoring during construction and operation of buildings and structures in areas of permafrost *Modern innovative technologies of survey, design and construction in the Far North* 62-67.

[23] Alekseev A G 2019 *Geotechnical monitoring on permafrost* (Moscow) 112

Assessment of the concentration and properties of dust in emissions into atmospheric air from the sources of the workshop for the aerated concrete structures production

T O Kondratenko*, A P Lapina

Don State Technical University, 162, Sotcialisticheskaya str., Rostov-on-Don, 344022, Russia

E-mail: tatkondr@rambler.ru

Abstract. Improving the environmental safety of manufacturing enterprises in construction industry remains an urgent task today. The need to expand the research in this area is dictated by a wide range of industrial enterprises, a range of products, as well as the varieties of bulk raw materials used in the manufacturing process. Exceeding the maximum permissible values for the content of fine dust PM_{10} and $PM_{2,5}$ in the atmospheric air, both on the territory of industrial sites and on the borders of sanitary protection zones, is due to the ineffective operation of the existing dust cleaning systems. Analysis of the dispersed composition of dust in emissions from the construction industry, as well as the study of dust properties depending on the type of enterprise, raw materials used, the characteristics of technological operations, are the basis for the development of measures optimal for a particular production to reduce dust emissions, improve the environmental situation in general.

Introduction

One of the stages in the effectiveness analysis of the existing systems for cleaning dust emissions at the construction industry enterprises, in particular for the aerated concrete blocks production, is the assessment of the fractional composition, the concentration of dust contained in the air. Dust control is carried out on the territory of the industrial site and at the border of the sanitary protection zone. The experimental studies were carried out in order to solve the problem of compliance with the standards for the content of harmful substances in the atmospheric air on the industrial site territory and the border of the sanitary protection zone of the enterprise for the autoclaved aerated concrete blocks production. In combination with the obtained results of the dispersed composition of dust, the results of studying the properties of dust in emissions into the atmospheric air will determine the directions for increasing the dust cleaning system efficiency [1-10].

Main part

The object of the study was the existing dust cleaning system of an enterprise for the production of building blocks from autoclaved aerated concrete. As noted by the author in [8-12], the conditions of the technological process for the manufacture of structures from aerated concrete are associated with the use of bulk materials (cement, quartz sand, lime, building gypsum, aluminum powder). Also, in the composition of technological shacks and, accordingly, technological operations, there is a shack for crushing the undercut layer, where dry shredding of production waste is performed.

In the course of the experimental studies of the dust emissions nature, based on the data obtained on the fractional composition of dust particles in emissions from the production line sources [8-12], the concentration of fine particles was calculated. The assessment of dust emissions into the air was carried out on the territory of the industrial site and at the border of the sanitary protection zone during the warm season. The following results were obtained on the content of fine dust particles (less than 10 microns) in the atmospheric air on the industrial site territory: for PM10 is from 12% till 40%, for PM2,5 from 0,3 % till 0,5 %. Disperse analysis of dust particles in the air at the sanitary protection zone border showed the content of PM10 particles of the order of 80% of the mass of dust, particles PM2,5 – 0,5%.

The median particle diameter is 11-15 microns [17-19]. Particle sizes vary from 1.5 to 20 microns [17-19].

Particle Concentration Results PM_{10} and $PM_{2,5}$, contained in the emissions into the atmospheric air from the sources of the workshop for the aerated concrete blocks production are presented in Table 1.

Table 1. Particle concentration PM10 and PM2,5 in emissions from the workshop sources

Emission source	General concentration, mg/m^3	PM_{10}		$PM_{2,5}$	
		content, %	concentration, mg/m^3	content, %	concentration, mg/m^3
Aspiration system from the conveyor of raw materials	160	83	133	10	16
Aspiration system from the crushing shack	109	40	43.6	0.5	0.55

The concentration determination of fine particles with sizes less than 10 microns and less than 2.5 microns was carried out using the approach proposed by JSC Scientific Research Institute for the Protection of Atmospheric Air (SRI Atmosfera) and LLC PTB Volgogradgrazhdanstroy.

Using the expressions below with the measured total dust concentration in emissions from the source C_T^E and data on the fractional composition of dust particles in emissions from the source, we calculate the fine particles concentration:

$$C_{\mathrm{PM}_{10}}^{E} = C_T^E D(d_p = 10\ \mu\mathrm{m}) \tag{1}$$

$$C_{PM_{2.5}}^{E} = C_T^E\ D(d_p = 2.5\ \mu\mathrm{m}) \tag{2}$$

The measurements of the total dust concentration in the atmospheric air in the summer period showed the following results: the average values of this value on the industrial site territory are 0.8 mg / m3; the average values of the total dust concentration at the border of the sanitary protection zone are 0.44 mg / m3.

Using the expressions (1) and (2) and the data of the results of assessing the power of dust emissions into the atmosphere, obtained earlier and contained in [8-17], we obtain the values of the PM10 and PM2.5 particles concentration in atmospheric air (Table 2).

Table 2. PM_{10} and $PM_{2,5}$ particle concentration in atmospheric air

Samples selection place	General concentration, mg/m^3	PM_{10}		$PM_{2,5}$	
		content, %	concentration, mg/m^3	content, %	concentration, mg/m^3
The industrial site territory	0.8	12-40	0.096-0.32	0.3-0.5	0.002-.004
The protective zone sanitary border	0.44	80	0.32	0.5	0.0022

The data obtained experimentally indicate that during the production of building blocks from aerated concrete, as a result of insufficient emissions dedusting, an excess of the content of PM_{10} particles is noted as the maximum permissible concentration, which for the air of populated areas is established as 0.3 mg/m^3.

According to the studies' results, the excess of the established standards for the $PM_{2,5}$ particles concentration was not established.

In addition to the dispersion analysis of the emitted dust, it is important to study the type and properties of dust for the equipment selection for a dust cleaning system.

In the course of the studies of the dispersed analysis of dust emissions into atmospheric air from the sources in the production of aerated concrete structures by the method of optical microscopy, the study of the dust particles' shape was also carried out [8-17].

The analysis of the results obtained showed that the dust particles formed during the production of aerated concrete are close to spherical or elongated, oblong and are characterized by the values of the geometric shape factor F_g and the sphericity coefficient (aspect ratio) from 1.061 to 1.564 and from 0.939 to 0.631, respectively [8-17].

According to the research results, it was determined that the apparent and bulk density of dust emitted during the production of aerated concrete and building blocks from it, ranges from 200-1710 kg / m^3 and 80-1100 kg / m^3, respectively [8-17].

The complex rheological properties of dust - flowability, aeration, arch formation - are characterized by an indirect indicator - the angle of repose, that is, the angle between the horizontal surface and the cone generatrix of the powdery material poured on it [8-17].

The angle at which the outer surface of the granular material is located when the particles fall onto the plane - the dynamic repose angle - was determined by the method of pouring powder onto a disk.

The inclination angle of the material surface resulting from the collapse of a layer of particles when the retaining wall is removed - the static angle of repose or the angle of collapse - was determined using a transparent rectangular container with an opening side wall [8-17].

According to the data obtained, the values of the slope of repose - dynamic and static - for the dust under investigation vary within 39-53 degrees and 44-73 degrees, respectively [8-17].

One of the dust properties, which characterizes the ability of a powdery product to become suspended, as well as the tendency of the product to aerate, is dust dispersibility or dusting ability.

Dust dispersibility, i.e., dusting ability - one of the properties of dust, which characterizes the ability of a powdery product to become suspended, as well as the tendency of the product to aerate.

Depending on the amount of dusting capacity, dust is divided into three groups: more than 70% - very dusty; 50-70% - dusty; 30-50% - slightly dusty [16].

To determine the dispersibility of the dust generated during the production of aerated concrete structures in the studied production conditions, the method of N.E. Pestov was used. [13, 16]. In accordance with this technique, the research was carried out as follows. A known mass of dust is directed in portions through a funnel into an air stream, which sprays the material. When determining the dusting capacity, the air flow speed is taken as 2 m / s. Part of the dust settles on the crystallizer. According to the experimental studies' results, it was established that the dust generated in the production of aerated concrete structures has a high dispersibility - from 45 to 88%. Thus, according to the experimental studies' results, it was found that the dust formed during the aerated concrete structures production, according to the existing classification, belongs to the dusty group.

Studies of the complex rheological properties of dust showed that the studied dust is characterized by:

- Good flowability - the ability of the material to form a discrete-continuous steady flow. Flowability depends on the size, density and moisture content of the dust particles. To assess flowability, it is recommended to use the dynamic slope of repose as the main indicator.
- Large aero capacity - the tendency of the material to form a fluidized system with an avalanche-like unstable outflow;
- Weak arching - the ability of dust to form a stable arch structure, causing the powdery material to hang over the outlet [8-15].

Summary

The conducted experimental studies showed the following results:

1. The main properties of dust particles entering the systems for localization and purification of emissions during the production of aerated concrete and building aerated concrete blocks have been investigated and generalized.
2. According to the results of the studies carried out, it was found that the studied dust has a high dust-forming ability and is characterized by good flowability. The studies have also shown that the dust generated in the considered production is polydisperse. In this case, the bulk of the particles has a shape close to spherical, or elongated, oblong.
3. The data obtained from the field studies' results indicate that dust enters the aspiration systems, from 25 to 100% the mass of which falls on the particles with sizes up to 20 microns. In this case, the median diameter varies from 5.5 μm to 28 μm, the proportion of particles with sizes less than 10 μm (PM10) is from 6% to 65%.
4. Analysis of the results obtained, characterizing the dispersed composition of dust particles, as well as the content of PM_{10} particles in emissions and in the atmospheric air, allows us to conclude that the fractional efficiency of dust collecting equipment used in dust cleaning systems is insufficient.
5. It was also found that when cyclones are used in dust control units for the production of aerated concrete and aerated concrete blocks, an excess of PM_{10} particles may be observed in atmospheric air.

References

[1] Azarov V N, Sergina N M, Kondratenko T O 2017 *Problems of protection of the urban air environment from pollution by industrial dust emissions MATEC Web of Conferences* **106 07017**. DOI:10.1051/matecconf/201710607017

[2] Kondratenko T O 2014 Environmental aspects of aerated concrete production *Problems of industrial and environmental protection: collection of materials and scientific works of environmental engineers, Volgograd* **6.**

[3] Nazarova M S 2014 Assessment of technological processes of aerated concrete production as sources of dust emission *Problems of industrial and environmental protection: collection of materials and scientific works of environmental engineers* **6.**

[4] Nikolenko M A 2014 Improvement of dust cleaning systems in the production of aerated concrete products *VIII International correspondence scientific and practical conference "Modern science: development trends.", Krasnodar,* 206-210.

[5] Nikolenko M A 2014 Improving the environmental safety of the production of aerated concrete products *XIV International Scientific and Practical Conference "Priority Research Areas: From Theory to Practice", Novosibirsk,* 106-111.
[6] Kondratenko T O, Saybel A V 2012 Assessment of the impact of construction production on the environment *Engineering Bulletin of the Don* **4 (2)**. Information on ivdon.ru/magazine/archive/n4p2y2012.
[7] Kondratenko T O, Saybel A V 2012 Environmental assessment when choosing building materials for new construction, reconstruction and restoration *Engineering Bulletin of the Don* **4 (2)**. Information on ivdon.ru/magazine/archive/n4p2y2012.
[8] Kondratenko T O 2014 Methods for ensuring environmental safety in the aerated concrete production *Scientific review* **11 (3)** 815-817.
[9] Sergina N M, Kondratenko T O, Nikolenko M A, Pushenko S L 2017 Principles of layout and assessment of the effectiveness of systems for protection against dust pollution of the air of the working area and atmospheric air *EMMFT 2017: International Scientific Conference Energy Management of Municipal Transportation Facilities and Transport EMMFT* 710-719. Information on https://link.springer.com/chapter/10.1007/978-3-319-70987-1_75
[10] Bogomolov A N, Sergina N M, Kondratenko T O 2016 On inertial systems for dust cleaning and dedusting of equipment and working areas in the production of aerated concrete with suction from the hopper of the VZP apparatus *Procedia engineering* **150** 2036-2041. DOI:10/1016/g.proeng.2016.07.290.
[11] Kondratenko T O 2014 Layout diagram of the dedusting system in the production of aerated concrete *Scientific Review* 11 (2) 518-520.
[12] Sergina N M, Nikolenko M A 2015 Experimental evaluation of the solution to reduce dust emissions into the atmosphere in the production of aerated concrete *Engineering Bulletin of Don* **1.** Information on http://www.ivdon.ru/ru/magazine/archive/n1y2015/2751.
[13] Azarov, V N, Sergina N M 2002 *Method for microscopic analysis of the dispersed composition of dust using PC* (Moscow) **1333.**
[14] Kondratenko T O, Danilova-Volkovskaya G M, Fursov V A, Kobalia T L 2019 The basic properties study of the dust particles entering the localization system and emissions cleaning in the aerated concrete production and building gas concrete blocks *IOP Publishing (CATPID-2019 IOP Conf. Series: Materials Science and Engineering; Kislovodsk, Russian Federation; 1-5 October 2019).* **698, 022075**. DOI: 10.1088/1757-899X/698/2/022075.
[15] Kondratenko T O 2020 Experimental studies to develop the measures for reducing dust emissions mass into the atmosphere from the sources of the aerated concrete structures shop *IOP Publishing, (CATPID-2020 IOP Conf. Series: Materials Science and Engineering; Nalchik; Russian Federation; 26-30 September 2020)* **913, 052047**. DOI: 10.1088/1757-899X/913/5/052047.
[16] Kouzov P A, L Ya Skryabina 1983 *Methods for determining the physical and chemical properties of industrial dusts* (Leningrad, Chemistry) 138.
[17] Kondratenko T O 2014 Research of the main properties of dust in the production of aerated concrete *III International scientific-practical conference "Problems and prospects of modern science." (*Stavropol) 67-73.

IOP Conf. Series: Materials Science and Engineering 1083 (2021) 012082 doi:10.1088/1757-899X/1083/1/012082

Geotechnical and architectural studies of the territory in the water protection zone of the Temernik River at the pre-design analysis stage

L A Zolotareva[1], S E Komarova[2*], D I Shesternina[3], N M Khansivarova[1]

[1]South Federal University, 105/42Bolshaya Sadovaya st., Rostov-on-Don, 344006, Russia

[2] Don State Technical University, 1 Gagarin sq., Rostov-on-Don, 344002, Russia

[3]ASA LLC, 97/243, Budennovsky avenue, Rostov-on-Don, 344011, Russia

E-mail: groppius@mail.ru

Abstract. The article describes the application of an integrated systematic approach to the organization of surveys at the architectural spatial environment pre-design studies stage in the water protection zone of the Temernik River. The authors propose to use the geological portal "Geobaza" when conducting engineering surveys on the territory to increase the economic feasibility of the developed architectural project and reduce the time for the pre-project engineering studies' execution.

Introduction

At present, especially in the economic crisis context, detailed and accurate engineering surveys of the territory are a rather costly research type, and it is sometimes difficult to perform in the conditions of difficult terrain and other geological features of the site. The study, to which the article is devoted, is intended to open up the possibility of quickly obtaining a fairly complete and reliable amount of information about the results of previously conducted surveys from an interactive map - the geological portal "Geobaza", developed by the General Director of LLC "Geoelement", Tryukhan V.A. jointly with the researchers from the Institute of Geosciences of the Southern Federal University.

Purpose of the study: Identification of factors, parameters and principles of creating a developing architectural spatial environment in the water protection zone of the Temernik River in Rostov-on-Don on the basis of a comprehensive survey using the geological portal "Geobaza", as well as on the basis of the results of the surveys to determine the architectural environment nature (a possible layout and development scheme territory).

Research object is the riverside area of the Temernik River.

Research subject is the environment of the land plot intended by the owner for the multifunctional residential complex construction.

Problem: lack of reliable forecasts of the negative processes development in the water protection zone and on the water body, as well as the results of monitoring the state of the bottom and banks, changes in morphometric features, the state and mode of the water protection zones use of the Temernik River in the city of Rostov-on-Don.

Research hypothesis: The ability to determine the development nature and the functional purpose of buildings based on the analysis of a comprehensive area study:

— engineering and geodetic surveys;
— environmental engineering surveys;
— geotechnical survey;
— engineering and hydrogeological surveys;
— experimental filtration studies in the water protection zone of the Temernik River;
— architectural pre-design analysis.

Methods of work and survey results

In accordance with clause 4 [1], the customer and the contractor determined the scope of work carried out in the course of engineering surveys of both basic and special types, their volume and method of implementation, taking into account the specifics of the relevant territory and the land plots located on them. According to clauses 1 and 5 of article 47 [2], in accordance with the design assignment and the survey execution programs, engineering surveys were performed to prepare the pre-design documentation for the capital construction object: engineering and geodetic surveys; environmental engineering; geological engineering; engineering and hydrological field and office work. In accordance with clause 4 of article 3 [3], during the construction of individual architectural objects, determined by the approved urban planning documentation, the architectural and planning assignment should be developed on the basis of mandatory pre-design studies.

According to [4], the implementation of the entire above-mentioned complex of works in close interaction of executive authorities, municipalities, scientific and educational institutions should lead to a significant improvement of the Temernik river basin, the creation of the necessary engineering and organizational prerequisites for the continuous ecological park formation in the riverside zone of the Temernik river on territories of the city of Rostov-on-Don, Aksaysky and Myasnikovsky districts of Rostov region.

In accordance with the current Land Use and Development Rules for the city of Rostov-on-Don [5], the projected territory is located in the following urban planning zone: MF-2/5/5 - a zone of multifunctional development along the Temernik River of the second type. The urban planning regulations of this territorial zone were developed to ensure the legal conditions for the formation, development and maintenance of the territories intended for the placement of multifunctional buildings with a high degree of landscaping, by reforming (transforming) the territory along the Temernik River.

Specialists of DonGeo LLC and Architectural and Construction Atelier LLC took part in the field and laboratory work with the involvement of students of the Southern Federal University (Academy of Architecture and Arts, Department of Structural Mechanics and Structures, Institute of Earth Sciences, Department of Engineering Geology) under an agreement of mutual participation.

Engineering and geodetic surveys

In order to establish on the ground points of the state geodetic network (SGN) on the territory of the urban district of the city Rostov-on-Don, as well as to determine the optimal schemes of geodetic networks, a reconnaissance of the area was carried out. And a request was submitted to the Office of the Federal Service for State Registration, Cadaster and Cartography in the Rostov Region to provide the coordinates and heights of geodetic points in the study area. During the reconnaissance survey of the survey area, the sites for laying points for satellite measurements were outlined and established, and in particular, the difficult places requiring more detailed survey were identified. The creation of the geodetic reference network was carried out on the technical assignment basis. The maximum errors in the survey justification position according to the adjustment sheet of the geodetic satellite network are in the plan - 0.020, in height - 0.029.

The work was carried out using two sets of dual-frequency satellite receivers S-Max GEO - №5726550876, №572755017. The processing of data from the field satellite measurements was carried out on the basis of a field party on personal computers using the Leica Geo Office (LGO) software product. As a result of the work, the coordinates and heights of the reference geodetic network Rp1 and Rp2 points in the coordinate system MSK-61 and the Baltic height system were obtained.

Figure 1. Work site scheme.

The survey of underground utilities was carried out with the involvement of the operating organizations' representatives.

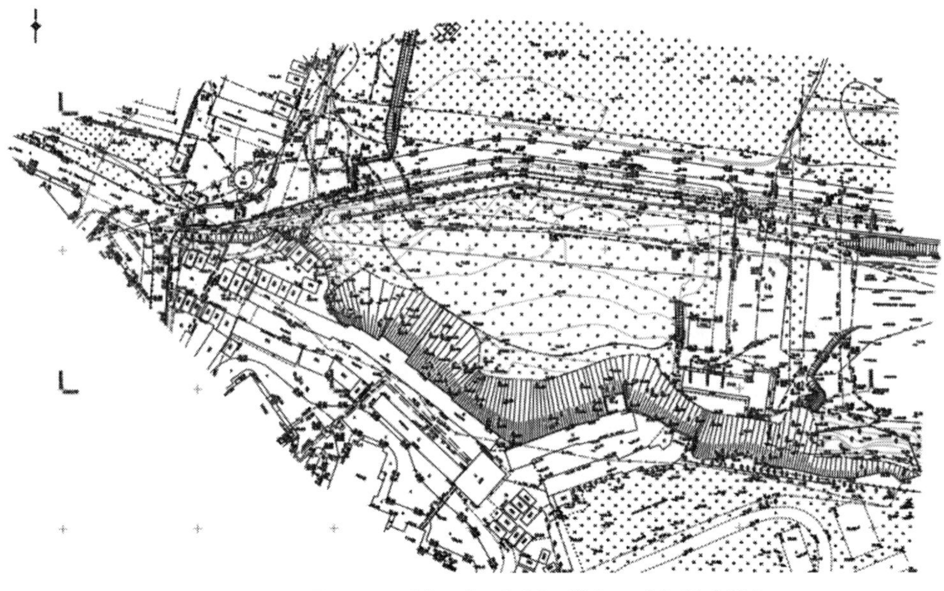

Figure 2. Topographic plan M 1: 500 on 20.01.2020.

The utilities' location was preliminarily determined by the operating organization according to the technical documentation, then by the locator to clarify the utilities' location and the occurrence depth. The routes of underground communications were marked with the landmarks on the ground and were taken into account in the process of topographic survey.

The topographic plans with applied underground utilities were subsequently sent to the operating organizations for approval. The agreements were obtained with the seals and signatures of all responsible authorities.

In the archival funds, a topographic survey was obtained by LLC YuzhGeoKom in April 2014, with the designation of zones with special conditions for the studied area territory use.

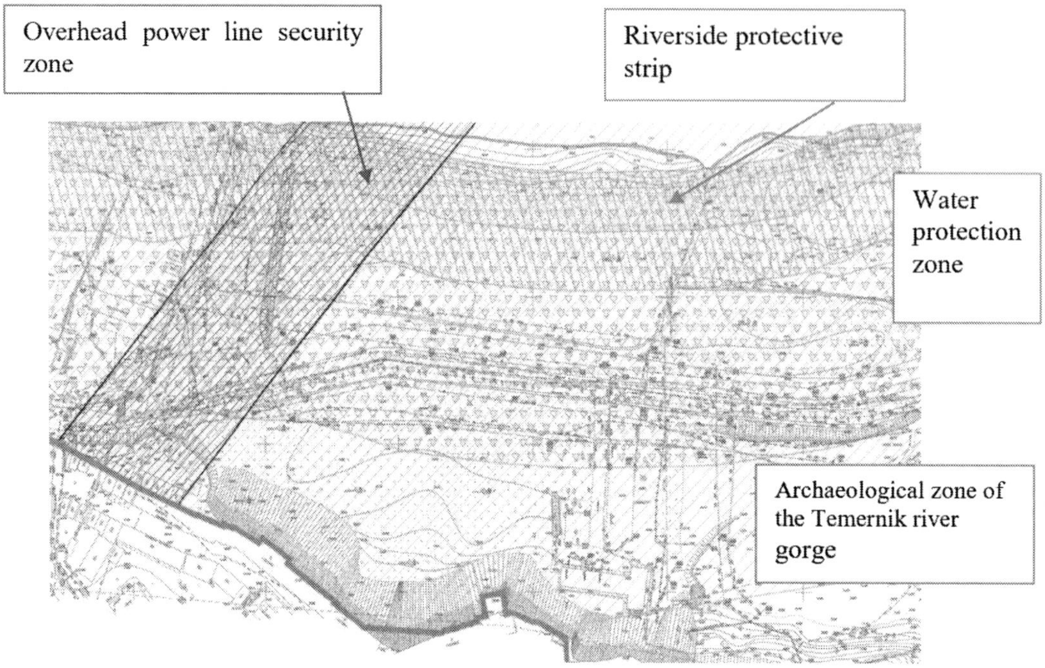

Figure 3. Topographic plan M 1: 500 with special areas of territory use on April 30, 2014.

Administratively, the survey site is located in the Oktyabrsky district of the urban district, the city of Rostov-on-Don, at 2 Pyatiletki street, 29, cadastral number 61: 44: 082201: 176. The area of the land plot is 1.35 hectares, the area of the adjacent residential buildings is 0.46 hectares. The site is located on the left bank of the Temernik River. The relief of the site is complex, with contour lines falling towards the Temernik River. Absolute surface elevations vary from 11.87 to 12.14 m in the Baltic system of heights. Along the southern border of the land plot, a slope with a drop in elevations towards the land plot along the border from 20.06 to 14.5 m at the top of the slope and from 13.28 to 12.07 m at the bottom of the slope in the Baltic system of heights, slope height difference from 3.0 m to 7, 0m.

The territory of the land plot is located within the following zones with special conditions of use:

- in the archaeological zone of the Temernik river gully;
- in the water protection zone of the Temernik river;
- the western part of the land plot within the boundaries of the PTL 110 kV security zone.

Engineering and environmental surveys

The purpose of the engineering and environmental surveys was to assess the current state and predict possible changes in environmental components under the anthropogenic load influence in order to prevent, minimize or eliminate harmful and undesirable environmental and related social, economic and other consequences.

The main tasks to be solved are as follows:

1. Assessment of the current ecological state of the natural environment individual components (atmospheric air, soil, vegetation and fauna) and geo- and ecosystems in general, their resistance to anthropogenic impacts and their ability to recover in the area of the studied object.
2. Assessment of socio-economic and medico-biological living conditions of the population in the area of engineering and environmental surveys based on stock materials.
3. Identification of soil pollution possible sources based on the current situation analysis and the territory use.

4. Assessment of physical factors of influence (noise from natural and man-made sources).
5. Development of proposals and recommendations for the environmental protection measures organization and industrial environmental control (monitoring).

At the work site, there are no hazardous natural and man-made processes affecting the relief. The vegetation is represented by tree plantations of oak, acacia and bushes, the bases are uneven, the tree trunks are inclined in different directions. The fourth part of the site surface is swampy. To drain groundwater, two drainage systems with concrete pipes and trays pass through the surface areas.

Physical factors of environmental impact - noise rarely generated by vehicles from the road passing along the site border.

Figure 1. View of the study area from North.

Figure 2. View of the study area from South.

According to the tasks set, the following necessary works were carried out during the implementation of engineering and environmental surveys. Collection, generalization and analysis of stock materials, information about natural conditions and the modern nature of economic development on the territory of the projected facility were carried out in specially authorized state bodies in the field of environmental protection and other specialized organizations performing medico-biological, hydrometeorological, landscape, soil research. Reconnaissance (route) survey of the territory was carried out in the course of field work during the construction site study.

During the soil samples' selection, the main visual signs were recorded: texture, moisture, color, structure, density, composition, neoplasms, inclusions, character of effervescence, transition character. Soil sampling was carried out for the purpose of their ecotoxicological assessment as the environment components. Geo-ecological sampling of the natural environment components was single. Taking into account the nature of the designed object, the range of controlled components was determined based on the potential contamination composition specifics. Laboratory studies of soil samples for the content of pollutants were carried out in specialized chemical-analytical centers and laboratories accredited in accordance with the established procedure: testing laboratory center FBEI "Center for Hygiene and Epidemiology in the Rostov Region", FSBU GTCAS "Rostov".

The assessment of vegetation and fauna was assessed using the stock data and directly visually at key sites of the surveyed area.

The background concentrations of pollutants were estimated based on the data presented by FSBI "North Caucasian UGMS" and climatic reference books. In addition, with the help of the multifunctional multicomponent portable gas analyzer Kometa-M, the content of the following gases in the atmospheric air was studied: oxygen, ammonia, carbon monoxide, hydrogen. The measurements were taken at the points coinciding with the places where soil samples were taken. The total number of measurements is 4. Measurements of the external gamma radiation dose rate (EGR) were carried out on a plot area of 6.7 hectares in constant listening-in mode using devices SRP-68-01, DRG-01T1, DKG-02У "Arbitr". EGR measurements were performed over the network of profiles. A total of 30 measurements were made. Contrasting EGR abnormal values were not established. Control gamma-spectrometric determination of radionuclides was recognized as inappropriate. Taking volumetric activity (VA) measurements of radon in the selected samples included: measurement of the residual activity of radon in the measuring chamber; mixing the sample between the sampler and the measuring chamber; VA radon samples dimension in the measuring chamber.

The sound (noise) level was estimated at the border with the residential area of the projected buildings. The measurements were made at two points in accordance with GOST R 53187-2008 "Noise monitoring of urban areas".

Cameral work included analysis of the current state of natural components based on processing the results of a route survey in the territory; laboratory data on soils; materials collected in the bodies for the environment protection and monitoring, the library of the Faculty of Geology and Geography. Working maps and diagrams were drawn up with digitization and preparation of cartographic material in electronic form, the correspondence of the identified parameters to the current standards was established, the natural complex value and current state were determined. The preparation of the final report on the engineering and environmental surveys results was carried out. Graphic applications are represented by a set of cards.

Engineering and geological surveys

According to engineering and geological conditions, the area of work belongs to the III category of complexity. The purpose of the geotechnical survey was to determine the feasibility of construction and to determine the possible construction costs. The tasks of engineering and geological surveys were: study of the geological and lithological structure and hydrogeological conditions of the construction site, the engineering and construction properties of soils, which will serve as a natural basis and environment for the designed object.

"One of the modern tools to achieve this goal is to use GIS-technologies. GIS capabilities for engineering-geological studies are significant: they allow accumulating, storing and processing large volumes of engineering-geological information, for example, forming an information resource about the composition and properties of the geological environment; allow solving the problems of typological engineering-geological zoning; architectural analysis of the design site, etc." [10]

Consolidation of efforts of specialists from survey organizations and researchers of the Southern Federal University, namely the Department of Structural Mechanics and Structures, the Department of Engineering and Construction Disciplines of the Academy of Architecture and Arts and the Department of General and Engineering Geology of the Institute of Geosciences, in accordance with the scientific plan, allowed to test the an interactive map [10, 11] of the geotechnical characteristics of soils in the territory of Rostov-on-Don and the Rostov region (electronic archive of information on the parameters and characteristics of soils of the buildings and structures foundations), which provides information from 2008 to 2020, including for the investigated territory. The information obtained helped to trace the changes' dynamics in the geological structure in the study area over 12 years.

Geomorphologically, the site belongs to the following geomorphological elements: the left bank slope of the Temernik River valley, its high floodplain and the first terrace above the floodplain. The relief of the study area surface has a slope in the north direction.

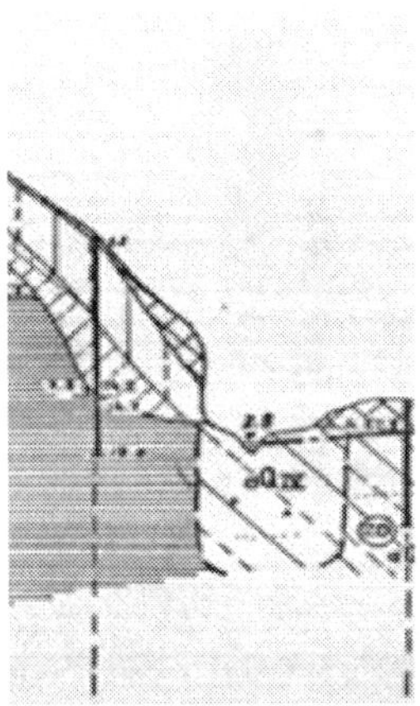

Figure 4. Geological section in the floodplain of the Temernik River 2008.

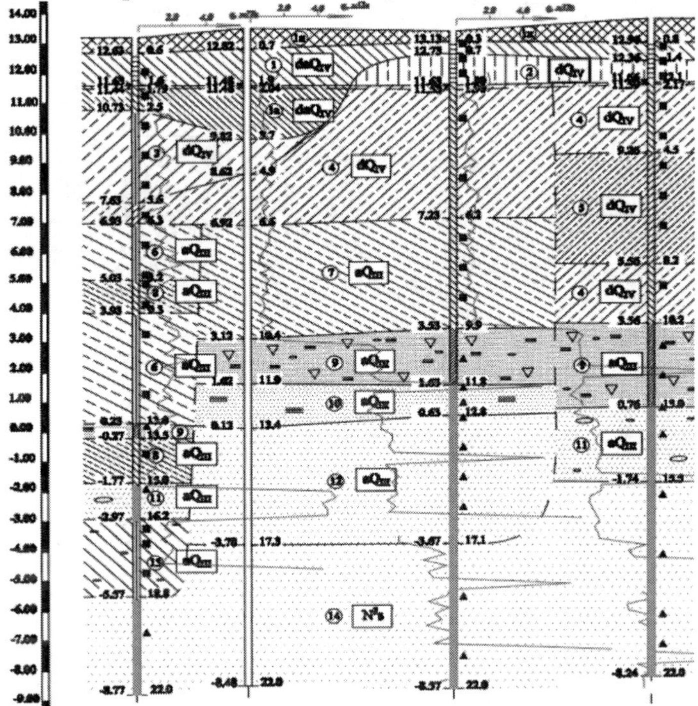

Figure 5. Geological section in the first terrace above the floodplain in 2020.

According to the archival materials of 2013, the geological structure of the site to a depth of 20.0 m involved Quaternary diluvial and alluvial loams, clays and sands, covered by technogenic soils, underlain by Neogene sediments, represented by clays and limestones from above. Groundwater is located at depths from 0.9 to 1.25 m (absolute marks 10.21 m - 12.15 m). Groundwater is hydraulically linked to the water level in the Temernik River. During flood periods, flooding of a low floodplain occurs, and in the areas of a high floodplain, groundwater may flow to the day surface.

According to the survey data of January 2020, the geological structure of the survey site is attended by diluvial deposits of the Quaternary age, overlapped from the surface by technogenic and humous soils. Diluvial loams are underlain by Quaternary alluvial clays, loams, gravel deposits and sands. Neogene sands occur down the section. When drilling wells in January 2020 the ground waters were discovered at depths of 0.4 m - 0.8 m (absolute marks 11.48 m - 12.10 m) in man-made loose loam. Under the prevailing hydrogeological conditions, without taking into account the amplitude of seasonal fluctuations, the structures foundations at the survey site will be periodically flooded.

Engineering and hydrogeological surveys

The main tasks of the research were: study of the geological section within the various geomorphological elements of the territory; clarification of the hydrogeological and engineering-geological conditions of the territory and the establishment of the ground water level position; obtaining the information about the geo-filtration parameters of soils necessary to substantiate the measures to prevent flooding and flooding phenomena and directions for subsequent hydrogeological monitoring and forecasts.

The scope of the work included: reconnaissance ecological and hydrogeological survey; manual drilling of mapping wells up to 7 meters deep in order to study the geological section and establish the groundwater level position; hydrogeological experimental work - single express pumping from wells.

The most important aspect of the study was the flooding and waterlogging processes in the territory. The study area is located on the first low left-bank above-floodplain terrace of the Temernik River. From the North it is bounded by a motor road, from the South by the slope of the second terrace above the floodplain. Temernik (Figures 3 and 4).

Figure 3. View of the western part of the slope slope the second above-floodplain Temernik terrace of the Temernik river.

Figure 4. To the eastern part of the second above the floodplain terrace of the river

Slope of the second terrace above the floodplain of the Temernik river, overgrown with trees, has a different steepness and is significantly disturbed by anthropogenic activity (remnants of a destroyed drainage system, soil dumps, construction and household waste). At the time of the survey, no groundwater outlets were found in the terrace body. Hazardous slope processes were not identified.

When examining the first lower terrace of the Temernik River, the phenomena of flooding its low sections, localized in the western part of the site, were established. It seems most likely that the anthropogenic impact on the relief and the drainage system disruption created the conditions for a very difficult water flow (backwater) and was the main reason for flooding, underflooding and swamping of low parts of the territory. The whole area is overgrown with weeds, bushes and reeds. Inspection of the drainage system, created to drain surface and lower the groundwater level and upper waters, indicated its unsatisfactory condition. Throughout the site, there are traces of its partial or complete destruction. Not all gutters and pipes were in working order.

Experimental filtration studies in the water protection zone of the Temernik River

Short-term pumping from wells is mainly performed to study the filtration characteristics of rocks with a relatively low permeability (0.01 <K <5, m / day). The calculated coefficient value, as a rule, depends on the nature of the level rise curve in the well. This curve is plotted in coordinates lgS_o/S from t and the filtration coefficient is determined only by the straight-line part of the graph passing through the maximum number of experimental points (Fig. 6). Reliable results have been obtained by means of the open hole tests. In this case, the mesh does not become clogged and the filter resistance does not affect the groundwater horizon filtration parameters' determination. The specificity of soils necessitates the random errors' elimination. For this, it was recommended to determine the filtration coefficient based on the results of the experiments carried out at several points of one site.

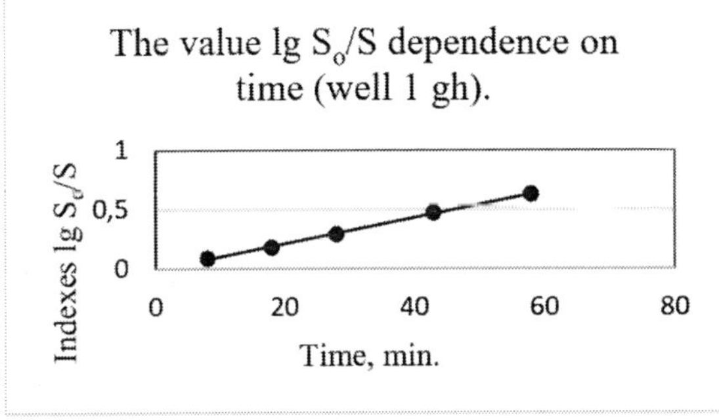

Figure 6. Dependence lgS_o/S from t.

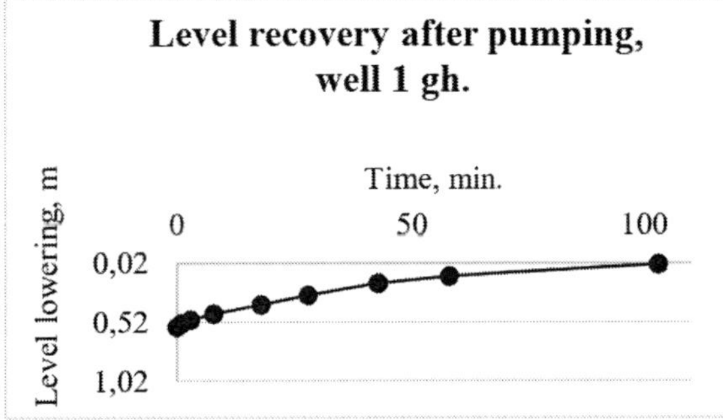

Figure 7. Level recovery curve No.1.

To conduct the experiment, the wells were drilled below the static water level by at least 40-60 cm. Water was pumped out or dumped out from them to the maximum possible depth, the restoration of the level of which was observed after 2-3 minutes at the beginning of the experiment and after 10-15 minutes - in the end. Based on the observation results, graphs of the dependence of the level h change on time was plotted. The filtration coefficient of watered homogeneous soils according to the experimental work results was determined according to the formula of K.Ya. Kozhanov.

$$k = \frac{\alpha(h_0 - r_0)}{h_0^2 \Delta t} r_0^2 \lg \frac{S_0}{S}$$

where

h_o – is the depth from static level to the bottom of the well ($h_o = h + S$), m;
r_o – is well radius, m;
Δt – is a calculated moment of time in days, determined from the graph in Figure 7.
S_o – is maximum lowering of the water level in the well from static, m;
S – is a decrease in water level at a point in time Δt, m.

In the production of instant pumping, on the recommendation of the author of this method K.Ya. Kozhanov, it is better to use the data obtained in the interval of (0,2-0,8) S_o.

According to the experiment results and the calculation results, the data given in Table 1 were obtained.

Table 1. Experimental results

Well No.	Well depth, m	Depth to static level to the bottom of the well, m	Well radius, r_o, m	Depth from static level to the bottom of the well h_o, m	The maximum decrease in the water level in the well from the static, S_o, m	Filtration coefficient, m/day
1	1.71	0.55	0.10	1.15	0.56	0.51
2	7.30	0.70	0.075	6.60	0.45	1.09
3	2.00	0.85	0.10	1.15	0.37	3.46
4	2.00	0.40	0.10	1.60	0.50	3.58

Based on the analysis of the field and laboratory work performed, the following conclusions were drawn:

1. Underground (ground) waters at the research site are widespread. They belong to non-pressurized groundwater, have a free surface, the position of which in the research course was recorded in wells at depths of 0.4 - 0.85 m (the first above-floodplain terrace). Water-bearing rocks are Quaternary deposits of various origins: technogenic soils, alluvial loams, clays; diluvial loams. Groundwater is fed over the entire area of its distribution due to the atmospheric precipitation infiltration. The aquifer is hydraulically connected with the waters of the Temernik River, into the channel of which groundwater is unloaded. The existing dilapidated drainage system has a certain effect on the groundwater discharge. In the relief depressions, groundwater came to the surface. The established facts indicate that under the existing conditions (climate, relief, conditions of surface and underground runoff) flooding, under-flooding and waterlogging of lowered areas in the territory occur.
2. The performed experimental filtration works make it possible to classify most of the soils composing the floodplain and the valley slopes as permeable (filtration coefficients 0.5 - 3.6 m / day). This testifies to the rather powerful flows of groundwater discharging into the Temernik river bed.

Architectural pre-design analysis

At the stage of the territory development architectural concept formation, it is necessary to carry out a pre-design analysis of the site and the surrounding area, which includes a study of the configuration and topography of the site, an exposure assessment of the relief and the degree of its complexity, as well as the presence of special geological conditions in the area under consideration.

As a part of a comprehensive pre-design architectural analysis, the following issues are also considered: the need for insolation and natural illumination calculations; the presence and possibility of reducing the sanitary protection zone and other especially dangerous protection zones; possible directions for the future territory development. The main purpose of pre-design studies is to collect and study information, criteria and factors to substantiate the possibility of placing a new construction object specified by the customer, taking into account the requirements of urban planning, historical, cultural, socio-economic, sanitary-hygienic and environmental regulations and the concept of the general city plan.

In the context of the research topic, to which this article is devoted, the central subject of the pre-project architectural analysis is the geometry of the site, on which the nature and form of the designed multifunctional residential complex directly depends. The relief of the site is complex, with the fall of the contours towards the Temernik River, the southern and southwestern borders are fenced off by a slope, the steepness of which reaches 45°, slope height - 7.0 m. The task of planning the territory is specific and is determined by the temporal and spatial development of the environment in the Temernik River water protection zone. A concomitant factor is the historical nature of the area development in the 60-90s of the last century - the land plot was previously a part of the territory entirely allotted for the operation of the "Nadezhda" sanatorium, where there were no main and citywide roads and streets, as well as urban transport. When designing buildings on a site with difficult terrain, a creative, individual approach is required, taking into account a set of technical, functional and compositional requirements, "... establishing the maximum connection between architecture and relief and other natural components is the most important condition for the successful solution of urban planning problems". [6]

Discussion

In modern conditions of constant shortage of sites for construction, the developers are forced to develop the territories that were previously considered unsuitable for urban construction. One of the parameters that have a direct impact on the future construction site development complexity degree is the site relief. Many specialists from different countries have devoted their scientific works to this problem solution. From the end of the 20th century to the present time, the international experience of using the territory located in difficult conditions has been studied. Thus, in the book "Urban Development on Slopes" under the general editorship of the candidate of architecture V.R. Krogius sets out the results of many years research by Soviet and foreign experts in this field: "... in the theory and practice of urban planning, the concept of "complex relief" has been formed, which in general terms characterizes the totality of forms of the Earth's surface, significantly affecting the functional, household, sanitary and hygienic, architectural and aesthetic and technical and economic characteristics of urban construction and economy and determining the use of special methods of planning, development and improvement of cities". [6]

The monograph "Comprehensive Assessment of the Territory in Urban Planning" under the general supervision of S.G. Sheina [12] is devoted to the GIS- systems application in the process of collecting and analyzing initial information for the economic assessment in the study area. The authors of the monograph defined an algorithm for financial assessment and formed a mathematical model for making calculations, as well as the monograph proposed ways to use the results of a comprehensive assessment in urban planning and management of the Rostov-on-Don city development. The quality of the Rostov-on-Don city territory suitability and investment attractiveness study was the subject of a scientific article by S.G. Sheina and Yu.V. Popova "Assessment of the suitability of the territories of Rostov-on-Don for the implementation of integrated development projects". [13]

In the article by Yu. V. Gorgorov, M. G. Sarkisyants, M. A. Sotnikov the principles of sustainable development of the architectural space of the Temernik River embankment in Rostov-on-Don and the issues of experimental design in the water protection zone of the linear park along the Temernik River are considered. [7]

The article by A.V. Ishchenko "Comprehensive analysis of built-up areas as a means of effective urban planning in landslide hazard zones (for example, Rostov-on-Don)" [10] analyzes the degree of environmental risk impact on landslide slopes, groundwater backwater, as well as contamination of the area. The author also proposes the construction of an electronic engineering-geological model of the Rostov-on-Don city, taking into account landslide processes, geotechnical analysis of the geological environment, detection and forecast of the vulnerable spots' formation with the maximum probability of the landslide processes formation.

L.A. Zolotareva, O.D. Alekseeva in an article "Isomorphism of the computational model of the object and the architectural concept" consider the main points of "the concept choice and the method of constructing a model of the future object, which depend on many factors - a set of functions of the object, the construction site, the requirements of the task for the project development, surrounding buildings, the regulatory framework, etc." [8]

In the articles of L.A. Zolotareva, N.M. Hansivarova, V.A. Tryukhan "Expansion of the functions of the portal of geological information for joint remote use by various structural divisions of SFedU in teaching students" and "Portal of geological information as a tool for remote practical training of bachelors of the training direction 05.03.01. "Geology" the components of the natural-technical system (NTS), their interaction and dynamics are described. "One of the most important economic and social tasks of our time is to create NTS, which is capable of meeting the needs of society to the maximum extent with minimal disruption of the existing ecological balance. Optimal design and NTS control can be provided only by joint activities of designers and geological engineers, since the decomposition aspects of any NTS are geotechnical and technical aspects.

"In the context of preventing the spread of a new coronavirus infection, a new format of industrial practices is relevant. The basis for its development was the geological portal "Geobaza". It is a constantly updated electronic archive of geological information, which was created by a graduate of the Faculty of Geology and Geography (since 2013 - the Institute of Geosciences) SFedU, General Director of LLC "Geoelement" V. A. Tryukhan. Geological portal contains the following information:

1. A map of factual material showing the points of the performed engineering-geological and environmental surveys in the territories of Russian cities.
2. Brief information about the survey objects.
3. Table of physical and mechanical properties of engineering and geological elements identified within the soil strata.
4. Data on the properties of groundwater [11]

Summary

The scientific novelty of the study lies in the study of a holistic picture of pre-design engineering and architectural studies required to analyze the main factors affecting the interaction process of a building with the geological environment. Thanks to the joint work of specialists from the survey organizations and researchers of the Southern Federal University, the data obtained by means of the geological portal "Geobaza" were used for the first time. Based on the results of the interactive map use of the factual material and the experimental and filtration studies carried out in the water protection zone of the Temernik River, the possible periodic flooding of the projected building foundations was accurately predicted, which even at the stage of pre-design development indicated the need for the individual design decisions to ensure safety, waterproofing and reliability of structures in special geological conditions. The results of the comprehensive studies carried out allowed the chief architect and chief engineer of the multifunctional residential complex project to most efficiently and effectively solve the fundamental planning structure of the projected facility.

Reflection of the results obtained in the course of various studies on the available maps of the geological portal makes it possible to compile the maps of various scales from the small sites of the planned construction to such large areas as the Rostov Region and the Southern Federal District, and in the future to cover the territory of the entire Russian Federation.

References

[1] *Decree of the Government of the Russian Federation of January 19, 2006* No. 20.

[2] *The Town Planning Code of the Russian Federation dated December 29, 2004*. No. 190-FL (as amended on December 27, 2019).

[3] *Federal Law "On Architectural Activity" dated 18.10.1995. as amended on July 19*, 2011.

[4] *Decree of the Government of the Rostov region dated 26.10.2018 No. 673* "On approval of the draft planning of riverside areas of the river Temernik, located in Rostov-on-Don, Myasnikovsky and Aksaysky districts of Rostov region (from the source of the Temernik river, located east of the village of Bolshiye Saly, Myasnikovsky district, to the confluence with the Don river) (stage 1)".

[5] *Decision No. 605 of the Rostov-on-Don City Duma of the sixth convocation of 12/21/2018*. as amended on 10/20/2020 on the approval of the rules for land use and development of the Rostov-on-Don city.

[6] Krogius V R, Abbott D, Pollit K and others 1988 *Urban planning on the slopes* (Moscow, Krogius, Stroyizdat) 328.

[7] Gorgorova Ju V, Sarkisyants M G and Sotnikova M A 2018 The principles of sustainable development of architectural space of the Temernik River embankment in Rostov-on-Don. International Conference on Sustainable Cities *IOP Publishing. IOP Conf. Series: Earth and Environmental Science* **177** 012035. doi: 10.1088 / 1755-1315 / 177/1/012035

[8] Zolotareva L A, Alekseeva O D 2017 *Isomorphism of the computational model of the object and the architectural design "Topical Issues of Science and Education"* (July 17, 2017, Warsaw, Poland) **1.**

[9] Zolotareva L A, Khansivarova N M, Tryukhan V A 2020 Expansion of the functions of the geological information portal for joint remote use by various structural divisions of SFedU in teaching students. Actual directions of development of the higher education system: Distance education - problems and advantages *Materials of the International scientific and practical conference (Rostov-on-Don, June 16-17, 2020) South Federal University. - Rostov-on-Don; Taganrog: Publishing House of the Southern Federal University* 302 p. DOI: 10.18522 / 801282720.

[10] Ishchenko A V 2012 Comprehensive analysis of built-up areas as a means of effective urban planning in landslide hazard zones (on the example of Rostov-on-Don) *Engineering Bulletin of the Don, North Caucasian Scientific Center of the Higher School of SFedU* **3**.

[11] Khansivarova N M, Zolotareva L A, Tryukhan V A 2020 Portal of geological information as a tool for remote practical bachelors' training of the training direction 05.03.01 "Geology" *Actual directions of development of the higher education system: Distance education - problems and advantages: Materials of the International scientific and practical conference (Rostov-on-Don, June 16-17, 2020)* South Federal University, Rostov-on-Don; Taganrog: Publishing House of the Southern Federal University, 302. DOI: 10.18522 / 801282720.

[12] Sheina S G, Babenko L L, Matveyko R B, Khamavo A A 2015 Comprehensive assessment of the territory in urban planning *Monograph. RSCEU* 103.

[13] Sheina S G, Popova Yu V 2018 Assessment of the territories' suitability in Rostov-on-Don for the integrated development projects' implementation *Construction and man-made safety* **13 (65)** 29-32.

Nitrogen oxide emissions reducing in cement production

T V Guseva[1], E N Potapova[2*], I O Tichonova[2] and K A Shchelchkov[1]

[1]Research Institute "Environmental Industrial Policy Centre", 42 Olimpijskij ave., Mytishchi, 141006, Russia
[2]D. Mendeleev University of Chemical Technology of Russia, 9 Miusskaya square, 125047, Moscow, Russia

E-mail: 55pen@mail.ru

Abstract. In the European Union, the Russian Federation, as well as in a number of other countries, cement production is regulated by legislation on the best available technologies. Mandatory requirements (technological parameters, BAT-AELs) are imposed, first of all, on the emissions limitation of inorganic dust, nitrogen and sulfur oxides. The majority of Russian cement plants are currently characterized by the nitrogen oxide emissions state-established technological indicators' excess. The technological indicators of the best available technologies and methods for reducing nitrogen oxide emissions operating in the Russian Federation and the EU countries are considered. It is demonstrated that the most efficient technologies are the staged fuel combustion and selective non-catalytic reduction (SNCR) of nitrogen oxides.

1. Introduction

A new environmental industrial policy based on the concept of integrated pollution prevention and control by introducing the best available technologies (BAT) has been developing and implementing in Russia since 2014 [1].

Technological rationing based on BAT principles means setting economically feasible goals for the introduction of resource-saving and waste-free industries, technological re-equipment, solving the problems of import substitution, the formation of a competitive industry that ensures the transition of the state's economy from the raw materials export to an innovative development type.

The transition process to a new state regulation in the field of environmental protection is based on the BAT use and normatively regulated reduction of technogenic effect on environment [2-7].

In 2019, the Russian Federation started the process of issuing the integrated environmental permits for large enterprises in key industries. Among the first enterprises to receive the integrated environmental permits in December 2019 were three cement plants: a branch LLC "HeidelbergCement Rus" in the village of Novogurovsky, Joint-stock company "Iskitimcement" and LLC "South Ural Mining and Processing Company".

Evaluation of technological, technical solutions, including the best available technologies, at these cement enterprises showed that there is an excess of nitrogen oxides emissions into the air (over the established industry technological indicators) at all enterprises, therefore, business entities have prepared the projects programs to improve environmental efficiency (PIEE), which were approved by the Interdepartmental Commission under the Ministry of Industry and Trade of the Russian Federation.

The problem of excess emissions of nitrogen oxides into the air currently exists at most Russian cement plants, in connection with which it is advisable to analyze the technological indicators and effective methods for reducing nitrogen oxides emissions in more detail.

2. Formation of nitrogen oxides NO_x when baking Portland cement clinker

Nitrogen oxides (with a share NO about 95 % and N_2O about 5 %) are the key pollutants emitted into the atmosphere during cement production.

There are three ways of nitrogen oxides formation during the Portland cement clinker baking [8]: thermal nitrogen oxides (thermal NO_x); fast nitrogen oxides (fast NO_x) and fuel nitrogen oxides (fuel NO_x).

Thermal NO_x can be formed at high temperatures (t>1200 °C) and high concentration of oxygen during oxidation of atmospheric nitrogen with oxygen during combustion. With an increase in the flame temperature and the excess air ratio in the furnace, the amount of generated thermal NO_x increases. Thermal NO_x can be mainly formed by burning gaseous fuels (natural gas and liquefied petroleum gas) and fuels that do not contain nitrogen-containing substances.

Fast NO_x can be formed in a burning fuel flame by complex reactions of radical interaction CH with nitrogen N_2 with HCN formation, which then is quickly disintegrated into NO_x, CO_2 and H_2O. Content of fast NO_x depends on the shape and temperature profile of the flame and increases during the combustion of enriched mixtures and during low-temperature combustion, reaching 25% of the total amount of the formed nitrogen oxides.

Fuel NO_x can be formed from nitrogen-containing compounds that are part of solid and liquid fuels, especially coal. NO_x formation mechanism consists in converting nitrogen-containing fuel compounds into ammonia NH_3 and HCN followed by additional oxidation to NO_x.

NO_x release greatly depends on the type of the process used. NO_x emissions' rates are rather high in long rotary kilns of the wet production process when the raw mixtures difficult to sinter are baked. NO_x emissions decline with increasing fuel moisture. In dry kilns with cyclonic heat exchangers and calciners, part of the fuel is burned in the calciner at the temperatures up to 950-1000 °C, which leads to a decrease in total NO_x emissions by reducing thermal NO_x formation.

3. NO_x emission levels in the kiln exhaust gases of rotary kilns, corresponding to BAT

Emission levels of nitrogen oxides in rotary kiln flue gases corresponding to BAT (BAT-Associated Emission Levels), are specified in the relevant regulatory documents (Table 1). In Russia, these levels are called sectoral technological indicators.

Table 1. Range of nitrogen oxide emissions values in Russia, the European Union and Germany.

Emissions of nitrogen oxides NO_x from the rotary kiln, mg/m^3	ITG 6-2015, Russia, [8,9]	BREF CLM EC, [10]	17. BImSchV, Germany, [11]
- for kilns with a cyclonic heat exchanger	≤ 500	<200-450	200
- for long kilns for wet production and Lepol type kilns	≤ 800	400-800	data unavailable

The values of nitrogen oxide emissions' technological indicators, given in Table 1, are established in the Russian ITG 6-2015, are close to the emission levels of cement plants in the European Union (BREF CLM). However, when using special technologies, which will be discussed below, and when using alternative fuels (waste incineration), nitrogen oxide emissions in dry-type furnaces with cyclonic heat exchangers should be no more than 200 mg/m^3 (17. BImSchV).

The cement plants' operation analysis, carried out in 2017-2019, showed that currently none of the 57 Russian cement plants uses special technologies, and the levels of nitrogen oxide emissions often exceed 1000 mg/m^3 [12-13].

4. Emissions reduction measures NO_x in kiln gas

To reduce oxide emissions during cement production, the use of both primary technical solutions integrated into the technological process and the special technologies or their combination with primary measures are considered to be the best available technologies (Table 2, Figure 1) [8, 10,14].

Table 2. Emission reduction solutions NOx, used in the cement production.

Technical solution	Emission reduction efficiency, %	Emission data, mg/nm^3
Primary technical solutions:		
• baking process optimization	25	1000
• flame cooling	0-35	less 500-1000
• use of burners with low NO_x emission	0-35	less 500-1000
• staged fuel combustion, combustion of fuel in the middle of the kiln	10-50	-
• use of mineralizers	10-15	-
Special technologies:		
• technology of selective non-catalytic reduction SNCR	30-90	200-500
• technology of selective catalytic reduction SCR	43-95	200-500

For both environmental and economic reasons, NO_x emission reductions can preferably be started with the implementation of the primary technical solutions integrated into the technological process. As it can be seen from Figure 1, staged fuel combustion is the most effective primary technical solution used in the EU countries.

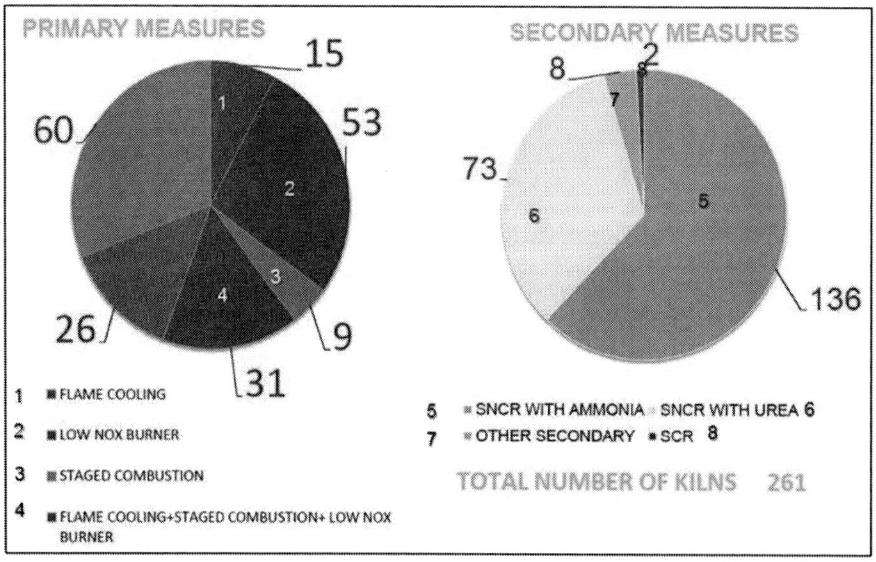

Figure 1. Various technical solutions' application to reduce nitrogen oxide emissions [14].

4.1. Staged fuel combustion

Staged fuel combustion is used in dry-process cement kilns (Figure 2). The first combustion stage takes place in a rotary kiln under optimal clinker firing conditions. The second combustion stage is in the burner at the entrance to the furnace, where the reducing atmosphere is formed. This leads to the decomposition of some of the nitrogen oxides accumulated in the firing zone. At a high temperature,

IOP Conf. Series: Materials Science and Engineering 1083 (2021) 012083 doi:10.1088/1757-899X/1083/1/012083

NO_x reduction reaction to elemental nitrogen takes place in this zone. At the third stage, fuel and tertiary air are fed to the calciner, which also results in a reducing atmosphere. At the same time, the number of NO_x, formed during fuel combustion, is reduced as well as the amount of NO_x, coming into the kiln from the outside. At the fourth final stage, the remaining tertiary air is fed to the top of the system for residual combustion.

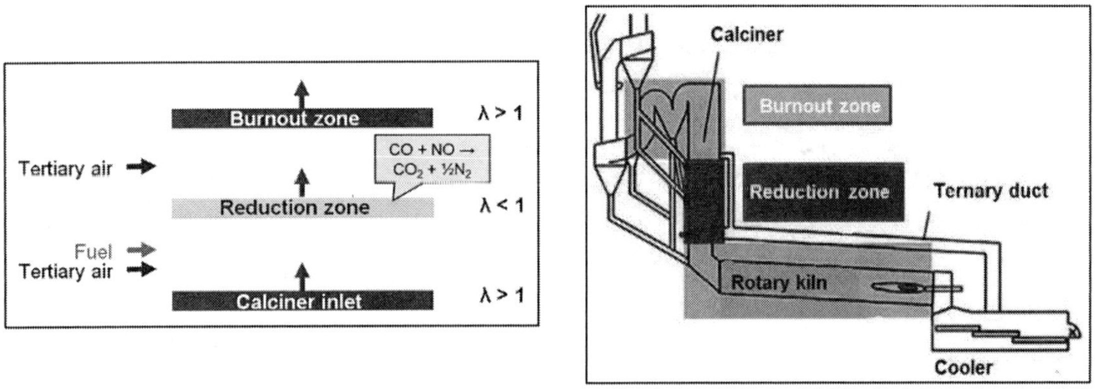

a b

Figure 2. Schematic representation of staged fuel combustion: *a* - the principle of staged fuel combustion; *b* - equipment location.

Combustion of lumpy fuel waste (for example, waste tires, tires, automobile tubes) is one of the options for the staged fuel combustion technology, while the combustion of fuel lumps is accompanied by the formation of a reducing atmosphere in the burnout zone. In the kilns equipped with baked heat exchangers and calciner, pieces of fuel are fed at the inlet to the kiln or calciner.

4.2. Selective non-catalytic reduction of nitrogen oxides (SNCR)

The technology of selective non-catalytic reduction of nitrogen oxides includes injection of an aqueous solution of ammonia into the flue gases (up to 25 % NH_3) or aqueous solutions of ammonia or urea compounds for reduction NO till N_2 (Figure 3):

urea $NH_2CONH_2 + 2\ NO + ½\ O_2 \rightarrow 2\ N_2 + CO_2 + 2\ H_2O$

ammonia $4\ NH_3 + 4\ NO + O_2 \rightarrow 4\ N_2 + 6\ H_2O$.

To ensure the sufficient contact time of the reducing agent with the flue gases, the optimum temperature should be 830 - 1050 °C.

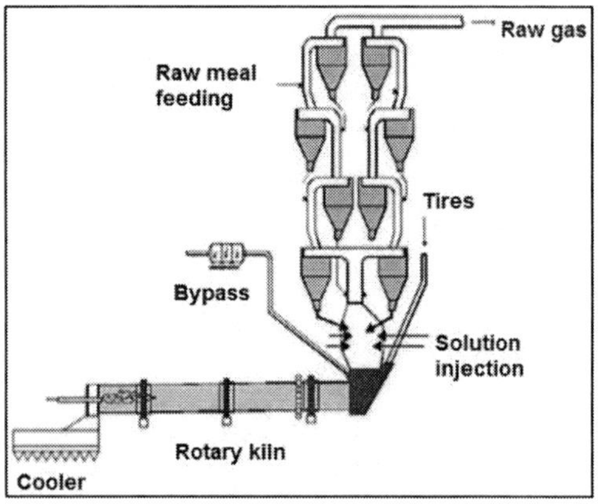

a

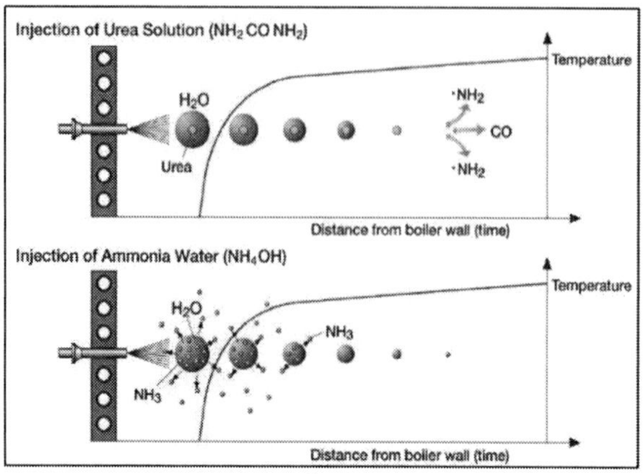

b

Figure 3. SNCR technology: *a* – reagent supply point; *b* - technology principle.

If the enterprise uses the technology of staged fuel combustion already, then to further reduce NOx emissions, the SNCR technology needs to be applied. With the simultaneous use of staged combustion and selective non-catalytic reduction of nitrogen oxides, it is necessary that the temperature, dwell time and gas atmosphere in the reaction section are consistent with each other.

4.3. Selective catalytic reduction of nitrogen oxides (SCR)

In SCR technology nitrogen oxides are reduced to N_2 by means of NH_3 and catalyst at a temperature of about 300-400 °C. This technology is widely used to reduce NO_x in other industries (for example, in thermal power plants with waste incineration), however, cement technology is practically not used.

SCR technology is used much less frequently, which is associated with a high cost (about 10 times higher than SNCR) and limited catalyst life.

In 2017, the authors of this report, together with the representatives of the German Cement Plants Association VDZ audited 3 Russian cement plants. The audit revealed NO_x emissions excess at all the enterprises. Based on the audit results, German colleagues presented a comparative assessment of effective technologies (Table 3).

Table 3. Comparison of technology options to reduce NO_x emissions [15].

Index	Staged combustion	SNCR	SCR
General technology assessment	Very cost effective sufficiently high potential for NO_x emissions reduction	Cost effective, medium / high NO_x emissions reduction potential	Expensive and complex but effective technology with low NO_x (~ 200 mg/m^3) and NH_3 emissions potential.
Applicability (according to [9])	For dry and wet processing production plants with an upgraded kiln	For dry process plants	For dry process plants
Additional information on applicability	Most effective for kilns with heat exchanger, calciner and tertiary duct. Built-in calciner required	Can be combined with staged combustion	Requires wide space for equipment
Investment costs	Low (can be high for complex modification projects)	Average	Very high

Operating costs	Current costs will not change	Medium / High	Average
NO_x emissions reduction rate	10–50%	30–90%	43–95%
Potential risks	Risk of increased CO emissions in the absence of optimization	Danger of significant NH_3 overshoots in some cases (for example, when high reduction rates are achieved) Risk of increased CO emissions combined with staged combustion and no optimization	Risk of increased NH_3emissions in the absence of optimization

5. Summary

Analysis of the technical solutions used to reduce nitrogen oxide emissions in cement production shows that the most preferable technologies are the staged fuel combustion and non-catalytic reduction of nitrogen oxides (SNCR). In the EU countries, especially in Germany, to reduce NO_x emissions to less 200 mg/m^3 these two technologies are applied simultaneously.

By 2025, all large industrial enterprises should have obtained the integrated environmental permits. Therefore, today it is necessary to start preparations for IEP receiving – to conduct an internal environmental audit - to analyze the application of the best available technologies at the enterprise, as well as to compare the achieved technological indicators with the industry indicators established by the order of the Ministry of Natural Resources and Ecology of the Russian Federation. Information and technical guide on BAT (ITG 6-2015) in this case, serves as a methodological basis for conducting an internal audit. The cement production enterprises with non-achieved BAT technological indicators, in particular, in terms of nitrogen oxide emissions, first of all, should take a set of measures to optimize the process and conditions for burning the raw mixture and consider the use of technologies for the staged fuel combustion, as well as selective non-catalytic reduction of nitrogen oxides.

References

[1] Guseva T, Skobelev D and Chechevatova O 2019 *SGEM* **19(5.1)**, 373-381. DOI: 10.5593/sgem2019/5.1/S20.047.
[2] Guseva T, Potapova E and Tichonova I 2018 *Jcement* **6,** 34-38.
[3] Best Available Techniques (BAT) for Preventing and Controlling Industrial Pollution, Activity 2: Approaches to Establishing Best Available Techniques Around the World. Environment, Health and Safety, Environment Directorate, OECD, 2018. Information on https://www.oecd.org/chemicalsafety/risk-management/best-available-techniques.htm#Activity2.
[4] Provis J L, Komljenovic M and Juenger M C G 2019 *Front. Mater.* **6 (298).** DOI: 10.3389/fmats.2019.00298.
[5] Skobelev D 2019 *SGEM* **19(5.3)** 291-298 DOI: 10.5593/sgem2019/5.3/S21.037.
[6] Potapova E N, Guseva T V, Tikhonova I O, Kanishev A S, Kemp R G 2020 *Construction Materials* **9** 15–20. DOI: https://doi.org/10.31659/0585-430X-2020-784-9-15-20.
[7] Guseva T, Vartanyan M, Tikhonova I and Shchelchkov K 2019 *Procedia Environmental Science, Engineering and Management* **6 (4)** 619-628.
[8] ITS 6-2015 *«Cement production»* 2015 (Moscow, Standartinform) 293.
[9] Order of the Ministry of Natural Resources of Russia dated 02.04.2019 N 210 «On approval of the normative document in the field of environmental protection «Technological indicators of the

best available cement production technologies».
[10] Integrated Pollution Prevention and Control (IPPC). Reference Document on Best Available Techniques in the Cement, Lime and Magnesium Oxide Manufacturing Industry 2013 (Seville: Institute for Prospective Technological Studies, European IPPC Bureau) http://eippcb.jrc.es/reference.
[11] Siebzehnte Verordnung zur Durchführung des Bundes-Immissionsschutzgesetzes (Verordnung über die Verbrennung und die Mitverbrennung von Abfällen - 17. BImSchV).
[12] Potapova E 2019 *Alitinform* **4 (57)** 13-21.
[13] Potapova E and Krivoborodov Yu 2020 *Technique and technology of silicates* **1 (27)** 2-6.
[14] Workshop to Promote the Ratification of Protocols of the UNECE Air Convention with Focus on Countries in the EECCA Region (14-16 May 2019 Berlin Germany) https://www.unece.org/index.php?id=51388.
[15] Technologies for reducing nitrogen oxide emissions (SCR, SNCR, staged combustion). Technical recommendations 2019. http://www.good-climate.com/materials.

The carcinogenic risk index of the Vladikavkaz city territory in GIS technologies is an indicator of natural and technological safety level

O G Burdzieva*, A S Kanukov

Geophysical Institute of the Vladikavkaz Scientific Center of the Russian Academy of Sciences, 93a,Markov str., Vladikavkaz, 362002, Russia

E-mail: olgaburdzieva@mail.ru

Abstract. Environmental pollution, in addition to harm to nature, leads to various diseases of the population living in the contaminated area, including cancer. In the Republic of North Ossetia-Alania, environmental pollution is mainly due to non-ferrous metallurgy enterprises and vehicles. The capital of the Republic - Vladikavkaz is the most polluted city in the region, where the main stationary sources of pollution and the largest number of vehicles are located. The material on cancer was selected from the data in various regions of the Vladikavkaz city. The main share of newly registered oncological disease falls on the age groups over 60 years. An analysis of cancer showed that in those areas of the republic, including the city of Vladikavkaz, where the environmental pollution level is higher, the rates of cancer are higher.The risk coefficient for public health under the influence of chemicals polluting the environment is calculated. For all the studied indicators, there is a significant excess of the maximum permissible values, with the exception of mercury. The data obtained are consistent with a high level of cancer incidence in the city, which indicates a close relationship between the incidence and the carcinogenic risk index.

1.Introduction

In the modern world, there are opportunities in a relatively short time to receive a huge amount of information about certain features of various studied objects, which can be processed fully only with the use of modern computer technologies. Due to the fact that the prevailing number of different factors have a spatial reference, it is necessary to develop not just informational, but geoinformation support. That is why geoinformation technologies and methods have become one of the main tools in conducting the environmental research [1], as well as in assessing and monitoring the environment and resources' state. The development and implementation of new methods for modeling the behavior of objects, moving or spreading a particular geophysical process, is an important task in determining the relationship and correlation of various physical fields' spatial distributions [2-6]. Modern equipment and methods for observing dangerous natural and technogenic processes [7] make it possible to get a huge amount of information, which gives an opportunity to simulate a particular process both at the macro [8] and at the micro level [9]. At the same time, it becomes possible to create the unified information systems combining information from various fields [10].

Environmental pollution is a complex and multidimensional problem, but the main one is the adverse effects on public health. To a greater extent, environmental pollution affects the health of the urban population [11-13].

As evidenced by the numerous studies, the adverse environmental factors, even with a low level of exposure can cause significant health problems. Environmental pollution, despite the relatively low concentration of substances, due to the long exposure duration, can lead to serious illnesses, including oncology [14].

The material was selected on cancer incidence in various regions of the Vladikavkaz city [15]. For processing and subsequent analysis of the obtained data, a special database was developed. The formative parameters of the base included location, sex, age of the patient, the lesion localization of the patient's body. The patients were divided into several groups by age: up to 20 years old, 20-29 years old, 30-39 years old, 40-49 years old, 50-59 years old, 60-69 years old and 70 years and older. Data types were strictly defined for the subsequent database connection to the map using modern GIS technologies. Connecting the database to a digital map in the geographic information system made it possible to perform the multiple data analysis due to the spatial reference of each record [16-17]. So, for example, it is possible to select the age categories of interest, filter by sex, a specific feature or localization of the disease. In GIS technologies, the distribution of the malignant tumors' incidence over the city area was plotted on a map, a city development chart. According to the generally accepted methodology of polyclinic territorial zoning along the polyclinic service boundaries, the incidence values for different years were obtained, as well as the average value for several years, which was depicted on the map [18] (Figure 1).

Environmental pollution in the Republic of North Ossetia-Alania comes from the stationary and mobile sources. Stationary sources include the industrial enterprises, which are represented in the republic by such industries as non-ferrous metallurgy, machine building and metalworking, electric power, fuel industry, forestry and woodworking industries, and the production of building materials [19-21]. The main stationary sources of atmospheric emissions are metallurgical enterprises [22-25]. A major contribution to environmental pollution is made by the mobile sources, in particular road transport, the amount of which is growing rapidly from year to year.

As it is known, the metal content in the soil is the content sum of the metals entering the soil due to natural processes and human activity [26-28]. It is obvious that the metals contribution from anthropogenic sources in the soil is higher than the contribution from natural ones. A significant increase in the metals content in soils is observed in the areas with a high level of industrial activity, where their accumulation can be several times higher than the average content in uninfected areas. In addition, in the areas remote from industrial centers, the elevated metal concentrations can also be observed as a result of long-distance transport by wind.

In 2014, the State Budgetary Institution "Directorate for the Implementation of Environmental Programs and Environmental Education" and the Open Joint-Stock Company "Sevosettingeoekomonitoring" conducted a "study of the soil pollution state including pesticides and radioactive substances and the preparation of environmental maps with the application of geopathogenic zones" to assess the soil pollution state, including pesticides and radioactive substances and the preparation of environmental maps with the application of geopathic zones. Based on these data, we calculated the index of carcinogenic risk to public health when exposed to chemicals polluting the environment.

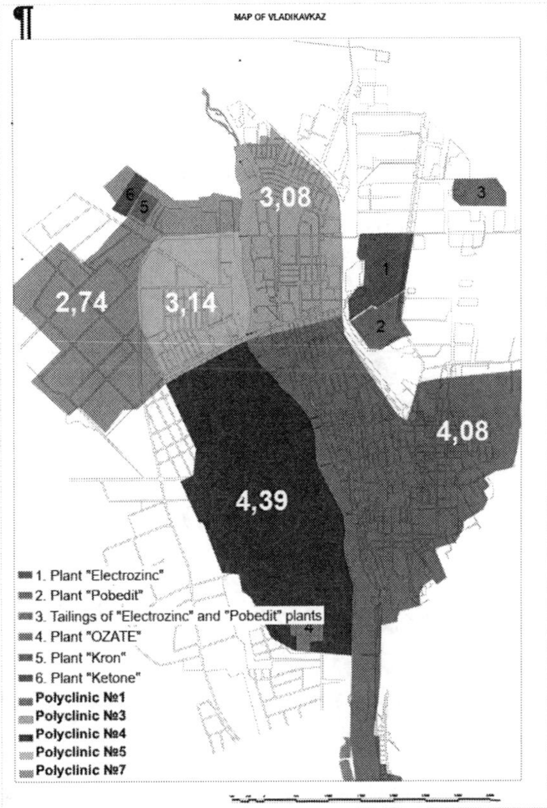

Figure 1. The incidence of neoplasms in urban polyclinics.

The standard formula was used to calculate the average daily dose for the inhalation exposure to chemicals entering the air from the soil (an explanation of all the parameters is given in Table 1):

I = (Ca x IR x ED x EF)/(BW x AT x 365)

Table 1. Standard formula for calculating the average daily dose for inhalation exposure to chemicals entering the air from the soil

Parameter	Characteristic	Standard value
I	Inhalation intake, mg / (kg / day)	-
Ca	The substance concentration in the air, mg/m3	Cs x (1/PEF + 1/EF)
Cs	Concentration in soil, mg/kg	-
PEF	Dust Particle Emission Factor, m3/kg	calculated value (Table 3.13)
VF	Soil Vapor Factor, m3/kg	calculated value (Table 3.14)
IR	In-flowRate, m3/day	20 m7day; children: 10 m3/day
EF	Exposure Frequency of, days / year	350 day/g
ED	Exposure Duration, years	30 years old; children: 6 years old
BW	Body Weight	70 kg; children: 15 kg
AT	AveragePeriod, years	For adults 30 years old; for children: 6 years old; carcinogens: 70 years

Reference guidelines for chronic inhalation exposure were taken from the guidelines for assessing public health risk from the exposure to chemicals polluting the environment (Table 2).

The carcinogenic risk index was calculated as the average daily dose ratio during the inhalation exposure to chemicals entering the air from the soil to the reference concentrations for chronic inhalation exposure. Values lower than 1 are considered to be safe. Based on the study results, the carcinogenic risk index maps were constructed for the Vladikavkaz city territory (Figures 2-3).

Table 2. Reference concentrations for chronic inhalation exposure

CAS	Substance	RFC mg/m5	Critical organs / systems
7440-43-9	Cadmium	$2*10^{-5}$	kidneys, respiratory system, hormone, cancer
7440-50-8	Copper	$2*10^{-5}$	respiratory system
7440-02-0	Nickel	$5*10^{-5}$	respiratory system, blood, immune, cancer, central nervous system
7439-97-6	Mercury	0,0003	Central nervous system, hormone, kidney
7439-92-1	Lead	0,0005	CNS, blood, development, reproductive syst., hormone., kidneys
7440-66-6	Zinc	0,0009	respiratory organs, immune (sens.), blood

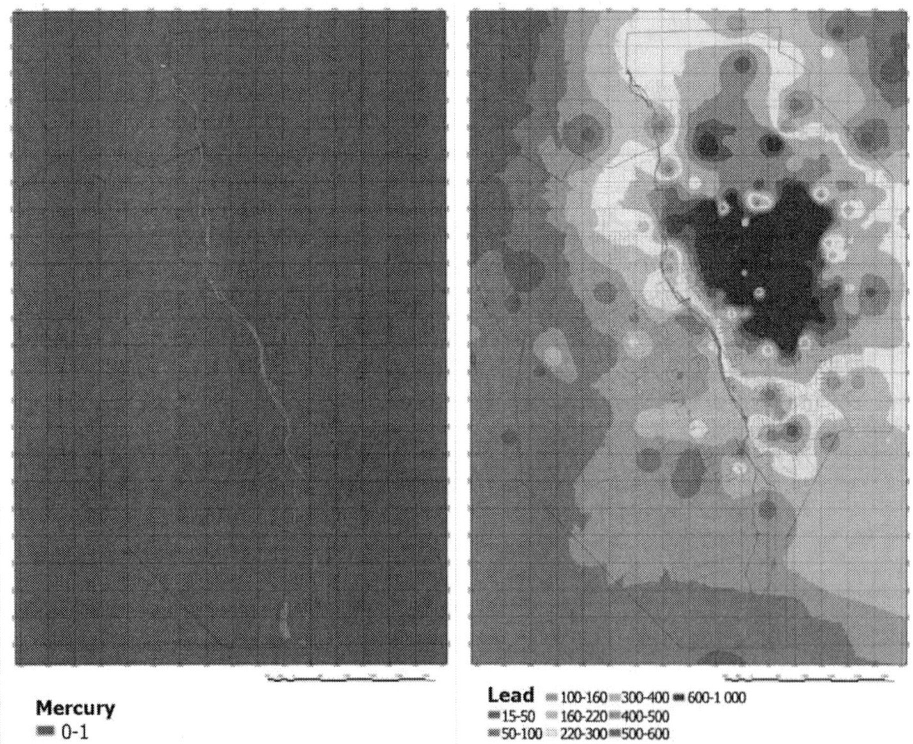

Figure 2. Carcinogenic risk index maps for the Vladikavkaz territory.

IOP Conf. Series: Materials Science and Engineering 1083 (2021) 012084 doi:10.1088/1757-899X/1083/1/012084

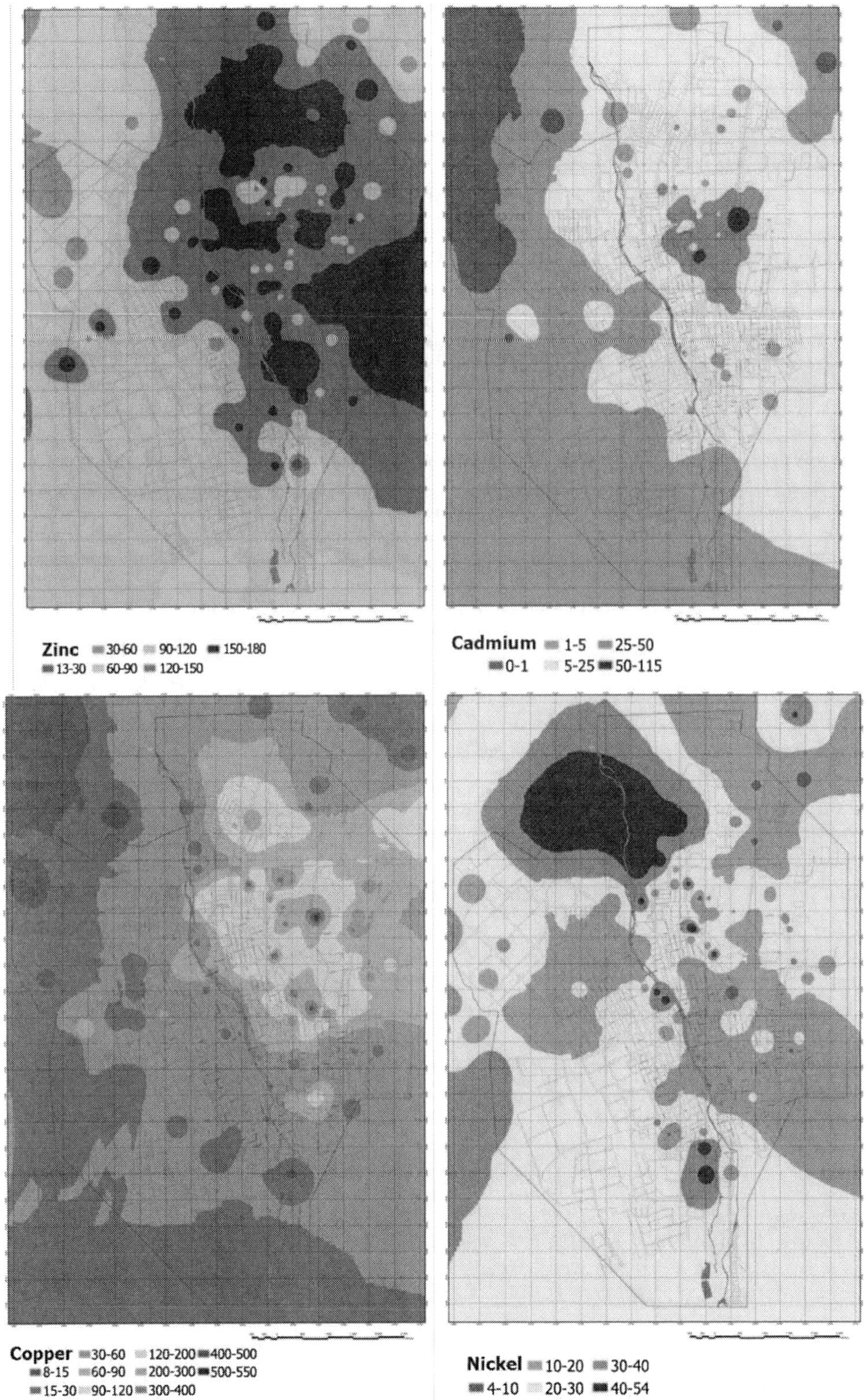

Figure 3. Carcinogenic risk index maps for the Vladikavkaz territory.

According to the constructed maps, for all investigated indicators, with the exception of mercury, there is a significant excess of the maximum permissible values. The data obtained are consistent with a high level of cancer incidence in the city, which indicates a close relationship between the incidence and the carcinogenic risk index.

Summary

- The capital of the republic, the Vladikavkaz city, is the most polluted city in the region, where the main polluting enterprises and the largest number of vehicles are located.
- The material was selected on the cancer incidence in various regions of the Vladikavkaz city. For processing and subsequent analysis of the data on cancer incidence of the population living in the region with intensive mining, a special database has been created.
- In the areas where environmental pollution by heavy metals is higher, the oncological morbidity level of the population is higher. It is established that the closer to industrial enterprises of non-ferrous metallurgy and the corresponding halos of soil contamination with heavy metals, the higher the oncological incidence of the population of the Vladikavkaz city.
- Based on the available data, the index was calculated and the carcinogenic risk maps of the population under the influence of lead, zinc, cadmium, copper and mercury (Pb, Zn, Cd, Cu, Ni, Hg) were calculated, which showed, with the exception of mercury, a multiple excess of the maximum permissible values.
- The data obtained are consistent with a high level of cancer incidence in the city, which indicates a close relationship between the incidence and the carcinogenic risk index.

References

[1] Zaalishvili V, Dzhgamadze A, Gogichev R, Dzeranov B, Burdzieva O 2018 *International Journal of GEOMATE***15(51)** 22-30
[2] Zaalishvili V, Melkov D, Kanukov A, Dzeranov B, Shepelev V 2016 *International Journal of GEOMATE***10(1)** 1670-1674
[3] Zaalishvili V, Nevskaya N, Melkov D 2014*Izvestiya. Physics of the Solid Earth***50(2)** 263-272
[4] Zaalishvili V, Melkov D, Dzeranov B, Morozov F, Tuaev G 2018 *International Journal of GEOMATE***15(47)** 158-163
[5] Zaalishvili V, Melkov D, Kanukov A, Dzeranov B 2016 *International Journal of GEOMATE***10(1)** 1656-1661
[6] Zaalishvili V, Melkov D, Kanukov A 2019 *Akustika***32** 279-283
[7] Zaalishvili V 2016 *Measurement Techniques***58(12)** 1297-1303
[8] Zaalishvili V, Melkov D 2014 *Izvestiya. Physics of the Solid Earth***50(5)** 707-718
[9] Grigorkina G, Ramonova A, Kibizov D, Kozyrev E, Zaalishvili V, Magkoev T, Fukutani K 2017 *Solid State Communications***257** 16-19
[10] Zaalishvili V, Kanukov A, Melkov D, Makiev V, Dzobelova L 2018 *International Journal of GEOMATE***15(51)** 160-166
[11] Alborov I, Burdzieva O 2017*An article in the collective monograph "Geological and geophysical studies of the deep structure of the Caucasus: geology and geophysics of the Caucasus: modern challenges and research methods"* 507 – 513
[12] Alborov I, Burdzieva O 2017*An article in the collective monograph "Geological and geophysical studies of the deep structure of the Caucasus: geology and geophysics of the Caucasus: modern challenges and research methods"* 514-518
[13] Beriev O and ZaalishviliV 2015*Geology and Geophysics of the South of Russia* **4** 47 – 36
[14] Beriev O, Zaalishvili V, Burdziev O, Sachs T, Kanukov A 2013*Geology and Geophysics of the South of Russia***3** 29-38
[15] Burdzieva O, Zaalishvili V, Beriev O, Sachs T and Kanukov A 2012*Earthquake-resistant construction. BuildingSafety***3** 62-67
[16] Zaalishvili V and Mel'kov D 2014 *Izvestiya, Physics of the Solid Earth* **50(5)** 707-718.

[17] Milyukov V, Yushkin V, Kopaev A, Zaalishvili V, Kanukov A and DzeranovB 2014*Measurement Techniques***56(10)** 1105-1110
[18] Burdzieva O, Zaalishvili V, Beriev O, Kanukov A and Maysuradze M 2016 *International journal of Geomate***10(19)** 1693-1697
[19] Zaalishvili V, Burdzieva O, Dzhgamadze A 2015 *Ecology, Environment and Conservation* **21** 145-149
[20] Borshevsky S, Kachurin N, Burdzieva O and Golik V 2017 *Sustainable Development of Mountain Territories* **9(1)** 81-91
[21] Zaalishvili V, Nevskaya N, Nevskii L and Shempelev A 2015 *Journal of Volcanology and Seismology* **9(5)** 333-338
[22] Shempelev A, Zaalishvili V and Kukhmazov S 2017 *Geotectonics* **51(5)** 479-488
[23] Gorbatikov A, Rogozhin E, Stepanova M, Kharazova Y, Andreeva N, Perederin F, Dzeboev B, Zaalishvili V, Melkov D, Dzeranov B, Gabaraev A 2015 *Izvestiya, Physics of the Solid Earth* **51(1)** 26-37
[24] Zaalishvili V, Beriev O, Burdzieva O, Sachs T, Kanukov A 2010*Proceedings of the international scientific-practical conference "Dangerous natural and technogenic processes in the mountain and foothill territories of the North Caucasus"* 433-438
[25] Burdziev O, Beriev O, Sachs T and Kanukov A 2016*Proceedings of the Kabardino-Balkarian Scientific Center of the Russian Academy of Sciences***6(74)** 102-108
[26] Golik V, Burdzieva O and Dmitrak Yu 2017*An article in the collective monograph "Geological and geophysical studies of the deep structure of the Caucasus: geology and geophysics of the Caucasus: modern challenges and research methods"* 542-247
[27] Golik V, Komashchenko V, Morkun V and Zaalishvili V 2015 *Metallurgical and Mining Industry***7(4)** 325-329
[28] Golik V and Zaalishvili V 2017 *Proceedings of the Tula States University-sciences of Earth***1** 52-67

Geological information database integration into a geographic information modeling system

A S Kanukov[*,1], **P G Ivanov**[2]

[1]Geophysical Institute of the Vladikavkaz Scientific Center of the Russian Academy of Sciences, 93a,Markova str., Vladikavkaz, 362002, Russia
[2]Geological Institute "Str. Dimitrov", Bulgarian Academy of Sciences, Acad. G. Bonchev st. bl.24, 1113 Sofia Bulgaria

Academy of Sciences
E-mail: akanukov@list.ru

Abstract. The article is devoted to the geological information database integration in the Vladikavkaz city into a specially developed system of geoinformation modeling. With the high technologies' development, various geographic information systems designed for collecting, storing, analyzing, and graphically visualizing spatial data and the related information about the objects represented in GIS have become widespread. A map of the engineering-geological zoning of the Vladikavkaz city territory has been created at the Geophysical Institute, covering the issues of the geological structure, hydrogeological conditions, lithology, morphology, tectonics and the distribution of various types of soils in the territory under consideration. There are areas on the territory characterized by different depths of pebbles or thicknesses of clay and loam on pebbles, which is the main parameter that determines the seismicity category according to Building Codes and Regulations-II-7-81*. Based on the physical and mechanical properties' analysis, the reliable differentiation of soil groups with various determining indicators characterizing the soils' category by their seismic properties has been carried out. This information is collected in a database of geological information generated in the form of a shape file with spatial reference of each well. The developed database is used as a basis in the geographic information modeling task. Based on epyopen source software, the integration of the geological information database in the Vladikavkaz city into the geographic information modeling system has been performed.

Introduction

Assessment of the seismic hazard in an urbanized area, based on modern scientific approaches and methods, is an essential component in determining the expected level of seismic risk in implementing a scenario-specific seismic impact for a given region and developing the specific risk management measures to reduce the territory vulnerability and ensure the safety of its population and further development.

The manifestation of a strong and destructive earthquake, in turn, is largely determined by the soil conditions of the territory [1]. In this regard, special attention is paid to the appropriate soil distribution schemes' creation for various types of soil in a given urbanized territory [2-7]. The detailed schemes of geological zoning of the settlements' territories in the Republic of North Ossetia-Alania were the basis of seismic microzoning [8-9]. In this case, a detailed geological zoning of soils for engineering and geological surveys in an urbanized area is carried out using modern geophysical methods [10-12]. A special place in the study and subsequent differentiation of soils with the allocation of their features and characteristics is occupied by the geoinformation modeling data [13-15].

Geoinformation modeling is currently still a relatively young area of scientific research, which covers an extensive range of questions on the creation and use of geoinformation systems (GIS systems) about objects and their characteristics, as well as on the mathematical methods and algorithms' application in these systems. GIS systems include DBMS (database management systems), editors of graphics presented in vector or raster format, various analytical tools, which allows them to be used in cartography, geology, geophysics, ecology and many other areas. Geoinformation modeling is widely used in geo-ecological [16-17], geophysical [18] and other [19] studies.

As a result of a number of studies, the Geophysical Institute created a map of the engineering and geological zoning of the Vladikavkaz city [8], covering the issues of its geological structure, hydrogeological conditions, lithology, morphology, tectonics, and the distribution of various types of soils in the territory under consideration. The areas of different pebble bedding depths and different thicknesses of clay and loamy cover on pebbles were identified on the territory, which are the main indicators that determine the seismicity category according to Building Codes and Regulations-II-7-81*.

Based on the soils' physical and mechanical properties analysis, the groups were differentiated according to their seismic properties (Building Codes and Regulations-7-II -81*).

The aim of the work was to compile a map of seismic microzoning based on the detailed geological zoning schemes.

A GIS project "Database of geological information of the Vladikavkaz city territory" was created.

The situational basis in the contours form of quarters, streets, the position of the Terek River and anthropogenic load is obtained from the existing specialized bases.

Three thematic maps created:

- well-location map;
- depth map and pebble roof depths;
- map of engineering and geological zoning.

Thematic maps are created with related databases.

There is only one layer of "mine workings" with fields on the well-location map:

- production type (well, trench, pit, hole);
- index (production number and type index), the field is rendered for signature on the card;
- conduit number.

The map of engineering-geological zoning also carries the information about the conduit. In addition, the following information is provided:

- slopes with steepness of more than 15 degrees (polygon layer, without attributive data);
- soil categories by seismic properties in accordance with existing Building Codes and Regulations requirements II-7-81*.

The subsequent development prospects of the created geocoded databases come down to the spatial analysis of attribute information. Based on it, it is possible:

- to create the repeating fields with the functions of refinement and information updating based on new received data (i.e., detailing);
- creating correction sheet;
- overlay with information about the hazardous industrial facilities;
- overlay of water horizons and lenses roof depths;
- solving other spatial problems.

The soils' description is given for all wells in the GIS project (Table 1). This information is collected in a database of geological information generated in the form of a shape file with spatial reference of each well.

Due to the fact that the developed database is used as a basis in the geoinformation modeling task, it had to be prepared beforehand in order to be able to work in the geoinformation system. In this regard, it should be noted that the materials in the geographic information system should be designed in the form of databases [14]. At the same time, any object put on the map should be linked to the materials related to it. Thus, it is necessary to combine databases and cartographic materials with spatial relationships. In this regard, the database management system should provide support for both text and graphical databases, and spatial. For example, PostGIS can serveas such a subsystem.

Table 1. Description of mining soils (example of database content)

Well numbers	Depth, m *from*	*till*	Description of soils	Absolute mark of the mouth, m	Depth of pebbles, m
4E2	0	0.1	Bulk layer	669	2.4
	0.1	0.3	Pebblesoillayer		
4E2	0.3	2.4	Brownish-brown loamy refractory humus with admixture of gravel and pebbles up to 10%	669	2.4
	2.4	12.0	Pebbles		

PostGIS is an extension of the free object relational database management system, which is designed to store the additional geographic attributes in the database - geometry. PostGis supports the OGC (OpenGeospatialConsortium) standard and has been released since 2001 by Refractions Research, competing with many commercial projects, while remaining a free software product that is an open source. The main advantage of PostGis is the ability to use the SQL language along with spatial functions and operators. At the same time, a fairly active development of the PostGis program is observed as PostgreSQL and the OpenSourceGeospatialFoundation project are developing, collecting the best of them. For example, due to the development of PostgreSQL in PostGis, it became possible to use such innovations as bitmap and Gist indexes.

Due to the alliance with OpenSourceGeospatialFoundation [14], PostGis has now the ability to support such programs as GEOS, Proj4, and JTS. In 2001, only one application - MapServer was able to display the data stored in PostGis. To date, almost all applications for working with cartographic data, both paid and distributed on a non-commercial basis, allow working with this data. Among the latest software are GRASS, QGIS, MapServer, uDig, GeoServer, GDAL / OGR, SharpMap, gvSIG, FeatureServer.

PostGIS provides the full functionality of spatial operations (respectively, to OGC), which make it possible to implement any spatial operations with geodata. Choosing PostgreSQL + PostGISgives a possibility to get the most fully functional system that will support the operations with spatial data, being a very powerful system when working with a database.

Due to this service, which stores geodata in a database and gives an opportunity to filter and select the necessary data using the standard SQL queries, it is possible search for the necessary objects, both by coordinates and by using other information stored in the attributes of each record.

Using such blocks makes it possible to store the already created GIS objects, create the new ones, and also perform any spatial operations with them.

GeoServer was used to visualize the contents of the spatial database. GeoServer is a web server that provides standard clients, such as web browsers and various GIS, with access to various maps and data that can be stored in almost any format, while the user does not need special knowledge about geographical data. In the simplest case, all that is required is a web browser that displays the maps in the exactly required form.

GeoServer represents a reference implementation of OGC standards: WFS (WebFeatureService), WCS (WebCoverageService) and is the main component GeospatialWeb.

It should also be noted that its distinguishing feature from a number of others (MapServer or FeatureServer) is a graphical system for managing settings and data description for the projects implemented in GeoServer. This system is implemented as a web service, with the interactive creation and modification of cartographic materials in the system.

In addition, GeoServer uses the StyledLayerDescriptor (SLD) description language to set the styles. Initially, it was created to work with WMS services [14]. It is possible to prepare a style file in the SLD language and transfer it to a third-party WMS server, and at the same time get a map designed in accordance with the user's requirements.

When solving such problems, it is also necessary to take into account the important fact that the database should be protected from unauthorized access. To do this, the geographic information system should have means for authenticating the users. Basic authorization is implemented in GeoServer, but is vulnerable to the authorization data interception attacks, making possible to view maps without registering on a web service. The way out of this situation is the use of the so-called https binding, in which all data between the client and server is encrypted.

Using this method, support for the https protocol can be introduced, even if the source software does not support it, which may occur in the case of a ready-made system. For these purposes, the pound program is suitable, which "listens" to a specific port and works through it using the secure https protocol. The data arriving at this port is decrypted and redirected to the internal port, which the program "listens to", and into which the protocol support https is entered.

For the pound service to work, a certificate should be obtained and installedto the system. The certificate consists of two keys - public and private. The public key is used to encrypt traffic from the client to the server in a secure connection, the private key is used to decrypt the encrypted traffic received from the client on the server. After generating the public and private keys, based on the public key, a certificate request is generated to the Certification Authority (CA), in response to which the CA sends a signed certificate. When signing, the CA checks the client, which allows him to guarantee that the certificate holder is who he claims to be. Each signed certificate has a validity period and a fee is usually charged for creating / renewing a signature.

However, it is possible to create such a certificate without contacting the Certification Authority. They can be created for the stations that run on Unix / Linux, which is also an advantage of this system. Such certificates are signed by themselves and therefore are called self-signed. If this certificate is not verified in any other way, then the use of this protocol can be subjected to a man-in-the-middle attack, that is, a man-in-the-middle. Its essence is that the intruders can connect to the channel through which a secure connection is established, and intercept all the requests that go between the server and the client. In this case, the intruder is presented to the client as a server, and to the server as a client. Since it is possible to confirm the fact that the server is what it claims to be only with a certificate issued by the CA, such attacks are easily feasible for the self-signed certificates. At the same time, the direct transmission of a self-signed certificate through a closed channel and installing it in the system as a trust that allows avoiding such attacks.

To display maps in an information system, a spatial data visualization system is required. The main requirement for this system is the support of layers provided by WMS (and WFS) servers, such as Mapserver, ArcIMS or Geoserver, and the data from Google map services.

As an example of a finished product, the OpenLayers library can be considered as the most functional and open of all the existing systems.

Using the OpenLayers library makes it possible to quickly create a web-service for viewing the cartographic materials, which can be presented in various formats and can be located on different servers. Due to OpenLayers, the developer has the opportunity to create his own map, the source materials for which can be stored on such servers as Mapserver, ArcIMS or Geoserver, as well as the data provided by Google services.

It should be noted that OpenLayers is a free software product, open source.

The following features are implemented in the OpenLayers system:

- Adding a navigation panel to the map that contains controls such as map navigation buttons and zooming.
- Moving the map with the help of the "mouse";
- Changing the map scale by scrolling the "mouse" wheel;
- Getting coordinates under the "mouse" pointer;
- Conclusion of the separate layers' visibility switches to the control panel;
- Selecting an object on the map and receiving all the related information;
- Changing the transparency of any of the layers presented on the map;
- Editing elements on the map by the user.

Thus, the open source software use made it possible to integrate the database of geological and geophysical (engineering-geological, hydrogeological, geomorphological) information on the Vladikavkaz city territory into a special geographic information system with significant functionality.

Summary

- As a result of a number of studies, a map of engineering-geological zoning of the Vladikavkaz city territory has been created, covering the issues of the geological structure, hydrogeological conditions, lithology, morphology, tectonics, the distribution of various types of soils in the territory under consideration.
- On the Vladikavkaz city territory,the plots with different depths of pebbles and thickness of clay and loam on pebbles were identified, which are the main indicators that determine the soils' seismicity category by theBuilding Codes and Regulations requirements-II-7-81*.
- Based on the physical and mechanical properties' analysis, including (if necessary), with the help of their additional study by geophysical methods, differentiation of soil groups by the seismic properties was carried out.
- A GIS project "Database of geological information of the territory of the city of Vladikavkaz" has been created, including the information on wells in the city with a detailed soil description.
- Based on the open source software, the integration of a geological information database in the Vladikavkaz city into a geographic information modeling system has been performed.

References

[1] Zaalishvili V, Kanukov A 2009*Proceedings of the III Caucasian International School of the seminar for young scientists "Seismic hazard and seismic risk management in the Caucasus"* 137–141.

[2] Zaalishvili V, Melkov D, Kanukov A, Dzeranov B, Shepelev V 2016 *International Journal of GEOMATE***10(1)** 1670-1674.

[3] Zaalishvili V, Melkov D, Dzeranov B, Morozov F, Tuaev G 2018 *International Journal of GEOMATE***15(47)** 158-163.

[4] Zaalishvili V, Melkov D, Kanukov A, Dzeranov B 2016 *International Journal of GEOMATE***10(1)** 1656-1661.

[5] Zaalishvili V 2016 *Measurement Techniques***58(12)** 1297-1303.

[6] Zaalishvili V, Nevskaya N, Melkov D 2014 *Izvestiya. Physics of the Solid Earth***50(2)** 263-272.

[7] Zaalishvili V, Melkov D, Kanukov A 2019 *Akustika***32** 279-283.

[8] Zaalishvili V, Dzhgamadze A, Gogichev R 2013*Geology and geophysics of the South of Russia***3** 20–28.

[9] Gorbatikov A, Rogozhin E, Stepanova M, Kharazova Y, Andreeva N, Perederin F, Dzeboev B, Zaalishvili V, Melkov D, Dzeranov B, Gabaraev A *2015 Physics of the Solid Earth***51(1)** 26-37.

[10] Shempelev A, Zaalishvili V, Kukhmazov S 2017 *Geotectonics***51(5)** 479-488.

[11] Zaalishvili V, Nevskaya N, Nevskii L, Shempelev 2015 *Journal of Volcanology and Seismology***9(5)** 333-338.

[12] Zaalishvili V, Dzhgamadze A, Melkov D, Dzeranov B, Gabaraev A, Gabeeva I, Dzeboev B, Kanukov A, Shepelev V 2015Geology and geophysics of the South of Russia**1** 5–20.
[13] Zaalishvili V, Kanukov A, Melkov D, Makiev V, Dzobelova L2018 *International Journal of GEOMATE***15(51)** 160-166.
[14] Milyukov V, Yushkin V, Kopaev A, Mironov A, Dem'Yanov G, Sermyagin R, Basmanov A, Popad'Ev V, Nasretdinov I, Zaalishvili V, Kanukov A, Dzeranov B 2014 Monitoring current vertical movements of the northern caucasus by absolute and relative gravimetry*Measurement Techniques***56(10)** 1105-1110.
[15] Burdzieva O, Zaalishvili V, Beriev O,Kanukov A, Maisuradze M 2016 *International Journal of GEOMATE***10(1)** 1693-1697.
[16] Zaalishvili V, Dzhgamadze A, GogichevR, Dzeranov B, Burdzieva O2018 *International Journal of GEOMATE***15(51)** 22-30.
[17] Zaalishvili V, Melkov D 2014 *Izvestiya. Physics of the Solid Earth***50(5)** 707-718.
[18] Grigorkina G, Ramonova A, Kibizov D, Kozyrev E, Zaalishvili V, Magkoev T, Fukutani K2017 *Solid State Communications***257** 16-19.

Features of the earthquakes' manifestation intensity formation in mountainous areas

V B Zaalishvili

Geophysical Institute of the Vladikavkaz Scientific Center of the Russian Academy of Sciences, 93a, Markova Geoscience str., Vladikavkaz, 362002, Russia

E-mail: vzaal@mail.ru

Abstract. Analysis of the destructive earthquakes' consequences engineering macro-seismic survey data shows that the earthquakes manifestation degree is in close correlation with the peculiarities of geomorphological conditions. Moreover, this dependence is not of a simple character: there is a clearly expressed multifactorial nature of correlations. The work shows that the earthquakes intensity is determined not only by the relief slope magnitude, but also by the specific location of the site in the mountain range hierarchy. The results of the experimental studies analysis (field and laboratory experiments) are presented. It is noted that from a physical point of view, the relief has a certain potential energy. To estimate the indicated energy, a phenomenological indicator was introduced in the form of the so-called "relief factor", which is the product of the relief inclination angle by the corresponding height. It is noted that upon reaching the value R=650 (degrees m) the intensity increment is one point. The use of a new relief parameter, which makes it possible to fully take into account the specific conditions of the location of the investigated area, increases the establishment validity of the mechanisms for the earthquake intensity formation in mountainous regions.

Introduction

In recent years, the list of studies, in which instrumental measurements are very actively used in various, very different fields of geophysics, has significantly expanded. This undoubtedly raised the quality of the results obtained as well as their reliability [1-10]. In this regard, it became possible to directly study such subtle effects as the relief influence on the amplification of vibrations under strong dynamic influences, which is clearly manifested during macro-seismic and other field surveys of various hazardous natural and man-made processes' manifestations. It is known that most of the seismically active zones, which are characterized by a large rugged relief and various engineering and geological conditions, are located in mountainous areas. The results of an engineering survey of the strong and destructive earthquakes' consequences show that it is precisely these features that largely determine the earthquakes intensity. The issues of relief influence adequate consideration and the engineering-geological structure of the area on the intensity of the destructive earthquakes' manifestation is the basis for carrying out the corresponding anti-seismic measures in the developed and newly developed territories. At the same time, traditionally, scientists and engineers pay more attention to the study of the influence of the engineering-geological structure of the area on the earthquake seismic effect.

Analysis of the engineering macro-seismic survey data of the destructive earthquakes consequences shows that the earthquakes manifestation degree is in close correlation with the peculiarities of geomorphological conditions. Moreover, this dependence is not of a simple character: there is a clearly expressed multifactorial nature of correlations. This circumstance imposes certain restrictions on the direct use of macro-seismic survey data, which is superficial to a certain extent. This is facilitated by the tight deadlines allotted for the direct study of various engineering and geological situations and reducing the earthquakes consequences analysis to a statement or descriptive form, etc. This leads to the need for special sorting of survey data based on corrective information. This approach will clarify the reasons for the change in the earthquakes' intensity in each case.

The seismic effect or the intensity of an earthquake is formed by a certain alignment of the factors considered above. This refers to the physical phenomena formed by a specific set of factors: strengthening or weakening of the incoming seismic waves by the soil layer, uneven precipitation, the phenomenon of resonance in the soil-structure system, etc.

The facts of the area geomorphological conditions' influence on the earthquake intensity manifestation are well known from the macro-seismic survey strong earthquakes data. In particular, during the Rachinsky earthquake (Georgia, 1991) in Ambrolaur and Onsky districts, a large number of rural-type buildings located at the base of the hills received minor damage (0-1 degrees) [14]. At the same time, the buildings located on the slopes exceeding 20 degrees were damaged significantly more (2-3 degrees). When vast territories are located on high plateaus, the seismic effect is often formed solely by the relief influence, but this is rarely paid attention to. At the same time, due to the proximity to the epicenter, partial and complete building collapses may prevail. For example, in the village of Bokva, during the 1991 Rachinsky earthquake damage of 4-5 degrees predominated, i.e. - partial and complete collapses of buildings and structures were observed. Here, when analyzing the reasons for the seismic effect formation, the influence of the relief, namely the building location height, was not taken into account.

Analysis of seismic norms of different countries in the world [13] shows that they take into account the seismic properties of soils, mainly due to their physical and mechanical properties. The influence of the relief is not even considered in them. An exception is the building codes of France [13]. In the early building codes of the former USSR (SNiP-II-A; 12-69), the influence of the relief was taken into account [11]. In subsequent Norms (SNiP - II -7-81), for the unknown reasons, this provision was excluded [12]. Moreover, the current Recommendations for seismic microzoning [16] recommend increasing the expected intensity of sites with a steepness of more than 15 degrees by one point. This simplified principle of intensity estimation is not at all typical for seismic microzoning. This is due to the lack of a reliable methodology to quantify such estimates. Hence, the question of assessing the contribution of relief features to the generated vibration intensity during earthquakes is of great scientific and practical interest.

Experimental research.

At the Institute, at different times, special studies were carried out to assess the influence of relief on the seismic effect of an earthquake. These studies were based on the materials of macro-seismic survey of a number of strong earthquakes. The studies were carried out on-site and on models. Field observations, in particular, were carried out in the area of the monastic cave complex David Gareji when solving the protection tasks of a historical and cultural monument near the military training ground (1988). In this case, the explosions were used as vibration sources [15].

In this work, as an example of the enhancement of the seismic effect caused by the features of the relief, the results of physical modeling are given on the example of the territory of Sukhumi [17]. A 3-D model of Sukhumi, the constancy of the wave propagation velocities, Poisson's ratio and soil density in the corresponding layers of nature – model was examined as the main conditions for simple similarity when creating a physical three-dimensional.

The territory of the city in the plan was 9x9 km. The scale of the model was M 1: 10000. The most dangerous vibrational frequencies for the city are the frequencies of 1-10 Hz, which for a model measuring 0.9 x 0.9 m corresponds to the frequencies 10 ÷100 kHz. The bedrock of the corresponding mountain range is represented by highly weathered from the surface, fractured limestone. Gypsum was used as a model material. The wave field of Rayleigh surface waves propagating from the earthquake source in the shelf zone near the city of Sukhumi was simulated.

The experiments were carried out on a setup consisting of a generator of ultrasonic vibrations with a period T=3÷20 m sec and the receivers (piezoelectric acceleration sensors TsTS-19). During the research, the frequency of the vibrator was 62 kHz. This corresponds to a natural vibration frequency of 6.2 Hz. When using a pulse with T = 15 msec, when the propagation speed of Rayleigh waves in gypsum was V_R = 1.21 mm / s, the vibration wavelength for the model was λ = 18 mm.

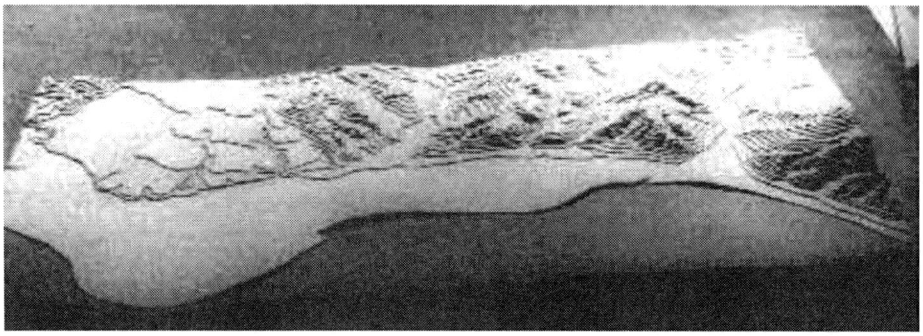

Figure 1. General view of the model

The source of ultrasonic vibrations was located on the model surface in the shelf area at the reference point. The surface of the model was divided into 96 squares and the wave field of Rayleigh waves was studied at 67 nodal points (Figure 3.). At each picket 7 measurements were made and the results were averaged. The experiments took into account the possibility of reflections from the side faces of the model (the phenomenon of reverberation).

The influence of the relief features on the Rayleigh wave passage was estimated relative to its amplitude on a reference circle, i.e., relative to vibration amplitudes at corresponding distances on a similar model with a homogeneous topography.

The increase coefficients in the amplitude level were determined by sequential movement along the profile on the model surface and compared with the amplitude at the reference point.

Data analysis showed that at an angle of inclination α ≥ 20 0 gain reached k=2.0, which corresponds to an increment of 1 point. Sometimes the increment reached 2 or more points. It should be noted that these points corresponded to detached peaks. In some cases, when α<<20 0 the gain, however, was large and also corresponded to 2 points. The analysis showed that this phenomenon is due to the location of these points between the dominant peaks. Hence, we can conclude that the inclination angle use as a characteristic of the relief is clearly insufficient. Moreover, this parameter does not always adequately correspond to the actual results of macro-seismic survey of the destructive earthquakes' consequences.

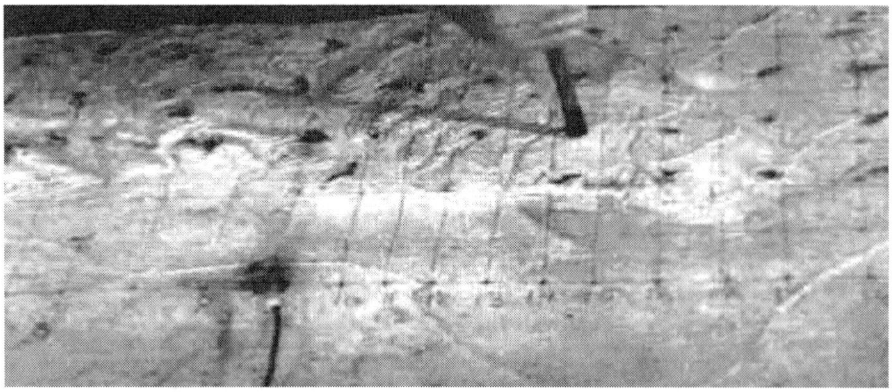

Figure 2. Carrying out the experiment

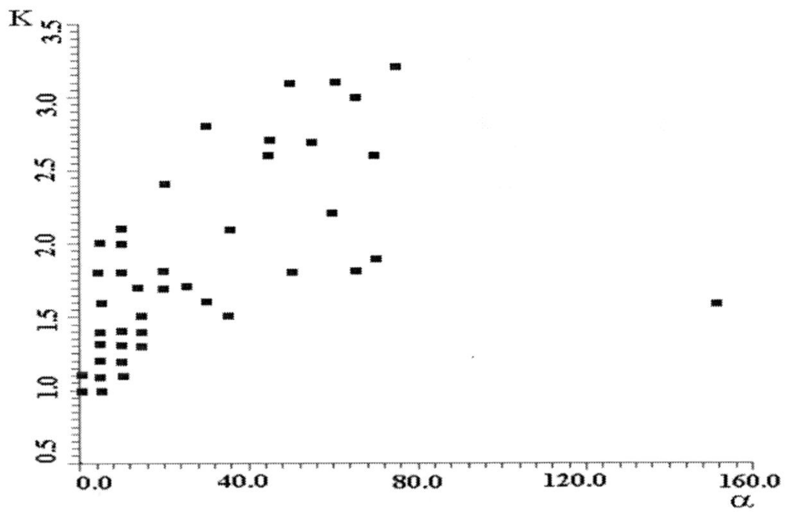

Figure 3. Dependence of the ground vibrations amplification on the relief inclination angle

Further, it was found that depending on the inclination angle, the intensity increase varies significantly (0 ÷ 3.0). On the other hand, the cross-sections (almost parallel to the coastline) show that, in addition to the observation angle, the amplification effect is significantly influenced by the observation height itself. Moreover, even in the absence of local tilt, the gain was nevertheless quite significant.

Analysis of the corresponding dependencies showed that to increase the reliability of observation results, it is necessary to take into account both factors simultaneously. This led to the introduction by the authors [15] the so-called coefficient of "relief" R in 1989 which makes it possible to take into account both factors.

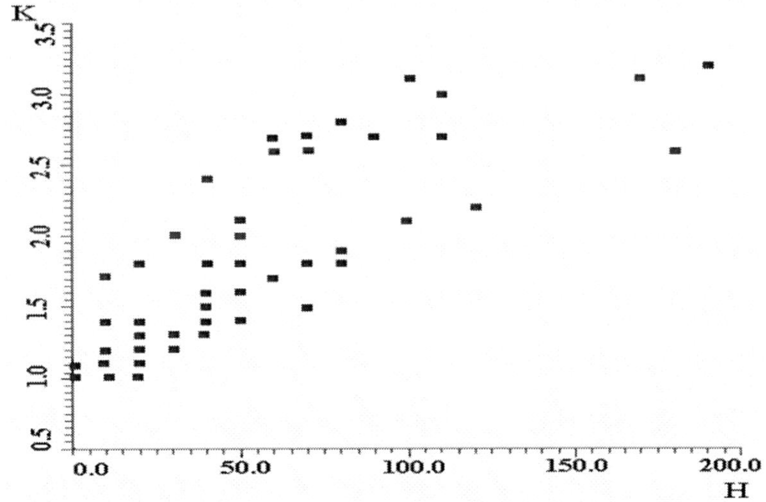

Figure 4. Dependence of the ground vibrations amplification on the observation height.

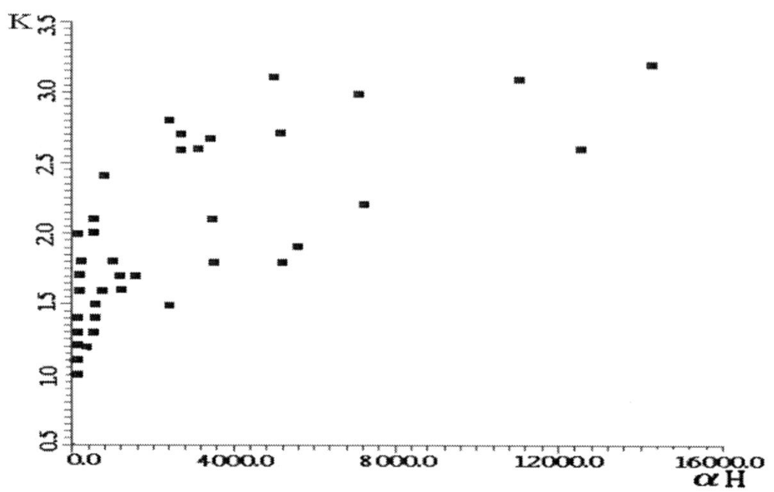

Figure 5. Dependence of the ground vibrations amplification on the relief factor

Analysis of the data obtained as a result of the experimental studies, as well as a result of the Rachinsky earthquake macro seismic examination showed that the relief influence on the seismic effect of an earthquake is most fully taken into account by the indicated indicator:

$$R = \alpha \times H \tag{1}$$

where α - is the relief inclination angle, degrees;
H – height, m.

For example, for $\alpha = 10^0$ and at height H =65 ÷ 70 m, i.e., at R=650-700 (degrees m) the increment in the intensity will be 1 point.

Summary

The experimental observations' results analysis shows that to characterize the influence of a terrain on the seismic effect of an earthquake, such indicators as the angle of inclination or only the height are clearly insufficient. At the same time, both factors undoubtedly form the seismic effect during earthquakes and dynamic impacts.

It is necessary to use an indicator that would allow taking into account both of these factors at the same time. From a physical point of view, the relief has a certain potential energy. To estimate the indicated energy, a phenomenological indicator was introduced in the form of the so-called "relief" coefficient which is the product of the angle by the corresponding height.

The work shows that the intensity is determined not only by the magnitude of the relief slope, but also by the specific location of the site in the mountain range hierarchy. The results of the analysis of previously conducted studies related to the need for practice, as well as the special experimental studies (field and laboratory experiments) are presented. The impact, in this case, was carried out in the form of explosions and monochromatic vibrations in the ultrasonic range, etc. Despite the significant differences in research conditions, the research results were generally characterized by a high degree of compliance.

Consideration of the geomorphological conditions' influence in the area should be carried out by using the so-called "relief" coefficient. It is noted that upon reaching the value R=650 (degrees m) the intensity reaches 1 point.

The use of the introduced parameter in the practice of earthquake-resistant construction makes it possible to fully take into account the specific conditions of the studied area location. The latter circumstance increases the establishment validity of the mechanisms for the earthquake intensity formation in mountainous regions.

References

[1] Zaalishvili V, Melkov D, Dzeranov B, Morozov F and Tuaev G 2018 *International Journal of GEOMATE* **15 (47)** 158-163.
[2] Zaalishvili V, Kanukov A, Melkov D, Makiev V and Dzobelova L 2018 *International Journal of GEOMATE* **15 (51)** 160-166.
[3] Zaalishvili V, Melkov D, Kanukov A and Dzeranov B 2016 *International Journal of GEOMATE* **10 (1)** 1656-1661.
[4] Zaalishvili V and Rogozhin E 2011 *Open Construction and Building Technology Journal* **5** 30-40.
[5] Zaalishvili V 2016 *International Journal of GEOMATE* **10 (2)** 1706-1717.
[6] Gorbatikov A, Rogozhin E, Stepanova M, Kharazova Yu, Andreeva N, Perederin F, Zaalishvili V, Melkov D, Dzeranov B, Dzeboyev B and Gabaraev A 2015 *Physics of the Earth* **1** 28.
[7] Zaalishvili V, Nevskaya N, Nevskii L and Shempelev A 2015 *Journal of Volcanology and Seismology* **9 (5)** 333-338.
[8] Zaalishvili V B, Melkov D A 2014 *Izvestiya. Physics of the Solid Earth* **50 (5)** 707-718.
[9] Khulelidze K K, Kondratyev Yu I, Zaalishvili V B, Betrozov Z S 2016 *Sustainable development of mountainous areas* **8 (1)** 46-51.
[10] Grigorkina G S, Ramonova A G, Kibizov D D, Kozyrev E N, Zaalishvili V B, Magkoev T T, Fukutani K 2017 *Solid State Communications* **257** 16-19.
[11] *SNiP II-A; 12-69 Building Codes and Regulations* (Stroyizdat, Moscow).
[12] *SNiP II-7-81 Building Codes and Regulations* (Gostroyizdat, Moscow).
[13] Paz M 1994 *International Handbook of Earthquake Engineering. Codes, Programs and Examples.*
[14] Gogmachadze S, Zaalishvili V, Kipiana D and Odisharia A 1991 *UNESCO report* 141-145.
[15] Zaalishvili V and Gogmachadze S 1989 *Report of ISMIS Academy of Sciences of Georgia* **№ state reg. 01.9.00.016412** 25-40.
[16] *Recommendations for seismic microzoning in engineering surveys for the construction*, Moscow, 1985 72.
[17] Gogmachadze S 1993 *Cand. Dissertation (ISMIS)* 156.

Vibrations of nonlinear, absorbing soil medium according to the seismic microzoning instrumental method

V B Zaalishvili

Geophysical Institute of the Vladikavkaz Scientific Center of the Russian Academy of Sciences, 93a, Markova str., Vladikavkaz, 362002, Russia

E-mail: vzaal@mail.ru

Abstract. In order to assess the nonlinear and absorbing properties of soils, the studies in typical areas of the Tbilisi territory were carried out using a powerful vibratory source SV-10/100. It was found that the nonlinearity indicators used in the previously obtained formulas for the increments of the intensity make it possible to evaluate the stress-strain state of soils (linear elasticity, nonlinear elasticity, inelasticity). The "pure" absorption, as the ratio of the normalized spectrum to the vibration frequency, is actually a loss factor. This will make it possible to widely use the easily calculated "pure" absorption index in solving many problems of engineering seismology and earthquake-resistant construction, as well as to increase the reliability of the data obtained. In particular, these indicators can be used in the energy calculation of buildings, structures, etc.

1. Introduction

In recent years, the list of studies, in which instrumental measurements are very actively used in various, very different fields of geophysics, has significantly expanded. This undoubtedly raised the quality of the results obtained as well as their reliability [1-3]. In this regard, it became possible to directly study such subtle effects as nonlinearity, which is clearly manifested during macro-seismic and other field surveys of various hazardous natural and man-made processes' manifestations. At the same time, it should be noted that the manifestations of physical nonlinearity of soils under strong impacts have been little studied. There were only a few works in this area [4]. This was explained by the complexity of the problem, on the one hand, and, on the other hand, by the fact that the contribution of nonlinearity of soils to the formation of earthquake manifestations was for a long time considered as insignificant [5]. At the same time, the analysis of the survey data for a number of strong earthquakes that have occurred in recent years has shown that the nonlinearity effect can be significant. This led to a sharp change in attitudes towards this problem [5, 6].

2. Non-linear and absorbing soil properties

In order to assess the nonlinear and absorbing properties of soils, the studies out on the territory of Tbilisi using a powerful vibration source SV-10/100 were carried. Based on geotechnical data, 6 plots with typical geotechnical conditions were selected.

With the help of the SV-10/100 source, sweep - signals in the range of 1-100 Hz were generated in the soil strata and seismic-type observations were carried out. The duration of exposure was 2s and 12s. Vibrations were recorded by the analog equipment (channel SM-3 - GB-III-3). Next, the vibration records were digitized, and the Fourier amplitude spectra were constructed. As the initial parameters, the maximum amplitudes of the spectrum, the weighted average frequency of vibrations and the area of the soil vibrations' real spectrum were considered (Table 1). In addition, the indicators of nonlinearity and "pure" nonlinearity, absorption as well as "pure" absorption indicators (Table 1), which are closely related to soil movement, were calculated [4].

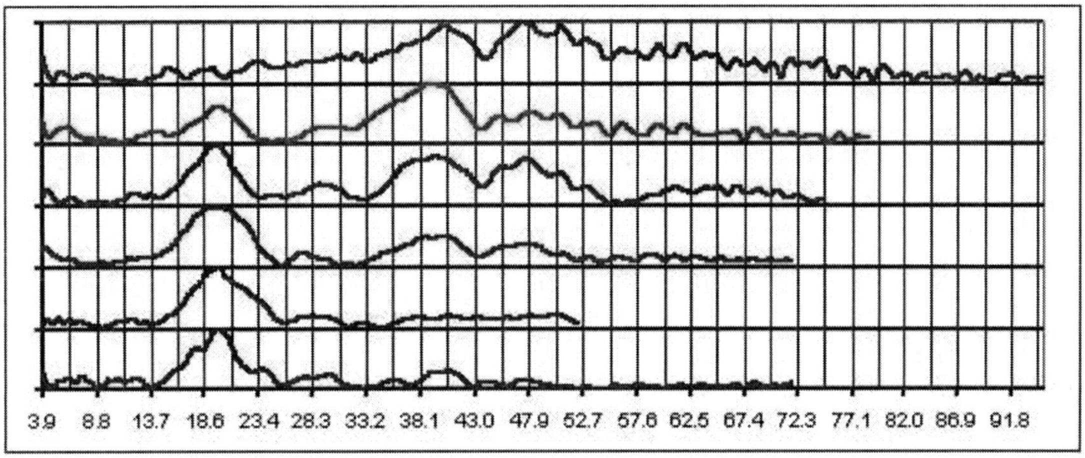

Figure 1. Sweep signal soil motion spectrum

$$\Delta I_1 = 2\lg\frac{S_{ri}}{S_{ro}} \quad (1)$$

$$\Delta I_2 = 3\lg\frac{A_i f_{cвi}}{A_o f_{cвo}} \quad (2)$$

$$\Delta I_3 = 2.4\lg\frac{S_{riб}S_{roд}}{S_{riд}S_{roб}} \quad (3)$$

$$\Delta I_4 = 3.3\lg\frac{(A_i f_{cвi})_б (A_o f_{cвo})_д}{(A_i f_{cвi})_д (A_o f_{cвo})_б} \quad (4)$$

Table 1. Indicators of soil vibrations under short-term exposure

Plot No.	Channel No.	Maximum spectrum vibration amplitude	Weighted average vibration frequency	Non-linearity index	Indicator of "pure" nonlinearity		Absorption rate	Indicator of "pure" absorption	Coefficient losses
		A	f_{wa}	S_r	A^*f_{wa}	S_r/f_{wa}	$S_r/A=S_n$	S_n/f_{wa}	η
1	1	590	54.4	10704.0	32096.0	196.8	18.1	0.3	0.3
	2	-	-	-	-	-	-	-	-
	3	73.4	63.6	1022.8	4668.2	16.1	14.0	0.2	0.2
	4	39.3	60.3	433.7	2369.8	7.2	11.0	0.2	0.2
	5	54.7	55.0	395.3	3008.5	7.2	7.2	0.1	0.2
	6	77.8	56.7	612.7	4411.3	10.8	7.9	0.1	0.1

2	1	894.0	39.0	29784.0	34866.0	763.7	33.3	0.9	0.7
	2	375.0	36.0	10709.5	13500.0	297.5	28.6	0.8	0.7
	3	130.0	30.3	3701.8	3939.0	122.2	28.5	0.9	0.4
	4	97.8	22.5	1354.2	2200.5	60.2	13.8	0.6	0.7
	5	55.1	22.3	708.8	1228.7	31.8	12.9	0.6	0.3
	6	34.6	20.3	412.3	702.4	20.3	12.0	0.6	0.2
3	1	772.0	42.1	22674.0	32501.2	538.6	29.4	0.7	0.6
	2	177.0	30.8	2792	5451.6	90.7	15.8	0.5	0.4
	3	57.2	23.4	584.6	1338.5	25.0	10.2	0.4	0.2
	4	46	25.4	510.4	1168.4	20.1	11.1	0.4	0.2
	5	87.7	23.7	879.3	2078.5	37.1	10.0	0.4	0.2
	6	87.1	23.0	654.4	2003.3	28.5	7.5	0.3	0.2
4	1	875.0	47.3	27690.0	40299.6	585.4	32.5	0.7	0.7
	2	211.5	39.8	4831.5	8417.7	121.4	22.8	0.6	0.4
	3	80.6	38.6	1903.2	3111.2	49.3	23.6	0.6	0.2
	4	51.8	32.9	921.0	1704.2	28.0	17.8	0.5	0.2
	5	50.6	26.2	584.7	1325.7	22.3	11.6	0.4	0.2
	6	33.6	27.8	387.6	934.1	14.0	11.5	0.4	0.2
6	1	1512	41.4	58568.0	62596.8	1414.7	38.7	0.9	0.8
	2	532.0	31.9	11758.0	16970.8	368.6	22.1	0.7	0.6
	3	115.6	24.8	1421.4	2866.9	57.3	12.3	0.5	0.2
	4	39.1	23.0	515.4	899.3	22.4	13.2	0.6	0.2
	5	37.9	24.4	419.2	924.8	17.2	11.1	0.5	0.1
	6	26.4	22.5	307.4	594.0	13.7	11.6	0.5	0.1

Table 2. Increments in soil intensity

Plot No.	Channel No.	Intensity increments, ΔI_1, intensity	Intensity increments, ΔI_2, intensity	Intensity increments, ΔI_3, intensity	Intensity increments, ΔI_4, intensity
1	1	-0.7	-0.02	-0.7	-1.2
	2	-	-		
	3	0.5	1.6		
	4	-0.1	0.9		
	5	-0.7	0.5		
	6	-0.1	1.0		
2	1	0.2	0.1	0.8	1.6
	2	1.2	1.2		
	3	1.6	1.4		
	4	0.8	0.8		
	5	-0.2	-0.7		
	6	-0.4	-1.4		
3	1	-	-	-	-
	2	-	-		
	3	-	-		
	4	-	-		
	5	-	-		
	6	-	-		

4	1	0.2	0.3	0.8	1.4
	2	0.5	0.6		
	3	1.0	1.1		
	4	0.5	0.5		
	5	-0.4	-0.6		
	6	-0.5	-1.0		
6	1	0.8	0.9	1.8	2.7
	2	1.2	1.5		
	3	0.8	1.0		
	4	0.01	-0.3		
	5	-0.6	-1.1		
	6	-0.7	-1.6		

In order to quantify the variations in nonlinearity, let us try to use their close relationship with the magnitude of the increments in the intensity. Using the formulas (1) and (2) [7], the increments in the intensity were calculated (Table 2). As a reference soil in this calculation, the soils of plot No. 3 were considered. At the same time, the indicators of soil vibrations in the near zone of the source were used as the compared data. Data analysis shows that the increment in the intensity in rocky soils (section 1), calculated by the formula (1), is - 0.7 points. At the same time, the increment obtained by the formula (2) was only 0.02 points. An even greater anomaly is observed for the soils of plot No. 2, presented in the form of lacustrine-bog deposits. Here, the increments are 0.1 and 0.2 points, respectively; the formulas (1) and (2) do not "work" here. With distance from the source, the increments in the intensity on the channel 2, calculated using both formulas, are 1.2 points, for the next measurement point (channel 3) - 1.6 and 1.4 points, and then the increments start decreasing to negative values. A similar picture is observed for other areas.

In the near field of the source, i.e., in the region with a high level of impact, the vibration spectrum "spreads" in the LF and HF spectral regions. With the distance from the source, the high-frequency component of soil vibrations is absorbed and the weighted average vibration frequency becomes less (Figure 1). This effect is hardly noticeable on rocky soils (section 1) due to low absorption and high frequency of its natural vibrations.

Thus, in the area adjacent to the source (Δ=1.5m), in non-linear soils, inelastic or permanent deformations prevail, which are not recorded by inertial geophones. With distance from the source, the amplitudes of the vibrations gradually increase, because inelastic vibrations are "transformed" into the elastic ones (while partially absorbed), reaching this state at a distance of 11.5 m (channel 3), where the intensity increment is maximum. Then the absorption process begins to dominate, which is higher in weak than in medium or rocky soils. In particular, the soils of section No. 6 in the form of macroporous sandy loam, characterized by a higher absorption index, reach a linear elastic state already at a distance of 6.5 m. Thus, when using the formulas (1) and (2), it is necessary to remove the geophones of the corresponding channels at distances of the order of 5-12 m.

The increments in the intensity can also be determined using the so-called formulas of "inelasticity" (3) and (4) [7]. They compare the indicators of soil movement in the close and remote zones of the source. The largest increase in the intensity is observed for the soils of plot No. 6. The overestimated value of the increment in intensity equal to 2.7 points (formula (4) for the "pure" nonlinearity indicator) is due to strong absorption in macroporous sandy loam for the close and remote zones of the source. In addition, the soils of plot No. 3 (clay p / t, 4.5 m thick, underlain by pebbles and rocky soils) are characterized in the remote zone by a high value of the "pure" nonlinearity index due to the emergence of waves refracted on the underlying rocky soils. However, the average value of the increment in the intensity obtained from both formulas is 2 points. For the soils of plots No. 2 and No. 4 - 1 point, etc.

Thus, when choosing a reference plot in the case of using powerful sources and dynamic indicators of movement, it is necessary to avoid the areas with heterogeneous soil conditions and underlying rocky soils close to the surface. At the same time, the use of both formulas of "inelasticity" makes it possible by averaging to obtain physically substantiated values even in such "unfavorable" conditions.

The analysis of variations in the "pure" absorption index is of particular interest. It can be seen that the indicated indicator for rocky soils corresponds to Sn/fwa = 0.1-0.3, for "medium" soils - Sn/fwa = 0.3-0.7 and for loose or "weak" soils, characterized by high nonlinearity, - Sn/fwa = 0.6-0.9. The maximum values Sn/fwa typical for the near-field zone of the source, i.e., energy absorption for all types of soil in the close zone is the highest. In this regard, it would be interesting to estimate the values of the loss coefficient η. Calculations of the loss factor for soils were carried out using the so-called "the resonant amplitude width" method [8]. Comparison of the data obtained (Table 1) shows that the "pure absorption" indicator is very close in value to the loss factor η and it is changed adequately to the latter. The conclusion suggests itself that the "pure" absorption rate is obviously the loss factor. This question undoubtedly requires additional research.

3. Summary

The nonlinearity indices used in the formulas for the intensity increments make it possible to evaluate the soils stress-strain state (linear elasticity, nonlinear elasticity, inelasticity). If we assume that the "pure" absorption rate is a loss rate, then it becomes possible to use the easily calculated "pure" absorption rate to evaluate many problems of engineering seismology and earthquake engineering, as well as increase the reliability of the data obtained. In particular, these indicators can be used in the energy calculation of buildings, structures, etc.

References

[1] Gorbatikov A V, Rogozhin E A, Stepanova M Yu, Kharazova Yu V, Andreeva N V, Perederin F V, Zaalishvili V B, Melkov D A, Dzeranov B V, Dzeboyev B A and Gabarayev A F 2015 Physics of the Earth **1** 28.
[2] Zaalishvili V B 2016 *International Journal of GEOMATE* **10 (2)** 1706-1717.
[3] Zaalishvili V B, Melkov D A 2014 *Physics of the Solid Earth* **50 (5)** 707-718.
[4] Nikolayev A V 1967 *Physics of the earth* **2** 23-31.
[5] Field E, Kramer S, Elgam A et al. 1998 *Seismological Research Letters* **69 (3).**
[6] Nikolayev A V and Galkin I N (ed.) *Problems of nonlinear seismic* (Moscow, Nauka) 288.
[7] Zaalishvili V B *Physical foundations of seismic microzoning* (Moscow, Institute of Physics of the Earth named after O. Yu. Schmidt of the Russian Academy of Sciences (IPE RAS) 367.
[8] Nashif A, Joage D and Henderson J 1988 *Vibration damping* (Moscow, Mir) 448.

IOP Conf. Series: Materials Science and Engineering 1083 (2021) 012088 doi:10.1088/1757-899X/1083/1/012088

The vibrations of a nonlinear, absorbing soil ground calculation by the method of the seismic microzoning multiple-reflected waves

V B Zaalishvili

Geophysical Institute of the Vladikavkaz Scientific Center of the Russian Academy of Sciences, 93a, Markova str., Vladikavkaz, 362002, Russia

E-mail: vzaal@mail.ru

Abstract. A nonlinear version of the program ZOND (nonlinear) has been developed based on the dependence of the shear and absorption modulus values on the seismic impact intensity, as well as on the introduction of ratios for calculating various indicators of soil movement (normal stress, deformation, etc.). The program algorithm is based on the multiple-reflected waves theory. An accelerogram of an earthquake, recorded on a rocky ground in Japan with earthquake characteristics similar to those for the territory of Tbilisi (magnitude, epicentral distance, spectral features, etc.), was chosen as an accelerogram set into the bedrock from the generated strong motion database AGESAS. The results of linear and nonlinear calculations with and without absorption are presented. It is clearly seen that with an increase in the intensity of the impact, the nonlinearity manifestation increases. At the same time, absorption increases. In this regard, the resulting movement at very high levels of impact may be below the initial level. This corresponds to the facts known from the analysis of the consequences of strong earthquakes that have occurred in recent years.

1. Introduction

In recent years, the list of studies, in which instrumental measurements are very actively used in various, very different fields of geophysics, has significantly expanded. This undoubtedly raised the quality of the results obtained and their reliability [1-6]. In this regard, it became possible to directly study such subtle effects as nonlinearity, which is clearly manifested during macro-seismic and other field surveys of various hazardous natural and man-made processes' manifestations. At the same time, it should be noted that the physical nonlinearity manifestations of the soils under strong impacts have been little studied. Accounting for the nonlinear soil properties in engineering seismology problems can be carried out in two ways: instrumental and calculated. Despite the fact that the instrumental method of seismic microzoning is the main one, nevertheless, very often it becomes necessary to solve such problems using the computational method.

2. Modeling nonlinear and absorbing soil properties

The use of the computational method makes it possible to simulate practically any conditions observed in nature. The solution of the problem of calculating the vibrations of the soil strata for the conditions of linear elastic deformations was implemented in different years by many authors [7-14]. The author and his students previously developed a linear version of the program ZOND (linear) [15].

At the same time, the needs of practice lead to the need to calculate soil vibrations for the conditions of nonlinear deformations (nonlinear elastic and inelastic deformations.). We believe that the elastic half-space is a linear elastic medium, and the soil overlying it exhibits strong nonlinear properties under intense seismic influences [16].

The solution of this nonlinear problem in an analytical form for soils is based on significant assumptions for such a complex medium as soil [16], and therefore the solution of nonlinear computational problems is based on the correlations established by the experimental studies. In other words, computational programs for solving nonlinear problems are essentially analytical-empirical. The most adequate programs have these peculiarities (SHAKE etc.) [17].

Let us have a seismic wave falling on the soil strata surface. Let us assume that the soil stratum is a nonlinear absorbing infinite medium with a density ρ and the propagation velocity of shear waves Vs. With small deformations, the shear modulus G will be maximum for a given soil:

$$G = G_{\max} = \rho V_s^2 \tag{1}$$

With an increase in deformation, the shear modulus initially remains constant, but upon reaching a certain value (specific for each material or soil), it is characterized by a noticeable change, i.e., the soil starts showing its nonlinear properties. With a continued increase in deformation, the growth of stresses slows down and then may not change until the destruction or hardening of the material, i.e., before the structural change.

The value of the plasticity number was chosen as the main indicator of the soil, characterizing its type and behavior under intense loads PI. On the empirical relations basis, the parameters necessary for the calculations are determined [18]:

$$k(\gamma, PI) = 0.5\left\{1 + \tanh\left[\ln\frac{0.000102 + n(PI)}{\gamma}\right]^{0.492}\right\} \tag{2}$$

where

$$n(P_I) = \begin{cases} 0.0 & \text{for } PI=0 \\ 3.37\times10^{-6} PI^{1.404} & \text{for } 0 < PI \le 15 \\ 7.0\times10^{-7} PI^{1.976} & \text{for } 15 < PI \le 70 \\ 2.7\times10^{-5} PI^{1.115} & \text{for } PI > 70 \end{cases}$$

$$d = 0.272\left\{1 - \tanh\left[\ln\left(\frac{0.000556}{\gamma}\right)^{0.4}\right]\right\} e^{-0.0145PI^{1.3}}$$

Next, the change in the shear modulus is determined based on the following relationship:

$$\frac{G}{G_{\max}} = k(\gamma, PI)(\sigma)^d \tag{3}$$

where G - is a current module,

σ - is normal voltage.

The seismic energy absorption is calculated by the formula [18]:

$$\xi = 0.333\frac{1+\exp\left(-0.0145PI^{1.3}\right)}{2}\left[0.586\left(\frac{G}{G_{\max}}\right)^2 - 1.547\frac{G}{G_{\max}} + 1\right] \tag{4}$$

On the basis of the indicated and introduced ratios to determine the required parameters (normal stress, deformation, etc.), a nonlinear version of the program ZOND (nonlinear) was developed. As an accelerogram set into the rock base from the database of strong movements AGESAS we have created [19], the earthquake accelerogram was chosen, recorded on a rocky ground in Japan, with characteristics similar to the territory of Tbilisi (magnitude, epicentral distance, spectral features, etc.) [20].

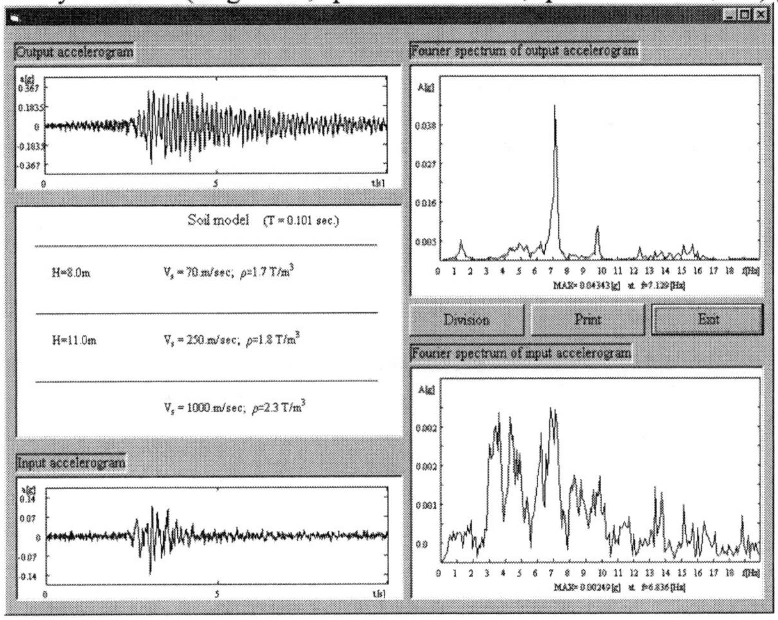

Figure 1. Plot 16. I=7. Non-linear calculation without absorption

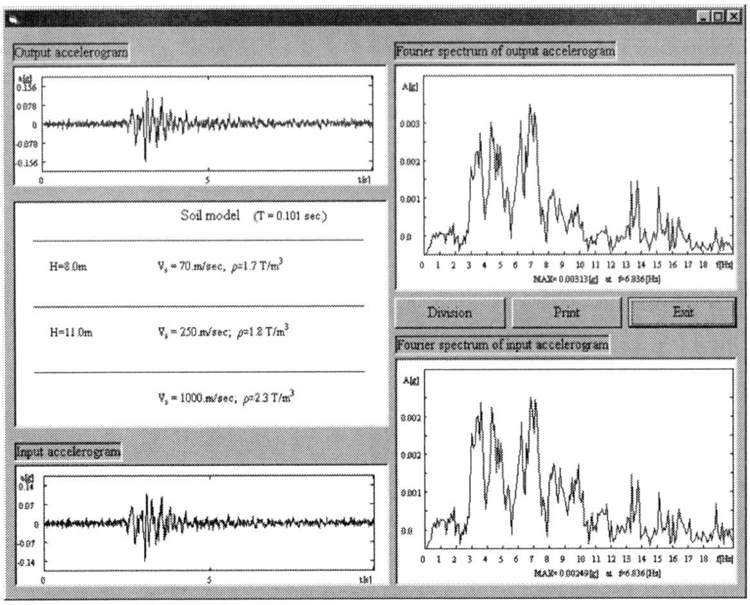

Figure 2. Plot 16. I=7. Nonlinear calculation with absorption

IOP Conf. Series: Materials Science and Engineering 1083 (2021) 012088 doi:10.1088/1757-899X/1083/1/012088

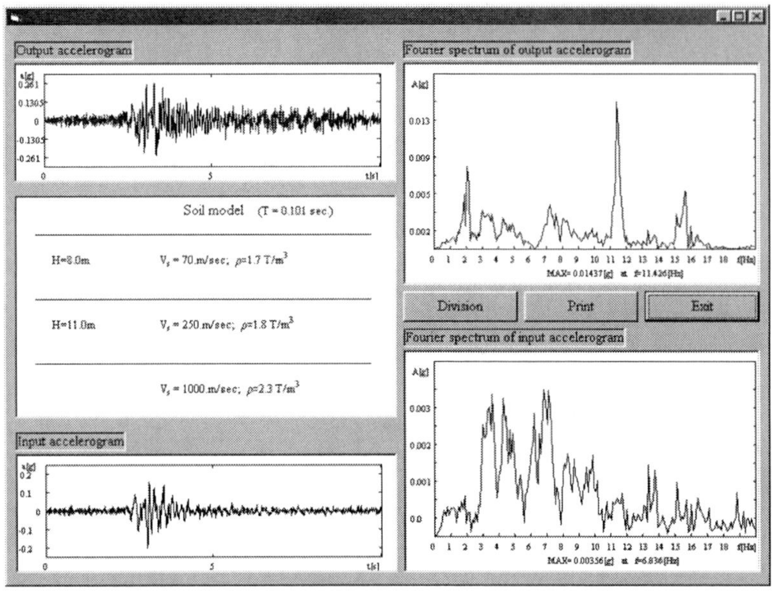

Figure. 3. Plot 16. I=8. Linear calculation without absorption

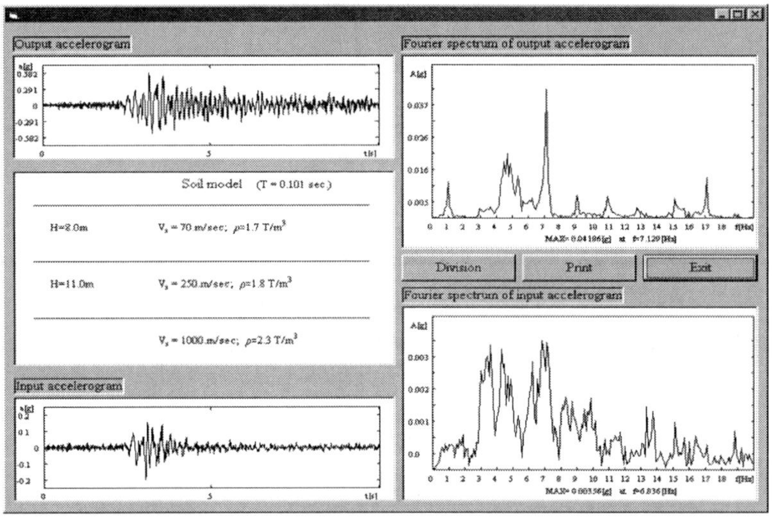

Figure. 4. Plot 16. I=8. Non-linear calculation without absorption

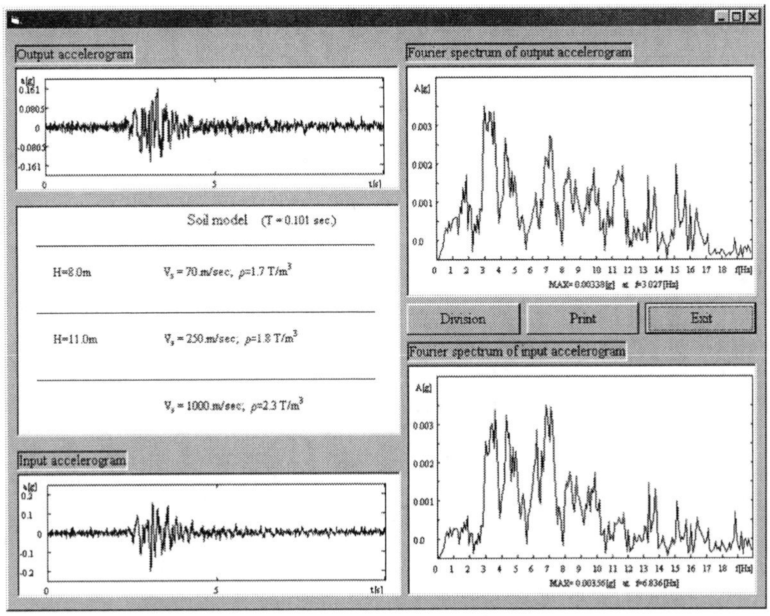

Figure. 5. Plot 16. I=8 . Linear calculation with absorption

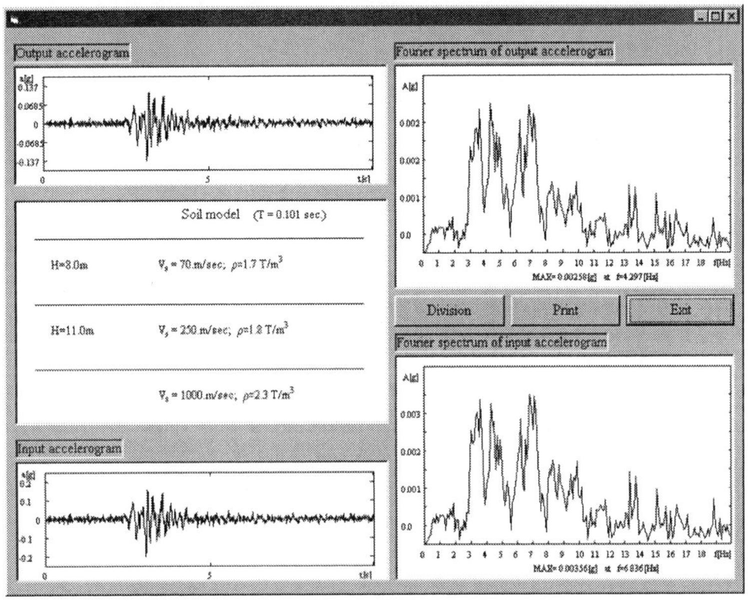

Figure. 6. Plot 16. I=8. Nonlinear calculation with absorption

3. Summary

The results' analysis of linear and nonlinear calculations with and without absorption for the models of specific sites in the territory of Tbilisi confirms the adequacy of the calculations to the physical phenomena observed in the soils under intense loads (Fig. 1-6). It is clearly seen that with an increase in the seismic impact intensity, the manifestation of nonlinearity increases (Fig. 2, Fig. 4). At the same time, absorption increases. Hence, the resulting movement at very high levels of impact may be below the initial level (Fig. 4). This corresponds to the facts known from the consequences analysis of the strong earthquakes that have occurred in recent years (for example, the Notridge earthquake, 1994).

References

[1] Zaalishvili V B, Melkov D, Kanukov A S and Dzeranov B V 2016 *International Journal of GEOMATE* **10 (1)** 1656-1661.
[2] Zaalishvili V B and Rogozhin E A 2011 *Open Construction and Building Technology Journal* **5** 30-40.
[3] Zaalishvili V B 2016 *International Journal of GEOMATE* T **10 (2)** 1706-1717.
[4] Gorbatikov A V, Rogozhin E A, Stepanova M Yu, Kharazova Yu V, Andreeva N V, Perederin F V, Zaalishvili V B, Melkov D A, Dzeranova B V, Dzeboev B A and Gabaraev A F 2015 *Physics of the Earth* **1** 28.
[5] Zaalishvili V B, Melkov D A 2014 *Izvestiya. Physics of the Solid Earth* **50(5)** 707-718.
[6] Grigorkina G S, Ramonova A G, Kibizov D D, Kozyrev E N, Zaalishvili V B, Magkoev T T, Fukutani K 2017 *Solid State Communications* **257** 16-19.
[7] Kanai K 1952 *Bull. Earthquake Res. Inst.* **30** 31-37.
[8] Kanai K, Tanaka T, Yoshizawa Sh. 1959 *Bull. of the earthquake Research Institute* **37.**
[9] Kanai K, Tanaka T 1961 *VIII Bull. Earthquake Research Institute* **39** 97-114.
[10] Shima E 1962 *Bull. of the Earthquake Research Institute* **40.**
[11] Kanai K, Tanaka T, Osada K and Suzuki T 1966 *Bull, Earthquake Res.Inst.* **44** 645-696.
[12] Jabauri G G 1979 *Messages of the Academy of Sciences of Georgia* (Moambe).
[13] Napetvaridze Sh G 1973(Metsniereba, Tbilisi).
[14] Napetvaridze Sh and Odisharia A 1970 *Proc. of a conf. Dinamic Waves in Civil Eng.* (London - New –York) 388.
[15] Zaalishvili V B, Otinashvilly M G and Timchenko I E 2000 *Structural theory and seismic resistance* (ISMIS Academy of Sciences of Georgia named after K.S. Zavrieva) **1** 67-71.
[16] Bonnet G, Heitz J 1994 *Proceedings of the 10th European Conference on Earthquake Engineering.Vienna* **1** 361-364.
[17] Schnabel B, Lysmer J, Seed H B 1972 *College of Engineering. University of California. Berkeley. Earthquake Engineering Centre* **EERC 72-12.**
[18] Ishibashi I and Zhang X 1993 *Soils and foundations* **33 (1)** 182-191.
[19] Zaalishvili V, Timchenko I, Lomidze G, Shengelia N 2000 *Third periodic report of INTAS.*
[20] Samkov B V, Jabauri G G, Datunashvili A S and Zaalishvili V B 1982 *Science* 127-131.

Rating assessment of an urbanized area for insurance purposes

V B Zaalishvili

Geophysical Institute of the Vladikavkaz Scientific Center of the Russian Academy of Sciences, 93a, Markova str., Vladikavkaz, 362002, Russia

E-mail: vzaal@mail.ru

Abstract. A rating assessment of the engineering-geological and geomorphological features of the territory has been carried out in the work. An area of 5 km x 4 km was allocated on the territory of Tbilisi. The area was divided into the squares with the size 100x100m. The minimum size of the plots was determined by the maps scale 1: 10,000. On the basis of these maps, each plot was assigned a rating according to the appropriate factor (soil type, layer thickness, groundwater level, relief, hazardous geological processes - landslide, avalanche, rockfall). Each factor was assigned a corresponding rating index (0-4). The best sites had the lowest index. Final maps of individual site ratings were compiled for various factors (soil type, thickness, etc.). On their basis, a summary rating map has been drawn up. All maps were made in GIS technology. The paper shows that the influence of these conditions can be significant in the absence of seismic impact. The results of the work can serve as a basis for assessments of any level, including for assessing seismic risk, as well as for insurance of buildings and structures located in the relevant territories.

1. Introduction

The list of studies, in which instrumental measurements are very actively used in various, very different fields of geophysics, has significantly expanded in recent years. This undoubtedly raised the quality of the results obtained as well as their reliability [1-5]. There is currently a concentration of population and basic means of production in the world in urbanized areas. From an economic, social and political point of view, this phenomenon is characterized by a high feasibility as it facilitates the management, organization, protection and communication of a concentrated population and industry. However, upon careful analysis of the situation, this fact turns out to be completely inappropriate in the conditions of a high earthquake probability, since this creates additional problems, especially if earthquakes occur in very large urban areas. This problem is very relevant for many cities in the world. Therefore, urbanized areas should be investigated in detail to establish all possible consequences of earthquakes or any other disasters.

2. Allocation of zones with different rating levels

One of the ways of a detailed study of an urbanized territory is a rating assessment of its engineering-geological and geomorphological features. For this purpose, a sufficiently large area was allocated on the territory of Tbilisi for relevant research. The allocated territory was 5 km long and 4 km wide, i.e., the total area of the study area was 20 km^2[6]. The region covers the western part of Tbilisi and completely includes the Vake region and most of the Saburtalo region. The choice of the indicated territory was based on the following considerations. Individual development occupies almost half of the territory here, and it is very concentrated. Industrial buildings take up a much smaller area. The population of the allocated area exceeds 200,000 people (this is more than 1/5 of the total population of the city). The allocated territory is one of the most important parts of the Tbilisi territory. Most of the offices and shops of Georgian and international firms and organizations are located here (Universities, UNO and UNESCO representative offices, the International Eurasia Foundation, Soros Foundation, etc.). At the same time, the Vake and Saburtalo districts (especially Vake) are considered prestigious districts of the city, and the new buildings are rapidly developing here. Most of the houses were built in the second half of the twentieth century. From an engineering-geological point of view, the area is very interesting, because a wide variety of soil types and variations in their thickness, groundwater levels and difficult terrain are presented here. In addition, there are areas where dangerous geological processes are possible.

The study area was divided into 100x100 m squares. Thus, the entire selected area was divided into 50x40 = 2000 squares. Each of these sites, in order to be able to compare them, was subject to an individual rating assessment. Information about the engineering-geological conditions of the selected area was obtained on the basis of studying the engineering-geological maps of Tbilisi with the scales of 1: 10000 and 1: 25000 [1-3]. The maps scale that determined the minimum size of the square areas was used. Each feature of the geotechnical map (soil type, layer thickness, groundwater levels, slope, etc.) was transferred to the corresponding model of the site in electronic form (into a computer).

The following factors were used to form the site rating:

- Soil types
- The thickness of the soil layers
- Groundwater level
- Relief (topography)
- The existence of hazardous geological processes

Each factor was assessed on a 4-point scale, where 0 corresponded to the least influence of this factor on the site status deterioration and 4 to the biggest. Based on the analysis of engineering-geological maps, the soils of the selected area were divided into several groups:

Unweathered and weathered rocky soils, conglomerates (A).

Semi-weathered clayey deposits and pebbles (B).

Clays and dense sandy loam (C).

Subsidence clays, loams and sandy loams (D).

Lacustrine-bog and silty soils (E).

This division was based on the fact that more or less homogeneous soil types behave differently at different degrees of watering or different slopes of the relief. When flooded (almost to the day surface), a relatively dense soil (for example, C soils) can become "weaker" than the initially "weak" but non-watered soil (E soils). The soils were ranked in the order of transition from dense soil to soft soil. At the same time, the soils were rated according to their types as follows: A soils were rated - 0, B - 1, C - 2, D - 3, E - 4 (Table 1 - 2.).

Table 1. Indicators of various features of soils on the territory of Tbilisi

Factor type	Factor type	Soil layer thickness	Ground water level	Relief	Hazardous geological processes
		m	*m*	*slope in degrees*	-
Symbol	*T*	*H*	*W*	*R*	*G*
					*Gm** *Gz**

A	Unweathered and weathered rocky soils and conglomerates	<5	1-3	1-10	-	-
		5-10	4-10	10-15	-	+
		10-20	10-20	15-20	-	+
		20-30	>20	20-30	-	+
		30-40		30-40	-	+
		>40		>40	-	+
B	Semi-rocky weathered clay deposits and pebbles	<5	1-3	1-10	-	-
		5-10	4-10	10-15	+	+
		10-20	10-20	15-20	+	+
		20-30	>20	20-30	+	+
		30-40		30-40	+	+
		>40		>40	+	+
C	Clays and loams, dense sandy loam	<5	1-3	1-10	-	-
		5-10	4-10	10-15	+	-
		10-20	10-20	15-20	+	-
		20-30	>20	20-30	+	-
		30-40		30-40	+	+
		>40		>40	+	+
D	Subsidence clays, loams and sandy loams	<5	1-3	1-10	-	-
		5-10	4-10	10-15	+	-
		10-20	10-20	15-20	+	-
		20-30	>20	20-30	+	-
		30-40		30-40	+	+
		>40		>40	+	+
E	Lacustrine-boggy sediments, silty and loose soils	<5	1-3	1-10	+	+
		5-10	4-10	10-15	+	+
		10-20	10-20	15-20	+	+
		20-30	>20	20-30	+	+
		30-40		30-40	+	+
		>40		>40	+	+
weight factor	4	4	4	4	4	

The soil layers' thickness was divided into 6 groups in the ranges: <5, 5 - 10, 10 - 20, 20 - 30, 30 - 40, >40 (m). For solid soil, power (for the selected area) does not matter. In this regard, the power effect for the corresponding rocky soil A is rated 0 points. For the soils B and C, the rating points were 0, 1, 1, 2, 2, 3. For the soils D and E, the rating was 0, 1, 2, 3, 3, 4, respectively. This indicates a gradual deterioration in the condition for "weak" soils (Table 1 - 2). With an increase in the power of "weak" soil, the number of points or its rating increases, i.e., the conditions worsen.

G_m - Landslides (soils B, C, D, E)

G_z - Rockfall or landslide (soils A, B); landslide (soils C, D, E) when the layers slope coincides with the relief slope.

The third factor was represented by the level of groundwater table. The levels are divided into groups 1 - 4, 4 - 10, 10 - 20, >20 (m) from the day surface. The influence of watering on the behavior of A soils is insignificant and it is 0. For E we have 1, 2, 3, 3, 4, 4. For all other soils, the index was 4, 3, 2 and 0 (Table 1-2).

The next factor is the terrain slope. The influence of the relief was estimated by the inclination angle: 1 - 10, 10 - 15, 15 - 20, 20 - 30, 30 - 40, > 40 (degree). For A soils, the index was 0, 0, 1, 1, 2 and 3. For B, C, D soils had: 0, 1, 1, 2, 3, 4. For E soils: 1, 2, 4, 4, 4, 4. Due to the fact that this type of soils, as a rule, absent on inclined terrain, they become very unstable in conditions of inclination. For "weak soils" this factor becomes very significant (Table 1 - 2).

Hazardous geological processes are associated with the relief and soil type. Three types of possible dangerous geological processes: landslides, rockfalls and landslides were taken into account. Landslides are possible for B, C, D soils. Rockfall or collapse is possible for A, B, C, D soils. Landslides in the absence of watering appear only when the slope of the layers coincides with the relief slope. A collapse is possible for soils C, D, E if the slope of the layers coincides with the slope of the relief. Soils E, as a rule, are free from these processes, because they do not exist on slopes. But if they exist there, then danger increases significantly. For these processes, the soils were assigned the following rating points A - 0, 0, 1, 2, 3, 4; B - 0, 0, 1, 2, 3, 4; C - 0, 1, 2, 2, 3, 4; D - 0, 1, 2, 2, 3, 4; E - 4, 4, 4, 4, 4, 4, respectively (Table 1- 2).

By summing the ratings of individual sites for various factors, the final rating maps of the selected area of the Tbilisi territory were compiled: a map of soil type ratings, a map of ratings by soil capacity, a map of groundwater levels, a map of terrain ratings, a map of hazardous geological processes ratings in GIS technology. This makes it possible to take into account the role of each factor in the formation of the vulnerability on the individual sites in the designated area of Tbilisi. Further, a summary map of the rating assessment of the allocated territory was built (Fig. 1). It is clearly seen that the most dangerous areas are occupied by the areas of dark colors. The areas with a low rating are characterized by a lighter color (up to white).

It should be noted that this study was based on the fact that there was no earthquake. The territory ranking was carried out to assess the impact of exclusively engineering-geological, hydro-geological and geomorphological conditions of the territory and can have the status of their "static" assessment. Taking into account the factor of the earthquake, the rating of the territory will acquire a “dynamic” form. In this case, a number of additional factors will appear, and some of the factors considered will manifest themselves in a different way.

The results obtained are the basis for assessments at any level: in assessing the engineering-geological hazard of urbanized territories, both for non-seismic and seismic regions. This, in turn, can be an important basis for assessing the seismic risk in a territory, as well as for insurance of buildings and structures located in the corresponding territories.

The analysis of the results obtained shows that the use of the considered concept for assessing the rating of one or another section of the soil strata makes it possible to create a very reliable methodology in its reliability. This methodology allows, when formalizing any indicators of the soil layer and its stress-strain state, to use it to assess the expected damage and vulnerability of any real natural and natural-man-made system. This, in turn, becomes the basis for the application of approaches to such a system for assessing risk, including economic and social losses, i.e., the insurance basis.

Table 2. Rating assessment of soil features in the territory of Tbilisi

Factor type	Factor type	Soil layer thickness	Ground water level	Relief	Hazardous geological processes
u		*m*	*m*	*degree*	-
			weight factor		

Symbol	*T*	*H*	*W*	*R*	*G*
				0	0
				0	0
A	0	0	0	1	1
				1	2
				2	3
				3	4
		0	4	0	0
		1	3	1	1
		1	2	1	2
B	1	2	0	2	2
		2	0	3	3
		3	0	4	4
		0	4	0	0
		1	3	1	1
C	2	1	2	1	2
		2	0	2	2
		2		3	3
		3		4	4
		0	4	0	0
		1	3	1	1
D	3	2	2	1	2
		3	0	2	2
		3		3	3
		4		4	4
		1	4	1	4
		2	3	2	4
E	4	3	2	4	4
		3	0	4	4
		4		4	4
		4		4	4

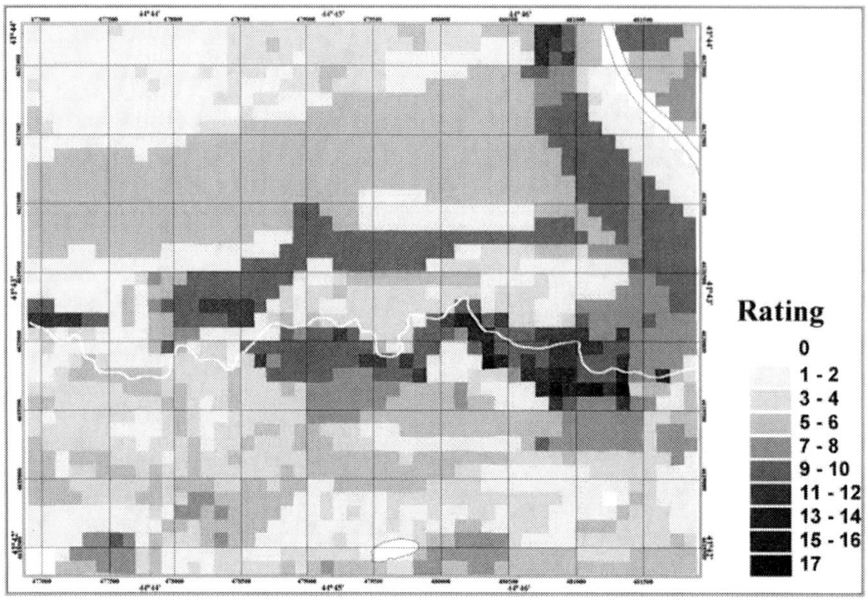

Figure 1. Summary map of the rating assessment of the allocated territory in Tbilisi

3. Summary

The rating assessment is the basis for assessments at any level: when assessing the engineering-geological hazard of urbanized territories, both for non-seismic and seismic regions. The analysis of the results obtained shows that the use of the considered concept for assessing the rating of one or another section of the soil strata makes it possible to create a very reliable methodology in its reliability. This methodology allows, when formalizing any indicators of the soil layer and its stress-strain state, to use it to assess the expected damage and vulnerability of any real natural and natural-man-made system. This, in turn, becomes the basis for the application of approaches to such system for assessing risk, including economic and social losses, i.e., the most important insurance basis both for the buildings and structures located in the relevant territories and for their residents.

References

[1] Zaalishvili V B 2016 *International Journal of GEOMATE* **10 (2)** 1706-1717.

[2] Gorbatikov A V, Rogozhin E A, Stepanova M Yu, Kharazova Yu V, Andreeva N V, Perederin F V, Zaalishvili V B, Melkov D A, Dzeranov B V, Dzeboyev B A and Gabarayev A F 2015 *Physics of the Earth* **1** 28.

[3] Zaalishvili V B, Melkov D A 2014 *Izvestiya Physics of the Solid Earth* **50 (5)** 707-718.

[4] Khulelidze K K, Kondratiev Yu I, Zaalishvili V B, Betrozov Z S 2016 *Sustainable development of mountainous areas* **8 (1)** 46-51.

[5] Grigorkina G S, Ramonova A G, Kibizov D D, Kozyrev E N, Zaalishvili V B, Magkoev T T, Fukutani K 2017 *Solid State Communications* **257** 16-19.

[6] Zaalishvili V B 2000 *Institute of Physics of the Earth named after O. Yu. Schmidt of the Russian Academy of Sciences (IPE RAS)* 367.

[7] Zaalishvili V B 2009 *Science* 365.

IOP Conf. Series: Materials Science and Engineering 1083 (2021) 012090 doi:10.1088/1757-899X/1083/1/012090

Accounting for nonlinear soil properties in the computational method of seismic microzoning

V B Zaalishvili

Geophysical Institute of the Vladikavkaz Scientific Center of the Russian Academy of Sciences, 93a, Markova str., Vladikavkaz, 362002, Russia

E-mail: vzaal@mail.ru

Abstract. The paper notes that in recent years, attention to the study of nonlinear phenomena in engineering seismology has sharply increased. The problem of accounting for nonlinear phenomena can be solved using instrumental and computational studies, as well as collaterally. In the latter case, the possible level of soil nonlinearity is estimated using the instrumental method, then the possible features of soil movements are calculated numerically. An algorithm for taking into account the nonlinearity of soils when calculating vibrations of the soil strata under intense seismic influences is proposed. The algorithm was developed on the basis of the known empirical relationships between the soil types and their indicators. The number of plasticity was chosen as an indicator of the soil type. Additionally, the necessary criteria and refinements have been introduced to correctly take into account the normal stress and calculate the deformation. The value of the shear wave propagation velocity in the soil, closely related to the physical state of the soil, is considered as a modulated indicator. The way of using the proposed algorithm for the method developed on the basis of the multiple-reflected waves theory and the method developed on the FEM basis is shown.

1. Introduction

In recent years, the list of studies, in which instrumental measurements are very actively used in various, very different fields of geophysics, has significantly expanded. This undoubtedly raised the quality of the results obtained and their reliability [1-9]. In this regard, it became possible to directly study such subtle effects as nonlinearity, which is clearly manifested during macro-seismic and other field surveys of various hazardous natural and man-made processes' manifestations. At the same time, it should be noted that the manifestations of physical nonlinearity of soils under strong impacts have been insufficiently studied. The influence of the soil conditions of the territory on the earthquakes' manifestation is well known. Many authors studied various aspects of the local conditions' influence: the dependence of the earthquakes' manifestation on the topographic, lithological, hydrogeological features of the territory. At the same time, the influence of each individual factor on the earthquake intensity formation is still taken into account rather formally. Relatively few works are known on the direct assessment of the soils' physical state influence on the intensity formation. With rare exceptions, there were almost no works devoted to the impact level influence on soils' behavior, i.e. – the studies of soil nonlinearity assessments [11]. On the one hand, this is explained by the greater complexity of the problem and, on the other hand, many leading scientists of the world believed that the nonlinearity influence on the earthquakes' manifestation features is insignificant [10]. Analysis of the survey data of strong earthquakes that have occurred in recent years has shown that the nonlinearity effect can be very significant. This led to a radical change in the attitude to this problem [10, 12].

This problem can be solved using the instrumental and computational studies, as well as collaterally. In the latter case, the soil nonlinearity level is assessed using the instrumental method.

Possible features of soil movements (the soil layer natural vibrations' frequency, etc.) are calculated by the calculation numerical methods.

2. Calculation method of seismic microzoning.

When carrying out seismic microzoning, it is necessary to simulate the real soil conditions, which are diverse. It is equally important to develop and use the principles of accounting for the nonlinear soil properties using the calculation method. The use of the computational method makes it possible to simulate any conditions observed in nature and evaluate the features of nonlinear phenomena deep in loose sediments. Such programs exist abroad. But, first, they are quite expensive, and second, the use of a ready-made program does not allow taking the results of our own research fully into account.

When solving such a problem, it seems that the half-space describing the rocky soil behaves like a linear elastic medium, and the loose soil overlying it, exhibits strong nonlinear properties under intense seismic or dynamic influences.

Half-space and loose soil can be characterized by the propagation velocity of the shear waves V_s, attenuation coefficient β and shear modulus G.

The nonlinear behavior of soils is most often assessed by the nonlinear relationship between stress τ and soil deformation γ [13]:

$$\tau = G(\gamma)\gamma \tag{1}$$

At present, on the basis of the existing modifications [5-12], the Institute has developed two numerical methods for calculating the soil layer vibrations: the first is based on the well-known theory of multiple-reflected waves (MRW) [14-21], and the second is the finite element method (FEM) [20]. Currently, the new modifications of both methods have been developed [22]. In these ways, the medium still appears to be linearly elastic. In addition, a method to take into account the nonlinearity of soils [22], based on high capabilities FEM, was developed. When implementing this method, it is possible to set any type of stress-strain relationship obtained from the laboratory research data or physical assumptions and calculate the corresponding reaction on the soil stratâ surface.

At the same time, when solving the scientific and practical problems, it is necessary to rely on the initial data obtained directly from the field observations (shear wave velocity in the soil, soil density, etc.), although if necessary, it is also possible to use this, well-known in earthquake-resistant design, approach.

2.1. Method of multiple - reflected waves of seismic microzoning

Let us have a seismic wave falling on the soil strata surface. Let us assume that the soil mass is a nonlinear, absorbing infinite medium with a density ρ and the propagation velocity of shear waves Vs. The shear modulus of the medium G can be determined from the relation:

$$G = G_{max} = \rho V_s^2 \tag{2}$$

In most practical problems, in particular, when drawing up a high-velocity section of a soil stratum, the values of the shear wave velocities obtained at small deformations are used. Hence, the shear modulus will correspond to the maximum value of the shear modulus for a given environment (soil).

In the process of applying a monotonically increasing seismic or dynamic load to the soil, several different stress-strain states of the latter will be observed. Initially, with the growth of small (for a given soil) deformations, the stresses in the soil increase in direct proportion to them. The process is linearly elastic. Then, upon reaching a certain value (specific for each material or soil), the material exhibits its nonlinear properties. The nature of the process becomes nonlinear elastic. In this case, there are no inelastic phenomena, the soil simply changes some of its elastic characteristics (for example, the prevailing frequency of the natural soil vibrations can go into the high-frequency region, etc.). With the nonlinear elastic nature of the relationship "stress-strain", as the deformations increase, the stresses slightly slow down the growth. But the process is still reversible. With a further increase in deformations, the process can become inelastic. In this case, stresses significantly reduce growth, may not change, or even have negative increments until the material is hardened or destroyed, respectively, i.e., before the change in physical condition.

In accordance with the noted stages of physical states, the propagation velocity of the shear seismic wave, which is a quantity closely related to the state and changes of the soil. From the formula (1) it can be seen that any, arbitrarily small, change in velocity leads to a noticeable change in the soil shear modulus. The change in soil density at the first stage can, in general, be neglected. At the same time, the soil density can be quite reliably determined for various types of soils directly from the shear wave velocity using the empirical dependences obtained by us for the conditions of Georgia [23]. Such connections are well known from the data of other authors. [24].

1. The behavior of soil under intense loads depends on the type of soil and the specific physical condition. The latter, in turn, is determined by a number of soil parameters: the degree of water cut, existing internal stresses, the past loads chronology, etc.

2. A very important factor is the magnitude of the normal stress caused by the lithostatic pressure of the overlying soil [12]. In this case, the normal stress can reach noticeable values. Thus, it is necessary to take into account the location of the soil in depth.

3. Next, we determine the expected deformations from the vibration velocity of soil particles in accordance with the corresponding level of intensity for a given area [23]. In the latter case, the intensity predicted for a given area can be used according to maps of general seismic zoning, detailed seismic zoning and even seismic microzoning.

If we estimate the indicated intensity directly from a seismic zoning map or detailed zoning, i.e., without taking into account the engineering-geological, hydrogeological and geomorphological conditions, then it should be attributed to the so-called medium soils. Such conditions should be carefully selected. Otherwise, the intensity and deformation will accordingly be underestimated or overestimated. The vibration velocity is determined from the tabular data for the corresponding intensity.

a) Further, in the ideal case, an accelerogram can be found from the database of strong movements, recorded on similar grounds and with the corresponding developed intensity. Then we recalculate the record from the strata surface onto the rocky ground (bedrock) The resulting accelerogram (or a set of them) will be the initial one for entering the soil sections into the bedrock for all areas in the study area. It is easier to select the records recorded directly on the rocky grounds with appropriate characteristics or create the synthetic accelerograms.

b) Another solution is as follows. We carry out the work using the instrumental method of seismic microzoning, determine the intensity level for the soils of the selected site (this also determines the level of possible nonlinearity) and estimate the vibration velocity according to the tabular data (Table 2) for the corresponding intensity [24]. Then the process is repeated.

When using the maps of seismic microzoning, the problem of the numerical method is somewhat narrowed and as a result, we study only the dependence of the vibrations frequency on the intensity. Then, if, for example, we have the soils in the 8-point zone of the city, on which the intensity calculated by the methods of seismic microzoning can be 9 points, then we use the corresponding value from the database or the specified table.

The calculation should take into account the characteristics of the soil type. This is realized on the basis of soil differentiation according to such indicators, which are especially characteristic for certain types of soils. Such indicators include: consistency indicator, porosity coefficient, plasticity number, deformation modulus, etc. By choosing the appropriate indicator, it is possible to reliably take into account the soil type and its expected behavior under intense loads.

It should be noted that the role of watering is to change the physical state of certain soil types. Therefore, the formal accounting of the groundwater level as one of the soils seismic properties indicators, which is still widely used in some modifications of seismic microzoning, can lead to errors [12]. On the other hand, watering is taken into account directly by the corresponding change in the physical soil state.

We define further, for a given level of deformations, the change in the shear modulus due to the above-mentioned factors, for example, $a_1 a_2 a_3 a_4$ etc. It should be borne in mind here that an increase in the number of factors that determine the soil type and its physical state, on the one hand, leads to greater concretization, but on the other hand, it seriously complicates the calculations due to the lack of appropriate analytical relationships. In this regard, in such studies, as a rule, the amount of soil deformation and one or two indicators of the soil itself are taken into account. To solve such problems, the results of special studies of the soils behavior under varying loads are used [12, 13, 26]:

$$\frac{G}{G_{\max}} = \prod(a_1 a_2 a_3 a_4 \gamma) \tag{3}$$

where G - is a current shear modulus.

7. In the considered method, one of the main parameters is the shear wave velocity. In this regard, we further determine the changes in the current velocity of the shear wave (ρ-represent constant or calculated from empirical dependence) for a given level and take into account in the corresponding calculations:

$$V_S = \sqrt{\frac{G}{\rho}} \tag{4}$$

8. If it is necessary to take into account absorption, its calculation is carried out according to the corresponding dependence:

$$\xi = \prod(a_1 a_2 a_3 a_4 \gamma) \tag{5}$$

Thus, the module G and absorption ξ are calculated based on the change of V_S.

$$G = \varphi(V_S)$$
$$\xi = \varphi(V_S) \tag{6}$$

The data obtained are the direct basis for taking into account nonlinear phenomena in the soil. At the same time, during the development, the well-known research results are used.

9. In the case of a layered soil stratum, a corresponding change in shear velocity is calculated for each layer. At the same time, seismic vibrations, already modulated by soil nonlinearity, will come to each subsequent layer from the depth. They are expressed by the corresponding accelerograms. As a result, on the soil layer surface, we obtain an accelerogram representing the integral effect of taking into account physical nonlinearity.

10. The algorithm was developed on the basis of the well-known empirical relationships between the soil types, the corresponding indicators and the indicators values' dependence on deformations. The number of plasticity and the corresponding dependence were chosen as an indicator of the soil

characteristics [26]. Additionally, the necessary criteria and refinements were introduced to take into account the normal stress and calculate the deformation.

2.2 Seismic microzoning FEM method

1. To make the calculations, it is necessary to know the values of Young's modulus E and Poisson's ratio μ, which are calculated according to well-known formulas. The main indicator is again the measured shear wave velocity. The values of the propagation velocities of longitudinal and transverse waves in the soil mass are determined without any special difficulties in field observations. In a methodical analysis based on the known shear wave velocity, the longitudinal wave velocity can be approximately determined. This is realized from the known ratios of the longitudinal and transverse waves' velocities [23, 27]. As noted above, the soil density can be determined from the empirical relationship between the shear wave velocity and the density obtained by us for the conditions of Georgia. Next, the Poisson's ratio (μ) and Young's dynamic modulus (E) are calculated. The calculations of these values are carried out using the longitudinal and transverse waves' velocities.

2. With a layered thickness, we calculate E and μ for each layer according to the above-depicted formulas. The calculation makes it possible to directly take into account the change in the soils' behavior due to their nonlinearity under intense loads.

3. In conclusion, it should be noted that the behavior of "weak" soils under intense loads is characterized by a significant difference in compression and tension. Moreover, non-cohesive soils (such as sand or gravel) near the earth surface have nearly zero Young's modulus and tensile shear modulus. This can and does lead to a decrease in the bearing capacity of the soil during multicyclic seismic vibrations, to its destruction (for example, the phenomenon of cavitation). The absence of a dataset does not allow to take into account directly the bi-modularity of a particular environment in the calculations at present. At the same time, if necessary, it is possible to take into account the bi-modularity of the environment based on the specified dependencies of the type:

$$G_{-} = \varphi(V_{S-}) \tag{7}$$

$$G_{+} = \varphi(V_{S+}) \tag{8}$$

Summary

A new algorithm has been developed based on the use of the method of multiple - reflected waves and the FEM method by taking into account the empirical relationships between the soil types, the corresponding indicators and the magnitude dependence of the indicators on deformations. The number of plasticity and the corresponding dependence were chosen as an indicator of the soil characteristics. The values μ of Young's modulus E and Poisson's ratio are also used. Additionally, the necessary criteria and refinements were introduced to take into account the normal stress and calculate the deformation.

References

[1] Zaalishvili V B and Rogozhin E A 2011 *Open Construction and Building Technology Journal* **5** 30-40.

[2] Zaalishvili V B, Kanukov A S, Melkov D A, Makiev V D and Dzobelova L V 2018 *International Journal of GEOMATE* **15 (51)** 160-166.

[3] Zaalishvili V B, Melkov D A, Kanukov A S and Dzeranov B V 2016 *International Journal of GEOMATE* **10 (1)** 1656-1661.

[4] Zaalishvili V B and Rogozhin E 2011 *Open Construction and Building Technology Journal* **5** 30-40.

[5] Zaalishvili V B 2016 *International Journal of GEOMATE* **10 (2)** 1706-1717.

[6] Gorbatikov A V, Rogozhin E A, Stepanova M Yu, Khazarova Yu V, Andreeva N V, Perederin F V, Zaalishvili V B, Melkov D A, Dzeranov B V, Dzeboev B A, Gabarayev A F 2015 *Physics of the Earth* **1** 28.

[7] Zaalishvili V B, Melkov D A 2014 **50 (5)** 707-718.

[8] Grigorkina G S, Ramonova A G, Kibizov D D, Kozyrev E N, Zaalishvili V B, Magkoev T T, Fukutani K 2017*Solid State Communications* **257** 16-19.
[9] Hulelidze K K, Kondratyev Yu I, Zaalishvili V B, Betrozov Z S 2016 *Sustainable development of mountainous areas* **8 (1)** 46-51.
[10] Field E H, Kramer S, Elgamal A W, Bray J D, Johnson P A, Cramer C, Roblee C, Wald D J, Bonilla L F, Dimitriu P P, Anderson J G 1998 *Seismoligical research letters* **69 (3)** 230-234.
[11] Nikolaev A 1967 *Physics of the Earth* **2** 23-31.
[12] Zaalishvili V B 2000 *Institute of Physics of the Earth named after O. Yu. Schmidt of the Russian Academy of Sciences (IPE RAS)*, (Moscow).
[13] Bonnet G, Heitz J F 1994 *Proceedings of the 10th European Conference on Earthquake Engineering* (Vienna) **1** 361-364.
[14] Kanai K 1952 *Bull. Earthquake Res.* Inst. **30** 31-37.
[15] Kanai K, Tanaka T and Yoshizawa Sh 1959 *Bull. of the earthquake Research Institute* **37.**
[16] Kanai K and Tanaka T 1961 *VIII Bull. Earthquake Res. Inst.* (Tokyo) **39** 97-114.
[17] Shima E 1962 *Bull. of the Earthquake Research Institute* (Tokyo) **40.**
[18] Napetvaridze Sh G 1973 *Some tasks of engineering seismology* (Tbilisi, Metsniereba) 162.
[19] Napetvaridze Sh and Odisharia 1970 *A Proc. of a conf. Dynamic Waves in Civil Eng* (London - New –York).
[20] Gogeliya T I 1976 *A numerical method for assessing the influence of local soil conditions on the intensity and spectral composition of seismic vibrations in the foundations of structures*. (Cand., Dissertation). Tbilisi, 1976.
[21] Jabauri G G 1979 *SA Georgia messages (*Moambe).
[22] Zaalishvili V, Timchenko I, Otinashvilly M and Javrishivili I 2000 *Report of ISMIS Academy of Sciences of Georgia.*
[23] Zaalishvili V 1986 *Auth. Cand. dissertation (*Tbilisi).
[24] Nazarov G N and Shemshurin V A 1977 *Seismic microzoning*
[25] Assessment of seismic hazard and seismic risk. *Institute of Physics of the Earth named after O. Yu. Schmidt of the Russian Academy of Sciences (IPE RAS)*, Moscow, 1997.
[26] Kramer S L 1996 *International Series in Civil Engineering and Engineering Mechanics.*
[27] Goryainov N N and Lyakhovitsky F M 1979 *Nedra.*

Research of abrasive properties separate dust to ensure the reliability of dusting systems in the production of building materials

B Ch Meskhi[1], A I Evtushenko[1*], N M Sergina[2], V N Azarov[2]

[1]Don State Technical University, 1, Gagarin sq., Rostov-on-Don, 344000, Russia
[2]Volgograd State Technical University, 1, Akademicheskaya str., Volgograd, 400074, Russia

E-mail: a.evtushenko@mail.ru

Abstract. The intensive development of the infrastructure of large cities, which took place in the second half of the twentieth century, led to the fact that enterprises specializing in the production of building materials and products were in close proximity to urban residential areas. This determines an increase in the requirements for the efficiency and reliability of dust removal systems designed, as is known, to ensure the standardized quality of the air environment of settlements and industrial premises. The dust generated in the production of building materials, in many cases, has abrasive properties, which leads to accelerated wear of the elements of dust cleaning systems. The experience of using such systems shows that dust collectors, as well as shaped parts of air ducts (bends, tees, etc.), through which the dust and gas mixture is transported to dust collectors, are most susceptible to abrasion. Abrasive wear of dust collectors during operation leads to an increase in dust leakage into the atmospheric air into the air of the working area. During operation of a dust-cleaning unit under pressure, damage to the walls of the air ducts also leads to knocking out the transported dust. The supercritical wear of the elements of the dust removal system leads to its failure. The article presents some results of studies of the dust properties of building materials that determine the process of abrasive wear.

1. Introduction

To ensure the reliability of dust removal systems, it is necessary at the design stage to predict the occurrence of failure as a result of abrasive wear of individual elements. As a rule, when mentioning the dust generated during the production of building materials, its abrasive properties are simply stated as a fact without specifying a quantitative indicator characterizing these properties - the abrasion coefficient, knowing which, one can predict the degree of abrasive wear of the dust removal system elements during its operation.

2. Formulation of the problem

Such a forecast can be made according to the results of calculations by the well-known method [1, 2], using data on the characteristics of the captured dust, which determine its abrasive properties (density, fractional composition, particle shape, etc.), as well as data on the value of the abrasion coefficient.

The values of the latter indicator are experimentally established only for several types of dust of building materials and are given in [3].

3. Objects and methods of research

As the objects of research, we selected systems designed for the localization, removal and cleaning of dust emissions from technological equipment in the workshops for the production of building expanded clay, concrete, reinforced concrete structures and products, as well as in the workshops for the mechanical processing of wood in the manufacture of wooden building products (floorboards, baseboards , window and doorways, etc.) at the enterprises of Volgograd and Rostov-on-Don. Schemes of some of the examined systems are shown in Figure 1.

To assess the abrasive properties of the studied dust, an experimental setup was used, the circuit of which is shown in Figure 2 [2].

The method is based on determining the degree of wear of the plate of the steel grade under study. In this case, the plate is located at an angle of 450 to the dust jet. The dust under investigation is fed from the hopper to the funnel and from it to the booster device, which is a tube with a diameter of 6 mm and a length of 150 mm. A sample cassette is installed at one end of the tube. The accelerating device is rotated by an electric motor in a horizontal plane. Dust particles are accelerated by centrifugal forces, hit the sample and through the opening of the cassette are carried out into the dust collector. By weighing the sample before and after the experiment, the mass wear of the sample was determined. The amount of dust falling asleep, the speed of rotation of the tube in all experiments were the same [2]. The abrasivity coefficient of the studied dust was determined by the results of experiments by the expression (1) [2].

$$K_a = A\Delta m, \ \mathrm{m}^2\cdot N^{-1} \qquad (1)$$

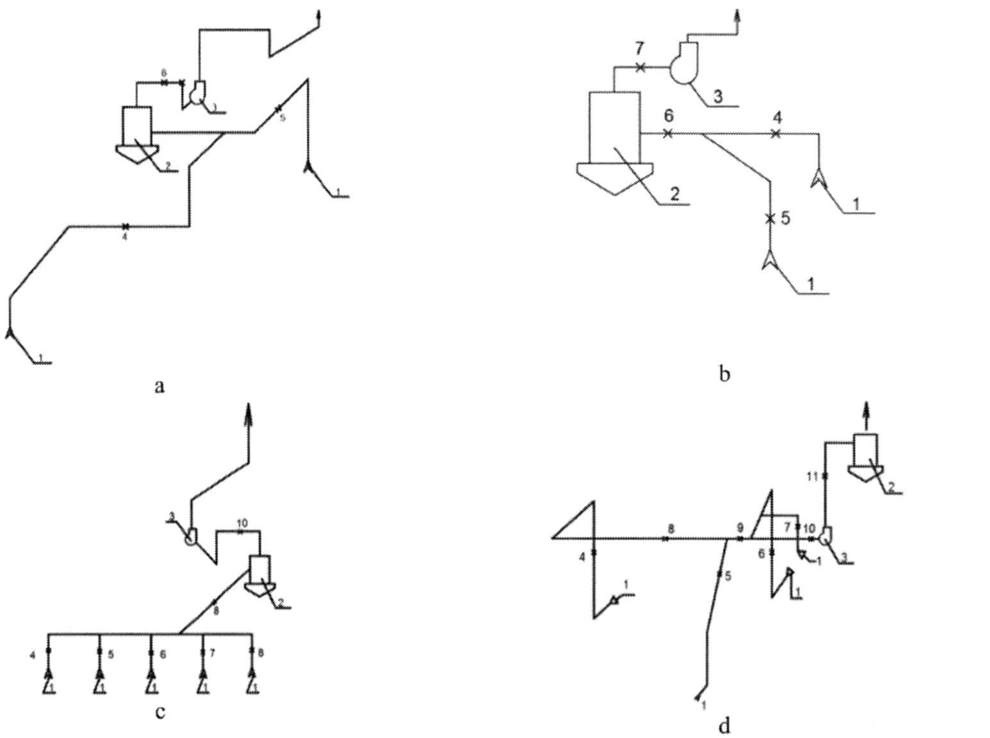

Figure 1. Diagrams of inspected dust removal systems: a - from the nodes of sand filling; b - from the sand conveyor; c - from the expanded clay conveyor; d - from woodworking machines. 1 - local suction; 2 - cyclone; 3 - fan; 4-11 - points of measurement

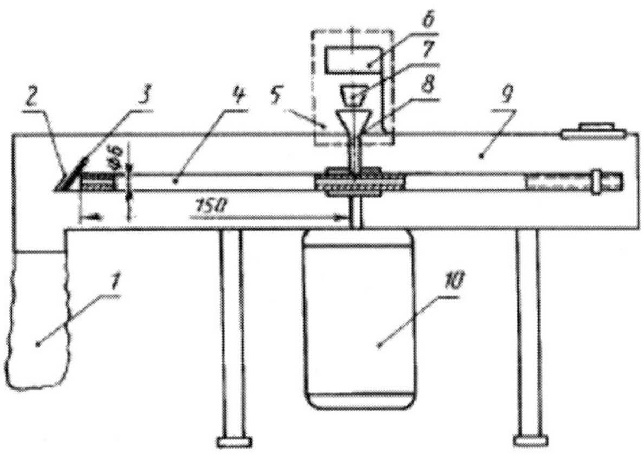

Figure 2. Scheme of the experimental setup for determining the coefficient of abrasiveness of the studied dust.
1 - dust collector; 2 - cassette for mounting the sample; 3 - sample; 4 - booster tube; 5 - vibration feeder; 6 - hopper; 7 - electromagnetic coil; 8 - funnel; 9 - case; 10 - electric motor [2]

Where:
A - installation constant (at an angular speed of the engine of 50s^{-1} and a tube length of the accelerating device 0.15 m, A = 1.0804·10-6 [2]), m^2/(N·kg);
Δm - is the change in mass of the steel sample, kg.

The arithmetic mean of the results of three main measurements was taken as the test result [2].

The obtained data on the value of the coefficient of abrasiveness were used to calculate the intensity of abrasive wear of cyclones when cleaning emissions from atmospheric dust from the studied dust. The calculations were carried out according to the expression (2) [2]

$$\tau = \frac{hq}{3600Ecv^2K_a}, \text{ч} \tag{2}$$

Where
τ - is the time of abrasive wear of an element of a dust removal unit to a given depth, h; h - linear wear, m;
c/q - mass concentration of the material, kg/s^2/m^2;
E - is the coefficient of probability of particles getting on the wearing surface, in fractions of one;
v- is the gas flow velocity, m/s;
K_a - coefficient of abrasiveness of the studied dust, m^2/kg.

Taking into account the results of the predicted calculation of the time of abrasive wear of cyclones for 2.5 years, full-scale studies were carried out every six months to assess the change in the amount of dust breakthrough depending on the operating time of dust removal plants.

To determine the density of the investigated dust, a manometric method based on the Boyle-Marriott law was used [2]. Evaluation of the fractional composition was carried out using the methods of microscopic analysis [2,4-7].

4. Results and discussion

Figure 3 in accordance with [2, 4-7] in the form of integral distribution functions shows the results of the analysis of the fractional composition of dust entering the aspiration system serving the expanded clay kiln [8, 9].

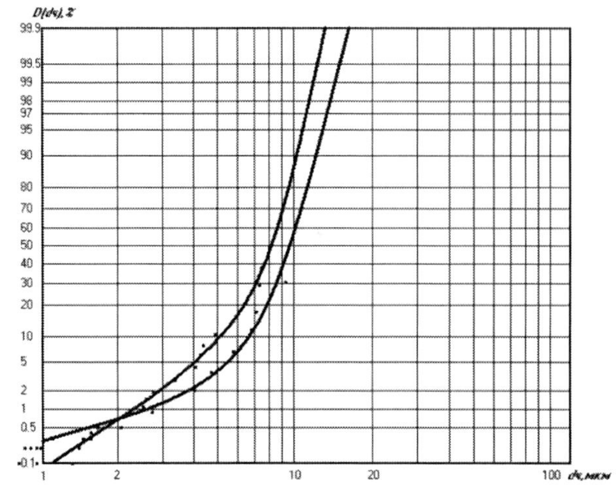

Figure 3. Integrated curves of the fractional distribution of particle mass in a probability-logarithmic grid for dust selected in a dust removal system serving a rotary expanded clay kiln

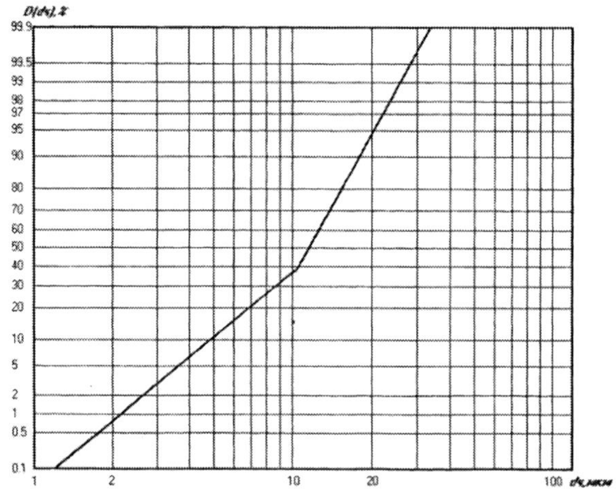

Figure 4. The results of the analysis of the fractional composition of sand dust entering the dust removal system

Analysis of the data showed that the particle sizes are: maximum - 16 microns, minimum - 0.3 microns, median diameter - 9 microns. The proportion of particles with particle sizes less than 10 microns - PM_{10} ranges from 60 to 90%. The content of particles with sizes less than 2.5 microns - $PM_{2.5}$ - is 1-1,8% [8, 9].

The sizes of sand dust particles vary in the range of 1.3–32 μm. Particles with sizes less than 10 microns account for 38%, the median diameter d_m is 12 microns (Figure 4) [10]. When machining pine wood, particles are formed with sizes from 5 microns to 35 microns. Particles with sizes less than 10 microns account for 9.5%, the median diameter d_m = 20 μm (Figure 5)

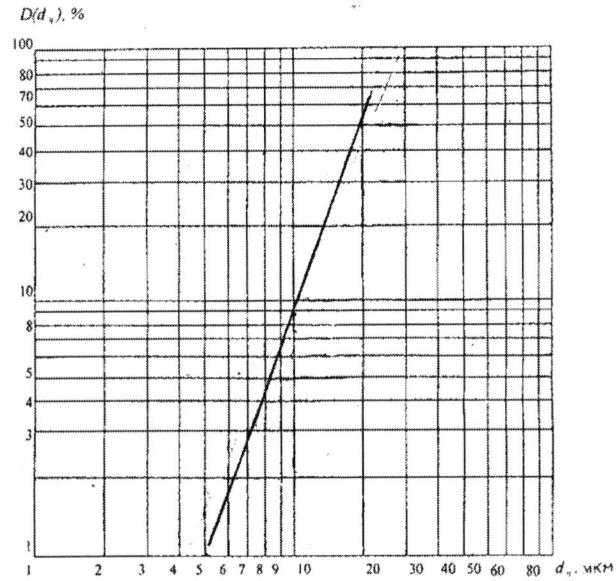

Figure 5. Perfractional distribution of mass over the diameters of dust particles entering the dedusting system generated during the mechanical processing of pine wood.

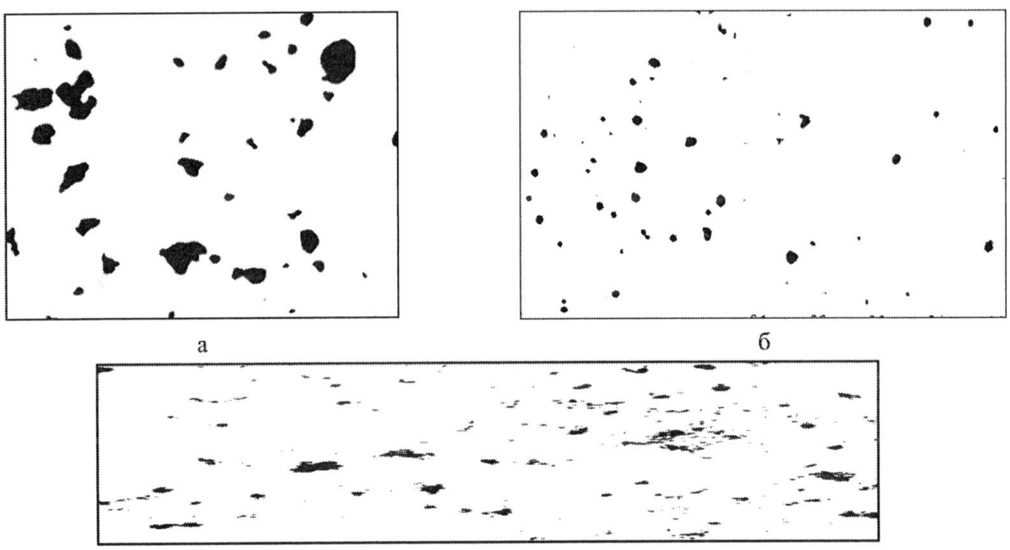

Figure 6. Microphotographs of particles of the studied dust entering the dust removal systems: a - expanded clay dust; b - sand dust; c - pine wood dust

In this case, the fractional composition of expanded clay dust and sand dust entering the dust removal systems obeys a truncated log-normal distribution [5]. The dispersed composition of dust generated during the mechanical processing of pine wood, as well as the dispersed composition of dust generated during the processing of wood of other tree species, is described by the Goden – Andreev – Schuman formula [5], which for the case under consideration takes the form:

$$D(d_p) = 0{,}0108 d_p^{2{,}84} \quad (3)$$

The study of the shape of the particles of the studied dust was carried out according to microphotographs obtained during the analysis of variance (Figure 6). Obviously, most sand particles have a shape close to spherical (Figure 6, b). Dust particles of pine wood are mainly characterized by an elongated rod-shaped or needle-shaped form (Figure 6, c). Expanded clay particles differ in the greatest variety of shapes - from lamellar, rod-shaped to rarely close to spherical (Figure 6, a).

The results of experimental studies to determine the density of the studied dust [2] showed that this average value is: for expanded clay - 2600 kg/m^3, sand - 1500 kg/m^3 [10-12], pine wood - 1300 kg/m^3. The experimentally obtained values of the coefficient of abrasivity for the studied dust are: expanded clay – $0{,}66 \cdot 10^{-10}$ m^2/kg, sand - $0{,}3 \cdot 10^{-10}$ m^2/kg, pine wood - $0{,}42 \cdot 10^{-10}$ m^2/kg.

Table 1 shows the results of a predictive calculation of the abrasive wear of cyclones in the examined dust removal systems, based on experimental data on the speed and concentration of dust in the cleaned air flows, as well as on the basis of experimentally obtained values of the abrasivity coefficients of the studied dust.

Table 1. The results of the predictive calculation of the abrasive wear of cyclones in the examined dust removal systems

Type of dust	Abrasion time, year
Sand dust	2.9
Expanded clay dust	1.8
Wood dust (pine)	1.9

The data of field studies to assess the change in the magnitude of dust breakthrough depending on the operating time of dust removal plants are presented in table 2 and in the figure 7-9.

Table 2. Change in the amount of dust leakage depending on the life of the dust removal systems

Type of dust	Duration of operation of the dust removal system, year				
	0.5	1	1.5	2	2.5
	Dust slip, %				
Sand dust	3.6	3.8	4	4.3	5
Sand dust	3.5	3.6	3.8	4.2	4.8
Sand dust	2.9	3.1	3.3	3.6	3.8
Expanded clay dust	11.5	12.0	14.9	16.8	
Expanded clay dust	10.2	10.4	13.8	16.2	
Wood dust (pine)	17.2	16.8	17.2	18.3	20.3
Wood dust (pine)	15.1	15.1	15.1	15.8	18.3

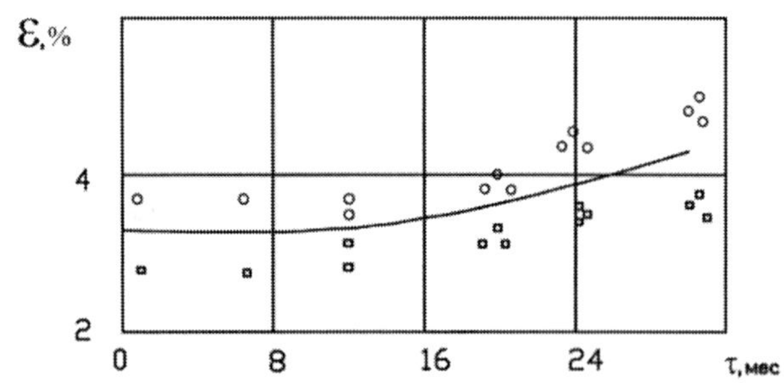

Figure 7. Change in the magnitude of the breakthrough when cleaning emissions from sand dust

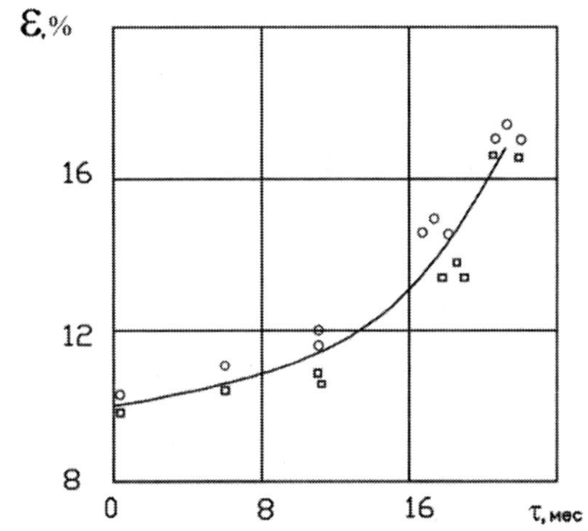

Figure 8. Change in the value of the breakthrough when cleaning emissions from expanded clay dust

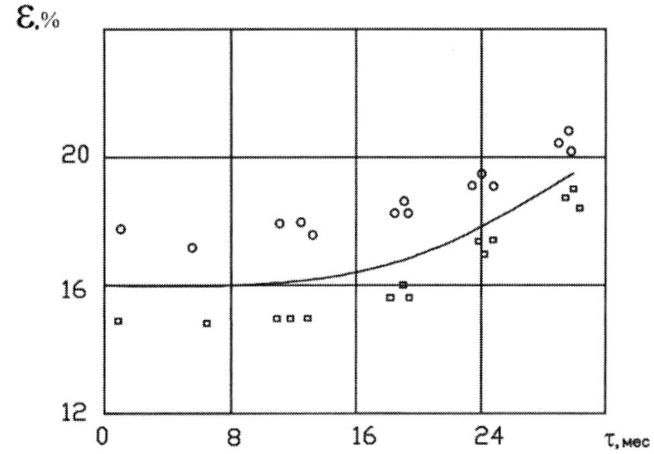

Figure 9. Change in the value of the breakthrough when cleaning emissions from dust generated by the mechanical processing of pine wood

These results indicate that during the operation of the sand dust emission cleaning system, breakthrough into the atmospheric air due to abrasive wear of cyclone elements increases on average by 28.6%, from wood dust - by 21%. The cyclones used for cleaning emissions from expanded clay dust are most subjected to abrasion. In this case, during operation, the dust leakage to the atmosphere increases on average by 69,8%.

5. Summary

- According to the results of experimental studies carried out using standard techniques, data on the properties (disperse composition, particle shape, density) that determine the abrasiveness of certain types of dust generated in the production of building materials and products were obtained.
- For wood dust, expanded clay and sand dust, the values of the coefficient of abrasivity were experimentally determined, data on the values of which are currently not available in the domestic reference literature.
- The obtained data on the values of the abrasiveness coefficients for the dust under study made it possible to predict the duration of the trouble-free operation of dust removal systems with cyclones.

- The results of field studies to assess the effect of abrasive wear of cyclones on the amount of dust leakage, conducted at the enterprises for the production of building materials, products and structures in Volgograd and Rostov-on-Don, confirmed the advisability of using vortex inertial dust collectors with counter-swirling flows in dust-control systems at enterprises in the industry (CDW) instead of cyclones, since CDW devices are not only characterized by greater efficiency, but also are more resistant to abrasion vnom wear.

References

[1] Aliev G M 1988 Installation and maintenance of gas cleaning and dust removal systems (Metallurgy, Moscow).

[2] Kouzov P A, Scriabin L Ya 1983 Methods for determining the physicochemical properties of industrial dusts (Leningrad: Chemistry).

[3] Scriabin L Ya 1981 Atlas of industrial dusts: overview information (Moscow).

[4] Samokhvalov N M, Zykova, Y A Improving the efficiency of dusting in production portland cement 10.36629/2686-777X-2018-1-12-93-96

[5] Kouzov P 1987 Fundamentals of the analysis of the dispersed composition of industrial dusts and ground materials (Leningrad: Chemistry).

[6] Azarov V N, Sergina N M 2002 Technique of microscopic analysis of the dispersed composition of dust using a PC *Moscow. - Dep. at VINITI* 07.15.2002 **1333** 9.

[7] Azarov V N, Yurkyan O V, Sergina N M 2004 Technique of microscopic analysis of the dispersed composition of dust using a personal computer (PC) *Legal and applied metrology* **1** 46-48.

[8] Kopytenkova O, Levanchuk A, Tursunov Z Health risk assessment for exposure to fine dust in production conditions *Russian Journal of Occupational Health and Industrial Ecology* 10.31089/1026-9428-2019-59-8-458-462

[9] Sergina N M, Semenova E A, Kislenko T A 2013 Dust removal system for expanded clay production *Engineering Journal of the Don* **4**. Information on ivdon.ru/magazine/archive/n4y2013.

[10] Sergina N [et al.] 2017 Study of some properties of certain types of dust of building materials *Modern Science and Innovations* **4** 148-150.

[11] Cherkasova N G 2020 How Adding Wood Dust Affects the Properties of Composite Construction Materials *E&ES* 459 (3) 032054.

[12] Pałubicki, Bartosz & Hlásková, Luďka & Rogozinski, Tomasz. 2020 Influence of Exhaust System Setup on Working Zone Pollution by Dust during Sawing of Particleboards *International Journal of Environmental Research and Public Health*. 17. 3626. 10.3390/ijerph17103626.

IOP Conf. Series: Materials Science and Engineering 1083 (2021) 012092 doi:10.1088/1757-899X/1083/1/012092

Innovative solutions to improve the purification method of chromium-containing waste water

D O Ignatkina[1], V P Batmanov[1], M Y Belogorodskaya[1] and G L Gizzatova[2*]

[1]Volgograd State Technical University (VSTU), 1 Akademicheskaya street, Volgograd, 400074, Russian Federation.
[2]Volgograd State Agrarian University, 26 University avenue, Volgograd, 400002, Russian Federation

E-mail: gulnaragizz@yandex.ru.

Abstract. The research results showed that the problem of resource saving during the chromium electroplating from standard electrolyte can be solved. Formulated requirements that ensure the implementation of the set tasks without costs for new equipment and chemical reagents. It has been proved that organic compounds of a certain structure introduced into the electrolyte influence the chromium electroplating in such a way that all technological characteristics, coating properties, physical and chemical characteristics of the solution change. The chrome plating process has been optimized using the experimental mathematical planning method, which made it possible to develop methods for obtaining chrome coatings protected by security documents on the basis of experimental data. All the proposed methods allow solving the problem of resource saving and environmental safety. The authors have consistently studied resource-saving and environmental indicators. In galvanic chromium plating technology, resource-saving indicators are: energy intensity; metal intensity; water intensity; and labour intensity. Environmental indicators: industrial emissions: gas emissions, waste water; occupational diseases.

Introduction

The main area of nature protection is the prevention of pollution directly in the technological cycle instead of catching it at treatment plants.

Preventing pollution is always economically beneficial. This is due not only to the reduction of raw material and energy costs in the product manufacturing but mainly to the reduction of waste processing costs.

The modern technical and technological base of industry makes it impossible to carry out deep air and water treatment at industrial enterprises due to the extremely high cost of these measures. The development of new technological processes with reference to which waste-free production can be created will ensure not only high technical and economic performance but also the integrated use of natural resources. However for technical and economic reasons the transition to zero-waste technology cannot be made immediately. The real way to greening technology is to gradually switch first to low-waste and then to waste-free closed cycles. In this way sustainable environmental management and protection can be achieved.

50% of available industrial waste and emissions can be prevented at their source by using technically sound, environmentally friendly and economically viable technologies [1-2].

The electroplating production of chromium is a rapidly growing and major part of a whole range of industries, which may reach a dead end if the main objective: the quality improving of the metal surface is not addressed comprehensively in the close interconnected development of the three areas:

- improving the environmental safety of production, and environmental protection;
- development of resource-saving processes that will increase the efficiency of electrochemical chrome plating by reducing energy and raw material costs not only by using new methods but also by optimizing the process;
- study of the chrome plating process mechanism in order to obtain coatings with predefined (improved) functional properties.

It is necessary to create a process chain of interrelated operations: efficient, environmentally friendly chrome plating and chrome-containing waste water treatment. Today the separation of these operations has resulted in a more complex expensive production process than the main one [3-4]. The treatment plant which is based on the recovery of chromium (VI) compounds to chromium (Hi) in the form of insoluble hydroxide slurry has become more complex.

Disposing of chromium-containing sludge into building materials is still problematic as there are no statistics yet available to confirm that there has been no change in human health across generations when living in houses whose building materials contain recycled chromium-containing waste.

The widespread use of galvanic chromium plating and the environmental problems it makes pose a pressing challenge: the creation of a new electrolytes generation that will provide a high level of process efficiency with minimal harmful effects on the environment.

Based on the analysis of gas chromium-containing emissions, the water effluents composition and methods of their treatment, the chromium electroplating process, the properties of chromium coatings and chromium alloys based on it the author considers it relevant to create a system of targeted development of resource-saving processes as the basis for environmentally friendly technologies of galvanic chromium plating from water environments.

The research results that are presented in the article are devoted to the search of the main factors influencing the intensification and ecological safety of chromium electroplating and its alloys based on aqueous solutions of chromium anhydride in order to create a method that comprehensively solves the problem of resource saving and environmental protection. Parameters are defined to predict more effective additives, which introduction into the electrolyte will ensure the implementation of environmental and technological objectives [5].

The research made it possible to formulate the requirements that an organic compound - an initiating additive - must meet:

- oxidize in a chromic acid solution to form an organochromic compound with chromium ions;
- regenerated at the anode;
- do not polymerize;
- have an aliphatic substituent with at least ten carbon atoms.

These requirements are met by substances with a cyclic structure:

- carbocyclic,
- alicyclic;
- aromatic with condensed benzene nuclei;
- aromatic with condensed benzene and five-member cycles;

heterocyclic (six or more carbon atoms, oxygen atom).

Main part

Natural oil is a promising substance that contains a sufficient number of such structures as the structures formed by the interaction of known organic additives with chrome plating electrolyte resemble the aromatic part of some oils first of all Anastasievskoye field. The maximum number of compounds capable to meet the above requirements is contained in a second vacuum chromium run at temperatures of 350-420°C (spindle distillate) [6].

Solving the resource saving problem of the chrome plating process first of all the conditions are determined under which it is possible to obtain quality coatings at minimum current densities and electrolyte temperature.

Chroming is the most energy-intensive of all galvanic processes and is sensitive to changes in temperature and current density.

The introduction of NF-VGISI additive into the standard chrome plating electrolyte makes it possible to obtain quality coatings at a wide range of current densities at 20 °C and deposition rates that are 2-2.5 times higher than standard. The higher current yield values of chromium in the electrolyte with NF-VHISI additive are explained by a significant increase in hydrogen overvoltage compared to the use of standard electrolyte (without NF-VHISI additive) but also by protonation of organic additive [7-8].

Intensity is an important indicator of resource saving energy. Chromium plating at low temperatures (without special heating) increasing the deposition rate significantly reduced energy consumption.

The NF-VGISI additive promotes the foam formation on the electrolyte surface, which dramatically reduces (by 80-85%) the gas content of chrome anhydride.

In the framework of this task, for the first time studies were conducted on the impact of the organic additive on the physical and chemical properties of the electrolyte solution: density, viscosity, electrical conductivity and surface tension.

It was established that the change point in physical and chemical characteristics of the solution so-called "critical concentration" corresponds to the composition of the solution containing 250 g/l of CrO_3 and 2-6 g/l of NF-VGISI. Sulphuric acid is excluded. The new structure is formed which facilitates the transport of electroactive particles with this composition and certain process modes. It was shown that the maximum current yield of chromium as well as optimal values of some physical characteristics of the obtained coatings are achieved at extreme (first or second derivative) points of functions of physical and chemical properties from the composition of chromium electrolyte solutions [9].

The resulting forecast model is expressed in the following way:

$$F[(\frac{d\varphi(C_i)}{dC_i} = 0 \vee \frac{d^2\varphi(C_i)}{dC_i^2} = 0) \Rightarrow C_i] = opt$$

where F is the predicted parameter (BT, microhardness and others),

C_i is the concentration of electrolyte component,

$\varphi(C_i)$ is the function of density, viscosity, surface tension, specific electrolyte solution conductivity,

opt - optimal value of the predicted parameter (for $BT\ opt = max$).

This expression does not allow the numerical value of the predicted parameter to be obtained but only indicates the area and direction in which it is searched for.

The mathematical process optimization with respect to the obtained data made it possible to recommend the electrolyte composition, g/l: 250 CrO_3, 4 NF-VGISI, electroplating from which will allow obtaining optimal technological indicators [10-12].

As can be seen, an improved composition of chrome plating electrolyte, which lacks inorganic catalyst (H_2SO_4) is the study result.

For the first time, the very strict work dependence of the electrolyte and the coatings quality on sulfuric acid was eliminated. Chromium electroplating from the resulting composition is less sensitive to changes in the concentration of chromium anhydride, chromium ions (III) and iron (III).

The total concentration of Cr^{3+} and Fe^{3+} ions critical for the electrolyte was increased from 7 g/l to 40 g/l.

The results listed in Table 1 allow us to reduce the number of operations performed by the corrector worker to analyze the electrolyte composition by a factor of 2; to facilitate the purification of chrome-containing effluents (the purification stage from SO_4^{2-} has been eliminated) [13].

Table 1. Resource-saving and environmental indicators of the chrome plating from water environment

Evaluation indicators						
Energy capacity	Metal capacity	Water capacity	Labour capacity	Gas emission	Water runoffs	Occupational diseases
Estimated parameters (components)						
Deposition rate increases 2-2.5 times Number of ventilation units reduce Electrolyte heating is excluded (t_{pct}=20-26°C) Periodical current	Electrolyte dissipation capacity increased by 40% Microhardness increased by 30-35% Wear resistance increased by 3-4 times Internal tension reduced by 30-45% Porosity reduced by 19-20 times Lead consumption reduced by 2.7 times	Electrolyte heating has been eliminated t_{pct}=20-25°C) Foam creation on the electrolyte surface Corrections frequency reduced by 1.8 times	Deposition rate increased by 2-2.5 times Dissipation capacity of the electrolyte increased by 40%	Electrolyte heating is excluded (t_{pct} =20-25°C) Foam creation on the electrolyte surface	Utilization of used electrolyte Exclusion of sulfate ions from aqueous effluents	Deposition rate increased by 2-2.5 times Electrolyte heating is excluded (t_{pct}=20-25°C) Foam creation on the electrolyte surface

In view of the foregoing conclusions are drawn which undoubtedly confirm innovative solutions for improving the treatment of chromium-containing waste water:

- the deposition rate of chromium is increased by a factor of 2-2.5 that reduced electricity consumption (energy consumption) by a factor of 2-2.5, the electroplaters number by a factor of 2 (from three-shift operation to one-shift operation), reduced the rate of solution removal from CrO_3, and reduced occupational diseases;
- electrolyte heating is eliminated, which reduced steam consumption, electrolyte evaporation, occupational diseases level by 100%, number of electrolyte adjustments by 1.5 times, amount of water consumed, number of electroplaters - adjusters, chemical reagents consumption;
- foam is formed on the electrolyte surface, which prevents electrolyte drift, reduces gas emissions by 70%, leads to the saving of chemical reagents, man-hours, electrolyte correction is reduced by 1.5 times that reduced the volume of industrial effluents;
- the dissipation electrolyte capacity is increased by 40% that made it possible to reduce the number of locksmiths producing special lead anodes and to save lead (100%);
- the concentration of Cr^{3+} and Fe^{3+} was increased to 40 g/l, thus increasing the electrolyte lifetime that reduces man-hours (for correction), chemical reagent consumption and the volume of chrome water effluent;
- microhardness increased, wear resistance (94-5 times) that increases the service life of the part and reduces metal consumption;

- internal stresses in coatings reduced (by 1.25 times);
- coating porosity reduced by 12-20 times (1-2 pores/cm^2);
- corrosion resistance increased by 3-4 times (acts in the annex) that made it possible to eliminate preliminary copper and nickel plating operations, thus reducing electricity consumption (by 1.2 times).

References

[1] Moskvicheva E V 1998 Resource-saving processes as a basis for environmentally friendly galvanic chrome plating technologies from water and non-water environment *Dokt, Diss.: 11.00.11, 05.17.03* Moscow 352.
[2] Furtatova O N 2004 Intensification of electrolytic chrome plating and neutralization of chromium-containing effluents *Cand.Diss: defended 30.03.2004, 05.17.03 / Oksana Nikolaevna Furtatova. Novocherkassk* 127.
[3] Elinany G A, et al. 1976 Polarography of metal-gallic complexes *Journal of Electroanalytical Chemistry and Interfacial Electrochemistry* **72.3** 363-369.
[4] Elynek T B 1992 The Successes of Electroplating Technology *An Assessment of the International Literature 1990-1991. Metal plating and surface treatment* **1 (3-4)** 7-26.
[5] Galkin Yu A, Lotosh V E 1990 Technology of sewage sludge disposal at machine building enterprises *Chemistry and water technology* **12 (6).**
[6] Libreich E 1934 *Theorie der Verchromung. Z. Elektrochem* **1** 73-87.
[7] Ignatiev V I, Ionicheva N S and Mareichev A V 1985 *Galvanic coatings in mechanical engineering* Handbook. In 2 volumes. Edited by Shluger M A.
[8] Kauspedens D V 1994 *Galvanotechnology and metal processing* **3** 43.
[9] Naidenko V V, Gubanov L I 1999 *Cleaning and utilization of electroplating industrial processes* (Nizhny Novgorod, Decon) 432.
[10] Unruch Y 1991 *Metaloberflaeche* **3** 107.
[11] Shkurikova E B 997 Environmental safety of electroplating production through reorganization of Consciousness *Electroplating and surface treatment* **(5) 1** 42-49.
[12] Kushnina K S, Vavilova A S 1990 *Wastewater treatment and disposal of electroplating sludge and sludge from hexavalent chromium production* (Moscow).
[13] Smirnov D N, Genkin V E 1980 *Wastewater treatment in metal processing processes* (Moscow, Metallurgy) 88.

IOP Conf. Series: Materials Science and Engineering 1083 (2021) 012093 doi:10.1088/1757-899X/1083/1/012093

Optimisation of environmental protection measures in the electrodeposition industry

D V Shchitov[1], Yu Yu Yuriev[2], A V Nesterchuk[1] and A V Moskvicheva[2*]

[1]Department of construction "North Caucasian Federal University" Institute of service, tourism and design (branch) NCFU in Pyatigorsk, 1 Pushkin street, Stavropol, 355009, Russian Federation

[2]Department of water supply and sewerage, Volgograd State Technical University (VSTU), 1 Akademicheskaya street, Volgograd, 40007, Russian Federation

E-mail: antonio1805@yandex.ru

Abstract. At present, the quality of wastewater treatment at machine-building enterprises and disposal of generated sludge does not meet modern technical, economic and ecological requirements. In the course of research into galvanic wastewater, the problem of its neutralization at galvanic production was analyzed. Ehe quality of industrial waste water and the structural composition of pollutant compounds molecules containing heavy metal ions were studied. Physical and chemical characteristics of real and model solutions were studied when different system parameters were changed: pH, temperature, oxidant impact (6% solution H_2O_2), reagent concentration). The negative impact of complex formation processes on the depth of electroplating methods was also studied and experimentally confirmed, which technical essence is due to the structure, size, charge size of pollutant particles (electrochemical, membrane, coagulation). The data obtained formed the basis of the methodology for determining the effectiveness of the treatment of these effluents using known methods.

Introduction

One of the most pressing problems in electrodeposition industry is the neutralization of industrial wastewater in the electrodeposition shops. The degree of water pollution in this production largely determines the environmental situation in water basins.

Unfortunately, in the European part of Russia it is difficult to find a reservoir that is not contaminated with heavy metal ions, which causes great environmental damage.

In Russia, most drinking water is taken from surface sources and its contamination with heavy metal ions is harmful to human health.

The required efficiency of wastewater treatment is determined by the conditions in which it is discharged. In Russia for heavy metal ions (Cd^{2+}, Cr^{3+}, Cu^{2+}, Ni^{2+}, Zn^{2+}, Fe^{3+}). The maximum permissible concentrations of these pollutants should not exceed 0.1; 2.7; 0.5; 1; 10; 0.5 mg/l respectively when delivering industrial wastewater to municipal biological treatment facilities. When discharging wastewater into a basin, the concentrations of these pollutants shall not exceed the following normative values for water basins of domestic use: 0.001; 0.5; 1; 0.1; 1; 0.3 mg/l or for fisheries: 0.005; 0.001; 0.001; 0.01; 0.01; 0.005 mg/l [1].

At present, the quality of wastewater treatment at machine-building enterprises and disposal of the resulting sludge does not meet modern technical, economic and environmental requirements (100 roubles of gross output are associated with waste generation for 2,700 roubles) (environmental damage) and water losses of up to 80%).

Water supply as a structure cannot be considered separately from other resource consuming systems in mechanical engineering therefore it is necessary to develop new or to upgrade existing technological solutions to eliminate existing problems at mechanical engineering enterprises.

Wastewater of electrodeposition industry is primarily polluted with electrolytes, whose chemical composition during electrodeposition must ensure the formation of quality metal coatings, which is achieved by using complex compounds in the presence of various organic additives [2].

The aim of the research was to study the qualitative composition of wastewater and the structural composition of the molecules of pollutant compounds containing heavy metal ions. Model and real solutions were studied over a 12-month period.

This article presents the study results aimed at identifying the causes of low efficiency of wastewater treatment by the reagent method.

The main attention was paid to the study of complex compounds formed in the wastewater of electrodeposition industry.

Main part

Solutions of complex compounds can be in four types of equilibrium: a) dissociation into complex and external ions; b) dissociation of complex ions (or substitution of ligands with solvent molecules); c) dissociation of ligands; d) redox equilibrium (the complex-forming agent exhibits a variable degree of oxidation).

The dissociation equilibriums of complex compounds into complex and external ions are called ionic equals

Ionic equilibriums are subject to patterns of behaviour in strong electrolyte solutions:

$$[Cr(H_2O)_4Cl_2]Cl \leftrightarrows Cr(H_2O)_4Cl_2]^{+} + Cl^{-}$$

$$[Cu(NH_3)_4](OH)_2 \leftrightarrows [Cu(NH_3)_4]^{2+} + 2OH^{-}$$

$$Na_2[Zn(OH)_4] \leftrightarrows 2Na^{+} + [Zn(OH)_4]^{2-}$$

$$K_3[Fe\ (CN)_6] \leftrightarrows 3K^{+} + [Fe\ (CN)_6]^{3-}$$

Under this scheme, only complex compounds - non-electrolytes - are not dissociated: $[Cd(NH_3)_3(N0_2)_3]$, $[Ni(NH_3)_2Cl_2]$, $[Zn(NH_3)_2Cl_2]$ and others.

Complex and external ions are easily detected by chemical analysis methods. Unfortunately, the constituents of complex ions are either detected by conventional chemical analysis methods using the most sensitive reagents or are not detected at all. The addition of potassium ferricyanide $K_3[Fe(CN)_6]$ potassium rhodanate or ammonium rhodanide to the solution does not cause the blood-red colouring inherent in $Fe(SCN)_3$ iron rhodanide, as the Fe^{3+} ions in the solution are complex and the reaction is virtually nonexistent. Also in the solution of potassium ferricyanide a complex ion $[Fe\ (CN)_6]^{3-}$ by the formation of ferrous ferricyanide $Fe_3[Fe(CN)_6]_2$ is easily detected in the presence of Fe^{3+} ions:

$$3FeSO_4 + 2K_3\ [Fe\ (CN)\ 6],\ Fe_3\ [Fe\ (CN)6] \downarrow + 3K_2SO_4.$$

To detect complex ions in a solution of $K_4\ [Fe\ (CN)_6]$ yellow prussiate of potash, one of the following reactions must be performed:

$$3K_4[Fe(CN)6]+4FeCl_3 = 12KCl + Fe_4[Fe(CN)6]_3\downarrow \text{ Blue residuum}$$

$$K_4[Fe(CN)_6]+2CuSO_4=2K_2SO_4 + Cu_2[Fe(CN)6]\downarrow \text{ Fallow residuum}$$

The nature of external ions and their quantity in a complex compound molecule can be determined by ion equilibriums. For example, the interaction of 1 mole of cobalt luteosalt $[Co(NH_3)_5Cl]Cl_2$ with silver nitrate produces 2 mole of silver chloride [3-4]:

$$[Co(NH_3)_5Cl]Cl_2 + 2AgNO_3 = [Co(NH_3)_5Cl](NO_3)_2 + 2AgCl\downarrow$$

This is confirmed by the presence of two chlorine ions (out of three) in the outer sphere.

The number of ions when a molecule of a given complex compound is dissociated can be determined by the size of its molecular electrical conductivity.

Solvatation equilibrium is an exchange reaction in which a certain number of molecules or ligand ions of the initial complex exchange with solvent molecules. Thus ion $[Cu(NH_3)_4]^{2+}$ in an aqueous solution exists in equilibrium with ion $[Cu(H2O)4]^{2+}$ and ammonia:

$$[Cu(NH_3)_4]^{2+} + 4H_2O \leftrightarrows [Cu(H_2O)_4]^{2+} + 4NH_3$$

This equilibrium can also be described as an exchange reaction under which a certain number of solvent molecules in a solvated elementary ion are exchanged for ions or ligand molecules.

Solvatation equilibrium is a particular case where one ligand is replaced by another. Such processes take place in stages and all intermediate compounds can be found in the solution:

$$[Cr(H_2O)_4Cl_2]^{+} + H_2O \leftrightarrows [Cr\,(H_2O)_5Cl]^{2+} + Cl^-$$

$$[Cr(H_2O)_5Cl]^{2+} + H_2O \leftrightarrows [Cr\,(H_2O)_6]^{3+} + Cl^-$$

The implementation of this equilibrium leads to the appearance in the solution of sometimes immeasurably small and sometimes significant amounts of ligands and hydrated elementary ions of the metal complexing agent [5].

The so-called elementary ions actually always exist in aqueous solutions in hydrated form (ion hydrates). Their presence can easily be proved by simple experiments observing a change of salts colour when the solvent is replaced. The chlorine and bromine copper solutions in acetone are herbaceous green and fallow respectively. Due to the formation of hydrated copper ions in the solution $[Cu(H2O)_4]^{2+}$ the water addition to these solutions causes blue colouring. Anhydrous cobalt chloride and its solution in alcohol are blue (colour of the Co^{2+} ion) and its aqueous solution is pink (colour of the $[Co(H_2O)_6]_{24}$ ion).

Based on the assumption that the ligands replacement in the complex ion by water molecules leads to the formation of a hydrated complex ion and ligands the solvatation equilibrium can be represented as the equilibrium of the electrolytic dissociation of the complex ion [6-7].

If more sensitive reagents are used it is possible to detect complexion-forming ions that cannot be determined by conventional qualitative reactions in saline.

The dissociation of complex ions usually occurs to an insignificant extent subject to the law of masses action and is characterized by the dissociation constant on the quantitative side. For equilibrium:

$$[Ag(NH_3)^2]^+ \leftrightarrows Ag^+ + 2NH_3$$

$$K_{dis} = [Ag^+\,NH_3]_2 / [Ag(NH_3)^2]^+ = 4\cdot 10^{-7}$$

Ion dissociation equation $[Ag(NH_3)_2]+$ must be written as: $[Ag(NH_3)^2]^+ + 2H_2O \leftrightarrows [Ag\,(H_2O)_2]^+ + 2NH_3$ and therefore an expression for the balance constant:

$$K = [Ag^+(H_2O)_2][NH_3]_2 / [Ag(NH_3)_2][H_2O]_2$$

However, the effective water mass in diluted solutions of the complexes is so large that one can neglect to calculate the relative values of dissociation constants by ion hydration.

For equilibrium:

$$[Ag(Cl^-)_2]^- \leftrightarrows Ag^+ + 2Cl^-$$

$$K_{dis} = [Ag^+][Cl^-]_2 / [Ag(Cl^-)_2]^+ = 1\cdot 10^{-21}$$

As the value of dissociation constants characterizes the stability of ions in the solution, they are called complex instability constants (Inco) (table 1). The less stable the complex in the solution the larger is the Inco.

Table 1. Instability constants of complex ions

Complex ion dissociation	Instability constant at room temperature
$[Ag(Cl^{-})2]^{+} \leftrightarrows Ag^{+} + 2Cl^{-}$	$1.0 \cdot 10^{-21}$
$[Ag(NH3)2]^{+} \leftrightarrows Ag^{+} + 2NH3$	$4.0 \cdot 10^{-7}$
$[AlF6]^{3-} \leftrightarrows Al^{3+} + \leftrightarrows 6F^{-}$	$2.0 \cdot 10^{-28}$
$[Cd(NH3)6^{2+} \leftrightarrows Cd^{2+} + 6NH3$	$8.0 \cdot 10^{-6}$
$[Cd(NH3)6]^{3+} \leftrightarrows Cd^{3+} + 6NH3$	$8.0 \cdot 10^{-36}$
$[Cu(NH3)_4]_2+ \leftrightarrows Cu^{2+} + 4NH_3$	$4.6 \cdot 10^{-14}$
$[Fe(Cl^{-})_6]_3- \leftrightarrows Fe^{3+} + 6Cl^{-}$	$1.0 \cdot 10^{-44}$
$[Fe(Cl^{-})_6]_4- \leftrightarrows Fe^{2+} + 6Cl^{-}$	$1.0 \cdot 10^{-37}$
$[PtCl_4]_2 \leftrightarrows Pt^{2+} + 4Cl^{-}$	$1.0 \cdot 10^{-16}$
$[StF_6]_2 \leftrightarrows StF_4 + 2F^{-}$	$6.3 \cdot 10^{-8}$
$[Zn(NH_3)_4]_2+ \leftrightarrows Zn^{2+} + 4NH_3$	$4.0 \cdot 10^{-10}$
$[Zn(CN)_4]_2- \leftrightarrows Zn^{2+} + 4Cl^{-}$	$6.3 \cdot 10^{-18}$
$[NH_4]^{+} \leftrightarrows NH_3 + H^{+}$	$6.0 \cdot 10^{-10}$

The dissociation of complex ions always runs in stages so the data in Table 1 only assess the total stability of the internal sphere of complexes [8].

All the constituent parts of a molecule can be easily detected in the solution of a complex compound with a high instability constant. For example, in alum solutions $KAI(SO_4)_2 \cdot 12H_2O$, $NH_4Fe(SO_4)_2 \cdot 12H_2O$ along with ions $[A1(SO_4)_2]^{-}$ and $[Fe(SO_4)_2]^{-}$ available in small concentrations, ions can be easily detected $A1^{3+}$, SO_4^{2-}, Fe^{3+}, K^{+} и $NH4^{+}$. Ions Ba^{2+} precipitate ions from these solutions SO_4^{2-}. Potassium or ammonium rhodanide causes encarnadine colouring of the iron alum solution. When alkali is added to the alum solution aluminium hydroxide sludge is precipitated and dissolved in excess alkali. When alkali is added to a solution of iron ammonium alum, fallow $Fe(OH)_3$ iron hydroxide sludge is precipitated and ammonia is released. When complex ions are dissociated the equilibrium is strongly shifted to the right:

$$[A1(SO_4)_2]^{-} \leftrightarrows A1^{3+} + 2SO_4^{2-}$$

$$[Fe(SO_4)_2]^{-} \leftrightarrows Fe^{3+} + 2SO_4^{2-}$$

The ions stability of the type $[A1(SO_4)_2]^{-}$ in crystals and their dissociation in the solution makes it possible to consider such substances as alum, ferrous ammonium sulfate and other double salts as complex compounds with a high value of the instability constant. Table 1 shows that for typical complex compounds the instability constant is small, i.e. the complex ion is a poorly dissociated particle, a weak electrolyte. For comparison, the values of dissociation constants at room temperature of weak acids and bases: K_{dis} of vinegar acid is 1.86-10-5, K_{dis} of carbon dioxide is $3\text{-}10^{-7}$, and K_{dis} of ammonium oxide hydrate is 1.75^{-10-5}.

According to Berthollet's rule, ion reactions are virtually irreversible if the ion interaction produces a substance that is removed from the chemical interaction sphere in the form of gas or vapour or if it is difficult to decompose at least one of the reaction products into the source ions (as a result of sludge deposition in the formation of a poor dissociating substance). From this point of view, the equilibrium shift of ionic reactions towards the formation of complex compounds becomes clear. The formation of complex compounds in the solution is the formation of complex ions - poorly dissociated particles. Using the example of a reaction between potassium cyanide and iron cyanide, which is practically irreversible due to the formation of a weak ion electrolyte, let us consider complex formation $[Fe(Cl)_6]^{4-}$

$$Fe(Cl)_2+4KCl = K_4[Fe(Cl)_6]$$

When potassium fluoride is added a yellow solution of iron chloride discolours due to the formation of a complex ion $[FeF_6]^{3-}$:

$$FeCl_3 + 6KF = K_3[FeF_6] + 3KCl$$
$$Fe^{3+} + 6F- = [FeF_6]^{3-}$$

When a concentrated solution of potassium rhodanide is added to a small amount of cobalt chloride solution a blue colouring appears due to the complex formation $[Cd(ClS)_4]^{2-}$

$$CdCl_2 + 4KClS = K_2[Cd(ClS)_4] + 2KCl$$

$$Cd^{2+} + 4ClS- = [Cd(ClS)_4]^{2-}$$

The balance shift of the reaction towards the formation of weak electrolytes is accompanied by the dissolution of the precipitate. Adding an ammonia solution to the silver chloride sludge results in the formation of soluble complex silver ammonia:

$$AgCl + 2NH_4OH = [Ag(NH_3)_2]^+ + Cl^- + H_2O$$

If a small amount of potassium iodide is added to the silver salt solution, yellow precipitate of silver iodide falls out:

$$AgNO_3 + KI = AgI\downarrow + KNO_3$$

$$Ag^+ + I^- = AgI$$

If there is excess potassium iodide, the silver iodide dissolves to form a complex:

$$AgI + KI = K[AgI_2]$$

$$AgI + I^- = [AgI_2]^-$$

Under the action of ammonia, the blue copper hydroxide sludge is transformed into a soluble complex base of blue:

$$Cu(OH)_2 + 4NH_4OH = [Cu(NH_3)_4](OH)_2 + 4H_2O$$

$$Cu(OH)_2 + 4NH_4OH = [Cu(NH_3)_4]^{2+} + 2OH^- + 4H_2O$$

The shift in the reaction equilibrium towards the formation of weaker electrolytes explains the destruction of some complexes and the formation of others with a lower instability constant. Therefore, if acid is used on the silver ammonia complex a reaction occurs with the formation of a new more stable complex - NH_4^+:ion:

$$[Ag(NH3)2]Cl + 2HNO3 = 2NH4NO3 + AgCl\downarrow$$

$$[Ag(NH_3)_2]^+ + Cl^- + 2H^+ + 2NO_3^- = 2NH_4^+ + +2NO_3^- + AgCl\downarrow$$

$$[Ag(NH3)2]^+ + Cl^- + 2H^+ = 2NH4^+ + AgCl\downarrow$$

The reaction is directed towards the formation of NH ion, since ion I_{nco} $[Ag(NH_3)_2]^+$ is 4-10^{-7} and I_{nco} ion $[NH_4]^+$ is $6\cdot10^{-10}$.

Resistance constants objectively characterize the strength of complex ions in the solution [9]. When assessing the complexes stability it is often not instability constants that are used for convenience but rather indices of instability constants equal to the negative, decimal logarithm of the instability constant, pC = – lgI_{nco}.

$$C_{Inco}\ of\ ion\ [Fe(Cl)_6]^{3-} = 1{,}0\cdot10^{-44},\ a\ pC = 44.$$

The higher the pC value of the complex the more stable the complex in the solution. The complex stability in the solution in relation to the action of the solvent primarily depends on the strength and nature of the bond between the ligands and the central atom. The strength of this bond depends on the nature of the central atom, the degree of its oxidation, the size and structure of its electronic shells and the nature of the ligands. The smaller the radius of the ion complex and the higher its charge, the more robust the complexes it is capable of forming. In this way the complex ion $Cd(NH_3)_6]^{2+}$ is reversibly broken down in water:

$$[Cd(NH_3)_6]^{2+} + 6H_2O \leftrightarrows [Cd(H_2O)_6]^{2+} + 6NH_3$$

While the complex ion $[Cd(NH_3)_6]^{3+}$ is virtually indestructible with water.

The influence of the central ion charge on the stability of complexes is reflected in the values of the instability constants of complex ions formed by central ions with close radii and similar electronic structures but differing in their charge size. For example one group of such complex-forming ions is Na^+, Ca^{2+}, I^{3+}, Th^{4+} (r = 1,04±0,06 A) and another group is Ag^+, Hg^{2+}, Tl^{3+} (r - 1,09±0,04 A) [10].

Table 2 shows the indicators values of instability constants of complex ions formed by the above-mentioned complex formers. These results were obtained by K.B. Yatsimirsky and V. P. Vasilyev.

Table 2 shows that the stability of complex ions increases with the charge of the complex ion-forming agent. The complexes strength in the solution is significantly affected by the ratio between the size of ligands and the central ion. Small cations form the strongest complexes with small anions while cations with a large radius form the strongest complexes with large anions. In the case of small single-charged anions (anion radius < 1.6A), the complex stability decreases with increasing cation size. In addition, in the case of large anions (nitrate, iodate, thiosulphate) the stability of the complexes they form increases with increasing cation size. If the central ion in the formation of a complex compound shows the maximum coordination number then a stronger complex is formed than in the case of a lower coordination number.

The stability of complex compounds in solutions depends on the size of the coordination number of the central atom and its electronic structure. The strongest complex compounds form transition metals that can be used to communicate with ligands, in addition to s- and p-orbitals of the outer shell, another d-orbitals of the lower electron shell. The formation of strong bonds between the central atom and the ligands is the result of this hybridisation.

Table 2. Indicators of instability complexes constants

Types of complex ions	Complexing ions			
	Na^+	Ca^{2+}	I^{3+}	Th^{4+}
RIO_3^{z-1}	—	0.9	—	2.9
RNO_3^{z-1}	—	0.3	—	0.6

RSO_4^{z-2}	0.7	2.3	3.5	4.1
$RC_2O_4^{z-2}$	—	3.0	7.3	—
	Ag^+	Hg^{2+}	Tl^{3+}	—
ROH^{z-1}	2.3	10.3	14.8	—
RNH_3^{z}	3.2	8.8	—	—
RCl^{z-1}	2.7	5.3	8.1	—
RBr^{z-1}	9	9.1	9.7	—

R is the central ion.

Z is the charge of the central ion.

The nature and properties of the ligands affect the stability of complex compounds in the same way as the nature and properties of central atoms. The strongest complexes form ligands with large charges and small radii. Strong complexes are also formed when the substituent has a large coordination capacity, i.e. when the inner sphere is cyclogenesis. At the same time five- and six-year cycles are the most stable. For example, C_{Inco} of complex ammonia cadmium $[Cd(NH_3)_4]^{2+}$ is equal to $3.3 \cdot 10^{-7}$, while C_{Inco} of Ethylene diamine cadmium complex is only $6.7 \cdot 10^{-13}$. This is explained by the fact that when ethylene diamine $H_2N–CH_2–CH_2—NH_2$ is coordinated, a five-member cycle is formed in the inner sphere. The strength of transition metal complexes also depends on the strength of the ligand crystal field [11].

When comparing the complexes stability, it is necessary to take into account the mutual influence of intraspherical substituents and their different mobility, which is determined by the law of transfluence.

This study is based on the quantitative, qualitative, wastewater composition, the structural composition of pollutant molecules containing heavy metal ions and the physical and chemical properties of the solutions under study (density, electrical conductivity, surface tension, viscosity). The study was conducted over a 12-month period on both the model solutions and the actual runoffs. It was found that electrodeposition wastewater is primarily polluted with electrolyte technological solutions, which chemical composition should ensure the formation of high-quality metal coatings during electrodeposition. This is achieved by introducing various organic additives into the solution. Therefore in electroplatings wastes heavy metal ions primarily for this reason are part of complex ions, as evidenced by the data obtained (see Table 1 and Table 2), which shows a discrepancy between the concentration values of metal ions taken for the implementation of the reagent method and the true ones, since pollutant ions are part of molecular complex structures in a related form, making it difficult for them to participate in the deposition process.

Evidence has also been obtained from an independent laboratory confirming the presence of three stable backgrounds: anionic, cationic and organic, which enhance the formation of complex compounds in the wastewater of electrodeposition industry [12].

Average composition (mg/l): average concentration value (12 measurements): pH = 6.66...8.5 (before treatment), pH = 7.07...7.48 (after treatment). t = 10...18 °C. Organic substances with concentrations > 1 %: SAS - sulphanols, soaps, inhibitors, dextrin, SAA -7, SAA -10, thiourea, PEPA, phthalenide. The total contaminants concentration is 17.0...19.0 g/l.

If there is no complex formation in the solution (solution No. 4) then the deposition results in concentrations close to the corresponding values of solution No. 3 (see Table 2). In solutions No. 1 and No. 2 (see Table 2) the change value in indicators is within 20% that is consistent with the values of the instability constant of complex ions, which include ions of heavy metal pollutants of the considered electroplatings wastes: pC $[Cu(P_2O_7)]^{2-}$ = 26,72; pC $[Cd(P_2O_7)]^{2-}$ = 18,4; pC $[Ni(H_2O)_4 \cdot SO_4]$ = 15.34;

pC $[ZnCl_4]^{2-}$ = 12.24; pC $[nCrO_3{\cdot}mH_2O]$ = 17.8. Their dissociation degree does not exceed 14...17%, which confirms the presence of complex ions. This conclusion can also be confirmed by the values of the solubility products of the corresponding ions: $Cu(OH)_2$, $Cd(OH)_2$, $Ni(OH)_2$, $Cr(OH)_2$, $Zn(OH)_2$.

The obtained data formed the basis of the determining methodology the treatment effectiveness of the considered effluents by the known methods: ion exchange, electrolysis, galvanic coagulation, sorption, etc. It was found that the efficiency of the method is related to the molecular mass of pollutants, surface tension value, electrical conductivity, density of the solutions being purified.

Summary

The obtained data on complexing, confirmed that the use of the reagent method for cleaning electroplatings wastes is not effective since there is a true values distortion of the heavy metal ions concentration; more than 80% of them are in the composition of weakly dissociating complex ions. Also experimentally confirmed the negative influence of complexation processes on the purification depth of electroplatings wastes by methods (electrochemical, membrane, coagulation), which technical essence is due to the structure, size, harge value of the pollutant particles.

References

[1] Olyansky Yu I, Bolev A A, Sakharova A A, Ignatkina D O, Yurin P F, Voytyuk A A 2013 Method of increasing the reliability functioning of water management systems *Internet Bulletin VSTU Series: Polythematic* **2 (27)**. Information on http://www.vestnik.vgasu.ru/

[2] Ryltseva T F, Ryltsev V V, Prikhodchenko A V, Boleev A A, Potolovsky R V 2012 On the issue of water quality system *Bulletin of the Volgograd State Architectural University. Series: Building and architecture* **26 (45)** 138-144.

[3] Moskvicheva E V, Komarov A Yu, Romanov V A, Sakharova N A, Shchelochkova A A 2010 Water saving in communal systems on the example of Volgograd *Internet Bulletin VSTU. Series: Polythematic* **2 (12) 5**. Information on http://www.vestnik.vgasu.ru/

[4] Nechaev E A, Pavlichenko V A, Belousova N A, Silina T F 1986 Relationship of instability constants of heavy metal complexes with half-wave potentials of oxidation of sulfur-containing reagents *Electrochemistry* **22 (3)** 320-324.

[5] Alykov N M, Alykova T V, Pashchenko K P 2002 Quantum-chemical cluster modeling of adsorption of aromatic hydrocarbons *Natural Sciences Journal of Basic and Applied Research* **4** 2006-2014.

[6] Kramer 1938 Die Hartvechromung *Metal Industry und Galvanotechnic* **33(19)** 411-42.

[7] Galkin Yu A, Lotosh V E 1990 Technology of utilization of sewage sludge from machine-building enterprises *Chemistry and technology of water* **12 (6)**.

[8] Moskvicheva E V 1972 Resource-saving processes as the basis for environmentally friendly technologies of galvanic chromium plating from aqueous and non-aqueous media *Dokt, Diss. Moscow.*

[9] Glikina F B 1972 Chemistry of complex compounds (Moscow, Education) 170

[10] Ignatkina D O, Voityuk A A, Moskvicheva A V, Moskvicheva E V, Seliverstov I V 2018 Device for wastewater treatment from multicomponent pollution *Patent RF no 178983.*

[11] Samokhina V N 1981 *Sewerage of settlements and industrial enterprises Revised and enlarged* (Moscow, Stroyizdat) 639

[12] Khan V A, Lerner M I, Myshkin F D, Tskhe A A 2013 Development of a complex for non-reagent water purification *Scientific journal KubGAU* **86 (02)**. Information on http://ej.kubagro.ru/2013/02/pdf/55.pdf. IDA 0861302055

Improvement of environmental and economic indicators of the electrolytic chromium plating process

E V Moskvicheva[1*], A G Timofeev[1], P A Sidyakin[2] and L N Chernoshchekov[3]

[1]Volgograd State Technical University (VSTU), 1 Akademicheskaya street, Volgograd, 40007, Russia
[2] "North Caucasian Federal University" Institute of service, tourism and design (branch) NCFU in Pyatigorsk, 1 Pushkin street, Stavropol, 355009, Russia
[3] Lomonosov Moscow State University, 1Leninskie Gory, Moscow, 119991, Russia

E-mail: antonio1805@yandex.ru

Abstract. In most cases, the improvement of technological processes is a beneficial measure to reduce anthropogenic impact of production on the environment. However, it is impossible to solve the problem of elimination of production emissions and wastes only by improving the technological process. That is why more and more integrated and complex measures are required to solve the problem, which include both improvement of technological processes and improvement of emissions purification schemes, waste utilization and recycling. At the same time, the measures should be both technically effective and economically feasible. The obtained results allowed to draw a conclusion about the increase of ecological and economic indicators: decrease of toxic chromium anhydride wear and tear by 2-2,7 times; decrease of water volume polluted in the process of technological cycle; decrease of expenses for lead anodes manufacturing and human resources; increase of operational characteristics of manufactured parts - wear resistance, microhardness.

Introduction

Currently, considerable attention is paid to environmental safety issues at hazardous facilities. This interest is stimulated by the unfavorable environmental situation in regions with a high concentration of production facilities that irrationally eliminate their waste as well as by the significant and gradually growing fines paid by companies for hazardous emissions into the atmosphere and industrial waste landfills.

The main solution to environmental problems is to reduce anthropogenic impact of industrial enterprises on the environment. The result can be achieved through the following activities:

- improvement of technological processes;
- improvement of emission purification methods and waste disposal.

Improvement of technological processes leads to improvement of technical indicators (for example reduction of energy intensity, amount reduction of raw materials or product quality improvement) [1].

In most cases, improvement of technological processes is an economically beneficial measure to reduce the anthropogenic production impact on the environment. However, it is impossible to solve the problem of elimination of production emissions and waste only by improving the technological process.

Therefore, its solution requires more complicated and complex measures, which include both improvement of technological processes and improvement of emissions purification schemes, utilization and waste processing. At the same time, the measures should be both technically effective and economically feasible.

Further analysis of decisions efficiency that reduce the anthropogenic impact will be carried out based on measures to improve electrolytic chromium plating production. [2]

At present, the technology of electrolytic chromium plating is improved by changing the composition of chromium plating electrolyte. Chroming electrolyte with thiophosphorus additive (TFA) implemented at the plant allowed:

- to reduce energy intensity of chroming process, as well as chromic anhydride entrainment in 2 - 3 times due to decrease of electrolysis temperature from 55 to 35 °C;
- eliminate the use of special anodes that repeat the surface of parts during their dimensional chroming;
- obtain chrome plates with improved physical and chemical properties (wear resistance increased by 4-5 times, microhardness increased by 1.1-1.2 as compared to currently used standard chrome electrolyte). Based on these data the environmental and economic indicators of the chroming process from the chroming electrolyte with TFA are comparable with the similar indicators of the chrome bath from standard electrolyte. Before conducting such calculation, it should be noted one more economic benefit from implementation of chroming electrolyte with TFA. There are three electroplating-chrome baths at this enterprise; each of them has 1000 liters volume. The baths are powered individually - each bath from its rectifier with a nominal current of 1500 A.

Main part

Originally, the project chromium plating area was designed for application of chrome plates on small parts (handles, caps for ball valves). Currently, the demand for such parts is not high, so there are only two baths of chromium plating in operation. Marketing research has shown that there is a demand for chrome plated ball valve plugs of a larger size (hereinafter - plugs). The plug of such a ball valve has an area of 1.08 m².

For chrome plating from standard electrolyte with minimum possible working current density of 45 A/dm², which provides obtaining chrome plates with high values of microhardness, the surge current (1.5 times higher than the working current) will make up:

$$J_t = 1.5 \bullet 45\ \mathrm{A/dm}^2 \bullet 1.08\ \mathrm{m}^2 \bullet 100 = 7300\ \mathrm{A}$$

The last value is 4 - 5 times higher than the nominal value of current strength, for which the current source available at the enterprise is designed. In addition, the bulk density of current in the chrome bath will be:

$$J_v = \frac{7300\ \mathrm{A}}{1.5 - 1000\ l} = 4.9\ \mathrm{A/l},$$

This will cause the electroplating-chrome bath to overheat, as the recommended volume current density for chrome plating is 1-1.5 A/l.

As a result, to chrome-plate a plug made of standard electrolyte it is necessary to replace both the power supply and the electroplating-chrome bath. However, such reconstruction as calculations show due to low production volumes (today the market demand is about 2100 plugs per year) is not profitable.

When chromium plating a plug made of chrome plating electrolyte with TFF, the operating current density can be reduced to 8 A/dm² therefore the surge current will be equal:

$J_t = 1.5 \bullet 8\ \mathrm{A/dm}^2 \bullet 1.08\ \mathrm{m}^2 \bullet 100 = 1300\ \mathrm{A}$, and the volume current density in the process of bath operation will be 0.86 A/l. Thus, the introduction of chrome plating electrolyte with TFA allows to chromium plating the plug without replacing the existing equipment. The calculation also shows that with the required thickness of chrome plates 20 microns, the electroplating-chrome bath will provide the required capacity of 2100 plugs per year, provided its continuous operation [3].

Specialists of the research department of the machine building plant for the implementation of chromium plating electrolyte with TFA made a number of changes in the design of the electroplating-chrome bath. First, they changed the mutual arrangement of lead anodes in the bath, which is necessary for chromium plating one part in it. Secondly, the bath lead cover with expired life was replaced by a by a vinyl plastics cover. For reasons of the low thermal vinyl conductivity, heating and cooling of the electrolyte began to be carried out with a coil of lead pipes lowered into the electrolyte and located on the bath walls. In such a lining not only eliminates the need for lining the bath with lead but also improves the coating uniformity as at least partially excluded the influence of the bath walls and bottom on the passage of current between the anode and cathode [4].

The carried out researches allowed drawing a conclusion that the process of chromium electrodeposition from electrolyte with TFA is most expedient at temperature 35 °C and current density 5-10 A/dm^2.

Table 1 presents properties of chromium plating electrolyte composition, (g/l): 250 CgO_3 + 2.5 H_2SO_4 + 1.75 TFA, and physical and mechanical properties of coatings obtained from it.

The conducted researches showed that the increase in current density up to 10 A/dm^2 causes the formation of burn-on along the chrome plates edges. Thus, apparently the optimal mode of chrome plating is electrolyte temperature of 35 °C and current density of 7-8 A/dm^2 [5].

The question of chromium plating electrolyte correction is important from practical point of view. Carried out researches have shown that TFA decomposes slowly at storage (and consequently at operation) of chromium plating electrolyte with formation of orthophosphoric acid. Calculations showed that chromium-plating electrolyte does not require correction of orthophosphoric acid during the whole period of its operation. Correction for TFA should be made in every 250 Ah/l on condition of continuous operation of electroplating-chrome bath [6].

Adjustment of the electrolyte for FTA is made by adding the necessary amount of "concentrate" to it. The concentrate is a liquid product of chromic anhydride reaction with diisobutyldithiophosphoric acid. To prepare the concentrate of diisobutyldithiophosphoric acid in the amount of 70 g / l is fed into an aqueous solution of chromium anhydride with a concentration of 500 g / l at a temperature of 55 °C then the reaction mass is maintained for 240 minutes at the same temperature to complete the reaction, after which it is cooled. Solid reaction products are separated from liquid ones and washed with water in the amount of 60-70% of the volume of chromic anhydride initial solution. Washing water is combined with liquid reaction products and as a result, concentrate is obtained. The carried out researches showed that the study of chromium plating electrolyte after its TFA correction can be not carried out, if TFA content in the bath before correction corresponds to the working interval of TFA concentrations [7].

The working interval of TFA concentrations depends on chromic anhydride and sulfuric acid concentrations. Decrease of chromic anhydride concentration in electrolyte from 250 g/l to 150 g/l leads to some reduction of working interval of TFA concentrations; increase of chromic anhydride concentration to 350 g/l does not change working interval of TFA concentrations. Thus, it can be considered that the optimal concentration of chromic anhydride in CgO_3 electrolyte is ~ 250 g/l. At this concentration of chromic anhydride the working interval of TFA concentrations, at which shiny chrome plates with microhardness of 11000 - 12600 MPa are obtained can be found from Fig.1, knowing the ratio between the concentration of sulfuric acid and chromic anhydride in the chromic electrolyte. At reception of the hardest chrome coverings (with microhardness not less than 12000 MPa) it is necessary to narrow the working interval of concentrations of TFA and H_2SO_4 (Figure 2) [8].

High wear resistance of chrome plates is explained by the fact that the irradiated chromium coatings have a high value of microhardness as well as are fine crystalline, which is confirmed by X-ray phase analysis presented in Figure 1.

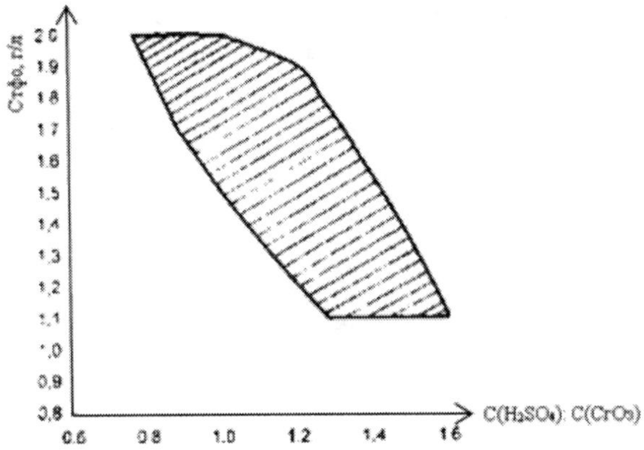

Figure 1. Working interval of TFA concentrations as a function of H_2SO_4 ratio: CrO_3, Anode - lead. $C(CrO_3)$ concentration = 250 ± 30 g/l. $C(H_2SO_4)$ concentration:100/$C(CrO_3)$ ratio = 0.76:1.6. Concentration of TFA 1.1 - 2.0 g/l. Electrolyte temperature is 35 °C. Cathode current density is 8 A/dm^2. Microhardness of chrome plates 11000 - 12600 MPa.

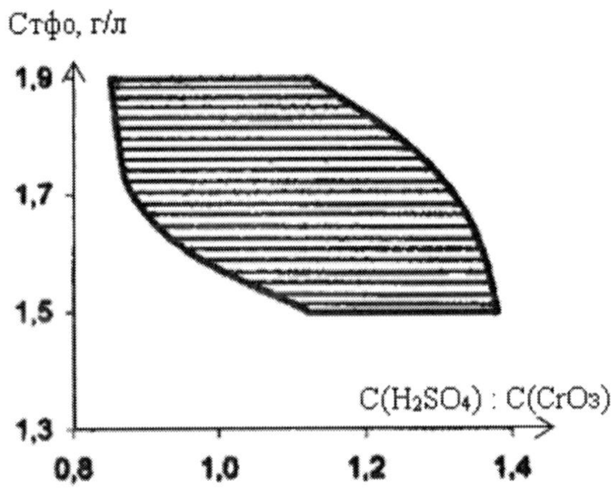

Figure 2. Operating interval of TFA concentrations as a function of H_2SO_4: CrO_3 ratio. CrO_3.concentration = 250 ± 30 g/l. $C(H_2SO_4)$:100/ $C(CrO_3)$ ratio = 0.85:1.38. Concentration of TFA 1.5 - 1.9 g/l. Anode - lead. Electrolyte temperature is 35 °C. Cathode current density is 8 A/dm. Microhardness of chrome plates 12000 - 12600 MPa.

Figures 1, 2 show that decrease (increase) of TFA concentration in chromium plating electrolyte can be compensated by increase (decrease) of sulfuric acid concentration within the working concentration interval.

Determination of physical-mechanical dependence properties of chrome plates and physical properties of chromium plating electrolytes depending on its composition allowed to reveal optimal composition of chromium plating electrolyte, g/l: CrO_3+ 2,5 H_2SO_4 + 1,75 TFA.

Figure 2 shows that in chromium plating electrolyte that contains 250 g/l of CrO_3, 2.5 g/l of H_2SO_4 the working interval of TFA concentrations is 1.6 - 1.9 g/l.

Operation experience of chromium plating electrolyte with TFA at the electrodeposition showed no difficulties connected with its correction in spite of narrow interval of working concentrations of TFA. The above mentioned achievement is a significant positive result with improvement of environmental and economic indicators. Electrolyte adjustment is accompanied not only by contamination of water effluents, but also by an increase in their volume, which has a very negative impact on the environmental safety of the production under consideration.

It is connected with the fact that low temperature of electrolysis leads to decrease of chromium plating electrolyte entrainment that stabilizes its composition and in the process of chromium plating electrolyte operation there is not only decrease of TFA concentration but also increase of $C(H_2SO_4)/C(CrO_3)$ ratio due to chrome anhydride concentration decrease [9].

This fact, taking into account Figure 2 allows to operate chromium plating electrolyte at wider concentration range of 1.5 - 1.9 g/l TFA.

Concentration of TFA STFA in g/l. In the chromium plating electrolyte is calculated, determining the total phosphorus content of $P_{comm.}$ in % and phosphorus content in the form of phosphate ions $P_{ph.}$ in % by the developed method.

$C(TFA)=324.3 - p - (P_{comm.} - P_{ph.})$;

Where p is the density of the analyzed chromium plating electrolyte, g/ml.

The concentration of sulfate ions in chromium plating electrolyte is determined by the developed method. Determination of chromium anhydride concentration is similar to its determination in standard chromium plating electrolyte. Adjustment of chromic anhydride and sulfuric acid electrolyte is performed in the same way as in standard chromium plating electrolyte. As in chromium plating electrolyte at correction according to TFA concentrate is added which contains chrome in recalculation to chromic anhydride in amount of 7,43 g CrO_3, it should be taken into account at the electrolyte correction according to the chromic anhydride [10].

Summary

Based on the obtained results, the conclusions were formed showing a demonstrative improvement in environmental and economic indicators of the production under consideration - electrodeposited chrome plating:

- removal of toxic chromium anhydride was reduced by 2-2.7 times;
- the volume of water contaminated during the technological cycle was reduced;
- reduced costs for lead anode production and human resources;
- improved performance of manufactured parts - wear resistance, microhardness.

As a result, the presented results have proved the expediency of the conducted research when the technogenic pollution can be reduced and the quality of the main products can be improved.

References

[1] Yuriev Yu, Sidyakin P A, Shchitov D M, Tatov A S 2020 Development of engineering decisions to reduce technological exposure to the environmental environment of oil flower *IOP Conference Series: Materials Science and Engineering. International Scientific Conference "Construction and Architecture: Theory and Practice of Innovative Development" (CATPID-2020). Part 1 (Nalchik, Russian Federation, 26-30 September 2020): International Scientific Conference on "Construction and Architecture: Theory and Practice of Innovative Development" (CATPID-2020)* **913** 8.

[2] Kramer 1938 Die Hartvechromung *Metal Industry und Galvanotechnic* **33 (19)** 411-42.

[3] Ignatkina D O, Voityuk A A, Fedulova E V 2019 Development of ecologically safe technology of the multi-component wastewater treatment for the food industry enterprises (in Russian) Actual problems of construction, housing and utilities and technosphere safety *Materials of VI All-Russia. Scientific technical conference of researchers, Volgograd, 22-27 April* 187-189.

[4] Matulis Yu 1980 Modern condition and prospects of development of electroplating equipment (in Russian) *Journal of the higher chemical school named after D.I.Mendeleev* **25 (2)** 122-126.

[5] Ignatkina D O, Voityuk A A, Moskvicheva A V, Moskvicheva E V, Seliverstov I V 2018 Device for wastewater treatment from multicomponent pollution. Patent RF no 178983.

IOP Conf. Series: Materials Science and Engineering 1083 (2021) 012095 doi:10.1088/1757-899X/1083/1/012095

Mud stream formation conditions and mud stream hazardous areas of the Chechen Republic

G A Sergeeva, V L Adamyan, E V Marinchenko, L A Seferyan*

Don State Technical University, 1, Gagarin sq., Rostov-on-Don, 344002, Russia

E-mail: Luiza.seferyan@mail.ru

Abstract. Mud streams in the mountainous zone are formed almost everywhere in the Chechen Republic: on the northern slope of the Main Ridge and its spurs, the Skalisty, Pastbischny, Lesisty, Terskiy, Sunzhensky ranges and in the depressions between them. The conditions for the mud streams formation and mud stream hazardous regions of the Chechen Republic are determined by orographic tectonic-geomorphological, geological-lithological, soil-plant, anthropogenic and especially hydrometeorological conditions and affect the mud streams with their indicators. The variety and complexity of the relief, as well as the temperature regime, are used to predict the mud stream threat and develop anti-mud stream measures. The authors of the article provide the following indicators: air temperature regime, also values of average monthly and annual air temperatures, monthly values of vertical air temperature gradients in the Chechen Republic territory, changes in annual precipitation (mm) with terrain altitude above the sea level on the northern slope of the Eastern Caucasus, schematic map of orohydrography and mud stream channels of the Chechen Republic on a scale -1: 500,000.

Introduction

In the Chechen Republic, mud streams are caused by orographic, tectonic-geomorphological, geological-lithological, soil-vegetation, anthropogenic and especially hydrometeorological conditions. Ehe latter directly affect mud streams by means of their following indicators: values and regime of air temperature, amount, composition, intensity of atmospheric precipitation, size and intensity of contemporary glaciation and snow cover melting, moisture content of the territory, presence of small rivers with large slopes of channels and their flood regime [1].

The variety and complexity of the relief, significant fluctuations in relative and absolute heights, combined with seasonal features of atmospheric circulation, create a great variety in the temperature regime on the territory of the Republic. In the high-altitude zone, the air temperature is also influenced by insignificant contemporary glaciation. In general, the typical air temperature is its increase from west to east, especially when considering its summer values, as well as its decrease with an increase in absolute heights for the territories under consideration (Tables 1, 2) [2].

Material and research methods

The intensity of the decrease in air temperature with altitude is not the same throughout the year. Throughout the territory under consideration, the smallest vertical air temperature gradients are observed in the cold period (X-IV), and the highest in the warm (V-IX); more significant gradients are observed in its eastern part (Table 3).

Table 1. Average monthly and annual air temperature (^{0}C) (western part of the Chechen Republic territory)

M Station	Height, m	I	II	III	IV	V	VI	VII	VIII
M Grozny	126	-3.5	-2.0	3.0	9.9	16.8	21.0	23.7	23.2
	500	-4.3	-2.4	2.8	8.3	14.8	18.8	21.1	21.0
	1000	-5.3	-2.9	2.6	6.3	12.3	16.0	17.8	18.2
	1500	-7.1	-5.1	-0.1	3.3	9.3	13.1	15.1	15.6
	2000	-8.9	-7.3	-2.8	0.3	6.3	10.2	12.4	13.1
	2500	-10.7	-9.5	-5.4	-2.7	3.3	7.3	9.7	10.6
	3000	-12.5	-11.7	-8.1	-5.7	0.3	4.4	7.0	8.1
	3500	-14.3	-13.9	-10.8	-8.7	-2.7	1.5	4.3	5.6
	4000	-26.1	-16.1	-13.4	-11.7	-5.7	-1.4	1.6	3.1

Table 1 extension.

M Station	Height, m	IX	X	XI	XII	X-IV	V-IX	Year
M Grozny	126	17.8	11.4	4.4	-0.6	3.2	20.5	10.4
	500	16.0	10.2	4.0	-0.6	2.6	18.3	9.1
	1000	13.8	8.7	3.5	-0.6	1.8	15.6	7.5
	1500	11.4	6.5	1.5	-2.1	-0.4	12.9	5.1
	2000	9.0	4.3	-0.5	-3.6	-2.6	10.2	2.7
	2500	6.6	2.1	-2.5	-5.1	-4.8	7.5	0.3
	3000	4.2	-0.1	-4.5	-6.6	-7.0	4.8	-2.1
	3500	1.8	-2.3	-6.5	-8.1	-9.2	2.1	-4.5
	4000	-0.6	-4.5	-8.5	-9.6	-11.4	-0.6	-6.9

Table 2. Average monthly and annual air temperature (^{0} C) (eastern part of the Chechen Republic territory)

M Station	Height, m	I	II	III	IV	V	VI	VII	VIII
	100	- 0.1	0.9	4.0	13.4	19.7	22.8	26.4	25.2
	500	- 0.9	0.5	3.8	11.8	17.7	20.6	23.8	23.0
M Botlikh	979	- 1.9	0.0	3.6	9.8	15.2	17.8	20.6	20.3
	1500	- 4.0	-2.5	0.8	6.6	11.8	14.6	17.5	17.4
	2000	- 6.0	-5.0	-2.0	3.4	8.4	11.4	14.4	14.6
	2500	- 8.0	-7.5	-4.8	0.2	5.1	8.2	11.3	11.6
M Sulak, high-mountain	2923	- 9.9	-9.7	-7.2	-2.6	2.2	5.2	8.5	8.9
	3000	-10.0	-10.0	-6.8	-3.0	1.8	5.0	8.2	8.8
	3500	-12.0	-12.5	-9.6	-6.2	-1.6	1.8	5.1	6.0
	4000	-14.0	-15.0	-12.4	-9.4	-5.0	-1.4	2.0	3.2

Table 2 extension

M Station	Height, m	IX	X	XI	XII	X-IV	V-IX	Year
	100	20.0	13.6	5.6	0.3	5.4	22.8	12.6

	500	18.2	12.4	5.2	0.3	4.7	20.7	11.4
M Botlikh	979	15.9	10.9	4.7	0.3	3.9	18.0	9.8
	1500	13.2	8.4	2.4	-1.6	1.4	14.9	7.1
	2000	10.5	6.0	-0.2	-3.5	-1.0	11.9	4.4
	2500	7.8	3.6	-2.4	-5.4	-3.5	8.8	1.7
M Sulak, high-mountain	2923	5.3	1.4	-4.0	-7.2	-5.6	6.0	-0.8
	3000	5.1	1.2	-4.6	-7.3	-5.8	5.8	-1.0
	3500	2.4	-1.2	-6.8	-9.2	-8.2	2.7	-3.7
	4000	-0.3	-3.6	-9.1	-11.1	-10.6	-0.3	-6.4

Table 3. Monthly values of vertical air temperature gradients on the Chechen Republic territory (° C / 100 m)

I	II	III	IV	V	VI	VII	VIII	IX	X	XI	XII	Year
Eastern North Caucasus at an altitude of 500-1000 m above the sea level, m												
-0.2	-0.1	0.05	0.40	0.50	0.55	0.65	0.55	0.45	0.30	0.10	0.00	0.30
Northern slope of the eastern part of the Greater Caucasus at heights of more than 1000 m (western part of the Republic territory)												
0.35	0.44	0.53	0.60	0.60	0.58	0.54	0.51	0.48	0.44	0.40	0.30	0.48
Northern slope of the eastern part of the Greater Caucasus at heights of more than 1000 m (eastern part of the Republic territory)*												
0.41	0.50	0.56	0.64	0.67	0.65	0.62	0.57	0.54	0.49	0.45	0.38	0.54

Note. * – These gradients were calculated by OVSI NC HMC according to M Botlikh - M Sulak, mountain region.

Research results and their discussion

The average annual air temperature decreases, respectively, from 8.6 ° C at an altitude of 1000 m, 6.2 ° C - 1500 m, 3.6 ° C - 2000 m, 1.0 ° C - 2500 m, -1.5 ° C - 3000 m, -4.1 ° C - 3500 m to -6.6 ° C at an altitude of 4000 m (Table 4). On average, the average annual air temperature for the territory under consideration is positive up to an altitude of 2800 m, above this altitude the average annual air temperatures are negative.

Average monthly positive air temperatures are observed up to an altitude of 4000 m in July-August. Above this altitude, the air temperature is negative throughout the year. During the cold period (X-IV), the average air temperature is positive up to an altitude of 1500-1600 m, above this altitude the average air temperature is negative. The average air temperature of the warm period (V-IX) is positive up to an altitude of 4000 m in July-August, and above this altitude it is negative [3].

Directly in the zone of contemporary glaciation, air temperatures are lower than at the same altitudes where there are no glaciers. When moving from rocky surfaces to a glacier, the air temperature drops sharply, that is, a "temperature jump" occurs, the magnitude of which is determined mainly by the size of the glacier. For small glaciers, such as the glaciers in the considered area, it is 0.5 ° C [4].

The transition of the air temperature through 0 ° C, i.e., the beginning of the mud stream dangerous period, observed at an altitude of 1000 m in the first decade of March, at an altitude of 2000 m is in the first decade of April and at an altitude of 3000 m in the second decade of May (Table 5).

Table 4. Average monthly and annual air temperature by altitude (°C) for the Chechen Republic territory

Height, m	I	II	III	IV	V	VI	VII	VIII	IX	X	XI	XII	X-IV	V-IX	Year
100	-1.8	-0.6	3.5	11.6	18.2	21.9	25.0	24.2	18.9	12.5	5.0	-0.2	4.3	20.9	11.5

500	-2.6	-0.1	3.3	10.0	16.2	19.7	22.4	22.0	17.1	11.3	4.6	-0.2	3.8	19.5	10.3
1000	-3.6	-1.4	3.1	8.0	13.8	16.9	19.2	19.2	14.8	9.8	4.1	-0.2	2.8	16.8	8.6
1500	-5.6	-3.8	0.4	5.0	10.6	13.8	16.3	16.5	12.3	7.4	2.0	-1.8	0.5	13.9	6.2
2000	-7.4	-6.2	-2.4	1.8	7.4	10.8	13.4.	13.8	9.8	5.2	-0.4	-3.6	-1.8	11.0	3.6
2500	-9.4	-8.5	-5.1	-1.2	4.2	7.8	10.5	11.1	7.2	2.8	-2.4	-5.2	-4.2	6.6	1.0
3000	-11.2	-10.8	-7.4	-4.4	1.0	4.7	7.6	8.4	4.6	0.6	-4.6	-7.0	-6.4	5.3	-1.5
3500	-13.2	-13.2	-10.2	-7.4	-2.2	1.6	4.7	5.8	2.1	-1.8	-6.6	-8.6	-8.7	2.4	-4.1
4000	-15.0	-15.6	-12.9	-10.6	-5.4	-1.4	1.8	3.2	-0.4	-4.0	-8.8	-10.4	-11.0	-0.4	-6.6

Table 5. Dates of average daily air temperatures' transition through 0 °C by heights

Height, m	M Station	Date	M Station	Date	Average date
668	Vladikavkaz	9.03	–	–	–
1000		25.03	Botlikh	14.02	5.03
2000		27.04		23.03	9.04
3000		31.05	Sulak, high-mountain	30.04	15.05
3659	Kazbegi, high-mountain	17.06		–	–

Due to the sharp dissection of the relief and large amplitudes of heights, atmospheric precipitation in the Chechen Republic is unevenly distributed. Their number varies from 480 mm at an altitude of 500 m, 630 mm at an altitude of 1000 m, 910 mm at an altitude of 2000 m and reaches a maximum of 1150 mm at an altitude of 3000 m. Above this altitude, the precipitation amount decreases and at an altitude of 4000 m it is 900 mm (Table 6).

Table 6. Change in annual precipitation (mm) with terrain height above the sea level on the northern slope of the Eastern Caucasus

Height, m	Annual precipitation, mm
500	480
1000	630
2000	910
2200	960
2400	1010
2600	1060
2800	1110
3000	1150
3200	1130
3400	1080
3600	1020
3800	960
4000	900

Table 7. Average monthly, annual for cold (X-IV), warm and mud stream (V-IX) periods precipitation (mm) by heights

Station	Height above sea-level, m	I	II	III	IV	V	VI	VII
M Grozny	123	18	16	25	32	53	70	57
M Bamut	345	19	21	33	53	101	122	90
M Vladikavkaz	668	22	24	37	69	129	154	115
M Vedeno	715	27	33	44	53	119	134	116
M Sulak, high-mountain	2923	34	41	67	128	158	169	135.

Table 7 extension

Station	Height above sea-level, m	VIII	IX	X	XI	XII	X-IV	V-IX	Year
M Grozny	123	49	41	33	28	20	152	270	422
M Bamut	345	69	58	37	31	21	215	440	655
M Vladikavkaz	668	85	75	46	34	24	256	558	814
M Vedeno	715	107	87	56	39	26	278	565	843
M Sulak, high-mountain	2923	110	96	77	49	28	424	668	1092

A small annual amount of precipitation is observed between the Skalisty and Glavny ridges, in the "rain shadow" zone, where its rate is 220-350 mm. [5].

Most of the annual precipitation falls on the warm, mud stream-prone period (V-IX) - 61-68% and only 32-39% in the cold (X-IV) with a maximum in May-July and a minimum in January-February. During the mud stream hazardous period, the territory under consideration receives an average of 65% of the annual precipitation (Table 7).

Accordingly, the vertical gradient of precipitation also changes with height, which up to an altitude of 3000 m is equal to 20-35 mm / 100 m. Above it, it is negative and as the absolute heights increase, it increases from -10 mm / 100 m at heights of 3000-3200 and up to -35 mm / 100 m at altitudes over 4000 m (Table 8).

Table 8. Vertical gradients of atmospheric precipitation on the Chechen Republic territory

Altitude zone, m	Vertical gradient, mm / 100 m	Altitude zone, m	Vertical gradient, mm / 100 m
500-1000	30	3000-3200	-10
1000-1200	35	3200-3400	-25
1200-1400	30	3400-4000	-30
1400-2800	25	more than 4000	-35
2800-3000	20		

Atmospheric precipitation in the territory under consideration falls in liquid, solid and mixed forms. At the same time, with height, the proportion of liquid and mixed precipitation decreases, and the proportion of solid precipitation increases. At an altitude of about 3000 m, solid precipitation makes up 100% in the period November-April, in other months their share decreases and in the mud stream-hazardous period (V-IX) it is only 23%. The share of liquid precipitation, on the contrary, increases and during this period they amount to 52% with the highest values in July-August (Table 9) [6].

Table 9. Average monthly amount of solid, liquid and mixed precipitation during the mud stream hazardous period (X-IX) in mm, % of their total amount

Station	Height, m	Months	solid		liquid		mixed		Total, mm
			мм	%	мм	%	мм	%	
Sulak, high-mountain	2923	V	95	60	16	10	47	30	158
		VI	25	15	85	50	59	35	169
		VII	1	1	119	88	15	11	135
		VIII	2	2	90	82	18	16	110
		IX	28	29	39	41	29	30	96
		V-IX	151	23	349	52	168	25	668

The greatest amount of precipitation, mainly in liquid form (heavy rains or steady downpours), during the mud stream-hazardous period falls at altitudes of about 3000 m, in the zone of non-sodded moraines. These sediments take an active part in the process of moraine material moistening and, consequently, in mud stream formation. Those precipitates that fall in a solid state during this period quickly melt and, like liquid ones, take part in the moistening of friable material, preparing it for movement. Due to the active development of convective clouds during the warm period of the year, steady downpours often fall on the territory under consideration. During the warm period in high-mountainous regions, heavy rainfall significantly prevails over steady downpours [7].

In high-water years, when the amount of precipitation during the cold period, which determines the soil moisture degree and the water content of rivers, exceeds their average long-term values and precipitation is observed in June - August above the norm, the rain mud streams probability increases sharply. The impetus for the mud stream process onset in these years can be a downpour with a precipitation of 25 mm / day. Also, for the years of average water content, showers of 50 mm / day, forming high floods, can serve as an impetus to the mud streams formation. Intense showers in the conditions of steep relief are instantly concentrated in temporary streams that can disturb the equilibrium state of friable material and landslide masses on the slopes. These movements, leading to the mud stream onset, coincide in time with the shower intensity maximums. The impulse to the mud stream process beginning is the maximum rainfall intensity of more than 0.1 mm / min. The critical rate of mud stream-forming precipitation in the area under consideration ranges from 25 to 50 mm per day. Heavy rainfall leads to the mud streams formation in the local areas of the Republic, and the frontal ones - on a significant territory. The precipitation amount for the previous months and the snow cover thickness characterize the moisture content of the territory. The climatic snow line here is 3980 m. It is the upper limit of the mud streams area [8].

Complex characteristics of mud stream phenomena

The Chechen Republic has 78 main mud stream channels, of which the following river basins are located: Fortanga - 5, Shalazha - 2, Gekhi - 1, Martan - 3, Goyt - 2, Sunzha - 2, Argun - 25, Sharoargun - 28, Dzhalka - 3, Khul-khulau - 4, Ansalta - 3. 58% of mud-and-stone streams, 33% of sedimentary and 9% of mud streams pass through them. Mud-stone mud streams play a leading role in high mountains 78%, and alluvial mud streams dominate in the middle mountains (62%) and low mountains (89%). Mud streams, which are the second most common in the middle mountains (29%), are inferior to mud streams in low mountains (7%) (Figure 1).

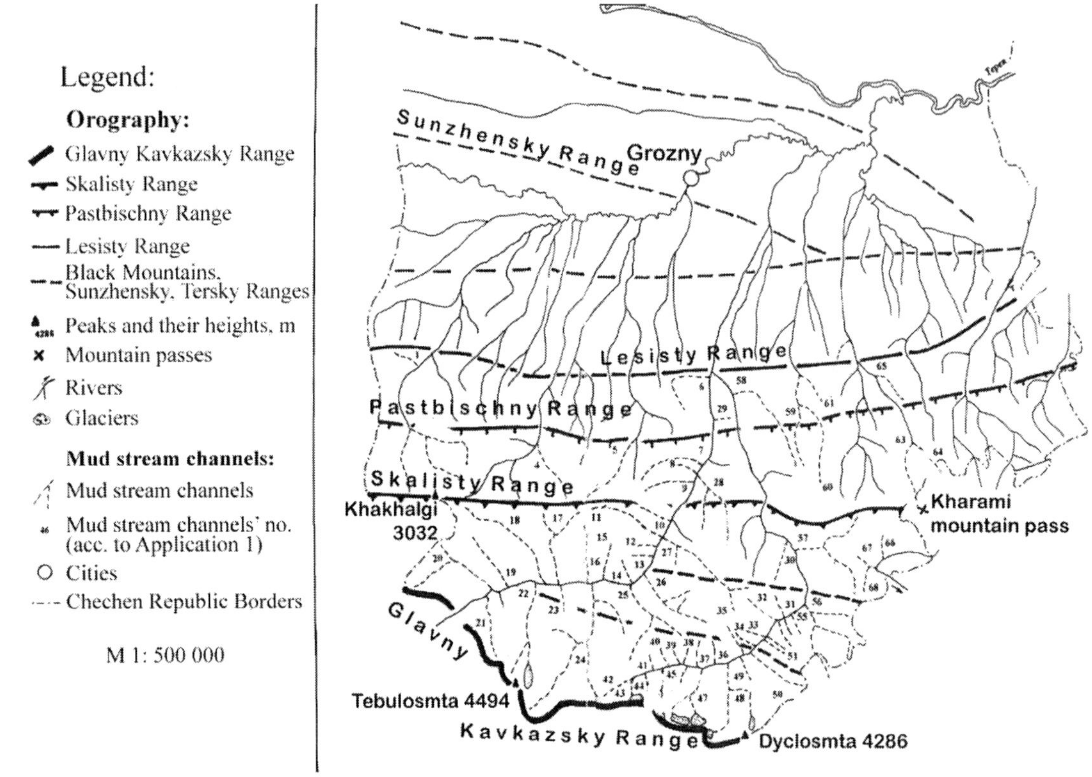

Figure 1. Schematic map of orohydrography and mud stream channels of the Chechen Republic

The solid component of mud streams in the highlands (more than 2000 m) are loose deposits of the rocky belt and fresh glacial moraines, in the middle mountains (2000-1000 m) - the material of ancient moraines, talus, avalanches, landslides, terrace alluvium, in the low mountains (less than 1000 m) - alluvial, diluvial-landslide, fine-detrital landslide deposits of erosional relief forms [9].

Mud streams of rain origin prevail (89%). There are no pure glacial mud streams, but mixed glacial-rain mud streams can form (8.5%). Snow mud streams are absent, snow-rain mud streams (1.5) and limnogenic mud streams (1%) are rarely formed.

Mud streams are formed at all altitude intervals, and their share increases from high mountains to low mountains. Glacial-rain mud streams follow the spread of contemporary glaciation, which in the Republic belongs to the scattered type, and is formed only in the highlands, due to glacial processes, mainly as a result of buried ice melting. Snow and rain mud streams are observed on the mid-mountain ranges slopes [10]. Limnogenic mud streams in the highlands are caused by the destruction of natural moraine dam and tarry moraine lakes, in the middle and low mountains - by the destruction of dams from landslide masses. Every year, the possibility of the anthropogenic mud streams formation, caused by the excess load on landscapes by irrational economic activities of a human and as a result of past military operations in the mountains, increases.

The debris flows volumes vary from tens to millions cubic meters. The most significant volumes are characteristic of mud streams that are formed in the regions of the Glavny Range northern slope and its spurs and in the North Jurassic depression, where in some cases they exceed 1 million m^3. Mud streams with volumes of no more than 10 thousand m^3 are observed in the areas of the forward ridges. Mud streams of small volumes (up to 10 thousand m^3) are formed almost annually, medium (10-100 thousand m^3) and significant (100 thousand m^3 - 1 million m^3), respectively, once every 1-5 years and once every 5-10 years old. Catastrophic mud streams (over 1 million m3), on average, are formed once every 15-20 years [11].

On the territory of the Chechen Republic, the mud stream hazardous period in the highlands lasts from May to September, and the period of greatest danger is July-August, which is associated with the maximum precipitation and intense melting of snow and ice in the highlands, in the middle mountains - from April to September, in the low mountains from March to September.

Mud stream processes cause significant damage to all economy sectors of the mountainous regions of the Republic. Currently, mud streams periodically threaten 69 objects of the economy, including: 50 settlements and 19 sections of highways [12].

Mud stream hazardous areas

On the territory of the Chechen Republic mud stream basins are distributed unevenly, but it is clearly possible to distinguish areas of all 4 categories of mud stream hazard (Fig. 2). Based on the available materials, the entire mountainous zone of the Chechen Republic is subdivided into mud stream hazardous areas of all four mud stream hazard categories.

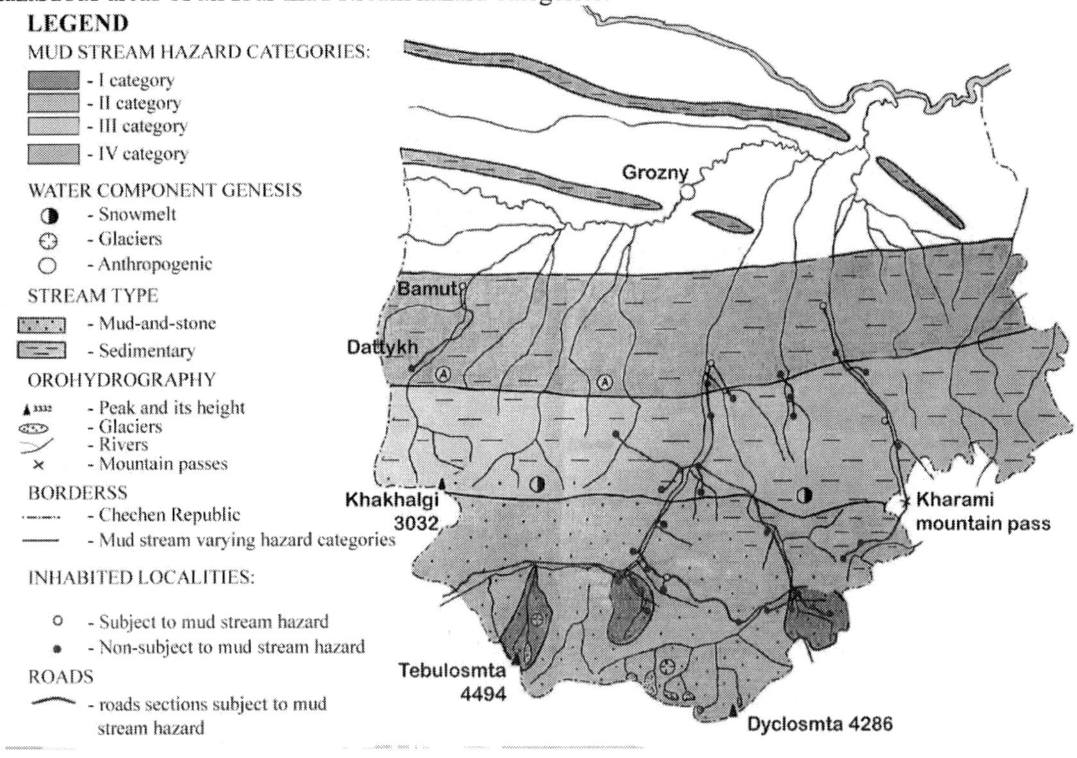

Figure 2. Schematic map of mud stream hazard areas of the Chechen Republic

I category region with high mud stream hazard is characterized by:

- the development of large mud stream centers, mainly mud stream incisions, more than 3 km long, with their catchment areas 5 km^2 and more, or very active with current-generating surfaces with the least filtration coefficient, less than 0.01 mm / min; morphologically, the centers of dispersed mud stream are characterized by a dense network of furrows, potholes, trays of micro mud stream, landslides-streams, developing mainly on the bodies of large ancient landslides, less often in diluvial deposits; rocky hearths are confined to the landslide-talus slopes of ancient kars and trogs;
- the highest availability of liquid precipitation;
- the potential for the ice-rain mud streams formation as a result of buried ice melting;
- catastrophic discharge of high-density mud streams capable of exceeding 250 m^3 / sec (for sediment mud streams);
- one-time volumes of removal is more than 1 million m^3 (for mud-and-stone streams).

Within the Chechen Republic, there are three local areas of high mud stream hazard, confined to the large tributaries of the Argun and Sharoargun rivers. The first region is the mud stream basin of the Maistikhi river. It is a right tributary of the river. Argun and has a basin area of 97 km^2. Mud-stone mud streams are formed in this basin. The water component is formed by rare but intense heavy rains. The main types of mud stream centers are: 1 rocky mud stream center, 2 incisions and 9 centers of diffuse mud stream formation. The second region is the mud stream basin of the Khacharoyakhkriver, its right tributary. Argun, with a basin area of about 100 km^2 [13]. The mud-and-stone type mud streams of rain origin are formed here. Only the most powerful mud streams reach the river valley. Argun, where they formed an alluvial cone up to 350 m wide and up to 500 m long. Within the basin there are 2 incisions and 2 centers of dispersed mud stream formation. The third region is the mud stream basin of the Kenkhi river. It is a right tributary of the river. Sharo-Argun and has a basin area of 46 km^3. Here mud-stone mud streams of rain genesis are formed. Partial accumulation of mud streams occurs at the mouth of the river Kenkhi, where a powerful fan is formed. Within the basin, there are 7 centers of dispersed mud stream, 5 incisions, 1 mud stream catchment.

In the main channels of the aforementioned mud stream basins, sedimentary mud streams can also be formed. The nature of the channels makes it possible to classify them as I category of mud stream hazard (basins of the Maistikhi and Khacharoyakhk rivers) and II category of mud stream hazard (basin of the Kenkhi river). The formation of sedimentary mud streams here is associated with the transformation of mud-stone flows as they move along the main channels.

Mud stream hazard period is May-September, with the greatest mud stream activity in July-August.

The recurrence rate of catastrophic mud streams is 1 time in 15 -20 years, large - 1 time in 5 years, small ones are formed annually.

II category region with medium mud stream hazard is characterized by:

- the presence of medium-sized mud stream centers: incisions with a length of 1-3 km, potholes with a length of 2-3 km, areas of their catchments 3-4 km^2, centers of dispersed mud streams, rocky centers, mud stream catchments, or runoff surfaces of medium activity with a filtration coefficient of 0.01 -0.05 mm / min;
- flow rates of mud streams 250-100 m^3 / sec.;
- one-time volumes of mud streams removal from 100 thousand m^3 to 1 million m^3.

The area is orographically located between the Glavny and Skalisty ridges and covers the North Jurassic depression between them. In this area, foci are formed in diluvial rubble-clayey and landslide-talus sandy-carbonate deposits. A large mass of loose material for the mud streams formation is supplied by landslide flows in the upper reaches of mud stream rivers and numerous mud streams along their channels [12].

There are 30 centers of dispersed mud stream formation in the Argun basin, 21 incisions, 14 rocky centers, 9 mud stream catchments, 5 potholes. Here the most mud stream-prone are the channels of the Heldikhoyerk, Dzumserk, and Verdyerk rivers. In the Sharoargun basin there are 21 centers of dispersed mud stream, 19 times, 7 rocky centers, 3 potholes, 1 mud stream catchment. The most mud stream hazardous is the Dayakhk river bed. And in the south-east of the Republic, the upper reaches of the Ansalta river are dangerous.

Here, both mudstone and sedimentary flows are formed. The genesis of mud streams is mainly rain. Snow and rain mud streams are formed rarely.

The mud stream hazardous period is April-September.

The average frequency of mud streams is once every 5-10 years.

III category region with low mud stream hazard is characterized by:

- development of predominantly small mud stream centers: mud stream catchments, potholes no more than 1 km long, rocky centers, centers of dispersed mud stream and single incisions, areas of their catchments of 2 km^2 or less, or weakly active drainage surfaces with a filtration coefficient of 0.05-0.1 m / min;
- discharge of sedimentary flows 100-10 m^3 / s;
- simultaneous volumes of mud-stone flows 100-10 thousand m^3.

The area is orographically confined to the Skalisty and Pastbizhny ridges and the Shatoi depression. Rocky mud stream centers, confined to limestone cliffs, and the centers of dispersed mud stream are mainly developed on the Skalisty and Pasture Ridges. The latter represent an extensive network of furrows, couloirs, and talus trays. Despite the abundance of loose material, the foci are short, mud streams are rarely formed and they are of low thickness, due to the large number of karst forms that absorb most of the rather abundant sediments in this area. The Shatoi Basin is characterized by mud stream catchments and the centers of dispersed mud stream formation. In the eastern part of the region, the formation of mud streams is often associated with landslides-streams.

The following river channels are most dangerous: the upper reaches of the Martanka river (and its tributary, the Meredzhi river), Gekhi, Martan; in the basin of the Argun river - Syuzha, Varanda; in the upper reaches of the Dzhalka river - Alakskakh and Bass rivers; in the Khulkhulau basin in its right tributary Bulk, and in its upper reaches its constituent rivers Kharachoy and Okholitlau.

Here, mainly sedimentary mud streams are formed, much less often - mud-stone, mud flows are characteristic of the eastern part of the region. The genesis of mud streams is mainly rain. Snow-rain mudslides are rarely formed, which are the characteristic of the western part of the district, and limnogenic mud streams associated with landslides-streams are the characteristic of the eastern part of the district.

Mud stream hazard period is March-September.

Average frequency of mud streams is rare - 1 time in 10-15 years.

IV category region with very low or potential mud stream hazard is characterized by:

- discharge of sedimentary flows less than 10 m^3 / s;
- volumes of mud-stone and mud flows carried out less than 10 thousand m^3.

The IV category of mud stream hazard also includes poorly studied areas or the areas where previously mud streams were not noted, but due to the ongoing changes in the natural environment, they may occur.

The area is orographically confined to the Lesisty (Black Mountains), Sunzhensky, Tersky ridges, where there are significant channel slopes, the presence of friable material or clayey, easily decaying rocks. In the eastern part of the Republic, within the Sunzhensky, Tersky ridges and the Black Mountains, there are valleys of permanent and temporary watercourses: gullies, ravines, ravines, potholes less than 1 km long, with widely developed landslide processes on clay slopes, where mud streams can form.

Basically, small sedimentary flows are formed in the area, caused by heavy rainfall. The formation of mud streams associated with anthropogenic activities in the areas of logging and construction is possible.

Mud stream hazard period is March-September.

The possibility of mud streams formation is very rare - once every 15 years or less.

Summary

The mud stream-hazardous period duration and mud stream activity at altitudes of more than 2000 m will increase in the coming decades in the North Caucasus, including in the Chechen Republic, due to climate warming, to an increase in the period duration of the positive air temperatures, an increase in the atmospheric precipitation amount, an increase snow line and increasing river flow. These phenomena will lead to an increase in the area of mud streams and the processes of frost weathering will increase the flow of significant masses of friable material necessary for the mud streams formation into the river beds. As a result of the landscape zones' shift to higher elevations, the economic load in the most elevated parts of the Republic will increase, which will intensify anthropogenic mud stream processes there. The research data was obtained by the team of authors on the basis of the studied scientific and technical reports, which are summarized in the documents and publications of the high-mountain hydrometeorological research of the North Caucasus HMC department.

References

[1] Sergeeva G A, Adamyan V L 2017 Distribution of anthropogenic mud streams in the North Caucasus region *Intelligent systems in production* **1 (32)** 114-117.

[2] Sergeeva G A, Volobueva L L, Krivosheeva E A 2015 Debris flow hazard of the Karachay-Cherkess Republic / Western Caucasus *Monograph. Palmarium Akademic Publishing* 269.

[3] Sheina S G, Girya L V 2017 Innovative methods of monitoring for deformations of man-made loaded areas *MATEC Web of Conferences, International scientific and technical conference "Modern directions and prospects for the development of processing technologies and equipment in mechanical engineering 2017"* (ISSN: 2261-236X, France).

[4] Sergeeva G A, Adamyan V L 2017 Serious areas in the Pshekha river basin *Ecology and life safety. Collection of articles of the XVII International Scientific and Practical Conference* 135-138.

[5] Sergeeva G A, Zhizhin K S 2018 Geological cataclysms of the North Caucasus *Scientific journal "Globus", materials of XXXIII Intern. scientific-practical conf. "Achievements and problems of modern science" St. Petersburg.*

[6] Sergeeva G A, Zhizhin K S 2016 Dynamics of snow avalanches in the South slope of the Western Cucasus *Innov: electronic scientific journal* **2 (27)** 1.

[7] Yazyeva S B, Seferyan L A, Oparina L A, Golubeva A Y 2018 Greening. technology organizing of multistorey buildings, when reconstructing architectural and planning decisions with the use of modern building materials *Materials Science Forum* **931** 883-888

[8] Seferyan L A, Kondrateva T N, Morozov V E, Leusenko I V 2019 Qualitative state improvement of public and courtyard areas in terms of Rostov-on-Don *IOP Conference Series: Materials Science and Engineering* **698 (5)** 055004.

[9] Sheina S G, Yudina K V 2018 Development technique of urban-environmental passport of the reconstruction territory *Materials Science Forum* **931** 822-826. Information on https://www.scientific.net/MSF.931.822

[10] Shvydenko N V, Axenov A A, Vinogradova E V, Seferyan L A 2020 Public private partnership as a housing development tool *IOP Conference Series: Materials Science and Engineering. – IOP Publishing* **913 (5)** 052021.

[11] Sheina S G, Chubarova K V, Girya L V, Seferyan L A 2020 The Use of Information Technology in Territorial Planning of a Subject of the Russian Federation *IOP Conference Series: Materials Science and Engineering, IOP Publishing* **913 (4)** 042025.

[12] Sergeeva G A, Adamyan V L, Seferyan L A, Reznikova A V 2020 The Debris-hazardous Areas' Research Methods (on the Republic of Ingushetia Example) *IOP Conference Series: Materials Science and Engineering, IOP Publishing*,**913 (4)** 042024.

Statistical estimation of the effect of precipitation redistribution using the historical regression method

S O Gekkieva*, A H Sherkhov

Federal State Budgetaru Instituion"High-Mountain Geophysical Instituion", 2, Lenin Ave., Nalchik, 360000, Russia

E-mail: sgekkieva@list.ru, fff.ddd.111@mail.ru

Abstract. Precipitation has significant natural variations in time and space. The calculation of the probability of increasing or decreasing the amount of precipitation against the background of their natural variability as a result of work on active influences is not an easy task. A statistical assessment of the effect of the redistribution of precipitation in the protected and control areas was carried out using the historical regression method.

Introduction

The water resources of our planet should be treated very carefully, considering water as a special kind of raw material, without which life on Earth is impossible. Water is used at all stages of human development and serves as one of the important products for each individual. Therefore, the fresh water consumed in economic activity by weight is an order of magnitude higher than all other raw materials in the aggregate. In The earth's biosphere, fresh water is 2.5%, but more than 99% of this water is in the form of snow and ice. Rivers and lakes contain 93 thousand cubic km of water at a time. Over the past 50 years, the consumption of fresh water has increased 4 times from 1000 cubic km to 4100 cubic km per year. 70% of the water is used for agriculture, and the remaining 30% is used in industry and utilities [4].

Materials and research methods

Currently, due to the acute shortage of fresh water, many countries of the world are forced to take certain measures to solve this problem. Therefore, active actions on clouds for the purpose of artificial increase in precipitation are one of the ways to solve this problem. But there is still no consensus on how these works affect the precipitation regime of neighboring territories. Moreover, this is difficult to estimate in the North Caucasus, since natural fluctuations in precipitation are extremely variable in space and time. This uneven distribution is due to the physical and geographical location of the region, where the Main Caucasian ridge runs on the one hand, and the black sea is close on the other. Figure 1 shows the time course of the average precipitation for the 15-year period before the start of work on artificial increase in precipitation, which fell on the territory of the Stavropol territory (PT) and its neighboring control (CT) and adjacent territories (AT).

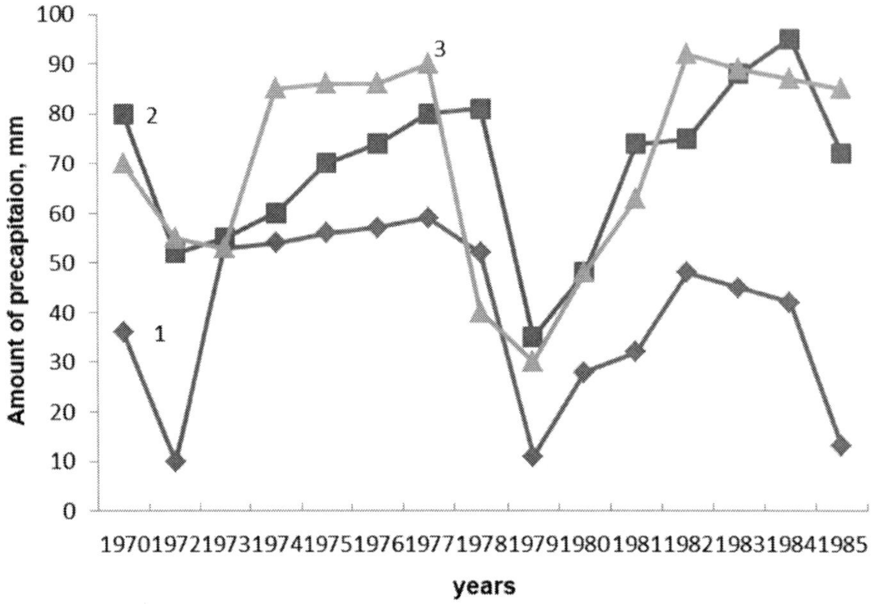

Figure 1. Time course of precipitation (may-august) in the control and protected areas: 1-CT, 2 – PT and 3-AT

The time course of precipitation over the entire region shows that precipitation over the entire territory has quasi-periodic fluctuations with cycles from 2 to 6 years. The years of maximum and minimum for all the studied areas coincide. The analysis of the time stroke also shows that in the period from 1971 to 1977 there is an increase in precipitation on the protected, control, adjacent and throughout the study area, and from 1977 to 1979 there has been a steady decreasing trend in average rainfall for the season (may - august) throughout. A certain synchronicity of this pattern is most likely due to the fact that the main amount of precipitation falls during frontal processes that take place throughout the region.

In this regard, a statistical assessment of the physical effect of the impact of artificial increase in precipitation in the Stavropol territory on the regime of liquid precipitation in neighboring territories was carried out using the historical regression method.

Research results and discussion

The presence of a high correlation of precipitation between the compared territories and the absence of influence of active impacts on the control territories allows using this method. The basic data on which the method is based are averaged precipitation data for the spring-summer season (may-august) from individual weather stations. Averaging precipitation makes it possible to reduce the coefficient of variation of precipitation. The calculation period was chosen from 1970 to 1985, since 1986 on the territory of the Stavropol territory, work was started on active effects on convective clouds in order to artificially increase precipitation, in connection with which the natural mode of precipitation was violated.

The data of statistical analysis of one-dimensional series for four compared pairs of territories (CT_1, CT_2, CT_3, CT_4 and AT_1, AT_2) are presented in table1. These are the values of the correlation coefficients (r) between the sedimentary series of the adjacent territory (AT) and the control territories (CT) with estimates of their calculation errors (S) for the period of time before active impacts, the values of the linear regression coefficients between the sedimentary series (x-y) and the resulting linear regression equation [2].

Table 1. Statistical analysis data

Statistical specifications	Compare pairs of precipitation series X_1–y_1	X_2–y_1	X_3–y_2	X_4–y_2
$r \pm S_r$	0.8 ±0.09	0.7±0.13	0.7±0,13	0.7±0.13
ρ	0.7±0.14	0.6±0.11	0.5±0,15	2.2±0.09
regression equation	0.7x + 38	0.6x+ 25.2	0.5x+ 21,3	2.2x+ 19.3
$\pm S$	±11.4	±11.4	±10.3	±7.9

An analysis of the data given in table1 allows us to use any regression ratio in calculating the effect of artificial increase in precipitation on the precipitation regime in the adjacent territory.

The regression relationship between CT and AT before the start of work on active impacts (1970 - 1985) and on active impacts (1986 - 1997) in the Stavropol territory is shown in Figure 1.

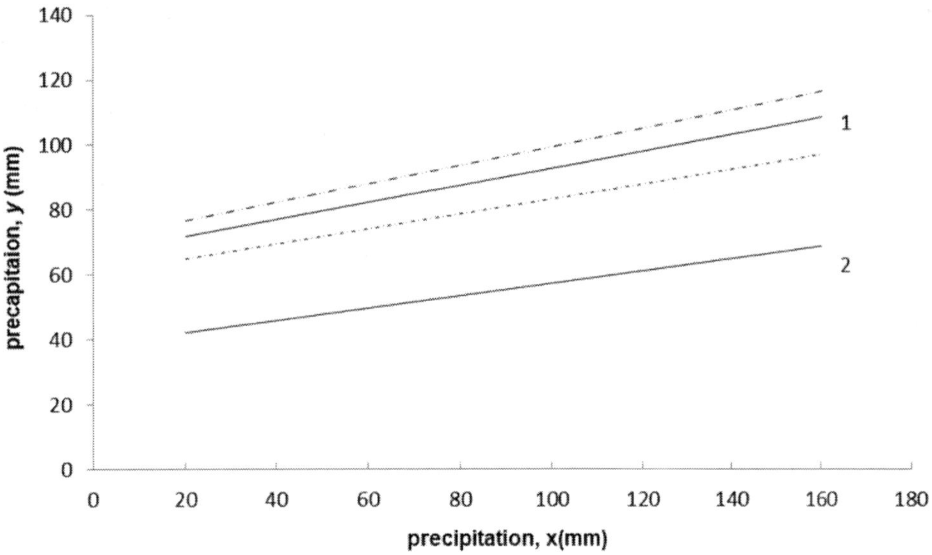

Figure 2. Regression relationship of precipitation of adjacent (y) and control (x) territories 1– in the period before the start of work, 2 - in the period of active impact

In figure 1, the confidence interval of the regression equation is indicated by dashed lines. Confidence interval for the regression equation obtained between the precipitation of the control (x) and adjacent (y) territories:

$$y = 0{,}7x + 38 \tag{1}$$

The correlation coefficient between precipitation CT and AT is equal to r = 0.8.

In order to test the hypothesis about the significance of the sample correlation coefficient, it is necessary, at a significance level of 0.05, to check the null hypothesis that the general correlation coefficient is equal to zero H_0: $r_r = 0$. A random variable is accepted as a criterion for testing the null hypothesis [3].

$$T = \frac{r\sqrt{n-2}}{\sqrt{1-r^2}} = \frac{0{,}8\sqrt{16-2}}{\sqrt{1-0{,}8^2}} = 4{,}93 \tag{2}$$

According to the significance level of 0.05 and the number of degrees of freedom k, we find from the student distribution table the critical point of the two-sided critical region t (0.05; 16) = 2.12

Since T > t, we reject the null hypothesis. Therefore, the sample correlation coefficient is significantly different from zero, i.e. precipitation in the adjacent and control areas are correlated. The error in determining the expected amount of precipitation at AT_1 by the regression equation (1) is equal to the standard deviation from the regression equation:

$$S = S_{AT}\sqrt{1 - r^2} \quad (3)$$

$$S = \pm 19{,}1\sqrt{1 - 0{,}8^2} = \pm 11{,}4 \text{ (mm)}$$

The confidence interval of the regression equation (1) is found by the formula:

$$y_{AT.est.} = \bar{y}_{AT} + a(x_{CT} - \bar{x}_{KT}) \pm t_{\alpha}\sigma_{AT}\frac{\sqrt{1-r^2}}{\sqrt{n-2}} \times \sqrt{1 + \left(\frac{x_{CT} - \bar{x}_{CT}}{\sigma_{CT}}\right)^2} \quad (4)$$

Calculations by formula (3) show that the error of the regression equation at a significance level of 10% is on average $\bar{s} = \pm 5{,}7$ mm. As can be seen from Figure 1, the actual amount of precipitation in the adjacent territory over the years of active impacts lies outside the confidence interval of the regression equation. In Figure 1, the confidence interval of the regression equation is indicated by dashed lines. Thus, in a statistical assessment by the method of historical regression of the effect of redistribution of precipitation, we obtain that the amount of additional precipitation for the season should be $\Delta y > 11.4$ mm, so that with a confidence probability of 68% we can talk about a significant redistribution effect. If we consider the entire period of active impact works (12 years), we find that the amount of additional precipitation of 5.7 mm is significant at a significance level of 0.5.

Summary

Based on studies of the climatic characteristics of the North Caucasus from 1958 to 2009 [1], it was found that over the past 50 years, a steady tendency has been observed throughout the region for changes in temperature and humidity, as well as precipitation and the climate as a whole. Therefore, in order to analyze the effect of redistribution of precipitation due to active impacts for each individual season with active impacts, it is necessary to determine which year it belongs to - with excess or deficit of precipitation.

References

[1] Abshaev M T, Abshaev A M 2017 Analysis of options for increasing rainfall in the local territory *Proceedings of HMGI* **100** 75 - 77.
[2] Ekba Ya A, Kaplan L G, Zakinyan R G 1992 On evaluation of physical efficiency of works on artificial increase of precipitation in Stavropol region. In proceedings: Physics of clouds and active influences *Proceedings of HMGI* **85** 71-72.
[3] Gekkieva S 2002 Ecological aspects of active impacts on clouds (Doct.Diss, Nalchik).
[4] Mitropolsky A K 1961 Technique of statistical calculations (Fizmatgiz, Moscow).
[5] Shipilov O 1983 The use of statistical methods to assess the effectiveness of work to increase precipitation. Hydrometeorology (VNIIGMI-MCD, Obninsk).

IOP Conf. Series: Materials Science and Engineering 1083 (2021) 012097 doi:10.1088/1757-899X/1083/1/012097

Researches of ice-forming efficiency of products of sublimation of pyrotechnic compositions consisting of silver iodide agi particles and zinc oxide nanotubes

B M Khuchunaev, Kh- M Kh Baysiev, S O Gekkieva*, A Kh Budaev

Federal State Budgetary Institution «High-Mountain Geophysical Institute», 2 Lenina avenue, Nalchik, 360030, The Russian Federation

E-mail: gekkieva@list.ru

Abstract. The article is a continuation of experimental research on the search for new, more effective compositions for active effects on clouds and fogs. The paper also presents the results of studies of the ice-forming efficiency of pyrotechnic compositions based on AgI and zinc oxide nanotubes. The dependence of the specific yield of ice-forming active particles for various zinc contents in the initial composition of AD-1 was revealed.

1. Introduction

Over the past decades, in many countries a huge number of theoretical, laboratory and field studies have been carried out on the means of influencing clouds and precipitation, answers to many questions have been received, and valuable diverse experimental material has been collected. However, the issue of developing new and improving existing compositions remains relevant today. AgI iodide silver pyrotechnic compositions are one of the most widely used reagents for active effects on clouds. It is believed that the crystal structure of AgI is similar to the structure of natural ice; therefore, it is an effective reagent for influencing supercooled cloud systems. The formation of the ice phase on AgI aerosol particles has been studied by many researchers. In early works [1, 2], it was assumed that ice crystals formed on AgI aerosol particles due to sublimation of water vapor upon saturation relative to ice. However, the development of condensed matter physics in recent years shows that the growth of a new phase on catalyst particles, such as the growth of ice on reagent particles, is a more complex process.

This paper presents the results of experimental studies on ways to increase the ice-forming efficiency of crystallizing reagents of anti-hail products. The main goal of the research was to determine the effect on the yield of ice-forming particles of various chemical components that increase the yield of ice-forming particles from one gram of reagent. As such a component, finely dispersed zinc powder was used in the experiments, with particle sizes from 0.01–0.05 mm, which were introduced into the initial pyrotechnic composition of AD-1 in relation to its total mass — 3%, 6%, and 9%, respectively. Zinc has a crystal lattice closer to the ice crystal lattice [3]. The pyrotechnic composition of AD-1 in relation to its total mass is 3%, 6%, and 9%, respectively.

2. Methods and materials

Structural and physical characteristics of zinc

Zinc crystals have a hexagonal packing of atoms, which are very similar to the structure of ice. Zinc is a brittle transition metal of a bluish-white color, which oxidizes in air, being covered with a thin layer of zinc oxide. It has a low melting point. At 100-150 ° C, zinc is plastic, at a temperature of 419 ° C it begins to melt, but if the boiling point is 913 ° C, it begins to turn into steam. It was shown in [4–6] that upon sublimation of zinc in the presence of graphite, zinc oxide nanotubes are formed. The formation of nanotubes occurs according to the well-known drip mechanism. The physical essence of the droplet mechanism lies in the fact that the formation of a closed interface of three phases adjacent to the singular face of the crystallization front, and the creation of conditions under which a drop of capillary pressure arises in the droplet or an uncompensated surface tension applied to the wetting perimeter provides a shift of the phase boundary in the direction of crystal growth. The role of the phase boundaries during the growth of whisker nanocrystals by the aforementioned dropping mechanism is to reduce the crystallization barriers due to the release (discharge) of a spheroidizing drop of excess free energy and lowering the saturation necessary for the growth of the singular vertex face of the nanowire at a given speed, as well as to ensure that the specified face steady equilibrium drops. The driving force of the process of moving the droplet is the excess of free energy, the source of which is the outer surface of the droplet. The crystallizing substance diffuses into this drop, precipitates at the liquid – crystal interface, and the drop remains on top. The growth rate under the drop is one to two orders of magnitude higher than other growth mechanisms.

In our experiments, at a high combustion temperature of the pyrotechnic composition of anti-hail products, zinc oxide nanotubes are also formed, in this case the catalyst (graphite) is part of AD-1.

The equipment and methodology for the experiment

Figure 1 shows a set of equipment for studying the ice-forming properties of reagents. To carry out the full cycle of the experiment, it is required: a 1m^3 sublimation chamber, a 9m^3 large cloud chamber, a 0.080 m^3 small cloud chamber, a device for creating a cloud environment, an Adventurer electronic balance, a reagent atomization device, a reagent sublimation device at high pressures, a microscope Motic for counting the number of crystals. In a large cloud chamber, the temperature can be maintained from 0 ° to –17 ° C, in a small cloud from 0 ° to –25 ° C [4-6].

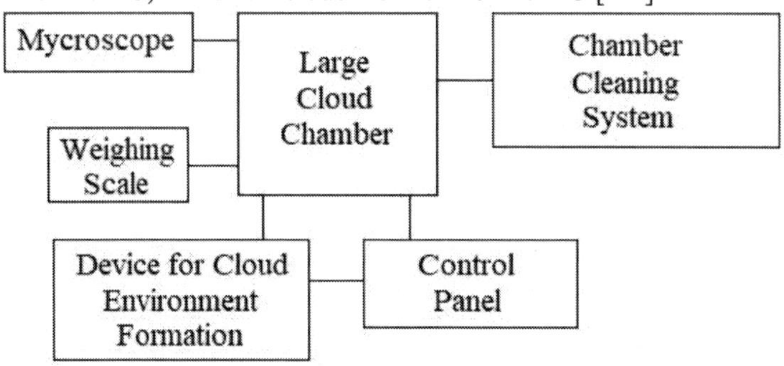

Figure 1. A Set of equipment for studying the ice-forming properties of reagents

3. Results

Laboratory experiments to determine the specific yield of ice-forming particles using AD-1 reagent with zinc nanotubes were carried out with different weight fractions (3, 6, 9% of the weight of AD-1). The determination of the yield of crystallization nuclei for each sample is calculated according to the generally accepted method [4-6].

Figure 2 shows the crystals deposited on the substrate, which were formed upon the introduction of pyrotechnic composition AD-1 and AD-1 with zinc additives.

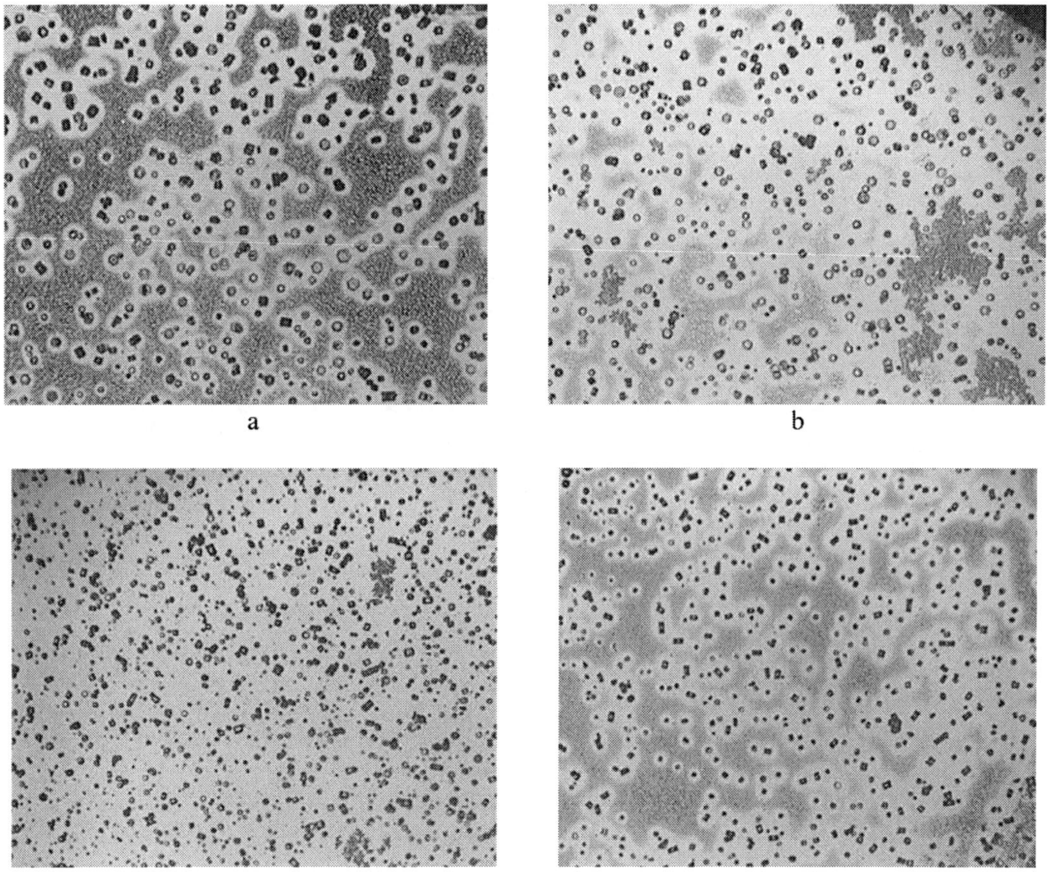

Figure 2. Ice crystals on a substrate when AD-1 and AD1 + zinc reagent is introduced into the cloud chamber: *a* - ice crystals when adding AD-1, *b* - ice crystals when adding AD-1 + zinc (3%), c - ice crystals when adding AD1 + zinc (6%), d - ice crystals when adding AD1 + zinc (9 %)

Table 1 presents the experimental data of the component of the reagent AD-1 without additives at temperatures from -5.9 ° C to -10.2 ° C.

Table 1. Average results of crystallization nuclei of pyrotechnic composition AD-1 with zinc additives at various temperatures

Temperature, °C	Mass of reagent AD-1, g	Specific output, g^{-1}
-13.4	0.01	4.4×10^{12}
-11.9	0.01	5.7×10^{12}
-10.6	0.01	1.2×10^{12}
-10.2	0.01	1.2×10^{12}
-9.7	0.01	6.4×10^{12}
-9.5	0.01	5.0×10^{12}
-9.1	0.01	6.2×10^{12}
-8.7	0.01	1.3×10^{12}
-8.2	0.01	3.3×10^{12}

-7.5	0.01	$3.2x10^{12}$
-7.2	0.01	$5.3x10^{12}$
-6.8	0.01	$6.5x10^{11}$
Temperature, °C	**Mass of reagent AD-1+Zn(3%), g**	**Specific output, g^{-1}**
-12.5	0.01	$1.7x10^{13}$
-12.3	0.01	$1.7x10^{13}$
-11.5	0.01	$9.4x10^{12}$
-10.5	0.01	$1.1x10^{13}$
-9.2	0.01	$2.0x10^{13}$
-9	0.01	$1.1x10^{13}$
-8.7	0.01	$8.6x10^{12}$
-8.5	0.01	$1.9x10^{13}$
-7.5	0.01	$9.9x10^{12}$
-5.5	0.01	$9.0x10^{12}$
-4.4	0.01	$1.1x10^{13}$
-2.2	0.01	$8.2x10^{12}$
Temperature, °C	**Mass of reagent AD-1+Zn(6%), g**	**Specific output, g^{-1}**
-13.3	0.01	$1.0x10^{14}$
-12.2	0.01	$1.43x10^{14}$
-10.4	0.01	$1.02x10^{14}$
-10	0.01	$6.0x10^{13}$
-9.6	0.01	$5.2x10^{13}$
-7.5	0.01	$5.4x10^{13}$
-5	0.00052	$1.8x10^{13}$
-4.2	0.00052	$1.2x10^{13}$
-3.4	0.00052	$1.0x10^{13}$
-2.2	0.00052	$1.3x10^{13}$
-1.6	0.00052	$4.7x10^{12}$
-1	0.00052	$6.6x10^{12}$
Temperature, °C	**Mass of reagent AD-1+Zn(9%), g**	**Specific output, g^{-1}**
-12.7	0.01	$3.5x10^{13}$
-12.1	0.01	$3.2x10^{13}$
-11.3	0.01	$1.9x10^{13}$
-10.7	0.01	$1.4x10^{13}$
-10	0.01	$2.1x10^{13}$
-9.5	0.01	$1.6x10^{13}$
-7.2	0.01	$1.3x10^{13}$
-7	0.01	$7.2x10^{12}$
-5.9	0.01	$1.3x10^{13}$
-5.2	0.01	$5.1x10^{12}$
-4.5	0.01	$3.9x10^{12}$
-4.3	0.01	$6.2x10^{12}$

In total, laboratory experiments confirm that the presence of finely dispersed zinc powder in the composition of the initial ice-forming fuel in relation to the total mass of 6% sharply increases the yield of ice-forming particles in the entire range of accepted temperatures, from zero to minus 14 °C.

With the addition of 9% zinc, there is a decrease in the specific yield of ice-forming particles. The physics of such a decline is still difficult to explain.

In figure 3 shows the dependence of the yield of ice-forming active particles (n) for various contents of finely divided zinc powder in the initial composition of AD-1. Curve 1 characterizes the yield of ice-forming particles from one gram of the initial AD-1 composition without zinc in it, curve 2 with a zinc content of 3%, curve 3 with a zinc content of 9% and curve 4— with a zinc content of 6%.

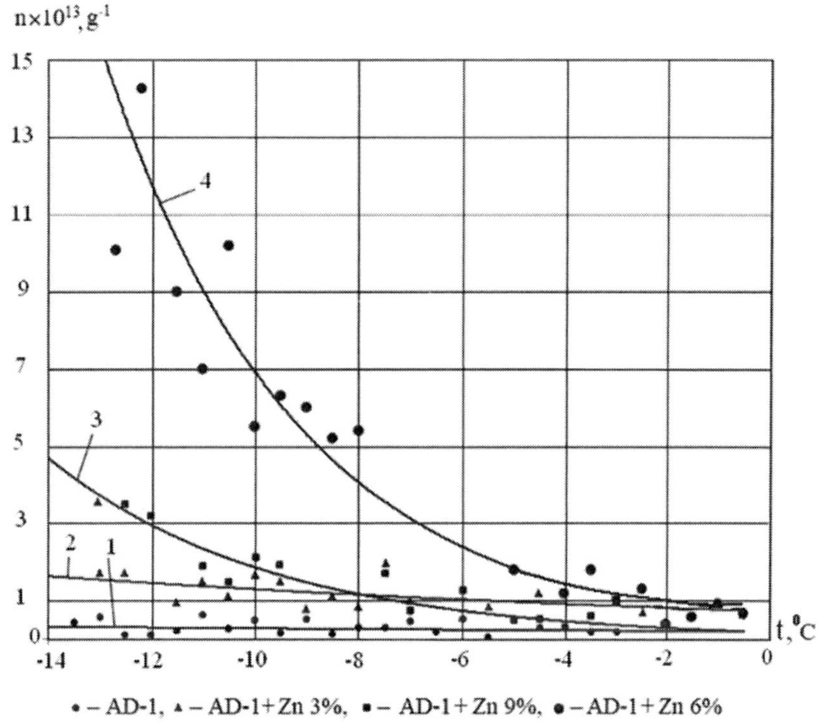

Figure 3. Output of active particles (n) for different concentration of dispersed zinc powder in AD-1 compound

The results of experiments conducted with various mass concentrations of zinc in the range from 0 to 9% are presented in table 2.

Table 2. The specific yield of active particles of the reagent AD-1 with the addition of zinc

t, °C	-11	-9	-7	-5
C_{Zn}, %	Specific output, g^{-1}	Specific output, g^{-1}	Specific output, g^{-1}	Specific output, g^{-1}
0	3.2×10^{12}	2.7×10^{12}	2.5×10^{12}	6.5×10^{11}
1	8.8×10^{12}	7.0×10^{12}	5.6×10^{12}	1.2×10^{12}
2	1.5×10^{13}	1.2×10^{13}	8.2×10^{12}	4.5×10^{12}
3	2.8×10^{13}	2.2×10^{13}	1.1×10^{13}	7.2×10^{12}
4	4.6×10^{13}	3.6×10^{13}	1.6×10^{13}	9.5×10^{12}
4.7	6.5×10^{13}	5.2×10^{13}	2.5×10^{13}	1.1×10^{13}
5	7.0×10^{13}	5.9×10^{13}	3.0×10^{13}	1.3×10^{13}
5.5	7.3×10^{13}	6.2×10^{13}	3.3×10^{13}	1.5×10^{13}
6	7.5×10^{13}	6.3×10^{13}	3.4×10^{13}	1.6×10^{13}
6.6	6.5×10^{13}	5.2×10^{13}	2.5×10^{13}	1.4×10^{13}
7	5.2×10^{13}	4.3×10^{13}	1.9×10^{13}	1.0×10^{13}

8	3.4×10^{13}	3.0×10^{13}	1.4×10^{13}	7.2×10^{12}
9	2.2×10^{13}	1.9×10^{13}	1.0×10^{13}	5.3×10^{12}

Note: the top line shows the specific yield of active particles of the AD-1 reagent at given temperatures.

The specific yield of active reagent particles depending on the concentration of powdered zinc is shown in figure 4. From the presented graph it is seen that the maximum yield of active ice-forming particles is provided in the range of zinc concentrations of 4.7-6.6 mass. % . This area in Fig. 4 is limited by vertical lines. From the above materials it follows that the presence of finely dispersed zinc powder in the initial ice-forming fuel in the ratio to the total mass of 6% sharply increases the yield of ice-forming particles in the entire range of accepted temperatures, from zero to minus 14 ^{0}C. This increase in the yield of active ice-forming particles in the temperature range under study is explained by the fact that zinc oxide nanotubes of various modifications and sizes that actively interact with a supercooled cloud medium in the accepted temperature range are formed at a high combustion temperature of the pyrotechnic composition. So, for example, at a temperature of minus 12 ^{0}C (curve 3 on the graph), the yield of ice-forming particles increases by almost an order of magnitude, and in the temperature range from minus from 2 ^{0}C to minus 4 ^{0}C - almost doubles, which makes it possible to affect warmer supercooled part of the cloud.

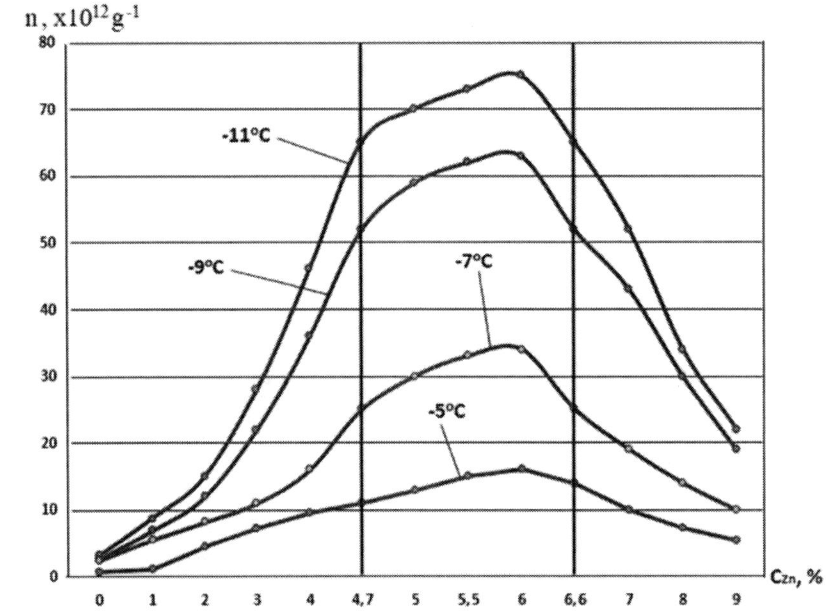

Figure 4. Dependence of specific output of ice-forming particles from zinc concentration in pyrotechnical compound AD-1 for different temperatures

Summary

1. The work presents the results of original studies, which showed that the presence of finely dispersed zinc powder in the composition of the initial ice-forming fuel AD-1 in relation to the total mass of 6% sharply increases the yield of ice-forming particles over the entire range of accepted temperatures.
2. The specific yield of ice-forming particles at a temperature of minus 12 ^{0}C increases almost by an order of magnitude, and in the temperature range from minus 2^{0}C to minus 4^{0}C - twice.
3. Based on the initial ice-forming fuel AD-1, a methodology has been developed for obtaining an effective pyrotechnic composition for equipping meteorological rockets designed for active impact

on thundercloud clouds with the aim of artificially increasing liquid precipitation and combating hail.

4. This composition can also be used to create promising ground-based generators for active impacts on supercooled mists, in order to provide favorable meteorological conditions for the functioning of spaceports, airports and road transport communications.

References

[1] Nikandrov V Ya 1959 *Artificial effects on clouds and fogs* (Leningrad, Gidrometeoizdat) 191.
[2] Plude N O 1966 To the question of the mechanism of crystallization of supercooled fog with an aerosol of silver iodide *Proceedings of the Main geophysical observatory* **186** 17.
[3] Koloskov B P, Korneev V P, Schukin G G 2012 Methods and means of modification of clouds, precipitation and fog *Monograph. SPb.: RGGMU* 36-38.
[4] Khuchunaev B M, Stepanova S I, Khuchunaev A B, Panaetov V P 2011 Investigation of the ice-forming properties of crystalline hydrates and zinc oxide nanotubes *Report of the All-Russian Conference on Cloud Physics and Active Impacts on Hydrometeorological Processes. Nalchik.* 396-402.
[5] Tapaskhanov V O, Khuchunaev B M, Tlisov M I, Khuchunaev A B, Kupovy G V 2014 Preliminary results of a study of the condensation ice-forming properties of nanotubes and zinc oxide *News of Universities. North Caucasus region. Natural Sciences* **6** 40-43.
[6] Khuchunaev B M, Kupovy G V, Tashilova A A, Teunova N V, Khuchunaev A B, Budaev A Kh 2018 Investigation of ice-forming properties of zinc oxide nanotubes *News of Universities. North Caucasus region. Natural Sciences* **4 (200)** 111-115.

Study of the dispersed composition of dust particles on the leaves of apricot trees (Prúnus armeníaca) in the residential area

I Glinyanova, V Azarov*

Institute of Architecture and Construction, Volgograd State Technical University, 1, st. Akademicheskaya, Volgograd, 400074, Russia

E-mail: azarovpubl@mail.ru

Abstract. The authors investigate and analyze the dispersion composition of dust particles on the leaves of apricot trees (Prúnus armeníaca) in the residential area of the Sredneakhtubinsky district of the Volgograd region near industrial enterprises in comparison with the dispersed composition of dust on the leaves of apricot trees (Prúnus armeníaca) growing on the territory in a conditionally clean zone (control) in the absence of anthropogenic pressure from industrial complexes in the SNT "Oroshenets" (Volgograd, Sovetsky district). The research points in the residential area were taken on the territory of social facilities (boarding school, hospital, kindergarten) and low-rise residential areas. The research material is the leaves of apricot trees (Prúnus armeníaca) in the residential area of the Sredneakhtubinsky district of the Volgograd region and in the SNT Oroshenets, Volgograd, Sovetsky district (conditionally clean zone, (control). Research methods: washing off dust from leaf blades into a glass with distilled water, filtering the suspension through an AFA-VP filter, obtaining a filtrate, natural drying of the filtrate at a temperature not higher than 30-40 ° C, placing the dried filtrate on a glass slide of an optical microscope, making measurements of dust particles, dust processing, study of the dispersed composition of dust (according to GOST R 56929-2016). Areas of environmental risk have been identified, these are the territories of a kindergarten, a hospital and other low-rise residential buildings, as well as places that are safe for the population (boarding school, etc.). These studies indicate environmental pollution, in the future, it is required to determine the chemical composition of dust particles and identify sources of pollution and take environmental measures.

Introduction

The study of the dispersed composition of dust in the atmospheric air of urban agglomerations makes it possible to predict the real ecological situation and, if necessary, develop a number of measures to improve the quality of life and reduce the risk of morbidity among the population. Until recently, it was believed that "dust particles larger than 5 microns do not pose a threat to human health, they are easily retained in the nasal cavity and do not pass into the body, while dust smaller than 5 microns is most dangerous for humans" [1], [2]. However, as shown by numerous studies of recent years, carried out by scientists in different countries of the world, both PM 2.5 and PM10 are sources of various kinds of human diseases. So, Wu RS; Zhong LJ; Huang XL; Xu HB and others found an association between fine dust (PM 2.5, PM 10) and mortality at age 65 from cardiovascular disease and respiratory diseases in China [3]. G. Cesar, A. Cristina, L. Nascimento demonstrated the effect of PM10 on hospitalization of children due to respiratory diseases [4]. Mehdi Mokhtari, Mohammad Miri, Hassan Khorsandi [etc.] and others noted high rates of premature mortality from cardiovascular and respiratory diseases associated with PM 10 and PM 2.5 in Yazd (Iran) [5]. T. Y. Kim, Tae-Young, H. Kim, S. Yi [etc.] have also confirmed the short-term impact of PM 2.5 and PM 2.5-10 on mortality among the population in large cities in South Korea [6].

R. Rumi, S. Ghosh, P.K. Padhy and others confirmed changes in hematological parameters as well as oxidative stress in rural women from tribal areas in northeastern India exposed to PM 10 and PM 2.5 [7]. G. Meng, B. Gufran, S. Shaojie [etc.] calculated annual changes in population mortality and life expectancy depending on the province due to the emissions of pollutants from power generation that generate PM2.5 into the environment [8]. T. Chan, Z. Zhang, B. Lin [etc.] investigated the relationship between long-term exposure to PM2.5 in the environment and chronic kidney disease in the population [9]. ZJ Andersen, M. Pedersen, G. Weinmayr [etc.] found compelling evidence for a link between long-term exposure to PM2.5 absorption from the ambient air, which is caused by traffic and the development of malignant brain tumors among urban populations in several European countries (Holland, Sweden, Italy, Austria, Denmark, etc.) [10].

Based on the analyzed works, the particle size has an important effect on the health of the population, therefore, the study of the dispersed composition of dust particles in the urban environment is undoubtedly an important and relevant topic. Monitoring of dust in the atmospheric air of urban areas is carried out traditionally by taking atmospheric air for a certain period of time and identifying fine dust and pollutants in air samples. Along with the traditional method of studying dust particles in the atmospheric air of an urban environment, there is a large number of experimental studies devoted to phytomonitoring of urban areas, namely, the study of dust particles and their characteristics collected on plant leaves. Analysis of works on phytomonitoring by domestic and foreign authors allows us to conclude that plant leaves are excellent indicators and passive biomonitors of fine dust in the atmospheric air of urban areas. Absorbing daily a large amount of dust-like particles floating in the atmospheric air, leaf blades of plants reflect the real ecological situation in the investigated area, since they have the property of accumulating dust on their surface over a certain period of time. The study of the dispersed composition of dust on plant leaves (for example, the leaves of apricot trees (Prúnus armeníaca)) can give an objective picture of the ecological situation in the residential area. The hypothesis of this scientific study is the following: the study of the dispersed composition on the leaves of trees (for example, the leaves of apricot trees (Prúnus armeníaca) of the residential zone in the Sredneakhtubinsky district of the Volgograd region) will identify areas of environmental risk and safe places to live in the residential zone.

The aim of the work is: to study the dispersion composition of dust particles on plant leaves (using the example of apricot trees (Prúnus armeníaca)) growing in the residential area of the Sredneakhtubinsky district of the Volgograd region in comparison with the conditionally clean zone (control) in the SNT Oroshenets.

Research objectives:

1) selection of the territory of the residential area, determination of sampling points and research material

2) collection of leaves from apricot trees (Prúnus armeníaca) in the residential area at the selected sampling points and in the conditionally clean area (control) in the SNT "Oroshenets", Sovetsky district of Volgograd

3) extraction of dust-like particles from leaves of apricot trees (Prúnus armeníaca) and study of the dispersed composition of dust

4) comparative analysis of the dispersed composition of dust particles collected from the leaves of apricot trees (Prúnus armeníaca) in the residential zone of the Sredneakhtubinsky district of the Volgograd region with the dispersed composition of dust particles in a conditionally clean zone (control)

5) conclusions and recommendations based on the research results.

Analysis of publications

Analysis of publications on the study of the dispersed composition of dust on plant leaves led to the conclusion that in Russia most of the works on this topic are devoted to the study of dust in the traditional way in an urban environment: sampling of atmospheric air in residential areas with subsequent analysis of dust particles. So, for example, Zaitseva N.V., May I.V., Maks A.A., Zagorodnov S.Yu. assessed the exposure of the population in the zones of influence of emissions from industrial stationary sources of enterprises of machine-building and metallurgical profiles in the study of the dispersed and component composition of the solid component of dust [11]. Golokhvast K.S. studied nano- and micro-sized particles of atmospheric suspensions and their ecological effect in the cities of the Far East [12]. Somewhat later, Golokhvast K.S., Revutskaya I.L. and others investigated the distribution of dust particles by seven classes in the snow of Birobidzhan and in the “Bastak” reserve zone. They showed that in the atmospheric suspensions of the city of Birobidzhan, in ecologically significant quantities (up to 72.4%), technogenic microparticles are contained. The authors noted that the city of Birobidzhan has a negative effect on the composition of atmospheric suspended matter in the Bastak reserve [13]. Prosviryakova I.A., Shevchuk L.M. presented the results of a hygienic assessment of the content of fine particulate matter in the ambient air in a residential area located in the zone of influence of vehicle emissions. The authors analyzed the fractional composition of solid particles, carried out a hygienic assessment of the degree of air pollution, and determined the levels of risk to public health caused by air pollution of residential areas with fine particles [14].

There are few results of analyzes of Russian researchers who studied dust on plant leaves, but without analyzing their fractional composition.

So, Chernyshenko O.V. studied the leaves of urban trees and shrubs (Populus balsamifera; Pópulus nígra; Symphoricarpos albus; Sambúcus nígra) according to the intensity of dust accumulation on them, but the dispersion of dust was not considered by the author [15]. Ageeva E.A., Kazantseva M.N. assessed the dust-holding capacity of leaves of trees and shrubs (Sorbus sibirica, Cotoneaster melanocarpu, etc.) in Tyumen. Analysis of the data obtained showed that the average amount of dust deposited by the leaves of urban plants is significantly higher than in the green zone of the city. The dispersion of dust was not considered by the authors [16]. L. I. Atkina, M. V. Ignatova studied the amount of dust trapped by leaves on trees of Malus baccata L., Sorbus aucuparia L., Acer negundo L., Crataegus sanguinea L. in urban plantings of Yekaterinburg, growing along the highway, nevertheless, they did not study the diameters of dust particles [17].

As for foreign scientists, in this direction there is a fairly large number of works devoted to the study of fine dust on plant leaves in an urban environment. Thus, A. Przybysz, A. Saebo, H. Hanslin [etc.] monitored the accumulation of solid particles and trace elements on the leaf blades of evergreen species (Taxus baccata L, Hedera helix L, and Pinus sylvestris L.) in urban and rural areas taking into account the amount of precipitation and time. The authors found the greatest accumulation of pollutants in the air in the foliage of plants protected from rain; in places prone to traffic pollution, and the least dust accumulation was found in the leaves of rural plants. Among the analyzed species, P. sylvestris had the greatest amount of precipitated fine particles and microelements. These results showed that when assessing the overall effect of vegetation on pollutant recovery, it is necessary to take into account both the dynamics of deposition and leaching of leaves by rains during the season [18]. Later, R. Popek, A. Lukowski, C. Bates [etc.] carried out a series of studies in five Polish cities to study the fractions of fine dust, heavy metals and PAHs on plant leaves (Tilia cordata Mill). The authors found that strong winds reduced the amount of fine dust on the leaves, especially the finest fractions, but they did not find any connection with precipitation [19]. G. Sgrigna, A. Saebo, S. Gawronski [etc.] studied fine dust of different fractions: from 0.2 to 2.5 microns and from 2.5 to 10 microns on the leaves of a plant - Quercus ilex, incl. and deposition of PM 2.5; PM 10 in four districts of Terni (Italy). In this work, changes were found in the deposition of fine dust, which correlated, according to the authors, “with the distance to the main roads and the position downwind relative to the industrial zone” [20]. Y. Song, B. Maher, F. Li [etc.] carried out studies of fine dust on the leaves of trees of five evergreen plants in Beijing (China), studying their morphology, size, elemental composition, incl. their mass concentration [21]. J. Shi, G. Zhang, H. An [etc.] studied the elemental composition, density and size fractions of fine particles on the leaf surfaces of 14 urban green plants (Salix matsudana, Euonymus japonicus, Magnolia denudate, Sophora japonica, etc.) from the overpass Xizhimen and to the Olympic Forest Park in Beijing (China) [22].

L. Mo, Z. Ma, Y. Xu [etc.] evaluated dust deposition on leaves and wax layer of 35 species (11 shrubs, 24 trees) in Beijing, China. Differences in the accumulation of fine dust between the species were found. Thus, Cephalotaxus sinensis, Euonymus japonicus, Broussonetia papyrifera, Koelreuteria paniculata, and Quercus variabilis were effective in capturing small particles. The results of this study can help to select species for urban green spaces, the purpose of which is to capture air pollutants and mitigate the adverse health effects of air pollution [23]. L. Lin, J. Yan, K. Ma [etc.] presented a new approach to the quantitative characterization of particulate matter deposited on the leaves of urban trees (Salix matsudana, Ailanthus altissima, Fraxinus chinensis), which consists in the accurate determination of the amount, size, shape and the spatial distribution of particles with different diameters on the leaves [24]. S. Janhall conducted an analytical review of studies devoted to the deposition rate of fine dust on the leaves of urban plants, the study of the density of PM2.5 and PM10, PM> 10 on the leaves; the relationship between aerodynamic drag and dispersion of dust particles, etc. and came to an interesting conclusion that the deposition of large particles on plant leaves is more effective at high wind speeds. In her opinion, urban vegetation affects air quality precisely through the effects of the deposition and dispersion of pollutants [25].

Y. Xu, W. Xu, L. Mo [etc.] studied the quantitative assessment of dust particles of three fractions: (0.2-2.5 microns), (2.5-10 microns), (> 10 microns), accumulated by 17 species of urban plants in Beijing (China). The authors concluded that the greatest accumulation of dust by weight on the leaves was accounted for by particles of the fraction:> 10 microns, while the accumulation of fractions: 0.2-2.5 microns and 2.5-10 microns was less [26]. U. Weerakkody, J. W. Dover, P. Mitchell [etc.] were also engaged in the quantitative assessment of dust particles on the leaves of plants of green walls in the city, where they found the largest number of particles of all sizes on the needles of Juniperus chinensis L. This study highlighted the importance of individual leaf size in capturing fine dust, regardless of their changing micromorphology [27]. S. Singh, P. Bhattacharya, N. Gupta studied the quantitative deposition of dust on the deciduous surface of urban trees: Thevetia peruviana, K. Schum in industrial, residential and other zones of Delhi (India). Their experiment gave a positive assessment to these plant species as tolerant bioindicators of atmospheric air pollution [28]. A. Przybysz, G. Nersisyan, S. Gawronski revealed the ability of evergreen conifers (Taxus baccata L. and Pinus nigra Arn.), As well as deciduous plants (Carpinus betulus L.) to remove fine dust from the atmospheric air with urban greenery in winter [29]. X.Sun, H. Li, X. Guo [etc.] determined the diameter of dust particles and the area they occupied on plant leaves (Euonymus japonicus, Pyracantha fortuneana, Ligustrum vicaryi, Amygdalus triloba, Ligustrum sinense, etc.). According to the authors, the structure of the leaf surface significantly influenced the ability of plants to retain fine dust: plants with a thick waxy layer or large and dense stomata, for example (E. japonicus), adsorbed more fine dust. This study provides a scientific basis for the ability of landscape plants to retain particles of different diameters [30]. Meanwhile, some researchers on dust on the leaves of urban plants emphasized the fact that this type of monitoring can be used as an alternative to traditional monitoring of air pollution. Thus, M. Zampieri, J. Sarkis, Pestana, C. Rafael [etc.], after conducting a study of dust particles on plant leaves in an urban environment, they concluded that, for example, the T. granulosa can be used as a passive biomonitor and a valuable alternative to air pollution monitoring and spatio-temporal assessment of fine dust [31]. T. Zhang, Y. Bai, X. Hong [etc.] measured the amount of fine dust and heavy metals deposited on the leaves of the Euonymus japonicus plant in Beijing (China) during the East Asian monsoon, noting that plants can be effectively used as biomonitors of environmental pollution [32]. L. Lin, G. Chen, J. Yan [etc.] conducted a factor analysis of the landscape metrics of dust particles deposited on leaf surfaces (Ailanthus, Ash, and Willow). As a result of the conducted research L. Lin, G. Chen, J. Yan [etc.] believe that it is possible to provide a high spatial monitoring of dust in the urban environment using indicators such as dust particles on the leaves of urban plants [33].

This work is aimed at studying the dispersion composition of dust particles on plant leaves (using the example of apricot trees (Prúnus armeníaca)) as passive biomonitors of atmospheric air quality and identifying areas of environmental risk in the residential zone of the Sredneakhtubinsky district of the Volgograd region.

Research materials and methods

In this case, the research material was the leaves of apricot trees (Prúnus armeníaca), which grow on this territory of the residential zone. This is a massive species found in the city of Volgograd and the Volgograd region with high frequency. The collection of material for the study was carried out at the end of June 2018 after the completion of intensive leaf growth. Leaf samples were taken at a height of 1.5–2 m of the crown (the height of the layer of atmospheric air inhaled by humans). The leaves were taken at 6 points at different distances from the industrial zone: from 150 m to 800 m, in different cardinal points, with different wind loads. When collecting leaves, attention was paid to the condition of the leaf blade: it should not be subject to changes, for example, traces of insect activity or the presence of bacterial necrosis or any other damage. 10 samples were made at each point. Each sample represents leaves from one species of apricot tree (Prúnus armeníaca). Leaf blades were selected in 10 replicates, using several apricot trees (Prúnus armeníaca) (up to 3 pcs.) At each point. At the same time, the leaves of an apricot tree (Prúnus armeníaca) were sampled in a conditionally clean zone (control) in the SNT "Oroshenets" on the bank of the Varvarovsky reservoir (Sovetsky district of Volgograd). All collected leaves for each sample were placed in paper bags. All packages were signed with the date and time of material collection, then they were delivered to the laboratory, avoiding shaking. The dust from the leaves of one sample was washed into a beaker with distilled water. The resulting suspension was filtered through the middle of the AFA-VP filter and dried at a temperature of no more than 30-40 ° C (the natural temperature of atmospheric air in summer in Volgograd and the Volgograd region). The dried filtrate (dust from leaves) was placed evenly on a glass slide, which was mounted on an optical microscope. Each sample contained a wash of dust from 10-15 leaves of one plant species (Prúnus armeníaca). At each point in the study, at least 1000 dust particles were collected.

At the second stage, a study of the dispersion composition of dust collected on the leaves of apricot trees (Prúnus armeníaca) in the residential area of the Sredneakhtubinsky district of the Volgograd region and on the leaves of apricot trees (Prúnus armeníaca) in the conditionally clean zone (control) in the SNT "Oroshenets" on the bank of the Varvarovsky reservoir was carried out (Soviet district of Volgograd). Measurements of dust particles, dust processing, study of the dispersed composition of dust were carried out in accordance with clauses 11-13 of GOST R 56929-2016 [34] using the SPOTEXPLORER computer program, which allows digital processing of black and white images in Windows Bitmap format (*. bmp), by the volume of a dusty particle, its equivalent diameter was calculated, and the number of particles of various sizes was determined. After scanning the photograph, the dispersed composition of the general population of dust was determined. The measurement results were drawn up in the form of tables, on the basis of which the differential and integral curves of the distribution of the mass of particles by size were constructed, plotting the values (particle diameter) on the abscissa axis, and on the ordinate axis - the distribution density of particles of the corresponding size in percent. Thus, not only the dispersed composition of the general population of dust was determined, but also its finely dispersed component according to GOST R 56929-2016.

At the 3rd stage, a comparison and analysis of the dispersion compositions of dust obtained as a result of sampling dust particles from the leaves of apricot trees (Prúnus armeníaca) in the residential area of the Sredneakhtubinsky district of the Volgograd region, from the leaves of apricot trees (Prúnus armeníaca) in a conditionally clean zone (control) in SNT "Oroshenets" on the bank of the Varvarovsky reservoir (Sovetsky district of Volgograd).

IOP Conf. Series: Materials Science and Engineering 1083 (2021) 012098 doi:10.1088/1757-899X/1083/1/012098

Results and their analysis

The studies were carried out in a residential area near residential buildings (points: No. 4, No. 5, No. 6) and social facilities (hospital - point No. 1, kindergarten - point No. 3, boarding school - point No. 2) in the Sredneakhtubinsky district of the Volgograd region.

The material for the study was the leaves of apricot trees (Prúnus armeníaca), which grow on the territory of the residential area at points : №1,2,3,4,5,6. Leaf blades were collected in dry weather at the end of June 2018. A total of 600 pieces were collected at six points. Apricot tree leaves (Prúnus armeníaca), 100 pcs. leaf blades at each point.

Dust was found on the leaf blades of apricot trees (Prúnus armeníaca). Dust particles on the leaves of common apricot (Prúnus armeníaca) were studied using a Versa 3D scanning microscope and are shown in Figure 1. The leaf blades (Prúnus armeníaca) shown in Figure 3, as an example, were collected at point 3 (Kavkazskaya street (1) in the Sredneakhtubinsky district of the Volgograd region.

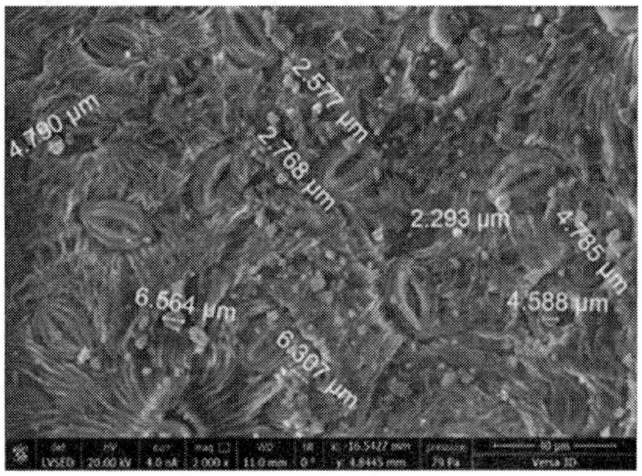

Figure 1. Photo of dust particles with indicated dimensions on a leaf blade of an ordinary apricot (Prúnus armeníaca)

Initially, studies were carried out of the dispersed composition of dust particles on the leaves of apricot trees (Prúnus armeníaca) in the conditionally clean zone (control) in the SNT "Oroshenets" of the Soviet district of Volgograd, which were subsequently compared with data from the residential zone. The distribution of the number of particles on the leaves of apricot trees (Prúnus armeníaca) from the conditionally clean zone by equivalent diameters (N (dh),%) according to PM2.5 was no more than 12.16%, PM10 - 18.73%; distribution of mass of particles over equivalent diameters (D (dp),%) according to PM 2.5 - 16.17%, and according to PM10 - 21.16%.

The research was carried out in 6 points of the residential zone of the Sredneakhtubinsky district of the Volgograd region.

Point number 1 - this is the territory of the hospital on the Hospital st. (Sredneakhtubinsky district of the Volgograd region). This is the western point and it is located approximately 660 meters from the industrial zone. The research results for this area are shown in Fig. 2,3. As an illustration, Fig. 2 shows the distribution of dust particles in the form of 10 integral curves constructed in a probabilistic-logarithmic grid based on 10 samples of dust particles taken from leaf blades (100 pcs.) Of apricot trees (Prúnus armeníaca) at each point under study, which contained at least 1000 pieces. Dust particles studied with an optical microscope. The abscissa axis (x) plotted the particle diameter (dp, µm), and the ordinate (y) - the mass fraction of particles (D (dp),%).

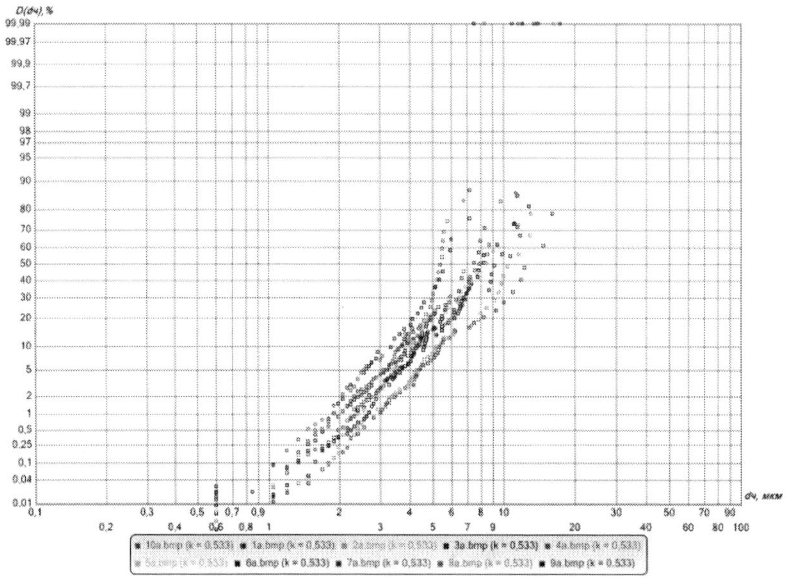

Figure 2. Integral curves of particle mass distribution over equivalent diameters in the probabilistic-logarithmic grid for dust sampled in the residential area on the street. Hospital (Sredneakhtubinsky district, Volgograd region) - Point number 1

Figure 3 shows diagrams of mass distribution (D (dp),%) and amount (N (dp),%) of dust particles by equivalent diameters along the Hospital st.

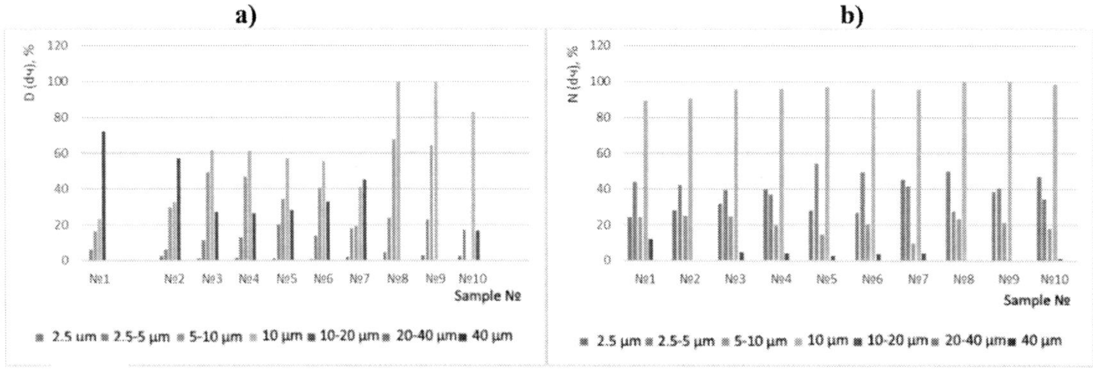

Figure 3. Distribution of mass and number of particles by equivalent diameters,% for dust sampled in a residential area on the street. Hospital, (Sredneakhtubinsky district, Volgograd region), point number 1: a) Diagram of particle mass distribution by diameter (D (dp),%) b) Diagram of the distribution of the number of particles by diameter (N (dp),%)

Point number 2 - this is the area around the boarding school (Vorovskogo street, Sredneakhtubinsky district, Volgograd region) and it is located in the southeast. The distance to the industrial zone is approximately 630 m. The research results for this territory are shown in Figure 4,5.

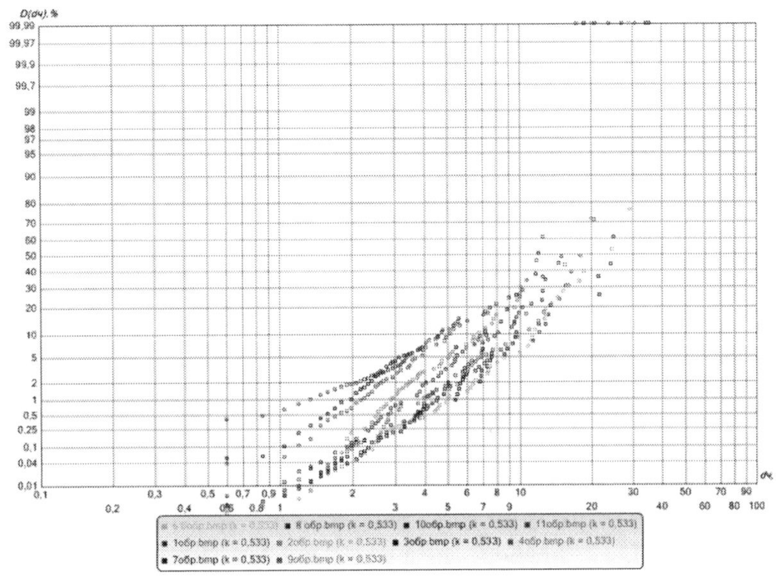

Figure 4 Integral curves of particle mass distribution over equivalent diameters in the probabilistic-logarithmic grid for dust sampled in the residential area on the street. Vorovskogo (Sredneakhtubinsky district, Volgograd region), point number 2

Figure 5 shows diagrams of mass distribution (D (dp),%) and number (N (dp),%) of particles by equivalent diameters along the Vorovsky st.

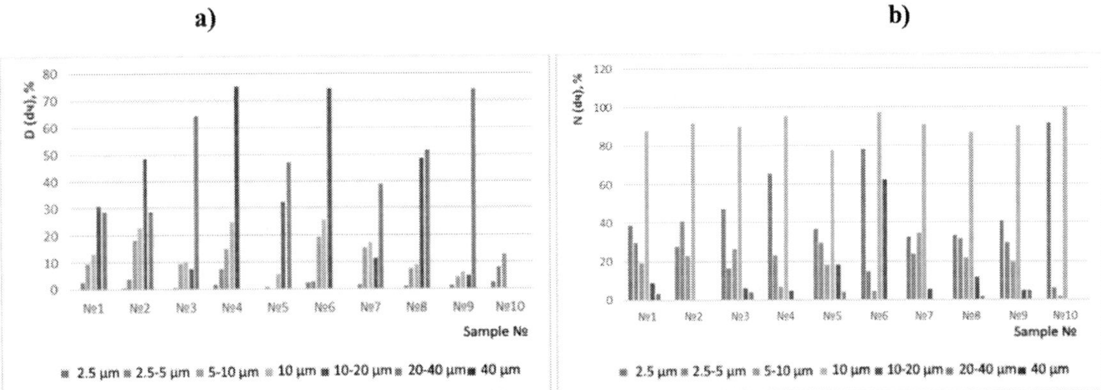

Figure 5. Distribution of mass and number of particles by equivalent diameters,% for dust sampled in a residential area on the street. Vorovskogo, (Sredneakhtubinsky district, Volgograd region), point number 2: a) Diagram of particle mass distribution by diameter (D (dp),%); b) Diagram of the distribution of the number of particles by diameter (N (dp),%)

Point number 3 - this is the territory of the kindergarten on the Kavkazskaya st. (1) in the Sredneakhtubinsky district of the Volgograd region. The research results for this area are shown in Fig. 6.

Figure 6 shows diagrams of particle mass distribution (D (dp), number of particles by (N (dp),%) by equivalent diameters.

a) b)

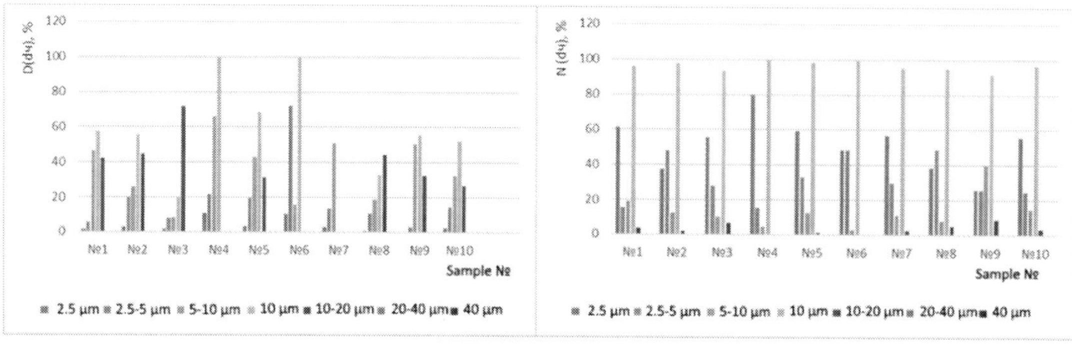

Figure 6. Distribution of mass and number of particles by equivalent diameters,% for dust sampled in a residential area on Kavkazskaya street (1), (Sredneakhtubinsky district, Volgograd region), point No. 3: a) Diagram of particle mass distribution by diameter (D (dp),%); b) Diagram of the distribution of the number of particles by diameter (N (dp),%)

Point number 4 - this is the territory of residential buildings, in the immediate vicinity of industrial enterprises, 300 meters from the main sources of emissions. The research results for this area are shown in Fig. 7.

Figure 7 shows the diagrams of the distribution of the mass of particles by diameter (D (dp),%) and the number of particles by diameter (N (dp),%).

a) b)

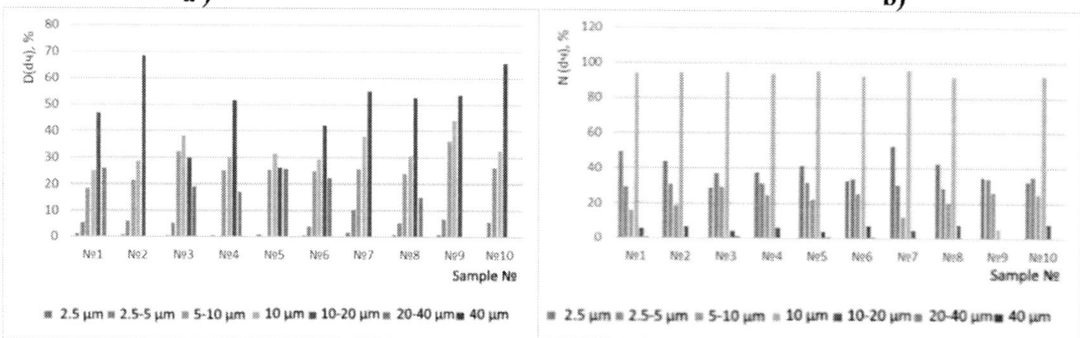

Figure 7. Distribution of mass and number of particles by equivalent diameters,% for dust sampled in a residential area on Kavkazskaya street (2), (Sredneakhtubinsky district, Volgograd region), point No. 4: a) Diagram of particle mass distribution by diameter (D (dp),%); b) Diagram of the distribution of the number of particles by diameter (N (dp),%)

Point number 5 - this is the territory of low-rise residential buildings in the north-west of the study area on the Nechayeva st. in the Sredneakhtubinsky district of the Volgograd region at a distance of less than 200 meters from the industrial zone. The research results for this territory are shown in Figure 8,9.

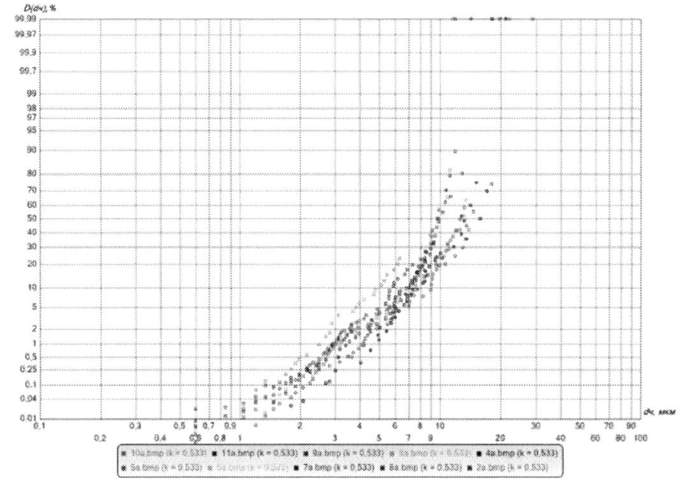

Figure 8. Integral curves of particle mass distribution over equivalent diameters in the probabilistic-logarithmic grid for dust sampled in a residential area on the Nechaev st. (Sredneakhtubinsky district, Volgograd region), point number 5

Figure 9 shows diagrams of particle mass distribution by equivalent diameters (D (dp),%) and the number of particles by diameter (N (dp),%).

a) **b)**

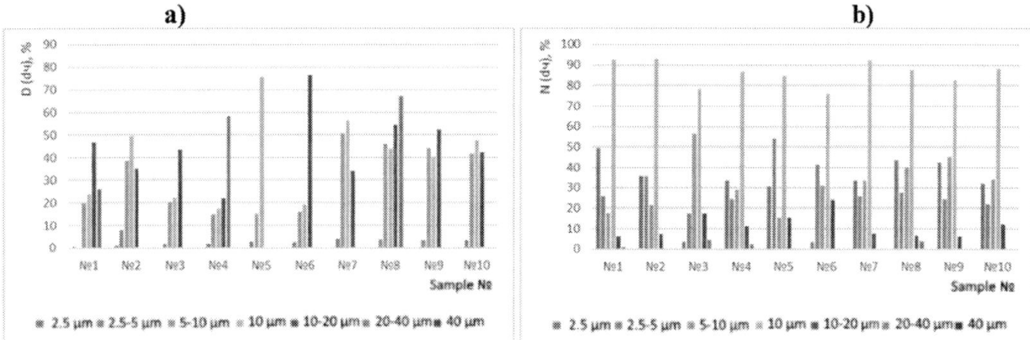

Figure 9. Distribution of mass and number of particles by equivalent diameters,% for dust sampled in a residential area on Nechaev st., (Sredneakhtubinsky district, Volgograd region), point 5: a) Diagram of particle mass distribution by diameter (D (dp),%); b) Diagram of the distribution of the number of particles by diameter (N (dp),%)

Point number 6 - this is the territory of low-rise residential buildings in the northeastern part of the investigated residential area on the Omskaya st. in the Sredneakhtubinsky district of the Volgograd region, 150 m from the industrial zone. The research results for this area are shown in Fig. 10,11.

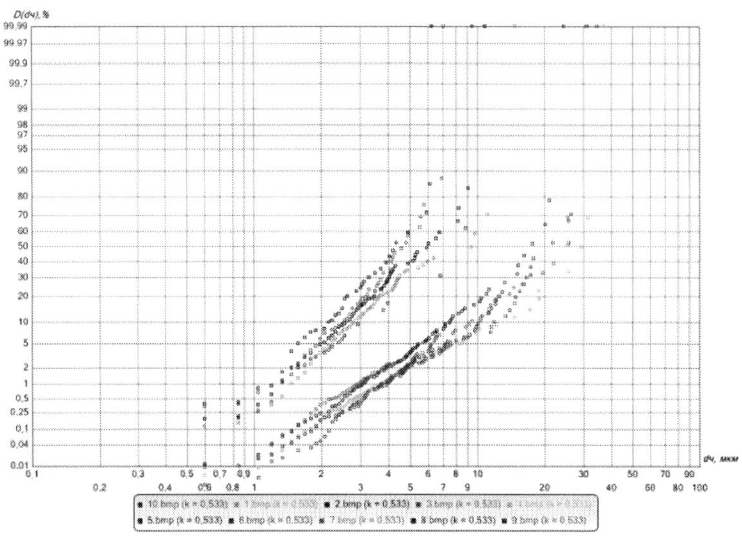

Figure 10. Integral curves of particle mass distribution over equivalent diameters in the probabilistic-logarithmic grid for dust sampled in a residential area on the Omskaya st. (Sredneakhtubinsky district, Volgograd region), point number 6

Figure 11 shows the diagrams of the distribution of the mass of particles (D (dp),%) and the number of particles (N (dp),%) by equivalent diameters.

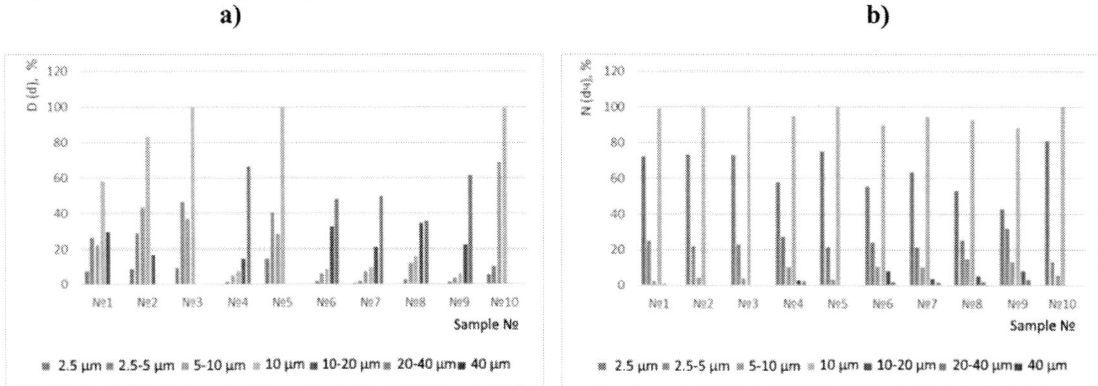

Figure 11. Distribution of mass and number of particles by equivalent diameters,% for dust sampled in a residential area on the Omskaya st., (Sredneakhtubinsky district, Volgograd region): a) Diagram of particle mass distribution by diameter (D (dp),%); b) Diagram of the distribution of the number of particles by diameter (N (dp),%)

From the presented figures, it can be concluded that the safest territory in the residential area is point No. 2 (boarding school on Vorovskogo street, Sredneakhtubinsiy district of the Volgograd region), heavier fractions were recorded on it, size: 10-20 microns and 20-40 μm, although the number of leaders is RM 10. This point is located at a distance of more than 600 m from the industrial zone.

Points No. 1,5,6 are located in the immediate vicinity of industrial enterprises, at a distance of about 200-250 m and represent zones of environmental risk for living, since the largest amount of fine dust (PM10) was recorded on the leaves of apricot trees, both in terms of quantity and weight. Dust particles of heavy fractions (20-40 microns) were found at point No. 6 in large quantities, but a fraction of 20-40 microns was not found at point No. 5. This indicates that the direction and speed of the wind have a great influence on the distribution of dust particles. It is also worth paying attention to point number 1, since at this point there is a social object - a hospital complex, and at this point the most fine dust was found - PM 10, both in quantity and in mass.

Points No. 3 also falls into the zone of environmental risk, although it is located at a distance of more than 800 m from the industrial zone, but PM 10 was found in it in large quantities, both in mass and in their number. A kindergarten is located at this point.

Points No. 4 occupies a borderline security position. On the one hand, a high percentage of fine dust (PM 10) was recorded in it in terms of quantity, however, in terms of mass, dust particles divide the percentage between PM 10 and a fraction of 10-20 microns, they in the samples at this point turned out to be practically in equal proportions.

The distribution of dust particles in the analyzed points is due to the absence of precipitation from April to July 2018 in the study area, but depended on the speed and direction of the wind and the number of days of wind loads, which are shown in Figure 12.

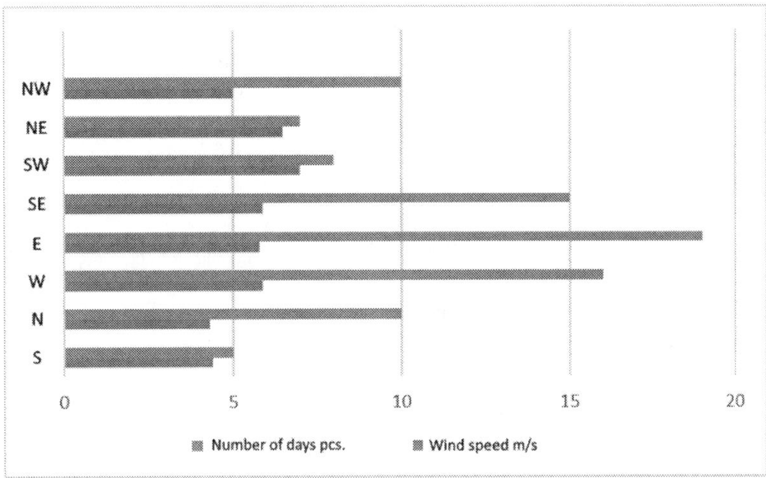

Figure 12. Diagram of wind direction and its speed versus number of days (March-June 2018)

It can be seen from the diagrams in Figure 16 that the prevailing winds: west, east, southeast did not affect the spread of dust particles, since the enterprise is located in the study area in its northern part, and the study of the residential zone took place south of the industrial zone.

However, the north wind, which prevailed for 10 days at a speed of 4.3 m / s, could have a significant effect on the spread of fine dust from the industrial zone to point No. 3, which can be attributed to the zone of environmental risk, there is a kindergarten, it is got into the zone of influence of the north wind, despite the fact that the territory of the kindergarten is 800 m away from the industrial zone. Nevertheless, it is known that fine dust can hover in the air for a long time, for 3-14 days or more, and in the absence of influence from precipitation, winds can be carried over long distances. The north wind could have affected the territories of points 4 and 5, where the largest amount of PM10 was found, up to 90%, and heavy fractions, which> 10 μm, predominate in weight.

The northeast wind, which blew for 7 days at a speed of 6.5 m / s, influenced point No. 6.2. The northwest wind stayed only 10 days during 3 months, with an average speed of 5 m / s and could also have an effect on point No. 1.

Summary

As a result of the studies, the dispersed composition of dust on the leaves of apricot trees (Prúnus armeníaca) in the residential area of the Sredneakhtubinsky district of the Volgograd region was revealed. Found fine particles: PM2.5, PM10, which in their values (N (dp),%), (D (dp),%) significantly exceed the data on fine dust in a conditionally clean zone (control) in SNT "Oroshenets" (Sovetsky district of Volgograd). This creates certain environmental risks for local residents. The excess of fine dust in the study area can be due to various factors: both natural and technogenic pollution of the environment. Therefore, in the future, simultaneously with studies of the dispersed composition of dust, it will be chemically analyzed in order to quickly detect sources of pollution in residential areas in Volgograd and the Volgograd region and develop environmental measures.

References

[1] Kouzov P A 1987 *Fundamentals of the analysis of the dispersion composition of industrial dust and grinded materials* (Chemistry, Leningrad).
[2] Belova T I, Shkrabak V S, Agashkov E M 2017 Results of a laboratory study of the dispersed composition of red beet food concentrate dust in the dust extraction system *Bulletin of the St. Petersburg Agrarian University* **4 (49)** 301-307.
[3] Wu R S , Zhong L J, Huang X L, Xu H B [etc.] 2018 Temporal variations in ambient particulate matter reduction associated short-term mortality risks in Guangzhou, China: A time-series analysis (2006-2016) *Science of the total environment* **645** 491-498.
[4] Cesar G, Cristina A, Nascimento L 2018 Coarse particles and hospital admissions due to respiratory diseases in children. An ecological time series study *Sao Paulo medical journal* **136(3)** 245-250.
[5] Mokhtari M, Miri M, Khorsandi H [etc.] 2015 Assessment of Air Quality Index and Health Impact of PM10, PM2.5 and SO2 in Yazd, Iran *Journal of Mazandaran University of Medical Sciences* 14-23.
[6] Kim T, Young T, H Kim S Yi [etc.] 2018 10th Asian Aerosol Conference (AAC): Korea Assoc Particle & Aerosol Res *South Korea aerosol and air quality research* **18 (7)** 1853-1862.
[7] Rumi R, Ghosh S, Padhy P 2018 Indoor air pollution in rural north-east India: Elemental compositions, changes in haematological indices, oxidative stress and health risks *Ecotoxicology and environmental safety* **165** 393-403.
[8] Meng G, Gufran B, Shaojie S [etc.] 2018 The impact of power generation emissions on ambient PM2.5 pollution and human health in China and India *Environment International* **121** 250-259.
[9] Chan T, Zhang Z, Lin B [etc.] 2018 Long-Term Exposure to Ambient Fine Particulate Matter and Chronic Kidney Disease: A Cohort Study *Environmental Health Perspectives* **126 10**.
[10] Andersen Z, Pedersen M, Weinmayr G [etc.] 2018 Long-term exposure to ambient air pollution and incidence of brain tumor: the European Study of Cohorts for Air Pollution Effects (ESCAPE) *Neuro-Oncology* **20 3** 420-432.
[11] Zaitseva N V, May I V, Max A A, Zagorodnov S Yu 2013 Analysis of the dispersed and component composition of dust for assessing the exposure of the population in the zones of influence of emissions from industrial stationary sources *Hygiene and Sanitation* **92 (5)** 19-23.
[12] Golokhvast K S 2013 Atmospheric suspensions of the cities of the Far East (FEFU Publishing House, Vladivostok).
[13] Golokhvast K S, Revutskaya I L et al. 2016 Nano- and microdimensional air pollution of the BASTAK reserve caused by the technogenic influence of the city of Birobidzhan *Bulletin of physiology and pathology of respiration* **61** 36-41.
[14] Prosviryakova I A, Shevchuk L M 2018 Hygienic assessment of the content of PM 10 and PM 2.5 particulate matter in the ambient air and the health risk of residents in the zone of influence of emissions from stationary sources of industrial enterprises Health risk analysis **2** 14–22.
[15] Chernyshenko O V 2012 Dust-filtering ability of woody plants *Lesnoy Vestnik* **3** 7-10.
[16] Ageeva E A, Kazantseva M N 2012 Assessment of the dust-holding capacity of leaves of trees and shrubs in the plantation of Tyumen *Actual problems of forestry* **31** 88-91.
[17] Atkina L I, Ignatova M V 2014 Features of dust-holding capacity of leaves of Malus baccata L., Sorbus aucuparia L., Acer negundo L., Crataegus sanguinea L. in urban plantings of Yekaterinburg *Forests of Russia and economy in them* **4 (51)** 79-82.
[18] Przybysz A, Saebo A, Hanslin H [etc.] 2014 Accumulation of particulate matter and trace elements on vegetation as affected by pollution level, rainfall and the passage of time *Science Of*

The Total Environment **481** 360-369.

[19] Popek R, Lukowski A, Bates C [etc.] 2017 Accumulation of particulate matter, heavy metals, and polycyclic aromatic hydrocarbons on the leaves of Tilia cordata Mill. in five Polish cities with different levels of air pollution *International Journal Of Phytoremediation* **19 (12)** 1134-1141.

[20] Sgrigna G, Saebo A, Gawronski S [etc.] 2015 Particulate Matter deposition on Quercus ilex leaves in an industrial city of central Italy *Environmental Pollution* **197** 187-194.

[21] Song Y, Maher B, Li F [etc.] 2015 Particulate matter deposited on leaf of five evergreen species in Beijing, China: Source identification and size distribution *Atmospheric Environment* **105** 53-60 .

[22] Shi J, Zhang G, An H [etc.] 2017 Quantifying the particulate matter accumulation on leaf surfaces of urban plants in Beijing, China *Atmospheric Pollution Research* **8 (5)** 836-842.

[23] Mo L, Ma Z, Xu Y [etc.] 2015 Assessing the Capacity of Plant Species to Accumulate Particulate Matter in Beijing, China *Plos One* **10 (10)** 8-18.

[24] Lin L, Yan J, Ma K [etc.] 2017 Characterization of particulate matter deposited on urban tree foliage: A landscape analysis approach *Atmospheric Environment* **171** 59-69.

[25] Janhall S 2015 Review on urban vegetation and particle air pollution - Deposition and dispersion *Atmospheric Environment* **105** 130-137.

[26] Xu Y, Xu W, Mo L [etc.] 2018 Quantifying particulate matter accumulated on leaves by 17 species of urban trees in Beijing, China *Environmental Science And Pollution Research* **25 (13)** 12545-12556.

[27] Weerakkody U, Dover J, Mitchell P [etc.] 2018 Quantification of the traffic-generated particulate matter capture by plant species in a living wall and evaluation of the important leaf characteristics *Science Of The Total Environment* **635** 1012-10-24.

[28] Singh S, Bhattacharya P, Gupta N 2018 Dust particles characterization and innate resistance for Thevetia peruviana in different land-use pattern of urban area *International Journal Of Environmental Science And Technology* **15 (5)** 1061-1072.

[29] Przybysz A, Nersisyan G, Gawronski S 2019 Removal of particulate matter and trace elements from ambient air by urban greenery in the winter season *Environmental Science And Pollution Research* **26 (1)** 473-482.

[30] Sun X, Li H [etc.] 2018 Capacity of six shrub species to retain atmospheric particulates with different diameters *Environmental Science And Pollution Research* **25 (3)** 2643-2650.

[31] Zampieri M, Sarkis J, Pestana R [etc.] 2013 Characterization of Tibouchina granulosa (Desr.) Cong. (Melastomataceae) as a biomonitor of air pollution and quantification of particulate matter absorbed by leaves *Ecological Engineerin* **61** 316-327.

[32] Zhang T, Bai Y, Hong X [etc.] 2017 Particulate matter and heavy metal deposition on the leaves of Euonymus japonicus during the East Asian monsoon in Beijing, China *Plos One* **12 (6)** 1-16

[33] Lin L, Chen G, Yan J [etc.] 2018 A factor analysis of landscape metrics of particles deposited on leaf surface *Environmental Science And Pollution Research* **25(28)** 28391-28402.

[34] GOST R 56929-2016. Emissions of pollutants into the atmosphere. Investigations of the fractional composition of dust by the optical method when normalizing air quality (Standartinform, M.).

The effectiveness of the dust particles of the flask in the apparatus of counter-swirling flows

A A Sakharova*, A Yu Sorokin, D V Azarov, V V Lupinogin

Institute of Architecture and Civil Engineering of Volgograd State Technical University, 1, Akademicheskaya st., Volgograd, 400074, Russia

E-mail: sax.nastya@yandex.ru

Abstract. In this research was studied gathering efficiency and breakthrough of flask dust particles in industrial plants. The experiments were carried out on devices with counter swirling flows with various modifications and diameters of 150, 400, 600, and 800 mm. Analyzed and graphically presented the dependences of particle breakthrough on tghe devices size. The breakthrough effect in small inertial VZP devices was investigated, also the construction of VZP device was designed and experimentaly tested under operating conditions in terms of productivity and energy losses close to standard devices, but excluding the breakthrough effect of larger particles. Recommendations are given for the device usage on counter swirling flows with the best characteristics for capturing flask dust particles.

1. Introduction

Dust is one of the unfavorable factors, which is affecting workers on industrial enterprises of various industries. The main measure to reduce its negative impact is the efficient and reliable work of dedusting ventilation systems. The usage of dust collectors on counter swirling flows is becoming more and more widespread in these systems, which, with the proper piping arrangement, provide sufficiently high values of cleaning efficiency, while maintaining operation reliability and design simplicity, characteristic to cyclone-type dust collectors. In addition, the following reasons for the low dedusting ventilation systems efficiency are observed: clogging of the cyclones inner surface and fabrics of bag filters with dust, outdated dust collectors, low efficiency of capturing fine particles, abrasive wear. This leads to a decrease in productivity, deterioration of the dust environment at workplaces and on the enterprise territory, urban environment pollution, as well as an increase in operating costs [1].

The first design of a vortex dust collector (VDC) (Fig. 1) was patented in 1953 by E. Schauffler and H. Zennek [2], a dusty stream was supplied into the device from below, and compressed air was supplied from above.

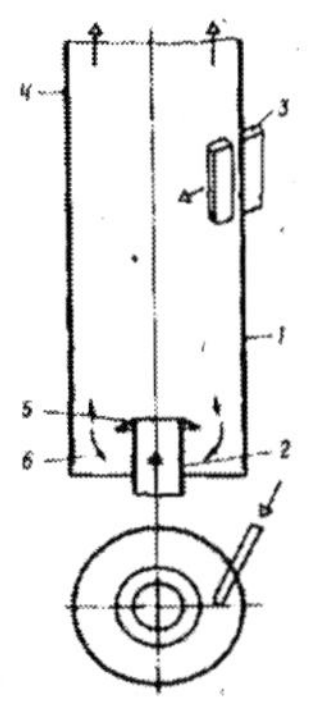

Figure 1. Dust collector made by Schauffler and Zennek: 1 - separation chamber; 2 - primary flow branch pipe; 3 - compressed air nozzle; 4 - exhaust; 5 – striker washer; 6 - bunker

In VZP devices, if the flows direction is in different directions, A.M. Kutepov and A.S. Latkin proposed to designate VZPp - (opposite twist), and in VZPc - i.e. with concurrent swirling flows [3].

In VZPc, the twist drops sharply due to oppositely swirling flows, meaning, the hydrodynamic resistance of the device decreases, the abrasion of the walls by the material, and most importantly, the collision of oppositely swirled flows creates a more developed turbulent structure than in the cyclone device, which should lead to an increase in the speed of the technological process. Such devices are used, for example, in drying techniques, but they are not considered in this research. The designation of the VZP corresponds to the designation of the VZPc. Currently, VZP devices are used in countries such as Germany, USA, France, Spain, Japan, Poland, etc. [4].

Back in the 90s of the last century at the Volgograd Chemical Plant in the shop for the calcium carbide production, a multistage aspiration system was introduced for the removal and cleaning of flue and strip gases of an electric furnace, the diagram of which is shown in Fig. 2. The main part of the cleaning system is the dust collectors blocks on counter swirling flows and concentrators. Dusty gases are fed to a dust concentrator installed at the inlet to the device. After separation of the solid and gaseous phases in the concentrator into two streams, the highly concentrated flow is directed through the air duct to the second concentrator of the dust collector on counter swirling flows with a diameter of 2200 mm, and the low-concentrated flow to the upper part of the apparatus [5].

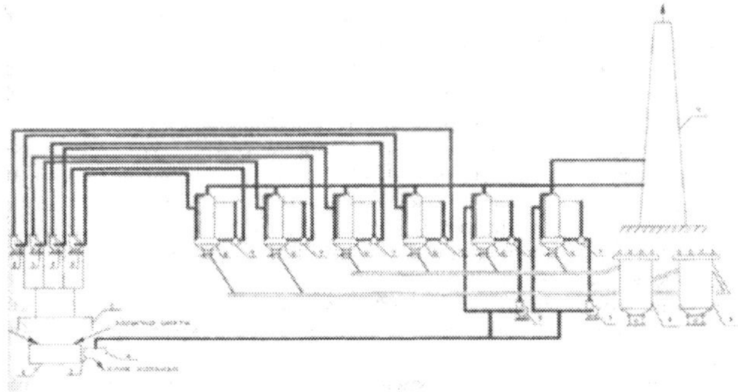

Figure 2. Diagram of a flue and strip gas purification unit in the calcium carbide production: 1 - three-phase carbide furnace; 2 - taphole; 3, 4 - umbrella; 3 - smoke exhauster DN-21.5; 6 - dust collector on counter swirling flows; 7- dust concentrator; 8 - hopper for receiving the captured dust; 9 - chimney

The development of vortex dust collectors blocks with separators-concentrators provided the development of schemes for capturing fine dust. There is a diagram of the aspiration system from the coke drying equipment in the calcium carbide production, which made it possible to ensure the dedusting efficiency within 88-94.5%, depending on the operating mode of the process equipment, is shown in Figure 3 [6].

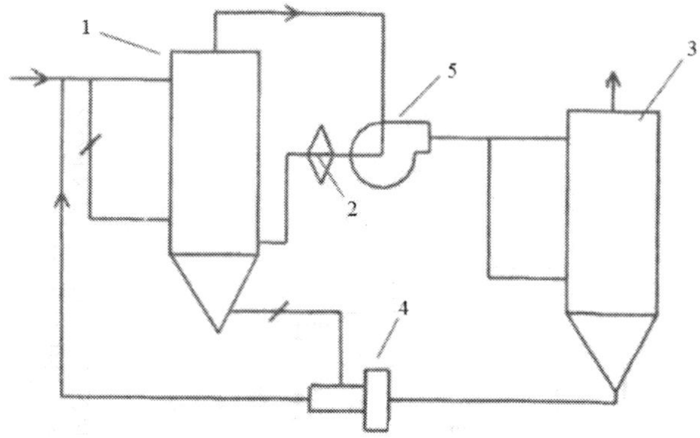

Figure 3. Diagram of the aspiration system from the coke drying department equipment in calcium carbide production: 1 - dust collector on counter swirling flows of the first stage VZP; 2 - small filter; 3 - dust collector on counter swirling flows of the second stage VZP, 4 - separator-concentrator; 5 - fan

It's obvious that with a decrease in the device diameter, its efficiency should increase. For small devices, this statement is fulfilled. However, it turned out that the breakthrough effect of the large particles takes place. It is expressed in the fact that a small number of large particles slips and the overall (total) capturing efficiency decreases, and the breakthrough function ceases to be decreasing in some area [7].

Due to the diversity of the VZP devices design, created for specific productions, the task to reduce the breakthrough of large particles appears.

Thus, researches aimed at improving dedusting ventilation systems by optimizing the structural elements of dust collectors for capturing flask dust particles are relevant.

2. Objective and tasks of research

The main purpose of the study is to improve the dedusting ventilation systems design with the use of counter swirling dust collectors (VZP) with an increase in reliability and efficiency [8,9].

Main tasks:

- selection of case studies: the efficiency of capturing large flask dust particles in small VZP devices;
- analysis of dust collection systems using devices with diameters from 150 to 800 mm and comparison of the breakthrough effect in these devices;;
- development of the VZP device design and experimental tests under operating conditions, excluding the breakthrough of larger particles in small devices;
- assessment of the application possibility in industry;
- increasing fractional efficiency of capturing, reliability and energy efficiency of dust removal systems;
- a description of the recommendations and conclusions.

3. Materials and methods

VZP devices with striker and dust concentrator (VZP_{OPK}-250 and VZP_{OPK}-600)

As an experimental setup, the authors have proposed a dust collector on counter swirling flows with a dust concentrator installed inside the dust collector body with a spiral partition and a hollow cylindrical baffle.

The vortex dust collector VZP_{OPK} consists of a cylindrical body with a conical dust collector, an upper axial outlet for the cleaned gas, an upper tangential inlet for dust contained gas and a lower tangential inlet for dust contained gas, a conical swirler with a striker washer, also the device has a dust concentrator, which separate dust contained gas flow on fractions with different dust content, which facilitates their further dedusting, as well as a spiral shutter and a hollow cylindrical baffle installed inside the dust collector body, creating an intensive swirling of dust contained gas flows and separating the emitted dust (Fig. 4)

The device has proven itself in industrial enterprises for the extraction and storage of flasks. Devices 250 and 600 are marked according to the diameters of the cylindrical body (Fig. 4, position 1) and corresponds to VZP-150 and VZP-400 in terms of performance and energy consumption.

The experiments consisted in conducting tests to study the functions of fractional flask dust particles breakthrough for two standard sizes of devices 250 and 600, as well as for the devices VZP_{OPK}-250 and VZP_{OPK}-600 developed by the authors.

In this work, for a comparative analysis, experimental studies of works [10 - 13] were carried out.

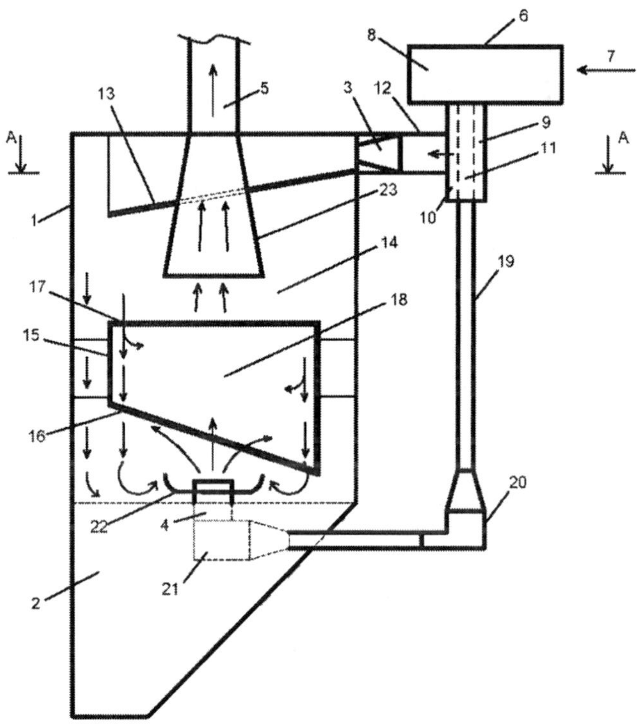

Figure 4. Device VZP_{OPK} dedusting flask production: 1 - cylindrical body; 2 - dust collector; 3 - upper tangential branch pipe; 4 - lower tangential branch pipe; 5 - purified gas outlet branch pipe; 6 - dust concentrator; 7 - tangential entrance; 8 - vortex chamber; 9 - axial tube; 10 - wall area; 11 - central zone; 12, 19 - air duct; 13 - spiral shutter; 14 - swirling zone of the dusty gas fraction; 15 - hollow cylindrical bump stop; 16 - lower entrance; 17 - upper entrance; 18 - separation zone; 20, 21 -swirler; 22 - annular striker washer; 23 - diffuser

To study the capturing efficiency, dust from the flask of the Kamenoyarsk and Astrakhan deposits was used. The dust density was ρ = 2200 - 2600 kg /m^3. Samples of $PM_{2.5}$ and PM_{10} dust were obtained in two stages: PM_{40} dust - by the sieve method through a sieve with a diameter of 40 microns and then by the $PM_{2,5}$ and PM_{10} sedimentation method. In this case, the integral distribution functions of the fractional composition, for all samples obtained from the original by cutting off larger particles [14], are no more than 6% for PM_{10} and 5% for $PM_{2,5}$.

To investigate the fractional efficiency of VZP dust collectors, were built the average integral distribution functions of the particles mass over equivalent diameters in the flask dust samples (Figure 5).

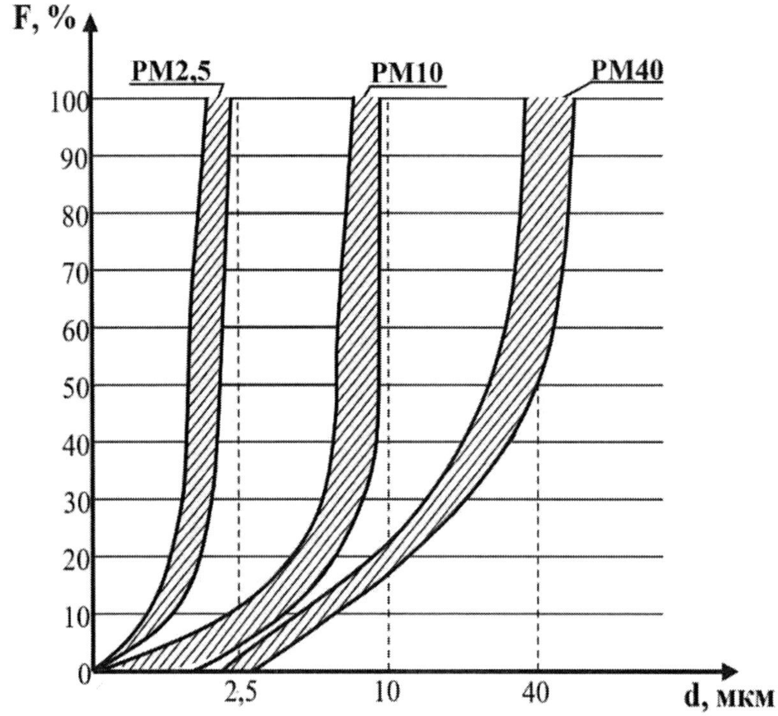

Figure 5. Variation ranges of integral functions particle mass distribution over equivalent diameters, %

4. Results and discussion

Equations of regression for a small dust breakthrough of different equivalent dust diameters ($PM_{2,5}$ и PM_{10}) in the VZP_{OPK}-600 device were obtained.

Similarly, according to the proven methodology, studies were carried out for the VZP_{OPK}-25 device (Figure 6). Comparison of the breakthrough in this device with the data obtained for other standard VZP devices by the authors themselves and other researchers shows that the VZP_{OPK}-250 device is relieved from the breakthrough effect of the large particles in small devices much more than in devices 400 and 800.

In this case, the data of other authors were selected for a particles close to the density of the flask, or the results were recalculated for a given density.

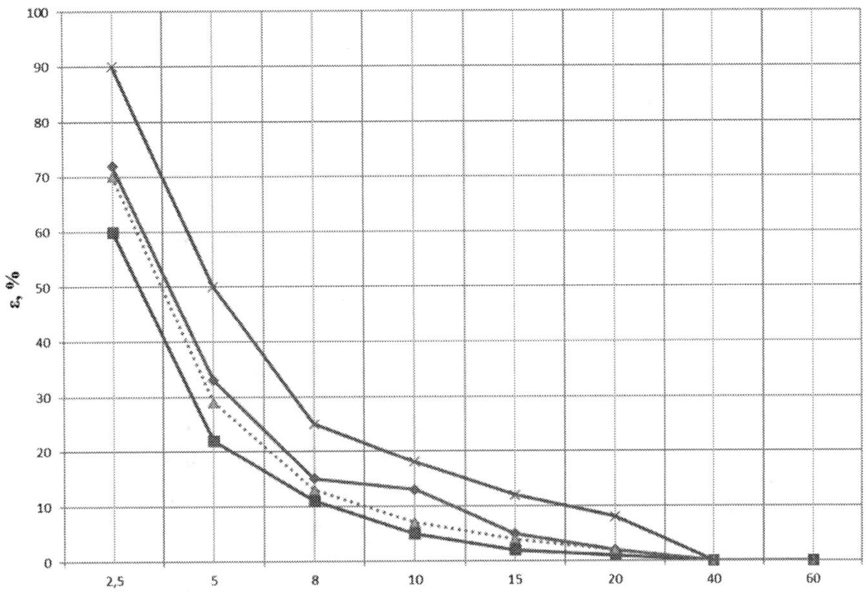

Figure 6. Fractional efficiency of dust capturing VZP devices

Thus, the VZP_{OPK} device for a small dust up to 10 microns has a breakthrough close to that of VZP-150, but for larger particles the breakthrough is significantly smaller. For example, for a particles 2,5 microns – in 1,1 times; 10 microns - in 1,5 times; 20 microns – in 3 times. Consequently, dust emissions into the atmosphere, for example, from the flask packing unit for the VZP_{OPK} device, will be 2 times less.

5. Summary

Based on the conducted research results, the following conclusions can be drawn:

1. The dispersed composition of flask dust in crushing and packing systems has been researched.
2. The possibility of creating dedusting installations using VZP devices has been analyzed on the example of a diagram of a flue and strip gas purification unit in the calcium carbide production (Fig.2) and a diagram of an aspiration system from the equipment for coke drying in the calcium carbide production (Fig.3).
3. The breakthrough effect of large solid flask particles in small inertial devices was investigated, using the example of VZP-150 and VZP-400.
4. The designs of the VZP_{OPK}-250 and VZP_{OPK}-600 devices have been developed in terms of performance and energy losses close to standard devices.
5. Experimental tests carried out under operating conditions showed that the breakthrough of larger particles in small devices VZP_{OPK}-250 and VZP_{OPK}-600 in 2 times less, and for particles with a size of 20 microns – in 3 times less than in standard devices.

References

[1] Novikov V T 2003 Equipment and design principles for environmental protection systems. Part 1. Dust, its properties and dust collection. Tutorial. - Tomsk: Ed. TPU 176 p.
[2] Blinov L A 1987 A. S. 1477479 USSR, MKI4 B 04 C 3/06. Vortex dust collector. -Stated on 28.04.1987; Publ. 28.04.1987. Bul. 17.
[3] Kutepov A M, Latkin A S 1999 Vortex processes for the modification of disperse systems (Nauka, M.).
[4] Azarov V N 2003 Dust collectors with counter swirling flows. Implementation experience (monograph, Volgograd: RPK "Polytechnic" VolgSTU).

[5] Azarov V N 2002 Aspiration systems for smoke and tap gases for the calcium carbide production *Building materials* **11** 20-21.
[6] Azarov V N, Volyntseva L N, Sergina N M, Yurkyan O V, Donchenko B T, Martyanov V N 1999 Dust collectors with counter swirling flows (Under. ed. V.N. Azarov, LLC "Association Volgogradekotekhzerno", Volgograd, (Review of inventions)).
[7] Sazhin B S, Gudim L I 1995 Vortex dust collectors (Chemistry, M.).
[8] Sazhin B S, Gudim L I 1982 Dust collectors with counter swirling flows. Environmental protection and rational use of natural resources (NIITEKhim, Issue. 1, M.).
[9] Shilyaev M I, Shilyaev A M, Grishchenko E P 2006 Methods for calculating dust collectors: textbook. allowance (Publishing house Tomsk. state architect-builds. University, Tomsk).
[10] Azarov D V, Lukanin D V, Ostaali M 2016 Estimation of emissions into the atmosphere from dedusting systems with dust collectors on counter swirling flows *Biosphere compatibility: man, region, technology* **3** 38-43.
[11] Chebanova S A, Azarov V N, Marinin N A, Vorobiev V I 2014 On the study of aerodynamic characteristics of ash dust in the cities air *Bulletin of the Volgograd State University of Architecture and Civil Engineering. Series: Building and architecture* **35 (54)** 207-211.
[12] Artyukhin A S, Baev A V 2007 On methods of calculating the efficiency of devices with counter swirling flows *Bulletin of VolgGASU. Ser. Natural science. Volgograd* **6 (23)** 121-127.
[13] Azarov V N, Lukanin D V, Borovkov D P, Redwan A M 2014 Experimental Study of Secondary Swirling Flow Influence on Flows Structure at Separation Chamber Inlet of Dust Collector with Counter Swirling Flows = [Experimental studies of the influence of the secondary swirling flow on the structure of flows] *International Review of Mechanical Engineering (IREME)* **8 (5)** 851-856.
[14] Azarov V N, Sakharova A A 2019 Aerodynamic Characteristics and Fractional Composition of the Flask Dust *All-Russian research-to-practice conference "Ecology and safety in the technosphere: current problems and solutions" (EST 2018, 22–24 November 2018, Yurga, Russian Federation). - Conference Series: Earth and Environmental Science. - Publishing Ltd IOP* **224 1** 8. Information on https://iopscience.iop.org/article/10.1088/1755-1315/224/1/012027/pdf.

The breakthrough of large particles in small devices of counter-swirling flows with top strikers

V V Lupinogin[1*], D V Azarov[2], A A Sakharova[1], D R Dobrinski

[1]Institute of Architecture and Civil Engineering of Volgograd State Technical University, 1, Akademicheskaya st., Volgograd, 400074, Russia
[2] JSC "Lukoilsevernefteprodukt", 55, 40 let VLKSM st., Volgograd, 400029, Russia

E-mail: vladlupinogin@gmail.com

Abstract. The work considers the breakthrough effect of large particles in a small diameter device on counter-swirling flows. The dependences of particle breakthrough on different diameter devices were built. An experiment was carried out on a device of 100, 200, 300 and 400 mm with top paddle strikers installed at the axial outlet. A comparative analysis is made with the calculated and experimental data of the researchers in this area. The results of devices with small diameter are obtained and presented in the visual graphs and conclusions.

Introduction

In terms of production a pollutant is emitted - dust. To prevent negative impact on environmental safety, dust collectors are installed at enterprises. The development and appearance of new devices operating according to the inertial principle on vortex swirling flows (VZP) has a positive effect on the operation of dedusting ventilation systems [1].

The general operation principle of a vortex dust collector with counter-swirling flows is that dust contained gas enters the device through two inlets simultaneously. The gas swirls under the centrifugal forces action of the two streams and the dust particles move to the cylindrical device walls. The interaction of the streams directs the pollutant into the dust collector. The cleaned gas is discharged through the axial outlet [2].

The creation history of the VZP dust collectors begins in Germany. In 1953, scientists E. Schauffler and H. Zenneck received the Federal Republic of Germany patent №1092281 "Vortex chamber for separating solid and liquid aerosol particles using an additional swirling gas flow", shown in Fig. 1 [3]. The main feature of this dust collector was that dust contained gas was supplied to the lower stream, and pure compressed gas to the upper stream.

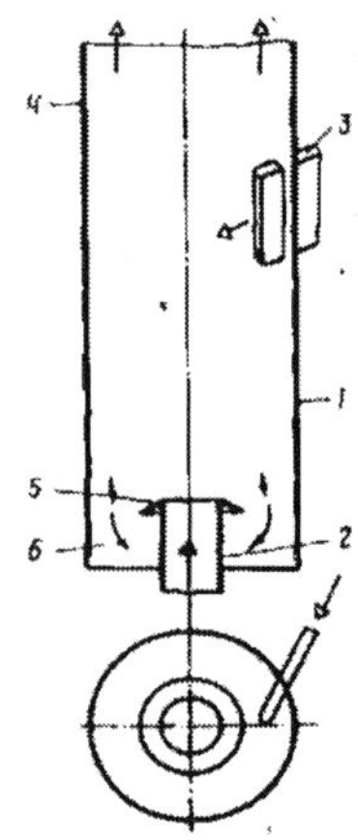

Figure 1. Dust collector by Schauffler and Zennek: 1 - separation chamber; 2 - primary flow branch pipe; 3 - secondary flow pipe; 4 - exhaust; 5 – striker washer; 6 - bunker

Next, the scientist H. Klein, taking the work of his German colleagues as a basis, presented the first experimental results and a description of the device operation [4].

During the operation of dust collectors with a gas flow, uncaught dust particles, especially fine impurities, can be carried away through the axial outlet.

Scientist L. A. Blinov in 1987 proposed a design in which fine impurities coagulate in a dust collector [5].

One year later, Soviet scientists B.S. Sazhin, L.I. Gudim. etc., the design of a dust collector was presented, where a nozzle is installed at the axial outlet and a conical shell is installed in the upper part of the body, forming an annular chamber for additional cleaning [6].

In 1997, the Ural TNP Joint Stock Company received a patent. The design feature is the presence of a filter made of fluffy threads, installed at the axial outlet of the purified gas. As the clogging progresses, the filter element is regenerated by shaking the sieve and blowing the filter element with compressed air [7].

A year later, a design of the vortex dust collector was protected by Samarskaya Yu.G, which provided an increase in the efficiency of dust collection due to the elimination of unwanted vortices in the separation chamber in the area of secondary gas flow formation [8].

Western scientists also contributed to the development of VZP devices. James G. C. and Burnard G. K. presented a patent for a device in which many small chambers are sprayed with liquid, and additional cleaning is carried out with a use of a wet filters complex [9].

At the moment, several authors have successfully defended their Ph.D. thesis on improving aspiration systems using devices on counter swirling flows [10-13].

In the history of VZP development, the problem of catching small dust particles was addressed. This problem was partially solved by operating devices of small sizes (up to 400 mm), but the next problem appeared - the breakthrough of large dust particles.

Objective and tasks of research

The main aim of the research is to analyse the calculated and experimental data of VZP devices, which are exists in the enterprises environmental safety system, and compare with experimental data obtained by the selected VZP for the study, based on the fractions breakthrough effect of the of large and small dust particles. Assessment and proposal of the optimal size with the top paddle strikers as alternatives for industrial use in order to prevent the large dust particles breakthrough.

Objective and tasks of research

This work consists of the next main parts:

- case studies selection: small diameter VZP devices;

- assessment of existing dust collection systems with a use of small diameters VZPs and results comparison in terms of breakthrough effect and device sizes; assessment of the possibility to use devices in industrial production;
- increasing the fractional efficiency of collection, reliability and energy efficiency of dedusting ventilation systems with dust collectors on counter-swirling flows by comparing VZP devices.
- results analysis, formulation of conclusions and recommendations.

Materials and methods

VZP device with small diameter and top paddle strikers

The authors have proposed a dust collector on counter swirling flows of small diameter with attachment to the axial outlet of the paddles to prevent the large particles breakthrough. The device has proven itself in industrial cement production plants.

The dust collector contains a cylindrical metal body, equipped with a conical dust collector in its lower part, and a straight pipe for dust contained air inlet, installed in the upper part of the body, as well as a tangential pipe for dust contained air inlet, installed in the lower part of the body, coaxial with the vertical axis of the device, and axial a branch pipe for the outlet of cleaned gas. On the lower part of the outlet pipe, there are fixated paddle strikers, designed to prevent large dust particles from entering the stream of cleaned air (Figure 2).

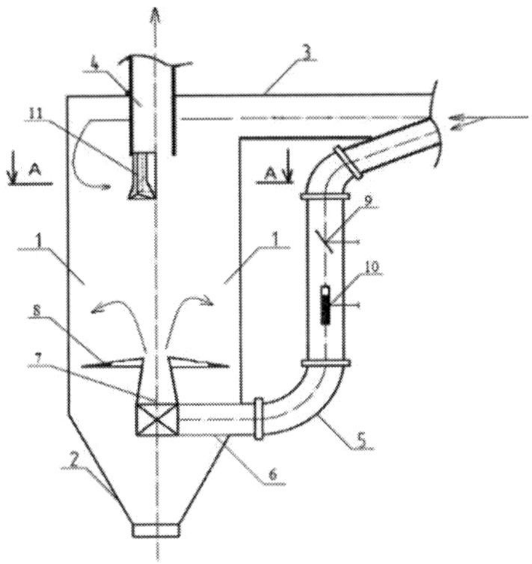

Figure 2. VZP device: 1 – body; 2 – dust collector; 3 - upper branch pipe; 4 – axial output; 5 – pipeline; 6 – lower branch pipe; 7 – swirler; 8 –striker washer; 9 – shutter; 10 – additional shutter; 11 – paddle strikers

The main list of studied parameters

The main list of the studied VZP devices with paddle strikers and the breakthrough effect of dust particles in them, is presented in Table 1.

Table 1. Device diameter and particle size used for research.

VZP diameter, [mm]	The size of researched dust,[μm]
100	2.5-50
200	
300	
400	

VZP devices with paddle strikers with diameters of 100, 200, 300 and 400 mm were researched, and a comparative analysis with the used dust collectors of these sizes without strikers was made.

In the work were used experimental datas from works. [14 - 15].

Resutls and discussion

The obtained experimental data for VZP devices with paddle strikers of 100, 200, 300 and 400 mm diameters after comparison with the calculated and experimental data of the authors [14 - 15], who studied VSF devices, are presented in the dependencies form (Figures 3 - 6).

The authors experimental data were made on the cement production dust.

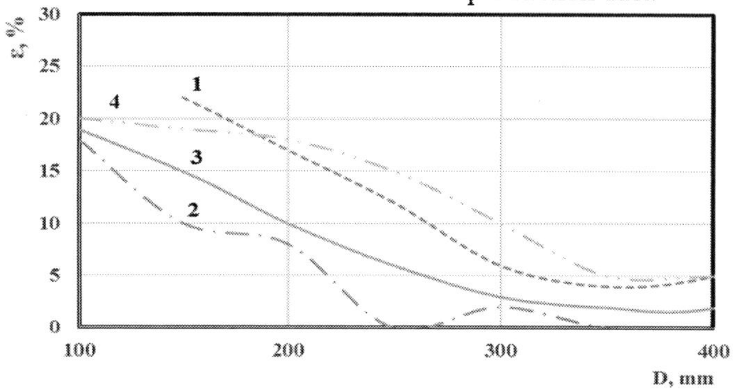

Figure 3. Dependence of the 50 μm dust particles breakthrough on the size of VZP device: 1 – authors experimental data; 2 – authors calculated data; 3 – experimental data of VZP with paddle strikers; 4 – experimental data of standard VZP

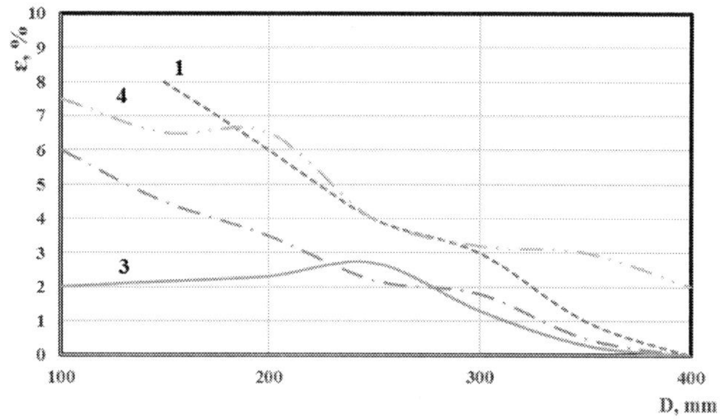

Figure 4. Dependence of the 10 μm dust particles breakthrough on the size of VZP device: 1 – authors experimental data; 2 – authors calculated data; 3 – experimental data of VZP with paddle strikers; 4 – experimental data of standard VZP

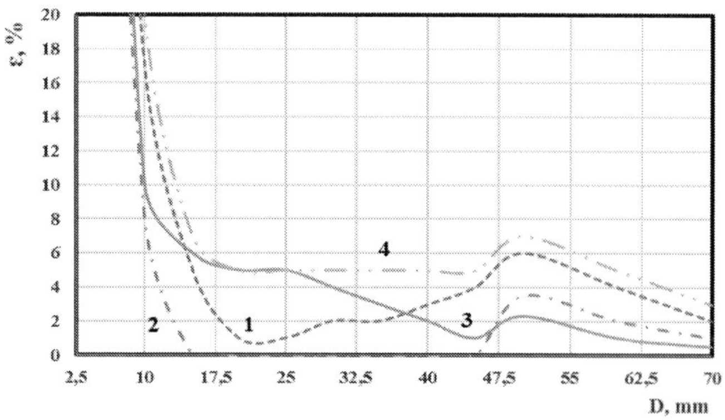

Figure 5. Dependence of the 2,5 μm dust particles breakthrough on the size of VZP device: 1 – authors experimental data; 2 – authors calculated data; 3 – experimental data of VSF with paddle strikers; 4 – experimental data of standard VZP

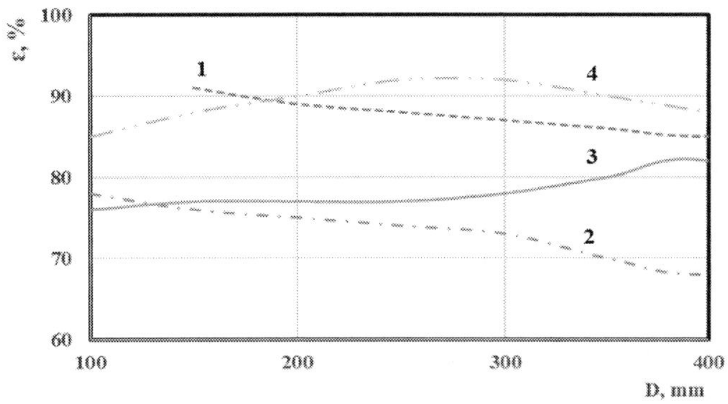

Figure 6. Dependence of the breakthrough on the dust particles size of the VZP – 200 device: 1 – authors experimental data; 2 – authors calculated data; 3 – experimental data of VZP with paddle strikers; 4 – experimental data of standard VZP

Summary

The research showed, that the particles of 2,5 microns are captured poorly by the devices, but at the same time VZP - 100 and VZP - 200 with paddle strikers show the results close to the calculated values.

The aim of the research was to identify the optimal size of the VZP device with top strikers, as a device with the high indicators of reliability and efficiency, and at the same time with less energy and resource consumption. Obtained results showed that the VZP device with a separation chamber of 200 mm diameter shows the highest operation quality. The emission of small fractions into the atmosphere is reduced by 12% for 2,5 microns, by 7% for 10 microns and by 4% for 50 microns.

Further research of dust particles in construction industries and modernization of their operating and design parameters is required.

Acknowledgements

This document was prepared with the support of the Volgograd State Technical University patent on a vortex paddle dust collector, application number № 2019109476 from 01.04.2019

References

[1] Sergina N M, Abduldzhalil M S A, Abramova L M 2015 Dust collectors with counter swirling flows in systems for cleaning dust emissions in the production of building materials *Engineering Bulletin of the Don* **37 (3)**.
[2] Azarov V N 2003 Dust collectors with counter swirling flows (Implementation experience: monograph, RPK "Polytechnic" VolgSTU, Volgograd).
[3] Schauffler E, Zenneck H 1953 Pat FRG № 1092281.
[4] Klein H 1963 Staub Reinhaltung der Luft [Text] **23 (4)**.
[5] Blinov L A 1987 A.S. 11477479 USSR, MKI4 B 04 C 3/06. Vortex dust collector / - Stated on 28.04.1987; Publ. 28.04.1987. Bul. No. 17.
[6] Sazhin B S, Goodim L I, Chumakov A G, Chumakov I A 1990 A.S. 1593708 USSR, MKI5 V 04 S 5/30. Centrifugal catcher /. - Stated 10.02.1988; Publ. 23.09.1990. Bul. No. 35.
[7] Pat. 2083295 Russia, MKI4 B 04 C 3/06. Dust collector O L Chernykh, O L Black. -Stated 19.05.1995; Publ. 10.07.1997. Bul. 19
[8] Dezhurov S V, Zhilnikov V B, Polonsky V M and others 1998 Pat. 2146562 Russia, MPK7 V 04 C 3/06, V 01 D 45/12. Vortex dust collector / - Stated 24.09.1998; Publ. 20.03.2000. Bul. No. 8.
[9] James G. C., Burnard G. K. Dust filter apparatus: pat. 4484937 USA. – 1984.
[10] Azarov D V 2004 Improvement of aspiration systems for carbide production with the use of vortex dust collectors [Text] (dis .. cand. tech. Sciences: 05.23.03, Volgograd).
[11] Artyukhin A S 2007 Improvement of the operating and design parameters of VZP devices in dust-free ventilation systems [Text] (dis. Cand. tech. Sciences: 05.23.03, 03.00.16 Volgograd).

IOP Conf. Series: Materials Science and Engineering 1083 (2021) 012100 doi:10.1088/1757-899X/1083/1/012100

[12] Borovkov D P 2004 Improvement of aspiration systems for enterprises in the building industry by means of flow swirling in air ducts [Text] (dis. ... Cand. tech. Sciences: 05.23.03, Volgograd).
[13] Ponomareva N S 2011 Improvement of dust protection systems using VZP devices at construction enterprises for the production of chalk [Text] (dis .. cand. tech. Sciences: 05.26.01, 05.23.19, Volgograd).
[14] Azarov D V, Lukanin D V, Ostaali M 2016 Estimation of emissions into the atmosphere from dedusting systems with dust collectors on counter swirling flows *Biosphere compatibility: man, region, technology* **3** 38-43.
[15] Sergina N M et al. 2016 On the theoretical evaluation of the efficiency of dust collectors with counter-swirling flows in the systems for cleaning emissions into the urban air environment *Modern Science and Innovations* **1** 112-117.

The forecast accumulation in vegetative and generative organs of black currants and plum with ^{90}Sr its location on the surface of the soil

A Gorodnichaya, I Glinyanova*

Kuban State Agrarian University, 13, Kalinina st., Krasnodar, 350044, Russia
Volgograd state technical university, 1, Akademicheskaya st. Volgograd, 400074, Russia

E-mail: ecoris@yandex.ru

Abstract. As a result of field studies, experimental material on the accumulation of ^{90}Sr in vegetative and generative organs of black currant and cherry plum in the link of the trophic chain «soil-plant» was obtained. Coefficients of transition (CT) from soil to bark, wood, leaves and fruits of black currant and coefficients of transition from soil to bark, wood, leaves and fruits of cherry plum which will allow to make the forecast of accumulation of radionuclide at any known density of pollution of the soil in the conditions of Krasnodar region on the studied soils are defined.

Introduction

Natural biogeochemical cycles on planet Earth have evolved throughout its evolution. Moreover, natural radionuclides were an integral part of them. However, as a result of human activity, technogenic radionuclides began to enter the environment, which were also included in natural biogeochemical cycles. Nuclear weapons testing, production, processing, use and disposal of radioactive substances, global redistribution of natural radionuclides (construction, production and use of natural resources) all lead to global redistribution and contamination of the environment with radioactive substances. As a result, radioactive contamination occurs not only in urban and natural areas, but also in agricultural land. The growing number of people on Earth against the background of a decrease in the area of fertile land for various reasons, including radioactive contamination, requires a solution to the issue of returning fertile territories for use in the economy, as well as territories for their possible use in construction.

Many scientific works have been devoted to the study of the migration of radionuclides along trophic chains in recent decades [1-6], however, the issue of reclamation of contaminated areas during construction work in the Krasnodar Territory remains open. This raises many questions, including the forecast of the possible use of radioactively contaminated areas for construction work. Therefore, the timeliness and relevance of such work in the conditions of the Krasnodar Territory is beyond doubt.

The Krasnodar Territory has always been distinguished by the variety of plants grown in agriculture. Cereals, vegetables, grapes are grown there and, of course, a certain area is set aside for growing orchards and obtaining fruit products from this.

In the urban conditions of the Krasnodar Territory, crops such as currants, gooseberries, raspberries, cherries, cherry plums, apple trees, pears, etc. are grown everywhere. Much attention is paid to the cultivation of crops such as cherry plum and black currant. These plants are distinguished by high productivity, quick entry into the season of fruiting, high nutritional value of fruits. Cherry plum fruits contain malic and citric acid, sugar, pulp contains pectin, tannins and minerals. Delicious syrups, compotes, preserves, and jam are prepared from the fruit; thanks to the presence of pectins in it, a fragrant golden jelly is obtained [7,8]. The fruits of black currant are rich in various useful substances: sugars, organic acids - citric, malic, salicylic, succinic. They contain many vitamins: B1, P, E, K, B12, PP, carotene.

The purpose of this work is based on experimental data on the accumulation of ^{90}Sr in cherry plum and black currant, with its surface location on the soil. Calculate the coefficients of transition to the vegetative and generative parts of them and make a forecast of the possible use of the territory for the needs of farming and construction work.

Research methodology

The experiments were carried out on leached chernozem, low-humus, super-powerful [9,10].

All work on the preparation of the site was carried out very carefully, observing the uniformity of conditions. For such experiments, plots with 6 plants are recommended and most often used in research work [11]. At the ends of the rows there are protective plants - 2 plants each. With that said, field trials are based on the standard method of placing options.

In 2008, an experimental site was laid in the field. On the plots of the first option, there were black currant plants. The food area is standard - 4x4m. The experiment was repeated 6 times. Cherry plum plants were located on the plots of the second option. The food area is standard - 6 × 4m. The experiment was repeated 6 times. The soil surface was contaminated in 1989 by the application of ^{90}SrCl2, the contamination level of the experimental site was 500 MBq / m2.

After sampling, the plants were divided into organs and parts, dried at a temperature of 105 ° C, weighed and milled in mills MRP-1 or EM-3A.

Tests of products on the basis of radioactive contamination were performed on the USK "Gamma Plus" device according to the method of measuring the activity of beta-emitting radionuclides in counting samples using the "Progress" software. The technique was developed by SE VNIIFTRI and approved by the State Standard of Russia on 05.05.1996. This technique is the main one in determining the values of the activity of beta-emitting radionuclides in a counting sample and allows you to calculate the error of each measurement. To register beta radiation from a counting sample, a beta-spectrometric path with a scintillation detection unit (SDU) is used. For exposure of counting samples, special aluminum cuvettes are used (Complex universal spectrometric "Gamma Plus", 1995).

When controlling the content of strontium-90 in soils and plants, methodological guidelines were applied (Guidelines for determining the content of strontium-90 and cesium-137 in soils and plants, TsINAO, 1985), GOST R 50801-95, as well as OST R 10070-95 Soils. Methods for the determination of strontium-90 in farmland soils (OST R 10070-95). The results obtained were processed by the methods of mathematical statistics according to B.A. Dospekhov [11].

CT = specific activity of plants (Bq / kg) / density of soil contamination (Bq / m2). When calculating the CT, the content of radionuclides in a plant is reduced to a unit of pollution density, which neutralizes the influence of the latter and makes it possible to reveal the dependence of the accumulation of radionuclides on soil-ecological conditions, species characteristics of plants and other factors.

Results and discussion

As a result of a field experiment when growing black currants on soil contaminated with ^{90}Sr when it is on the soil surface, the coefficients of the nuclide transfer to bark and wood in the soil-plant trophic chain were calculated (Figure 1).

In the studied variant of the experiment, the contact of the root system of black currant with a radionuclide located on the soil surface was close and prolonged.

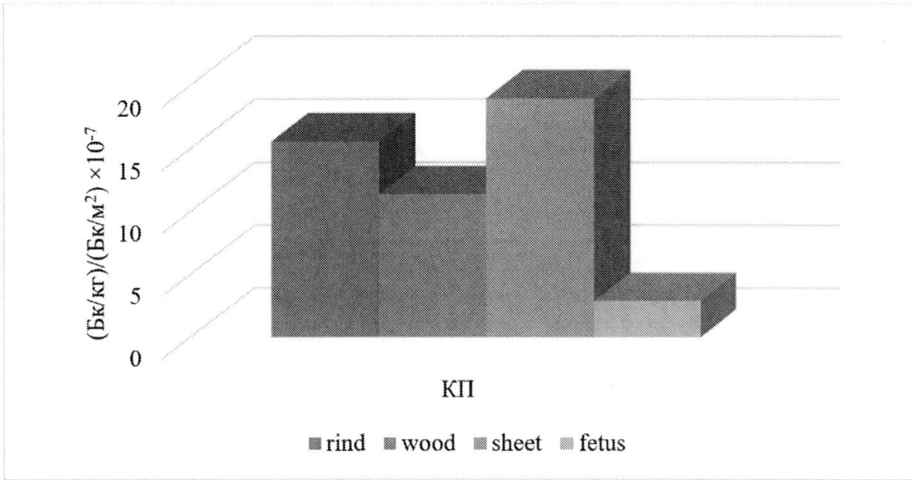

Figure 1. Coefficients of transfer from soil to plant (black currant) ^{90}Sr when located on the soil surface

When the radionuclide is located on the soil surface, the leaf of the studied plant has the highest accumulation coefficient. The difference in CT between black currant leaves and bark is 1.2 times. The leaves of black currant have the greatest pollution due not only to the intake of a nuclide from the soil, but also due to their external pollution due to falling dust torn from the surface of the field. Through the stomata from the surface of the leaf blade, the nuclide can penetrate into the leaf and increase its radioactive contamination.

In general, the CT for bark and leaves of black currant is 1.7 and 1.4 times higher than for wood, respectively. Radioactive contamination of black currant wood occurs only by the root route, while the KP is 11.4 × 10-7 (Bq / kg) / (Bq / m2).

The obtained coefficients of transfer of ^{90}Sr from soil to plants for vegetative organs will not only make it possible to make the right decisions about the advisability of growing this crop in a specific contaminated site, but also to develop options for the disposal of contaminated plant parts.

The lowest CT of ^{90}Sr from soil to plants at the studied location of the nuclide on the soil was determined for black currant fruits. The difference in CT in the fruits of black currant and bark, wood, foliage of this plant, respectively, was 5.4; 3.9; 6.6 times. Due to the fact that the fruits of black currant are used for food, the established fact is important for making a forecast and recommendations on the further use of the contaminated area for the needs of the national economy. In general, after the performed experiment and the calculated CT of ^{90}Sr from soil to plants, it can be argued that this berry shrub can be used when growing in an urban zone.

When making a forecast about the option of using a radioactively contaminated area, one should take into account the CT of the nuclide from soil to plants. But it can only be applied to the environmental and soil conditions in which it was obtained. For other regions, more research should be done.

When growing cherry plum on soil contaminated with ^{90}Sr when it was on the soil surface, experimental material was obtained and the coefficients of the transition of the nuclide into bark and wood in the "soil-plant" trophic chain were calculated (Figure 2).

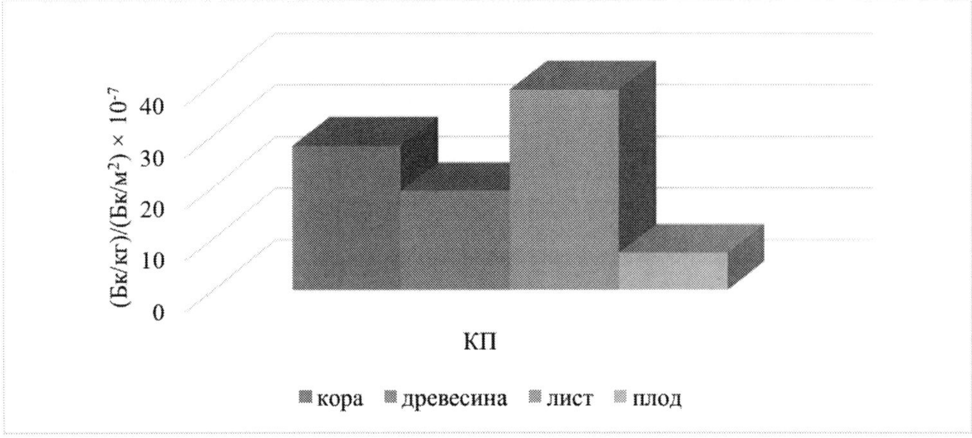

Figure 2. Coefficients of transfer from soil to plant (cherry plum) ^{90}Sr at its location on the soil surface

In the studied version of the experiment, the contact of the cherry plum root system with the radionuclide located on the soil surface was at first close, but later the close contact decreases, since the root system penetrates deep into the depths, and the nuclide remains on the soil surface.

With the surface location of the radionuclide on the soil in cherry plum, the highest transfer coefficient was recorded for leaves. The difference in CT between cherry plum leaves and bark is 1.4 times. Cherry plum leaves are located in the crown of a tree, nevertheless, their contamination can occur due to the ingress of radioactive dust torn from the surface of the contaminated soil. This additional variant of leaf contamination should be taken into account and, using various options for agricultural technology, reduce its likelihood.

In general, the CT of ^{90}Sr from soil to plants for the bark and leaves of cherry plum is 1.5 and 2.0 times higher than for wood, respectively. Radioactive contamination of cherry plum wood occurs only by the root route, while the KP is 19.2 × 10-7 (Bq / kg) / (Bq / m2)

The smallest CT of ^{90}Sr from soil to plants at the studied location of the nuclide on the soil was determined for cherry plum fruits. The difference in CT in cherry plum fruits and bark, wood, foliage of this plant, respectively, was 3.9; 2.7; 5.4 times. Cherry plum fruits are grown in order to be used for food or to prepare offal. Therefore, the established fact on the CT is important for making a forecast and recommendations on the further use of the contaminated area for growing this woody plant.

The obtained transfer coefficients for vegetative organs will not only make it possible to make the right decisions about the advisability of growing this crop on a specific contaminated site and develop options for the disposal of contaminated parts of the plant, but also to give recommendations on the possible use of the territory for industrial and civil construction.

As a result of the experimental studies performed and the calculated CT, it became possible to determine the difference in the coefficient of ^{90}Sr transfer from the soil to the vegetative and generative organs of the studied plants (Fig. 3).

Coefficients of transfer of ^{90}Sr from soil to vegetative and generative organs of cherry plum are higher than for black currant. The difference in CT for bark between cherry plum and black currant was 1.8 times, for wood - 1.7 times, for foliage - 2.4 times and for fruits - 2.5 times.

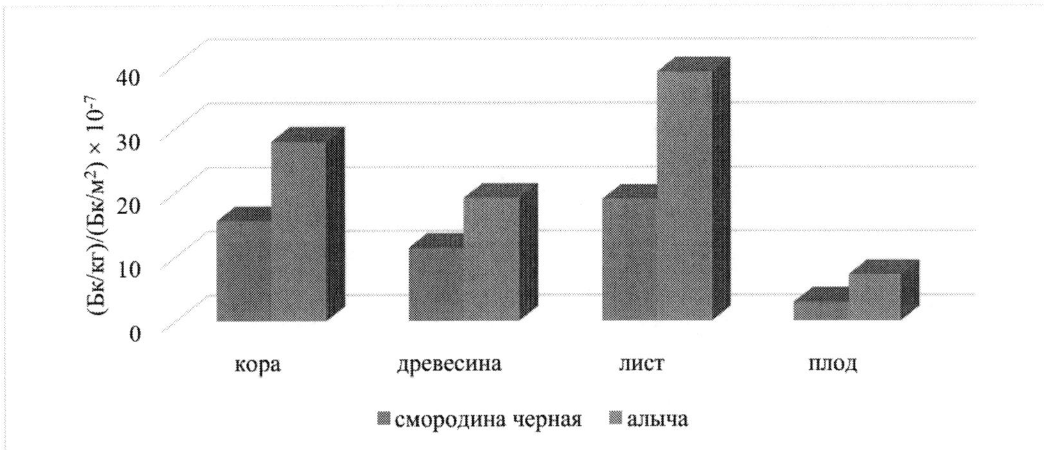

Figure 3. Coefficients of transfer from soil to studied plants ^{90}Sr at its location on the soil surface

As a result of the experiment performed, it was found that over the entire research period (8 years) in the first years of research in the vegetative and generative organs of the cherry plum an intense accumulation of a nuclide occurs, then in the dynamics of observations it decreases.

The content of the nuclide in the vegetative and generative organs of black currant also gradually increases during the first part of the study period, but later on, part of the roots goes deep into the soil, therefore, the intensity of accumulation of the nuclide decreases.

Cherry plum and black currant are grown for the purpose of obtaining fruits. As a result of the experiment, it was found that when the nuclide is located on the soil surface, the intensity of its accumulation in cherry plum fruits is higher than in black currant. The calculated CT of ^{90}Sr from the soil to the fruits of cherry plum and black currant, when the radionuclide is located on the soil surface, are shown in Figure 3.

Summary

1. The calculated transfer coefficients (CT) for the studied plants allow us to predict the pollution of their vegetative and generative organs at a known density of radioactive contamination of the territory.
2. For the leaves of black currant and cherry plum, the highest ^{90}Sr CT from soil to plants was determined in comparison with other vegetative organs.
3. The smallest CT of ^{90}Sr from soil to plants was determined for fruits of black currant and cherry plum.
4. In general, for the vegetative and generative organs of cherry plum, CT of ^{90}Sr from soil to plants is higher than for black currant.
5. With a superficial location of the nuclide on the soil, cherry plum plants are more suitable for land reclamation than black currants.

References

[1] Ipatiev V A 1999 *Forest Person* (Chernobyl, Gomel) 454.

[2] Melchenko A I, Pogorelova V A 2016 Dependence of the migration of radionuclides in the fruit cenosis in the soil horizon of leached chernozem on the depth of their occurrence *Bulletin of the Altai State Agrarian University* **8 (142)** 54-60.

[3] Melchenko A I, Melchenko V A, Melchenko E A 2012 Efficiency of cleaning agricultural plants from radioactive contamination depending on the method of irrigation with water containing radionuclides *Proceedings of the Kuban State Agrarian University* **1 (34)** 166-173.

[4] Melchenko A I, Melchenko V A, Melchenko E A 2011 Accumulation of radionuclides in agricultural crops depending on the time of their contact with plants *Proceedings of the Kuban State Agrarian University* **4 (31)** 157-162.

[5] Melchenko A I, Melchenko V A, Melchenko E A 2012 Accumulation of radionuclides in agricultural plants depending on the physicochemical properties of radionuclides *Proceedings of the Kuban State Agrarian University* **1 (34)** 91-99.
[6] Shcheglov A I 2000 *Biogeochemistry of technogenic radionuclides in forest ecosystems* (Moscow) 268.
[7] Yakushev V I 1982 *Fruit growing* (Moscow, Kolos) 415.
[8] Kolesnikov V A 1979 *Fruit growing* (Moscow, Kolos) 415.
[9] Simakin A I 1988 *Fertilizer, soil fertility and harvest* (Krasnodar) 270.
[10] Tarasenko B I 1981 *Improving the fertility of soils in the Kuban* (Krasnodar) 146.
[11] Dospekhov B A 1968 *Field experiment technique* (Moscow, Kolos) 336.

IOP Conf. Series: Materials Science and Engineering 1083 (2021) 012102 doi:10.1088/1757-899X/1083/1/012102

Management of organizational and technological design in construction based on information modeling (BIM-technologies)

Yu V Gushchina*, Yu N Nikolaev, A A Chernikov

Volgograd State Technical University, 28 Lenin ave., Volgograd, 400005, Russia

E-mail: jgushina@mail.ru

Abstract. The article discusses the modern application aspects of information modeling technologies for buildings and structures (BIM-technologies) in the construction projects' management. The introduction of information modeling technologies into the investment and construction project management process will allow not only using automated tools for performing various types of analysis and verification, visual planning and optimization of the construction process, but also provide regulated access to the object data to all interested parties in a single information environment.

1. Introduction

The construction industry is currently undergoing major changes, in part due to the introduction of Building Information Modeling (BIM) and the integration of new technologies. By combining the BIM modeling provisions with the efficient production principles, there is potential to improve the productivity and quality of organizational and technological design in construction. This tool gives an opportunity to divide the construction projects into work zones, get a fully automated product, and offers 4D construction color modeling for short term project planning process. This will make it possible to organize the construction planning systematic assessment and analysis in terms of labor productivity, labor distribution, taking into account the process of short-term planning, and will contribute to the continuous improvement of planning for the forthcoming construction of the facility [1].

In today's unstable socio-economic conditions, the global construction industry is going through a process of serious transformation associated with the abandonment of traditional design and construction methods with the transfer of project information in paper form in favor of the implementing projects' innovative ways. With the Building Information Modeling (BIM) introduction the construction industry is undergoing a significant transformation. Construction projects are extremely information-rich in nature. Their growing complexity, the lack of the necessary information to make decisions at the right time, the increasing pressure on timing in the conditions of traditional methods of their implementation partly explain the insufficiently high efficiency of the industry as a whole. Uncertainty typical for the construction activities organization is associated with changing factors of the external and internal environment and manifests itself in the form of various risks that reduce the implementation efficiency of the investment and construction projects and can lead to the disruption of contractual obligations stipulated in the general contract with the customer [2].

The gradual widespread transition to information modeling technologies for buildings and structures (BIM-technologies) was a response to the need to collect, record and process significant amounts of information in the design, construction and operation of capital construction facilities, subsequent (sometimes multiple) data correction during the project [3]. The possibility of such a transition is due to the intensive development of information technologies and the emergence of specialized software products aimed at creating a digital information model of a construction object, including all the necessary information about it. The presence of such a construction object model allows not only the use of automated tools for carrying out various types of analysis and checks, issuing design and working documentation, visual planning and optimization of the construction process, estimating the estimated cost, obtaining other data, but also provides regulated access to the data about the object for all interested parties in a unified information environment [4]. With the Building Information Modeling (BIM) introduction, construction industry is undergoing a significant transformation.

2. Integration of organizational and technological design and information modeling in construction

The advantages of information modeling technology for buildings and structures lead to its widespread introduction into the world design practice and construction management practice. The processes of BIM implementation are also going on in Russia, but at the moment there is some lag behind the developed countries, where the practice of using BIM is already widespread and makes it possible to draw the conclusions about the achievement of high speed, volume and quality of construction in combination with an economic efficiency increase [5].

Construction as a type of activity is a material production, and therefore has a significant impact on the development of a number of related industries and requires a fairly high volume of investment. Through the investment and construction projects' implementation, significant volumes of investment resources, in connection with which the accuracy of assessing their effectiveness is of particular importance, are under development process [6]. In the process of developing investment resources, the independent organizations, which sometimes have completely different goals and works, take part. Most often, the interests of the investor and the construction organization coincide. So, the investor's main goal is to get the maximum income, so the main task is to build and commission the facility as soon as possible with minimal financial investment. The construction organization, being a commercial enterprise, seeks to get the maximum profit, therefore it needs such an organization of construction production that will reduce the costs and construction time without compromising the quality of the facility being built. Thus, both the investor and the construction organization have the same desire to reduce costs and construction time, as well as in the ratio of price and quality indicators [7].

Analysis of projects implemented by the construction companies using BIM technologies revealed various effects. The main ones are:

- significant cost savings during the construction phase;
- potential savings in operating costs;
- improving planning accuracy and transparency;
- reduction of time losses for intercompany approvals;
- well-coordinated teamwork;
- the ability to use innovative design solutions;
- ensuring a unified vision of the goals of the project by all its participants [8].

Thus, the introduction of information modeling technologies in the process of managing organizational and technological design in construction will not only allow the use of automated tools for various types of analysis and inspections, visual planning and optimization of the construction process, but will also provide regulated access to the data about the facility to all interested parties, to the participants in the construction cycle in a unified information environment. Construction projects today are complex, burdened with many uncertainties and subject to various changes in the plan and course of their implementation. Scheduling uses activity analysis and scheduling optimization to identify and remove constraints. Optimization is used to remove constraints and ensure that prerequisites are available according to actual demand. During the screening process, work can be grouped into three different task states:

(1) "the task is ready" if all constraints are removed and the task is ready for execution,

(2) "the task may be ready" if the task is indeed limited, but will most likely, over time, all restrictions will be lifted, and

(3) "The task cannot be ready" if the task does not have its limitations, is deleted and will not be completed as planned [9].

The design visual drawing is in line with the construction work area schedule for project implementation and schedule management. After the actual working time has been entered, the work area will show the project completion status and other completion status will be displayed in different colors. In this way, schedule management makes it possible to track the progress of each production job and accurately capture the status of the components. The actual completion status is compared to the planned completion status of the construction project [10]. When the progress problem is found, the relevant data should be sent to the support-cloud, notify the person in charge to deal with them in time. After the problem is resolved, the relevant data should be sent to the cloud for the convenience of project and request management [11].

3. Management of organizational and technological design based on BIM technologies in construction

The planned tasks in the management of organizational and technological design in construction must meet the following four quality criteria: certainty, validity, consistency and scope. The learning phase describes the completed work, which is a tool to optimize by tracking the performance of the short-term planning process to improve productivity and project efficiency.

Critical principles with the highest concentration of unique interactions include:

a) get the right quality the first time (reduce product variability);

b) focus on improving upstream variability (reducing variability production);

c) Shorter production cycle times [12].

BIM functionality:

a) aesthetic and functional assessment;

b) multi-user viewing of combined or separate multidisciplinary models;

c) 4D visualization of construction schedules;

d) prompt communication of product and process information.

Considering that the BIM functions a), b) and d) are mainly related to the field of construction management, it can be argued that the integration of an effective management approach with the technical capabilities of BIM technologies will bring benefits to the overall production efficiency and labor productivity. Tekla software, BIM-sight and Bentley are some of these automated systems [13]. Any other project planning software that exports its planning data in XML format files (such as Primavera P6 or Power Project) can also be used.

Factors of work productivity in construction and duration of tasks play an essential role in the assessment. Construction companies primarily rely on experience with previous projects. However, reliability and labor productivity factors are often weak, and during project implementation, as a rule, construction intensity is usually not monitored and not used for adjustments in construction planning [14]. 3D models can be widely used as visualization and communication, as the use of 4D visualization provides a more intuitive understanding of the construction process compared to traditional 2D drawings and separate information on the schedule [15]. By checking the collisions of the 3D model, problems in building structures can be found in advance to reduce their number, design changes and avoid rework. A reasonable solution to this problem, which reduces the possibility of loss of labor, material losses and rework caused by mistakes or inefficiency during the construction process can be proposed before construction. The 3D model solves more serious problems the design of various specialties, low production efficiency, low construction efficiency, high construction costs and high costs of various resources in the construction process [16]. Based on dynamic construction modeling, it is also possible to automatically generate capital and resource consumption curve, which will allow all major participants in the construction process to intuitively respond to the consumption of funds and resources, as well as perceive changes in demand for funds and resources in real time, to take appropriate measures to ensure balance of resource provision [17].

Summary

In the construction industry, there is a change in benchmarks, which is associated with the increased requirements for the equipment used, the technologies used and economic proposals. The complexity scale and level of the investment and construction projects being implemented are increasing, which requires special attention to the management organization, the timing of construction work, and the construction products' quality. The construction industry has been experiencing slow growth for several decades. Over the past few years, labor productivity has dropped by almost 20%. Some reasons include: lack of realistic information on current design and construction techniques and technologies and poor short-term planning and work plans. The development of the BIM system and other technologies such as augmented reality, drones and advanced building materials that have reached market maturity have enormous potential for improving quality, organization efficiency as well as construction production technology [18].

The introduction of BIM technologies will allow solving one of the key problems, which is to attract all the specialists involved in the construction project to participate in the project. BIM modeling also provides the ability to optimally analyze and control time-dependent spatial collisions through simulated 4D construction progress, which leads to more efficient execution and is critical, especially for the large construction sites where several contractors are working on a construction project [19].

The development of a multidimensional information model for the construction of an object is an effective solution to the urgent tasks of the construction industry and determines the integrated management of organizational and technological design using BIM technologies, which also opens up the opportunities for further research [20]. For example, a comprehensive project planning tool will automate the entry of work times into a planning software (e.g., MS Project) that is automatically assessed using the labor productivity construction coefficients.

Thus, the implementation of the main organizational and technological indicators (expected work, finished work and especially, the percentage of the work schedule completion), which serve as key indicators for monitoring the continuous process of improving operational planning, is an important step in increasing the effectiveness of monitoring in the short-term planning. In the future, it is necessary to conduct additional research in order to gain more practical experience and the amount of information obtained in mobile applications and rendering of 3D information models on tablets or other mobile devices, since this area has enormous potential for solving the problems of increasing the reliability and efficiency of organizational and technological design in construction.

References

[1] Bolotin S A, Drozdov A D, Nefedova V K 2018 Pedagogical parallels *Materials of the V International Scientific conference* 71-77.

[2] Petrov K S, Kuzmina V A, Fedorova K V 2017 *Engineering herald of the Don* **2.** Information on ivdon.ru/ru/magazine/archive/N2y2017/4057.

[3] Fil O A 2015 *Materials of the XI International scientific and practical conference, Trends of modern science. Volume 5. Economic science. Sheffield. Science and education* 92-96.

[4] Grakhov V P, Mokhnachev S A, Ishtryakov A Kh 2015 Development of BIM-design systems as an element of competitiveness *Modern problems of science and education* **1-1.**

[5] Burlachenko O V, Elfimov K A, Bunin D V 2018 Information support of life cycle management of construction objects in the BIM concept *Bulletin of the Volgograd State Architectural and Construction University. Series: Construction and architecture* **54 (73)** 217-227.

[6] Dixit M K, Venkatraj V, Ostadalimakhmalbaf M, Pariafsai F, Lavy S 2019 Integration of facility management and building information modeling (BIM) *A review of key issues and challenges (2019) Facilities* **37 (7-8)** 455-483.
[7] Dmitriev A N, Cygankova A A, Papikjan L M 2018 Materials of the VIII International Scientific conference *Modern problems of projects and investments' management in construction industry. Moscow* 186-195.
[8] Eastman Ch, Teicholz P, Sacks R, Liston K 2011 BIM Handbook: a guide to building information modeling for owners, managers, designers, engineers and contractors *Information Modeling Guide. New York: John Wiley and Sons* 640.
[9] Heigermoser D, García de Soto B, Abbott E L S, Chua D K H 2019 BIM-based Last Planner System tool for improving construction project management *Automation in Construction* **104** 246-254.
[10] Ji S, Hu T 2019 Construction Simulation and Real-time Monitoring Research of Concrete Dam Based on BIM *IOP Conference Series: Earth and Environmental Science* **340 (5)** 052056.
[11] Chen J, Wu J, Qu Y 2020 Monitoring Construction Progress Based on 4D BIM Technology *IOP Conference Series: Earth and Environmental Science* **455 (1)** 012034
[12] Wang X, Jiang R, Wu W, Ma B, Zhu J 2019 Research and Application of Dynamic Construction Management based on BIM5D *IOP Conference Series: Earth and Environmental Science* **234 (1)** 012007.
[13] Li M, Huang L, Yu H 2019 The researches on construction management teaching reform based on BIM talent training *Conference Proceedings of the 7th International Symposium on Project Management, ISPM* 1003-1008.
[14] Hu S J, Wang J P, Xu B 2019 Research on the risk factors of BIM application in construction enterprises *Proceedings of 22nd International Conference on Advancement of Construction Management and Real Estate, CRIOCM* 201-208.
[15] Yu Z, Peng H, Zeng X, Sofi M, Xing H, Zhou Z 2018 Smarter construction site management using the latest information technology *Proceedings of the Institution of Civil Engineers: Civil Engineering* **172 (2)** 89-95.
[16] Mandičák T, Mesároš P, Tkáč M 2018 Impact of management decisions based on managerial competencies and skills developed through BIM technology on performance of construction enterprises *Pollack Periodica* **13 (3)** 131-140.
[17] Zhang J 2019 BIM-based Fine Cost Management of the Whole Process *IOP Conference Series: Earth and Environmental Science* **267 (4)** 042009.
[18] Huang J, Mao Y, Wang D, Zhu J, Zhou Y 2019 Construction schedule-cost joint control of RCC dams based on BIM *Advances in Science and Technology of Water Resources* **39 (5)** 66-72 and 88.
[19] Yi S, Yi L 2018 Computer aided architectural design and construction management based on BIM *Journal of Advanced Oxidation Technologies* **21 (2)** 201812807.
[20] Li Y-W, Cao K 2020 Establishment and application of intelligent city building information model based on BP neural network model *Computer Communications* **153** 382-389.

Clustering and its key features' formation in construction

A V Ischenko, D V Shishkunova, N P Guryanov

National Research Moscow State University of Civil Engineering (NRU MGSU), 26, Yaroslavskoe highway, Moscow, 129337, Russia

E-mail: Ishchenko.AB@yandex.ru

Abstract. The purpose of this study is to substantiate the terminology "Clustering in construction" used in the regulation of construction organizations in a single system of interaction, taking into account the organizational and technological model of the new and existing buildings' formation reliability. When implementing such projects, a set of joint tasks of the organizational and technological configuration is outlined and a number of indicators (comfortable housing with deliberately used free space; a combination of residential development with other objects of social, educational and entertainment infrastructure) are taken for consideration, as well as the basic principles of organizational and technological clustering criteria in construction. The most important scientific theory is that with the use of cluster modeling methods, competitiveness indicators can be increased, the range of material and technical base of construction production can be increased, and the investment durability of the construction organization can be formed. As a result of the study, the characteristic properties of the construction cluster were established for use in an integrated model.

Materials and methods

The term "cluster" means "set, bundle, totality, concentration". In a general interpretation, this is a set of components, which, in turn, have specific functions for solving the established issues.

A cluster is a form of economic ties' association. It was primarily used to increase competitiveness. However, in the approved direction of improving the economy, organization with modern economy features, the cluster started to be used when developing an increasingly wide range of tasks, in particular:

– in the study of the state, region, industry competitiveness;
– as the basis of a nationwide production course;
– when preparing regional formation programs;
– as a basis for promoting innovative activity;
– as a base for cooperation between small and medium businesses, etc.

However, the term "cluster" can be considered from a different point of view - it can be applied to management. In management process, various techniques and means, which should be closely interconnected and act as a whole unit are used, the form of this whole unit should be characterized by the cluster method. In this case, the "Cluster System" [1] of the organization can be analyzed not only locally, but also more extensively. In this meaning, "cluster" is understood as an amalgamation of coordinated applications, consolidated by one business process (technological process), which covers various mutually influencing fragments, they guarantee detailed information about the work and allow to make appropriate administrative decisions based on the acquired information.

The hierarchical form of the methods system seems to be a distinctive feature of the cluster (as opposed to the systematic approach, in which the organization methods are analyzed as an interconnected set). With the cluster method in the administration organization, each category of tools performs a specific functional meaning in conjunction with other tools [2].

Currently, the modern construction association is fragmented, consists of many separate economic elements that have specific characteristics. In terms of economic content, the construction association has a diversity, manifests itself as a system of organizations and enterprises that guarantee the formation of new and modernization of the existing basic assets for production and non-production purposes. The activity of the construction complex is divided into interrelated operations that are performed by various structures.

In construction industry, there are a lot of cooperation ties with organizations and companies, both in the current industry and in other production branches. This is due to the difficulty, consumption of materials and the variety of construction products. In construction, unlike in industry, not a single construction and installation organization is able to carry out the entire range of work on the formation of existing construction products with the available capacities. Therefore, for the implementation of individual work, all kinds of specialized organizations (subcontractors) are involved. All cooperation ties in the organization of finished construction products can be divided into resource and production.

In those areas where construction clusters are organized, the volume of house construction is much higher. The presence in the cluster of a service part in the form of banks, the legal state administration, the cluster system stability makes it possible to increase the existing indicator by supplying the population with affordable mortgage loans with guaranteed terms for the housing commissioning.

The role of communities of entrepreneurs and social organizations in the issue of the construction branch independent regulation is most noticeably increasing in Russia today [3]. This is due to the fact that the basic part of its subjects is in private ownership, and the investments are off-budget. Therefore, based on the logic of the presence and formation of the market, the state task is to determine the systematization of regulations, norms, that is, institutions for private initiative, the emergence of investment markets, contract work, production and supply of material and technical resources, etc. Social associations are designed to control connections within all the markets between the market and the state.

Cluster types

Clusters function as semi-independent parts of the project under the overall coordination of the contractor or project manager, replicating the logic of single-point responsibility throughout the supply chain. The relationship between clusters and the overall project can change as the authority delegation level increases: clusters can be endowed with functional powers [4]. They can grant wider discretionary powers over technical and commercial agreements within a dedicated cost target, or they can fully delegate responsibility for a work portion, including direct management of all contractual and commercial relationships with the suppliers involved in activities in different typologies of development.

Types of buildings

The choice of building types is most often the result of such economic considerations as the land plot cost, its availability and the infrastructure cost. However, each type of building and combination of different types interacts with local environmental conditions in different ways, which affects both the microclimate around the building and the microclimate inside it.

The types of buildings are presented in Table 1.

Table 1. Types of buildings.

Type	Description

Free standing / Semi-detached	Detached and adjoining houses have a high open shell-to-floor ratio, this increases heat transfer between the building and the environment and provides more options for passive heating, cooling and better use of adjacent open spaces.
Courtyard / Patio	Backyard and patio homes are quite common in hot or maritime climates. In both cases, the building completely or partially covers an open space - a closed or attached, semi-closed open space. Such spaces are commonly referred to as climate modifiers. They are believed to have better climatic conditions than surrounding open areas, and is expected to have a positive effect on the building surrounding volume microclimate. This has a beneficial effect under certain conditions, providing solar access to all parts of the building and providing better ventilation of the interior. However, the specific design of the complex (building and adjacent open space) should always be considered taking into account the specific climatic conditions of the area, hours of operation and design details, as well as the cultural, conservation and religious aspects of the societies within which this typology is developed.

Access types

Row buildings (townhouses) and high-rise buildings can be subdivided according to the form of access into corridor and staircase / elevator types (Table 2). A single congested corridor can provide solar exposure and can provide good cross ventilation if properly designed. Two-story corridor buildings will almost always suffer from poor solar access and cross-ventilation, at least in some of the blocks. Buildings with a staircase / lift core serving two blocks per floor may have full solar access and cross ventilation. [5,6]. When more than two units are served by one core of this type, the use of sun and wind in air conditioning the space becomes difficult, although the ratio of open shell area to floor area per unit becomes less.

Table 2. Access types.

Type	**Description**
Row (Townhouse) Buildings	Row buildings, three to four stories high, can serve as windbreakers to protect adjacent low-rise areas. The use of common walls between the blocks minimizes the open envelope of the building and thus contributes to energy savings. However, this can also have some disadvantages in certain climatic conditions, and this requires appropriate design of the unit layout to ensure cross ventilation.
High building	High-rise buildings are very common in urban centers, where they allow more intensive land plot exploitation. Their internal and adjacent spatial microclimate is ambivalent. Good ventilation can be provided for most units by positioning the openings on both the upwind and leeward sides. While intermediate blocks have a smaller open shell, edge blocks (especially on upper floors) can be exposed to adverse wind conditions. High-rise buildings cast long shadows, thus creating unpleasant adjacent open spaces in winter and causing solar access difficulties when designing nearby buildings. In addition, their volume and height create localized air movements that are difficult to control. Such tasks can be solved at the ground level by integrating the protruding shelves located close to the ground level into high-rise buildings. They can be used to distract and control air movement. Wind speed at the ground level near tall buildings can be 300% higher than in the nearby areas with lower buildings. Such phenomena may be undesirable in specific cases, but can also have positive microclimatic effects in dense urban centers suffering from poor ventilation, as well as in the areas with high air pollution recommendation that different types of buildings should be included in the cluster design to allow access to the sun, as well as wind protection. Row buildings can be used as wind deflectors, while high-rise

	buildings can aid ventilation in dense open spaces. Low-rise buildings should be designed and positioned in such a way that they take full advantage of their connection to open spaces, but should be treated to avoid excessive heat exchange with the environment.

Results and discussions

The implementation of the cluster approach, the creation on the state scale of the requirements for the construction clusters existence will provide the prospect of increasing the industry level, and the social influence from clustering will affect the increase in the degree of accessibility of the housing stock for the state residents [7]. From this it follows that consolidation into a cluster of construction organizations with a successful outcome will eliminate a large number of issues.

The formation of a construction cluster will make it possible to implement such cluster initiatives, the **tasks** of which are:

1. Improving the interactions approaches between the construction organizations of the state regions, which cooperate within the framework of sectoral and cross-sectoral value chains;
2. Promotion of the construction industry production area;
3. Creation of working tools for outsourcing construction companies in the regions;
4. Formation of joint plans within the boundaries of a certain region to improve the spatial and industrial infrastructure;
5. Formation of collegial projects within the framework of the personnel course creation and functioning within the framework of innovations;
6. Formation of joint management companies to regulate and control the projects being implemented [8].

The variability of a construction cluster in the form of a clustering object is caused by the following conditions:

1. The formation of a large volume of large investment projects aspires to the construction of new enterprises and industrial territories.
2. Providing residents with a home fund within the boundaries of various government initiatives reduces the public component of investment risks and supports the development of user and production capabilities.
3. The existing unresolved issues of the construction branch negatively affect the socio-economic formation of the country's regions and, in total, on the investment environment.
4. The production association is characterized by a large increase in investment and helps the formation of other branches of various industries [9].

It is necessary to pay attention to some significant approaches characterized by the implementation of the cluster method in construction.

The strategic approach is manifested in the fact that the cluster is a working model of cooperation between the government and entrepreneurship, an apparatus for bringing the diversity of the regional economy and contributes to the formation of an effective tool for increasing the production competitive ability.

Economic approach:

– the cluster model of the construction formation and the creation of building materials has a significant opportunity to improve the consolidated production characteristics, for example, as an increase in gross domestic product, the creation of new job positions, an increase in work efficiency, etc.;

– along with an increase in the construction branch productivity, there is a stimulation of the economy adjacent branches' formation: the mechanical branch, the production of apparatus and technical equipment, logistics, etc.;

– the cluster approach will increase the rate of formation of small and medium-sized businesses in the construction branch [10,11].

Technological approach:
– the cluster approach contributes to the construction industry improvement through the formation of stable transmission lines of key skills and the exchange of skills within the boundaries of the common technical and information zone;

– the emergence of an extensive distribution and management of new effective quality standards by increasing competitiveness, stimulating the process development by introducing new techniques to the market.

Public approach:
– manifests itself in the form of an increase in the citizens' working capacity degree, an increase in the living standarts quality due to an increase in the productivity of the activities of individual organizations within the cluster.

As a result, the complex impact of the work of organizations in the cluster is contained in the consolidation of the manufacturing companies' competitive advantages. Manufacturers of building materials in a highly competitive environment within the framework of the cluster method increase the choice, the quality characteristics of the finished product increase, there is a desire to reduce the production costs, increase the efficiency of activities using effective resource-saving and energy-saving techniques in construction. The shipping cost of production materials also decreases, production costs decrease, which leads to a decrease in the price of the housing stock, making it higher quality and understandable for society.

Thus, the creation and growth of the cluster method in construction guarantees:

1. Implementation of a profitable and effective construction scheme, understandable for the citizens on the principle of modern technologies, which provide an opportunity to implement the first-class materials at low cost;
2. Durability of building units and rational use of resources;
3. Increase in the products and services' export due to the entry of organizations into strategic cooperation networks and deliberate advertising of goods on the world market;
4. Increasing the investment attractiveness of the regions due to the specific identification of the sphere economic type and, thereby, the supply of the involved local and foreign investments in the economy [12];
5. Innovative activity increase of the companies in the cluster structure, due to increased competence on new techniques;
6. Productivity increase of the order of training professionals due to the specific establishment of prescriptions and responsibilities for the specialty of employees, which the cluster member organizations are looking for and the availability of educational approaches for builders and specific academic platforms that improve the quality of construction and installation activities;
7. Management positions in construction science, design and conduct of surveys and formations in the construction branch;
8. Formation of small and medium-sized businesses in a position of public competitive competition, by increasing the need for the latest research and attracting investment in the economy [13,14];
9. Revenues increase to the budgets of various sizes, an increase in the number of jobs and salaries due to an increase in exports and a number of products and services by the organizations participating in the cluster, continuously increasing competitiveness in the development of innovations and the implementation of working formation policies.

Summary
Thus, the formation of the cluster aspect combines the sources in the production of building materials and will contribute to the introduction of innovative resource-saving techniques and progressive approaches in construction, as well as allow the formation of favorable requirements for solving the housing problems.

The carried-out analysis of the construction industry suggests that it is necessary to increase the indicators for improving the development of this industry, however, there are a sufficient number of reasons that rightly hold back the formation of construction organizations. To achieve a favorable existing environment, the task of which is to stimulate the industry development and the constant participation of the government in this matter, it is rational to assert that an effective partnership between the construction organizations is possible only on the basis of the cluster method implementation.

Having analyzed the essence of the cluster method, we can conclude that the current event reduces costs leads to the growth of the domestic market and access to the international level, increases the wages level, increases the quality of the goods and its level of competition. Steady and solid production, which has achieved high results and is considered the basis of the cluster, "pull up" small and medium-sized organizations, obliging them to improve the quality of the equipment supplied to them. The formation of clustered webs in construction can facilitate the immediate release of a new set of goods and services. In addition, cluster cooperation of firms is promising in investment development, since for an investor, be it government or an individual investor, it is not private objects that are more attractive, but the systemic regulations, in the implementation of which the companies' complex coordination is required.

References

[1] Voronin A V 2007 Cluster approach to the management of the investment and construction complex of the region *Regional economy: theory and practice.*
[2] Porter M E, Stern S 2001 Innovation: Location Matters *MIT Sloan Management Review* **4.**
[3] Ivanchenko E A, Kosilov M S Don State Technical University, Rostov-on-Don, Russian Federation. Building problems in modern urban planning.
[4] Matveeva O A, Vasilenok V L 2014 Implementation of the cluster approach in construction **1**
[5] Kiyanenko K V 2005 Architecture and Social Modeling of Houses.
[6] Ischenko A V, Shishkunova D V 2017 Analysis of the sustainable development principles of the territories and their relationship *Innovative development of construction and architecture: a look into the future* 12-15.
[7] Asaul A N 2004 Construction cluster - a new regional production system *Construction Economics* **6**.
[8] Topchiy D V 2015 Assessment of organizational, technological and economic parameters when moving enterprises outside the city limits *Technology and organization of construction production* **4(1).**
[9] Khunagov R D, Popov R A Clusters as a form of organization and construction technology in a volatile environment: a regional aspect
[10] Enright M 2000 Regional clusters and multinational enterprises: independence, dependence or interdependence? *International Studies of Management and Organization* **30(2)**.
[11] Ischenko A V, Shishkunova D V 2018 Methodology for constructing an imitation model for planning the territory of a constituent entity of the Russian Federation taking into account the principles of sustainable development *Investments, construction, real estate as a material basis for modernization and innovative development of the economy* 155-158.
[12] Animitsa E G Strategic planning of the territorial development: essence, basic principles, problems
[13] Oleinik P P, Brodsky V I 2015 Organizational forms of construction *Scientific publication Textbook* 84.
[14] Lapidus A A, Telichenko V I, Tumanov D K, Ershov M N, Oleinik P P, Feldman O A, Ishin A V 2014 Technology and organization of construction production **2** 10.

IOP Conf. Series: Materials Science and Engineering 1083 (2021) 012104 doi:10.1088/1757-899X/1083/1/012104

Numerical and analytical calculation of longitudinal bending of prismatic elastic rods under the action of axial compressive load taking into account the authorized weight

Azimli Aziz Shakhin Oglu[1], A Yu Nesterov[1], S B Yazyev[2], A S Chepurnenko[2], L G Babahova[2]

[1]Samara University, 34, Moskovskoye shosse, Samara, 443086, Russia
[2]Don State Technical University, 1, Gagarin sq., Rostov-on-Don, 344002, Russia

E-mail: ps62@mail.ru

Abstract. The article is devoted to the problem of calculating the stability of compressed rods taking into account their own weight. A resolving differential equation of the second and fourth order with respect to the dimensionless deflection is obtained. A technique for numerically analytical and numerical solution of the obtained equation using the finite difference method is proposed. As a result, the problem was reduced to a generalized eigenvalue problem. Comparison with the solution of other authors is given.

The problem of determining the critical values for the rods is posed as the problem of finding the domain of these values depending on the boundary conditions and the coefficients α and β. The objective function is a converging power series with unknown coefficients. The solution of the prismatic rod stability problem is carried out in the MatLab environment. The values of α and β, for which the critical load takes a minimum value, are found.

To test the methodology, a number of problems were solved and compared with known solutions.

The proposed technique, in contrast to analytical solutions, allows solving the problem with arbitrary fixation of the ends of the rod. It is also suggested to take into account the length-varying stiffness and heterogeneity of the rod. Test problems showed good agreement with literature data. In the future, it is planned to develop a calculation method taking creep into account.

The higher quality of the analytical solution is shown in comparison with the existing methods.

Introduction

The problems of calculating stability compressed rods are studied by many authors, and from the latest publications on this problem it is worth noting the following works [1-19]. In the articles [1-3], the solution of this problem is considered taking into account creep using the finite difference method. In the works [4-5] the energy method in the form of Ritz-Timoshenko is applied. The finite element method in the form of the Bubnov-Galerkin method [6-9] can also be used to calculate the compressed rods, including creep. The problems of dynamic stability of rods are discussed in articles [10-16]. In the articles [17-19] for calculating the rods stability, the apparatus of the probability theory is connected. However, none of these publications dealt with self-weight sustainability. The existing analytical solutions [20-24] are applicable only for specific fixings of the rod ends.

Consider a homogeneous, rectilinear, vertically positioned rod with arbitrary fixing of its ends, experiencing the specified load (compressive force and dead weight).

Let us select some infinitesimal section of the rod that lags behind the origin at a distance t within the interval $[0, x]$ and measuring dt (Figure 1).

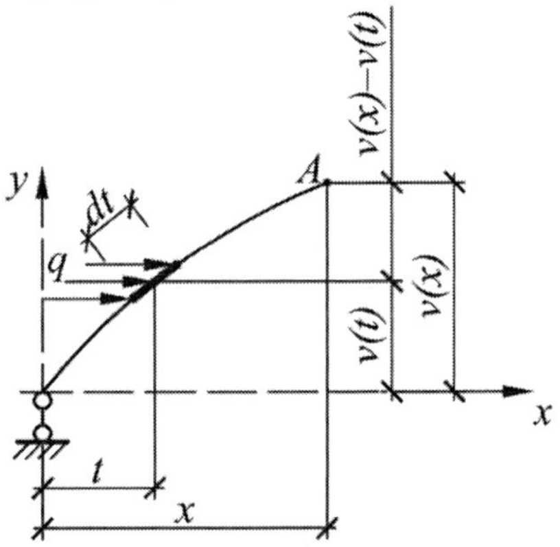

Figure 1. Element dt of the infinitely small rod section.

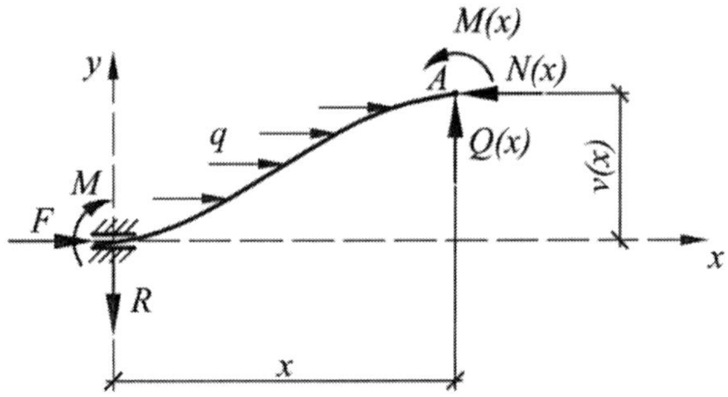

Figure 2. The adopted scheme of forces when drawing up an expression for a bending moment in a section

Then within the considered element $v(x) = const$. Here $v(x)$ — is the amount of deflection at the center of an infinitesimal element dt, and $[v(x) - v(t)]$ — defines the load arm q, acting on an infinitesimal element dt relative to point A. Then the moment relative to this point will be:

$$M_A = q \cdot \int_0^x [v(x) - v(t)]dt = qv(x)x - q \int_0^x v(t)dt$$

To take into account all the fixing options, we consider a rigid fixing, movable along the axis x (Figure 1).

In this case, in the section A moment will be equal:

$$M(x) = F \cdot v(x) + \left\{ qv(x)x - q \int_0^x v(t)dt \right\} + Rx - M \quad (1)$$

On the other hand, in the case of small deformations, the differential equation of the curved axis will be as follows:

$$\frac{d^2v(x)}{dx^2} = -\frac{M(x)}{EI_z} \quad (2)$$

Here E — is a Young's modulus of the material, I_z — is an axial moment of inertia, $M(x)$ — is a bending moment.

Then the integro-differential equation of the curved axis of the rod takes the form:

$$\frac{d^2v(x)}{dx^2} = -\frac{1}{EI_z}\left[F \cdot v(x) + \left\{qv(x)x - q\int_0^x v(t)dt\right\} + Rx - M\right] \tag{3}$$

Further, it is convenient to consider the resolving equations in dimensionless quantities. Let us make the following replacement:

$$x = l\varphi;\ t = l\xi;\ \begin{cases} v(x) = lv(\varphi) \\ v(t) = lv(\xi) \end{cases} \tag{4}$$

$$\frac{d^2v(x)}{dx^2} = \frac{1}{l^2}\frac{d^2v(x)}{d\varphi^2} \xrightarrow{v(x)\to v(\varphi)} \frac{1}{l^2}\frac{l \cdot d^2v(\varphi)}{d\varphi^2} = \frac{1}{l}\frac{d^2v(\varphi)}{d\varphi^2} \tag{5}$$

We rewrite the term in relation (3) in curly brackets with the integral in relative coordinates:

$$\frac{q}{EI_z}\int_0^x v(t)dt = \frac{q}{EI_z}\int_0^\varphi lv(\xi)\cdot ld\xi = \frac{ql^2}{EI_z}\int_0^\varphi v(\xi)d\xi \tag{6}$$

The expression (3), taking into account (5), can be rewritten in dimensionless coordinates:

$$\frac{1}{l}\frac{d^2v(\varphi)}{d\varphi^2} + \frac{Fl}{EI_z}v(\varphi) + \frac{ql\cdot v(\varphi)}{EI_z}l\cdot\varphi - \frac{ql^2}{EI_z}\int_0^\varphi v(\xi)d\xi = -\left(\frac{R\varphi l}{EI_z} - \frac{M}{EI_z}\right) \tag{7}$$

Multiplying the expression (7) by l, finally, in dimensionless coordinates, we represent:

$$\frac{d^2v(\varphi)}{d\varphi^2} + \alpha^2 v(\varphi) + \beta^3\varphi v(\varphi) - \beta^3\int_0^\varphi v(\xi)d\xi = m - r\varphi \tag{8}$$

here

$$\alpha^2 = \frac{Fl^2}{EI_z};\quad \beta^3 = \frac{ql^3}{EI_z};\quad m = \frac{Ml}{EI_z};\quad r = \frac{Rl^2}{EI_z}; \tag{9}$$

Since the solution will also be presented numerically, for the integration convenience under various boundary conditions, we differentiate (8) twice:

$$\frac{d^4v(\varphi)}{d\varphi^4} + \alpha^2\frac{d^2v(\varphi)}{d\varphi^2} + \beta^3\left[\frac{dv(\varphi)}{d\varphi} + \varphi\frac{d^2v(\varphi)}{d\varphi^2}\right] = 0. \tag{9a}$$

Materials and methods

The solution (8a) will be sought in the form of a power series:

$$v(\varphi) = \sum_{k=0}^{n} A_k\varphi^k; \tag{10}$$

Below are the first to fourth derivatives $v(\varphi)$:

$$\frac{dv(\varphi)}{d\varphi}=\sum_{k=1}^{n}kA_k\varphi^{k-1}; \quad \frac{d^2v(\varphi)}{d\varphi^2}=\sum_{k=2}^{n}k(k-1)A_k\varphi^{k-2}; \quad \frac{d^3v(\varphi)}{d\varphi^3}=\sum_{k=3}^{n}k(k-1)(k-2)A_k\varphi^{k-3}; \quad \frac{d^4v(\varphi)}{d\varphi^4}=\sum_{k=4}^{n}k(k-1)(k-2)(k-3)A_k\varphi^{k-4}; \tag{10a}$$

For the same degree at (φ^{k-4}), as, for example, for the 4th derivative, we write (10a) differently:

$$\frac{dv(\varphi)}{d\varphi}=\sum_{k=4}^{n+1}A_{k-3}(k-3)\varphi^{k-4}; \quad \frac{d^2v(\varphi)}{d\varphi^2}=\sum_{k=4}^{n+2}A_{k-2}(k-3)(k-2)\varphi^{k-4}; \quad \varphi\frac{d^2v(\varphi)}{d\varphi^2}=\sum_{k=2}^{n}A_kk(k-1)\varphi^{k-1}=\sum_{k=5}^{n+3}A_{k-3}(k-4)(k-3)\varphi^{k-4}; \quad \frac{d^4v(\varphi)}{d\varphi^4}=\sum_{k=4}^{n}k(k-1)(k-2)(k-3)A_k\varphi^{k-4}; \tag{11}$$

After setting the corresponding derivatives in the equation (8a) and canceling by $(k-3)$, we get:

$$\sum_{k=4}^{n}k(k-1)(k-2)A_k\varphi^{k-4}=-\alpha^2\sum_{k=4}^{n+2}A_{k-2}\varphi^{k-4}-\beta^3\left[\sum_{k=4}^{n+1}A_{k-3}\varphi^{k-4}+\sum_{k=5}^{n+3}A_{k-3}(k-4)\varphi^{k-4}\right]=0$$

Or, equating the coefficients at φ^{k-4} , get:

$$A_k=-\frac{[\alpha^2A_{k-2}(k-2)+\beta^3A_{k-3}(k-3)]}{k(k-1)(k-2)} \tag{12}$$

Let us determine the coefficients at φ^0:

$$v(\varphi)=A_0+A_1\varphi+A_2\varphi^2+A_3\varphi^3+\sum_{k=4}^{n}A_k\varphi^k;$$
$$\frac{dv(\varphi)}{d\varphi}=A_1+2A_2\varphi+3A_3\varphi^2+4A_4\varphi^3+5A_5\varphi^4+6A_6\varphi^5\ldots;$$
$$\frac{d^2v(\varphi)}{d\varphi^2}=2A_2+6A_3\varphi+12A_4\varphi^2+20A_5\varphi^3+30A_6\varphi^4\ldots; \tag{13}$$
$$\varphi\frac{d^2v(\varphi)}{d\varphi^2}=2A_2\varphi+6A_3\varphi^2+12A_4\varphi^3+20A_5\varphi^4+30A_6\varphi^5\ldots;$$
$$\frac{d^3v(\varphi)}{d\varphi^3}=6A_3+24A_4\varphi+60A_5\varphi^2+120A_6\varphi^3\ldots; \quad \frac{d^4v(\varphi)}{d\varphi^4}=24A_4\varphi+120A_5\varphi+360A_6\varphi^2\ldots;$$

We substitute the corresponding terms for φ^0 in (13):

$$A_4=-\frac{[\alpha^2 2\cdot A_2+\beta^3A_1]}{24} \tag{14}$$

We substitute the corresponding terms for φ^1 в (2.13):

$$A_5 = -\frac{[\alpha^2 6 \cdot A_3 + \beta^3 4 \cdot A_2]}{24} \quad (15)$$

Let us consider the boundary conditions for various fixings of the rod:

1. Hinge- Hinge

$$v(0) = 0;\ \frac{d^2v(0)}{d\varphi^2} = 0;\ v(1) = 0;\ \frac{d^2v(1)}{d\varphi^2} = 0; \quad (16)$$

2. Hinge-Pinch

$$v(0) = 0;\ \frac{d^2v(0)}{d\varphi^2} = 0;\ v(1) = 0;\ \frac{dv(1)}{d\varphi} = 0; \quad (17)$$

3. Pinch-Hinge

$$v(0) = 0;\ \frac{dv(0)}{d\varphi} = 0;\ v(1) = 0;\ \frac{d^2v(1)}{d\varphi^2} = 0; \quad (18)$$

4. Pinch-Pinch

$$v(0) = 0;\ \frac{dv(0)}{d\varphi} = 0;\ v(1) = 0;\ \frac{dv(1)}{d\varphi} = 0; \quad (19)$$

The study results. model problems solution

Task 1. Variant of the Hinge-Hinge rod fastening

Below the schemes of tasks to be solved and in accordance with the graphics will be given.

Let us consider in more detail the problem for a rod, taking into account its own weight with the option of "Hinge-Hinge" fixing (Figure 3).

From the boundary conditions:

$$v(0) = 0;\ \to A_0 = 0;\ \frac{d^2v(0)}{d\varphi^2} = 0;\ \to A_2 = 0;$$

$$v(1) = 0;\ \to \sum_{k=0}^{n} A_k = 0\ \to \qquad k_{11}\,A_1 + k_{13}\,A_3 = 0;$$

To define a series k_{11}, we use $\sum_{k=0}^{n} A_k$, but we assume in this amount $A_1 = 1,\ A_3 = 0$;

To determine k_{13}, we use $\sum_{k=0}^{n} A_k$, but we assume in this amount $A_1 = 0,\ A_3 = 1$;

$$\frac{d^2v(1)}{d\varphi^2} = 0;\ \to \sum_{k=4}^{n+2} A_{k-2}\,(k-3)(k-2) = 0 \to \quad k_{31}\,A_1 + k_{33}\,A_3 = 0;$$

To determine k_{31}, we use $\sum_{k=4}^{n+2} A_{k-2}\,(k-3)(k-2)$;, but we assume in this amount $A_1 = 1,\ A_3 = 0$;

To determine k_{13} , we use $\sum_{k=4}^{n+2} A_{k-2}\,(k-3)(k-2)$;, but we assume in this amount $A_1 = 0,\ A_3 = 1$;

We write the obtained relations

$$\begin{cases} k_{11}A_1 + k_{13}A_3 = 0 \\ k_{31}A_1 + k_{33}A_3 = 0 \end{cases}$$

The condition for the nonzero solutions existence

$$D = \begin{Vmatrix} k_{11} & k_{13} \\ k_{31} & k_{33} \end{Vmatrix} = 0$$

The sequence of numerical implementation is as follows: we assume $\alpha = 0; 0{,}1; 0{,}2 \dots \dots . . \pi$ from the condition $D = 0$, and determine the values β by solving the nonlinear equation in Matlab using the function *fzero*. The results are shown in Table 1.

Table 1. Dependence of the coefficients β from α « Hinge-Hinge»

α	β	α	β	α	β
0	2.648057	1.1	2.540819	2.2	2.135201
0.1	2.647209	1.2	2.519243	2.3	2.071442
0.2	2.644669	1.3	2.495281	2.4	1.999977
0.3	2.640422	1.4	2.468781	2.5	1.919087
0.4	2.634448	1.5	2.439619	2.6	1.826341
0.5	2.626722	1.6	2.407531	2.7	1.71802
0.6	2.617205	1.7	2.372278	2.8	1.587946
0.7	2.605857	1.8	2.333558	2.9	1.424479
0.8	2.592618	1.9	2.291001	3.0	1.200214
0.9	2.577424	2.0	2.244158	3.1	0.803247
1.0	2.560195	2.1	2.196462	3.141593	0.0

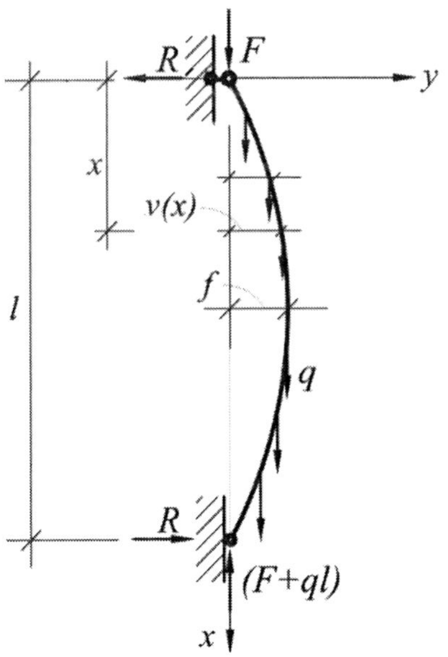

Figure 3. Design scheme "Hinge-Hinge"

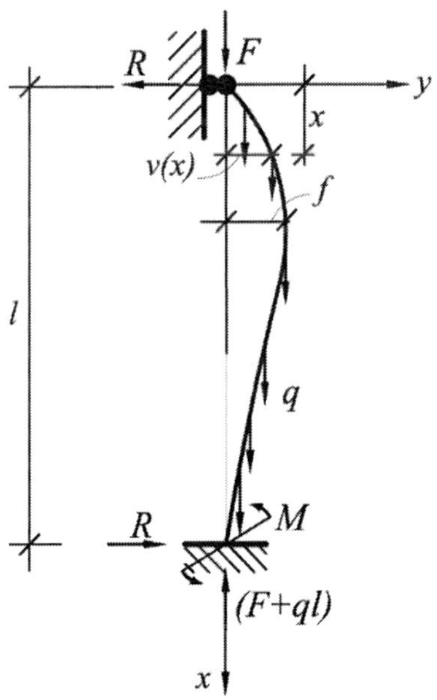

Figure 4. Design scheme "Hinge-Pinch"

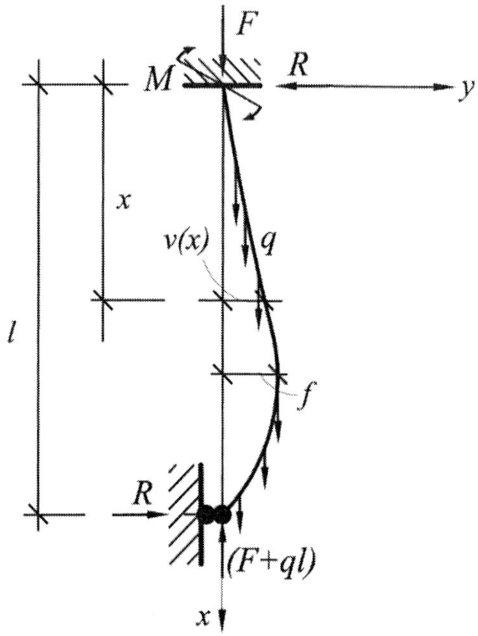

Figure 5. Design scheme "Pinch-Hinge"

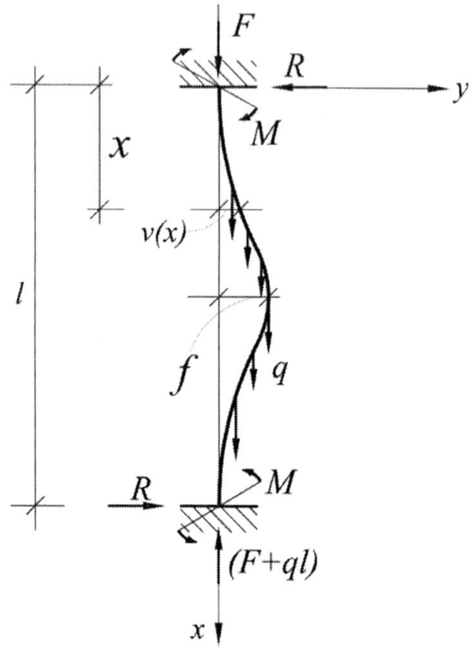

Figure 6. Design scheme "Pinch-Pinch"

Figure 6. Design scheme "Pinch-Pinch" α, β has the form (Figure 7).

By $\alpha = \pi$ and $\beta = 0$, according to (2.9), we obtain L. Euler's formula [25], and for $\alpha = 0$ and $\beta = 2{,}648057$, according to (2.9), we obtain the formula of A.N. Dinnik [22]

$$l_{kp} = 2{,}65\sqrt[3]{\frac{EI_z}{q}} \tag{20}$$

The dependence (Figure 7) can be approximated by the following curve [24]:

$$\alpha^2 + a\beta^3 = b \tag{21}$$

Having defined constant a and b at the values α and β on the axes, we get:

$$\alpha^2 + 0{,}532\beta^3 = \pi^2 \tag{22}$$

or taking into consideration (9)

$$F + 0{,}532ql = \frac{\pi^2 EI_z}{l^2} \tag{23}$$

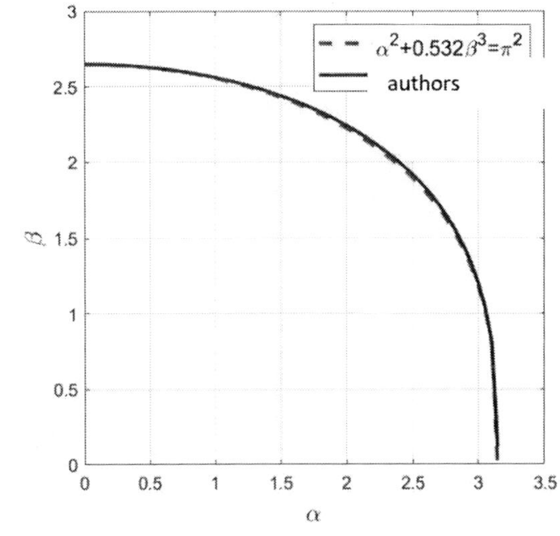

Figure 7. Stability region for a rod hinged at the ends

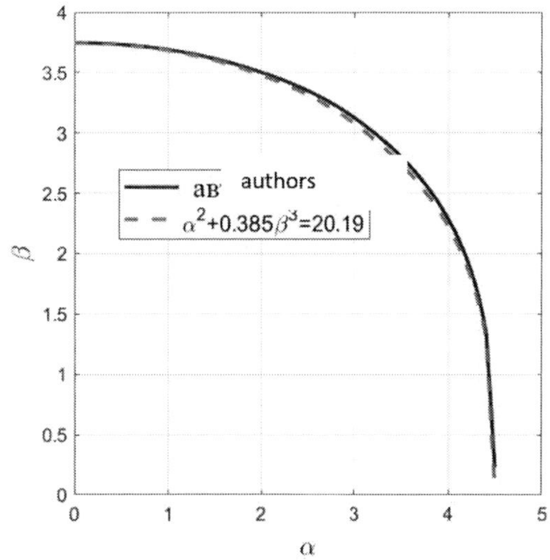

Figure 8. Stability region for a rod, the upper end of which is hinged and the lower end is rigidly clamped.

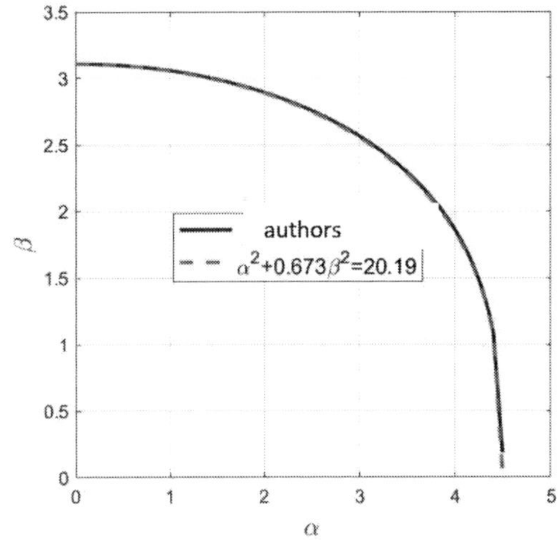

Figure 9. The stability region for a rod, the upper end of which is rigidly clamped, and the lower end is hinged.

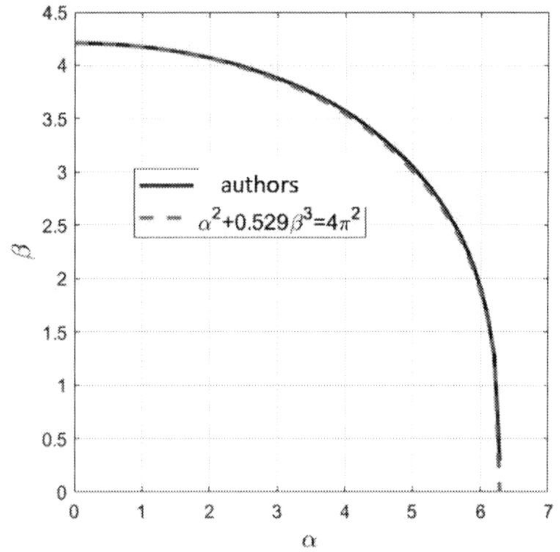

Figure 10. The stability region for a rod with rigidly clamped ends

Summarizing the above-mentioned, we note that the values and β , determined by the formula (22) i.e.

$$\beta = \sqrt[3]{\frac{(\pi^2 - \alpha^2)}{0{,}532}} \tag{24}$$

lie inside the stability region, which goes into the stability margin. This dependence in (Figure 3) is shown with a dotted line.

We find the section where the deflection takes the greatest value. It is worth recalling that we are considering the hinge-hinge case taking into account its own weight.

The border conditions:

$$v(1) = 0; \rightarrow \sum_{k=0}^{n} A_k \qquad k_{11} A_1 + k_{13} A_3 = 0;$$

To define a series k_{11}, we use $\sum_{k=0}^{n} A_k$, but we suppose $A_1 = 1,\ A_3 = 0$ in it ;

To determine k_{13}, we use $\sum_{k=0}^{n} A_k$, but we suppose $A_1 = 0,\ A_3 = 1$ in it ;

$$\frac{dv(\mu)}{d\varphi} = 0; \quad \rightarrow \sum_{k=4}^{n+1} A_{k-3}(k-3)\varphi^{k-4}; \qquad k_{31}^{*} A_1 + k_{33}^{*} A_3 = 0;$$

To determine k_{31}^{*}, we use $\sum_{k=4}^{n+1} A_{k-3}(k-3)\varphi^{k-4}$, but we suppose $A_1 = 1,\ A_3 = 0$ in it ;

To determine k_{33}^{*} , we use $\sum_{k=4}^{n+1} A_{k-3}(k-3)\varphi^{k-4}$, but we suppose $A_1 = 0,\ A_3 = 1$ in it ;

From it:

$$D_1 = \begin{Vmatrix} k_1 & k_r \\ k_{31}^{*} & k_{33}^{*} \end{Vmatrix} = 0$$

We carry out a numerical search using the found relationship between α and β (Figure 2.3), i.e., at $\alpha = 0$ and $\beta = 2{,}648057$ we find $\mu = 0{,}542750$, and for and $\beta = 0 \ \rightarrow \ \mu = 0{,}5$

So, the maximum deflection lies within:

$$0{,}5l < x_0 < 0{,}542750l \tag{25}$$

where $x_0 = 0{,}5l$ can be when $q = 0$, a $x_0 = 0{,}542750l$ by $F = 0$

It should be noted that the problem was solved under the assumption that the material of the rod does not receive plastic deformations under compression, i.e., stresses arising at critical loads are less than the proportional limit. Hence, with the flexibility of the rod λ greater λ_{lim} formula (23) is applicable and vice versa.

In case of plastic deformations, the formula (23) takes the following form:

$$F + 0{,}532ql = \frac{\pi^2 E_{lim} I_z}{l^2} \tag{26}$$

where

$$E_{lim} = \frac{I_1 E_k + I_2 E}{I_z} \tag{27}$$

Here I_1 and I_2 — are the moments of reloading and unloading zones inertia relative to the zero line, I_z — is a moment of inertia of the entire section about the central axis, E_k — is a tangent module.

Task 2. Variant of fastening the rod "Hinge-Pinch"

Let us consider a rod with an articulated upper end and a pinched lower end (Figure 4). According to boundary conditions (17), we determine the coefficients A_0, A_2 , and the numeric search is similar to the previous one. Dependency between α and β shown in Figure 8.

When $\alpha = 4{,}493407$ and $\beta = 0$, taking into account (9), we obtain the F.S. Yasinsky formula [23],

$$F_{kp} = \frac{20{,}19EI_z}{l^2} \tag{28}$$

When $\alpha = 0$ and $\beta = 3{,}744452$, taking into account (9), we obtain the formula of A.N. Dinnik [22],

$$l_{kp} = 3{,}74\sqrt[3]{\frac{EI_z}{q}} \quad (29)$$

Having defined the constant a and b at the values α and β on the axes in (8), the approximate dependence in this case has the following form:

$$\alpha^2 + 0{,}385\beta^3 = 20{,}19$$

or taking into account (9)

$$F + 0{,}385ql = \frac{20{,}19EI_z}{l^2} \quad (30)$$

Let us note that the values α and β , determined by an approximate dependence, also lie within the stability region (Figure 8), and the greatest deflection at

$$0{,}4l < x_0 < 0{,}46l \quad (31)$$

where $x_0 = 0{,}4l$ can be when $q = 0$, a $x_0 = 0{,}46l$ when $F = 0$

Task 3. Variant of fastening the rod "Pinch-Hinge"

Let us consider a rod which upper end is pinched and the lower end is hinged (Figure 5). According to the boundary conditions 2.18, we determine the coefficients A_0, A_2 , and the numeric search is similar to the previous one.

The dependency between α and β is shown graphically in Figure 9. When $\alpha = 4{,}4934070$ and $\beta = 0$, taking into account (9), we obtain the formula (28) of F.S. Yasinsky [23],

When $\alpha = 0$ and $\beta = 3{,}107555$, taking into account (9), we refine the formula of A.N. Dinnik [22], which coefficient is 3.09.

$$l_{kp} = 3{,}11\sqrt[3]{\frac{EI_z}{q}} \quad (32)$$

The approximate dependence (21) in such a fixing sequence has the following form:

$$\alpha^2 + 0{,}673\beta^3 = 20{,}19$$

or taking into account (9)

$$F + 0{,}673ql = \frac{20{,}19EI_z}{l^2} \quad (33)$$

Let us note that the values α and β, determined by an approximate relationship, also underestimate the exact values. The greatest deflection (Figure 5) at

$$0{,}6l < x_0 < 0{,}63l \quad (34)$$

where $x_0 = 0{,}6l$ can be when $q = 0$, and $x_0 = 0{,}63l$ obtained with $F = 0$

Task 4. Variant of fastening the rod "Pinch-Pinch"

Let us consider a rod in which the upper and lower ends are pinched (Figure 6). According to the boundary conditions 19, we determine the coefficients A_0, A_2 , and the numeric search is similar to the previous one. The dependency between α and β is shown in Figure 10.

When $\alpha = 6{,}283186$ and $\beta = 0$, taking into account (2.9), we obtain the F.S. Yasinsky formula [23],

$$F_{kp} = \frac{4\pi^2 EI_z}{l^2} \tag{35}$$

When $\alpha = 0$ and $\beta = 4{,}210175$, taking into account (2.9), we correct the formula of A.N. Dinnik [22], which coefficient is 4,19.

$$l_{kp} = 4{,}21\sqrt[3]{\frac{EI_z}{q}} \tag{36}$$

The approximate dependence (2.21) in this fixing sequence has the following form:

$$\alpha^2 + 0{,}529\beta^3 = 4\pi^2$$

or taking into account (2.9)

$$F + 0{,}529ql = \frac{4\pi^2 EI_z}{l^2} \tag{37}$$

It is necessary to note that the values α and β, determined by an approximate dependence, also lie within the stability region (Figure 10), and the greatest deflection (Figure 6) at

$$0{,}5l < x_0 < 0{,}545l \tag{38}$$

where $x_0 = 0{,}5l$ can be when $q = 0$, and only force acts F, and at $x_0 = 0{,}545l$ is obtained from the action of its own weight.

All the problems presented above are also solved numerically, in particular, by the finite difference method. To implement the grid method, the interval $\xi \in [0;1]$ breaks down into N_ξ parts with the step $\Delta\xi$, and the derivatives in the equation (8a) and in the boundary conditions are replaced by the grid derivatives.

The problems are reduced to a system of linear algebraic equations of the form:

$$([A] + \alpha^2[B] + \beta^3[C])\{X\} = 0 \tag{39}$$

where $\{X\} = \{v_1\ v_2\ v_3 \ldots v_{n-1}\ v_n\}$ — is the nodes' dimensionless displacements vector.

Here the corresponding matrices$[A], [B], [C]$ have the forms:

$$[A] = \frac{1}{\Delta\varphi^4}\begin{bmatrix} a_{22} & -4 & 1 & 0 & 0 & 0 & \ldots & 0 & 0 & 0 & 0 & 0 & 0 \\ -4 & 6 & -4 & 1 & 0 & 0 & \ldots & 0 & 0 & 0 & 0 & 0 & 0 \\ 1 & -4 & 6 & -4 & 1 & 0 & \ldots & 0 & 0 & 0 & 0 & 0 & 0 \\ 0 & 1 & -4 & 6 & -4 & 1 & \ldots & 0 & 0 & 0 & 0 & 0 & 0 \\ \ldots & \ldots & \ldots & \ldots & \ldots & \ldots & \ldots & \ldots & \ldots & \ldots & \ldots & \ldots & \ldots \\ 0 & 0 & 0 & 0 & 0 & 0 & \ldots & 1 & -4 & 6 & -4 & 1 & 0 \\ 0 & 0 & 0 & 0 & 0 & 0 & \ldots & 0 & 1 & -4 & 6 & -4 & 1 \\ 0 & 0 & 0 & 0 & 0 & 0 & \ldots & 0 & 0 & 1 & -4 & 6 & -4 \\ 0 & 0 & 0 & 0 & 0 & 0 & \ldots & 0 & 0 & 0 & 1 & -4 & a_{nn} \end{bmatrix}$$

$$[B] = \frac{1}{\Delta\varphi^2}\begin{bmatrix} -2 & 1 & 0 & 0 & 0 & \ldots & 0 & 0 & 0 & 0 \\ 1 & -2 & 1 & 0 & 0 & \ldots & 0 & 0 & 0 & 0 \\ 0 & 1 & -2 & 1 & 0 & \ldots & 0 & 0 & 0 & 0 \\ \ldots & \ldots & \ldots & \ldots & \ldots & \ldots & \ldots & \ldots & \ldots & \ldots \\ 0 & 0 & 0 & 0 & 0 & \ldots & 1 & -2 & 1 & 0 \\ 0 & 0 & 0 & 0 & 0 & \ldots & 0 & 1 & -2 & 1 \\ 0 & 0 & 0 & 0 & 0 & \ldots & 0 & 0 & 1 & -2 \end{bmatrix},$$

$$[C]=\frac{1}{\Delta\varphi^2}\begin{bmatrix} -2\varphi_2 & \varphi_2+\frac{\Delta\varphi}{2} & 0 & 0 & 0 & \dots & 0 & 0 & 0 & 0 \\ \varphi_3-\frac{\Delta\varphi}{2} & -2\varphi_3 & \varphi_3+\frac{\Delta\varphi}{2} & 0 & 0 & \dots & 0 & 0 & 0 & 0 \\ 0 & \varphi_4-\frac{\Delta\varphi}{2} & -2\varphi_4 & \varphi_4+\frac{\Delta\varphi}{2} & 0 & \dots & 0 & 0 & 0 & 0 \\ \dots & \dots & \dots & \dots & \dots & \dots & \dots & \dots & \dots & \dots \\ 0 & 0 & 0 & 0 & 0 & \dots & 1 & -2\varphi_{n-2} & \varphi_{n-2}-\frac{\Delta\varphi}{2} & 0 \\ 0 & 0 & 0 & 0 & 0 & \dots & \varphi_{n-2}-\frac{\Delta\varphi}{2} & \varphi_{n-1}-\frac{\Delta\varphi}{2} & -2\varphi_{n-1} & \varphi_n+\frac{\Delta\varphi}{2} \\ 0 & 0 & 0 & 0 & 0 & \dots & 0 & 0 & \varphi_n-\frac{\Delta\varphi}{2} & -2\varphi_n \end{bmatrix}$$

where $\varphi_i = (i-1)\Delta\varphi$

The coefficient a_{22} in the matrix$[A]$ equals to 5 if the upper end is hinged and 7 if it is rigidly clamped. The coefficient for a_{nn} for the lower end is calculated in a similar way.

The system (39) is homogeneous and has a nonzero solution if its determinant is zero:

$$|[A]+\alpha^2[B]+\beta^3[C]|=0 \tag{40}$$

For the known α, the problem of finding $\lambda=\beta^3$ is a generalized eigenvalue problem. In the same way, knowing β, it is easy to find α. The discrepancy with the numerical analytical solution is less than 0.01%.

The proposed technique, in contrast to analytical solutions, allows solving the problem with arbitrary fixation of the rod ends. It is also possible to take into account the length-varying stiffness and heterogeneity of the rod. Test problems showed good agreement with the literature data. In the future, it is planned to develop a calculation method taking into account creep deformations.

References

[1] Litvinov S V, Klimenko E S, Kulinich I I, Yazyeva S B 2015 Longitudinal bending of polymer rods with account taken of creep strains and initial imperfections *International Polymer Science and Technology* **42 (2)** 23-25.

[2] Andreev V I, Tsybin N Y 2015 On the stability of rod with variable cross-section *Procedia Engineering* **111** 42-48.

[3] Andreev V I, Rodmenkova E V 2014 Iterative method of optimization of stress state of column under eccentric compression *Procedia Engineering* **91** 20-25.

[4] Yazyev S B, Kozelskaya M Yu, Strelnikov G P, Litvinov S V 2017 Energy method in solving the problems of stability for a viscoelastic polymer rods *MATEC Web of Conferences* **129** 05010. Information on http://www.matec-conferences.org/articles/matecconf/pdf/2017/43/matecconf_icmtmte2017_05010.pdf

[5] Andreev V I, Yazyev B M, Chepurnenko A S 2014 Energy method in the calculation stability of compressed polymer rods considering creep *Advanced Materials Research* **1004-1005** 257-260.

[6] Yazyev B M, Yazyev S B, Grinev A P, Britikova E A 2018 The Definition of a Critical Deflection of Compressed Rods with the Creep by the Method of Bubnov-Galerkin *Material Science Forum* **931** 127-132.

[7] Danilova-Volkovskaya G M, Chepurnenko A S, Begak A, Savchenko A A 2017 Calculation of the bending of electromechanical aircraft element made of the carbon fiber *IOP Conference Series: Earth and Environmental Science* **90**. Information on iopscience.iop.org/article/10.1088/1755-1315/90/1/012046/pdf

[8] Veremeenko A A, Chepurnenko A S, Shvetsov P A, Zorchenko L A, Yazyev S B 2017 Finite-element modeling of loading of spring from an orthotropic material *IOP Conference Series: Earth and Environmental Science* **90**. Information on http://iopscience.iop.org/article/10.1088/1755-1315/90/1/012099

[9] Jan Walczak J, Sieniawski J, Bathe K 1983 On the analysis of creep stability and rupture *Computers & Structures* **17** 783-792.
[10] Aslan T A, Noori A R, Temel B 2018 Dynamic response of viscoelastic tapered cycloidal rods *Mechanics Research Communications* **92** 8-14.
[11] Chirkunov Yu A 2015 Non-linear longitudinal oscillations of a viscoelastic rod in Kelvin's model *Journal of Applied Mathematics and Mechanics* **79** 506-513.
[12] Eratlı N, Argeso H, Çalım F F, Temel B, Omurtag M H 2014 Dynamic analysis of linear viscoelastic cylindrical and conical helicoidal rods using the mixed FEM *Journal of Sound and Vibration* **333** 3671-3690.
[13] Khudayarov B A, Turaev F Zh 2016 Mathematical simulation of nonlinear oscillations of viscoelastic pipelines conveying fluid *Applied Mathematical Modelling* **66** 662-679.
[14] Zhe Ding Li Li, Jianyi Kong, Li Qin 2018 A modal projection-based reduction method for transient dynamic responses of viscoelastic systems with multiple damping models *Computers & Structures* **194** 60-73.
[15] Yan Liu, Mingbin Wang 2015 A model and numerical study for coiling of Kelvin-type viscoelastic filament *Mechanics Research Communications* **70** 17-23.
[16] Lang H, Arnold M 2012 Numerical aspects in the dynamic simulation of geometrically exact rods *Applied Numerical Mathematics* **62** 1411-1427.
[17] Potapov V D 1992 Stability of a viscoelastic rod subject to a random stationary longitudinal force *Journal of Applied Mathematics and Mechanics* **56** 90-95.
[18] Potapov V D, Marasanov A Y 1997 The investigation of the stability of elastic and viscoelastic rods under a stochastic excitation *International Journal of Solids and Structures* **34** 1367-1377.
[19] Deng J W, Xie C, Pandey M D 2014 Higher-order stochastic averaging to study stability of a fractional viscoelastic column *Journal of Sound and Vibration* **333** 6121-6139.
[20] Volmir A S 1967 Stability of deformable systems (Fizmatlit, Moscow).
[21] Timoshenko S 1929 Stabilitätsprobleme der Elastizitat *Handb. der Phys. und techn. Mech.* 81–145.
[22] Dinnik A N 1912 Limparticular bend lim and the action of its own weight *Bulletinof the Alekseev Don Polytechnic institute* **1** 19–46.
[23] Yasinsky F S 1952 Selected works on the stability of compressed rods (Gostekhizdat, Moscow – Leningrad).
[24] Belousov V P 1982 Stability of the plane bending of rods of variable transverse section (In the book: Dynamics of a rigid body. Alma-Ata).
[25] Euler L 1744 Methodus inveniendi lineas eurvas maximi minimive propietate gaudentes sive solution problematic isoperimetric lentissimo sensu accept *Lausanne et Genevae* 245–250.

IOP Conf. Series: Materials Science and Engineering 1083 (2021) 012105 doi:10.1088/1757-899X/1083/1/012105

Building life cycle information modeling at the operation stage

S G Sheina, L V Girya*, A S Livitchuk

Don State Technical University, 1, Gagarin, Sq., Rostov-on-Don, 344000, Russia

E-mail: Giryal@inbox.ru

Abstract. The purpose of this article is to simulate the technical condition of a building under various operating modes. The article proves the necessity of using information modeling to control and plan the real estate objects' maintenance.
BIM technologies application for collecting and analyzing information about a building at all stages of the life cycle has been investigated. The programs that increase the accuracy of designing complex building structures and the timing of the formation of documentation for planning repairs at the stage of building operation are described.
The modeling of the technical condition of a residential building is shown. An example of using the IAS Housing and Communal Services program, which makes it possible to predict the technical condition at any stage of the object's operation and determine the type of recommended repairs under various operating modes, is given.

Introduction

In the modern world, when information technologies are rapidly being introduced, at all stages of the life cycle from pre-design to demolition, a large amount of information about the object is collected and analyzed in different programs, it is necessary to look for new ways of processing data and presenting them in a "one window" mode. [1-3] Information Modeling using BIM technologies helps to solve these problems.

The information model is a repository of all the necessary information about a construction project. With ease of access and a simplified asset management data structure, documentation can be found quickly and maintenance can be accelerated throughout the entire lifecycle of a building. It can serve as a basis for urban planning management of territories within the municipality urban environment sustainable development framework [4-5]

Throughout the world, the use of BIM technology is becoming an important and even mandatory process to ensure that planning, design and construction of buildings are highly efficient. BIM is a process that encompasses the creation and management of the physical and functional information for a project. The result of this process is the construction of information models, which are ultimately digital files that describe every aspect of the project and support decision-making throughout the entire project cycle. This model is based on information collected and updated at the key stages of the project. The use of information modeling of buildings can improve the quality of work, speed and productivity, as well as reduce costs at the stages of buildings' design, construction and operation [6, 7].

Materials and Methods

BIM provides a collaborative environment for the digital data exchange between the investment and construction project participants. During the construction phase, the ideas, intentions and decisions made are transformed to reality. Later, after the construction phase, a BIM model can be used for efficient management and maintenance of the building. Using BIM, it is possible to develop an annual maintenance plan or easily find parts that need maintenance. Implementing BIM in the lifecycle of a construction project brings together accurate data for building management and delivers more business value during the operational phase. With BIM, it is now possible to document and manage every element of a building and use these models to control the system functioning. The fact that not only the building but also the systems are accurately documented takes building management to a new level and provides significant benefits in terms of financial management [6-11].

Building Information Modeling (BIM) implementation tools are Revit, Allplan and others. Such programs allow the use of dynamic information in intelligent models, which increases the accuracy of the complex building structures design and the formation of documents in a short time during the buildings' technical maintenance. The BIM model can be used at any stage of the object's life cycle, in spatial analysis related to its components, data on the time and costs of construction and reconstruction of the building.

Building information models (BIMs) are characterized by providing up-to-date information about the buildings that can be used to support efficient operation and maintenance of the existing buildings [12-16].

Results and Discussion

Solving the life cycle problems at the stage of building operation requires a systematic approach and should include a planned-mandatory system of repair work. The methodology for modeling the life cycle of an object includes five stages, each of which has a certain result and is aimed at solving the specific problems. The proposed methodology makes it possible to choose the most efficient option for building operation. [14-23]

The tool of which is the IAS Housing and Utilities Program (Figure 1), which allows storing, processing and studying the building technical condition examining results [23-25], and making a forecast of the technical condition at any stage of the facility operation. It is necessary to simulate the operational phase of the facility, on the basis of which to make effective management decisions on the facility operation throughout its life cycle.

The conceptual diagram of the numerical modeling technique is shown in Figure 2.

In the software product IAS Housing and Communal Services, a methodology is proposed, on the basis of which the physical and moral wear of the capital construction object and its cost estimate are analyzed. The criterion for determining the needs of capital construction objects is the coefficient "K" by the value of which one or another type of repair work is determined (Figure 4), it is defined as the ratio of the repair work cost to eliminate physical and moral wear to the replacement cost of the building. (Figure 3).

On the basis of the research carried out, the data and factors forming the information model of the operational phase of the object's life cycle are described (Figure 4).

As an example of the life cycle modeling methodology implementation, it is proposed to consider the building at the address: Orbital str., 115 in Rostov-on-Don, taking into account various operating options. The object under study is a 21-storey residential building, built in 2019, capital group -I (Figure 5).

Let us take an example of a construction object's life cycle numerical modeling stages.

1st stage.

Formation of an electronic technical passport in the IAS Housing and Communal Services

We fill out an electronic technical passport in the IAS Housing and Communal Services.

2nd stage

Modeling the life cycle of a construction object under various operating modes.

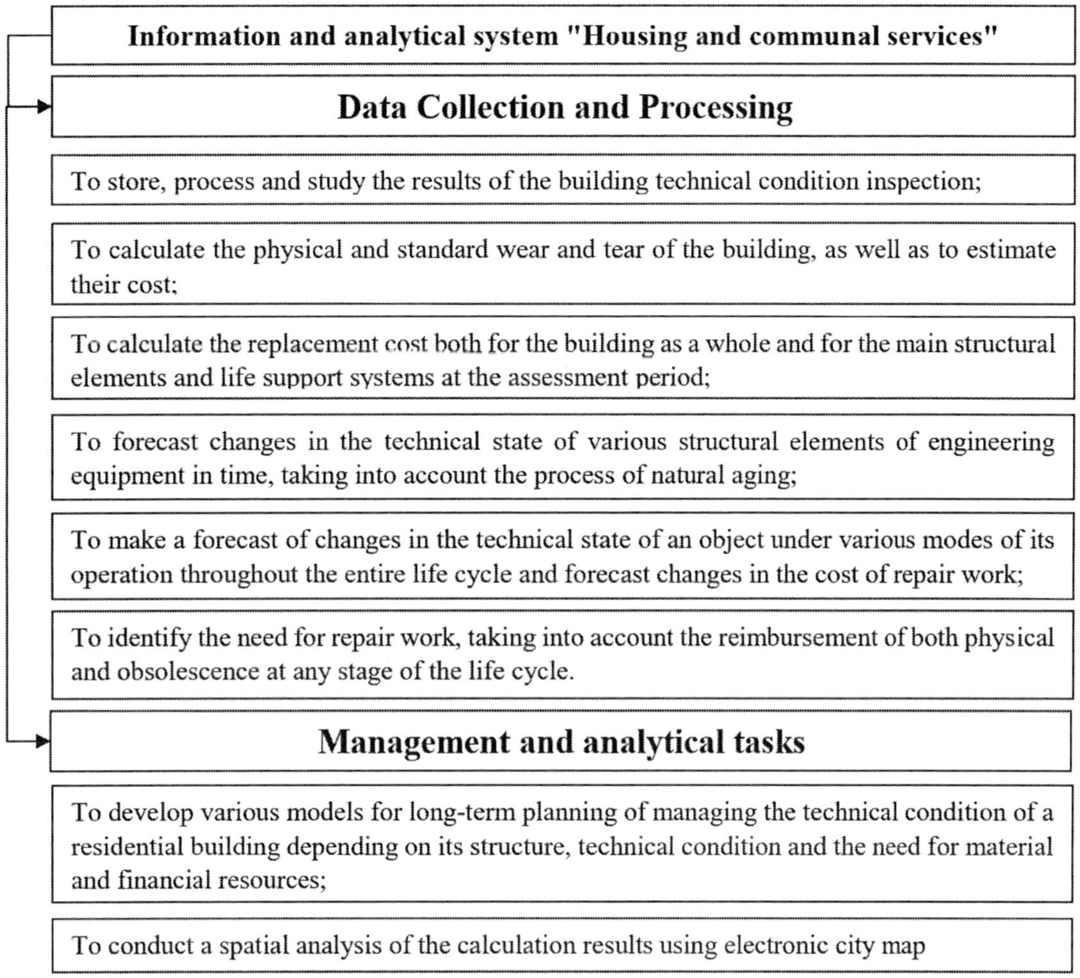

Figure 1. Capabilities of the information and analytical system "Housing and Communal Services"

At the second stage, the life cycle of a construction object is modeled according to three options under different operating modes.

The first option is the 1st model - taking into account natural aging. Modeling is carried out until the requirements of complete replacement of one or more elements $k \geq 1$ are achieved (Figure 5). The results of modeling the technical state of an object during natural aging (model 1) are presented.

Results of modeling the technical state of an object during natural aging (model 1)

The technical condition of structural elements (physical wear (%) in different periods of the building's life cycle during operation of model 1is given in Table 1.

After modeling, the following data were obtained, presented in Figures 6 and 7.

The replacement cost by 2048 for the building as a whole is 112,464,538 rubles, the cost estimate of physical wear is 37,546,927 rubles. The ratio coefficient of carrying out repair work for the building as a whole is 0.35.

For the main structural elements, the replacement cost is 16 183 647 rubles, the cost estimate of physical wear is 16 108 016 rubles. The efficiency factor of the repair work for the main structural elements is 1.01.

After 30 years, all major structural elements need to be completely replaced for the further operation of the building.

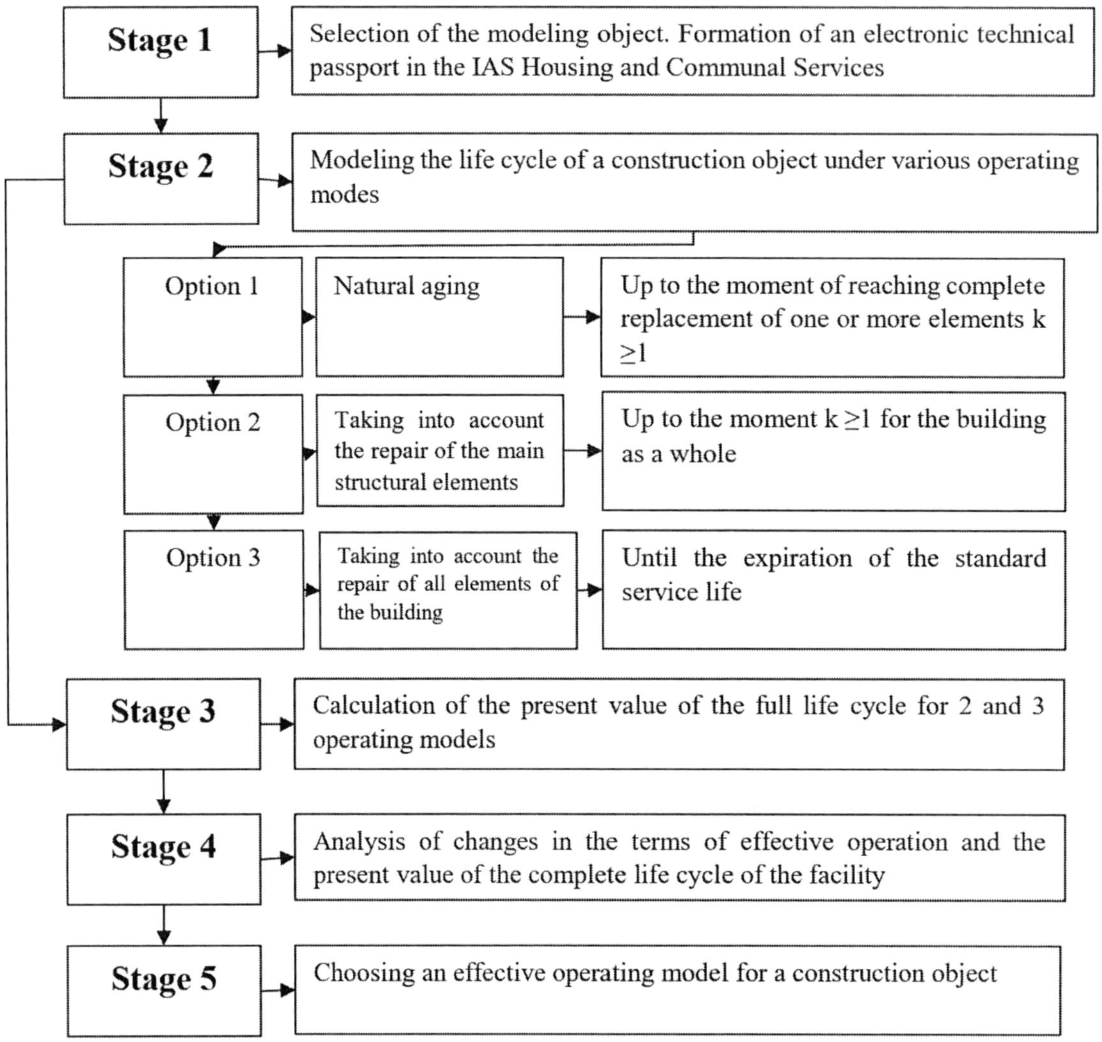

Figure 2. Technique for numerical modeling of a construction object's life cycle

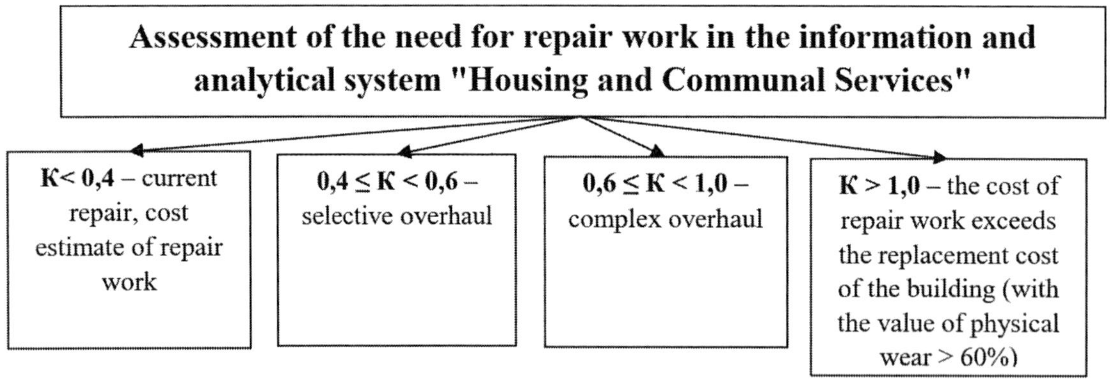

Figure 3. Assessment of the need for repair work in the information and analytical system "Housing and Communal Services"

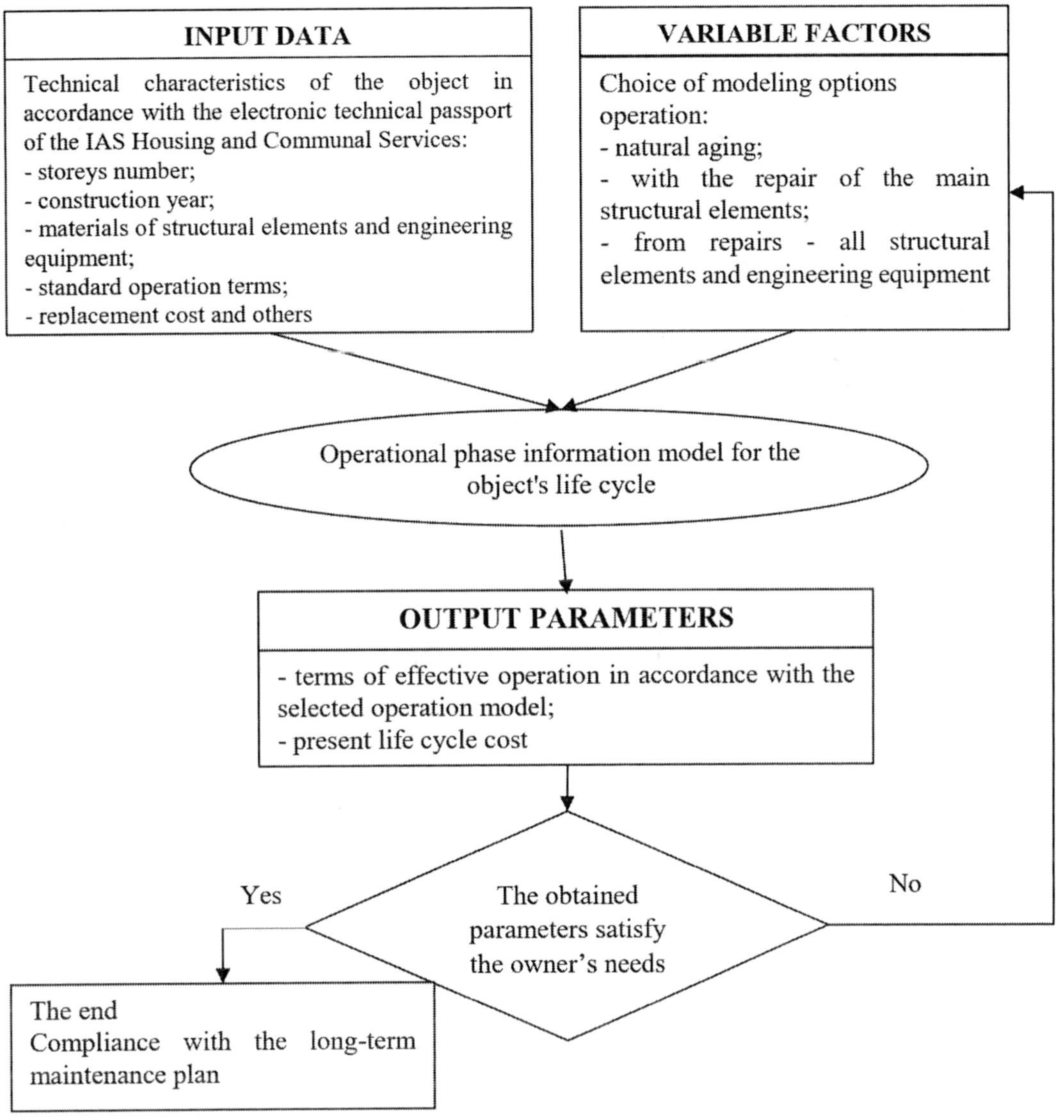

Figure 4. Information model of the object's life cycle operational phase

The recommended type of repair is current for the building as a whole, and reconstruction, modernization for the main structural elements and the life support systems.

The second option is the 2nd model of operation - taking into account the repair of the main structural elements. The 2nd model was developed taking into account the repair of the main structural elements that ensure the object's service life. These include 6 replaceable elements: roofing, cold and hot water supply, sewerage, heating and other things (balconies, drains, canopies). Modeling is carried out up to the moment $k \geq 1$ for the building as a whole. (Figures 8, 9).

By 2162, the replacement cost for the building as a whole will be 416,118,790 rubles, the cost estimate of physical wear is 410,450,393 rubles. The efficiency factor of the repair work for the building as a whole is 1.00 (Figure 8).

Recommended type of repair:
- for the building as a whole - reconstruction, modernization,
- for the main elements - complex overhaul.

Table 1. Physical and standard wear during operation for model 1

Year	Building wear in general, %	Average wear, %	Main construction elements' wear, %	Natural wear estimate cost for the building as a whole, rub.
1	**2**	**3**	**4**	**5**
2021	2.18	4.40	5.35	40 428
2022	4.25	6.45	10.32	90 988
2023	6.22	8.40	15.12	293 472
2024	8.06	10.55	19.28	619 373
2025	9.77	12.60	22.65	1 095 102
2026	11.40	14.08	25.76	1 705 220
2027	12.95	16.70	27.49	2 807 465
2028	14.48	17.75	30.31	4 109 387
2029	16.03	18.10	33.02	5 593 103
2030	18.79	19.25	36.98	6 572 984
2031	19.56	20.88	38.46	7 892 563
2032	20.16	21.93	39.28	9 198 516
2033	21.23	22.99	40.93	10 739 635
2034	22.24	23.98	41.51	12 836 490
2035	23.19	24.92	43.49	13 080 341
2036	24.09	25.81	45.89	14 729 738
2037	24.95	26.67	46.98	16 383 043
2038	25.78	27.51	47.23	18 899 500
2039	26.59	28.34	48.76	20 734 470
2040	27.40	29.14	49.63	21 769 230
2041	28.22	30.95	50.08	23 992 084
2042	29.07	32.79	52.21	25 899 655
2043	29.97	33.67	53.35	27 993 492
2044	30.92	35.62	55.72	29 645 805
2045	31.96	37.65	57.09	31 749 570
2046	32.91	39.60	59.43	33 406 258
2047	33.86	40.55	61.02	35 858 731
2048	34.89	41.85	64.87	37 546 927

When the building is operated according to model 2, the effective service life is 143 years or 95.3% of the standard service life.

Total, the cost estimate of the physical wear of the building as a whole by 2162 will be 410,450,393 rubles. The building according to the second model will be effectively operated for 143 years.

Option three is the 3rd model - taking into account the repair of all building elements.

Modeling is carried out until the expiration of the standard service life, depending on the capital group. Modeling the technical condition of the facility during the repair of all elements is shown in Figures 10 and 11.

The object's technical condition modeling during all elements' repair is shown in Figure 1.9. The replacement cost for the building as a whole is 1,354,508,513 rubles, the cost estimate of physical wear is 415,436,965 rubles.

The efficiency factor of the repair work for the building as a whole is 0.31, for the main structural elements of life support - 0.45.

Monitoring of the technical condition taking into account repair work is shown in Figure 11.

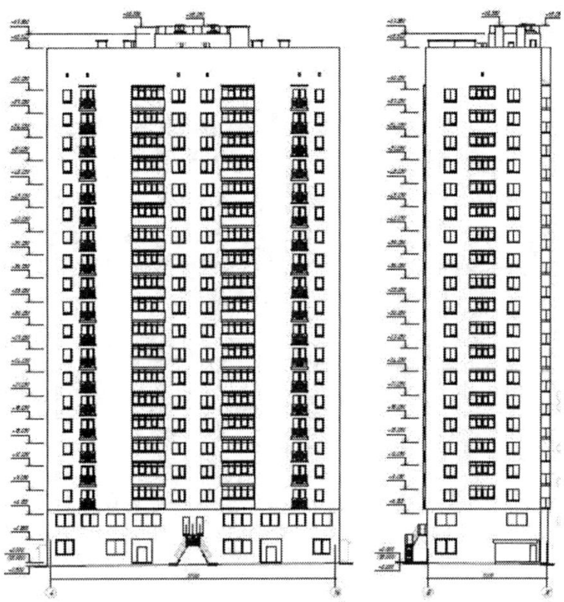

Figure 5. A residential building model

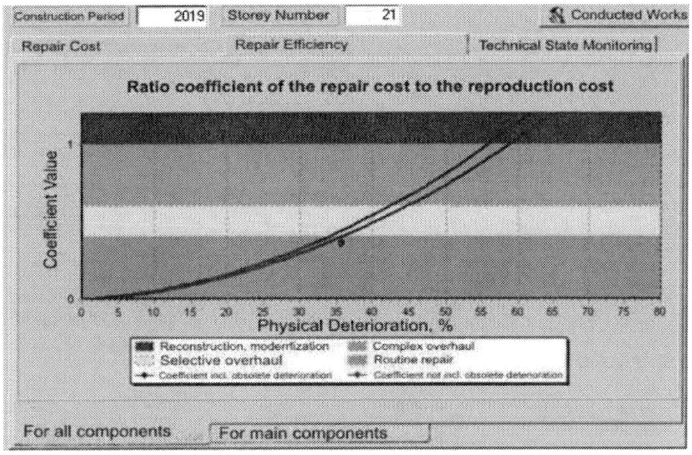

Figure 6. The effectiveness of repair work for model 1

Calculation of costs (in current prices) for carrying out repair work according to 3 models using the IAS Housing and Communal Services program on the basis of automatically generated acts, showed that for the entire period of operation the cost of repairs will amount for the 3rd operation building model will be 1,354,508,513 rubles, which is significantly more than the costs for the 2nd model - 416 118 790 rubles.

3rd stage.

Calculation of the present value of the full life cycle for the operating models 2 and 3

The forecast of the reduced costs for carrying out repair work is carried out during the operational period of the object's life cycle (the standard operation period is 150 years)

Present value is the future amount of money reduced to the present. To determine the present value, the increment coefficient formula is used:

$$K = (1 + R) n \quad (1)$$

where R is the discount rate;

n - denotes years.

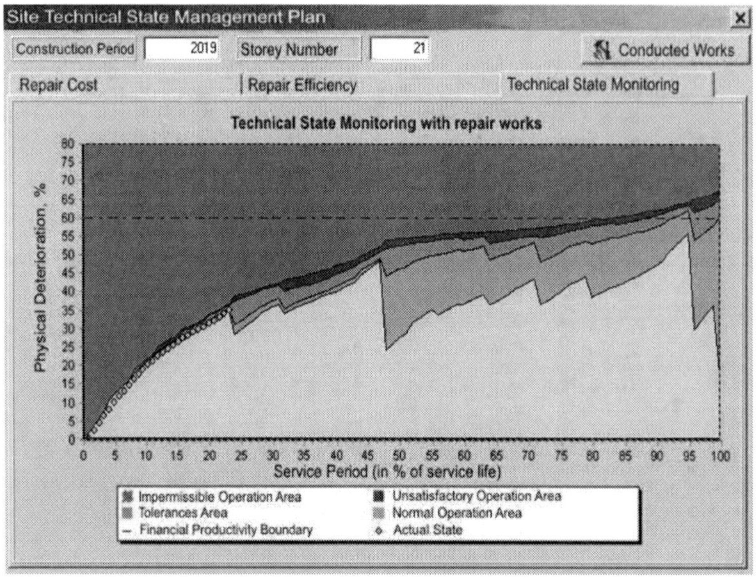

Figure 7. Technical condition monitoring

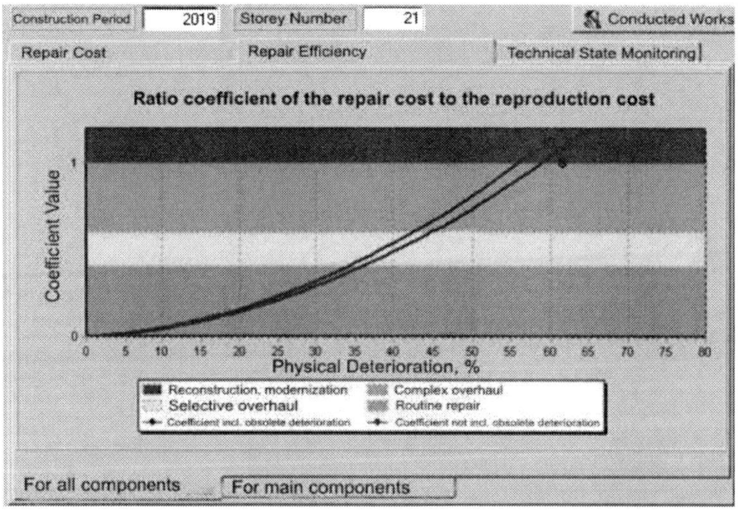

Figure 8. Efficiency of repair work (for all structural elements) for model 2

Due to the impossibility of presenting an official inflation forecast for the entire period of the building's standard operation (150 years), when calculating the reduced costs, we will take two forecast horizons - 50 years and 100 years. The cost of construction of the building for 2019 according to the EIRC is 416.75 million rubles. Using the formula for the increment coefficient, we will make forecast calculations of the building operating costs.

According to model 1, the main structural elements and life support systems of the building by 2048 will reach physical wear and tear equal to 64.87%, and further operation of the building becomes technically impossible due to failures of structural elements and engineering equipment. The term of effective operation of the building for 1 model was 30 years or 20% of the standard.

According to model 2, the cost estimate of the physical wear of the building as a whole by 2162 will amount to 410.4 million rubles. The building according to model 2 will be effectively operated for 143 years.

Calculation of the costs (in current prices) for carrying out repair work for three operation models using the IAS Housing and Utilities Program on the basis of automatically generated acts, showed that for the entire period of operation, the cost of repair work for the building will amount to 1.3 billion rubles for 3 operation models, which is much more than the costs for the 2 model - 416 million rubles. Comparison of costs is carried out by calculating the ratio of the repair work cost C_{rw} to the cost of new construction $C_{constr.}$

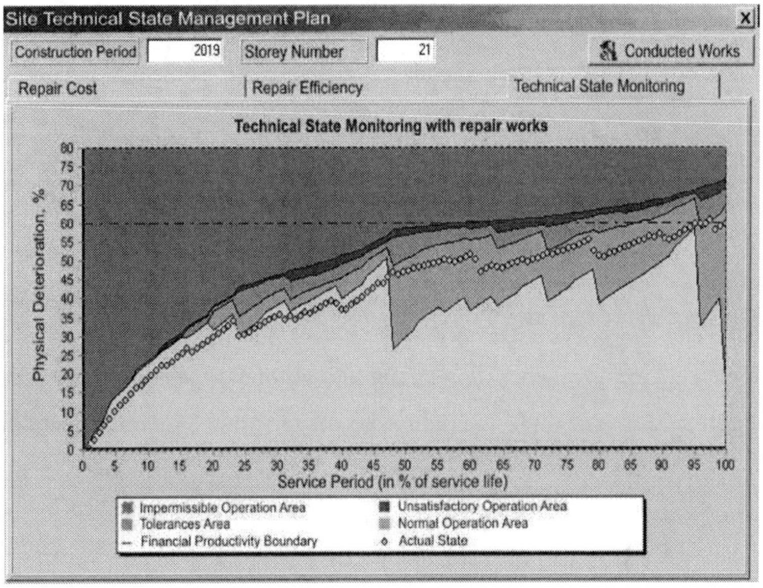

Figure 9. Technical condition monitoring taking into account repair work

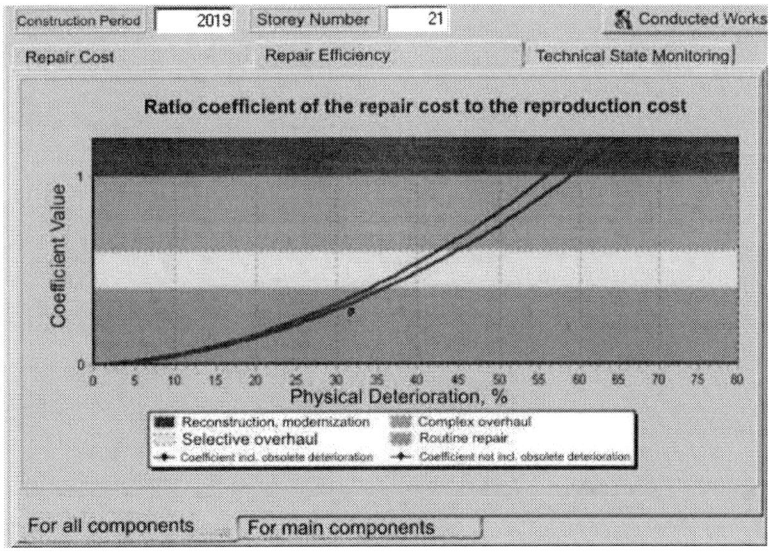

Figure 10. Repair work efficiency (for all structural elements) for the model 3

4th stage

Analysis of changes in the terms of effective operation and carrying out the cost of the full life cycle for a period

Analysis of changes in the terms of effective operation and carrying out the cost of the full life cycle for a period is given in graphical form (Figures 10-11).

IOP Conf. Series: Materials Science and Engineering 1083 (2021) 012105 doi:10.1088/1757-899X/1083/1/012105

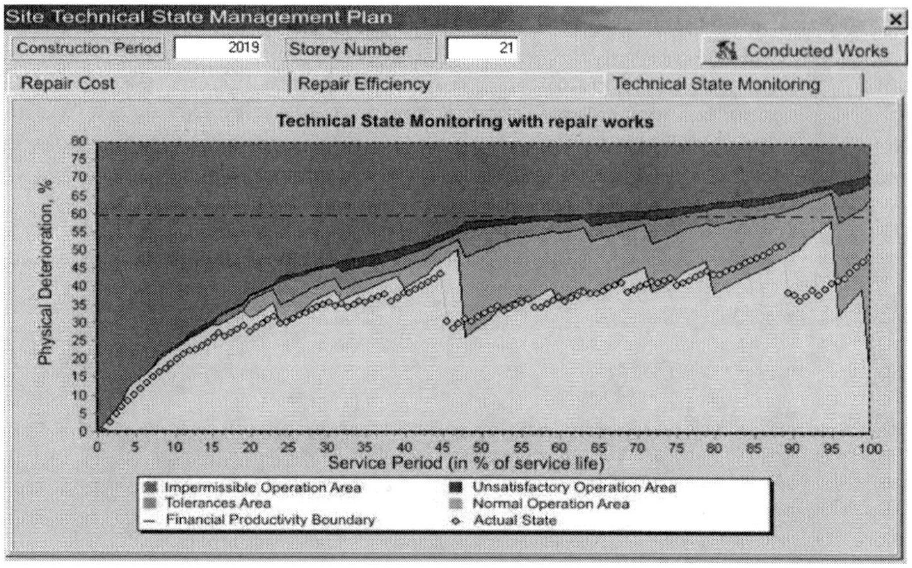

Figure 11. Monitoring technical condition taking into account repair work according to 3 models

Table 2. Comparison of the reduced costs for 2 and 3 operating models

Year	For model 2	For model 3
0	$C_{rw}/C_{constr}=0$	$C_{rw}/C_{constr}=0$
50	$C_{rw}/C_{constr}=1.4$	$C_{rw}/C_{constr}=7.2$
100	$C_{rw}/C_{constr}=10.4$	$C_{rw}/C_{constr}=112.5$

Repairs relative cost growth forecast to the replacement cost of the building for 2 and 3 operating models

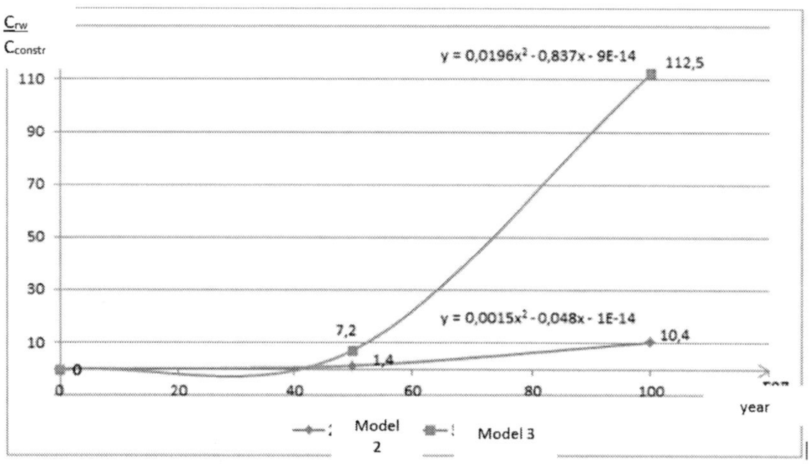

Figure 12.

5th stage

Choosing an effective operating model for a construction object

The choice of an effective operation model for a construction object is carried out by comparing the terms of effective operation of the building and the values of the reduced costs for carrying out repair work (Figure 11), Table 3.

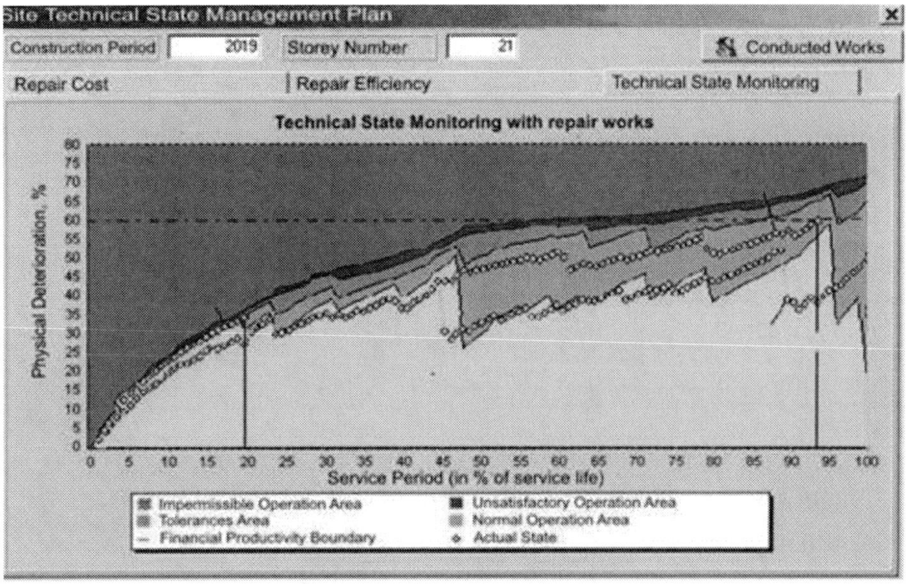

Figure 13. General combined schedule for monitoring the technical condition of the building, taking into account repair work according to three operating models (without repairs - natural aging, with the repair of the main structural elements, with the repair of all elements)

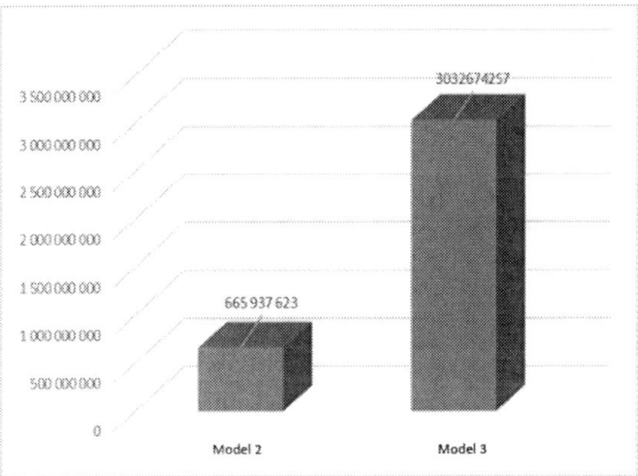

Figure 14. Costs for repair work with a forecasting horizon of 50 years for 2 and 3 operating models (in rubles)

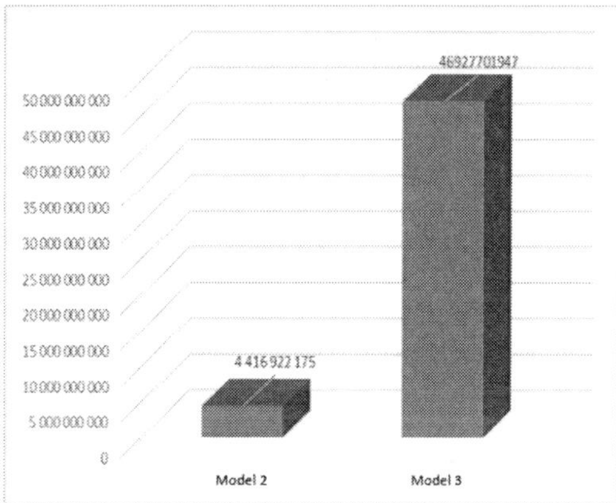

Figure 15. Costs of repair work with a forecasting horizon of 100 years for 2 and 3 operating models (in rubles)

Based on the performed numerical modeling and the presented graphs, it can be concluded that the most rational is model 2 - with the repair of the main structural elements, since the effective service life in this case is 143 years or 95.3% of the standard service life, and the cost of carrying out repair work is significantly (almost 10 times) lower than for the third model.

Summary

Thus, the article reveals the relevance of using BIM technologies for managing the technical condition of buildings. The necessity of using information modeling for monitoring and planning the maintenance of real estate objects has been substantiated. An example of using the IAS Housing and Communal Services program for developing an effective plan for managing the technical condition of a building according to three operation models using information technologies is given.

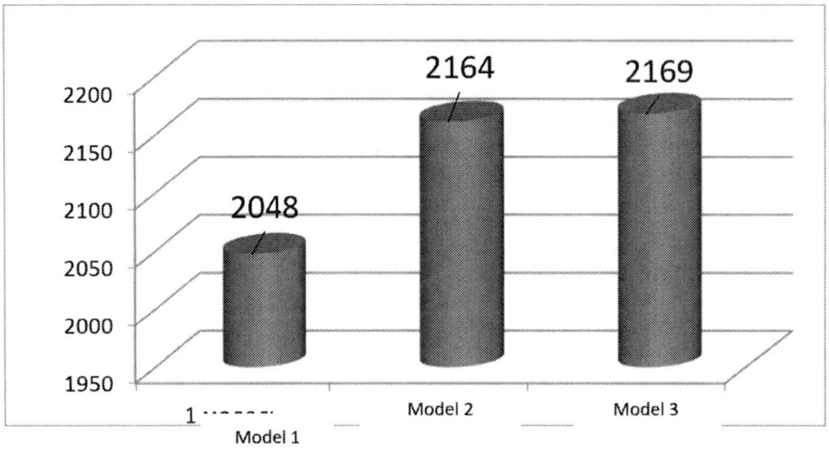

Figure 16. Comparison of the effective operation terms of the building (years)

Table 3. Comparison of terms and reduced costs for different forecast horizons for 2 and 3 operation models

Model	Effective operation period, years	Costs for the RW performed on the EIRC, rubles	The given value of the costs for conducting the RW by EIRC, rubles
Model 2	2164	416 118 790	8 066 848 534
Model 3	2169	1 354 508 513	75 454 184 804

References

[1] Mayilyan L R 2009 *Directory of a modern engineer of housing and communal services* (Rostov on Don, Phoenix) 380 [1] p., Ill. (Construction and design)

[2] Maria Fernanda Silva Rodrigues, A Costa, Raquel Matos 2018 Conservation level of residential buildings: Methodology evolution *Construction and Building Materials* 2018. Information on https://www.researchgate.net/publication/324533738

[3] Telichenko V I, Shcherbina E V 2019 Social-Natural-Technogenic System of Sustainable Environment of Vital Activity *Industrial and Civil Engineering* **6** 5-12. (in Russian). DOI: 10.33622/0869-7019.2019.06.5-12.

[4] Shcherbina E, Gorbenkova E, Belal A 2018 *Rural Areas: Critical Drivers for Sustainable Development* **51 (30)** 786-790. Information on https://doi.org/10.1016/j.ifacol.2018.11.195

[5] Kasyanov V F, Silin R V 2019 Method for multi-criteria evolution of urban planning *IOP confidence serious materials science and engineering* **687** 055040.

[6] Gusakova E A 2004 Life cycle of a construction object *System engineering of construction* (Moscow, ASV) 45-47.

[7] Gusakova E A 2004 System engineering for organizing the life cycle of a construction object *Monograph. Moscow: New Millennium Foundation* 350.

[8] Zilberova I Y, Mailyan V D, Mailyan A L 2019 Organizational and technological stages of the forensic construction and technical expert examination when determining the compliance of a building facility with regulatory and technical documentation *IOP Conference Series: Materials Science and Engineering* **698 (2)**. Information on https://doi.org/10.1088/1757-899X/698/2/022085

[9] Zilberova I Y, Nikulina O V, Novoselova I V 2019 Main principles of choice of multi-purpose design solutions for construction operations when conducting judicial construction and technical expertise (JCTE) *IOP Conference Series: Materials Science and Engineering* **698 (5)**. Information on https://doi.org/10.1088/1757-899X/698/5/055012.

[10] Zilberova I Y, Mailyan V D, Novoselova I V 2019 Modern methods for evaluating the technical and organizational-technological solutions for repair and construction production *IOP Conference Series: Materials Science and Engineering* **698 (5).** Information on https://doi.org/10.1088/1757-899X/698/5/055013

[11] Zilberova I Yu, Tomashuk E A and Bobkina V A 2019 Program-target approach as a basis for issuing a backlog in construction *IOP Conference Series: Materials Science and Engineering* **698 (5).** Information on https://doi.org/10.1088/1757-899X/698/5/055021

[12] Levtsev A P, Lapin E S, Zhang T S 2019 Increase of efficiency of heat transfer of sectional radiators in heat supply systems of buildings *Construction engineering magazine* **8 (92)** 63–75. DOI: 10.18720 / MCE.92.5

[13] Rybakov V A, Ananeva I A, Pichugin E D, Garifullin M 2020 Heat protective properties of enclosure structure from thin-wall profiles with foamed concrete *Construction engineering magazine* **94 (2)** 11–20. DOI: 10.18720/MCE.94.2

[14] Sheina S G, Girya L V, Fedyayeva P 2016 Methods of the Numerical Modeling of Residential Buildings Lifecycle for Evaluating the Effectiveness of Energy-Saving Solutions *Advanced Engineering Forum* **17.**

[15] Sheina S G, Girya L V, Fedyayeva P 2015 The Procedure of Substantiation of selection of the energy-efficient design solutions for residential buildings *International Journal of Applied Engineering Research* **10 (8).**

[16] Sheina S G, Girya L V, Seraya E S, Matveyko R B 2019 Intelligent municipal system and sustainable development of the urban environment: conversion prospects *IOP Conference Series: Materials Science and Engineering* **698**, doi:10.1088/issn.1757-899X, Online ISSN: 1757-899X Print ISSN: 1757-8981

[17] Sheina S, Girya L 2017 Innovative methods of monitoring for deformations of man-made loaded areas *Matec Web of Conferences* (ISSN: 2261-236X, France).

[18] Sheina S G, Yudina K V 2018 Development technique of urban-environmental passport of the reconstruction territory *Materials Science Forum* **931** 822-826. Information on https://doi.org/10.4028/www.scientific.net/MSF.931.822

[19] Strazhnikov A M 2002 Monitoring the quality of the housing stock (Moscow, Scientific work) 388.

[20] Payne G, Majale M 2004 *The urban housing manual* (London, Earthscan) 134.

[21] Mohamed Elkaftangi, Bassem Eid Mohamed 2019 Rethinking the Old Residential Sector in the UAE: Methodology for Sustainable Retrofitting *International Journal of Service Technologies and Management* **25 (1)** 4 – 17. DOI: 10.1504 / IJSTM.2019.096600

[22] *Price Waterhouse Financial and Cost Management Team* 1997 CFO: Architect of Corporation's Future. New York: John Wiley and Sons.

[23] Kashevarova G G, Tonkov Yu L 2020 Technical diagnostics of reinforced concrete structures with the use of intelligent systems *Engineering and construction journal* **1 (93)** 13–26. DOI: 10.18720 / MCE.93.2

[24] Rusanov A E, Baiburin A Kh, Baiburin D A, Bianco V 2020 Heat loss from defects of hinged facade systems of buildings *Magazine of Civil Engineering* **95 (3)** 57–65. DOI: 10.18720/MCE.95.6

[25] Beskopylny A N, Vernezi N L, Veremeenko A A and Arakelyan R M 2019 Experience non-destructive testing of the metal of bridge structures *Materials Science and Engineering*. Information on https://doi.org/10.1088/1757-899X/698/6/066001 (date of application: 05.06.2020).

Historical transformations of early medieval basilica churches of the North-Eastern Black Sea region in the light of the new data method based on lime mortars

V V Pishchulina*, V D Kotlyar

Don State Technical University, 1 Gagarin square, Rostov-on-Don, 344002, Russia

E-mail: viktvlad@mail.ru

Abstract. The article is devoted to the study of the medieval temples' architecture historical transformations in Musser, Tsandripsh, Loo in comparison with the data of the temples' mortar analysis by the X-ray phase method. The scientific novelty of the article is due to the fact that a comprehensive study of historical transformations and volumetric-spatial solutions of the temples made it possible to clarify their dating and construction periods.

Introduction

The relevance of this article is due to the lack of consensus today regarding the dating and construction periods of the temples presented in the article. The Mussera temple has repeatedly attracted the researchers' attention. However, the assumptions differ significantly on its dating. So, L. D. Rcheulishvili calls the end of the VI-VII centuries. [6], I.V. Berdzenishvili - VII-VIII centuries, [1], E. Neubauer inclines to VIII-IX centuries [8], V.P. Pachulia dates the temple to the 9th – 12th centuries [5], A.Yu. Vinogradov and D.V. Beletsky called the date 970 [2]. Yu.N. Voronov refers the construction to the X-XI centuries [3], L.G. Khrushkova does not name the estimated date [9], V.V. Pishchulina and E.M. Kishkinova proposed the date of the construction beginning as the 6th century, the second construction period - the 10th century [4]. L.G. Khrushkova identified two construction stages in the course of archaeological excavations of the Tsandripshsky temple: the 6th century - the main structure and the 10th century – its western porch [7]. Temple in Loo based on the results of archaeological excavations by B.B. Ovchinnikova dates the X-XI centuries, noting, nevertheless, the features of an earlier structure in its western part - including the absence of a walled narthex [3].

The purpose of this article is to clarify the construction periods of the studied temples. The tasks are: to study the architectural features of the temples in Musser, Tsandripsh, Loo in comparison with the data of a new method of dating medieval objects using mortars [9].

Material and methods

All three temples are located on the Black Sea coast. Mussera Church is located in the Gudauta region of Abkhazia, on the territory of the Pitsunda-Mussera Biosphere Reserve, on the left bank of the Ambara River, which flows into the Black Sea.

Located on a hill, the Musser Temple was once located inside a fortress, the ruined walls of which can be seen around it. In plan, the temple is a rectangle measuring 22.5x18.6 m, elongated in the west-east direction. The entrance from the west is located almost in the center of the facade, with a slight offset to the north, and leads to the narthex. In the lower tier, the narthex has two window openings on the northern and southern sides. The embrasures of all windows of the monument are trapezoidal and widen inward. From the narthex, two openings lead to the central nave. It significantly surpasses the side walls in height and width and ends with an apse, semicircular in plan from the inside and not protruding beyond the outer walls in the exterior, cut through by one window. Wide aisles lead from the central nave to the side ones. On its longitudinal walls there are two pairs of pilasters, and in the eastern part, near the apse, one more pair of pilasters with ledges form asymmetrical "shoulders". The central apse is connected with the side rooms by narrow passages.

The side aisles contain two rooms and are not symmetrical. The northern nave is divided into a small room on the west, with one window, and a longer one on the east, with three windows, with the rooms connected to each other. The eastern room ends with an apse with a window, which has a drop-shaped outline in the plan, curved towards the central axis of the building. From this extension, a narrow corridor leads to a small room - a "pocket", which has a lenticular outline of an irregular oval shape. The apse of the north nave is connected with the central apse by a stepwise widening passage. The southern nave, on the other hand, is divided into a short section from the east and an extended section from the west, and they are not connected to each other. A short section, which has one window from the east, that is, in the apse, and from the south, is connected with the central apse by a step-expanding passage. A short corridor leads from it to a room between the apses, which has a shape close to a triangle in plan. The extended western section of the south nave, completed with an apse, opens onto the facade with three arched doorways. The placement of the entrance from the south is known to be typical for the Syrian tradition.

The temple above the Loo village has similar constructional composition peculiarities, it was examined in detail by B.B. Ovchinnikova [3]. A feature of the temple in Loo village is the presence of a pentahedral central apse with semicircular lateral ones. This combination is characteristic of numerous temples of this period in the Eastern provinces of Byzantium, for example, in the Crimea, Bulgaria and Trebizond, the basilicas of the Northern Black Sea coast of the 6th century (for example, the basilica in Tsandripsh), but in the cross-domed churches of the region, we see a similar reception only in the Anakopia church of Simon the Canaanite. The performed architectural and archaeological investigations made it possible to clarify the dimensions of the temple, facades and individual details of the building. The total length of the temple with pilasters and apses is 21 m, width - 12.25 m, thickness of the walls - 1.1 m. This is a three-nave, three-apse building elongated in length. Local features of the temple are the masonry originality of the walls with transverse white stone inserts forming pilasters, as well as the presence of sacred trees near the temple. Six pillars are symmetrically located in the center of the temple, only their bases have survived, the size of the dome square side is 3.5 m. The distance between the eastern and central pairs of pillars is equal to the width of the central nave (3.5 m), the pillars form a square in plan. This allowed B.B. Ovchinnikova to suggest that these supports were domed. However, no parts of the cross-domed system were found among the ruins of the temple.

The middle apse is somewhat elongated. Its floor is raised above the floor of the central nave by 0.30 m. The western part of the temple is separated from the prayer hall not by a wall, but by a pair of pillars. This feature distinguishes the temple in Loo from many religious buildings of the X-XI centuries, similar in composition and plan, in Chersonesos, Anakopia, Bzybi, Lykhny. There was no nartex, that is the similarity of this temple with the basilica temples of the early Middle Ages. The temple was illuminated by narrow (0.4 m) window openings with a wedge-shaped arched end made of lime slabs and shell rock slabs. From the outside, in the decoration of the temple walls, in a number of places, the remains of the facing have been preserved. Another feature of the temple is the dissection of the inner facades of the southern, northern and western walls by pilasters, which could turn into arched semicircles. A facing block with individual letters of the Greek alphabet was found [3, p. 14].

Tsandripshsky temple is located 17 km north-west of the Gagra city, on the territory of a large seaside village. Nearby, at the mouth of the Khashupsa River, the remains of an ancient fortification have been preserved, which are included in the new buildings' composition. The monument is surrounded by the graves of a modern cemetery, which did not allow exploring the adjacent territory. In the Khashupsy gorge, 30 kilometers from the sea, on a high cliff, there are the ruins of a large fortress. It controlled the pass to the North Caucasus. The fortress has been little explored, according to the construction technique and analogies of L.G. Khrushkova was attributed it to the 6th century, which was confirmed by the results of the analysis of the solution by the X-ray phase method in 2020, performed by the authors of this article. Under the cover of this fortress, the Tsandripshsky temple existed. Tsandripshsky temple is also a three-aisled basilica with three protruding apses, a narthex and an adjoining small portico, which has been uncovered by excavations. Overall external dimensions 27.5x16.6 m (without the southern portico, which belongs to a later time). Very often in basilicas, the altar barrier was moved into the nave to increase the area of the altar. The situation was different in Tsandripsh, here the visitors from both the baptismal room and the memorial chapel could not move to the altar space, which was intended only for the clergy. They moved to the eastern part of the central nave, separated from the altar space by a fence. The area of the altar is increased due to the bema. There is no integral three-part altar space, united by doorways and intended for the liturgy of the Eucharist. These are the three isolated cells, with its own special functions each. There is no altar, no deacon. Tsandripsh materials confirm the opinion of many researchers, which emphasize the illegality of the name altar and deacon premises next to the altar in early Christian churches. The functions of these premises (we join the opinion of L.G. Khrushkova) may be different [9]. Basilica with its dynamic focus on the altar corresponded well with the early Christian "liturgy processions". Even D. Kraufut noted that the Palestinian and Syrian three-apse churches, built no later than the second quarter of the 6th century, are not associated with the new Constantinople liturgy. The Syrian origins of the liturgical space organization in Tsandripsh are reflected in other features. In Syria, a memorial chapel is often located in the southern room of a three-part sanctuary. A similar solution in Tsandripsh. The baptismal font in the apse is another characteristic Syrian feature.

Discussion

The dating of the Musser Temple is controversial. So, L. D. Rcheulishvili dated this monument to the turn of the 6th - 7th centuries and considered it earlier than the Kiach temple. A. Yu. Vinogradov and D. V. Beletsky, as well as L. G. Rcheulishvili, compare the composition of the western wall of the Myusser temple with the Lykhny temple and Kvelatsminda in Gurjaani, as well as with the Tsandripsh basilica, and the door at the organ loft level is attributed to the influence of the cathedral in Mokvi. Highlighting two stages in the construction history of the temple, these researchers date the monument to the X century [2, p. 153-155].

Comparing the Myusser temple with the temple in the village. Kondamiani (Georgia), dating back to the 7th century, it should be noted the similarity of the altar rooms separated by a wall, as well as the three-arch solution of the southern facade. Apparently, the Mussera narthex was added later, possibly in the 10th century, which explains the presence of a wall delimiting it from the main hall, and the "closed" organ lofts characteristic of the 10th century monuments. There is also a noticeable similarity in the nature of the masonry, as well as the design of the two churches' facades - in Musser and Kondamiani. The reconstruction of the Mussera temple in the 10th century also affected its southern part - an apse appeared there in the southern gallery, which turned the temple into a three-church basilica. Such techniques for arranging the apse in the gallery, after the apse in the pastoforium, are characteristic of the X century (for example, the temple in Nekresi), which allows us to conclude about the time of the building's reconstruction [4].

The solutions' analysis confirmed the estimated date of the construction beginning of the Mussera temple - VI century. The main facade, an altar inside and the remains of the ancient building from the north belong to this time. By the VII century the north wall and galleries had been completed. The assumption was also confirmed that in the X century the temple underwent reconstruction - the western wall with the organ loft, the temple facing were altered. This explains the similarity of the western wall with the organ lofts with similar monuments of the 10th century, noted by the researchers who dated the temple to the 10th century. Outside, the altar was renovated in the 12th century. (Table 1).

Throughout its centuries-old life, the Tsandripsh Basilica underwent restructuring. The first reconstruction, which greatly changed the temple's appearance, took place around the 8th-9th centuries. The most significant alterations were the elimination of the narthex with an adjoining portico and the construction of stone vaults instead of wooden floors above the first floors of the side naves and the central (except for the apse and bema). The vault of the central nave rested on new additional pillars attached to the original pillars, which in fact turned out to be pilasters, passing at the top into supporting arches. The thickness of the vault was 0.9 m, the main building material was well-fired rectangular bricks, tiles were used in backing. The foundations of the new pillars are only 0.6 m deep. Three pillars were installed on the tombs of the central nave. In connection with the vaults' arrangement, the organ lofts in the side aisles were eliminated. The altar barrier was expanded and slightly shifted to the east. In the X-XI centuries, the appearance of the temple changed again. On the western side, at the entrance to the church, there was a vestibule-vestibule, resting on four corner pillars, and at the southern entrance - a small portico made of broken limestone. A new floor was made, with brick paving and a layer of loose lime concrete.

The solutions' analysis of the Tsandripshsky temple confirmed the date of the beginning of its construction at the end of the 5th - beginning of the 6th century. The southern wall was completed by the 7th century. The porch, according to the analysis results, was added at the end of the 10th - beginning of the 11th centuries (Table 1).

Table 1.

No	Regist-ration number	Location, object name	Sampling location	The intensity of the main peak of calcite crystallinity (ARLXTRA device)	Ave-rage	Dated by the main peak intensity of calcite crystallinity, century ARLXTRA	Intensity of the main peak of calcite crystal-linity (Dron device)	Dated by the main peak intensity of calcite crystal-linity, century Drone	Comm on dating, century	Note
1.	**EA-20-TS-1**	**Tsand-ripsh**	**pillar**	5200	4800	5-6			6	Khrush-kova
2.	EA-20-TS-2	Tsand-ripsh	porch	3050-2020 3150-2019		11			10	Khrush-kova
3.	EA-20-TS-3	Tsand-ripsh	altar	5000		6	382	The beginning of the 6th century	6	Khrush-kova
4.	EA-20-TS-4	Tsand-ripsh	Southern wall	4200		7			10	Khrush-kova
5.	EA-20-TS-5	Tsand-ripsh	arch	4700		6			6	Khrush-kova
6.	**EA-20-MU-12**	**Musser**	Neighbor construc-tions	4950		5-6			-	
7.	EA-20-MU-4	Musser	Northern porch	4900	4900	5-6			6-7	Rcheu-lishvili Berdzeni-shvili
8.	EA-20-MU-10	Musser	Main facade	4900-2020 4750-2019		6	371	The beginning of the 6th century	10	Vinog-radov Voronov
9.	**EA-20-MU-3**	**Musser**	Northern wall	4000		7			6-7	Rcheu-lishvili Berdzeni-shvili

10.	EA-20-MU-8	Musser	Altar northern apse	4300		7			-	
11.	EA-20-MU-11	Musser	gallery	4100		7			-	
12.	EA-20-MU-7	Musser	Organ loft coating	3500		10			-	
13.	EA-20-MU-5	Musser	Altar outside	2500		12			-	
14.	EA-20-MU-9	Musser	Western wall with organ lofts	3200	3300	10			10	Vinog-radov, Voronov
15.	EA-20-MU-13	Musser	Northern gallery facing	3550		10			10	Vinog-radov, Voronov
16.	**C-20-LO-1**	**LOO**	Southern portico	4070		7			10-11	Ovchin-nikova
17.	C-20-LO-2	LOO	Of a fallen piece at the entrance	3570		10			11	Ovchin-nikova
18.	C-20-LO-3	LOO	column	3100		11			11	Ovchin-nikova
19.	C-20-LO-4	LOO	altar	4050		7			11	Ovchin-nikova
20.	C-20-LO-5	LOO	Of the wall with inserts	4600		7			11	Ovchin-nikova

Regarding the rebuilding of the Loos Temple, B.B. Ovchinnikova notes that the samples of solutions from the syntron in the central apse and the annex on the inner side of the southern wall differ in their lower strength, slightly different percentage of filler and mineral composition. According to her investigations, they were erected later, when miscalculations in the initial selection of the construction site became obvious and the southern wall deviated from the central axis of the temple. However, later this wall collapsed.

The solutions' analysis of the Loo temple gave interesting conclusions. The altar, the South wall and the portico in front of it, according to the mortar, can be dated to the end of the 6th - beginning of the 7th century. The fallen fragment of the western wall dates to the 10th century, and the inserts into the columns - to the 11th century. All this suggests that the Lossky temple was reconstructed into a cross-domed one from an older structure in the X - XI centuries. Its pillars were reinforced, which matches the findings of B.B. Ovchinnikova.

Summary

Comparison of the temples' architecture in Musser, Tsandripsh and Loo shows that the beginning of their construction can be attributed to the 6th-7th centuries. Their architectural features point to Syrian influence. In the 10th century, all three churches were rebuilt. It was then that a wall with an organ loft appeared in Musser and the northern gallery was rebuilt in the western part. The temple in Tsandripsh received a vestibule from the west. And the Loo temple was rebuilt into a cross-domed one. Thus, in our dating we join the conclusions of the majority of archaeological researchers.

Acknowledgments

The research is supported by grant RFBR №19-012-00402\20

References

[1] Berdzenishvili I V 2006 The issue of the spread of Christianity in Abkhazia according to archeology (IV – VIII centuries) *Author's diss. ... cand. East. Sciences, Tbilisi*, 25.

[2] Vinogradov A Yu, Beletsky D V 2015 *Church architecture of Abkhazia in the era of the Abkhaz kingdom. VIII-X century* (Moscow, Indrik) 372.

[3] Ovchinnikova B B 1997 Results of field research of the Loos archaeological expedition of the Ural State University named after A.M. Gorky *Materials of the USU conference: Archeology, architecture and ethnographic processes of the North-Western Caucasus* (Yekaterinburg, USU) 7 - 31.

[4] Pishchulina V, Kishkinova E 2020 To the question of age determination of the medieval temple in Miuseri (Abkhazia) *IOP Conf. Ser.: Mater. Sci. Eng.* **913** 032028.

[5] Pachulia V P 1968 *Historical Monuments of Abkhazia. Their value and protection* (Moscow, The main editorial office of oriental literature of the Nauka publishing house) 168.

[6] Rcheulishvili L D 1988 *Dome architecture of the 8th – 10th centuries in Abkhazia* (Tbilisi, Metsniereba) 90.

[7] Hrushkova L G 2018 *Eastern Black Sea region in the Byzantine era. History. Architecture. Archeology* (Kaliningrad-Moscow, ROS-DOAFK Publishing House) 480.

[8] Neubauer E 1978 Abchazische Architektur im Spannungsfeld zwischen Georgien und Byzanz (6. bis 11. Jh.), Byzantinische Kunstexport, Halle 70—79.

[9] Pishchulina V, Kotlyar V, Argun A 2020 *New data on chronology of medieval architectural objects of northern provinces of the Byzantine oykumena* Nordsci Conference on Social Sciences.

AUTHOR INDEX